ADVANCED MATHEMATICAL METHODS FOR SCIENTISTS AND ENGINEERS

INTERNATIONAL SERIES IN PURE AND APPLIED MATHEMATICS

E. H. Spanier and G. Springer
Consulting Editors

AHLFORS: Complex Analysis
BENDER AND ORSZAG: Advanced Mathematical Methods for Scientists and Engineers
BUCK: Advanced Calculus
BUSACKER AND SAATY: Finite Graphs and Networks
CHENEY: Introduction to Approximation Theory
CHESTER: Techniques in Partial Differential Equations
CODDINGTON AND LEVINSON: Theory of Ordinary Differential Equations
CONTE AND DE BOOR: Elementary Numerical Analysis: An Algorithmic Approach
DENNEMEYER: Introduction to Partial Differential Equations and Boundary Value Problems
DETTMAN: Mathematical Methods in Physics and Engineering
GOLOMB AND SHANKS: Elements of Ordinary Differential Equations
HAMMING: Numerical Methods for Scientists and Engineers
HILDEBRAND: Introduction to Numerical Analysis
HOUSEHOLDER: The Numerical Treatment of a Single Nonlinear Equation
KALMAN, FALB, AND ARBIB: Topics in Mathematical Systems Theory
LASS: Vector and Tensor Analysis
MCCARTY: Topology: An Introduction with Applications to Topological Groups
MONK: Introduction to Set Theory
MOORE: Elements of Linear Algebra and Matrix Theory
MOURSUND AND DURIS: Elementary Theory and Application of Numerical Analysis
PEARL: Matrix Theory and Finite Mathematics
PIPES AND HARVILL: Applied Mathematics for Engineers and Physicists
RALSTON AND RABINOWITZ: A First Course in Numerical Analysis
RITGER AND ROSE: Differential Equations with Applications
RUDIN: Principles of Mathematical Analysis
SHAPIRO: Introduction to Abstract Algebra
SIMMONS: Differential Equations with Applications and Historical Notes
SIMMONS: Introduction to Topology and Modern Analysis
STRUBLE: Nonlinear Differential Equations

ADVANCED MATHEMATICAL METHODS FOR SCIENTISTS AND ENGINEERS

Carl M. Bender
Professor of Physics
Washington University

Steven A. Orszag
Professor of Applied Mathematics
Massachusetts Institute of Technology

McGRAW-HILL BOOK COMPANY

New York St. Louis San Francisco Auckland Bogotá Düsseldorf
Johannesburg London Madrid Mexico Montreal New Delhi
Panama Paris São Paulo Singapore Sydney Tokyo Toronto

ADVANCED MATHEMATICAL METHODS FOR SCIENTISTS AND ENGINEERS

1234567890 FGRFGR 78321098

Library of Congress Cataloging in Publication Data

Bender, Carl M.
Advanced mathematical methods for scientists and engineers.

(International series in pure and applied mathematics)
Bibliography: p.
Includes index.
1. Differential equations—Numerical solutions.
2. Difference equations—Numerical solutions.
3. Engineering mathematics. 4. Science—Mathematics.
I. Orszag, S., joint author. II. Title.
QA371.B43 519.4 77-29168
ISBN 0-07-004452-X

This book was set in Monophoto Times Mathematics.
The editor was Rose Ciofalo and the production supervisor was Jeanne Selzam.
Fairfield Graphics was printer and binder.

To Our Wives

JESSICA AND REBA

and Sons

MICHAEL AND DANIEL

and

MICHAEL, PETER, AND JONATHAN

CONTENTS†

† Each section is labeled according to difficulty: **(E)** = easy, **(I)** = intermediate, **(D)** = difficult. A section labeled **(T)** indicates that the material has a theoretical rather than an applied emphasis.

PREFACE

The triumphant vindication of bold theories—are these not
the pride and justification of our life's work?

—Sherlock Holmes, *The Valley of Fear*
Sir Arthur Conan Doyle

The main purpose of our book is to present and explain mathematical methods for obtaining approximate analytical solutions to differential and difference equations that cannot be solved exactly. Our objective is to help young and also established scientists and engineers to build the skills necessary to analyze equations that they encounter in their work. Our presentation is aimed at developing the insights and techniques that are most useful for attacking new problems. We do not emphasize special methods and tricks which work only for the classical transcendental functions; we do not dwell on equations whose exact solutions are known.

The mathematical methods discussed in this book are known collectively as asymptotic and perturbative analysis. These are the most useful and powerful methods for finding approximate solutions to equations, but they are difficult to justify rigorously. Thus, we concentrate on the most fruitful aspect of applied analysis; namely, obtaining the answer. We stress care but not rigor.

To explain our approach, we compare our goals with those of a freshman calculus course. A beginning calculus course is considered successful if the students have learned how to solve problems using calculus. It is not necessary for a student to understand the subtleties of interchanging limits, point-set topology, or measure theory to solve maximum-minimum problems, to compute volumes and areas, or to study the dynamics of physical systems. Asymptotics is a newer calculus, an approximate calculus, and its mathematical subtleties are as difficult for an advanced student as the subtleties of calculus are for a freshman. This volume teaches the new kind of approximate calculus necessary to solve hard problems approximately. We believe that our book is the first comprehensive book at the advanced undergraduate or beginning graduate level that has this kind of problem-solving approach to applied mathematics.

The minimum prerequisites for a course based on this book are a facility with calculus and an elementary knowledge of differential equations. Also, for a few advanced topics, such as the method of steepest descents, an awareness of complex variables is essential. This book has been used by us at Washington University and at M.I.T. in courses taken by engineering, science, and mathematics students normally including juniors, seniors, and graduate students.

We recognize that the readership of this book will be extremely diverse. Therefore, we have organized the book so that it will be useful to beginning students as well as to experienced researchers. First, this book is completely self-contained. We have included a review of ordinary differential equations and ordinary difference equations in Part I for those readers whose background is weak. There is also an Appendix of useful formulas so that it will rarely be necessary to consult outside reference books on special functions.

Second, we indicate the difficulty of every section by the three letters E (easy), I (intermediate), and D (difficult). We also use the letter T to indicate that the material has a theoretical as opposed to an applied or calculational slant. We have rated the material this way to help readers and teachers to select the level of material that is appropriate for their needs. We have included a large selection of exercises and problems at the end of each chapter. The difficulty and slant of each problem is also indicated by the letters E, I, D, and T. A good undergraduate course on mathematical methods can be based entirely on the sections and problems labeled E.

One of the novelties of this book is that we illustrate the results of our asymptotic analysis graphically by presenting many computer plots and tables which compare exact and approximate answers. These plots and tables should give the reader a feeling of just how well approximate analytical methods work. It is our experience that these graphs are an effective teaching device that strengthens the reader's belief that approximation methods can be usefully applied to the problems that he or she need to solve.

In this volume we are concerned only with functions of one variable. We hope some day to write a sequel to this book on partial differential equations.

We thank our many colleagues, especially at M.I.T., for their interest, suggestions, and contributions to our book, and our many students for their thoughtful and constructive criticism. We are grateful to Earl Cohen, Moshe Dubiner, Robert Keener, Lawrence Kells, Anthony Patera, Charles Peterson, Mark Preissler, James Shearer, Ellen Szeto, and Scot Tetrick for their assistance in preparing graphs and tables. We are particularly indebted to Jessica Bender for her tireless editorial assistance and to Shelley Bailey, Judi Cataldo, Joan Hill, and Darde Khan for helping us prepare a final manuscript. We both thank the National Science Foundation and the Sloan Foundation and one of us, S. A. O., thanks the Fluid Dynamics Branch of the Office of Naval Research for the support we have received during the preparation of this book. We also acknowledge the support of the National Center for Atmospheric Research for allowing us the use of their computers.

Carl M. Bender
Steven A. Orszag

PART ONE

FUNDAMENTALS

I am afraid that I rather give myself away when I explain.
Results without causes are much more impressive.

Sherlock Holmes, *The Stock-Broker's Clerk*
Sir Arthur Conan Doyle

Part I of this book is a synopsis of exact methods for solving ordinary differential equations (Chap. 1) and ordinary difference equations (Chap. 2). Since our primary emphasis in later parts is on the approximate solution of such equations, it is important to review those exact methods that are currently known.

Our specific purpose with regard to differential equations is to refresh but not to introduce those concepts that would be learned in a low-level undergraduate course. Although Chap. 1 is self-contained in the sense that it begins with the most elementary aspects of the subject, the language and pace are appropriate for someone who has already had some experience in solving elementary differential equations. Our approach highlights applications rather than theory; we state theorems without proving them and stress methods for obtaining analytical solutions to equations.

The beauty of differential equations lies in their richness and variety. There is always a large class of equations which exhibits a new behavior or illustrates some counterintuitive notion. Unfortunately, many students, rather than enjoying the abundance of the subject, are confounded and appalled by it. To those who view the subject as an endless collection of unrelated methods, rules, and tricks, we say that the collection is actually finite; apart from transform methods (see the References), it is entirely contained in Chap. 1. The reader who masters the material in Chap. 1 will be fully prepared for any problems he or she may encounter. And to those mathematicians who prefer to study the general properties of a forest without having to examine individual trees, we are pleased to say that as we progress toward the approximate study of differential equations in Parts II, III, and IV our approach becomes far more general; approximate methods apply to much wider classes of equations than exact methods.

Our presentation in Chap. 2 is more elementary because most students are unfamiliar with difference equations. Our treatment of the subject emphasizes the parallels with differential equations and again stresses analytic methods of solution.

CHAPTER

ONE

ORDINARY DIFFERENTIAL EQUATIONS

Like all other arts, the Science of Deduction and Analysis
is one which can only be acquired by long and patient study,
nor is life long enough to allow any mortal to attain the
highest possible perfection in it. Before turning to those
moral and mental aspects of the matter which present the
greatest difficulties, let the inquirer begin by mastering
more elementary problems.

Sherlock Holmes, *A Study in Scarlet*
Sir Arthur Conan Doyle

(E) 1.1 ORDINARY DIFFERENTIAL EQUATIONS

An nth-order differential equation has the form

$$y^{(n)}(x) = F[x, y(x), y'(x), \ldots, y^{(n-1)}(x)], \tag{1.1.1}$$

where $y^{(k)} = d^k y/dx^k$. Equation (1.1.1) is a *linear* differential equation if F is a linear function of y and its derivatives (the explicit x dependence of F is still arbitrary). If (1.1.1) is linear, then the *general* solution $y(x)$ depends on n independent parameters called constants of integration; all solutions of a linear differential equation may be obtained by proper choice of these constants. If (1.1.1) is a nonlinear differential equation, then it also has a general solution which contains n constants of integration. However, there sometimes exist special additional solutions of nonlinear differential equations that cannot be obtained from the general solution for any choice of the integration constants. We omit a rigorous discussion of these fundamental properties of differential equations but illustrate them in the next three examples.

Example 1 *Separable equations.* Separable equations are the easiest differential equations to solve. An equation is *separable* if it is first order and the x and y dependences of F in (1.1.1) factor. The most general separable equation is

$$y'(x) = a(x)b(y). \tag{1.1.2}$$

Direct integration gives the general solution

$$\int^y \frac{dt}{b(t)} = \int^x a(s)\, ds + c_1, \tag{1.1.3}$$

where c_1 is a constant of integration. [The notation $\int^x a(s)\, ds$ stands for the antiderivative of $a(x)$.]

Linear equations have a simpler and more restricted range of possible behaviors than nonlinear equations, but they are an important class because they occur very frequently in the mathematical description of physical phenomena. Formally, a linear differential equation may be written as

$$Ly(x) = f(x), \tag{1.1.4}$$

where L is a linear differential operator:

$$L = p_0(x) + p_1(x)\frac{d}{dx} + \cdots + p_{n-1}(x)\frac{d^{n-1}}{dx^{n-1}} + \frac{d^n}{dx^n}. \tag{1.1.5}$$

It is conventional, although not necessary, to choose the coefficient of the highest derivative to be 1. If $f(x) \equiv 0$, the differential equation (1.1.4) is *homogeneous*; otherwise it is *inhomogeneous*.

Example 2 *Solution of a linear equation.* The general solution of the homogeneous linear equation

$$y'' - \frac{1+x}{x}y' + \frac{1}{x}y = 0 \tag{1.1.6}$$

is $y(x) = c_1 e^x + c_2(1 + x)$, which shows the explicit dependence on the two constants of integration c_1 and c_2. Every solution of (1.1.6) has this form.

Nonlinear equations have a richer mathematical structure than linear equations and are generally more difficult to solve in closed form. Nevertheless, the solution of many difficult-looking nonlinear equations is quite routine.

Example 3 *Solutions of nonlinear equations.* Two nonlinear differential equations which can be routinely solved (see Secs. 1.6 and 1.7) are the Riccati equation

$$y' = \frac{A^2}{x^4} - y^2, \qquad A \text{ is a constant,} \tag{1.1.7}$$

whose general solution is

$$y(x) = \frac{1}{x} + \frac{A}{x^2}\frac{c_1 - e^{2A/x}}{c_1 + e^{2A/x}} \tag{1.1.8}$$

and the equidimensional equation

$$y'' = yy'/x \tag{1.1.9}$$

whose general solution is

$$y(x) = 2c_1 \tan(c_1 \ln x + c_2) - 1. \tag{1.1.10}$$

There is a special solution to (1.1.9), namely $y = c_3$, where c_3 is an arbitrary constant, which cannot be obtained from the general solution in (1.1.10) for any choice of c_1 and c_2. (See Prob. 1.2.)

The rest of this chapter gives a brief theoretical discussion of the existence and uniqueness of solutions to initial- and boundary-value problems and surveys the elementary techniques for obtaining closed-form solutions of differential equations like those in the above three examples.

Systems of First-Order Equations

The general nth-order differential equation (1.1.1) is equivalent to a system of n first-order equations. To show this, we introduce the n dependent variables $y_k(x) = d^k y/dx^k$ ($k = 0, 1, 2, \ldots, n - 1$). These variables satisfy the system of n first-order equations

$$\frac{d}{dx} y_k(x) = y_{k+1}(x), \qquad k = 0, \ldots, n - 2,$$

$$\frac{d}{dx} y_{n-1}(x) = F[x, y_0, y_1, y_2, \ldots, y_{n-1}(x)].$$

Conversely, it is usually true that a system of n first-order equations

$$\frac{d}{dx} z_k = f_k(x, z_1, z_2, \ldots, z_n), \qquad k = 1, 2, \ldots, n, \tag{1.1.11}$$

can be transformed to a single equation of nth order. To construct an equivalent nth-order equation for $z_1(x)$, we differentiate (1.1.11) with respect to x, using the chain rule and (1.1.11) for dz_k/dx. We obtain

$$\frac{d^2}{dx^2} z_1 = \frac{\partial f_1}{\partial x} + \sum_{k=1}^{n} \frac{\partial f_1}{\partial z_k} f_k \equiv f_1^{(1)}(x, z_1, \ldots, z_n).$$

Repeating this process $(n - 1)$ times we obtain n equations of the form

$$\frac{d^j}{dx_j} z_1 = f_1^{(j)}(x, z_1, \ldots, z_n), \qquad j = 1, \ldots, n, \tag{1.1.12}$$

where $f_1^{(0)} = f_1$ and $f_1^{(j+1)} = \partial f_1^{(j)}/\partial x + \sum_{k=1}^{n} (\partial f_1^{(j)}/\partial z_k) f_k$. If these n equations can be solved simultaneously to eliminate $z_2, z_3, \ldots, z_n$ as functions of x, z_1, dz_1/dx, $d^2 z_1/dx^2, \ldots, d^{n-1} z_1/dx^{n-1}$, then the system (1.1.11) has been transformed to an nth-order equation for z_1. Can you construct an example in which the equations (1.1.12) cannot be solved for $z_2, \ldots, z_n$?

(E) 1.2 INITIAL-VALUE AND BOUNDARY-VALUE PROBLEMS

A solution $y(x)$ to a differential equation is not uniquely determined by the differential equation alone; the values of the n independent constants of integration must also be specified. These constants of integration may be specified in several quite disparate ways. In an *initial-value problem* one specifies y and its first $n - 1$ derivatives, $y', \ldots, y^{(n-1)}$, at *one* point $x = x_0$:

$$\begin{aligned} y(x_0) &= a_0, \\ y'(x_0) &= a_1, \ldots, \\ y^{(n-1)}(x_0) &= a_{n-1}. \end{aligned} \tag{1.2.1}$$

The numbers $a_0, \ldots, a_{n-1}$ are the n constants of integration of the initial-value problem. In a *boundary-value problem* a total of n quantities are specified at two or more points. For example, for a fifth-order differential equation, one might impose the conditions: $y(x_1) = a_1$, $y'(x_2) = a_2$, $y'(x_3) = a_3$, $y''(x_3) = a_4$, $y'''(x_1) + [y(x_2)]^2 = a_5$. It may be that some or all of the points $x_1, x_2, \ldots$ are interior to and not on the boundary of the region in which $y(x)$ is to be found. In any case, the quantities a_i are still the constants of integration for the problem.

The many theorems which deal with the existence and uniqueness of the solutions to initial-value and boundary-value problems provide the basis for two levels of applied analysis. *Local* analysis consists of investigating the solution in the neighborhood of a point; the results are valid when the neighborhood is sufficiently small. *Global* analysis, which is much more difficult, attempts to determine the nature of the solution over a region of finite extent. This distinction provides the organizational structure of this book. For example, asymptotic series (see Part II) are only valid locally, whereas WKB approximations (see Part IV) are global.

Initial-value problems are intrinsically much simpler than boundary-value problems. Initial-value problems may be subjected to local analysis to determine whether a unique solution $y(x)$ exists in a sufficiently small neighborhood of x_0. For example, the differential equation (1.1.1) may be differentiated repeatedly (if this is possible) to compute the coefficients in the Taylor expansion of $y(x)$ about the point x_0. (A series expansion is a typical tool of local analysis.) In fact, there is a standard existence theorem for initial-value problems which applies whether or not this Taylor series exists or converges. In general, if F, $\partial F/\partial y$, $\partial F/\partial y'$, ..., $\partial F/\partial y^{(n-1)}$ are all continuous functions of $x, y, y', \ldots, y^{(n-1)}$ in some neighborhood of $x_0, a_0, a_1, \ldots, a_{n-1}$, then a unique solution $y(x)$ satisfying (1.1.1) and (1.2.1) exists in some interval containing x_0. In Prob. 1.3 we examine the proof of this theorem for first-order equations using a technique that may be generalized to nth-order equations.

Example 1 *Initial-value problems having unique solutions.* Because they satisfy the conditions for the existence and uniqueness theorem, the following five initial-value problems all have unique solutions which exist for x sufficiently near 0:

(*a*) $y' = \sin(xy)$, $\quad y(0) = 1$;
(*b*) $y' = (x + y)x^2y^2$, $\quad y(0) = 1$;
(*c*) $y' = e^x + x/y$, $\quad y(0) = 1$;
(*d*) $y'' = y^2 + e^x$, $\quad y(0) = y'(0) = 0$;
(*e*) $y''' = e^{xyy'}$, $\quad y(0) = y'(0) = y''(0) = 0$.

Unfortunately, the solutions to these problems have at best a metaphysical existence because we do not know how to express any of them in terms of known functions.

If F is not a smooth function of its arguments, then the existence and uniqueness theorem may not hold.

Example 2 *Differential equation having a nonunique solution.* For the initial-value problem $y' = y^{1/3}$ [$y(0) = 0$] it is clear that F is continuous near $x = 0$ and $y = 0$, but that $\partial F/\partial y$ is not. It is therefore not surprising that there are two solutions to this problem: $y = 0$, $y = (2x/3)^{3/2}$. Notice that the general solution to the first-order equation $y' = y^{1/3}$ is $y(x) = \eta(2x/3 + c_1)^{3/2}$, where $\eta = \pm 1, 0$ and c_1 is arbitrary. This general solution depends on more than the single constant of integration c_1; it also depends on the discrete parameter η. In general, if F in (1.1.1) is not smooth, then the solution $y(x)$ may depend on more than n constants of integration.

Example 3 *Differential equation having a unique solution.* The initial-value problem $y' = x^{-1/2}$ $[y(0) = 1]$ does have a unique solution $y = 2x^{1/2} + 1$, even though F is discontinuous at $x = 0$. But this does not contradict the uniqueness theorem which gives a sufficient, but not necessary, condition on F for existence and uniqueness.

If it can be shown that the initial-value problem has a solution near x_0, the next step is to determine the global properties of this solution. In particular, one must determine the extent of the interval containing x_0 on which the local solution exists. This is relatively easy to do for linear equations. For example, it can be shown (see Prob. 1.4) that, independent of the initial conditions, the interval is at least as large as the largest interval containing no discontinuities of $p_0(x)$, $p_1(x)$, $p_2(x), \ldots, p_{n-1}(x)$. Additional analysis may establish that the interval is even larger.

Example 4 *Interval of existence of solution to a linear equation.* The result in Prob. 1.4 ensures that the solution to $y' = (\tan x)y + 1$ $[y(0) = 1]$ exists in the interval in which $\tan x$ is continuous, namely $(-\pi/2, \pi/2)$. However, the solution $y(x) = (1 + \sin x)/\cos x$ actually exists for $-3\pi/2 < x < \pi/2$.

Global analysis of initial-value problems for nonlinear differential equations is much more difficult. Solutions of nonlinear equations may develop spontaneous singularities whose locations depend on the initial conditions (see Chap. 4). Hence, the region in which the solution exists depends on the initial conditions as well as on the equation. Observe that the positions of the singularities of solutions to the nonlinear equations in Example 3 of Sec. 1.1 change as the initial conditions vary.

Boundary-value problems are inherently global. Existence and uniqueness theorems for solutions must be proved for an interval large enough to include *all* the points $x_1, x_2, \ldots$. Local analysis of the solution near any of these points is insufficient. The global properties of linear differential equations are relatively easy to obtain, but it is very difficult to state rigorous criteria for existence and uniqueness of solutions to boundary-value problems for nonlinear differential equations. Boundary-value problems for nonlinear differential equations may have no solution, a finite number k of solutions, or even infinitely many solutions. The same possibilities are true for linear equations except that $k = 1$.

(TE) 1.3 THEORY OF HOMOGENEOUS LINEAR EQUATIONS

In this section we review the elementary aspects of the elegant theory of initial-value and boundary-value problems for homogeneous linear differential equations.

Linear Independence of Solutions

The general solution of an nth-order homogeneous linear equation

$$y^{(n)} + p_{n-1}(x)y^{(n-1)} + \cdots + p_0(x)y = 0 \tag{1.3.1}$$

has the particularly simple form

$$y(x) = \sum_{j=1}^{n} c_j y_j(x), \tag{1.3.2}$$

where the c_j are arbitrary constants of integration and $\{y_j(x)\}$ is a linearly independent set of functions, each satisfying (1.3.1). There are always exactly n linearly independent solutions to (1.3.1) in any region where the coefficient functions $p_0(x)$, $p_1(x), \ldots, p_{n-1}(x)$ are continuous.

[A set of functions $\{y_j(x)\}$ is said to be *linearly dependent* if it is possible to find a set of numbers $\{c_j\}$ which are not all zero and which satisfy $\sum_{j=1}^{n} c_j y_j(x) = 0$. If it is not possible to find such a set, then $\{y_j(x)\}$ is *linearly independent.*]

The concept of linear independence is important because it enables one to decide whether any solution of the form (1.3.2) is indeed the general solution. If the n functions $\{y_j(x)\}$ are not linearly independent, then at least one y_j is a linear combination of the others and it is necessary to search for more solutions of the differential equation. Note that it does not make sense to discuss linear independence at a point; linear independence is a global concept. A set of functions is always said to be linearly dependent or independent throughout an interval.

The Wronskian

There is a simple test for the linear dependence of a set of differentiable functions. The Wronskian $W(x)$ is defined as the determinant

$$W(x) = W[y_1(x), y_2(x), \ldots, y_n(x)]$$

$$\equiv \det \begin{vmatrix} y_1 & y_2 & \cdots & y_n \\ y_1' & y_2' & \cdots & y_n' \\ \cdots & \cdots & \cdots & \cdots \\ y_1^{(n-1)} & y_2^{(n-1)} & \cdots & y_n^{(n-1)} \end{vmatrix} \tag{1.3.3}$$

$W(x)$ vanishes identically over an interval if and only if $\{y_j(x)\}$ is a linearly dependent set of functions (see Prob. 1.5). If $\{y_j(x)\}$ is linearly independent over some interval, then $W(x)$ does not vanish, except possibly at isolated points.

Example 1 *Linear dependence.* Since $W[e^x, e^{-x}, \cosh x] \equiv 0$ for all x, $\{e^x, e^{-x}, \cosh x\}$ is a linearly dependent set of functions.

Example 2 *Linear independence.* To verify that the solution $y(x) = c_1 e^x + c_2(1 + x)$ of $y'' - y'(1 + x)/x + y/x = 0$ in Example 2 of Sec. 1.1 is in fact the general solution, we evaluate the Wronskian: $W[1 + x, e^x] = xe^x$. Note that xe^x vanishes only at $x = 0$ (because e^x and $1 + x$ are tangent there). Thus, for all x, $\{1 + x, e^x\}$ is a linearly independent set.

Homogeneous linear equations have a remarkable property: the Wronskian $W(x)$ of any n solutions of (1.3.1) satisfies the simple first-order equation

$$W'(x) = -p_{n-1}(x)W(x). \tag{1.3.4}$$

(For the derivation of this equation, see Prob. 1.7.) The solution of (1.3.4) is known as Abel's formula:

$$W(x) = \exp\left[-\int^{x} p_{n-1}(t)\,dt\right]. \tag{1.3.5}$$

Thus, we have the surprising result that $W(x)$ can be computed before any of the solutions of the differential equation are known. The indefinite integral in Abel's formula means that $W(x)$ is determined up to an arbitrary multiplicative constant. Choosing a new set of n solutions (which, of course, will all be linear combinations of the old set) merely alters the constant.

Let us now use these theoretical results to discuss the wellposedness of initial-value and boundary-value problems. We define a *well-posed* problem here as one for which a unique solution exists. (Wellposedness is a concept which is usually associated with partial differential equations, but it is also appropriate here.)

Initial-Value Problems

To solve an initial-value problem one must choose the c_j in (1.3.2) so that the initial conditions in (1.2.1) are satisfied. The c_j are determined by a set of n simultaneous algebraic equations $\sum_{j=1}^{n} c_j y_j^{(i)}(x_0) = a_i$ $(i = 0, 1, \ldots, n-1)$. But, according to Cramer's rule, these equations have a unique solution only if

$$\det\,[y_j^{(i)}(x_0)] = W(x_0) \neq 0. \tag{1.3.6}$$

Thus, the Wronskian appears naturally in the study of initial-value problems. It actually has two related but distinct diagnostic applications. First, it may be used *globally* to determine whether a solution of the form in (1.3.2) is in fact the general solution of (1.3.1) by testing whether $\{y_j\}$ is a linearly independent set. In fact, the exponential form of the Wronskian implies that the general solution in one region remains the general solution in any region which can be reached without passing through singularities of the coefficient functions. [The exponential in (1.3.5) can never vanish except possibly at a singularity of $p_{n-1}(x)$.]

Second, and less importantly, the Wronskian may be used *locally* to spot an ill-posed initial-value problem without actually solving the differential equation by simply evaluating (1.3.5) and referring to (1.3.6). (A problem is ill posed if it has no solution or if the solution is not unique.) A homogeneous initial-value problem is ill posed if the initial conditions are given at a point x_0 for which the Wronskian, as calculated by Abel's formula, vanishes; either there is no solution at all or else there are infinitely many solutions.

Example 3 *Ill-posed problem—vanishing Wronskian.* The Wronskian for the differential equation in Example 2 vanishes only at $x = 0$. Thus, it is not surprising that the initial-value problem

$y'' - y'(1 + x)/x + y/x = 0$ $[y(1) = 1, y'(1) = 2]$ is well posed. When these initial conditions are replaced with $y(0) = 1$, $y'(0) = 2$, the problem is ill posed because there is no solution. When these initial conditions are replaced with $y(0) = 1$, $y'(0) = 1$, the problem is still ill posed because the solution $y(x) = e^x + c(e^x - 1 - x)$ is not unique.

Unfortunately, a nonvanishing Wronskian does not imply that the problem is well posed.

Example 4 *Ill-posed problem—nonvanishing Wronskian.* The initial-value problem

$$y'' - 6y/x^2 = 0$$

$[y(0) = 6,\ y'(0) = 6]$ is ill posed because the general solution $y(x) = c_1x^3 + c_2x^{-2}$ is either infinite or vanishing at $x = 0$. Abel's formula (1.3.5) tells us that the Wronskian is constant everywhere, including $x = 0$, and that this constant does not vanish because $y(x)$ is the general solution. Thus, unfortunately, the nonvanishing of $W(x)$ at $x = x_0$ in (1.2.1) is a necessary but not a sufficient condition for the wellposedness of an initial-value problem. Of course, this initial-value problem is clearly suspect because the initial conditions are given at a discontinuity of $p_0(x)$.

When the Wronskian is infinite at x_0, the initial-value problem may or may not be well posed.

Example 5 *Singular Wronskian.* The general solution of $(x^2 + x)y'' + (2 - x^2)y' - (2 + x)y = 0$ is $y = c_1/x + c_2 e^x$ and the Wronskian is $W(x) = W[c_1/x, c_2 e^x] = c_1 c_2 e^x(x + 1)/x^2$. The initial conditions $y(0) = 0$, $y'(0) = 1$ give an ill-posed problem because no solution exists, but the initial conditions $y(0) = 1$, $y'(0) = 1$ give a well-posed problem whose unique solution is $y(x) = e^x$.

Thus, although the Wronskian is a theoretically interesting object, it is sometimes impractical as a tool for local analysis of initial-value problems. The best advice is to beware of initial conditions that are given at a discontinuity of the coefficient functions; one should never worry about initial conditions given at a point where the coefficient functions in (1.3.1) are continuous.

Boundary-Value Problems

The solution to a boundary-value problem is determined by conditions given at two or more distinct points. In contrast with initial-value problems, continuity of the coefficient functions is not sufficient to guarantee a solution to a boundary-value problem:

Example 6 *Boundary-value problem with no solution.* Consider the boundary-value problem $y'' + y = 0$ $[y'(0) = 0,\ y(\pi/2) = 1]$. The general solution to the differential equation is $y(x) = c_1 \sin x + c_2 \cos x$. The Wronskian $W(\sin x, \cos x) = -1$ is nonvanishing and the coefficient functions of the differential equation are continuous. Nevertheless, the boundary-value problem has no solution; the condition $y'(0) = 0$ implies that $c_1 = 0$ and the remaining solution $y(x) = c_2 \cos x$ cannot satisfy $y(\pi/2) = 1$ for any value c_2 because $\cos(\pi/2) = 0$.

The previous example may suggest that boundary-value problems rarely have solutions. However, exactly the opposite is true; most boundary-value problems have a unique solution. Suppose we consider the general second-order linear differential equation $y'' + p_1(x)y' + p_0(x)y = 0$ and impose the boundary conditions $y'(x_0) = a$, $y(x_1) = b$. If $y_1(x)$ and $y_2(x)$ are two linearly independent solutions of the differential equation, then the general solution can be written as $y(x) = Ay_1(x) + By_2(x)$. The boundary conditions require that $Ay_1'(x_0) + By_2'(x_0) = a$, $Ay_1(x_1) + By_2(x_1) = b$. These equations can be solved for unique A and B in terms of a and b, provided that

$$\det \begin{vmatrix} y_1'(x_0) & y_2'(x_0) \\ y_1(x_1) & y_2(x_1) \end{vmatrix} \neq 0.$$

This condition, which is a generalization of the Wronskian condition in (1.3.6), is usually satisfied, guaranteeing a unique solution to the boundary-value problem. However, if the points x_0 and x_1 are chosen so that $y_1'(x_0)y_2(x_1) - y_2'(x_0)y_1(x_1) = 0$, then the boundary-value problem is ill posed; either there is no solution or there are an infinite number of solutions. The latter situation may occur if $by_1'(x_0) = ay_1(x_1)$ and $y_1'(x_0) \neq 0$, $y_1(x_1) \neq 0$. When this happens the boundary-value problem has infinitely many solutions which differ from each other by arbitrary constant multiples of $z(x) = y_1(x)y_2'(x_0) - y_2(x)y_1'(x_0)$. Because $z(x)$ satisfies the boundary conditions $z'(x_0) = z(x_1) = 0$, $y(x) + \alpha z(x)$ (α constant) satisfies the same boundary conditions as $y(x)$.

(E) 1.4 SOLUTIONS OF HOMOGENEOUS LINEAR EQUATIONS

Here we turn to the more practical aspects of the equation $Ly = 0$; namely, how to solve it. A first-order equation of this type is easy to solve because it is separable (see Example 1 of Sec. 1.1). However, when the order n is ≥ 2, exact closed-form solutions exist only rarely. We briefly classify and discuss some of the equations which are soluble.

Constant-Coefficient Equations

Constant-coefficient equations are characterized by having $p_0, p_1, \ldots, p_{n-1}$ independent of x. One seeks solutions of the form $y(x) = e^{rx}$. Substituting this trial function into the differential equation gives $L[e^{rx}] = e^{rx}P(r)$, where

$$P(r) = r^n + \sum_{j=0}^{n-1} p_j r^j$$

is an nth-degree polynomial. The solutions to $Ly = 0$ corresponding to *distinct* roots $r_1, r_2, \ldots$ of $P(r) = 0$ are

$$y = e^{r_1 x}, e^{r_2 x}, \ldots. \tag{1.4.1}$$

However, if there are repeated roots, (1.4.1) is not a complete set of solutions. To construct the remaining solutions, assume that r_1 is an m-fold repeated root. Then

$$L[e^{rx}] = e^{rx}(r - r_1)^m Q(r), \tag{1.4.2}$$

where Q is a polynomial of degree $n - m$. The function in square brackets is a solution of the differential equation if it makes the right side of (1.4.2) vanish. Indeed, letting $r = r_1$ shows that $e^{r_1 x}$ is a solution. To generate more solutions, we simply take derivatives with respect to r and set $r = r_1$. This process yields the m solutions

$$y = e^{r_1 x},\ xe^{r_1 x},\ x^2 e^{r_1 x},\ \ldots,\ x^{m-1} e^{r_1 x}. \tag{1.4.3}$$

A linear combination of all the solutions in (1.4.1) and (1.4.3) constitutes a general solution to the differential equation.

Example 1 *Constant-coefficient equations.*

(*a*) Substituting $y = e^{rx}$ into the equation $y'' - 5y' + 4y = 0$ gives the quadratic equation $(r - 1)(r - 4) = 0$ that must be satisfied by r. The general solution is therefore

$$y = c_1 e^x + c_2 e^{4x}.$$

(*b*) Substituting $y = e^{rx}$ into the equation $y''' - 3y'' + 3y' - y = 0$ gives a cubic polynomial equation for r with a triple root at $r = 1$. The general solution is therefore

$$y(x) = c_1 e^x + c_2 xe^x + c_3 x^2 e^x.$$

Equidimensional Equations

Equidimensional (or Euler) equations are so named because they are invariant under the scale change $x \to ax$. The coefficients have the form $p_j(x) = q_j/x^{n-j}$, where q_j is independent of x. Equidimensional differential equations may be solved by transforming them into constant-coefficient equations in t by the change of variables

$$x = e^t, \qquad x\frac{d}{dx} = \frac{d}{dt}. \tag{1.4.4}$$

Alternatively, equidimensional equations may be solved by directly substituting the trial function $y = x^r$ into the differential equation. This substitution gives $L[x^r] = P(r)x^{r-n}$ where $P(r)$ is a polynomial of degree n. Thus, the solutions have the form

$$y = x^{r_1},\ x^{r_2},\ \ldots, \tag{1.4.5}$$

when $r_1, r_2, \ldots$ are distinct roots of $P(r)$. When $P(r)$ has a repeated root r_1, a complete set of solutions is derived by differentiating the relation $L[x^r] = P(r)x^{r-n}$ with respect to r and then setting $r = r_1$. A complete set of solutions has the form

$$y = x^{r_1},\ x^{r_1}\ln x,\ x^{r_1}(\ln x)^2,\ \ldots, \tag{1.4.6}$$

when r_1 is a repeated root of $P(r)$. The roots of $P(r)$ are called *indicial exponents*.

Example 2 *Equidimensional equation.* If we substitute $y = x^r$ into the equidimensional equation $y'' + y/4x^2 = 0$, we obtain the polynomial equation $(r - \frac{1}{2})^2 = 0$. The general solution is therefore $y(x) = c_1\sqrt{x} + c_2\sqrt{x}\ln x$.

Exact Equations

An exact equation is a derivative of an equation of lower order: $Ly = (d/dx) \times (My) = 0$. This equation is simplified by integrating with respect to x: $My = c_1$. (The resulting equation is no longer homogeneous.)

Example 3 *Exact equation.* The equation $y'' + xy' + y = 0$ may be rewritten as $(d/dx) \times (y' + xy) = 0$. Thus, $y' + xy = c_1$, which is easily solved (see Sec. 1.5):

$$y = \left(c_1 \int_0^x e^{t^2/2}\, dt + c_2\right) e^{-x^2/2}.$$

An *integrating factor* is a function of x and y which, when multiplying a differential equation, makes it exact.

Example 4 *Integrating factor.* The differential equation $y'' + y'(1 + x)/x + y(x - 1)/x^2 = 0$ is not exact, but it becomes exact when multiplied by the integrating factor e^x:

$$e^x y'' + \frac{1 + x}{x} e^x y' + \frac{x - 1}{x^2} e^x y = \frac{d}{dx}\left(e^x y' + \frac{e^x}{x} y\right).$$

Thus, $y' + y/x = c_1 e^{-x}$, which is easily solved (see Sec. 1.5):

$$y(x) = -c_1(1 + x)e^{-x}/x + c_2/x.$$

Reduction of Order

Reduction of order is a technique for simplifying a linear differential equation to one of lower order by factoring off a solution that one has been lucky enough to find. Let $y_1(x)$ be a solution of $Ly = 0$. One then seeks further linearly independent solutions of the form

$$y(x) = u(x)y_1(x). \tag{1.4.7}$$

Clearly, substituting this expression for $y(x)$ into $Ly = 0$ gives a new equation for $u(x)$ of the form $Mu = 0$. The beauty of this substitution is that Mu has no term of the form $p_0(x)u$. Thus, $Mu = 0$ is a linear homogeneous equation of order $(n - 1)$ for $v(x) = u'(x)$.

Example 5 *Reduction of order.* We observe that the sum of the coefficients of the differential equation $y'' - y'(1 + x)/x + y/x = 0$ is 0. It follows that one solution is $y_1(x) = e^x$. Substituting $y(x) = u(x)e^x$ gives $u'' + u'(x - 1)/x = 0$, which is a first-order equation for $u'(x)$. The general solution for $y(x)$ is given in Example 2 of Sec. 1.1.

Transformation to a Known Equation

If the other techniques fail, it is sometimes possible to transform the differential equation into one of the classical equations of mathematical physics. Some well-analyzed equations that appear frequently in this text are the Airy equation

$$y'' = xy, \tag{1.4.8}$$

the parabolic cylinder (Weber-Hermite) equation

$$y'' + (\nu + \tfrac{1}{2} - \tfrac{1}{4}x^2)y = 0, \tag{1.4.9}$$

and the Bessel equation

$$y'' + \frac{1}{x}y' + \left(1 - \frac{\nu^2}{x^2}\right)y = 0. \tag{1.4.10}$$

Some properties of the solutions to these and other classical differential equations are given in the Appendix.

(E) 1.5 INHOMOGENEOUS LINEAR EQUATIONS

Inhomogeneous linear differential equations are only slightly more complicated than homogeneous ones. This is because the difference of any two solutions of $Ly = f(x)$ is a solution of $Ly = 0$. As a result, the general solution of $Ly = f(x)$ is the sum of *any* particular solution of $Ly = f(x)$ and the general solution of $Ly = 0$.

Example 1 *General solution to an inhomogeneous equation.* Suppose $y = x$, $y = x^2$, and $y = x^3$ satisfy the second-order equation $Ly = f(x)$. Can you find the general solution without knowing the explicit form of L and f? The differences $x - x^2$ and $x^2 - x^3$ are both solutions of $Ly = 0$. These functions are linearly independent, so the general solution of $Ly = 0$ is $y(x) = c_1(x - x^2) + c_2(x^2 - x^3)$. Hence, the general solution of $Ly = f(x)$, which must contain two arbitrary constants of integration, is $y(x) = c_1(x - x^2) + c_2(x^2 - x^3) + x$.

All first-order linear inhomogeneous equations are soluble because it is always possible to find an integrating factor which is a function of x only. The integrating factor $I(x)$ for

$$y'(x) + p_0(x)y(x) = f(x) \tag{1.5.1}$$

is $I(x) = \exp\left[\int^x p_0(t)\,dt\right]$. Multiplying by $I(x)$ gives $I(x)y'(x) + p_0(x)y(x)I(x) = (d/dx)[I(x)y(x)] = f(x)I(x)$. So the solution of (1.5.1) is

$$y(x) = \frac{c_1}{I(x)} + \frac{1}{I(x)}\int^x f(t)I(t)\,dt, \tag{1.5.2}$$

Example 2 *First-order inhomogeneous equation.* The equation $y'(x) = y/(x + y)$ is not linear in y, but *is* linear in x! To demonstrate this, we simply exchange the dependent variable y with the independent variable x:

$$\frac{d}{dy}x(y) = \frac{x(y) + y}{y}.$$

An integrating factor for this equation is $I(y) = 1/y$. Multiplying by $I(y)$ gives $(d/dy) \times (x/y) = 1/y$ or $x(y) = y \ln y + c_1 y$.

The technique of exchanging the dependent and independent variables is essential for the solution of Prob. 1.22. A generalization of this method to partial differential equations is called the *hodograph transformation.*

There are several standard techniques for solving higher-order inhomogeneous linear equations.

Variation of Parameters

The only new complication in solving an inhomogeneous equation if the associated homogeneous equation is soluble is finding one particular solution. The method of *variation of parameters* is a general and infallible technique for determining a particular solution. The method could be classified as a super reduction of order.

We illustrate with a second-order equation. Let $y_1(x)$ and $y_2(x)$ be two linearly independent solutions of the homogeneous equation $Ly = 0$, where $L = d^2/dx^2 + p_1(x)\, d/dx + p_0(x)$. We seek a particular solution of $Ly = f(x)$ having the symmetric form

$$y(x) = u_1(x)y_1(x) + u_2(x)y_2(x). \tag{1.5.3}$$

Of course, u_1 and u_2 are underdetermined so we have the freedom to impose a constraint which simplifies subsequent equations. We choose this constraint to be

$$u_1'(x)y_1(x) + u_2'(x)y_2(x) = 0. \tag{1.5.4}$$

Next, we differentiate (1.5.3) twice, substitute into $Ly = f(x)$, and remember that $Ly_1 = Ly_2 = 0$. Using (1.5.4) we have

$$u_1'(x)y_1'(x) + u_2'(x)y_2'(x) = f(x). \tag{1.5.5}$$

The solution of the simultaneous equations (1.5.4) and (1.5.5) for $u_1'(x)$ and $u_2'(x)$ is

$$\begin{aligned} u_1'(x) &= -\frac{f(x)y_2(x)}{W(x)}, \\ u_2'(x) &= \frac{f(x)y_1(x)}{W(x)}, \end{aligned} \tag{1.5.6}$$

where $W(x) = W[y_1(x), y_2(x)]$ is the Wronskian. Observe that the denominators W do not vanish because $y_1(x)$ and $y_2(x)$ are assumed to be linearly independent solutions of $Ly = 0$.

Integrating (1.5.6) gives the final expression for the particular solution in (1.5.3):

$$y(x) = -y_1(x) \int^x \frac{f(t)y_2(t)}{W(t)}\, dt + y_2(x) \int^x \frac{f(t)y_1(t)}{W(t)}\, dt. \tag{1.5.7}$$

Example 3 *Variation of parameters.* To solve $y'' - 3y' + 2y = e^{4x}$ by variation of parameters, we first determine that two solutions of the associated homogeneous equation are $y_1 = e^x$ and $y_2 = e^{2x}$. Next we compute the Wronskian: $W(e^x, e^{2x}) = e^{3x}$. Substituting into (1.5.7) gives

$$\begin{aligned} y(x) &= -e^x \int^x dt\, e^{4t}e^{2t}e^{-3t} + e^{2x} \int^x dt\, e^{4t}e^{t}e^{-3t} \\ &= c_1 e^x + c_2 e^{2x} + \tfrac{1}{6}e^{4x}, \end{aligned}$$

which is the general solution to the inhomogeneous differential equation.

Variation of parameters for nth-order equations is discussed in Prob. 1.15.

Green's Functions

There is another general method for constructing the solution to an inhomogeneous linear differential equation which is equivalent to variation of parameters. This method represents the solution as an integral over a *Green's function.*

To define a Green's function it is necessary to introduce the Dirac delta function $\delta(x-a)$. This function may be thought of as a mathematical idealization of a unit impulse; it is an infinitely thin spike centered at $x = a$ having unit area.† The δ function has two defining properties. First,

$$\delta(x-a) = 0, \qquad x \neq a. \tag{1.5.8a}$$

Second,

$$\int_{-\infty}^{\infty} \delta(x-a)\,dx = 1. \tag{1.5.8b}$$

From these properties we have the crucial result (see Prob. 1.16) that

$$\int_{-\infty}^{\infty} \delta(x-a)f(x)\,dx = f(a) \tag{1.5.9}$$

if $f(x)$ is continuous at a.

There are many ways to represent the δ function. It may be expressed (nonuniquely) as the limit of a sequence of functions:

$$\delta(x-a) = \lim_{\varepsilon \to 0+} F_\varepsilon(x), \tag{1.5.10a}$$

where

$$F_\varepsilon(x) = \begin{cases} 0, & x < a - \frac{1}{2}\varepsilon, \\ 1/\varepsilon, & a - \frac{1}{2}\varepsilon \le x \le a + \frac{1}{2}\varepsilon, \\ 0, & a + \frac{1}{2}\varepsilon < x; \end{cases}$$

or

$$\delta(x-a) \equiv \lim_{\varepsilon \to 0+} \frac{\varepsilon}{\pi[(x-a)^2 + \varepsilon^2]}; \tag{1.5.10b}$$

or

$$\delta(x-a) \equiv \lim_{\varepsilon \to 0+} (\pi\varepsilon)^{-1/2} e^{-(x-a)^2/\varepsilon}; \tag{1.5.10c}$$

or

$$\delta(x-a) \equiv \lim_{L \to +\infty} \frac{1}{2\pi} \int_{-L}^{L} e^{i(x-a)t}\,dt. \tag{1.5.10d}$$

(The notation $\varepsilon \to 0+$ means that ε approaches 0 through positive values only.) It is easy to verify (see Prob. 1.17) that the formulations in (1.5.10) satisfy (1.5.8).

Alternatively, $\delta(x-a)$ may be viewed as the derivative of a discontinuous function. If $h(x-a)$ is the Heaviside step function defined by

$$h(x-a) \equiv \begin{cases} 0, & x < a, \\ \frac{1}{2}, & x = a, \\ 1, & x > a, \end{cases}$$

then (see Prob. 1.18)

$$\delta(x-a) \equiv \frac{d}{dx} h(x-a). \tag{1.5.11}$$

† Technically, the δ function is not really a function; it is a distribution (see References).

Notice that integration is a *smoothing* operation but that differentiation is unsmoothing. For example, the Heaviside function, which is the integral of the δ function,

$$h(x-a) = \int_{-\infty}^{x} \delta(t-a)\,dt,$$

has a finite jump discontinuity while the δ function has an infinite jump discontinuity. Similarly, the ramp function, which is the integral of the Heaviside function,

$$r(x-a) = \int_{-\infty}^{x} h(t-a)\,dt$$

$$= \begin{cases} 0, & x \le a, \\ x-a, & a \le x, \end{cases}$$

is continuous everywhere.

Next, we define the Green's function. The Green's function $G(x, a)$ associated with the inhomogeneous equation $Ly = f(x)$ satisfies the differential equation

$$LG(x, a) = \delta(x-a). \tag{1.5.12}$$

Once $G(x, a)$ is known, it is easy to represent the solution to $Ly = f(x)$ as an integral

$$y(x) = \int_{-\infty}^{\infty} da\, f(a)G(x, a). \tag{1.5.13}$$

To verify that $y(x)$ in (1.5.13) solves $Ly = f$, we differentiate under the integral:

$$\begin{aligned} Ly(x) &= L\int_{-\infty}^{\infty} da\, f(a)G(x, a) \\ &= \int_{-\infty}^{\infty} da\, f(a)LG(x, a) \\ &= \int_{-\infty}^{\infty} da\, f(a)\delta(x-a) \\ &= f(x), \end{aligned}$$

where we have used (1.5.12) and (1.5.9) in turn.

The only remaining problem is to solve (1.5.12) for $G(x, a)$. But this is easy once the solutions to the associated homogeneous equation $Ly = 0$ are known. To illustrate we solve the second-order equation

$$LG(x, a) = \left[\frac{d^2}{dx^2} + p_1(x)\frac{d}{dx} + p_0(x)\right]G(x, a) = \delta(x-a). \tag{1.5.14}$$

We denote two linearly independent solutions to $Ly = 0$ by $y_1(x)$ and $y_2(x)$. Then, when $x \ne a$, the right side of (1.5.14) vanishes and we have

$$G(x, a) = A_1 y_1(x) + A_2 y_2(x), \qquad x < a,$$

$$G(x, a) = B_1 y_1(x) + B_2 y_2(x), \qquad x > a.$$

In order to relate the solution for $G(x, a)$ for $x < a$ to the solution for $x > a$, we argue that $G(x, a)$ is continuous at $x = a$ and that $\partial G/\partial x$ has a finite jump discontinuity of magnitude 1 at $x = a$. To show this, we observe that the most singular term on the left side of the Green's function equation (1.5.14) must be $\partial^2 G/\partial x^2$ because differentiation is an unsmoothing operation; if G or $\partial G/\partial x$ had an infinite jump discontinuity at $x = a$ like that of a δ function, then $\partial^2 G/\partial x^2$ would be even more singular than a δ function and (1.5.14) could not be satisfied. Thus, (1.5.14) implies that $\partial^2 G/\partial x^2 - \delta(x - a)$ must be less singular than a δ function at $x = a$. Therefore, integrating $\partial^2 G/\partial x^2 - \delta(x - a)$ from $-\infty$ to x gives a function which is continuous even at $x = a$: $\partial G/\partial x - h(x - a)$ is continuous everywhere. Hence the discontinuity in $\partial G/\partial x$ at $x = a$ is the same as that of the Heaviside function $h(x - a)$:

$$\lim_{\varepsilon \to 0+} \left[\left. \frac{\partial G}{\partial x} \right|_{x=a+\varepsilon} - \left. \frac{\partial G}{\partial x} \right|_{x=a-\varepsilon} \right] = 1. \tag{1.5.15}$$

Finally, since $\partial G/\partial x$ has only a finite jump discontinuity, its indefinite integral $G(x, a)$ must be continuous at $x = a$.

Continuity of $G(x, a)$ at $x = a$ gives the condition

$$A_1 y_1(a) + A_2 y_2(a) = B_1 y_1(a) + B_2 y_2(a).$$

Also, (1.5.15) requires that

$$B_1 y_1'(a) + B_2 y_2'(a) - A_1 y_1'(a) - A_2 y_2'(a) = 1.$$

Using these relations and solving for $B_1 - A_1$ and $B_2 - A_2$, we obtain

$$B_1 - A_1 = -\frac{y_2(a)}{W[y_1(a), y_2(a)]}, \tag{1.5.16}$$

$$B_2 - A_2 = \frac{y_1(a)}{W[y_1(a), y_2(a)]}. \tag{1.5.17}$$

Observe the strong parallel between these equations and (1.5.6).

We have now completed the solution of the Green's function equation (1.5.14). However, A_1 and A_2 are still arbitrary because $G(x, a)$ is only determined by (1.5.14) up to a solution of the homogeneous equation. Choosing $A_1 = A_2 = 0$ and using (1.5.16) and (1.5.17) to determine B_1 and B_2, we obtain

$$G(x, a) = \begin{cases} \dfrac{-y_2(a)y_1(x) + y_1(a)y_2(x)}{W[y_1(a), y_2(a)]}, & x \ge a, \\ 0, & x < a. \end{cases} \tag{1.5.18}$$

Substituting this formula for $G(x, a)$ into (1.5.13) reproduces exactly the variation of parameters result in (1.5.7).

The Green's function approach has a distinct advantage over the method of variation of parameters when it is necessary to solve a differential equation $Ly = f$ where L and the boundary conditions are fixed but f ranges over a wide variety of functions. (Why?) The analysis is particularly simple when the boundary conditions are homogeneous.

Example 4 *Solution of a boundary-value problem by Green's functions.* The Green's function for the boundary-value problem $y'' = f(x)[y(0) = 0, y'(1) = 0]$ is defined by the equations

$$(\partial^2 G/\partial x^2)(x, a) = \delta(x - a), \qquad G(0, a) = 0, \qquad (\partial G/\partial x)(1, a) = 0.$$

Notice that we have chosen G to satisfy the same *homogeneous* boundary conditions as y. The solution for $G(x, a)$ is

$$G(x, a) = \begin{cases} -x, & x < a, \\ -a, & x \geq a, \end{cases}$$

when $0 < a < 1$. For any $f(x)$, $y(x)$ can then be represented as $y(x) = \int_0^1 G(x, a) f(a)\, da$ $(0 \leq x \leq 1)$. Note that we do not integrate from $-\infty$ to $+\infty$. Why?

Example 5 *Solution of a boundary-value problem by Green's functions.* The Green's function for the boundary-value problem $y'' - y = f(x)[y(\pm\infty) = 0]$ is defined by the equations $\partial^2 G/\partial x^2 - G(x, a) = \delta(x - a)$, $G(\pm\infty, a) = 0$. The solution for $G(x, a)$ is $G(x, a) = -\frac{1}{2}e^{-|x-a|}$. Thus, for any $f(x)$, $y(x) = -\frac{1}{2}\int_{-\infty}^{\infty} e^{-|x-a|} f(a)\, da$.

Reduction of Order

For the sake of completeness, it is important to state that *reduction of order* reduces the order of inhomogeneous as well as homogeneous equations. Thus, sincc all first-order linear equations are soluble, reduction of order is especially useful for second-order linear equations.

Example 6 *Reduction of order for an inhomogeneous equation.* One solution of the homogeneous equation $a(x)y'' + xy' - y = 0$ is $y_1(x) = x$. Therefore, to solve the inhomogeneous equation $a(x)y'' + xy' - y = f(x)$ by reduction of order, we seek a solution of the form $y(x) = y_1(x)u(x) = xu(x)$. Substituting gives a first-order equation for $u'(x)$ which is easy to solve: $xa(x)u'' + [2a(x) + x^2]u' = f(x)$.

Method of Undetermined Coefficients

There is another technique for determining a particular solution to $Ly = f(x)$ called the method of *undetermined coefficients,* which we discuss briefly. This method is really little more than organized guesswork, but when it works it is faster than variation of parameters. Its application is usually limited to constant-coefficient equations where $f(x)$ is an additive or multiplicative combination of e^x, $\sin x$, $\cos x$, and polynomials in x, or equidimensional equations where $f(x)$ is a polynomial in x.

Example 7 *Method of undetermined coefficients.*

(*a*) To solve $y''' + y = e^x \sin x$ we guess a particular solution of the form $y = ae^x \sin x + be^x \cos x$ and determine the "undetermined coefficients" a and b by substituting into the differential equation. The results are $a = -\frac{1}{5}$ and $b = -\frac{2}{5}$.

(*b*) To solve $y'' - y = e^x$ we guess a particular solution of the form $y = axe^x$ because e^x already solves the homogeneous equation. The result is $a = \frac{1}{2}$.
(*c*) To solve $y'' - y/x^2 = x^4 + x^3$ we guess a particular polynomial solution of the form $y = ax^6 + bx^5$. We find that $a = \frac{1}{29}$ and $b = \frac{1}{19}$.
(*d*) To solve $y'' + xy' + 2y = 1$ or x^4 we guess particular polynomial solutions of the form $y = a$ or $y = ax^4 + bx^2 + c$, respectively. The results are $y = \frac{1}{2}$ or $y = x^4/6 - x^2/2 + \frac{1}{2}$.

(E) 1.6 FIRST-ORDER NONLINEAR DIFFERENTIAL EQUATIONS

Although most nonlinear differential equations are too difficult to solve in closed form, it is important to be able to recognize those equations which are soluble and to know the appropriate techniques for obtaining a solution. For first-order equations the usual procedure is to make a substitution which converts the equation into one that is either linear or exact.

Bernoulli Equations

Bernoulli equations have the form

$$y' = a(x)y + b(x)y^P, \tag{1.6.1}$$

where $a(x)$ and $b(x)$ are arbitrary functions of x and P is any number. This equation has two elementary cases: when $P = 0$ the equation is linear and when $P = 1$ the equation is separable. For all other values of P, dividing (1.6.1) through by y^P suggests the substitution

$$u(x) = [y(x)]^{1-P}. \tag{1.6.2}$$

The new differential equation for $u(x)$,

$$u'(x) = (1 - P)a(x)u(x) + (1 - P)b(x), \tag{1.6.3}$$

is soluble because it is linear in $u(x)$.

Example 1 *Bernoulli equation.* The differential equation $y'(x) = x/(x^2y^2 + y^5)$ is *not* a Bernoulli equation in y. However, exchanging the dependent and independent variables gives $(d/dy)x(y) = xy^2 + y^5/x$ which *is* a Bernoulli equation in x $(P = -1)$. The solution is $x(y) = \pm(c_1e^{2y^3/3} - \frac{1}{2}y^3 - \frac{3}{4})^{1/2}$.

Riccati Equations

Riccati equations are quadratic in $y(x)$:

$$y'(x) = a(x)y^2(x) + b(x)y(x) + c(x). \tag{1.6.4}$$

There are two elementary cases: when $a(x) = 0$ the equation is linear and when $c(x) \equiv 0$ the equation is a Bernoulli equation. Unfortunately, apart from these special cases, there is no general technique for obtaining a solution. This is not at all surprising because the substitution

$$y(x) = -\frac{w'(x)}{a(x)w(x)} \tag{1.6.5}$$

converts the Riccati equation into a second-order linear equation for $w(x)$:

$$w''(x) - \left[\frac{a'(x)}{a(x)} + b(x)\right] w'(x) + a(x)c(x)w(x) = 0. \tag{1.6.6}$$

This transformation also goes in reverse. There is a Riccati equation for every second-order homogeneous linear equation. Thus, a general closed-form solution for all Riccati equations (say, like the one for Bernoulli equations) would be equivalent to a quadrature solution for all linear second-order equations, which has never been discovered!

Nevertheless, many Riccati equations *can* be solved. For these equations the procedure is to guess just *one* solution $y = y_1(x)$, no matter how trivial, and then to use this solution to reduce the Riccati equation to a Bernoulli equation by an *additive* kind of reduction of order. Specifically, one seeks a general solution of the form

$$y(x) = y_1(x) + u(x). \tag{1.6.7}$$

The resulting Bernoulli equation for $u(x)$ is

$$u'(x) = [b(x) + 2a(x)y_1(x)]u(x) + a(x)u^2(x).$$

This equation is soluble. The transformation in (1.6.5) which connects Riccati and second-order linear equations motivates this beautiful substitution, which replaces one Riccati equation with another that is lacking a $c(x)$ term. Reduction of order for linear equations requires the multiplicative substitution $y = y_1(x)u(x)$ [see (1.4.7)]. The transformation in (1.6.5) converts (1.4.7) to an *additive* substitution of the form in (1.6.7).

Example 2 *Riccati equation.* It is not hard to see that a solution of $y'(x) = y^2 - xy + 1$ is $y_1(x) = x$. This is not the general solution which must contain an arbitrary integration constant and which is much too difficult to guess; it is merely *one* solution.

Now let $y = x + u(x)$; the equation for $u(x)$ is $u' = u^2 + xu$. The solution of this Bernoulli equation is

$$u(x) = \frac{e^{x^2/2}}{c_1 - \int_0^x e^{t^2/2}\,dt}.$$

So the general solution of the Riccati equation is

$$y(x) = x + \frac{e^{x^2/2}}{c_1 - \int_0^x e^{t^2/2}\,dt}.$$

Example 3 *Difficult Riccati equation.* It requires some fiddling to discover that a solution of the Riccati equation (1.1.7) $y'(x) = A^2/x^4 - y^2$ is $y_1(x) = 1/x - A/x^2$. However, once this $y_1(x)$ has been found it is routine to solve the resulting Bernoulli equation for $u(x) = y(x) - y_1(x)$:

$$u' + 2u/x - 2uA/x^2 = -u^2.$$

The solution to this equation gives $y(x)$ in (1.1.8).

Example 4 *Factoring.* Riccati equations arise from the attempt to solve a second-order linear equation by factoring. Factoring a linear equation means rewriting the nth-order linear operator L as a product of first-order linear operators:

$$Ly = \left[\frac{d^n}{dx^n} + p_{n-1}(x)\frac{d^{n-1}}{dx^{n-1}} + \cdots + p_1(x)\frac{d}{dx} + p_0(x)\right] y(x)$$

$$= \left[\frac{d}{dx} + a_1(x)\right]\left[\frac{d}{dx} + a_2(x)\right][\cdots]\left[\frac{d}{dx} + a_n(x)\right] y(x). \tag{1.6.8}$$

Observe that for nonconstant $a_1(x) \neq a_2(x)$ the factors may not commute. For example,

$$y'' + \left(x + \frac{1}{x}\right)y' + 2y = \left(\frac{d}{dx} + \frac{1}{x}\right)\left(\frac{d}{dx} + x\right)y \neq \left(\frac{d}{dx} + x\right)\left(\frac{d}{dx} + \frac{1}{x}\right)y.$$

Also, the factorization may not be unique (!)

$$y'' - y = \left(\frac{d}{dx} + 1\right)\left(\frac{d}{dx} - 1\right)y = \left(\frac{d}{dx} + \tanh x\right)\left(\frac{d}{dx} - \tanh x\right)y.$$

If an nth-order linear equation $Ly = f$ is factored, it is as good as solved because it is merely a sequence of first-order linear equations. To illustrate, we consider (1.6.8) and define

$$w_1(x) \equiv \left[\frac{d}{dx} + a_2(x)\right]\left[\frac{d}{dx} + a_3(x)\right]\cdots\left[\frac{d}{dx} + a_n(x)\right] y(x),$$

$$w_2(x) \equiv \left[\frac{d}{dx} + a_3(x)\right]\cdots\left[\frac{d}{dx} + a_n(x)\right] y(x),$$

$$w_3(x) \equiv \left[\frac{d}{dx} + a_4(x)\right]\cdots\left[\frac{d}{dx} + a_n(x)\right] y(x),$$

and so on. Then $w_1(x)$ satisfies the first-order linear equation $(d/dx)w_1 + a_1 w_1 = f(x)$. The solution to this equation has one integration constant. $w_2(x)$ solves the first-order linear equation $w_2'(x) + a_2(x)w_2(x) = w_1(x)$ and contains two integration constants. Continuing this process $(n-1)$ times gives an easy-to-solve equation for $y(x)$:

$$y'(x) + a_n(x)y(x) = w_{n-1}(x). \tag{1.6.9}$$

The solution of (1.6.9) for $y(x)$ contains all n integration constants and is therefore the general solution of $Ly = f$.

The catch, of course, is that it is often very difficult to factor a linear operator. To illustrate, we try to factor the second-order operator $L = p_0(x) + p_1(x)d/dx + d^2/dx^2$ by force. We write

$$Ly = \left[\frac{d}{dx} + a_1(x)\right]\left[\frac{d}{dx} + a_2(x)\right] y.$$

Multiplying out the factors gives $Ly = y'' + (a_1 + a_2)y' + (a_2' + a_1 a_2)y$, and allows us to identify $a_1 + a_2 = p_1$ and $a_2' + a_1 a_2 = p_0$. Eliminating a_1 from these equations gives a Riccati equation for a_2:

$$a_2' = a_2^2 - p_1 a_2 + p_0;$$

this is just as difficult to solve as the original second-order linear equation (although in rare cases it is easier to spot a solution to a Riccati equation). Indeed, the transformation (1.6.5) for linearizing a Riccati equation, $a_2(x) = -y'(x)/y(x)$, gives $Ly = y'' + p_1 y' + p_0 y = 0$, which is exactly the original second-order linear operator we tried to factor by force!

Exact Equations

First-order *exact* equations can be written in the form

$$M[x, y(x)] + N[x, y(x)]y'(x) = \frac{d}{dx} f[x, y(x)] = 0$$

and the solution of this equation is $f[x, y(x)] = c_1$. A necessary and sufficient condition for exactness is that (see Prob. 1.24)

$$\frac{\partial}{\partial y} M(x, y) = \frac{\partial}{\partial x} N(x, y). \tag{1.6.10}$$

Example 5 *Exact equation.* To check that the equation $y'(x) = (x^2 - y)/(y^2 + x)$ is exact, we identify $M = y - x^2$ and $N = y^2 + x$ and observe that $\partial M/\partial y = \partial N/\partial x = 1$. The solution of the exact equation is $y^3 + 3xy - x^3 = c_1$.

Example 6 *Separable equations.* Separable equations are exact because they have the form $M(x) + N(y)y'(x) = 0$. Thus, $\partial M/\partial y = \partial N/\partial x = 0$.

Example 7 *Integrating factor.* The equation $(1 + xy + y^2) + (1 + xy + x^2)y'(x) = 0$ is not exact because $\partial M/\partial y \neq \partial N/\partial x$. However, it becomes exact upon multiplying through by the integrating factor $I = e^{xy}$. Once this integrating factor has been guessed, it is easy to rewrite the equation as $(d/dx)[(x + y)e^{xy}] = 0$ and to obtain the solution $(x + y)e^{xy} = c_1$.

The existence theorem for solutions tells us that *every* nonlinear equation can be made exact. However, the previous example suggests that it may be quite difficult to find the integrating factor. A brute force approach is useless. For example, let us try to solve the equation $a(x, y) + b(x, y)y'(x) = 0$ by multiplying by an unspecified integrating factor $I(x, y)$ and demanding that the resulting equation be exact. The condition for exactness (1.6.10) gives a linear partial differential equation for $I(x, y)$:

$$a(x, y)\left(\frac{\partial}{\partial y}\right) I(x, y) - b(x, y)\left(\frac{\partial}{\partial x}\right) I(x, y) = \left(\frac{\partial b}{\partial x} - \frac{\partial a}{\partial y}\right) I(x, y).$$

The usual way to solve such an equation is by the method of characteristics, which is a standard technique of partial differential equations. It is not surprising that the characteristics are given by $dy/dx = -a/b$ which is precisely the differential equation we originally set out to solve! A more delicate approach which makes explicit use of the equation to be solved is required.

Substitution

Sometimes, it is possible to find a *substitution* which converts a nonlinear equation to one that is directly solvable. Linear substitutions should be considered first because they are easiest to spot.

Example 8 *Substitutions.*

(*a*) The substitution $u = x + y$ makes $y'(x) = \cos(x + y)$ separable: $u' = 1 + \cos u$.

(*b*) The linear transformation $x = av + bw + c$, $y = dv + ew + f$, with a suitable choice of a, b, c, d, e, and f, converts $y'(x) = (Ax + By + C)/(Dx + Ey + F)$ into a separable equation for $w(v)$.

(*c*) It is not immediately obvious that the solutions of $y' = (y - x)/(y + x)$ are logarithmic spirals until the substitution

$$x = r\cos\theta, \qquad dx = \cos\theta\, dr - r\sin\theta\, d\theta,$$

$$y = r\sin\theta, \qquad dy = \sin\theta\, dr + r\cos\theta\, d\theta,$$

is made. The new equation for $r(\theta)$ is separable, $r'(\theta) = -r$, and its solution is $r = c_1 e^{-\theta}$.

(*d*) Substituting $u = x + y$ changes the equation $y' = y/x + 1/(y + x)$ into one that is a Bernoulli equation: $u' = u/x + 1/u$. What is the solution for $y(x)$? [See Prob. 1.31(*a*).]

(*e*) When an equation has the form $y'(x) = F(y/x)$, the substitution $u = y/x$ gives a separable equation for u: $u'(x) = [F(u) - u]/x$. This substitution applies directly to the differential equation in part (*c*) of this example.

(*f*) Although the substitution $u = xy$ looks natural and promising, it is ineffective against the formidable differential equation $y' = \cos(xy)$. The new equation in terms of u is $u' = x\cos u + u/x$, which is no easier to solve than the original equation in x and y. As a rule, multiplicative substitutions like $u = xy$, $u = x^2y$, or $u = xy^2$ are ineffective for solving equations of the form $y' = F(xy)$, $F(x^2y)$, or $F(xy^2)$.

(I) 1.7 HIGHER-ORDER NONLINEAR DIFFERENTIAL EQUATIONS

In this section we give a brief summary of the techniques for solving nth-order nonlinear differential equations. Most of these techniques try to reduce the order of the differential equation; the lower the order of the differential equation, the higher the probability of ultimately finding a closed-form solution.

Autonomous Equations

An *autonomous* equation is one whose independent variable does not appear explicitly. The following are autonomous equations: $y'' + y' + y = 0$, $y''' + yy' = 0$, $yy' = y''y'''$. Autonomous equations are invariant under the translation $x \to x + a$.

An nth-order autonomous equation can always be replaced by a nonautonomous equation of order $(n - 1)$. The standard trick is to express $u \equiv y'(x)$ as a function of y and to find an equation satisfied by u and its derivatives. We thus let

$$y'(x) = u(y), \tag{1.7.1a}$$

$$y''(x) = \frac{du}{dx} = \frac{du}{dy}\frac{dy}{dx} = u'(y)u(y), \tag{1.7.1b}$$

$$y'''(x) = \frac{d}{dx}[u'(y)u(y)] = u(y)[u'(y)]^2 + [u(y)]^2u''(y), \tag{1.7.1c}$$

and so on. The independent variable in the new differential equation is y. Note that the highest derivative of u with respect to y in the new equation is always one less than the highest derivative of y with respect to x in the original equation.

Example 1 *Reduction of order of an autonomous equation.*

(*a*) The substitution $y'(x) = u(y)$ simplifies the equation $y''' + yy' = 0$ to the second-order equation $u'^2 + uu'' + y = 0$, after dividing through by u.
(*b*) The equation $yy' = y''y'''$ may be replaced by the second-order equation $y = uu'^3 + u^2u'u''$.

Equidimensional-in-x Equations

An equation is said to be *equidimensional in* x if the scale change $x \to ax$ leaves the equation unchanged. The following equations are equidimensional in x: $y'' + 17y'/x + 101y/x^2 = 0$, $y'' = yy'/x$, $y'' = y'''y'x^2$.

All equations which are equidimensional in x can be transformed into autonomous equations of the same order. The necessary change of variable was already given in (1.4.4):

$$x = e^t, \tag{1.7.2a}$$

$$x\frac{d}{dx} = \frac{d}{dt}, \tag{1.7.2b}$$

$$x^2\frac{d^2}{dx^2} = \frac{d^2}{dt^2} - \frac{d}{dt}, \tag{1.7.2c}$$

and so on.

Example 2 *Conversion of an equidimensional-in-x equation to an autonomous equation.* The equidimensional equation (1.1.9) $y'' = yy'/x$ may be rewritten as $x^2y''(x) = yxy'(x)$. The change of variables in (1.7.2) transforms this equation into the autonomous equation $y''(t) - y'(t) = y(t)y'(t)$.

The substitution in (1.7.1) reduces this equation to one that is first order: $uu' - u = yu$. This equation implies that either $u = 0$ so that $y(x) = c_3$, a constant, or else $u'(y) = y + 1$, whose solution is $u(y) = y^2/2 + y + c_1$.

If we now recall that $u(y) = y'(t)$ and that $x = e^t$, then we must finally solve the separable equation $y'(t) = y^2/2 + y + c_1$ for $y(x)$. The final solution is given in (1.1.10). (See also Prob. 1.2.)

Scale-Invariant Equations

A differential equation is *scale invariant* if there is a value of p for which the scale transformation

$$x \to ax, \qquad y \to a^p y \tag{1.7.3}$$

leaves the original differential equation unchanged.

Example 3 *Scale-invariant equations.*

(*a*) The Thomas-Fermi equation $y'' = y^{3/2}x^{-1/2}$ is scale invariant under the transformation $x \to ax$, $y \to a^{-3}y$.

(b) $u'^2 + uu'' + y = 0$, from part (a) of Example 1, is scale invariant under the transformation $y \to ay$, $u \to a^{3/2}u$.
(c) $x^2y'' + 3xy' + 2y = x^{-4}y^{-3}$ is scale invariant under the transformation $x \to ax$, $y \to a^{-1}y$.
(d) The first-order equation $y'(x) = F(y/x)$ considered in part (e) of Example 8 of Sec. 1.6 is scale invariant under the transformation $x \to ax$, $y \to ay$.

Recognizing that a differential equation is scale invariant is major progress toward its solution because all scale-invariant equations may be transformed into equations which are equidimensional in x by substituting

$$y(x) = x^p u(x). \tag{1.7.4}$$

Example 4 *Conversion of a scale-invariant equation to an equidimensional-in-x equation.*

(a) The Thomas–Fermi equation in part (a) of the previous example may be made equidimensional by substituting $y = x^{-3}u$. The resulting equation $x^2u'' - 6xu' + 12u = u^{3/2}$ is indeed equidimensional, so the substitution $x = e^t$ makes it autonomous: $u''(t) - 7u'(t) + 12u = u^{3/2}$. This equation is equivalent to the first-order equation $ww'(u) - 7w + 12u = u^{3/2}$, where $w(u) = u'(t)$. Unfortunately, this first-order equation is too difficult to solve in closed form; approximate rather than exact analytical methods are appropriate for understanding the Thomas–Fermi equation. A discussion of the approximate solution is given in Example 7 of Sec. 4.3.
(b) The equation in part (b) of the previous example may be converted to an equidimensional equation by the substitution $u(y) = y^{3/2}v(y)$. The resulting equation is $3v^2 + y^2v'^2 + 6yvv' + y^2vv'' + 1 = 0$. The subsequent substitution $y = e^t$ makes this equation autonomous: $3v^2(t) + [v'(t)]^2 + 6v(t)v'(t) + v(t)v''(t) - v(t)v'(t) + 1 = 0$. Finally, this equation may be reduced to first order by letting $w(v) = v'(t)$:

$$3v^2 + w^2 + 6vw + vw\frac{dw}{dv} - vw + 1 = 0.$$

Again the final equation is very complicated, but it is the *first-order* equivalent of the third-order equation $y''' + yy' = 0$. The reduction through two orders has followed the sequence of transformations autonomous (order 3) → scale invariant (order 2) → equidimensional (order 2) → autonomous (order 2) → complicated mess (order 1).
(c) Not all equations become too complicated to solve when their order is reduced. The equation in part (c) of the previous example simplifies beautifully when the transformation suggested by scale invariance is made. This transformation $y = u(x)/x$ reduces the scale-invariant equation

$$x^2y'' + 3xy' + 2y = x^{-4}y^{-3}$$

to the equidimensional equation

$$x^2u'' + xu' + u = u^{-3}.$$

The exponential substitution $x = e^t$ reduces this equation to one that is autonomous: $u'' + u = u^{-3}$. The equivalent first-order equation is easy to solve and the final closed-form solution for $y(x)$ is

$$y(x) = \pm\frac{1}{x}\sqrt{\cosh c_1 + (\sinh c_1)\sin(2\ln x + c_2)},$$

where c_1 and c_2 are constants of integration.

Equidimensional-in-*y* Equations

If an equation is invariant under the scale change $y \to ay$, then the equation is said to be *equidimensional in y*. When an equation is equidimensional in y, the transformation

$$y(x) = e^{u(x)} \tag{1.7.5}$$

always reduces the order of the equation by one.

Example 5 *Reduction of order of equidimensional-in-y equations.*

(*a*) All homogeneous linear equations are equidimensional in y. However, for these equations the substitution in (1.7.5) does not usually bring about progress. Letting $y(x) = e^{u(x)}$ in $y''(x) + p_1(x)y'(x) + p_0(x)y(x) = 0$ gives the Riccati equation

$$v' + v^2 + p_1(x)v + p_0(x) = 0,$$

where $v(x) = u'(x)$. This first-order Riccati equation is no easier to solve than was the original second-order equation.

(*b*) The equation $x^2yy'' + xy'y'' + yy' = 0$ is equidimensional in y. The transformation $y = e^u$ reduces this equation to one that is first order in $v(x) = u'(x)$:

$$v' + v^2 + 1/x - 1/(x + v) = 0.$$

This equation is still too difficult to solve in closed form, but we have certainly reduced its order by one!

(E) 1.8 EIGENVALUE PROBLEMS

An eigenvalue problem is a boundary-value problem that has *nontrivial* solutions only when a parameter E that enters the problem has special values called *eigenvalues.*

Example 1 *A simple eigenvalue problem.* Consider the boundary-value problem

$$y'' + Ey = 0, \qquad y(0) = y(1) = 0. \tag{1.8.1}$$

For every value of the parameter E, there is a trivial solution to this problem: $y(x) = 0$ for $0 \le x \le 1$. However, for special values of E there are additional nonzero solutions to (1.8.1):

$$E = (n\pi)^2, \qquad y(x) = A_n \sin n\pi x, \qquad n = 1, 2, 3, \ldots, \tag{1.8.2}$$

where A_n are arbitrary constants. The numbers $E = \pi^2, 4\pi^2, 9\pi^2, 16\pi^2, \ldots$ are the eigenvalues and the corresponding nontrivial solutions $\sin \pi x$, $\sin 2\pi x$, $\sin 3\pi x$, $\sin 4\pi x$, ... of (1.8.1) are called *eigenfunctions.*

It is easy to show that there are no other eigenvalues of (1.8.1). The general solution to $y'' + Ey = 0$ is $y(x) = A \sin (x\sqrt{E}) + B \cos (x\sqrt{E})$, so $y(0) = 0$ implies that $B = 0$, and $y(1) = 0$ requires that $A \sin \sqrt{E} = 0$. If $A = 0$ then $y(x) \equiv 0$, which is a trivial solution. Thus, the condition for an eigenvalue is $\sin \sqrt{E} = 0$ or $E = (n\pi)^2$ $(n = 1, 2, 3, \ldots)$. Notice that $E = 0$ is not an eigenvalue because it gives a trivial solution $y(x) \equiv 0$.

When E is an eigenvalue of a homogeneous linear boundary-value problem, the solution to the boundary-value problem is not unique; in addition to the solution $y(x) \equiv 0$ there are an infinite number of other solutions which are constant multiples of an eigenfunction. On the other hand, when E is not an eigenvalue the trivial solution $y(x) \equiv 0$ is unique.

Example 2 *Another simple eigenvalue problem.* Consider the eigenvalue problem

$$y'' + Ey = 0, \qquad y(0) = 0, \qquad y'(1) = 0. \tag{1.8.3}$$

Again, $y(x) \equiv 0$ is a trivial solution for all E; the eigenvalues are determined by the condition that there exist nontrivial solutions. Since the general solution of $y'' + Ey = 0$ that satisfies $y(0) = 0$ is $y(x) = A \sin(x\sqrt{E})$, $y'(1) = 0$ requires that $A\sqrt{E} \cos \sqrt{E} = 0$. Therefore, the eigenvalues are $E = (\frac{1}{2}\pi)^2, (\frac{3}{2}\pi)^2, (\frac{5}{2}\pi)^2, \ldots$.

Example 3 *Eigenvalue problem on an infinite domain.* The eigenvalue problem

$$y'' + (E - \tfrac{1}{4}x^2)y = 0, \qquad -\infty < x < \infty, \tag{1.8.4a}$$

$$y(\pm\infty) = 0, \tag{1.8.4b}$$

is known as the quantum harmonic oscillator. For any E this problem has the trivial solution $y(x) \equiv 0$, but for special E there are nontrivial solutions. The eigenvalues of (1.8.4) are $E = \frac{1}{2}, \frac{3}{2}, \frac{5}{2}, \frac{7}{2}, \ldots$; when $E = n + \frac{1}{2}$ the corresponding eigenfunction is

$$y(x) = A \operatorname{He}_n(x) e^{-x^2/4}, \tag{1.8.5}$$

where A is an arbitrary constant and $\operatorname{He}_n(x)$ is the Hermite polynomial of degree n [$\operatorname{He}_0(x) = 1$, $\operatorname{He}_1(x) = x$, $\operatorname{He}_2(x) = x^2 - 1, \ldots$]. It will be shown in Example 9 of Sec. 3.8 that $E = n + \frac{1}{2}$ are the only eigenvalues of (1.8.4).

The homogeneous boundary conditions (1.8.4*b*) will be shown (see Example 4 of Sec. 3.5) to be equivalent to the seemingly weaker inhomogeneous constraints

$$y(x) \text{ bounded as } x \to \pm\infty.$$

For the differential equation (1.8.4*a*) boundedness of $y(x)$ as $|x| \to \infty$ implies that $y \to 0$ as $|x| \to \infty$.

Example 4 *Eigenvalue problem having transcendental eigenvalues.* Consider the eigenvalue problem

$$y'' + (E - x)y = 0 \ (0 < x < \infty), \qquad y(0) = 0, \qquad y(\infty) = 0. \tag{1.8.6}$$

It will be shown in Example 5 of Sec. 3.5 that the general solution to (1.8.6) that vanishes as $x \to +\infty$ is

$$y(x) = A \operatorname{Ai}(x - E), \tag{1.8.7}$$

where $\operatorname{Ai}(t)$ is an Airy function [see (1.4.8)]. The boundary condition $y(0) = 0$ gives the eigenvalue condition $\operatorname{Ai}(-E) = 0$. The Airy function $\operatorname{Ai}(-E)$ is a transcendental function whose zeros may be computed numerically. A computer calculation gives the infinite discrete sequence of eigenvalues $E_0 \doteqdot 2.338$, $E_1 \doteqdot 4.088$, $E_2 \doteqdot 5.521$, $E_3 \doteqdot 6.787$, $E_4 \doteqdot 7.944$, $E_5 \doteqdot 9.023, \ldots$. The graph of the Airy function in Fig. 3.1 may be used to determine the approximate values of the first few zeros. Asymptotic methods give accurate approximations to the larger zeros (see Sec. 3.7).

Example 5 *Eigenvalue problem having a finite number of eigenvalues.* Consider the eigenvalue problem

$$y'' + (E + v \operatorname{sech}^2 x)y = 0, \qquad y \to 0 \text{ as } |x| \to \infty. \tag{1.8.8}$$

There are only a finite number of discrete eigenvalues: $E = -\frac{1}{4}(2n + 1 - \sqrt{1 + 4v})^2$ $(0 \le n \le V)$, where $V = (\sqrt{1 + 4v} - 1)/2$ and n is an integer (see Prob. 1.38).

Example 6 *Sturm-Liouville eigenvalue problem.* The eigenvalue problem

$$a(x)y''(x) + b(x)y'(x) + c(x)y(x) + d(x)Ey(x) = 0, \tag{1.8.9}$$

subject to the boundary conditions $y(\alpha) = y(\beta) = 0$, can always be transformed to *Sturm-Liouville* form (see Prob. 1.39):

$$\frac{d}{dx}\left[p(x)\frac{dy}{dx}\right] + [q(x) + Er(x)]y = 0, \qquad y(\alpha) = y(\beta) = 0. \tag{1.8.10}$$

Equation (1.8.10) is a Sturm-Liouville problem. It is a mathematical property of Sturm-Liouville problems which we do not prove here that when

$$p(x) > 0, \qquad q(x) \le 0, \qquad r(x) > 0, \qquad \alpha \le x \le \beta,$$

there are an infinite number of eigenvalues $E = E_0, E_1, E_2, \ldots$ which are all real and positive. Moreover, the eigenfunctions y_n $(n = 0, 1, 2, \ldots)$ associated with the eigenvalues E_n can be normalized so that they are *orthonormal* with respect to the *weight function* $r(x)$:

$$\int_\alpha^\beta dx\,r(x)y_n(x)y_m(x) = \begin{cases} 1, & \text{if } n = m, \\ 0, & \text{if } n \ne m. \end{cases}$$

It is often useful to expand a given function in terms of an orthonormal set of functions. For a discussion of the convergence of these expansions and their applications see the References.

In Example 5 of Sec. 10.1 we show how to use WKB theory to find approximate formulas for the eigenvalues E_n and eigenfunctions $y_n(x)$.

Example 7 *Schrödinger eigenvalue problem.* If we choose $V(x)$ to be a positive function which satisfies $V(x) \to +\infty$ as $|x| \to \infty$, then the Schrödinger eigenvalue problem

$$-y''(x) + V(x)y(x) = Ey(x), \qquad y(\pm\infty) = 0, \tag{1.8.11}$$

has an infinite number of real positive eigenvalues $E_0, E_1, E_2, \ldots$. Physically, the eigenvalues are the allowed energy levels of a particle in the potential $V(x)$. The eigenvalue problem in (1.8.4) is an example of a Schrödinger eigenvalue problem.

A closed-form solution to (1.8.11) exists only for very special choices of $V(x)$. However, in Sec. 10.5 we show how to find an approximate formula for the eigenvalues E_n and eigenfunctions $y_n(x)$ of a Schrödinger eigenvalue problem using WKB theory.

(TE) 1.9 DIFFERENTIAL EQUATIONS IN THE COMPLEX PLANE

Until now we have used the letter x to denote the independent variable of a differential equation and have considered x to be real. However, in some of our later analysis we will be concerned with the properties of differential equations in the complex plane where the independent variable is written as z. When a differential equation is generalized from the real-x axis to the complex-z plane, it is necessary to interpret all derivatives in the sense of complex derivatives. This leads to an important conclusion which does not hold for functions of a real variable; in regions where the complex derivative $y'(z)$ exists, the function $y(z)$ is analytic. This conclusion restricts the kinds of differential equations that can be formulated in the complex plane.

Example 1 *First-order differential equation.* The differential equation $dy/dx = |x|$ makes sense on the real axis. Its solution is $y(x) = \frac{1}{2}x|x| + c_1$. However, the differential equation $dy/dz = |z|$ does not make sense in the complex plane because it is understood that when we write $y'(z)$ we mean that $y(z)$, and therefore $y'(z)$, are analytic. But the differential equation sets $y' = |z|$ which is nowhere analytic.

In general, in order for the complex differential equation $dy/dz = g(z)$ to make sense, it is necessary that there be some region in the complex-z plane where $g(z)$ is analytic.

Example 2 *Second-order differential equation.* If two linearly independent solutions to the differential equation $y'' + p_1(z)y' + p_0(z)y = 0$ exist in some region, then $p_0(z)$ and $p_1(z)$ are restricted to be analytic (except possibly at isolated points). To see why, suppose that y_1 and y_2 are two linearly independent solutions (which must be analytic because their complex derivatives exist). Then (see Prob. 1.40)

$$p_0(z) = \frac{y_1''y_2' - y_2''y_1'}{y_1'y_2 - y_1y_2'}, \qquad p_1(z) = \frac{y_2''y_1 - y_1''y_2}{y_1'y_2 - y_1y_2'}, \tag{1.9.1}$$

where the denominators can vanish only at isolated points.

It is possible for a second-order complex differential equation with nonanalytic coefficients to have a solution, but the solution may not contain two independent arbitrary parameters. The differential equation $y'' - |z|y' + (|z| - 1)y = 0$ has the solution $y = c_1 e^z$, but no other solutions exist.

Aside from these restrictions, there are no other major distinctions between real and complex differential equations. The methods that we have introduced in this chapter for solving real differential equations, such as reduction of order, variation of parameters, and scaling, work equally well on complex differential equations.

PROBLEMS FOR CHAPTER 1

Section 1.1

(E) **1.1** Solve the separable equations:

(*a*) $y' = e^{x+y}$;

(*b*) $y' = xy + x + y + 1$.

(I) **1.2** Show that in addition to the special solution $y = c_3$ to the differential equation in (1.1.9) there is another special solution of the form $y = -2/(c_4 + \ln x) - 1$, where c_4 is arbitrary. The derivation of the general solution to (1.1.9) is given in Example 2 of Sec. 1.7. Explain how this special solution can arise in the derivation of the general solution. Also show how this special solution can be interpreted as a singular limit of the general solution in (1.1.10).

Section 1.2

(TI) **1.3** Existence and uniqueness theorem for initial-value problems:

(*a*) Consider the initial-value problem

$$\frac{dy}{dx} = F(x, y), \qquad y(x_0) = a_0. \tag{*}$$

Show that if F and $\partial F/\partial y$ are bounded continuous functions of x and y for x near x_0 and y near a_0 then the *Lipschitz* condition $|F(x, y) - F(x, z)| \le L|y - z|$ holds for some L and for $|x - x_0| \le r$, $|y - a_0| \le R$, and $|z - a_0| \le R$.

(*b*) Prove the existence of a solution to (*) by constructing the sequence of functions

$$y_0(x) = a_0,$$

$$y_n(x) = a_0 + \int_{x_0}^{x} F[t, y_{n-1}(t)]\, dt, \qquad n = 1, 2, \ldots. \tag{**}$$

From the Lipschitz condition prove that there exists a number $r_1 > 0$ such that

$$|y_{n+1}(x) - y_n(x)| \le \frac{1}{2} \max_{|t-x_0| \le r_1} |y_n(t) - y_{n-1}(t)|, \qquad |x - x_0| \le r_1.$$

From this result deduce that

$$\max_{|t-x_0| \le r_1} |y_{n+1}(t) - y_n(t)| \le \frac{1}{2^n} \max_{|t-x_0| \le r_1} |y_1(t) - y_0(t)|$$

so the sequence $y_n(t)$ converges uniformly for $|t - x_0| \le r_1$ as $n \to \infty$. The limit

$$y(t) = \lim_{n\to\infty} y_n(t) = y_0(t) + \lim_{n\to\infty} \sum_{m=1}^{n} [y_m(t) - y_{m-1}(t)]$$

exists by the comparison test for convergence of infinite series. Since $y(x)$ is the uniform limit of continuous functions, $y(t)$ is continuous and, since the Lipschitz condition implies that

$$\lim_{n\to\infty} F(t, y_n) = F(t, y),$$

the limit of (**) is

$$y(x) = a_0 + \int_{x_0}^{x} F[t, y(t)]\, dt$$

for $|x - x_0| \le r_1$. This integral equation is *equivalent* to the initial value problem (*) (why?), so the existence of a solution $y(x)$ for $|x - x_0| \le r_1$ is assured.

(*c*) Prove that the solution $y(x)$ is unique by supposing that there exist two solutions, $y(x)$ and $z(x)$, to the above integral equation. Use the Lipschitz condition to show that if this were true then

$$|y(x) - z(x)| \le \tfrac{1}{2} \max_{|t-x_0| \le r_1} |y(t) - z(t)|.$$

If $y(x)$ and $z(x)$ were not identical, this inequality would be violated for that x which maximizes $|y(x) - z(x)|$. This shows that a unique solution to (*) exists for x sufficiently close to x_0.

(*d*) Now we generalize to higher-order equations. Show that the second-order differential equation $y'' = F(x, y, y')$ may be replaced by a system of first-order differential equations $y' = F_1(x, y, u)$, $u' = F_2(x, y, u)$. Explain how to generalize the existence and uniqueness theorem of parts (*a*) to (*c*) to cover this system of equations.

(*e*) How does one generalize the existence and uniqueness theorem to *n*th-order equations?

(TI) **1.4** (*a*) Assume that $F(x, y)$ is a linear function of y and that $F(x, y)$ is a continuous function of x on the interval $a \le x \le b$. Show that the solution to the initial-value problem $y' = F(x, y)$, $y(x_0) = y_0$ $(a < x_0 < b)$ exists and is unique on the interval $a < x < b$.

(*b*) Show that the solution to an initial-value problem for an *n*th-order linear differential equation exists and is unique on the largest interval in which the initial conditions are specified and on which the coefficient functions $p_0(x), p_1(x), \ldots, p_{n-1}(x), f(x)$ are all continuous.

Section 1.3

(TI) **1.5** (*a*) Show that if $W(y_1, y_2, y_3) \equiv 0$, then there are numbers c_1, c_2, c_3 such that $c_1 y_1 + c_2 y_2 + c_3 y_3 = 0$.

(*b*) Prove that $W(y_1, y_2, \ldots, y_n)$ vanishes identically over an interval if and only if $y_1, y_2, \ldots, y_n$ are linearly dependent on that interval.

(TE) **1.6** Let $\{y_j(x)\}$ be n linearly independent functions. Show that the functions $y_j(x)$ solve a *unique* nth-order homogeneous linear equation $Ly = 0$, with L having the form in (1.1.5). In particular, show that

$$Ly = \frac{\det \begin{vmatrix} y_1 & y_1' & y_1'' & \cdots & y_1^{(n)} \\ y_2 & y_2' & y_2'' & \cdots & y_2^{(n)} \\ \cdots & \cdots & \cdots & \cdots & \cdots \\ y_n & y_n' & y_n'' & \cdots & y_n^{(n)} \\ y & y' & y'' & \cdots & y^{(n)} \end{vmatrix}}{W(y_1, y_2, \ldots, y_n)}.$$

(TE) **1.7** Verify (1.3.4).
Clue: Differentiate $W(x)$ in (1.3.3) once with respect to x and use the differential equation (1.3.1).

(E) **1.8** (*a*) Show that the initial-value problem $yy' = 1$, $y(0) = 0$ is not well posed. Find two solutions.
(*b*) Is $d^4y/dx^4 - 4y''' + 3y'' + 4y' - 4y = 0$ $[y(0) = 1, y(+\infty) = 0]$ well posed?

Section 1.4

(TE) **1.9** Assume that $y_1(x)$ is a solution of $y'' + p_1(x)y' + p_0(x)y = 0$. From the definition of the Wronskian and Abel's formula, derive an expression for another linearly independent solution. Observe that the second solution contains $y_1(x)$ as a factor. This result motivates the reduction-of-order substitution (1.4.7).

(E) **1.10** Use reduction of order to obtain the repeated root solution of $y''' - 3y'' + 3y' - y = 0$.

(I) **1.11** Solve $y'' + (x + 2)y' + (1 + x)y = 0$.

(TE) **1.12** Show that a substitution identical to that used in reduction of order $y(x) = u(x)f(x)$ can be used to eliminate the $y^{(n-1)}(x)$ term from an nth-order homogeneous linear differential equation. (When the one-derivative term has been eliminated from a linear second-order differential equation, the resulting equation is a *Schrödinger* equation.)

(E) **1.13** (*a*) Show that if a is a constant and $b(x)$ is a function, then

$$y'' + \frac{b'(x)}{b(x)}y' - \frac{a^2}{[b(x)]^2}y = 0$$

has a pair of linearly independent solutions which are reciprocals; find them.
(*b*) $y(x)$ and $[y(x)]^2$ are both solutions of $y'' + p(x)y' + 2y = 0$. Find $y(x)$.

(I) **1.14** Find the general solution to $x^{2n}(d/dx - a/x)^n y = ky$.

Section 1.5

(TI) **1.15** Formulate the method of variation of parameters for a third-order linear equation. How does it work for an nth-order equation?

(TE) **1.16** Prove (1.5.9) using (1.5.8).

(I) **1.17** Verify that the representations of $\delta(x - a)$ given in (1.5.10*a* to *d*) are valid by showing that each one satisfies (1.5.8).

(E) **1.18** Verify (1.5.11).

(TI) **1.19** Find the formula for the Green's function of a third-order inhomogeneous linear equation. Generalize this formula to the nth order.

(TI) **1.20** In Examples 4 and 5 of Sec. 1.5 we use Green's functions to solve inhomogeneous differential equations subject to homogeneous boundary conditions. How do we generalize to the case where the boundary conditions are inhomogeneous?

(E) **1.21** By reduction of order, find the general solution of $x^2y'' - 4xy' + 6y = x^4 \sin x$ after observing that $y_1 = x^2$ is a solution of the associated homogeneous equation.

(I) **1.22** There is a marvelous problem, the Snowplow Problem of R. P. Agnew, which reads, "One day it started snowing at a heavy and steady rate. A snowplow started out at noon, going 2 miles the first hour and 1 mile the second hour. What time did it start snowing?" (*Answer:* 11:23 A.M.) We also recommend a more sophisticated variation by M. S. Klamkin called the Great Snowplow Chase: "One day it started snowing at a heavy and steady rate. Three identical snowplows started out at noon, 1 P.M., and 2 P.M. from the same place and all collided at the same time. What time did it start snowing?" (*Answer:* 11:30 A.M.)

Clue: The speed of the plow is inversely proportional to the height of the snow.

Section 1.6

(E) **1.23** A man stands atop a mountain whose altitude is given by $z = e^{-(x^4+4y^2)}$ and pours boiling oil upon the climbers below him. What paths do the rivulets of oil follow? [Assume that these paths are orthogonal to the contour lines (level curves) of the mountain.]

(TI) **1.24** Show that a necessary and sufficient condition that $M(x, y) + N(x, y)\, dy/dx = 0$ be exact is $\partial M/\partial y = \partial N/\partial x$.

(TI) **1.25** Use the method of factoring in Example 4 of Sec. 1.6 to derive Abel's formula (1.3.5).

(I) **1.26** Show that the differential equation $xy'' + (cx + a)y' + cby = f$ can be solved explicitly in closed form provided that either (*a*) b is an integer or (*b*) $b - a$ is an integer.

Clue: This second-order equation can be factored into a product of first-order operators by using the novel trick of first converting the equation into a higher-order equation that can be easily factored. To do this introduce the operator notation $\delta = x\, d/dx$ so that the equation becomes $\delta(\delta + a')y + c(\delta + b')xy = xf$ where $a' = a - 1$, $b' = b - 1$. First show that if $b' > 0$ is an integer and we set $y = (\delta + 1)(\delta + 2) \cdots (\delta + b')z$, then the equation factors into the product of first-order operators $\delta(\delta + 1) \cdots (\delta + b')(\delta + a' + cx)z = xf$. If $b' < 0$ show that application of the operator $(\delta - 1) \cdots (\delta + b' + 1)$ to the equation gives the factorization

$$(\delta + a' + cx)\delta(\delta - 1) \cdots (\delta + b' + 1)y = (\delta - 1) \cdots (\delta + b' + 1)xf.$$

Use similar tricks to solve the differential equation when $b - a$ is an integer. For a similar approach to linear difference equations see Prob. 2.25.

(E) **1.27** Four caterpillars, initially at rest at the four corners of a square centered at the origin, start walking clockwise, each caterpillar walking directly toward the one in front of him. If each caterpillar walks with unit velocity, show that the trajectories satisfy the differential equation in part (*c*) of Example 8 in Sec. 1.6.

(E) **1.28** Discuss the existence and uniqueness of solutions to the initial-value problem $y' = \sqrt{1 - y^2}$ $[y(0) = a]$, for all initial values a. Is there a unique solution if $a = 1$?

(E) **1.29** Find a differential equation having the general solution $y = c_1(x + c_2)^n$.

(D) **1.30** (*a*) At $t = 0$, a pig, initially at the origin, runs along the x axis with constant speed v. At $t = 0$, a farmer, initially 20 yd north of the origin, also runs with constant speed v. If the farmer's instantaneous velocity is always directed toward the instantaneous position of the pig, show that the farmer never gets closer than 10 yd from the pig.

(*b*) Now suppose that the pig starts over again from $x = 0$, $y = 0$ at $t = 0$ and starts running with the speed v. The farmer still starts 20 yd north of the pig but can now run at a speed of $\frac{3}{2}v$. The farmer is assisted by his daughter who starts 15 yd south of the pig at $t = 0$ and can run at a speed of $\frac{4}{3}v$. If both the farmer and the farmer's daughter always run toward the instantaneous position of the pig, who catches the pig first?

(*c*) At $t = 0$, a pig initially at $(1, 0)$ starts to run around the unit circle with constant speed v. At $t = 0$, a farmer initially at the origin runs with constant speed v and instantaneous velocity directed toward the instantaneous position of the pig. Does the farmer catch the pig?

(I) **1.31** Solve the following differential equations:

(*a*) $y' = y/x + 1/y$;

(*b*) $y' = xy/(x^2 + y^2)$;

(*c*) $y' = x^2 + 2xy + y^2$;

(*d*) $yy'' = 2(y')^2$;
(*e*) $y' = (1 + x)y^2/x^2$;
(*f*) $x^2y' + xy + y^2 = 0$;
(*g*) $xy' = y(1 - \ln x + \ln y)$;
(*h*) $(x + y^2) + 2(y^2 + y + x - 1)y' = 0$, using an integrating factor of the form $I(x, y) = e^{ax+by}$;
(*i*) $-xy' + y = xy^2$ $[y(1) = 1]$;
(*j*) $y'' - (1 + x)^{-2}(y')^2 = 0$ $[y(0) = y'(0) = 1]$;
(*k*) $2xyy' + y^2 - x^2 = 0$;
(*l*) $y'' = (y')^2e^{-y}$ (if $y' = 1$ at $y = \infty$, find y' at $y = 0$);
(*m*) $y' = |y - x|$ [if $y(0) = \frac{1}{2}$, find $y(1)$];
(*n*) $xy' = y + xe^{y/x}$;
(*o*) $y' = (x^4 - 3x^2y^2 - y^3)/(2x^3y + 3y^2x)$;
(*p*) $(x^2 + y^2)y' = xy$, $y(e) = e$;
(*q*) $y'' + 2y'y = 0$ $[y(0) = y'(0) = -1]$;
(*r*) $x^2y'' + xy' - y = 3x^2$ $[y(1) = y(2) = 1]$;
(*s*) $y^3(y')^2y'' = -\frac{1}{2}$ $[y(0) = y'(0) = 1]$;
(*t*) $xy' = y + \sqrt{xy}$;
(*u*) $(xy)y' + y \ln y = 2xy$ [try an integrating factor of the form $I = I(y)$];
(*v*) $(x \sin y + e^y)y' = \cos y$;
(*w*) $(x + y^2x)y' + x^2y^3 = 0$ $[y(1) = 1]$;
(*x*) $(x - 1)(x - 2)y' + y = 2$ $[y(0) = 1]$;
(*y*) $y' = 1/(x + e^y)$;
(*z*) $xy' + y = y^2x^4$.

(D) **1.32** Find a closed-form solution to the following Riccati equations:
(*a*) $xy' + xy^2 + x^2/2 = \frac{1}{4}$;
(*b*) $x^2y' + 2xy - y^2 = A$;
(*c*) $y' + y^2 + (\sin 2x)y = \cos 2x$;
(*d*) $xy' - 2y + ay^2 = bx^4$;
(*e*) $y' + y^2 + (2x + 1)y + 1 + x + x^2 = 0$.

(I) **1.33** Under what conditions does the differential equation $y' = f(x, y)$ have an integrating factor of the form $I(xy)$?

(E) **1.34** Express the solution of the initial-value problem

$$x\frac{d}{dx}\left(x\frac{d}{dx} - 1\right)\left(x\frac{d}{dx} - 2\right)\left(x\frac{d}{dx} - 3\right)y(x) = f(x), \qquad y(1) = y'(1) = y''(1) = y'''(1) = 0,$$

as an integral.

(I) **1.35** An Abel equation has the general form $y' = a(x) + b(x)y + c(x)y^2 + d(x)y^3$. Solve the particular equation $y' = dy^3 + ax^{-3/2}$, where d and a are constants.

Section 1.7

(I) **1.36** Reduce the order and, if possible, solve the following equations:
(*a*) $y'' + y'(y + 1)/x + y/x^2 = 0$;
(*b*) $y''' + yy'' = 0$ (the Blasius equation);
(*c*) $yy'' + y'^2 - yy'/(1 + x) = 0$;
(*d*) $y' + (2y - y^2)/x = 0$;
(*e*) $y'' + y'^2/y - y'y/x = 0$;
(*f*) $y'' + y'(2y + 1)/x = 0$;
(*g*) $x^2y'' - (1 + 2y/x^2)\,xy' + 4y = 0$;
(*h*) $y'' + [(x - 2y)/x^2](y/x - y') = 0$;
(*i*) $y'' - y'(x^2 + 2y)/x^3 + 4y/x^2 = 0$;
(*j*) $xy' + 3y - xy^2 = 0$;
(*k*) $xyy'' = yy' + xy'^2$;
(*l*) $y'' + 3yy' + y^3 = 0$.

(I) **1.37** Consider the equation $y'' + 2y'/x + y^n = 0$ $[y(\infty) = 0]$ for $n \neq 0, 1$. Using the methods of Sec. 1.7 show that the equation is soluble in terms of elementary functions when $n = 5$ and solve it.

Sections 1.8 and 1.9

(D) **1.38** Verify the claim in Example 5 of Sec. 1.8 that there are only a finite number of eigenvalues E for the eigenvalue problem $y'' + (E + v \operatorname{sech}^2 x)y = 0$ $(y \to 0$ as $|x| \to \infty)$.

Clue: Solve the differential equation in terms of hypergeometric functions. The *Bateman Manuscript Project* or the *Handbook of Mathematical Functions* are good references on hypergeometric functions.

(TI) **1.39** Show that any equation of the form (1.8.9) can be transformed to Sturm-Liouville form (1.8.10).

(E) **1.40** Verify (1.9.1).

CHAPTER

TWO

DIFFERENCE EQUATIONS

From a drop of water a logician could infer the possibility of an Atlantic or a Niagara without having seen or heard of one or the other. So all life is a great chain, the nature of which is known whenever we are shown a single link of it.

—Sherlock Holmes, *A Study in Scarlet*
Sir Arthur Conan Doyle

This chapter is a summary of the elementary methods available for solving difference equations. Difference equations are used to compute quantities which may be defined recursively, such as the nth coefficient of a Taylor series or Fourier expansion or the determinant of an $n \times n$ matrix which is expanded by minors. Difference equations arise very frequently in numerical analysis where one attempts to approximate continuous systems by discrete ones.

Difference equations are the discrete analog of differential equations. The solution of a difference equation is a function defined on the integers; the discrete index n replaces the continuous independent variable x of differential equations. Many of the analytical methods developed for differential equations in Chap. 1 (reduction of order, variation of parameters, integrating factors) are also applicable to difference equations. Therefore, the presentation of this chapter will closely parallel that of Chap. 1.

(E) 2.1 THE CALCULUS OF DIFFERENCES

The study of difference equations rests on the notions of discrete calculus which we review here. A function defined on the integers associates with each integer n the number $a(n)$, usually denoted by a_n. The discrete derivative of a function a_n of the integers is defined as

$$Da_n \equiv a_{n+1} - a_n.$$

The second derivative $D^2 a_n$ is the derivative of the derivative; that is,

$$D^2 a_n = (a_{n+2} - a_{n+1}) - (a_{n+1} - a_n) = a_{n+2} - 2a_{n+1} + a_n.$$

The kth derivative is defined by taking k derivatives sequentially.

The discrete antiderivative or integral b_n of the function a_n is defined as the sum $b_n = \sum_{j=n_0}^{n} a_j$.

Example 1 *Discrete derivative of a sequence.* The integer function that corresponds to the continuous function $f(x) = x^k$ is the discrete function $f_n = n(n+1) \cdots (n+k-1)$, which also has k factors. The continuous derivative of $f(x)$ is $f'(x) = kx^{k-1}$, while the discrete derivative of f_n is

$$\begin{aligned} f_{n+1} - f_n &= (n+1) \cdots (n+k) - (n) \cdots (n+k-1) \\ &= (n+1) \cdots (n+k-1)[(n+k) - (n)] \\ &= k(n+1) \cdots (n+k-1), \end{aligned}$$

which has $(k-1)$ factors.

Example 2 *Discrete integral of a sequence.* The integral of $f(x) = x^k$ is $x^{k+1}/(k+1) + c_0$, where c_0 is an integration constant. The discrete integral of $f_n = (n) \cdots (n+k-1)$ is

$$\begin{aligned} \sum_{j=n_0}^{n} f_j &= \sum_{j=n_0}^{n} (j) \cdots (j+k-1) \\ &= \sum_{j=n_0}^{n} (j) \cdots (j+k-1) \frac{[(j+k) - (j-1)]}{k+1} \\ &= \frac{1}{k+1} \sum_{j=n_0}^{n} [(j) \cdots (j+k-1)(j+k) - (j-1)(j) \cdots (j+k-1)] \\ &= \frac{1}{k+1} n(n+1) \cdots (n+k) + c_0, \end{aligned}$$

where c_0 is a summation constant whose value is $-[(n_0 - 1)(n_0) \cdots (n_0 + k - 1)]/(k+1)$. Observe that to compute such a sum we convert the summand to a discrete derivative. In this form the summand is a "ladder" whose rungs cancel in pairs: if $f_n = g_{n+1} - g_n$ then $\sum_{j=n_0}^{n} f_j = g_{n+1} - g_{n_0}$, which is the discrete analog of the fundamental theorem of integral calculus $\int_a^b g'(x)\, dx = g(b) - g(a)$.

Example 3 *Discrete integral of a sequence.* The discrete function which corresponds to the continuous function $f(x) = (x+A)^{-k}$ $(k > 1)$ is $f_n = 1/[(n+A)(n+A+1) \cdots (n+A+k-1)]$. The integral $\int_x^a f(t)\, dt = (x+A)^{1-k}/(k-1) + c_0$. Correspondingly, the discrete integral of f_n is

$$\begin{aligned} &\sum_{j=n}^{n_0} \frac{1}{(j+A) \cdots (j+A+k-1)} \\ &\quad = \frac{1}{1-k} \sum_{j=n}^{n_0} \left[\frac{1}{(j+A+1) \cdots (j+A+k-1)} - \frac{1}{(j+A) \cdots (j+A+k-2)} \right] \\ &\quad = \frac{1}{k-1} \frac{1}{(n+A)(n+A+1) \cdots (n+A+k-2)} + c_0, \qquad n < n_0, \end{aligned}$$

where c_0 is the summation constant $[(1-k)(n_0+A+1)(n_0+A+2) \cdots (n_0+A+k-1)]^{-1}$.

(E) 2.2 ELEMENTARY DIFFERENCE EQUATIONS

Recalling the definition of an Nth-order differential equation from Sec. 1.1, it is natural to define an Nth-order difference equation as

$$a_n^{(N)} = F[n, a_n, a_n^{(1)}, a_n^{(2)}, a_n^{(3)}, a_n^{(4)}, \ldots, a_n^{(N-1)}],$$

where $a_n^{(j)}$ is the jth discrete derivative of a_n. However, it is better to simplify this definition to avoid repetition of $a_n, a_{n+1}, \ldots$, in the arguments of F. Thus, the conventional definition of an Nth-order difference equation is

$$a_{n+N} = G[n, a_n, a_{n+1}, \ldots, a_{n+N-1}]. \tag{2.2.1}$$

Clearly, if one is given a difference equation of this form, it is easy to rewrite it in a form which makes the discrete derivatives explicit.

The general solution of the Nth-order difference equation (2.2.1) depends on N independent parameters $c_1, c_2, \ldots, c_N$. These parameters appear as constants of summation.

Example 1 *Factorial function.* The solution of the first-order linear homogeneous difference equation $a_{n+1} = na_n$ is $a_n = c_1(n-1)!$. The arbitrary multiplicative constant c_1 is the constant of summation. The factorial function is closely related to the gamma function Γ (see Example 2).

Example 2 *Gamma function* $\Gamma(z)$. The gamma function is the most widely used of all the higher transcendental functions. It is usually discussed first in reference texts on higher transcendental functions because the function $\Gamma(z)$ appears in almost every integral or series representation of the other advanced mathematical functions.

The gamma function $\Gamma(z)$ is the generalization to complex z of the factorial function, which is defined only for integers: when $z = n$ is a positive integer, $\Gamma(n) = (n-1)!$. More generally, when Re $z > 0$, $\Gamma(z)$ can be defined as the integral

$$\Gamma(z) = \int_0^\infty e^{-t} t^{z-1}\, dt \tag{2.2.2}$$

which converges for all z in the half plane Re $z > 0$. This integral cannot be evaluated in terms of elementary functions except when z is an integer or a half integer. Other integral representations hold for Re $z \le 0$ (see Prob. 2.6).

$\Gamma(z)$ does *not* solve a simple differential equation but it is a solution of the difference equation

$$\Gamma(z+1) = z\Gamma(z). \tag{2.2.3}$$

Using integration by parts it is easy to show that the integral in (2.2.2) satisfies (2.2.3).

This difference equation for $\Gamma(z)$ can be used to evaluate $\Gamma(z)$ for Re $z \le 0$ from the integral representation (2.2.2). $\Gamma(z)$ is a single-valued analytic function for all complex values of z except for $z = 0, -1, -2, -3, -4, \ldots$. At these points $\Gamma(z)$ has simple poles (see Prob. 2.6 and Fig. 2.1). For additional properties of the Γ function see Prob. 2.6.

Example 3 *General first-order linear homogeneous difference equations.* The solution of

$$a_{n+1} = p(n)a_n \tag{2.2.4}$$

is a product over the function $p(n)$:

$$a_n = a_1 \prod_{j=1}^{n-1} \frac{a_{j+1}}{a_j} = a_1 \prod_{j=1}^{n-1} p(j). \tag{2.2.5}$$

In this formula a_1 is an arbitrary constant.

To derive this result another way, we take the logarithm of both sides of (2.2.4): $\ln a_{n+1} = \ln p(n) + \ln a_n$. Letting $b_n = \ln a_n$ gives $b_{n+1} - b_n = \ln p(n)$. The left side of this equation is an exact discrete derivative. Therefore, the solution b_n is obtained by computing the discrete integral: $b_n = b_1 + \sum_{j=1}^{n-1} \ln p(j)$. Exponentiating this equation gives the result in (2.2.5) with $a_1 = e^{b_1}$.

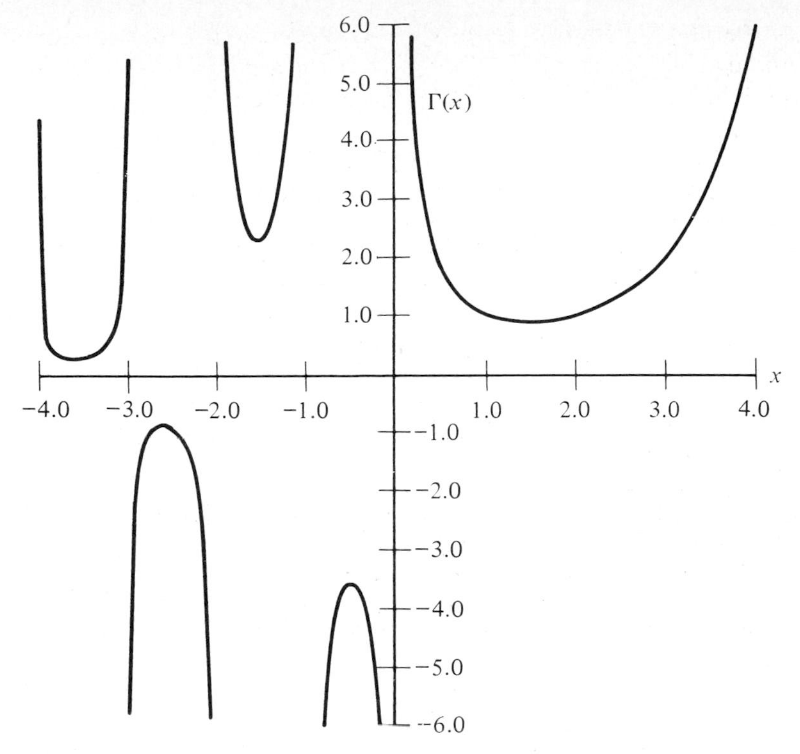

Figure 2.1 A plot of $\Gamma(x)$ for $-4 \le x \le 4$.

Observe the close similarity between this method of solution and the way in which we solve the differential equation $y' = p(x)y$: $y'/y = p(x)$, $\ln [y/y(0)] = \int_0^x p(t)\,dt$, $y(x) = y(0) \exp [\int_0^x p(t)\,dt]$.

Example 4 *General first-order linear inhomogeneous difference equations.* To solve

$$a_{n+1} = p(n)a_n + q(n), \tag{2.2.6}$$

we use a "summing factor" which is the difference equation analog of an integrating factor. The summing factor is $[\prod_{j=1}^{n} p(j)]^{-1}$. Multiplying both sides of (2.2.6) by this summing factor gives $a_{n+1}/\prod_{j=1}^{n} p(j) - a_n/\prod_{j=1}^{n-1} p(j) = q(n)/\prod_{j=1}^{n} p(j)$, in which a_n appears as an exact discrete derivative. Summing both sides from 1 to $n-1$ gives the solution to (2.2.6):

$$a_n = \prod_{j=1}^{n-1} p(j) \left[\sum_{k=1}^{n-1} \frac{q(k)}{\prod_{j=1}^{k} p(j)} + a_1 \right], \qquad n \ge 2. \tag{2.2.7}$$

As a special case of (2.2.6), consider $a_{n+1} = na_n/(n+1) + n$. To solve this equation we multiply by the summing factor $[\prod_{j=1}^{n} j/(j+1)]^{-1} = n+1$. The resulting equation is $(n+1)a_{n+1} - na_n = n(n+1)$. Summing this equation from 1 to $n-1$ gives $na_n - a_1 = \frac{1}{3}(n-1)n(n+1)$ (see Example 2 of Sec. 2.1). Therefore, $a_n = a_1/n + \frac{1}{3}(n^2 - 1)$ $(n \ge 1)$.

Example 5 *First-order linear difference equation with boundary condition.* The requirement that the solution to $a_{n+1} = na_n/(n+1) + (n-1)/[(n+1)^2(n+2)(n+3)]$ be finite at $n = 0$ is a boundary condition that uniquely determines a_n for all n. Multiplying by the summing factor

$n + 1$ and summing the resulting equation from 1 to $n - 1$ gives the general solution $a_n = (a_1 + \frac{1}{6})/n - 1/[(n+1)(n+2)]$ $(n > 0)$. (Verify this!) If a_0 is finite, then a_1 must be $-\frac{1}{6}$. (Why?) Thus, the unique solution to this problem is $a_n = -1/[(n+1)(n+2)]$ $(n \geq 0)$.

There is a simple differential analog of this example. Suppose we require that the solution to $y' + (\cot x)y = -1$ be finite at $x = 0$. The general solution to this equation is $y(x) = (\cos x + K)/\sin x$. The condition that $y(0)$ be finite requires that $K = -1$ and uniquely determines the solution.

Nonlinear difference equations are usually harder to solve than nonlinear differential equations. However, here are some examples that can be solved in closed form.

Example 6 $a_{n+1} = a_n^2$. To solve this equation we take the logarithm of both sides: $\ln a_{n+1} = 2 \ln a_n$. The solution to this equation is easy to find because it is a linear equation for $\ln a_n$. The solution is $\ln a_n = 2^{n-1} \ln a_1$. Thus, $a_n = a_1^{(2^{n-1})}$.

Example 7 $a_{n+2} = a_{n+1}^2/a_n$. This equation can also be solved by taking logarithms of both sides: $\ln a_{n+2} - 2 \ln a_{n+1} + \ln a_n = 0$. Thus, the second discrete derivative of $\ln a_n$ is zero. Two consecutive summations give $\ln a_n = c_1 n + c_2$ so that $a_n = e^{c_1 n + c_2}$, where c_1 and c_2 are arbitrary summation constants.

Example 8 $a_{n+1} = 2a_n^2 - 1$. This equation is soluble because the transcendental functions $\cos x$ and $\cosh x$ satisfy the functional relation $f(2x) = 2[f(x)]^2 - 1$. If $|a_1| \leq 1$, we substitute $a_n = \cos \theta_n$ and if $|a_1| \geq 1$ we substitute $a_n = \cosh \theta_n$. In either case, the resulting equation for θ_n is linear: $\theta_{n+1} = 2\theta_n$. Thus, $\theta_n = 2^{n-1}\theta_1$ and

$$a_n = \begin{cases} \cos(2^{n-1} \cos^{-1} a_1), & \text{if } |a_1| \leq 1, \\ \cosh(2^{n-1} \cosh^{-1} a_1), & \text{if } |a_1| \geq 1. \end{cases}$$

(I) 2.3 HOMOGENEOUS LINEAR DIFFERENCE EQUATIONS

In this section we survey the methods that are commonly used to solve higher-order homogeneous linear difference equations. Our discussion will emphasize the strong parallel between differential and difference equations.

A *homogeneous* linear difference equation of order N has the general form

$$a_{n+N} + p_{N-1}(n)a_{n+N-1} + p_{N-2}(n)a_{n+N-2} + \cdots + p_1(n)a_{n+1} + p_0(n)a_n = 0, \qquad (2.3.1)$$

where $p_0, p_1, \ldots, p_{N-1}$ are arbitrary functions of n. As is the case with differential equations, the general solution a_n to (2.3.1) is an arbitrary linear combination of N linearly independent solutions.

Constant-Coefficient Equations

To illustrate this property we consider the easiest-to-solve class of higher-order homogeneous linear difference equations, the constant-coefficient equations. These equations take the form (2.3.1) in which the coefficients $p_0, p_1, \ldots, p_{N-1}$ are independent of n.

We recall from our treatment of constant-coefficient differential equations in Sec. 1.4 that solutions usually have the form of exponentials: $y(x) = \exp(rx)$. For an Nth-order differential equation there are N solutions of the form $\exp(r_1 x)$, $\exp(r_2 x), \ldots, \exp(r_N x)$, unless the polynomial in r that results from substituting $y = \exp(rx)$ has repeated roots. If r_1 is a double root, then solutions have the form $\exp(r_1 x)$ and $x \exp(r_1 x)$; if r_2 is a triple root, then solutions have the form $\exp(r_2 x)$, $x \exp(r_2 x)$, and $x^2 \exp(r_2 x)$; and so on.

Using differential equations as a guide, we seek exponential solutions to (2.3.1) of the form $a_n = r^n$. Substituting this into (2.3.1) gives a polynomial equation for r:

$$r^N + p_{N-1} r^{N-1} + p_{N-2} r^{N-2} + p_{N-3} r^{N-3} + \cdots + p_1 r + p_0 = 0. \qquad (2.3.2)$$

In general, if the roots of (2.3.2) are all distinct then solutions to (2.3.1) have the form $r_1^n, r_2^n, r_3^n, \ldots$, and if r_1 is a multiple root then solutions have the form $r_1^n, nr_1^n, n^2 r_1^n, \ldots$. (See Prob. 2.9.)

Example 1 *Constant-coefficient equations.*

(*a*) To solve $a_{n+2} + 3a_{n+1} + 2a_n = 0$ we substitute $a_n = r^n$ and obtain the polynomial equation $r^2 + 3r + 2 = 0$. The roots of this equation are -1 and -2. Thus, the general solution is a linear combination of $(-1)^n$ and $(-2)^n$: $a_n = c_1(-1)^n + c_2(-2)^n$.

(*b*) To solve $a_{n+3} - 6a_{n+2} + 12a_{n+1} - 8a_n = 0$ we substitute r^n for a_n. The resulting equation for r is $r^3 - 6r^2 + 12r - 8 = 0$, which has a triple root at $r = 2$. Thus, the general solution has the form $a_n = (c_1 + c_2 n + c_3 n^2)2^n$.

(*c*) To solve $a_{n+3} + a_{n+2} - a_{n+1} - a_n = 0$ we substitute $a_n = r^n$. The resulting polynomial in r is $r^3 + r^2 - r - 1 = 0$, whose roots are -1, -1, and 1. Thus, the general solution for a_n is $a_n = (c_1 + nc_2)(-1)^n + c_3$.

Linear Independence and Wronskians

Since a complete solution to a homogeneous linear difference equation is a linear combination of N *linearly independent* solutions, it is important to be able to test the linear independence of a set of N functions of n. We define a finite set of functions $a_n, b_n, \ldots$ to be linearly independent for n in the interval $n_1 \le n \le n_2$ if the only solution to $k_1 a_n + k_2 b_n + \cdots = 0$ $(n_1 \le n \le n_2)$ is $k_1 = k_2 = \cdots = 0$. (We always assume that the number of functions is smaller than $n_2 - n_1 + 1$.)

In the study of functions of a continuous variable x the Wronskian provides a simple way to examine the linear independence of a set of functions. There is also a Wronskian W_n (sometimes called a Casoratian) for functions of the discrete variable n. If there are N functions $a_n, b_n, c_n, \ldots$, then W_n is defined as the $N \times N$ determinant:

$$W_n = W(a_n, b_n, c_n, \ldots) \equiv \det \begin{vmatrix} a_n & b_n & c_n & \cdots \\ a_{n+1} & b_{n+1} & c_{n+1} & \cdots \\ a_{n+2} & b_{n+2} & c_{n+2} & \cdots \\ \cdots & \cdots & \cdots & \cdots \end{vmatrix}. \qquad (2.3.3)$$

It is a theorem that $W_n = 0$ for all $n_1 \le n \le n_2$ if and only if $a_n, b_n, c_n, \ldots$ constitute a linearly dependent set of functions. For the general proof of this theorem see Prob. 2.11. We prove this theorem for the simple case of two functions in the next example.

Example 2 *Wronskian for two functions.* Two functions a_n and b_n are linearly dependent for $n_1 \le n \le n_2$ if b_n is a constant multiple of a_n. To see whether the Wronskian provides a test of the linear dependence of two functions we compute

$$W_n = \det \begin{vmatrix} a_n & b_n \\ a_{n+1} & b_{n+1} \end{vmatrix} = a_n b_{n+1} - b_n a_{n+1}.$$

Suppose first that b_n is indeed a constant multiple of a_n in some domain: $b_n = ca_n$ $(n_1 \le n \le n_2)$. Then the Wronskian vanishes in this domain: $W_n = a_n a_{n+1}(c - c) = 0$ $(n_1 \le n \le n_2)$. Now consider the converse. Suppose that $W_n = 0$ for $n_1 \le n \le n_2$. Then, assuming that $a_n \ne 0$ and $b_n \ne 0$, we have $a_{n+1}/a_n = b_{n+1}/b_n$ $(n_1 \le n \le n_2)$. Hence, $\prod_{j=n_1}^{n-1} a_{j+1}/a_j = \prod_{j=n_1}^{n-1} b_{j+1}/b_j$, so $a_n = cb_n$ $(n_1 \le n \le n_2)$, where $c = a_{n_1}/b_{n_1}$. Thus the vanishing of W_n implies the linear dependence of a_n and b_n. (How do we alter this argument if $a_n = 0$ or $b_n = 0$ for some n?)

As is the case with differential equations, we can compute the Wronskian of all N solutions to an Nth-order homogeneous linear difference equation, even if the solutions are not known explicitly. This is because the Wronskian satisfies a first-order difference equation. For the proof of this result see Prob. 2.12. In the next example we consider the special case of a second-order difference equation.

Example 3 *Wronskian of the solutions to a second-order homogeneous linear difference equation.* Suppose that a_n and b_n are two solutions to the second-order homogeneous linear difference equation

$$a_{n+2} + p_1(n)a_{n+1} + p_0(n)a_n = 0. \tag{2.3.4}$$

If we multiply (2.3.4) by b_{n+1} and subtract the equation obtained by interchanging a and b, the result is $a_{n+2}b_{n+1} - b_{n+2}a_{n+1} + p_0(n)(a_n b_{n+1} - b_n a_{n+1}) = 0$. But, a glance at (2.3.3) shows that this equation is a first-order difference equation for W_n: $W_{n+1} = W_n p_0(n)$. Thus,

$$W_n = W_{n_0} \prod_{j=n_0}^{n-1} p_0(j), \tag{2.3.5}$$

where W_{n_0} is an arbitrary constant. Equation (2.3.5) actually holds for difference equations of any order (see Prob. 2.12). It is the equivalent of Abel's formula for differential equations (1.3.5).

From (2.3.5) we can see that if W_{n_0} is nonzero, then W_n $(n \ge n_0)$ is also nonzero in regions in which $p_0(n) \ne 0$ $(n \ge n_0)$. [Note that $p_0(n)$ is nonvanishing whenever the difference equation (2.3.4), and also (2.3.1), functions as a backward as well as a forward difference equation; i.e., if we can compute a_n for successively smaller values of n as well as for successively larger values of n.] Thus, so long as $p_0(n) \ne 0$, the difference equation preserves the linear independence of solutions.

Example 4 *Wronskian for a constant-coefficient equation.* Equation (2.3.5) predicts that the Wronskian of two solutions to $a_{n+2} - 5a_{n+1} + 6a_n = 0$ is proportional to 6^n: $W_n = c\ 6^n$. To check this result we note that two linearly independent solutions, a_n and b_n, to this equation are $a_n = c_1 2^n$ and $b_n = c_2 3^n$. Thus, $W_n = a_n b_{n+1} - b_n a_{n+1} = c_1 c_2 6^n$.

Initial-Value and Boundary-Value Problems

The general solution to an Nth-order homogeneous linear difference equation is an arbitrary superposition of N linearly independent solutions. There are two possible ways to determine the N arbitrary parameters. In an *initial-value problem* on the domain $n_1 \le n \le n_2$ we specify N discrete derivatives of a_n at one point n_0 $(n_1 \le n_0 \le n_2 - N)$ or, equivalently, we specify N successive values of a_n starting with a_{n_0}. Either $a_{n_0}, a_{n_0}^{(1)}, a_{n_0}^{(2)}, \ldots, a_{n_0}^{(N-1)}$ is given or $a_{n_0}, a_{n_0+1}, a_{n_0+2}, \ldots, a_{n_0+N-1}$ is given. (Why are these two formulations of the initial-value problem equivalent?) Sufficient conditions for this initial-value problem to be well posed (i.e., for a unique solution to exist) on the interval $n_1 \le n \le n_2$ are:

(*a*) the Wronskian at n_0 does not vanish and
(*b*) $p_0(n)$ does not vanish on the interval $n_1 \le n \le n_0$.

It is not necessary that $p_0(n)$ be nonvanishing for $n_0 \le n \le n_2$. (Why?) What happens if $p_0(n) = 0$ for $n_1 \le n \le n_2$?

In a *boundary-value problem* on the domain $n_1 \le n \le n_2$ we specify N values of a_n or any of its discrete derivatives at any values of n scattered throughout the interval $n_1 \le n \le n_2$: for example, $a_{n_1}, a_{n_1+7}^{(1)}, a_{n_2}, a_{n_2-1}^{(3)}, \ldots$. The question of how to determine whether a boundary-value problem is well posed is difficult; there is no simple criterion like that for initial-value problems. See Prob. 2.13.

Reduction of Order

Abel's formula (2.3.5) for the Wronskian implies that if one solution to an Nth-order difference equation is known, then we can find an $(N-1)$th-order equation for the remaining unknown solutions. For example, if $N = 2$ and one solution a_n is known, a second solution b_n satisfies the first-order equation $a_n b_{n+1} - b_n a_{n+1} = W_n$.

There is a general procedure, called *reduction of order*, for lowering the order of any difference equation once one solution A_n of the equation is known. The object is to seek additional solutions a_n in the form of a product

$$a_n = A_n x_n \tag{2.3.6}$$

in which x_n is to be found. In general, $x_n^{(1)} = Dx_n$ satisfies a homogeneous linear difference equation of one lower order.

Let us examine reduction of order for second-order equations. Let A_n be one solution of $a_{n+2} + p_1(n)a_{n+1} + p_0(n)a_n = 0$. Substituting $a_n = A_n x_n$ gives $A_{n+2} x_{n+2} + p_1(n)A_{n+1} x_{n+1} + p_0(n)A_n x_n = 0$. But, since A_n is already a solution, we have $p_1(n)A_{n+1} = -A_{n+2} - p_0(n)A_n$. Hence, we can eliminate $p_1(n)A_{n+1}$ from the equation for x_n:

$$A_{n+2}(x_{n+2} - x_{n+1}) = p_0(n)A_n(x_{n+1} - x_n).$$

Note that this equation is a first-order homogeneous equation for $x_n^{(1)}$.

Example 5 *Reduction of order for a second-order equation.* The equation $(n+4)a_{n+2} + a_{n+1} - (n+1)a_n = 0$ is difficult to solve because it is not a constant-coefficient equation. However, by careful inspection you may be lucky enough to spot the solution: $A_n = 1/[(n+1)(n+2)]$. (Check that this is indeed a solution!)

To find a second solution linearly independent of A_n we substitute $a_n = x_n/[(n+1)(n+2)]$: $x_{n+2}/(n+3) + x_{n+1}/[(n+2)(n+3)] - x_n/(n+2) = 0$. After rearranging terms, we have $(x_{n+2} - x_{n+1})/(n+3) + (x_{n+1} - x_n)/(n+2) = 0$. Thus $x_n^{(1)}$ satisfies the first-order equation $x_{n+1}^{(1)}/(n+3) + x_n^{(1)}/(n+2) = 0$, whose solution is $x_n^{(1)} = c_1(n+2)(-1)^n$, where c_1 is an arbitrary constant. Now we must solve for x_n, which satisfies $x_{n+1} - x_n = c_1(n+2)(-1)^n$. Summing this equation from 0 to $n-1$ gives $x_n = c_1(-1)^{n+1}(2n+3)/4 + c_2$, where c_2 is an arbitrary constant.

Thus, the general solution to the original second-order difference equation is

$$a_n = \frac{c_1(-1)^{n+1}(2n+3)}{4(n+1)(n+2)} + \frac{c_2}{(n+1)(n+2)}.$$

The use of reduction of order for difference equations of order greater than 2 is discussed in Prob. 2.17.

Euler Equations

You will recall from Sec. 1.4 that, in addition to constant-coefficient differential equations, there is another kind of equation, called an Euler equation, which can be solved in general. Euler differential equations take the form

$$x^N \frac{d^N}{dx^N} y + p_{N-1} x^{N-1} \frac{d^{N-1}}{dx^{N-1}} y + \cdots + p_1 x \frac{dy}{dx} + p_0 y = 0, \tag{2.3.7}$$

where $p_0, p_1, \ldots, p_{N-1}$ are constants.

To construct the difference-equation analog of Euler equations we recall from Example 1 of Sec. 2.1 that there are simple discrete analogs of powers of x: $x \leftrightarrow n$, $x^2 \leftrightarrow n(n+1)$, $x^3 \leftrightarrow n(n+1)(n+2), \ldots, x^N \leftrightarrow n(n+1)(n+2) \cdots (n+N-1)$.

It is advantageous to adopt a simpler way of writing these lengthy formulas. We do so in terms of the Γ function:

$$\frac{\Gamma(n+N)}{\Gamma(n)} = n(n+1)(n+2) \cdots (n+N-1). \tag{2.3.8}$$

The gamma function is defined in Example 2 of Sec. 2.2.

Using this new notation, we define an *Euler difference equation* by

$$\frac{\Gamma(n+N)}{\Gamma(n)} a_n^{(N)} + p_{N-1} \frac{\Gamma(n+N-1)}{\Gamma(n)} a_n^{(N-1)} + \cdots + p_1 n a_n^{(1)} + p_0 a_n = 0. \tag{2.3.9}$$

Let us now recall how to solve the Euler differential equation (2.3.7). The substitution

$$y(x) = x^r \tag{2.3.10}$$

reduces (2.3.7) to an algebraic equation for r:

$$r(r-1)(r-2) \cdots (r-N+1) + p_{N-1} r(r-1) \times (r-2) \cdots (r-N+2) + \cdots + p_1 r + p_0 = 0. \tag{2.3.11}$$

If the roots $r_1, r_2, \ldots, r_N$ of (2.3.11) are all distinct, then the N linearly independent solutions to (2.3.7) are $y(x) = x^{r_1}, x^{r_2}, \ldots, x^{r_N}$. If (2.3.11) has repeated roots then solutions involve powers of $\ln x$. For example, if (2.3.11) has a triple root at r_1, then three solutions to (2.3.7) have the form $y(x) = x^{r_1}, x^{r_1}(\ln x), x^{r_1}(\ln x)^2$.

Apparently, to solve the discrete Euler equation (2.3.9) we must substitute the discrete analog of the function x^r, where r is a real number. We must therefore generalize (2.3.8) to noninteger values of N. We define the discrete analog of x^r to be $\Gamma(n + r)/\Gamma(n)$. Note that one discrete derivative of this is

$$\frac{\Gamma(n + r + 1)}{\Gamma(n + 1)} - \frac{\Gamma(n + r)}{\Gamma(n)} = \frac{r}{n}\frac{\Gamma(n + r)}{\Gamma(n)},$$

which is the discrete analog of rx^r/x. Two derivatives give $r(r - 1)\Gamma(n + r)/[\Gamma(n)n(n + 1)]$, which is the discrete analog of $r(r - 1)x^r/x^2$. N derivatives of $\Gamma(n + r)/\Gamma(n)$ give

$$\left[\frac{r(r - 1)(r - 2) \cdots (r - N + 1)}{\Gamma(n + N)/\Gamma(n)}\right]\left[\frac{\Gamma(n + r)}{\Gamma(n)}\right],$$

the discrete analog of $r(r - 1) \cdots (r - N + 1)x^r/x^N$.

It is now clear that substituting $a_n = \Gamma(n + r)/\Gamma(n)$ into (2.3.9) gives a polynomial equation for r:

$$r(r - 1) \cdots (r - N + 1) + p_{N-1}r(r - 1) \cdots (r - N + 2) + \cdots + p_1 r + p_0 = 0. \qquad (2.3.12)$$

The roots of this equation determine the solutions to (2.3.9).

Example 6 *Discrete Euler equations.*

(*a*) To solve the Euler equation

$$n(n + 1)a_n^{(2)} + na_n^{(1)} - \tfrac{1}{4}a_n = 0 \qquad (2.3.13)$$

we substitute $a_n = \Gamma(n + r)/\Gamma(n)$. This gives the polynomial equation $r^2 - \frac{1}{4} = 0$, which has two distinct solutions: $r_1 = \frac{1}{2}$, $r_2 = -\frac{1}{2}$. Thus, the general solution to (2.3.13) is $a_n = c_1\Gamma(n + \frac{1}{2})/\Gamma(n) + c_2\Gamma(n - \frac{1}{2})/\Gamma(n)$.

(*b*) To solve the Euler equation

$$n(n + 1)a_n^{(2)} + \tfrac{1}{4}a_n = 0 \qquad (2.3.14)$$

we again substitute $a_n = \Gamma(n + r)/\Gamma(n)$. However, now we obtain the polynomial equation $r^2 - r + \frac{1}{4} = 0$, which has a double root at $r = \frac{1}{2}$. Thus, we have obtained only *one* solution to (2.3.14): $a_n = \Gamma(n + \frac{1}{2})/\Gamma(n)$.

To find a second linearly independent solution we use reduction of order (see Prob. 2.18) and obtain $a_n = \varphi(n)\Gamma(n + \frac{1}{2})/\Gamma(n)$, where $\varphi(n) = \sum_{k=1}^{n} 1/(k - \frac{1}{2})$. Note that $\varphi(n)$ is the discrete analog of $\ln(x - \frac{1}{2}) = \int_{3/2}^{x} dt/(t - \frac{1}{2})$. Thus, the general solution of (2.3.14) is

$$a_n = c_1\frac{\Gamma(n + \frac{1}{2})}{\Gamma(n)} + c_2\varphi(n)\frac{\Gamma(n + \frac{1}{2})}{\Gamma(n)}. \qquad (2.3.15)$$

Discrete Euler equations are very similar to Euler differential equations. Whenever (2.3.12) has repeated roots, the solution to (2.3.7) contains discrete logarithms. These logarithm terms can be generated by differentiating with respect to r (see Prob. 2.19). The procedure is similar to that used in Sec. 1.4 for Euler differential equations.

Generating Functions

To facilitate the solution of a difficult difference equation, it is sometimes advisable to transform the difference equation into a differential equation. This transformation is accomplished by means of a generating function $F(x)$, which depends on the continuous variable x. The hope is that if the solution a_n to a difference equation is taken to be the nth coefficient in the Taylor expansion of $F(x)$, then $F(x)$ will satisfy an easy-to-solve differential equation.

Example 7 *Solution of a difference equation by means of a generating function.* To solve the difference equation

$$(n+1)(n+2)a_{n+2} - 2(n+1)a_{n+1} - 3a_n = 0, \qquad n \geq 0, \tag{2.3.16}$$

subject to the initial conditions $a_0 = 2$, $a_1 = 2$, we take a_n to be the nth coefficient in the Taylor expansion of a generating function $F(x)$: $F(x) \equiv \sum_{n=0}^{\infty} a_n x^n$. Differentiating this series term by term gives $F'(x) = \sum_{n=0}^{\infty} na_n x^{n-1} = \sum_{n=0}^{\infty} (n+1)a_{n+1} x^n$ and $F''(x) = \sum_{n=0}^{\infty} n(n-1)a_n x^{n-2} = \sum_{n=0}^{\infty} (n+2)(n+1)a_{n+2} x^n$. Hence, the difference equation (2.3.16) implies that $F(x)$ satisfies the constant-coefficient differential equation $F'' - 2F' - 3F = 0$, subject to the initial conditions $F(0) = 2$, $F'(0) = 2$. This initial-value problem is easy to solve and a_n is recovered by taking derivatives of $F(x)$ according to Taylor's formula; $F(x) = e^{3x} + e^{-x}$ so

$$a_n = \frac{1}{n!}\frac{d^n}{dx^n} F(x)\bigg|_{x=0} = \frac{1}{n!}[3^n + (-1)^n].$$

A generating function is particularly helpful when it replaces the difference equation by a differential equation of lower order.

Example 8 *Generating function as a means of reducing the order of a linear equation.* The second-order difference equation

$$a_{n+2} = a_{n+1} + (n+1)a_n, \tag{2.3.17}$$

subject to the initial conditions $a_0 = a_1 = 1$, can be transformed to a first-order differential equation for the generating function $F(x)$ if we define $F(x) \equiv \sum_{n=0}^{\infty} a_n x^n/n!$. To obtain the differential equation for $F(x)$ we multiply (2.3.17) by $x^{n+1}/(n+1)!$ and sum from $n = 0$ to ∞. The first term gives

$$\begin{aligned}
\sum_{n=0}^{\infty} \frac{a_{n+2}x^{n+1}}{(n+1)!} &= \frac{d}{dx}\sum_{n=0}^{\infty}\frac{a_{n+2}x^{n+2}}{(n+2)!} \\
&= \frac{d}{dx}\sum_{n=2}^{\infty}\frac{a_n x^n}{n!} \\
&= \frac{d}{dx}[F(x) - a_0 - a_1 x] \\
&= F'(x) - 1.
\end{aligned}$$

The second term gives

$$\sum_{n=0}^{\infty} \frac{a_{n+1} x^{n+1}}{(n+1)!} = \sum_{n=1}^{\infty} \frac{a_n x^n}{n!} = F(x) - 1.$$

The third term gives

$$\sum_{n=0}^{\infty} \frac{x^{n+1} a_n}{n!} = xF(x).$$

Combining these three results gives the equation for $F(x)$: $F'(x) = (x+1)F(x)$, $F(0) = 1$. Thus, $F(x) = e^{x^2/2+x}$, and the formula for a_n is

$$a_n = \frac{d^n}{dx^n} e^{x^2/2+x} \Big|_{x=0}$$

For further practice in solving difference equations using generating functions see Probs. 2.22 to 2.24.

Eigenvalue Problems

As is the case with differential equations, there are difference-equation boundary-value problems which determine eigenvalues and corresponding eigenfunctions. We conclude this section with one example of such a problem.

Example 9 *Difference-equation eigenvalue problem.* Let us consider the difference equation

$$(n+8)(n+2)a_{n+2} - (n+1)(2n+11)a_{n+1} + (n+1)(n+2)a_n = Ea_{n+1}, \tag{2.3.18}$$

subject to the two boundary conditions

$$a_{-1} \text{ is finite,} \tag{2.3.19}$$

$$S = \sum_{n=0}^{\infty} a_n \text{ is finite.} \tag{2.3.20}$$

We will see that (2.3.18) to (2.3.20) cannot hold simultaneously except for special values of the parameter E called eigenvalues.

The boundary condition in (2.3.19) implies that the elements of the sequence $a_0, a_1, a_2, \ldots$ are all multiples of a_0. To see why, we simply evaluate (2.3.18) for $n = -1$: $a_1 = Ea_0/7$. If we let $n = 0$ in (2.3.18), we obtain $a_2 = (E^2 + 11E - 14)a_0/112$. Using a_0 and $a_1 = Ea_0/7$ as initial values we can continue in this manner and calculate the entire sequence $a_0, a_1, a_2, \ldots$ from (2.3.18). Since (2.3.18) is linear and homogeneous, all elements of this sequence are multiples of a_0. For simplicity we choose $a_0 = 1$. Now, a_n is uniquely determined.

Next we examine the boundary condition in (2.3.20). Is this constraint already satisfied for any E? In Sec. 5.2 we show how to determine the approximate large-n behavior of solutions to linear difference equations. Using the techniques described there it is relatively easy to verify [see Prob. 5.6(f)] that there are two linearly independent solutions to (2.3.18) which for large n behave like

$$c_1(E)n^{-3+\sqrt{2+E}} \quad \text{and} \quad c_2(E)n^{-3-\sqrt{2+E}}, \tag{2.3.21}$$

where $c_1(E)$ and $c_2(E)$ depend on E. From these approximations we can test the convergence of the series in (2.3.20). (A series of the form $\sum a_n$ will converge if a_n approaches 0 like $n^{-\alpha}$ with $\alpha > 1$ as $n \to \infty$.) It is easy to see that when $E < 2$ this series always converges and when

$E \geq 2$ the series will not converge, except possibly for discrete (isolated) values of E for which $c_1(E)$ is zero.

Next we use a trick which gives one eigenvalue E. We simply sum the difference equation (2.3.18) from $n = -1$ to ∞. The right side gives ES, where S is the sum in (2.3.20). The left side simplifies dramatically:

$$\sum_{n=-1}^{\infty} (n+8)(n+2)a_{n+2} - \sum_{n=-1}^{\infty} (n+1)(2n+11)a_{n+1} + \sum_{n=-1}^{\infty} (n+1)(n+2)a_n$$

$$= \sum_{n=0}^{\infty} n(n+6)a_n - \sum_{n=0}^{\infty} n(2n+9)a_n + \sum_{n=0}^{\infty} (n+1)(n+2)a_n$$

$$= 2\sum_{n=0}^{\infty} a_n = 2S(!).$$

Thus, so long as the sum S exists,

$$2S = ES. \tag{2.3.22}$$

Now we must consider two possibilities. If S is nonzero, then we can conclude that $E = 2$; for all other finite values of E, S must vanish.

If we set $E = 2$ in (2.3.18), then we find the relatively simple solution

$$a_n = \frac{6!}{(n+2)(n+3)(n+4)(n+5)(n+6)} \tag{2.3.23}$$

which satisfies the conditions that a_{-1} be finite, that S exist, and that $a_0 = 1$. You should check that (2.3.23) really does satisfy (2.3.18) with $E = 2$ and that $S = \frac{3}{2}$. Note also that as $n \to \infty$, a_n goes to 0 like n^{-5}; this is precisely what is predicted in (2.3.21) with $E = 2$ and $c_1 = 0$. We say that (2.3.23) is the eigenfunction which corresponds with the eigenvalue $E = 2$. Can you use reduction of order to find another linearly independent solution to (2.3.18) with $E = 2$? This other solution will go to 0 like $1/n$ as $n \to \infty$ according to (2.3.21) and the sum S will not converge.

We know that for all $E < 2$ there are solutions a_n to (2.3.18) which satisfy the boundary conditions in (2.3.19) and (2.3.20). However, most of these solutions are very complicated functions of n. To find simple solutions we note that according to (2.3.21), when $E = -1$ and when $E = -2$, a_n approaches 0 as $n \to \infty$ like an *integral* power of $1/n$. Therefore, these solutions may well be simpler than those for other values of E. Indeed, for these two values of E, there are solutions which are rational functions of n. For $E = -1$,

$$a_n = \frac{(1 - 3n/2)\,6!}{(n+2)(n+3)(n+4)(n+5)(n+6)} \tag{2.3.24}$$

and for $E = -2$,

$$a_n = \frac{(n^2 - 13n + 6)\,5!}{(n+2)(n+3)(n+4)(n+5)(n+6)}. \tag{2.3.25}$$

You should verify that a_n in (2.3.24) and (2.3.25) satisfy the boundary conditions that a_{-1} be finite and that $a_0 = 1$, and you should prove by direct calculation that $S = \sum_0^\infty a_n = 0$. Can you find other simple eigenfunctions?

In many differential-equation eigenvalue problems eigenfunctions corresponding with different eigenvalues are *orthogonal*. The same is true here. It is easy to prove directly from (2.3.18) that if a_n and b_n are eigenfunctions corresponding with eigenvalues E_a and E_b then

$$\sum_{n=0}^{\infty} (n+2)(n+3)(n+4)(n+5)(n+6)a_n b_n = 0 \tag{2.3.26}$$

if $E_a \neq E_b$ (see Prob. 2.28). You should verify by direct calculation that a_n in (2.3.23) to (2.3.25) satisfy the orthogonality condition in (2.3.26).

Additional examples of eigenvalue problems may be found in Probs. 2.29 and 2.30.

(I) 2.4 INHOMOGENEOUS LINEAR DIFFERENCE EQUATIONS

The general solution to an inhomogeneous linear difference equation takes the form of a particular solution added to an arbitrary linear combination of solutions to the associated homogeneous linear difference equation. As with differential equations, once the homogeneous equation has been solved, it is relatively easy to find the general solution of the inhomogeneous equation. The three standard methods that are used have the same names as those used for differential equations: variation of parameters, reduction of order, method of undetermined coefficients. In this section we explain and illustrate these three techniques.

Variation of Parameters

The method of variation of parameters is used to construct a solution to the Nth-order inhomogeneous difference equation

$$a_{n+N} + p_{N-1}(n)a_{n+N-1} + \cdots + p_1(n)a_{n+1} + p_0(n)a_n = q(n) \tag{2.4.1}$$

in terms of a set of N linearly independent solutions to the associated homogeneous equation

$$a_{n+N} + p_{N-1}(n)a_{n+N-1} + \cdots + p_1(n)a_{n+1} + p_0(n)a_n = 0. \tag{2.4.2}$$

To explain the method we show how to solve second-order equations here. The case of Nth-order equations is treated in Prob. 2.31. We assume that A_n and B_n are two known linearly independent solutions to

$$a_{n+2} + p_1(n)a_{n+1} + p_0(n)a_n = 0. \tag{2.4.3}$$

To solve

$$a_{n+2} + p_1(n)a_{n+1} + p_0(n)a_n = q(n), \tag{2.4.4}$$

we seek a solution a_n of the form

$$a_n = A_n x_n + B_n y_n. \tag{2.4.5}$$

Substituting a_n in (2.4.5) into (2.4.4) gives

$$A_{n+2}x_{n+2} + B_{n+2}y_{n+2} + p_1(n)A_{n+1}x_{n+1} + p_1(n)B_{n+1}y_{n+1} + p_0(n)A_n x_n + p_0(n)B_n y_n = q(n).$$

We can eliminate A_{n+2} and B_{n+2} from this equation using the fact that A_n and B_n solve (2.4.3). We obtain:

$$-x_{n+2}p_1(n)A_{n+1} - x_{n+2}p_0(n)A_n - y_{n+2}p_1(n)B_{n+1} - y_{n+2}p_0(n)B_n + p_1(n)A_{n+1}x_{n+1} + p_1(n)B_{n+1}y_{n+1} + p_0(n)A_n x_n + p_0(n)B_n y_n = q(n). \tag{2.4.6}$$

Until now we have treated x_n and y_n as two independent functions of n. We are free to impose a constraint which in effect determines y_n in terms of x_n but leaves x_n arbitrary. It is useful to impose the constraint

$$A_{n+1}(x_{n+2} - x_{n+1}) + B_{n+1}(y_{n+2} - y_{n+1}) = 0 \tag{2.4.7}$$

because it simplifies (2.4.6) dramatically:

$$-A_n p_0(n)(x_{n+2} - x_{n+1}) - p_0(n)B_n(y_{n+2} - y_{n+1}) = q(n). \tag{2.4.8}$$

We can solve (2.4.7) and (2.4.8) simultaneously for $x_{n+1} - x_n$ and $y_{n+1} - y_n$. The result is

$$x_{n+2} - x_{n+1} = \frac{-q(n)B_{n+1}}{p_0(n)W_n},$$

$$y_{n+2} - y_{n+1} = \frac{q(n)A_{n+1}}{p_0(n)W_n},$$

where $W_n = A_n B_{n+1} - B_n A_{n+1}$, the Wronskian of the two solutions of the homogeneous equation. We therefore obtain for a_n, the general solution of the inhomogeneous equation,

$$a_n = c_1 A_n + c_2 B_n - A_n \sum^{n-2} \frac{q(j)B_{j+1}}{W_{j+1}} + B_n \sum^{n-2} \frac{q(j)A_{j+1}}{W_{j+1}}, \tag{2.4.9}$$

where we have left the lower summation limits arbitrary. In (2.4.9) c_1 and c_2 are arbitrary constants and we have used the fact that $W_{j+1} = p_0(j)W_j$ [see (2.3.5)]. Equation (2.4.9) is the final result and is the difference-equation version of (1.5.7).

Example 1 *Variation of parameters for a second-order equation.* Here we use variation of parameters to find a particular solution to the second-order inhomogeneous difference equation

$$(n + 4)a_{n+2} + a_{n+1} - (n + 1)a_n = 1. \tag{2.4.10}$$

Comparing this equation with that in (2.4.4) in which the coefficient of a_{n+2} is 1, we identify the coefficient functions $p_1(n) = 1/(n + 4)$, $p_0(n) = -(n + 1)/(n + 4)$, $q(n) = 1/(n + 4)$. Recall that two linearly independent solutions to the associated homogeneous equation are already known (see Example 5 of Sec. 2.3):

$$A_n = \frac{1}{(n + 1)(n + 2)},$$

$$B_n = \frac{(-1)^{n+1}(2n + 3)}{4(n + 1)(n + 2)}.$$

Using these formulas for A_n and B_n we compute W_n:

$$W_n = A_n B_{n+1} - B_n A_{n+1}$$

$$= \frac{(-1)^n}{(n + 1)(n + 2)(n + 3)}.$$

Thus, we have

$$\frac{B_{n+1}}{W_{n+1}} = -\frac{(n + 4)(2n + 5)}{4}, \qquad \frac{A_{n+1}}{W_{n+1}} = (-1)^{n+1}(n + 4).$$

Next, we substitute these results into (2.4.9) and obtain a formula for a particular solution:

$$a_n = -A_n \sum^{n-2} \frac{q(j)B_{j+1}}{W_{j+1}} + B_n \sum^{n-2} \frac{q(j)A_{j+1}}{W_{j+1}}$$

$$= \frac{1}{4(n+1)(n+2)} \left[\sum^{n-2} (2j+5) + (-1)^n(2n+3) \sum^{n-2} (-1)^j \right]$$

$$= \frac{1}{4(n+1)(n+2)} \left[\sum_{j=0}^{n} (2j+1) + (-1)^n(2n+3) \sum_{j=0}^{n} (-1)^j \right],$$

where in the last step we have for definiteness chosen the lower limit of summation to be 0.

Evaluating the indicated sums gives

$$a_n = \frac{1}{4(n+1)(n+2)} \left[(n+1)^2 + (2n+3)\frac{1+(-1)^n}{2} \right].$$

Note that we may discard the $(-1)^n(2n+3)$ term in the square brackets because it is proportional to B_n, a solution of the homogeneous equation. Also, we may simplify further by dropping constant (n-independent) terms in the square brackets because they are proportional to A_n. We are left with $a_n = n(n+3)/[4(n+1)(n+2)]$. It is easy to check that this is a particular solution to (2.4.10).

The general solution is obtained by adding a linear combination of solutions to the associated homogeneous equation:

$$a_n = c_1 \frac{1}{(n+1)(n+2)} + c_2 \frac{(-1)^{n+1}(2n+3)}{4(n+1)(n+2)} + \frac{n(n+3)}{4(n+1)(n+2)}.$$

Reduction of Order

Recall from Sec. 2.3 that if one solution to an Nth-order homogeneous linear difference equation is known, reduction of order may be used to divide out the known solution and reduce the order of the difference equation to $N-1$. Reduction of order also works for inhomogeneous equations. Thus, if the order of the original inhomogeneous equation is 2, then reduction of order gives a first-order inhomogeneous equation which is solvable (see Sec. 2.2).

In the case of the second-order equation

$$a_{n+2} + p_1(n)a_{n+1} + p_0(n)a_n = q(n), \tag{2.4.11}$$

we assume that A_n is a known solution of the associated homogeneous equation. As in (2.3.6) we seek a general solution to the inhomogeneous equation of the form

$$a_n = x_n A_n. \tag{2.4.12}$$

Substituting (2.4.12) into (2.4.11) gives

$$A_{n+2}(x_{n+2} - x_{n+1}) - p_0(n)A_n(x_{n+1} - x_n) = q(n). \tag{2.4.13}$$

This is a *first-order* inhomogeneous equation for $x_n^{(1)} = x_{n+1} - x_n$.

Example 2 *Reduction of order for a second-order inhomogeneous equation.* Reduction of order may be used to find the general solution of the inhomogeneous equation $a_{n+2} - (2n+1)a_{n+1}/n + na_n/(n-1) = n(n+1)$ once one solution A_n to the associated homogeneous equation is known. We assume that we have already discovered that $A_n = n - 1$.

Using $A_n = n - 1$, $p_0(n) = n/(n-1)$, and $q(n) = n(n+1)$, (2.4.13) becomes $(n+1) \times (x_{n+2} - x_{n+1}) - n(x_{n+1} - x_n) = n(n+1)$. But this equation was already solved in Example 4 of Sec. 2.2. The result is $x_{n+1} - x_n = c_1/n + (n^2 - 1)/3(n \geq 1)$, where c_1 is an arbitrary constant. We solve for x_n by summing from 1 to $n - 1$: $x_n = c_1\phi(n) + c_2 + n(n+1)(2n-5)/18$ where c_2 is another arbitrary constant and $\phi(n) = \sum_{j=1}^{n-1} 1/j$. Finally, we substitute x_n into (2.4.12) and obtain the general solution to the inhomogeneous difference equation:

$$a_n = c_1\phi(n)(n-1) + c_2(n-1) + (n-1)(n)(n+1)(2n-5)/18.$$

Method of Undetermined Coefficients

In contrast with variation of parameters and reduction of order, the method of undetermined coefficients is not a general method. It produces particular solutions to Nth-order equations of the form in (2.4.1) for which (*a*) $p_0, \ldots, p_{N-1}$ are *constants* and (*b*) $q(n)$ is a sum of terms of the form $c^n P(n)$, where c is a constant and $P(n)$ is a polynomial in n. Only very rarely does the method work for more complicated kinds of equations. We illustrate the method in the following example.

Example 3 *Method of undetermined coefficients.*

(*a*) To find a particular solution to $a_{n+2} + a_{n+1} + a_n = n$, we seek a solution of the form $a_n = An + B$, where A and B are undetermined coefficients. Since the right side is a polynomial of degree 1, it is clear that there must be a solution of this form.

Substituting into the difference equation gives $3An + 3A + 3B = n$. This equation is satisfied for all n if $A = \frac{1}{3}$ and $B = -\frac{1}{3}$. Thus, a particular solution is $a_n = (n-1)/3$.

Note that the method of undetermined coefficients gives only a particular solution without even a clue to the general solution of the inhomogeneous equation. We emphasize that there is much more information contained in one solution to the associated homogeneous equation than in a particular solution to the inhomogeneous equation. (Why?) In fact, knowing one homogeneous solution is like knowing *two* different particular solutions. (Why?)

(*b*) To find a particular solution to $a_{n+3} - a_{n+2} + a_{n+1} - a_n = 2^n n^2$, we seek a solution of the form $a_n = 2^n(An^2 + Bn + C)$. We obtain $5An^2 + (36A + 5B)n + (58A + 18B + 5C) = n^2$.

This equation is satisfied for all n if $A = \frac{1}{5}$, $B = -\frac{36}{25}$, $C = \frac{358}{125}$, so a particular solution is

$$a_n = 2^n(25n^2 - 180n + 358)/125.$$

(*c*) To find a particular solution to $a_{n+2} + 2a_{n+1} + a_n = \cos n$, we rewrite the inhomogeneous term as a sum of two terms $(e^{in} + e^{-in})/2$ and find a particular solution for each term. For example, we solve $a_{n+2} + 2a_{n+1} + a_n = e^{in}/2$ by seeking a particular solution of the form $a_n = Ae^{in}$. Substituting this form into the equation, we obtain

$$A = \frac{1}{2(e^{2i} + 2e^i + 1)} = \frac{e^{-i}}{2(2\cos 1 + 2)},$$

$$a_n = \frac{1}{4}\frac{e^{i(n-1)}}{1 + \cos 1}.$$

Adding this solution to its complex conjugate gives the complete solution to the problem $a_n = [\cos(n-1)]/[2(1 + \cos 1)]$.

(*d*) The method of undetermined coefficients happens to give a particular solution of $5a_{n+2} + na_{n+1} + (n-1)a_n = n^2$ even though the associated homogeneous equation does *not* have constant coefficients. If we look for a solution of the form $a_n = An + B$, then we obtain

$2An^2 + (5A + 2B)n + 10A + 4B = n^2$, which is satisfied for all n if $A = \frac{1}{2}$ and $B = -\frac{5}{4}$. Thus, a particular solution is $a_n = (2n - 5)/4$.

We must regard the success of the method of undetermined coefficients for the above equation as sheer luck. If we try to solve $Ka_{n+2} + na_{n+1} + (n-1)a_n = n^2$, where $K \neq 5$ and $K \neq 0$, then the above approach fails because it is equivalent to solving three simultaneous equations in the two unknowns A and B.

(E) 2.5 NONLINEAR DIFFERENCE EQUATIONS

There are very few general techniques beyond those already mentioned in Sec. 2.2 for solving nonlinear difference equations in closed form. In the following examples we summarize very briefly those which are most widely applicable; namely, substitution, use of known nonlinear functional relations, and introduction of generating functions.

Example 1 *Use of substitution to solve a nonlinear equation.* The equation $a_{n+1} = 2a_n(1 - a_n)$ may be solved by substituting $a_n = (1 - b_n)/2$. The new equation for b_n is $b_{n+1} = b_n^2$, whose solution was already given in Example 6 of Sec. 2.2 as $b_n = b_1^{(2^{n-1})}$. Thus,

$$a_n = [1 - (1 - 2a_1)^{(2^{n-1})}]/2.$$

Example 2 *Use of a nonlinear functional relation to solve a nonlinear equation.* The equation $a_{n+1} = 4a_n^3 - 3a_n$ ($|a_0| \leq 1$) can be solved by substituting $a_n = \cos\theta_n$. The equation simplifies to $\cos\theta_{n+1} = \cos(3\theta_n)$ because $\cos\theta$ satisfies the nonlinear functional relation $\cos(3\theta) = 4\cos^3\theta - 3\cos\theta$. Thus, the solution is $a_n = \cos(3^n \cos^{-1} a_0)$. Can you solve this difference equation if $|a_0| > 1$?

Example 3 *Use of a generating function to solve a nonlinear equation.* The convolution equation

$$a_{n+1} = K \sum_{j=0}^{n} a_j a_{n-j}, \qquad a_0 = 1, \tag{2.5.1}$$

is best solved by introducing the generating function $f(x) = \sum_{n=0}^{\infty} a_n x^n$ because $[f(x)]^2 = \sum_{n=0}^{\infty} x^n \sum_{j=0}^{n} a_j a_{n-j}$. If we multiply (2.5.1) by x^n and sum from $n = 0$ to ∞, we obtain an algebraic equation for $f(x)$: $[f(x) - 1]/x = K[f(x)]^2$. The solution to this equation is $f(x) = (1 \pm \sqrt{1 - 4xK})/2xK$ and we must choose the minus sign to insure that $f(0) = a_0 = 1$. Expanding the above expression into a Taylor series in powers of x gives

$$f(x) = \sum_{n=0}^{\infty} \frac{(4xK)^n \Gamma(n + \frac{1}{2})}{(n+1)!\,\Gamma(\frac{1}{2})}.$$

Thus, a_n, the coefficient of x^n, is given by

$$a_n = \frac{(4K)^n \Gamma(n + \frac{1}{2})}{(n+1)!\,\Gamma(\frac{1}{2})}.$$

Further examples of solvable nonlinear difference equations are given in the problems for Sec. 2.5.

PROBLEMS FOR CHAPTER 2

Section 2.1

(E) **2.1** Compute $\sum_{j=1}^{n} a_j$, where $a_j = j^2$ and j^3.

(I) **2.2** (*a*) Compute

$$\sum_{j=1}^{\infty} \frac{2j+3}{j^2(j+1)^2(j+2)^2(j+3)^2}.$$

(*b*) Compute

$$\sum_{j=1}^{\infty} \frac{2j+3}{j(j+1)^2(j+2)^2(j+3)}.$$

(E) **2.3** Compute

$$\sum_{n=1}^{35} \frac{1}{(n+1)(n+2)(n+3)}.$$

(E) **2.4** Show that $a_{n+k} = (D+I)^k a_n$ where $Ia_n = a_n$ and $Da_n = a_{n+1} - a_n$.

Section 2.2

(E) **2.5** Solve the following equations:

(*a*) $a_{n+1} = a_n(n+1)/n$;
(*b*) $a_{n+1} - na_n = n^2$;
(*c*) $a_{n+1} + a_n = n$;
(*d*) $a_{n+1} - a_n = 2^n$;
(*e*) $a_{n+1} - na_n/(n+1) = 1/n$.

(I) **2.6** The Γ function is defined in (2.2.2) as $\Gamma(z) = \int_0^\infty e^{-t} t^{z-1}\, dt$ (Re $z > 0$). Show that:

(*a*) $\Gamma(1) = 1$, $\Gamma(2) = 1$, $\Gamma(\frac{1}{2}) = \sqrt{\pi}$.

(*b*) $\Gamma(z+1) = z\Gamma(z)$.

Clue: Use integration by parts.

(*c*) Given that the difference equation in (*b*) is valid for all z, deduce the result that $\Gamma(z)$ has simple poles when $z = 0, -1, -2, -3, \ldots$. Find the residue of $\Gamma(z)$ at each of these poles.

(*d*) Derive the duplication formula $\sqrt{\pi}\,\Gamma(2z) = 2^{2z-1}\Gamma(z)\Gamma(z+\frac{1}{2})$ and the reflection formula $\Gamma(z)\Gamma(1-z) = \pi/\sin(\pi z)$.

Clue: It is difficult to derive these two formulas from the integral representation (2.2.2). See the references on the Γ function for useful ideas.

(*e*) Given that $\Gamma(\frac{1}{4}) \doteq 3.62561$ and $\Gamma(\frac{1}{3}) \doteq 2.67894$, compute $\Gamma(\frac{2}{3})$, $\Gamma(\frac{3}{4})$, $\Gamma(\frac{1}{6})$, $\Gamma(\frac{5}{6})$.

(*f*) Show that the integral representation $2\pi i/\Gamma(z) = \int_C t^{-z} e^t\, dt$ is valid for all z. Here C is any contour in the complex-t plane that begins at $t = -\infty - ia$ ($a > 0$), encircles the branch cut that lies along the negative real axis, and ends up at $-\infty + ib$ ($b > 0$) (see Fig. 6.12).

Clue: When Re $z < 1$, use the reflection formula stated in (*d*) to show the equivalence of this integral representation with that in (2.2.2).

(I) **2.7** Consider the partial difference equation $0 = (j+1)(2j+1)c_{n,j+1} - 2jc_{n,j} + c_{n-1,j-2} - \sum_{k=1}^{n} c_{k,1} c_{n-k,j}$, where $c_{0,0} = 1$, $c_{-1,j} = c_{0,j} = 0$ ($j > 0$), $c_{n,j} = 0$ ($j > 2n$). Use this partial differential equation to derive the following results:

(*a*) $c_{n,2n} = 1/4^n n!$;
(*b*) $c_{n,2n-1} = (4n+5)/[3 \cdot 4^n(n-1)!]$;
(*c*) $c_{n,2n-2} = (16n^2 + 64n + 82)/[18 \cdot 4^n(n-2)!]$.

(I) **2.8** Solve the partial difference equation $a_{n,m} = a_{n-1,m} + a_{n,m-1}$ ($n \geq 1, m \geq 1$) if $a_{0,m} = a_{n,0} = 1$ ($n \geq 0$, $m \geq 0$).

Section 2.3

(TE) **2.9** Show that if (2.3.2) has a multiple root r_1 of order p, then there are p linearly independent solutions to (2.3.1) of the form $r_1^n, nr_1^n, \ldots, n^{p-1}r_1^n$.

Clue: As in Sec. 1.4 for constant-coefficient differential equations, treat r as a continuous variable and differentiate with respect to r to generate new solutions to the difference equation.

(E) **2.10** Solve:

(a) $a_{n+2} - 5a_{n+1} + 4a_n = 0$ $(a_0 = 0,\ a_1 = 1)$;
(b) $a_{n+2} - 4a_{n+1} + 4a_n = 0$ $(a_0 = 0,\ a_1 = 1)$;
(c) $a_{n+3} - 3a_{n+2} + 3a_{n+1} - a_n = 0$ $(a_0 = 1,\ a_1 = 2,\ a_2 = 5)$;
(d) $a_{n+3} - a_{n+2} + a_{n+1} - a_n = 0$ $(a_0 = 0,\ a_1 = 1,\ a_2 = 2)$;
(e) $a_{n+2} - a_{n+1} + 2a_n = 0$ $(a_0 = 1,\ a_1 = 1)$.

(TI) **2.11** (a) Prove that if $W(a_n, b_n, c_n) = 0$ for all $n_1 \le n \le n_2$ where $n_2 - n_1 \ge 3$ and W is defined in (2.3.3), then a_n, b_n, and c_n are linearly dependent. Also prove the converse.

(b) Generalize the above result to N sequences $a_n, b_n, c_n, \ldots$.

(TI) **2.12** Derive (2.3.5) for difference equations of arbitrary order N.

(E) **2.13** Show that the solution to the boundary-value problem $\sqrt{2}\, a_{n+1} - (1 + \sqrt{3})a_n + \sqrt{2}\, a_{n-1} = 0$ $(2 \le n \le 6)$, $2\sqrt{2}\, a_2 = (1 + \sqrt{3})a_1$, $a_7 = 0$ is not unique.

(E) **2.14** One solution to $2na_{n+2} + (n^2 + 1)a_{n+1} - (n + 1)^2 a_n = 0$ is clearly $a_n = 1$. Find the general solution.

(E) **2.15** One solution to $n^2 a_{n+2} - n(n + 1)a_{n+1} + a_n = 0$ is $a_n = n$. Find the general solution.

(TE) **2.16** Use reduction of order to rederive the result stated in Prob. 2.9.

(TI) **2.17** Explain the method of reduction of order for difference equations of order larger than 2.

(E) **2.18** Use reduction of order to obtain a second solution to (2.3.14) given that one solution is $\Gamma(n + \frac{1}{2})/\Gamma(n)$. [The general solution is given in (2.3.15).]

(TI) **2.19** Show how to determine the structure of the general solution to an Euler difference equation when (2.3.12) has repeated roots. In particular, show that differentiating with respect to r gives rise to discrete logarithm terms as in (2.3.15).

(E) **2.20** Solve the following Euler difference equations:

(a) $n(n + 1)a_{n+2} - n(2n - 1)a_{n+1} + (n^2 - 2n + 2)a_n = 0$;
(b) $n(n + 1)a_{n+2} - 2n(n - 1)a_{n+1} + (n^2 - 3n + 4)a_n = 0$.

(E) **2.21** Solve:

(a) $n(n + 1)(n + 2)a_n^{(3)} - n(n + 1)a_n^{(2)} + na_n^{(1)} - a_n = 0$;
(b) $n(n + 1)(n + 2)(n + 3)a_n^{(4)} - a_n = 0$.

(I) **2.22** Legendre polynomials $P_n(z)$ satisfy the difference equation $(n + 1)P_{n+1} - (2n + 1)zP_n + nP_{n-1} = 0$, with $P_0(z) = 1$, $P_1(z) = z$.

(a) Define the generating function $f(x, z)$ by $f(x, z) = \sum_{n=0}^{\infty} P_n(z)x^n$. Show that $f(x, z) = (1 - 2xz + x^2)^{-1/2}$.

(b) If $g(x, z) = \sum_{n=0}^{\infty} P_n(z)x^n/n!$, show that $g(x, z) = e^{xz}J_0(x\sqrt{1 - z^2})$, where J_0 is a Bessel function.

Clue: $J_0(t)$ satisfies the differential equation $ty'' + y' + ty = 0$ with $J_0(0) = 1$, $J_0'(0) = 0$.

(I) **2.23** (a) The Bessel functions $J_n(z)$ satisfy the difference equation $J_{n+1}(z) - 2nJ_n(z)/z + J_{n-1}(z) = 0$ $(-\infty < n < \infty)$, with $J_0(0) = 1$ and $J_n(0) = 0$ $(n \ne 0)$. Define the generating function $f(x, z)$ by $f(x, z) = \sum_{n=-\infty}^{\infty} x^n J_n(z)$. Show that $f(x, z) = \exp[\frac{1}{2}z(x - 1/x)]$.

(b) Show that $J_{-n}(z) = J_n(-z) = (-1)^n J_n(z)$.

(c) Show that $1 = J_0(z) + 2\sum_1^{\infty} J_{2n}(z)$, $z = 2\sum_0^{\infty} (2n + 1)J_{2n+1}(z)$.

(I) **2.24** Hermite polynomials $\text{He}_n(z)$ satisfy the difference equation $\text{He}_{n+1}(z) = z\,\text{He}_n(z) - n\,\text{He}_{n-1}(z)$, with $\text{He}_0(z) = 1$, $\text{He}_1(z) = z$. Define the generating function $f(x, z)$ by $f(x, z) = \sum_{n=0}^{\infty} [x^n\, \text{He}_n(z)/n!]$. Show that $f(x, z) = e^{xz - x^2/2}$.

(I) **2.25** Solve in closed form the second-order linear difference equation $\delta(\delta + a)y_n + c(\delta + b)\xi y_n = f_n$, where b or $b - a$ is an integer. Here the difference operators δ and ξ are defined by $\delta y_n = n(y_{n+1} - y_n)$, $\xi y_n = ny_{n+1}$.

Clue: Show that the difference equation can be factored by converting it first to a higher-order difference equation with explicit first-order factors. To do this follow step-by-step the clue given in Prob. 1.26 and perform the analogous operations on the difference equation. In this connection, it is helpful to first show that $\delta\xi^N y_n = \xi^N(\delta + N)y_n$ where $\xi^N = \xi \cdots \xi$ with N factors.

(I) **2.26** (a) A tridiagonal matrix A has nonzero entries $a_{i,j}$ only when $|i - j| \le 1$. Define d_n as the determinant of the submatrix consisting of the first n rows and columns of A. Show that d_n satisfies a three-term difference equation.

(b) Use this result to find the eigenvalues of the $n \times n$ matrix whose entries are $a_{i,i} = 0$, $a_{i,i-1} = 1$, $a_{i,i+1} = 1$ for all i.

(TI) **2.27** Let A be a real symmetric tridiagonal matrix whose nonzero entries are $a_{i,i} = a_i$, $a_{i,i\pm 1} = b_i \neq 0$. Let A_n be the $n \times n$ matrix consisting of the first n rows and columns of A. Prove that the eigenvalues of A_n and those of A_{n-1} interlace. That is, show that if the eigenvalues of A_{n-1} and A_n in order of increasing size are $e_1, e_2, \ldots, e_{n-1}$ and $f_1, f_2, \ldots, f_n$, respectively, then $f_1 < e_1 < f_2 < e_2 < f_3 < \cdots < e_{n-1} < f_n$.

Clue: Construct the three-term difference equation satisfied by d_n, the determinant of A_n, and note the signs of the coefficients in this relation. The proof may be done by induction.

(E) **2.28** Verify (2.3.26).

(D) **2.29** Consider the eigenvalue problem

$$\frac{(n+3)^3(n-1)}{(n+1)(2n+3)}a_{n+1} + \left[\frac{16}{n(n+1)} - n(n+1)\right]a_n + \frac{(n-2)^3(n+2)}{n(2n-1)}a_{n-1} = Ea_n, \qquad n = 2, 3, 4, \ldots,$$

subject to the boundary conditions that a_1 be finite and $a_n \to 0$ as $n \to \infty$.

(a) Show that $a_n = (2n+1)/[(n+2)(n+1)(n)(n-1)]^2$ and

$$a_n = (2n+1)[2n(n+1) - 13]/[(n+2)(n+1)(n)(n-1)]^2, \qquad n \geq 2,$$

are eigenfunctions corresponding to the eigenvalues $E = -3$ and $E = -7$, respectively.

(b) Show that these eigenfunctions are orthogonal with respect to some appropriate norm.

(c) Find more eigenfunctions and eigenvalues.

(D) **2.30** Consider the eigenvalue problem $(n+1)(n+3)a_{n+1} - 2n(n+2)a_n + n^2a_{n-1} = Ea_n$, with a_{-1} finite and $a_n \to 0$ as $n \to \infty$. One eigenfunction is $a_n = 1/[(n+1)(n+2)]$, $n \geq 0$, and its associated eigenvalue is $E = 1$. Find another eigenfunction.

Section 2.4

(TI) **2.31** Assume that N linearly independent solutions to (2.4.2) are known. Find a particular solution to (2.4.1) using variation of parameters.

(E) **2.32** Find a particular solution to

(a) $a_{n+2} + a_{n+1} + a_n = ne^n + n^2e^{2n}$;

(b) $a_{n+2} - a_{n+1} + a_n = \sinh n$;

(c) $4a_{n+3} + 3a_{n+2} + 2a_{n+1} + a_n = n$.

(I) **2.33** Find particular solutions to $a_{n+2} - 2a_{n+1} + a_n = n + 1$ and $a_{n+2} - (n-1)a_{n+1} + na_n = n + 1$.

(I) **2.34** Note that $a_n = 1$ is a solution to the homogeneous part of $na_{n+2} - (n+1)a_{n+1} + a_n = 1/(n+1)$. Find the general solution.

(I) **2.35** Note that $a_n = n$ is a solution to the homogeneous part of $n^2a_{n+2} - n(n+1)a_{n+1} + a_n = 2^n$. Find the general solution.

(I) **2.36** Find the general solution to $a_{n+2} - 4a_{n+1} + 4a_n = 2^n$.

Section 2.5

(I) **2.37** Solve $a_{n+1} = 2a_n^2/(1 - 2a_n^2)$.

(I) **2.38** Solve $a_{n+1} = 4a_n(1 - a_n)$ and $a_{n+1} = 4a_n(1 + a_n)$.

(E) **2.39** Solve $a_{n+1} = [(a_n)^3(a_{n-1})^3]^{1/2}$.

(I) **2.40** The Fibonacci numbers a_n satisfy the linear recursion relation $a_{n+2} = a_{n+1} + a_n$. The first few numbers are $a_0 = 0$, $a_1 = 1$, $a_2 = 1$, $a_3 = 2$, $a_4 = 3$, $a_5 = 5$. Prove that $\sum_{j=0}^{n} a_j a_{n-j} = (n-1)a_n/5 + 2na_{n-1}/5$.

(E) **2.41** Solve $a_{n+1}a_n - p(n)(a_{n+1} - a_n) + 1 = 0$.

Clue: Let $a_n = \tan b_n$.

(E) **2.42** Solve $a_{n+1}a_n + \sqrt{1 - a_{n+1}^2}\sqrt{1 - a_n^2} = p(n)[|p(n)| \le 1]$.
Clue: Let $a_n = \cos b_n$.

(I) **2.43** Solve $a_n(a_{n+1} - a_n) = n(a_{n+1} - a_n)^2 + 1$.
Clue: Let $b_n = a_{n+1} - a_n$, apply the difference operator D to the resulting equation, and show that b_n satisfies either $b_n = b_{n+1}$ or $(n + 1)b_n b_{n+1} = 1$.

(E) **2.44** (*a*) Nonlinear difference-equation analogs of the Bernoulli differential equations $y'(x) = [y(x)]^{-1}$ and $y'(x) = [y(x)]^{-2}$ which can be solved in closed form are $a_{n+1} - a_n = 2/(a_{n+1} + a_n)$ and $a_{n+1} - a_n = 3/(a_{n+1}^2 + a_n a_{n+1} + a_n^2)$. Solve these equations.
(*b*) Find a solvable analog of $y'(x) = [y(x)]^{-N}$.

PART TWO

LOCAL ANALYSIS

My mind rebels at stagnation. Give me problems, give me work, give me the most abstruse cryptogram, or the most intricate analysis, and I am in my own proper atmosphere.

—Sherlock Holmes, *The Sign of the Four*
Sir Arthur Conan Doyle

When the methods of Part I do not yield an exact closed-form solution of a differential or difference equation or when the exact solution is too complicated to be useful, then one should try to ascertain the approximate nature of the solution. The first step toward an approximate solution is called *local analysis.*

The purpose of local analysis is to represent the solutions of equations which cannot be solved in closed form as simple expressions in terms of elementary functions. The results of a local analysis are valid in a sufficiently small neighborhood of a point. Ultimately, a uniform approximation to the behavior of the solution over an entire interval may be found by piecing together neighborhoods in which the local behavior is known. This piecing-together process uses the techniques of *global analysis* which are discussed in Part IV.

Chapter 3 introduces the basic tools of local analysis for linear differential equations. It begins by classifying the possible local behaviors of linear equations.

The techniques introduced in Chap. 3 are a prototype for all the asymptotic and perturbative analysis given in later chapters. The chapter establishes our attitude toward asymptotic analysis. Aside from stating a few general facts, our approach will be intuitive and heuristic. As in many areas of applied mathematics, it is more urgent to get the answer, the local behavior in this case, than it is to justify rigorously the means of getting it. Of course, the local behavior can usually be checked by substituting it back into the differential equation. If the differential equation is approximately satisfied, one may argue strongly for the validity of the approximations!

Chapter 4 extends the discussion to nonlinear differential equations, where the results are more limited because the techniques are less general. We will see that it is not always obvious whether the solution of a nonlinear differential equation is regular (analytic) or singular. Local analysis helps to predict which of these two behaviors occurs and, if the solution is singular, local analysis determines the nature of the singularity. Nonlinear equations can have an astonishing variety of singularities.

In Chap. 5 we draw on the close similarity between differential and difference equations to develop techniques, analogous to those in Chaps. 3 and 4, to obtain approximate solutions of linear and nonlinear difference equations.

Chapter 6 introduces techniques for the local analysis of integrals. The methods discussed here find frequent application because solutions of many mathematical and physical problems are expressible as integrals which are too difficult to evaluate exactly.

Part II of this book is designed to enlighten those mathematicians who only know how to seek exact results and who are often seen idly turning the pages in dusty books on special functions and thick tables of integrals. A practitioner of local analysis has an arsenal of useful and relatively easy things to do; local analysis is a handy first-aid kit for difficult equations.

CHAPTER

THREE

APPROXIMATE SOLUTION OF LINEAR DIFFERENTIAL EQUATIONS

Singularity is almost invariably a clue.

—Sherlock Holmes, *The Boscombe Valley Mystery*
Sir Arthur Conan Doyle

The theory of linear differential equations is so powerful that one can usually predict the local behavior of the solutions near a point x_0 without knowing how to solve the differential equation. It suffices to examine the coefficient functions of the differential equation in the neighborhood of x_0.

For example, the solutions to the formidable fourth-order differential equation $d^4y/dx^4 = (x^4 + \sin x)y$ are not expressible in terms of known mathematical functions. Nevertheless, the behavior of $y(x)$ as $x \to \infty$ is well approximated by a linear combination of the four elementary functions $x^{-3/2}e^{\pm x^2/2}$, $x^{-3/2} \sin (x^2/2)$, and $x^{-3/2} \cos (x^2/2)$. You will be pleased to know that after reading this chapter you will be able to derive this nifty result on the back of a postage stamp.

Even when the solution to a differential equation can be expressed in terms of the common higher transcendental functions, the techniques of local analysis are still very useful. For example, saying that the solutions to $y'' = x^4y$ are expressible in terms of modified Bessel functions of order $\frac{1}{6}$ does not convey much qualitative information to someone who is not an expert on Bessel functions. On the other hand, an easy local analysis of the differential equation shows that solutions behave as linear combinations of $x^{-1}e^{\pm x^3/3}$ as $x \to +\infty$.

Some of the examples in this chapter involve differential equations that can be solved exactly in terms of higher transcendental functions. Although the emphasis is on asymptotic analysis and not on the properties of these special functions, by the end of this chapter we will have accumulated a large number of results concerning special functions. For the reader's convenience, we have included an appendix which summarizes the most important formulas.

(E) 3.1 CLASSIFICATION OF SINGULAR POINTS OF HOMOGENEOUS LINEAR EQUATIONS

In this section we begin the process of local analysis by classifying a point x_0, which may be complex, as an *ordinary point*, a *regular singular point*, or an *irregular singular point* of a homogeneous linear equation. This classification gives the first indication of the nature of the solutions near x_0 and suggests the appropriate route for further systematic analysis.

In this chapter we continue to represent homogeneous linear differential equations by the form used in Chap. 1:

$$y^{(n)}(x) + p_{n-1}(x)y^{(n-1)}(x) + \cdots + p_1(x)y^{(1)}(x) + p_0(x)y(x) = 0, \qquad (3.1.1)$$

where $y^{(k)}(x) = d^k y/dx^k$. The classification scheme we are about to describe assumes that $p_0, \ldots, p_{n-1}$ have been defined for complex as well as for real values of their arguments.

Ordinary Points

The point x_0 ($x_0 \neq \infty$) is called an *ordinary point* of (3.1.1) if the coefficient functions $p_0(x), \ldots, p_{n-1}(x)$ are all analytic in a neighborhood of x_0 in the complex plane.

Example 1 *Ordinary points.*

(*a*) $y'' = e^x y$. Every point $x_0 \neq \infty$ is an ordinary point because e^x is entire.
(*b*) $x^5 y''' = y$. Every point x_0 except for $x_0 = 0$ and ∞ is an ordinary point.
(*c*) $y' = |x| y$. There are *no* ordinary points in the complex-x plane because $|x|$ is nowhere analytic. (See Sec. 1.9.)

Fuchs proved in 1866 that *all n linearly independent solutions of (3.1.1) are analytic in a neighborhood of an ordinary point.* Moreover, he proved that if any solution is expanded in a Taylor series about the ordinary point x_0, $y(x) = \sum_{n=0}^{\infty} a_n(x - x_0)^n$, then the radius of convergence of this series is at least as large as the distance to the nearest singularity of the coefficient functions in the complex plane (see Prob. 3.9). The location of a singularity of a solution must coincide with the location of a singularity of a coefficient function. The solution of a linear equation cannot have singularities at any other points.

Example 2 *Taylor series at an ordinary point.* The equation $(x^2 + 1)y' + 2xy = 0$ has an ordinary point at 0. The solution $y = c(1 + x^2)^{-1}$ can be expanded in a Taylor series whose radius of convergence is 1; this is the distance to the coefficient singularities at $x = \pm i$ when the differential equation is written in the form of (3.1.1).

Regular Singular Points

The point x_0 ($x_0 \neq \infty$) is called a *regular singular point* of (3.1.1) if not all of $p_0(x), \ldots, p_{n-1}(x)$ are analytic but if all of $(x - x_0)^n p_0(x)$, $(x - x_0)^{n-1} p_1(x), \ldots,$ $(x - x_0)p_{n-1}(x)$ *are* analytic in a neighborhood of x_0.

Example 3 *Regular singular points.*

(*a*) $(x-1)y''' = y$ has a regular singular point at 1.
(*b*) $x^2y'' + xy' = y$ has a regular singular point at 0.
(*c*) $x^3y' = (x+1)y$ does *not* have a regular singular point at 0.

A solution of (3.1.1) *may be analytic at a regular singular point. If it is not analytic, its singularity must be either a pole or an algebraic or logarithmic branch point.* Fuchs showed that there is always at least one solution of the form

$$y = (x - x_0)^\alpha A(x), \tag{3.1.2}$$

where α is a number called the *indicial exponent* and $A(x)$ is a function which is analytic at x_0 and which has a Taylor series whose radius of convergence is at least as large as the distance to the nearest singularity of the coefficient functions in the complex plane (see Prob. 3.23).

Example 4 *Taylor series at a regular singular point.* The equation $y' = y/\sinh x$ has a regular singular point at 0. The solution $y(x) = c \tanh (x/2)$ is analytic at $x = 0$ but has poles at $x = \pm i\pi$. Thus, the radius of convergence of the Taylor expansion of $y(x)$ is π, the distance to the nearest singularities in the complex plane of the coefficient function $1/\sinh x$.

If (3.1.1) is of order $n \geq 2$, then there is a second linearly independent solution having one of two possible forms:

$$y = (x - x_0)^\beta B(x) \tag{3.1.3}$$

or

$$y = (x - x_0)^\alpha A(x) \ln (x - x_0) + C(x)(x - x_0)^\beta. \tag{3.1.4}$$

Equations (3.1.3) and (3.1.4) are generalizations of the solutions of an Euler equation for nonrepeated and repeated indicial exponents (see Sec. 1.4). $B(x)$ and $C(x)$ are new functions which are also analytic at x_0 and which have radii of convergence at least as large as the distance to the nearest singularity of the coefficient functions. $A(x)$ is the same function that appears in (3.1.2).

In general, for each new linearly independent solution there is a new analytic function of x and either a new indicial exponent or another power of $\ln (x - x_0)$. Thus, the form of the nth solution is at worst

$$y(x) = (x - x_0)^\gamma \sum_{i=0}^{n-1} [\ln (x - x_0)]^i A_i(x), \tag{3.1.5}$$

where all the functions $A_i(x)$ are analytic at x_0. Conversely, Fuchs showed that *if all n solutions have the forms* (3.1.2) *to* (3.1.5), *then x_0 is at worst a regular singular point of the equation* (see Probs. 3.14 to 3.17).

Irregular Singular Points

The point x_0 ($x_0 \neq \infty$) is called an *irregular singular point* of (3.1.1) if it is neither an ordinary point nor a regular singular point. There is no comprehensive theory of irregular singular points, but we can say that at an irregular singular point at least one solution is *not* of the forms (3.1.2) to (3.1.5). Typically, *at an irregular singular point, all solutions exhibit an essential singularity, often in combination with a pole or an algebraic or logarithmic branch point. But this is not always the case. Some of the solutions may not have an essential singularity and may even be analytic at* x_0.

Classification of the Point $x_0 = \infty$

We have completed the classification of points x_0 in the finite complex plane, but it is also useful to classify the point $x_0 = \infty$. We do this by analytically mapping the point at infinity into the origin using the inversion transformation

$$x = \frac{1}{t},$$

$$\frac{d}{dx} = -t^2 \frac{d}{dt}, \tag{3.1.6}$$

$$\frac{d^2}{dx^2} = t^4 \frac{d^2}{dt^2} + 2t^3 \frac{d}{dt},$$

and so on, and then classifying the point $t = 0$. *The point* $x_0 = \infty$ *is called an ordinary, a regular singular, or an irregular singular point if the point at* $t = 0$ *is correspondingly classified.*

Illustrative Examples

The remainder of this section is a collection of elementary examples which further explain and illustrate how to classify points of differential equations. In the following examples observe how solutions behave in the vicinity of ordinary, regular singular, and irregular singular points.

Example 5 *Comparison of ordinary, regular singular, and irregular singular points.* Consider the three equations

$$\frac{dy}{dx} - \frac{1}{2} y = 0, \tag{3.1.7}$$

$$\frac{dy}{dx} - \frac{1}{2x} y = 0, \tag{3.1.8}$$

$$\frac{dy}{dx} - \frac{1}{2x^2} y = 0. \tag{3.1.9}$$

The transformation $x = 1/t$ gives three new equations which are respectively $dy/dt + y/2t^2 = 0$, $dy/dt + y/2t = 0$, $dy/dt + y/2 = 0$. Every point of (3.1.7) is an ordinary point except for ∞ which is an irregular singular point. As expected, the solution $y(x) = ce^{x/2}$ is analytic except for an essential singularity at $x = \infty$. Every point of (3.1.8) is an ordinary point except for 0 and ∞ which are regular singular points. The solution $y(x) = cx^{1/2}$ is analytic except for branch points at $x = 0$ and $x = \infty$. Every point of (3.1.9) is an ordinary point except for 0 which is an irregular singular point. The solution $y(x) = ce^{-1/2x}$ is analytic in the extended plane except for an essential singularity at $x = 0$.

Example 6 *Taylor series about an ordinary point.* The equation $y' + y/(x - 1) = 0$ has regular singular points at 1 and ∞. The solution $y(x) = c/(1 - x)$ has a pole at $x = 1$ and is analytic at ∞. The Taylor series of the solution about $x = 0$, $y(x) = c \sum_{n=0}^{\infty} x^n$, has radius of convergence 1, which is the distance to the regular singular point at 1.

Example 7 *Taylor series solutions which converge beyond the nearest singular point of the differential equation.* The equation $(x - 1)(2x - 1)y'' + 2xy' - 2y = 0$ has regular singular points at $\frac{1}{2}$, 1, and ∞. One solution of this equation is $y(x) = 1/(x - 1)$. A Taylor series expansion of this solution about the ordinary point at 0 converges beyond the first singular point at $\frac{1}{2}$ but ceases to converge at $|x| = 1$. A linearly independent solution is $y(x) = x$ whose Taylor series about $x = 0$ converges for all x.

Example 8 *Essential singular behavior near an irregular singular point.* The equation $y'' + 3y'/2x + y/4x^3 = 0$ has an irregular singular point at 0 and a regular singular point at ∞. Two linearly independent solutions are $y(x) = \sin x^{-1/2}$, $y(x) = \cos x^{-1/2}$. Both of these solutions have an essential singularity at the origin. The first of these also has branch points at $x = 0$ and $x = \infty$. The second solution has no branch cut and is analytic at $x = \infty$.

Example 9 *Analytic solutions near singular points.* At regular or irregular singular points, one or even several linearly independent solutions may be analytic. The equation $y'' + (1 - x)y'/x - y/x^2 = 0$ has a regular singular point at 0 and an irregular singular point at ∞. One solution, $y(x) = (e^x - 1 - x)/x$, is analytic at $x = 0$ but has an essential singularity at $x = \infty$. A linearly independent solution, $y(x) = (1 + x)/x$, has a pole at $x = 0$ but is analytic at $x = \infty$.

The equation in Example 2 of Sec. 1.3,

$$y'' - \frac{1 + x}{x} y' + \frac{1}{x} y = 0, \tag{3.1.10}$$

again has a regular singular point at 0 and an irregular singular point at ∞. *Both* linearly independent solutions

$$y(x) = e^x, \qquad y(x) = 1 + x \tag{3.1.11}$$

are analytic at $x = 0$. The first has an essential singularity at $x = \infty$ and the second has a pole at $x = \infty$. In general, all linearly independent solutions may be analytic at a regular singular point but at least one solution must be singular at an irregular singular point.

Sometimes it is possible to alter the character of a singular point by a transformation of the independent or dependent variable.

Example 10 *Removing a singularity by transforming the independent variable.* The irregular singular point at 0 of $y' - \frac{1}{2}x^{-1/2}y = 0$ disappears if we introduce a new independent variable $t = x^{1/2}$. The resulting equation, $dy/dt - y = 0$, has an ordinary point at $t = 0$.

Example 11 *Removing a singularity by transforming the dependent variable.*

$$y'' + \frac{2}{x} y' - y = 0 \tag{3.1.12}$$

has a regular singular point at 0 and an irregular singular point at ∞. The singular point at 0 may be removed by the transformation $y(x) = w(x)/x$, where $w(x)$ satisfies an equation which still has an irregular singularity at ∞: $w'' - w = 0$.

Example 12 *Removing a singularity by converting to a linear system.* Ostensibly the reason why the singularity of (3.1.12) is removable is that the two solutions have the same kind of singularity at $x = 0$: $y = e^x/x$, $y = e^{-x}/x$. However, it is difficult to find a transformation which eliminates the regular singular point at 0 of (3.1.10), $y'' - (1 + x)y'/x + y/x = 0$, even though both solutions $y = e^x$ and $y = 1 + x$ are analytic at $x = 0$! The only way to eliminate this singular point is to convert (3.1.10) into a linear *system* of equations of the form

$$\begin{aligned} y'(x) &= a(x)y(x) + b(x)z(x), \\ z'(x) &= c(x)y(x) + d(x)z(x). \end{aligned} \tag{3.1.13}$$

If the general solution of a second-order differential equation is analytic at $x = 0$, then it is possible to find an equivalent homogeneous linear system of equations of the form (3.1.13) whose coefficients are analytic at $x = 0$. This fact may be especially helpful to an unhappy numerical analyst who is trying to solve a differential equation which has a singular point where the solutions are all known to be analytic. An appropriate system for (3.1.10) is $y' = y + xz$, $z' = 0$.

Can you find *another* system of the form in (3.1.13) with analytic coefficients which is equivalent to (3.1.10)?

(E) 3.2 LOCAL BEHAVIOR NEAR ORDINARY POINTS OF HOMOGENEOUS LINEAR EQUATIONS

The examples in Sec. 3.1 illustrate various kinds of local behavior that solutions to homogeneous linear differential equations may exhibit in the neighborhood of a point x_0. However, those examples all have simple closed-form solutions. In the rest of this chapter we show how to represent the local behavior in terms of infinite series expansions when exact closed-form solutions are not known. The general procedure consists of first classifying the point x_0 as an ordinary, a regular singular, or an irregular singular point and then selecting a suitable form for the series based on this classification.

In general, it is very useful to know how to obtain answers to hard problems in terms of infinite series. This mode of attack can reduce an otherwise tough analysis to a sequence of simple operations for generating the terms in the series. Infinite series techniques are a good psychological restorative because they can give a scientist who is stalled on a problem a sense of forward progress as he or she computes the terms. Moreover, the first few terms of the series are usually sufficient to give a very accurate approximation to the local behavior of the solution to a differential equation.

A local series expansion about x_0 of a solution to a differential equation is an example of a *perturbation* series. A perturbation series is a series in powers of a small parameter. In this instance the small parameter is the distance between x and x_0. Since $y(x)$ is analytic near x_0 if x_0 is an ordinary point, the perturbation expansion is a Taylor series in powers of $x - x_0$. Here the nth approximant (the sum of the first n terms of the perturbation expansion) to the local behavior near an ordinary point becomes a more accurate approximation to the solution as $|x - x_0|$ becomes smaller, or as n increases, or both. A general treatment of perturbation series is given in Chap. 7.

To obtain a series solution about an ordinary point we substitute the Taylor series $y(x) = \sum_{n=0}^{\infty} a_n(x - x_0)^n$ into the differential equation and determine the coefficients a_n by solving a recursion relation.

Example 1 *Taylor series solution of a first-order differential equation.* Consider the initial-value problem $y' = 2xy$ $[y(0) = 1]$. Since $x = 0$ is an ordinary point, we may seek a solution in the form of the Taylor series $y(x) = \sum_{n=0}^{\infty} a_n x^n$. Substituting into the differential equation and differentiating term by term gives $\sum_{n=0}^{\infty} na_n x^{n-1} = 2x \sum_{n=0}^{\infty} a_n x^n$. After a shift of indices on the right side, we find that

$$\sum_{n=0}^{\infty} na_n x^{n-1} = 2 \sum_{n=2}^{\infty} a_{n-2} x^{n-1}.$$

We now resort to an important formal procedure that is used frequently. Since the coefficients of a Taylor expansion of a function are unique, it follows that if two functions are equal, their Taylor series coefficients must agree. This argument justifies equating the coefficients of x^{n-1} in y' and $2xy$ and leads to the recursion relation of a_n:

$$na_n = 0, \qquad n = 0, 1,$$
$$na_n = 2a_{n-2}, \qquad n = 2, 3, \ldots,$$

which can be solved in closed form. From the first equation with $n = 0$ we conclude that a_0 is an arbitrary constant and from the first equation with $n = 1$ we conclude that $a_1 = 0$. The solutions to the second equation are $a_{2n} = a_0/n!(n \geq 1)$, $a_{2n+1} = 0$ $(n \geq 1)$. Thus, the *general* solution to $y' = 2xy$ is

$$y(x) = a_0 \sum_{n=0}^{\infty} x^{2n}/n! = a_0 e^{x^2}.$$

This result may be verified by substituting it into the differential equation.

Example 2 *Local analysis of the Airy equation at* $x = 0$. To find the local behavior of the solutions to the Airy equation (1.4.8) $y'' = xy$ near $x = 0$ we substitute the series $y(x) = \sum_{n=0}^{\infty} a_n x^n$ and differentiate term by term. The result is

$$\sum_{n=0}^{\infty} a_n n(n-1)x^{n-2} = \sum_{n=0}^{\infty} a_n x^{n+1}.$$

After a shift of indices, the right side becomes $\sum_{n=3}^{\infty} a_{n-3} x^{n-2}$.

Equating the coefficients of x^{n-2} in y'' and xy gives

$$a_n n(n-1) = 0, \qquad n = 0, 1, 2,$$
$$a_n n(n-1) = a_{n-3}, \qquad n = 3, 4, \ldots,$$

which can be solved in closed form. The first equation is already satisfied for $n = 0$ and $n = 1$. Hence, a_0 and a_1 are *arbitrary* constants. Also, $a_2 = 0$. The solutions of the second equation are

$$\begin{aligned} a_{3n} &= \frac{a_0}{3n(3n-1)(3n-3)(3n-4)\cdots 9\cdot 8\cdot 6\cdot 5\cdot 3\cdot 2} \\ &= \frac{a_0}{3^n n!\, 3^n(n-\frac{1}{3})(n-1-\frac{1}{3})(n-2-\frac{1}{3})\cdots(\frac{5}{3})(\frac{2}{3})} \\ &= \frac{a_0\Gamma(\frac{2}{3})}{3^n n!\, 3^n\Gamma(n+\frac{2}{3})}, \\ a_{3n+1} &= \frac{a_1}{(3n+1)(3n)(3n-2)(3n-3)\cdots 10\cdot 9\cdot 7\cdot 6\cdot 4\cdot 3} \\ &= \frac{a_1}{3^n n!\, 3^n(n+\frac{1}{3})(n-1+\frac{1}{3})(n-2+\frac{1}{3})\cdots(\frac{7}{3})(\frac{4}{3})} \\ &= \frac{a_1\Gamma(\frac{4}{3})}{3^n n!\, 3^n\Gamma(n+\frac{4}{3})}, \\ a_{3n+2} &= 0. \end{aligned}$$

[The gamma function $\Gamma(z)$ is a function of complex z which satisfies $\Gamma(z+1) = z\Gamma(z)$. It is defined so that $\Gamma(n+1) = n!$ when n is a positive integer. $\Gamma(z)$ is discussed in Example 2 of Sec. 2.2.] If we now define $c_1 = a_0\,\Gamma(\frac{2}{3})$ and $c_2 = a_1\,\Gamma(\frac{4}{3})$, then the *general* solution of the Airy equation is

$$y(x) = c_1 \sum_{n=0}^{\infty} \frac{x^{3n}}{9^n n!\,\Gamma(n+\frac{2}{3})} + c_2 \sum_{n=0}^{\infty} \frac{x^{3n+1}}{9^n n!\,\Gamma(n+\frac{4}{3})}.$$

We have obtained *two* linearly independent solutions, each multiplied by an arbitrary constant of integration, even though we started with only one series!

The general solution is approximated for all finite x in terms of c_1 and c_2 by a rapidly convergent Taylor expansion. As $|x|$ gets smaller the expansion converges faster; fewer terms are required for any desired accuracy. The radii of convergence of these series are infinite because the Airy equation has no singular points in the finite x plane. Note, however, that c_1 and c_2 must be known before $y(x)$ can be approximated. Ordinarily, c_1 and c_2 would be determined from initial conditions given at $x = 0$, $c_1 = \Gamma(\frac{2}{3})y(0)$, $c_2 = \Gamma(\frac{4}{3})y'(0)$, or computed numerically from initial conditions given at some other point. If c_1 and c_2 are not known, then the expansion of $y(x)$ is not known either. This is typical in perturbation theory (see Chap. 7): to compute the nth term in a perturbation expansion, one must know the previous terms. In this problem, once the first two terms are known (which requires that c_1 and c_2 be given) then the rest of the expansion is relatively easy to obtain.

It is conventional to define two special linearly independent solutions of the Airy equation by

$$\text{Ai}\,(x) \equiv 3^{-2/3} \sum_{n=0}^{\infty} \frac{x^{3n}}{9^n n!\,\Gamma(n+\frac{2}{3})} - 3^{-4/3} \sum_{n=0}^{\infty} \frac{x^{3n+1}}{9^n n!\,\Gamma(n+\frac{4}{3})}, \tag{3.2.1}$$

$$\text{Bi}\,(x) \equiv 3^{-1/6} \sum_{n=0}^{\infty} \frac{x^{3n}}{9^n n!\,\Gamma(n+\frac{2}{3})} + 3^{-5/6} \sum_{n=0}^{\infty} \frac{x^{3n+1}}{9^n n!\,\Gamma(n+\frac{4}{3})}. \tag{3.2.2}$$

Ai (x) and Bi (x) are called Airy functions. When x is not too large, the Taylor series in (3.2.1) and (3.2.2) give good approximations to the Airy functions with only a few retained terms. For an error estimate, see Prob. 3.7.

In Fig. 3.1 we have plotted Ai (x) and Bi (x) for $-10 \le x \le 2$. These functions are of primary importance in perturbative analysis of differential equations (see Part IV). In (3.2.1) and (3.2.2) c_1 and c_2 are chosen so that Ai (x) is exponentially decreasing as $x \to +\infty$ and that Bi (x) oscillates 90° out of phase from Ai (x) as $x \to -\infty$ (see Fig. 3.1). It is not trivial to derive these properties from the series (3.2.1) and (3.2.2); these properties are established in Example 10 of Sec. 3.8 and Probs. 6.75 and 6.76.

(I) 3.3 LOCAL SERIES EXPANSIONS ABOUT REGULAR SINGULAR POINTS OF HOMOGENEOUS LINEAR EQUATIONS

We have seen in Sec. 3.2 that Taylor series are a good way to represent the local behavior of a solution to a differential equation near an ordinary point. What happens if we try to represent the local behavior of a solution near a regular singular point by a Taylor series?

Example 1 *Breakdown of a Taylor series representation at a regular singular point.* If we seek a solution to $y'' + y/4x^2 = 0$ in the form of a Taylor series about $x = 0$, the formal expansion procedure which works for ordinary points is not fruitful here. To wit, if we substitute the Taylor series $y = \sum_{n=0}^{\infty} a_n x^n$ into this equation and differentiate term by term, we obtain the following sequence of equations for a_n: $(n - \frac{1}{2})^2 a_n = 0$ $(n = 0, 1, 2, \ldots)$. The solution to these equations is $a_n = 0$ for all n. Thus, we obtain only the trivial solution $y(x) = 0$. This is not progress.

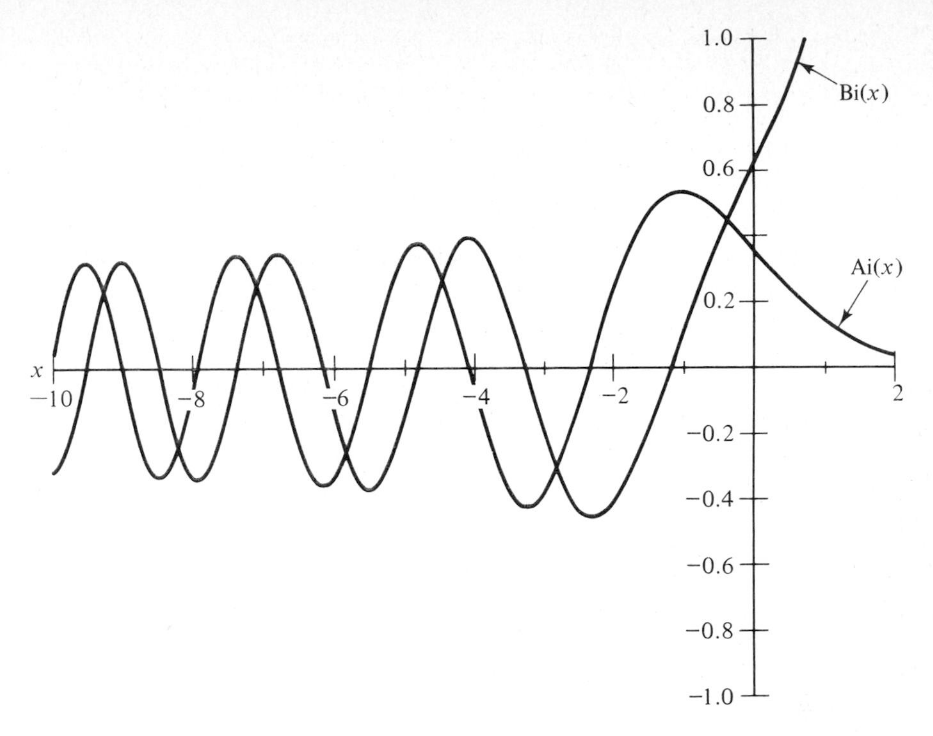

Figure 3.1 A plot of the Airy functions Ai(x) and Bi(x) for $-10 \le x \le 2$. Both functions are oscillatory for negative x; Bi(x) grows exponentially and Ai(x) decays exponentially as $x \to +\infty$.

Taylor series expansion failed in Example 1 because Taylor series are not general enough to describe the local behavior of solutions near regular singular points. Fortunately, the result of Fuchs stated in Sec. 3.1 suggests a more general structure than Taylor series. In particular, we learned that if x_0 is a regular singular point of a linear homogeneous differential equation, at least one solution must have the form (3.1.2): $y(x) = (x - x_0)^\alpha A(x)$, where $A(x)$ is analytic at x_0. This solution has an algebraic branch point at x_0 if α is nonintegral and a pole at x_0 if α is a negative integer. $y(x)$ is analytic at x_0 if $\alpha = 0, 1, 2, \ldots$. The other $n - 1$ linearly independent solutions of an nth-order equation have the form (3.1.3), (3.1.4), or (3.1.5), so they may also exhibit logarithmic branch points.

Since $A(x)$ is analytic, it can be expanded in a Taylor series:

$$y(x) = (x - x_0)^\alpha A(x) = (x - x_0)^\alpha \sum_{n=0}^{\infty} a_n (x - x_0)^n. \tag{3.3.1}$$

We call the right side of (3.3.1) a *Frobenius* series and we call the number α an *indicial exponent*. It is conventional to assume that $a_0 \neq 0$ in a Frobenius series, which is ensured by a proper choice of α. A Taylor series is a special case of a Frobenius series.

Example 2 *Local analysis at a regular singular point.* If we try a Frobenius series of the form (3.3.1) with $x_0 = 0$ in the differential equation of Example 1, we find that $[(n + \alpha) \times (n + \alpha - 1) + \frac{1}{4}]a_n = 0$ $(n = 0, 1, 2, \ldots)$. Since we are assuming that $a_0 \neq 0$, it follows that

$\alpha(\alpha - 1) + \frac{1}{4} = 0$ so $\alpha = \frac{1}{2}$. With $\alpha = \frac{1}{2}$, the remaining equations for a_n give $a_1 = a_2 = a_3 = \cdots = 0$. Thus, the Frobenius series takes the simple form $y(x) = a_0\sqrt{x}$, where a_0 is arbitrary. This one-term Frobenius series is an exact solution of the differential equation $y'' + y/4x^2 = 0$.

From our study of Euler equations in Sec. 1.4, we know that a second solution to this differential equation is $y(x) = b\sqrt{x} \ln x$, which is not in the form (3.3.1) of a Frobenius series. The general theory only guarantees that there be one solution in the form of a Frobenius series. The other solutions can be slightly more complicated.

In this section we review a systematic procedure advanced by Fuchs and Frobenius for calculating the indicial exponent α and the expansion coefficients a_n. This procedure can also be used to find series expansions of these solutions having the form (3.1.4) and (3.1.5) in which logarithms appear. A central part of the procedure is a recipe for deciding which of the forms (3.1.2), (3.1.3), (3.1.4), or (3.1.5) are appropriate to represent the other solutions. The method is first worked out in detail for second-order equations; the necessary generalizations to nth-order equations are indicated at the end of the section and further developed in the problems.

Frobenius Method for Second-Order Equations

If the differential equation

$$y'' + \frac{p(x)}{x - x_0} y' + \frac{q(x)}{(x - x_0)^2} y = 0 \tag{3.3.2}$$

has a regular singular point at x_0, then $p(x)$ and $q(x)$ are analytic at x_0. Thus, we may expand $p(x)$ and $q(x)$ in Taylor series about x_0: $p(x) = \sum_{n=0}^{\infty} p_n(x - x_0)^n$, $q(x) = \sum_{n=0}^{\infty} q_n(x - x_0)^n$. We substitute these Taylor expansions into (3.3.2) and obtain a solution $y(x)$ in the form of the Frobenius series (3.3.1) by equating the coefficients of $(x - x_0)^{n+\alpha-2}$ for $n = 0, 1, 2, \ldots$:

$$(x - x_0)^{\alpha-2}: \qquad [\alpha^2 + (p_0 - 1)\alpha + q_0]a_0 = 0, \tag{3.3.3a}$$

$$(x - x_0)^{n+\alpha-2}: \qquad [(\alpha + n)^2 + (p_0 - 1)(\alpha + n) + q_0]a_n = -\sum_{k=0}^{n-1} [(\alpha + k)p_{n-k} + q_{n-k}]a_k, \qquad n = 1, 2, \ldots. \tag{3.3.3b}$$

By assumption a_0 is nonzero, so (3.3.3a) requires that α must be a root of the indicial polynomial $P(\alpha)$, where

$$P(\alpha) = \alpha^2 + (p_0 - 1)\alpha + q_0. \tag{3.3.4}$$

Given these values of α we must then solve the recursion relation (3.3.3b) for a_n in terms of a_0. The constant a_0 is arbitrary and will ultimately appear as an overall multiplicative factor in the solution $y(x)$. However, the recursion relation (3.3.3b) can be solved for a_n in terms of a_k for $k < n$ only if $P(\alpha + n) \neq 0$ because the left side of (3.3.3b) is precisely $P(\alpha + n)a_n$. If this condition holds for all positive integers n, then it may be shown that the series (3.3.1) converges in a circle whose radius is at least as large as the distance to the nearest complex singularity of $p(x)$ or $q(x)$ (see Prob. 3.23).

Let α_1 and α_2 denote the two roots of the indicial polynomial $P(\alpha)$ which are ordered so that $\text{Re}\ \alpha_1 \geq \text{Re}\ \alpha_2$. If we let $\alpha = \alpha_1$ then $P(\alpha_1 + n) \neq 0$ for $n = 1, 2, \ldots$ because α_2 is the only other root of P. Thus, the recursion relation (3.3.3b) can be solved for a_n in terms of a_0 for all n. This explains why there is always at least one solution in the form of a Frobenius series.

Example 3 *Modified Bessel equation of order ν.* The modified Bessel equation

$$y'' + \frac{1}{x}y' - \left(1 + \frac{\nu^2}{x^2}\right)y = 0 \tag{3.3.5}$$

has a regular singular point at $x = 0$. Substituting the Frobenius series $y = \sum_{n=0}^{\infty} a_n x^{\alpha+n}$ into (3.3.5) gives

$$\sum_{n=0}^{\infty} (\alpha + n)(\alpha + n - 1)a_n x^{\alpha+n-2} + \sum_{n=0}^{\infty} (\alpha + n)a_n x^{\alpha+n-2} - \sum_{n=0}^{\infty} a_n x^{\alpha+n} - \nu^2 \sum_{n=0}^{\infty} a_n x^{\alpha+n-2} = 0.$$

Rewriting $\sum_{n=0}^{\infty} a_n x^{\alpha+n}$ as $\sum_{n=2}^{\infty} a_{n-2} x^{\alpha+n-2}$ and equating coefficients of powers of x to zero gives

$$\begin{aligned} x^{\alpha-2}&: & (\alpha^2 - \nu^2)a_0 &= 0, \\ x^{\alpha-1}&: & [(\alpha + 1)^2 - \nu^2]a_1 &= 0, \\ x^{\alpha+n-2}&: & [(\alpha + n)^2 - \nu^2]a_n &= a_{n-2}, \qquad n = 2, 3, \ldots. \end{aligned} \tag{3.3.6}$$

Since $a_0 \neq 0$, α must be a root of the indicial polynomial $P(\alpha) = \alpha^2 - \nu^2$. Therefore, $\alpha = \pm\nu$. Since ν appears as ν^2 in the Bessel equation, we may assume that $\text{Re}\ \nu \geq 0$ and denote $\alpha_1 = +\nu$ and $\alpha_2 = -\nu$. Thus $P(\alpha_1 + n) \neq 0$ for $n = 1, 2, \ldots$ and $a_1, a_2, a_3, \ldots$ may be easily determined from (3.3.6). The results are $a_1 = a_3 = a_5 = \cdots = 0$ and

$$\begin{aligned} a_{2n} &= \frac{a_{2n-2}}{2^2 n(\nu + n)} = \frac{a_{2n-4}}{2^4 n(n-1)(\nu + n)(\nu + n - 1)} = \cdots \\ &= \frac{a_0 \Gamma(\nu + 1)}{2^{2n} n!\, \Gamma(\nu + n + 1)}. \end{aligned} \tag{3.3.7}$$

Thus,
$$y(x) = a_0 \Gamma(\nu + 1) x^{\nu} \sum_{n=0}^{\infty} \frac{(\frac{1}{2}x)^{2n}}{n!\, \Gamma(\nu + n + 1)}.$$

It is conventional to set $a_0 = 2^{-\nu}/\Gamma(\nu + 1)$ in the above equations; the result is the Frobenius series expansion of the modified Bessel function $I_\nu(x)$:

$$I_\nu(x) = \sum_{n=0}^{\infty} \frac{(\frac{1}{2}x)^{2n+\nu}}{n!\, \Gamma(\nu + n + 1)}. \tag{3.3.8}$$

The ratio test shows that this series has an infinite radius of convergence; we could have predicted this result by noting that the only singular point of (3.3.5) in the finite complex plane is at $x = 0$.

A second solution to the modified Bessel equation may be obtained by choosing $\alpha = -\nu$ whenever ν is nonintegral. In fact, if 2ν is also nonintegral then $P(-\nu + n) \neq 0$ for all positive integers n, so the second Frobenius series exists. However, so long as 2ν is not an even integer, a_n in (3.3.7) is determined by a_0 and the series (3.3.8) for $I_{-\nu}(x)$ converges. The solution $I_{-\nu}$ is clearly linearly independent of I_ν when ν is nonintegral because their series start with different powers of x. Thus, we conclude that the Frobenius method gives a complete set of linearly independent solutions so long as $\nu \neq 0, 1, 2, 3, \ldots$. It is more difficult to find a second solution when ν is an integer. The appropriate method will be explained later in this section.

In general, one can always construct a pair of linearly independent solutions to (3.3.2) by solving the recursion relation (3.3.3*b*) with $\alpha = \alpha_1$ and $\alpha = \alpha_2$, provided that $\alpha_1 - \alpha_2$ is not an integer. If $\alpha_1 = \alpha_2$ the solution obtained by choosing $\alpha = \alpha_2$ is identical to the first solution obtained with $\alpha = \alpha_1$ except for a multiplicative factor a_0. Thus, to find a second solution requires more analysis. If $\alpha_1 - \alpha_2 = N$, a positive integer, then the recursion relation (3.3.3*b*) with $\alpha = \alpha_2$ and $n = N$ reads

$$0a_N = -\sum_{k=0}^{N-1} [(\alpha + k)p_{N-k} + q_{N-k}]a_k. \tag{3.3.9}$$

There are now two possibilities:

1. If the right side of (3.3.9) is nonzero, then a_N does not exist; thus, to find a second linearly independent solution requires further analysis.
2. If the right side of (3.3.9) conspires to vanish, then (3.3.9) reduces to the identity $0 = 0$ and does not determine a_N. It follows that a_N is an arbitrary constant. The recursion relation has successfully jumped the hurdle at $n = N$. We may now calculate the rest of the coefficients $a_{N+1}, a_{N+2}, a_{N+3}, \ldots$ as functions of the two arbitrary constants a_0 and a_N and thereby determine a second linearly independent solution to the differential equation.

We may summarize the above discussion as follows:

CASE I $\alpha_1 \neq \alpha_2$; $\alpha_1 - \alpha_2 \neq$ integer. There are two linearly independent solutions in Frobenius form.

CASE II $\alpha_1 - \alpha_2 = N = 0, 1, 2, \ldots$. This case must be subdivided into two:

(*a*) $\alpha_1 = \alpha_2$. There is just one solution in Frobenius form. We explain shortly how to construct a second solution.
(*b*) $\alpha_1 - \alpha_2 = 1, 2, 3, \ldots$. This case must be further subdivided into two:

(i) The right side of (3.3.9) is nonzero. There is just one solution in Frobenius form. We explain shortly how to construct a second solution.
(ii) The right side of (3.3.9) is zero. There are two linearly independent solutions in Frobenius form.

We dispose of the special case II(*b*)(ii) first because all solutions may be represented as Frobenius series.

Discussion of case II(*b*)(ii) This case is characterized by the vanishing of the right side of (3.3.9). This ensures that a Frobenius series with indicial exponent α_2 can be found.

Example 4 *Frobenius series at an ordinary point.* Although it is improbable that the right side of (3.3.9) is zero, this case is not hard to illustrate. One sure way to observe case II(b)(ii) is to use Frobenius theory at an ordinary point of a differential equation. If x_0 is an ordinary point then $p_0 = q_0 = q_1 = 0$. Thus, the indicial equation is $P(\alpha) = \alpha(\alpha - 1) = 0$, and the indicial exponents are $\alpha_1 = 1$, $\alpha_2 = 0$. Accordingly, the solution with $\alpha_1 = 1$ has the form of the Taylor series $y(x, \alpha_1) = \sum_{n=0}^{\infty} a_n(x - x_0)^{n+1}$. This solution satisfies the initial conditions $y(x_0) = 0$, $y'(x_0) = a_0 \neq 0$.

A linearly independent solution of the form (3.3.1) with $\alpha = \alpha_2 = 0$ and a_0 and a_1 arbitrary is guaranteed to exist. Thus, the second solution is also a Taylor series.

Example 5 *Modified Bessel equation with order* $\nu = \frac{1}{2}, \frac{3}{2}, \ldots$. If we choose ν in the modified Bessel equation (3.3.5) to be a half-odd integer, $\nu = \frac{1}{2}, \frac{3}{2}, \frac{5}{2}, \ldots$, then the indices α_1 and α_2 differ by an integer: $\alpha_1 - \alpha_2 = 1, 3, 5, \ldots$. Here, $p_0 = 1$, $q_0 = -\nu^2$, $q_2 = -1$ and all other p_n and q_n are zero. It can be shown (see Prob. 3.25) that the right side of (3.3.9) vanishes. Thus, both solutions have Frobenius series expansions: the Frobenius series expansions for $I_\nu(x)$ and $I_{-\nu}(x)$ are given explicitly by (3.3.8).

In cases II(a) and II(b)(i) the second solution does not have the form of a Frobenius series; a local expansion of the second solution involves the function $\ln(x - x_0)$.

Discussion of case II(a) We will find a second linearly independent solution to (3.3.2) when $\alpha_1 = \alpha_2$ by differentiating the Frobenius series (3.3.1) with respect to the indicial exponent α. This procedure is an extension of the trick which was introduced in Sec. 1.4 for generating a second solution to a linear equidimensional equation. To prepare for differentiating with respect to α, we leave α arbitrary for the moment by ignoring (3.3.3a) and solve the recursion relation (3.3.3b) for a_n as a function of a_0 *and* α. We denote the resulting Frobenius series by $y(x, \alpha)$:

$$y(x, \alpha) = (x - x_0)^\alpha \sum_{n=0}^{\infty} a_n(\alpha)(x - x_0)^n. \tag{3.3.10}$$

Of course, $y(x, \alpha)$ is not a solution of (3.3.2) unless $\alpha = \alpha_1$.

It is convenient to define the shorthand notation

$$L \equiv \frac{d^2}{dx^2} + \frac{p(x)}{x - x_0}\frac{d}{dx} + \frac{q(x)}{(x - x_0)^2}.$$

Any solution $y(x)$ of (3.3.2) satisfies the equation $Ly(x) = 0$.

Since $a_n(\alpha)$ by construction satisfies the recursion relation (3.3.3b), $y(x, \alpha)$ in (3.3.10) is almost, but not quite, a solution to the differential equation (3.3.2) when $\alpha \neq \alpha_1$. Instead of satisfying $Ly = 0$, $y(x, \alpha)$ satisfies

$$Ly(x, \alpha) = a_0(x - x_0)^{\alpha - 2}P(\alpha). \tag{3.3.11}$$

If at this point we choose $\alpha = \alpha_1$ then the right side of (3.3.11) vanishes; this shows that $y(x, \alpha_1)$ is just the Frobenius series solution of $Ly = 0$. However, if we differentiate both sides of (3.3.11) with respect to α and then let $\alpha = \alpha_1$, the right side again vanishes because $P(\alpha) = (\alpha - \alpha_1)^2$ when $\alpha_1 = \alpha_2$:

$$\frac{\partial}{\partial\alpha} a_0[(x-x_0)^{\alpha-2}(\alpha-\alpha_1)^2]\bigg|_{\alpha=\alpha_1}$$

$$= a_0[2(\alpha-\alpha_1)(x-x_0)^{\alpha-2} + (x-x_0)^{\alpha-2}(\alpha-\alpha_1)^2 \ln(x-x_0)]\bigg|_{\alpha=\alpha_1} = 0.$$

Thus, $(\partial/\partial\alpha)y(x, \alpha)\,|_{\alpha=\alpha_1}$ is a new solution to (3.3.2) because

$$L\left[\frac{\partial}{\partial\alpha} y(x, \alpha)\bigg|_{\alpha=\alpha_1}\right] = 0.$$

If we differentiate (3.3.10) with respect to α, we see that the form of this new solution is that given in (3.1.4):

$$\frac{\partial}{\partial\alpha} y(x, \alpha)\bigg|_{\alpha=\alpha_1} = y(x, \alpha_1)\ln(x-x_0) + \sum_{n=0}^{\infty} b_n(x-x_0)^{\alpha_1+n}, \tag{3.3.12}$$

where

$$b_n = \frac{\partial}{\partial\alpha} a_n(\alpha)\bigg|_{\alpha=\alpha_1}. \tag{3.3.13}$$

We have now obtained series expansions for *two* linearly independent solutions to (3.3.2) near a regular singular point. The new expansion in (3.3.12) has a radius of convergence at least as large as the distance to the nearest singular point of the differential equation (3.3.2).

Example 6 *Modified Bessel equation of order* $\nu = 0$. When $\nu = 0$ in (3.3.5) the indicial polynomial $P(\alpha) = \alpha^2 - \nu^2$ has a double root at $\alpha = 0$. One solution (3.3.8) is

$$I_0(x) = \sum_{n=0}^{\infty} \frac{(\frac{1}{2}x)^{2n}}{(n!)^2}. \tag{3.3.14}$$

A linearly independent solution must have the form (3.3.12). Thus, it is necessary to evaluate the coefficients b_n from (3.3.13) by performing the indicated differentiation. From equations (3.3.6) with $\nu = 0$ it follows that $a_0(\alpha) = a_0$, $a_{2n+1}(\alpha) = 0$ $(n = 0, 1, 2, \ldots)$, and

$$a_{2n}(\alpha) = \frac{a_0}{(\alpha+2n)^2(\alpha+2n-2)^2\cdots(\alpha+2)^2}, \qquad n = 1, 2, \ldots.$$

Thus, $b_0 = (\partial/\partial\alpha)a_0(\alpha)|_{\alpha=0} = 0$, $b_{2n+1} = 0$ $(n = 0, 1, 2, \ldots)$, and

$$b_{2n} = \frac{\partial}{\partial\alpha} a_{2n}(\alpha)\bigg|_{\alpha=0} = \frac{-a_0}{2^{2n}(n!)^2}\left(1 + \frac{1}{2} + \frac{1}{3} + \cdots + \frac{1}{n}\right), \qquad n \ge 1,$$

Substituting this result into (3.3.12) gives a solution which is linearly independent of I_0:

$$\frac{\partial}{\partial\alpha} y(x, \alpha)\bigg|_{\alpha=\alpha_1} = a_0 \ln x \sum_{n=0}^{\infty} \frac{(\frac{1}{2}x)^{2n}}{(n!)^2} - a_0 \sum_{n=1}^{\infty} \frac{(\frac{1}{2}x)^{2n}}{(n!)^2}\left(1 + \frac{1}{2} + \cdots + \frac{1}{n}\right).$$

It is conventional to construct a special second solution $K_0(x)$ as follows. We take $a_0 = -1$ in the above equation. Then we add to this a constant multiple of I_0, where the constant is $\ln 2 - \gamma$; $\gamma = 0.5772\cdots$ is an irrational number called Euler's constant (see Prob. 6.5). Thus,

$$K_0(x) = -(\ln \tfrac{1}{2}x + \gamma)I_0(x) + \sum_{n=1}^{\infty} \frac{(\frac{1}{2}x)^{2n}}{(n!)^2}\left(1 + \frac{1}{2} + \frac{1}{3} + \cdots + \frac{1}{n}\right). \tag{3.3.15}$$

It is, of course, not obvious why the function $K_0(x)$ is defined in this fancy way (see Probs. 3.65 and 3.66). The reason is that $K_0(x)$ has a beautifully simple behavior as $x \to \infty$, to be discussed later in Secs. 3.5 and 3.8 and again in Example 3 of Sec. 6.4.

Discussion of case II(*b*)(i) To find a second linearly independent solution to (3.3.2) when $\alpha_1 - \alpha_2 = N$, a positive integer, and the right side of (3.3.9) is nonvanishing we try to follow the same procedure as in case II(a). Let us differentiate the equation (3.3.11) with respect to α and see what differential equation $(\partial/\partial\alpha)y(x, \alpha)|_{\alpha=\alpha_1}$ satisfies. This time we are not so lucky: $(\partial/\partial\alpha)y(x, \alpha)|_{\alpha=\alpha_1}$ is a particular solution of the *inhomogeneous* equation

$$L\left[\frac{\partial}{\partial\alpha}y(x, \alpha)\bigg|_{\alpha=\alpha_1}\right] = a_0 P'(\alpha_1)(x - x_0)^{\alpha_1 - 2}$$

$$= a_0 P'(\alpha_1)(x - x_0)^{\alpha_2 + N - 2}. \tag{3.3.16}$$

The right side of this equation does not vanish as it does in case II(a) because $P(\alpha) = (\alpha - \alpha_1)(\alpha - \alpha_2)$ does not have a double root at $\alpha = \alpha_1$.

We really want to construct a solution to the homogeneous part of (3.3.16). One way to do this is to construct a *second* particular solution to the inhomogeneous equation (3.3.16) and subtract it from the first. The difference will be a solution to the associated homogeneous equation (3.3.2). It is nice to discover that the second particular solution has an ordinary Frobenius expansion

$$\sum_{n=0}^{\infty} c_n(x - x_0)^{\alpha_2 + n}. \tag{3.3.17}$$

Substituting this series into (3.3.16) and equating coefficients of $(x - x_0)^{\alpha_2 + n - 2}$ gives

$$(x - x_0)^{\alpha_2 - 2}: \qquad P(\alpha_2)c_0 = 0, \tag{3.3.18a}$$

$$(x - x_0)^{\alpha_2 + n - 2}: \qquad P(\alpha_2 + n)c_n + \sum_{k=0}^{n-1} [(\alpha_2 + k)p_{n-k} + q_{n-k}]c_k = 0, \qquad n \neq 0, N, \tag{3.3.18b}$$

$$(x - x_0)^{\alpha_2 + N - 2}: \quad P(\alpha_2 + N)c_N + \sum_{k=0}^{N-1} [(\alpha_2 + k)p_{N-k} + q_{N-k}]c_k = a_0 P'(\alpha_1). \tag{3.3.18c}$$

Since $P(\alpha_2 + N) = 0$, the relation (3.3.18c) does not determine c_N; rather it relates the value of a_0 to the coefficients $c_0, c_1, \ldots, c_{N-1}$:

$$a_0 = \frac{1}{P'(\alpha_1)} \sum_{k=0}^{N-1} [(\alpha_2 + k)p_{N-k} + q_{N-k}]c_k. \tag{3.3.19}$$

Note that a_0 does not vanish because the right side of (3.3.9) is nonzero.

We can now construct a second solution $y(x)$ to the homogeneous equation (3.3.2) by subtracting the two solutions to the inhomogeneous equation (3.3.16):

$$y(x) = \sum_{n=0}^{\infty} c_n(x - x_0)^{\alpha_2 + n} - \frac{\partial}{\partial\alpha}y(x, \alpha)\bigg|_{\alpha=\alpha_1}. \tag{3.3.20}$$

The coefficients c_n in (3.3.20) satisfy the recursion relation (3.3.18*b*); c_0 and c_N are arbitrary, and the constraint (3.3.19) determines a_0. Solutions with different values of c_N differ by constant multiples of $y(x, \alpha_1)$. The expansions in (3.3.20) converge in a complex disk whose radius is at least as large as the distance to the nearest singularity of $p(x)$ and $q(x)$.

Example 7 *Modified Bessel equation for integer order 1.* When $\nu = 1$ one solution of the modified Bessel equation (3.3.5) is $I_1(x)$, which is given by the Frobenius series (3.3.8). A second solution of the form (3.3.20) may be determined by the method just explained. It is conventional to choose the arbitrary constants c_0 and c_2 as follows: $c_0 = 1$, $c_2 = \frac{1}{2}\gamma - \frac{1}{2}\ln 2 - \frac{1}{4}$, where γ is Euler's constant. These choices define the modified Bessel function $K_1(x)$ (see Prob. 3.65):

$$K_1(x) = (\gamma + \ln \tfrac{1}{2}x)I_1(x) + \frac{1}{x} - \frac{x}{4} - \sum_{n=1}^{\infty} \frac{(\frac{1}{2}x)^{2n+1}}{n!\,(n+1)!}\left(1 + \frac{1}{2} + \frac{1}{3} + \cdots + \frac{1}{n} + \frac{1}{2n+2}\right). \tag{3.3.21}$$

This completes our exposition of the Frobenius method for second-order equations.

Frobenius Series for Higher-Order Equations

The Frobenius method extends easily to the general nth-order homogeneous linear differential equation at a regular singular point x_0:

$$\frac{d^n y}{dx^n} + \frac{q_{n-1}(x)}{x - x_0}\frac{d^{n-1}y}{dx^{n-1}} + \frac{q_{n-2}(x)}{(x - x_0)^2}\frac{d^{n-2}y}{dx^{n-2}} + \cdots + \frac{q_0(x)}{(x - x_0)^n}y = 0,$$

where $q_0(x), \ldots, q_{n-1}(x)$ are analytic at x_0. The indicial equation for α is

$$\alpha(\alpha - 1)\cdots(\alpha - n + 1) + q_{n-1}(x_0)\alpha(\alpha - 1)\cdots(\alpha - n + 2) \\ + q_{n-2}(x_0)\alpha(\alpha - 1)\cdots(\alpha - n + 3) + \cdots + q_0(x_0) = 0.$$

If the n roots α do not differ by integers, then there will be n linearly independent solutions of the form (3.3.2); otherwise, the form of the solution must be generalized to (3.1.4) or (3.1.5). Enumeration of special cases is left to Probs. 3.15 to 3.17.

(E) 3.4 LOCAL BEHAVIOR AT IRREGULAR SINGULAR POINTS OF HOMOGENEOUS LINEAR EQUATIONS

Our analysis of ordinary and of regular singular points is fundamentally different from the asymptotic analysis to be used throughout the remainder of this book. The approach of the last two sections has been prosaic and heavy-handed: the series were convergent, the manipulations were mechanical and rigorous, and the treatment was thorough to the point of being devastatingly boring (to us). This mummified style reflects the completeness of the theory; there was no room for imaginative mathematics.

From this point on, our style changes. To analyze the local behavior of solutions near irregular singular points, we will be forced to develop entirely new mathematical tools which as a whole comprise a calculus of approximations. Although it is possible to justify the use of these tools on a rigorous level, such a justification makes for slow reading and is contrary to the spirit of this book. Our intention is to omit time-consuming rigor and to emphasize careful problem solving. In contrast to the methods of the last two sections which we would describe as exact, rigorous, systematic, limited in scope, and deadly, these new methods are approximate, intuitive, heuristic, powerful, and fascinating.

In this section we will see that the formulas which express the local behavior of a solution near irregular singular points are generalizations of Frobenius series in much the same way that Frobenius series are generalizations of Taylor series. Let us begin our analysis by discovering why Frobenius series are insufficient to describe behavior near irregular singular points. We know that if *all* solutions to a linear differential equation in the neighborhood of a point x_0 can be expanded in Frobenius series, then x_0 is a regular singular point. Thus, at an irregular singular point, at least one solution must not have a Frobenius series representation. Let us observe explicitly how Frobenius series fail.

Example 1 *Irregular singular point at which there are no solutions of Frobenius form.* The differential equation $y' = x^{1/2}y$ has an irregular singular point at $x = 0$. The solution $y(x)$ may be represented as a convergent series about $x = 0$: $y(x) = a_0 e^{2x^{3/2}/3} = a_0 \sum_{n=0}^{\infty} (2x^{3/2}/3)^n/n!$; but do not mistake this series for a Frobenius series. Apart from an overall factor of x^α, a Frobenius series involves only integral powers of x.

Example 2 *Irregular singular point at which there are no solutions of Frobenius form.* What happens if we try to expand a solution of the differential equation

$$x^3 y'' = y \tag{3.4.1}$$

in a Frobenius series about the irregular singular point at 0? If a solution of Frobenius form $y = \sum_{n=0}^{\infty} a_n x^{n+\alpha}$ with $a_0 \neq 0$ exists, then $\sum_{n=0}^{\infty} (n+\alpha)(n+\alpha-1)a_n x^{n+\alpha+1} = \sum_{n=0}^{\infty} a_n x^{n+\alpha}$. Equating coefficients of x^α gives $a_0 = 0$, which is a contradiction. Therefore, no solution of Frobenius form exists about $x = 0$.

In Example 2 the assumption of a Frobenius series led to an immediate contradiction. However, in the next example the difficulty with assuming a Frobenius series occurs on a more subtle level.

Example 3 *Irregular singular point at which there are no solutions of Frobenius form.* The differential equation

$$x^2 y'' + (1 + 3x)y' + y = 0 \tag{3.4.2}$$

has an irregular singular point at 0. What happens if we try to expand $y(x)$ in the Frobenius series $y = \sum a_n x^{n+\alpha}$? Substituting this series gives

$$x^2 \sum_{n=0}^{\infty} (n+\alpha)(n+\alpha-1)a_n x^{n+\alpha-2} + (1+3x) \sum_{n=0}^{\infty} (n+\alpha)a_n x^{n+\alpha-1} + \sum_{n=0}^{\infty} a_n x^{n+\alpha} = 0.$$

Equating coefficients of powers of x to zero, we get

$$x^{\alpha-1}: \qquad \alpha a_0 = 0,$$

$$x^{n+\alpha}: \qquad (n+\alpha+1)a_{n+1} + (n+\alpha+1)^2 a_n = 0, \qquad n = 0, 1, \ldots.$$

Since $a_0 \neq 0$ in a Frobenius series, the indicial equation is just $\alpha = 0$. It follows that $a_{n+1} = -(n+1)a_n$ $(n = 0, 1, 2, \ldots)$. Thus, $a_n = (-1)^n n!\, a_0$ and the solution $y(x)$ is

$$y(x) = a_0 \sum_{n=0}^{\infty} (-1)^n n!\, x^n. \tag{3.4.3}$$

We have encountered no formal difficulty in constructing the nontrivial series solution (3.4.3). However, a closer examination of this series shows that its radius of convergence is zero! The series diverges for all $x \neq 0$. Since, by assumption, a Frobenius series has a nonvanishing radius of convergence, we have again reached a contradiction; there are no Frobenius series solutions.

Although the result of the above example may appear worthless, it is not. The divergent series (3.4.3) is a new kind of series expansion that is often encountered in applied mathematics. It is called an *asymptotic series* and has the remarkable property that although it is divergent, it can be used to obtain accurate approximations to the solution $y(x)$ of the differential equation. This example is typical of what is usually found when we perform local analysis about an irregular singular point. Specifically, if we seek a solution involving an infinite power series, the series is usually divergent.

Notice also that (3.4.3) gives just one of the two linearly independent solutions in Example 3. The missing solution does not have a formal (convergent or divergent) power series expansion about $x = 0$. The rest of this section develops a heuristic procedure for constructing series representations of *all* the solutions in the neighborhood of an isolated irregular singular point.

Brief Introduction to Asymptotics

The asymptotic methods we are about to introduce are best understood if the reader first masters the mechanical aspects, which are explained in this section and which are no more difficult than those used to obtain Taylor or Frobenius series. However, since the formal techniques are more difficult to justify mathematically, we have postponed a more intensive but still nonrigorous discussion until Secs. 3.7 and 3.8.

We must introduce two new symbols which express the relative behavior of two functions. The notation

$$f(x) \ll g(x), \qquad x \to x_0,$$

which is read "$f(x)$ is much smaller than $g(x)$ as x tends to x_0," means

$$\lim_{x \to x_0} f(x)/g(x) = 0.$$

Second, the notation

$$f(x) \sim g(x), \qquad x \to x_0,$$

which is read "$f(x)$ is asymptotic to $g(x)$ as x tends to x_0," means that the relative error between f and g goes to zero as $x \to x_0$:

$$f(x) - g(x) \ll g(x), \qquad x \to x_0,$$

or, equivalently,

$$\lim_{x \to x_0} f(x)/g(x) = 1.$$

Note that if $f(x) \sim g(x)$ $(x \to x_0)$ then $g(x) \sim f(x)$ $(x \to x_0)$.

Example 4 *Asymptotic relations.*

(*a*) $x \ll 1/x$ $(x \to 0)$.
(*b*) $x^{1/2} \ll x^{1/3}$ $(x \to 0+)$, where the limit $x \to 0+$ means that x approaches 0 through positive values only.
(*c*) $(\log x)^5 \ll x^{1/4}$ $(x \to +\infty)$.
(*d*) $x^{1/2} \sim 2$ $(x \to 4)$.
(*e*) $e^x + x \sim e^x$ $(x \to +\infty)$, but note that the difference between the left and right sides of this relation, x, goes to ∞ as $x \to +\infty$. Thus, even if two functions are asymptotic, they may not be approximately equal.
(*f*) $x^2 \not\sim x$ $(x \to 0)$ because x^2 and x approach 0 at different rates. In this case, even though x^2 and x are approximately equal to 0 as $x \to 0$, they are not asymptotic.
(*g*) It is a common mistake to assert that a function is asymptotic to zero. For example, the equation $x^3 \sim 0$ $(x \to 0)$ is wrong; by definition, no nonzero function can ever be asymptotic to zero.
(*h*) $x \ll -10$ $(x \to 0+)$, even though the signs are different.

Behavior Near Irregular Singular Points

Let us set aside any further discussion of the properties of asymptotic relations until Sec. 3.7. For the moment, we will not be concerned with the validity of adding, multiplying, integrating, or differentiating asymptotic relations. It is more urgent for us to return to the two examples at the beginning of this section whose behaviors we were unable to find using a Frobenius series.

Each of the second-order differential equations (3.4.1) and (3.4.2) has two linearly independent solutions. Yet, only one solution out of those four could be expressed as a series in powers of x, and this one was the divergent series (3.4.3). We will shortly formulate a procedure for discovering the local behavior of the other solutions near $x = 0$, but since it is always easier to derive a result that is already known, let us peek at the answers. One solution to (3.4.1) exhibits the behavior

$$y(x) \sim c_1 x^{3/4} e^{2x^{-1/2}}, \qquad x \to 0+; \tag{3.4.4a}$$

the other solution has the behavior

$$y(x) \sim c_2 x^{3/4} e^{-2x^{-1/2}}, \qquad x \to 0+. \tag{3.4.4b}$$

Also, the missing solution to (3.4.2) exhibits the behavior

$$y(x) \sim c_2 x^{-1} e^{x^{-1}}, \qquad x \to 0+. \tag{3.4.5}$$

Observe that these behaviors all involve exponentials of functions which become singular at the irregular singularity of the differential equation. Thus, these three functions have essential singularities at $x = 0$. This is not surprising; most of the solutions to the differential equations in the examples of Sec. 3.1 also have essential singularities at irregular singular points.

The asymptotic behaviors (3.4.4) and (3.4.5) are actually the first terms of infinite series representations of the local behaviors of the solutions, whose terms will be derived later in this section. We will refer to the first term in such a series as the *leading behavior* of the series. We will also refer to the most rapidly changing component of the leading behavior in the limit $x \to x_0$ as the *controlling factor*. For example, the controlling factor of the leading behavior given in (3.4.4*a*) is $e^{2x^{-1/2}}$, in (3.4.4*b*) it is $e^{-2x^{-1/2}}$, and in (3.4.5) it is $e^{x^{-1}}$. Also, the leading behavior of the series (3.4.3) is a_0. For the Frobenius series $\sum a_n x^{n+\alpha}$, the leading behavior is $a_0 x^\alpha$ and the controlling factor is x^α.

The first step in deriving the leading behavior of a solution near an irregular singular point consists of identifying the controlling factor. Since the controlling factor is usually in the form of an exponential, this suggests the substitution

$$y(x) = e^{S(x)}, \tag{3.4.6}$$

originally proposed by Carlini (1817), Liouville (1837), and Green (1837). Although it is not immediately apparent why this substitution should facilitate the calculation, it is, in fact, a beautiful trick. It reduces an nth-order linear differential equation to an approximate *first*-order differential equation for $S(x)$ which is usually valid in a neighborhood of the irregular singular point.

To see how this exponential substitution (3.4.6) works, let us use it to solve for the controlling factor of the solutions of the second-order differential equation

$$y'' + p(x)y' + q(x)y = 0 \tag{3.4.7}$$

near an assumed irregular singular point at x_0. Substituting $y = e^S$ gives

$$S'' + (S')^2 + p(x)S' + q(x) = 0. \tag{3.4.8}$$

This equation is just as difficult to solve as (3.4.7), but if x_0 is an irregular singular point, it is usually true that

$$S'' \ll (S')^2, \qquad x \to x_0. \tag{3.4.9}$$

For example, suppose the controlling factor of the behavior of $y(x)$ has the form $\exp[a(x - x_0)^{-b}]$, where $b > 0$ so that $y(x)$ has an essential singularity at x_0. Then $(S')^2 \sim a^2b^2(x - x_0)^{-2b-2}$ and $S'' \sim ab(b + 1)(x - x_0)^{-b-2}$. Thus, the comparison (3.4.9) is indeed valid because $b > 0$. In this argument we have differentiated an asymptotic relation; we discuss the correctness of this procedure later.

The asymptotic differential equation

$$(S')^2 \sim -p(x)S' - q(x), \qquad x \to x_0, \tag{3.4.10}$$

that is obtained by dropping S'' is very easy to solve and its solutions may be checked to see if the approximation (3.4.9) is valid. [Notice that we have taken the trouble to move two terms to the right side of (3.4.10) to avoid asserting that a quantity is asymptotic to 0!] Some examples in which the assumption in (3.4.9) is *not* valid are given in Probs. 3.32 and 3.46. Also, the assumption (3.4.9) does not hold if x_0 is an ordinary or regular singular point. Why?

Example 2 (revisited) *Controlling factor near the irregular singular point at 0.* Let us return to the differential equation (3.4.1), $x^3y'' = y$, and find the controlling factor of the solutions near the irregular singular point at 0. For this problem the asymptotic differential equation (3.4.10) is $(S')^2 \sim x^{-3}$ $(x \to 0+)$. The two possible solutions are $S'(x) \sim \pm x^{-3/2}$ $(x \to 0+)$, and, therefore,

$$S(x) \sim \pm 2x^{-1/2}, \qquad x \to 0+. \tag{3.4.11}$$

We have identified the controlling factor already given in (3.4.4). We are now considering only the one-sided limit $x \to 0+$ to avoid complications arising from imaginary values of $S(x)$.

Observe that to solve for $S(x)$ we had to integrate an asymptotic relation of the form $f(x) \sim g(x)$ $(x \to x_0)$. If we were integrating an equality, the integrals would differ by an integration constant c. When we have an asymptotic relation, the integrals differ by a function $C(x)$ whose derivative is small compared with f and g as $x \to x_0$. The two integrals $\int^x f\,dx$ and $\int^x g\,dx$ are asymptotic as $x \to x_0$ only if $C(x) \ll \int^x f\,dx$ or $\int^x g\,dx$. There are pathological cases for which $C(x)$ is *not* smaller than $\int^x f\,dx$ or $\int^x g\,dx$ (see Probs. 3.54 and 3.58). However, if $f \sim a(x - x_0)^{-b}$ as $x \to x_0$, where $b > 1$, then

$$\int^x f\,dx \sim [a/(1-b)](x - x_0)^{1-b} \qquad \text{as } x \to x_0;$$

if $b < 1$, then $\int^x f\,dx \sim c$ as $x \to x_0$, where c is some constant; if $b = 1$, then $\int^x f\,dx \sim a \ln|x - x_0|$ as $x \to x_0$ (see Prob. 3.28).

Example 2 (revisited) *Leading behavior near the irregular singular point at 0.* We have found two solutions (3.4.11) for the controlling factor. This is not surprising because the differential equation (3.4.1) is second order and must have two linearly independent solutions. Let us focus our attention on the solution $S \sim 2x^{-1/2}$ $(x \to 0+)$. The treatment of the other solution $S \sim -2x^{-1/2}$ $(x \to 0+)$ is analogous (see Prob. 3.29).

We can improve upon (3.4.11) by estimating the integration function $C(x)$, where

$$S(x) = 2x^{-1/2} + C(x), \qquad C(x) \ll 2x^{-1/2}, \qquad x \to 0+. \tag{3.4.12}$$

Again, we cannot hope to calculate $C(x)$ exactly because that would be equivalent to solving the original differential equation (3.4.1). Instead, we seek an asymptotic estimate of $C(x)$.

To obtain this estimate, we substitute (3.4.12) into the differential equation (3.4.8) with $p = 0$ and $q = -x^{-3}$ and obtain $\frac{3}{2}x^{-5/2} + C'' - 2x^{-3/2}C' + (C')^2 = 0$. This equation may be approximated by a soluble asymptotic differential equation for $C(x)$ obtained by using (3.4.12). From this it follows that $S' \sim -x^{-3/2}$ $(x \to 0+)$ or, equivalently, $C' \ll x^{-3/2}$ $(x \to 0+)$. Therefore, $(C')^2 \ll x^{-3/2}C'$ $(x \to 0+)$, which gives the asymptotic differential equation

$$\tfrac{3}{2}x^{-5/2} + C'' \sim 2x^{-3/2}C', \qquad x \to 0+.$$

This linear equation is soluble, but for practice let us simplify it further by making another asymptotic approximation. Since $C' \ll x^{-3/2}$ $(x \to 0+)$, then (if differentiation of this order relation is permissible) $C'' \ll x^{-5/2}$ $(x \to 0+)$. Therefore,

$$\tfrac{3}{2}x^{-5/2} \sim 2x^{-3/2}C', \qquad x \to 0+, \tag{3.4.13}$$

whose solution is $C \sim \frac{3}{4}\ln x$ $(x \to 0+)$. Note that differentiating (3.4.13) gives $C'' \sim -\frac{3}{4}x^{-2}$ $(x \to 0+)$, so $C'' \ll x^{-5/2}$ $(x \to 0+)$ as assumed. Substituting $C(x)$ into (3.4.12) gives

$$S(x) = 2x^{-1/2} + \tfrac{3}{4}\ln x + D(x), \tag{3.4.14}$$

where $D(x)$, the arbitrary function which arises from integrating (3.4.13), satisfies

$$D(x) \ll \ln x, \qquad x \to 0+. \tag{3.4.15}$$

Let us attempt to refine our asymptotic analysis even further by computing $D(x)$. Following the procedure used to determine the asymptotic behavior of $C(x)$, we substitute (3.4.14) into (3.4.8) with $p = 0$, $q = -x^{-3}$:

$$-3x^{-2}/16 + D'' + (D')^2 - 2x^{-3/2}D' + 3x^{-1}D'/2 = 0.$$

As usual, this equation can be replaced by a much simpler asymptotic equation. We make the following estimates. First, since $x^{-1} \ll x^{-3/2}$ $(x \to 0+)$,

$$3x^{-1}D'/2 \ll 2x^{-3/2}D', \qquad x \to 0+.$$

Second, $D' \ll x^{-1}$ $(x \to 0+)$ by differentiating (3.4.15), so

$$(D')^2 \ll x^{-1}D', \qquad x \to 0+.$$

Finally, differentiating (3.4.15) twice gives

$$D'' \ll x^{-2}, \qquad x \to 0+.$$

These estimates enable us to replace the above exact five-term differential equation for $D(x)$ by the two-term asymptotic differential equation $-2x^{-3/2}D' \sim 3x^{-2}/16$ $(x \to 0+)$ or

$$D' \sim -3x^{-1/2}/32, \qquad x \to 0+.$$

If we now compute the indefinite integral of this asymptotic differential equation, we find that

$$D(x) - d \sim -3x^{1/2}/16, \qquad x \to 0+,$$

where d is some constant. In other words,

$$D(x) = d + \delta(x), \tag{3.4.16}$$

where $\delta(x) \sim -3x^{1/2}/16$ as $x \to 0+$.

We will return to a more detailed analysis of the function $\delta(x)$ later in this section. But first, let us observe that we have determined all the contributions to $S(x)$ which do not vanish as $x \to 0+$; the last correction $\delta(x)$ vanishes as $x \to 0+$.

We can now make our definition of leading behavior more precise. *The leading behavior of $y(x)$ is determined by just those contributions to $S(x)$ that do not vanish as x approaches the irregular singularity.* Specifically, for this problem the leading behavior of $y(x)$ is $y(x) \sim \exp(2x^{-1/2} + \frac{3}{4}\ln x + d)$ $(x \to 0+)$ or $y(x) \sim c_1 x^{3/4} e^{2x^{-1/2}}$ $(x \to 0+)$, where $c_1 = e^d$. This result is precisely (3.4.4*a*). To obtain the other leading behavior (3.4.4*b*), one repeats the above arguments starting with $S(x) \sim -2x^{-1/2}$ $(x \to 0+)$ in (3.4.11).

Example 2 (revisited) *Discussion and numerical verification.* Before continuing with our asymptotic analysis of the local behavior of the solutions to (3.4.1), let us pause to examine empirically the results that we have already obtained. If the reader has not encountered the kind of asymptotic analysis just used to derive the leading behaviors in (3.4.4), he will find such an examination most worthwhile. For this purpose it is convenient to set $e^{\delta(x)} = 1 + \varepsilon(x)$, where $\varepsilon(x) \to 0$ as $x \to 0+$.

We have solved (3.4.1) numerically taking as initial values $y(1) = 1$, $y'(1) = 0$ to verify that $\varepsilon(x)$ really does approach zero as $x \to 0+$. The solution of this initial-value problem is a linear combination of solutions, one decaying exponentially to 0 as $x \to 0+$ [as in (3.4.4*b*)] and the other growing exponentially to ∞ as $x \to 0+$ [as in (3.4.4*a*)]. With the above initial conditions for our numerical integration the coefficient of the growing solution is nonzero. Since the growing solution overwhelms the decaying solution as $x \to 0+$, the behavior of the growing solution dominates the behavior of the solution to this initial-value problem.

We can test our prediction of the leading behavior

$$y(x) = c_1 x^{3/4} e^{2x^{-1/2}}[1 + \varepsilon(x)] \tag{3.4.17}$$

by fitting it to the numerical solution. To test the accuracy of this fit, we plot in Fig. 3.2 the computed values of $y(x)/(x^{3/4}e^{2x^{-1/2}})$ versus x for $0 < x < 1$. Observe that this ratio approaches the limit $c_1 \doteqdot 0.1432$ as $x \to 0+$. The distance between this limiting value and the plotted curve is proportional to the function $\varepsilon(x)$. This graph illustrates the power of asymptotic analysis. For the plotted values of x, the function $y(x)$ varies by many orders of magnitude. Yet, the leading

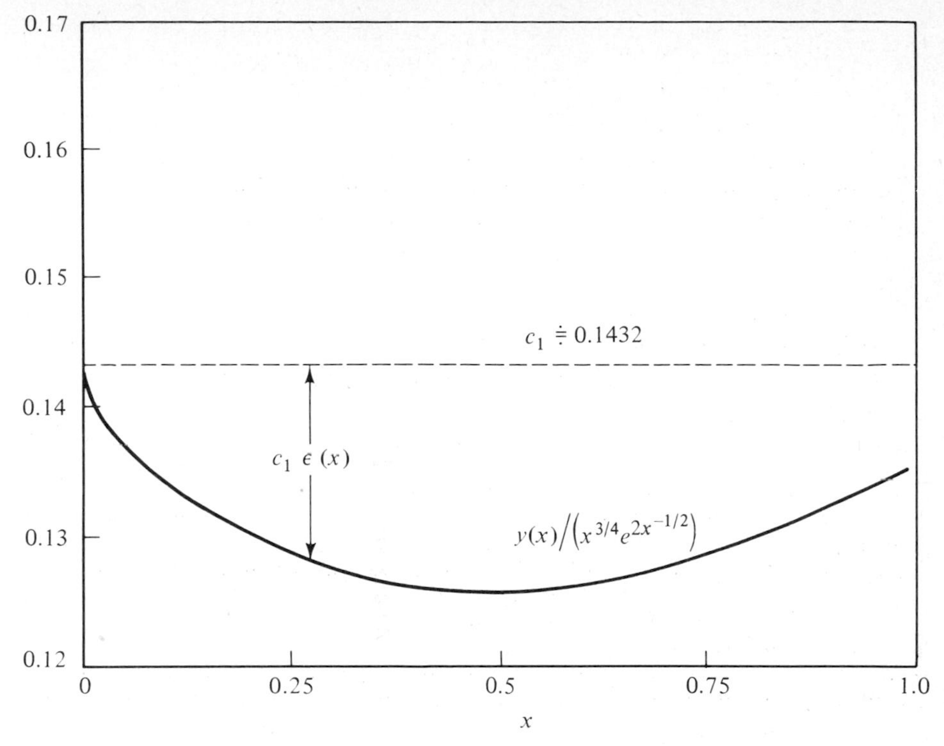

Figure 3.2 A plot of the ratio of $y(x)$, the numerical solution of (3.4.1) with $y(1) = 1$ and $y'(1) = 0$, to $x^{3/4} \exp(2x^{-1/2})$ for $0 \le x \le 1$. Asymptotic analysis [see (3.4.17)] shows that this ratio approaches a constant c_1. For these initial conditions, $c_1 \doteq 0.1432$. The difference between the plotted ratio and c_1 is $c_1\varepsilon(x)$.

behavior of $y(x)$ is a good estimate of $y(x)$ over the whole interval $0 < x \le 1$. Figure 3.2 shows that the relative error $\varepsilon(x)$ is at most 15 percent over this interval. Observe that the leading behavior becomes more accurate as $x \to 0+$. Figure 3.2 also shows that if we could estimate $\varepsilon(x)$, we could improve the approximation to $y(x)$.

Method of Dominant Balance

Before completing the analysis of Example 2 it is important to review the approach we have been using in Example 2 to solve differential equations approximately; namely, the *method of dominant balance*. The method of dominant balance is used to identify those terms in an equation that may be neglected in an asymptotic limit [in the way, for example, that we neglected S'' in (3.4.8)]. The technique of dominant balance consists of three steps:

1. We drop all terms that appear small and replace the exact equation by an asymptotic relation.
2. We replace the asymptotic relation with an equation by exchanging the $\sim$ sign

for an = sign and solve the resulting equation exactly (the solution to this equation automatically satisfies the asymptotic relation although it is certainly not the only function that does so).

3. We check that the solution we have obtained is consistent with the approximation made in (1). If it is consistent, we must still show that the equation for the function obtained by factoring off the dominant balance solution from the exact solution itself has a solution that varies less rapidly than the dominant balance solution. When this happens, we conclude that the *controlling factor* (and not the leading behavior) obtained from the dominant balance relation is the same as that of the exact solution.

The dominant balance argument that we have just outlined may appear circular, and indeed it is! Nevertheless, it is the most general and powerful procedure available for finding approximate solutions to equations.

Example 2 (revisited) *Corrections to the leading behavior near the origin.* Our analysis of the behavior of solutions to (3.4.1) is still incomplete because it is possible to find a sequence of approximations which are better and better estimates of $y(x)$ as $x \to 0+$. Our plan of attack will be to calculate successive approximations to $\varepsilon(x)$ in (3.4.17). We will first find the differential equation that $1 + \varepsilon(x)$ satisfies. Then, we will find a solution to this equation as a formal series of powers of x,

$$1 + \varepsilon(x) = 1 + a_1 x^\alpha + a_2(x^\alpha)^2 + a_3(x^\alpha)^3 + \cdots, \tag{3.4.18}$$

where α is a positive number. When the coefficients of this series are determined, the local analysis of (3.4.1) as $x \to 0+$ will be complete. We will have arrived at a representation of $y(x)$ as a series of elementary functions.

When the formal series (3.4.18) is substituted into (3.4.17), the resulting expansion is a generalization of a Frobenius series in three respects. First, it contains an exponential function which has an essential singularity. Second, it is not a series in integral powers of x (α will turn out to be $\frac{1}{2}$). Third, the series (3.4.18), like the series (3.4.3), is divergent for all $x \neq 0$. The interpretation of this divergent series is postponed to Sec. 3.5.

Let us now proceed to calculate the coefficients a_n for $n = 1, 2, \ldots$ and α. To do this, we first find the equation satisfied by $w(x) = 1 + \varepsilon(x)$ by substituting (3.4.17) into (3.4.1). The result is

$$w'' + \left(\frac{3}{2x} - \frac{2}{x^{3/2}}\right) w' - \frac{3}{16x^2} w = 0. \tag{3.4.19}$$

By replacing the dependent variable y by w we have divided or peeled off the leading behavior. This technique of peeling off the most rapidly varying behavior is frequently used in asymptotic analysis. The purpose of peeling off the leading behavior is to construct an equation having one solution that behaves asymptotically like a constant. It would be naive to expect that peeling off the leading behavior would eliminate the irregular singular point from the differential equation. There will still be one solution of (3.4.19) whose leading behavior has an essential singularity at $x = 0$, and (3.4.19) does indeed have an irregular singular point at $x = 0$. It follows from (3.4.4) and (3.4.17) or by direct local analysis of (3.4.19) that the possible leading behaviors of $w(x)$ are proportional to 1 and $e^{-4x^{-1/2}}$.

It is best to begin our calculation of $\varepsilon(x)$ by obtaining its leading behavior. We substitute $w(x) = 1 + \varepsilon(x)$ into (3.4.19) and obtain

$$\varepsilon'' + (3/2x - 2/x^{3/2})\varepsilon' - 3/16x^2 - 3\varepsilon/16x^2 = 0.$$

Following our usual procedure for finding leading behaviors, we replace this equation by a simpler asymptotic differential equation by observing that $\varepsilon \ll 1$ as $x \to 0+$, so

$$3\varepsilon/16x^2 \ll 3/16x^2, \qquad x \to 0+.$$

Also,

$$3x^{-1}/2 \ll 2x^{-3/2}, \qquad x \to 0+.$$

This gives the three-term asymptotic differential equation

$$\varepsilon'' - \frac{2}{x^{3/2}}\varepsilon' \sim \frac{3}{16x^2}, \qquad x \to 0+, \tag{3.4.20}$$

which we must solve subject to the condition that $\varepsilon \to 0$ as $x \to 0+$.

To solve this equation, we use the method of dominant balance. We argue that as $x \to 0+$, any one term in this equation cannot be much larger than the others without violating the conditions that (3.4.20) be an asymptotic relation. Therefore, there are four cases, which we investigate in turn. We will see that only case (*d*) is mathematically consistent and that, as already implied by (3.4.16), $\varepsilon(x) \sim -3x^{1/2}/16$ as $x \to 0+$:

(*a*) $\varepsilon'' \sim 3x^{-2}/16$ and $2x^{-3/2}\varepsilon' \ll 3x^{-2}/16$ as $x \to 0+$. Integrating the former asymptotic relation twice gives $\varepsilon(x) \sim -3 \ln x/16$ $(x \to 0+)$. This is inconsistent with the condition that $\varepsilon \to 0$ as $x \to 0+$, and it also violates the assumption that $2x^{-3/2}\varepsilon' \ll 3x^{-2}/16$ $(x \to 0+)$. Therefore, possibility (*a*) is excluded.

(*b*) $\varepsilon''/\varepsilon' \sim 2x^{-3/2}$ and $3x^{-2}/16 \ll 2x^{-3/2}\varepsilon'$ as $x \to 0+$. The integral of the first asymptotic relation is $\ln \varepsilon' \sim -4x^{-1/2}$ $(x \to 0+)$. This result violates the assumption that $3x^{-2}/16 \ll 2x^{-3/2}\varepsilon'$ as $x \to 0+$. Therefore, possibility (*b*) is also excluded.

(*c*) None of the three terms of (3.4.20) is negligible compared with the other two as $x \to 0+$. It is left for the reader to show in Prob. 3.30 that possibility (*c*) is excluded.

(*d*) $-2x^{-3/2}\varepsilon' \sim 3x^{-2}/16$ and $\varepsilon'' \ll 2x^{-3/2}\varepsilon'$ as $x \to 0+$. The integral of the first asymptotic relation is $\varepsilon(x) \sim -3x^{1/2}/16$ $(x \to 0+)$, where we have excluded the constant of integration because $\varepsilon \to 0$ as $x \to 0+$. This result is consistent because $\varepsilon'' \ll 2x^{-3/2}\varepsilon'$ as $x \to 0+$. Possibility (*d*) is the only consistent balance in (3.4.20).

We have determined that $w(x) - 1 \sim -3x^{1/2}/16$ $(x \to 0+)$, or that

$$w(x) = 1 - 3x^{1/2}/16 + \varepsilon_1(x),$$

where $\varepsilon_1 \ll 3x^{1/2}/16$ as $x \to 0+$.

Let us now return to the numerical results plotted in Fig. 3.2. You may have been surprised to see that $\varepsilon \to 0$ as $x \to 0+$ with vertical slope. The reason for this behavior is now clear; the slope of $3x^{1/2}/16$ is infinite at $x = 0$. Figure 3.3 shows that including the leading behavior of $\varepsilon(x)$ improves the approximation to the numerical solution $y(x)$ as $x \to 0+$.

We could now determine the leading behavior of $\varepsilon_1(x)$ by an argument like the one just given. We would find that

$$\varepsilon_1(x) = -15x/512 + \varepsilon_2(x), \qquad \varepsilon_2 \ll 15x/512, \qquad x \to 0+.$$

Evidently we can continue this process forever and generate a series representation for $w(x)$. However, since the formal structure of this series is now becoming apparent (it is a series in powers of $x^{1/2}$), it is much more efficient to substitute the full formal series representation (3.4.18) with $\alpha = \frac{1}{2}$, $w(x) = \sum_{n=0}^{\infty} a_n x^{n/2}$ $(a_0 = 1)$, directly into (3.4.19). We find that

$$\sum_{n=1}^{\infty} \frac{n}{2}\left(\frac{n}{2} - 1\right) a_n x^{n/2-2} + \frac{3}{2}\sum_{n=1}^{\infty} \frac{n}{2} a_n x^{n/2-2} - 2\sum_{n=1}^{\infty} \frac{n}{2} a_n x^{n/2-5/2} - \frac{3}{16}\sum_{n=0}^{\infty} a_n x^{n/2-2} = 0.$$

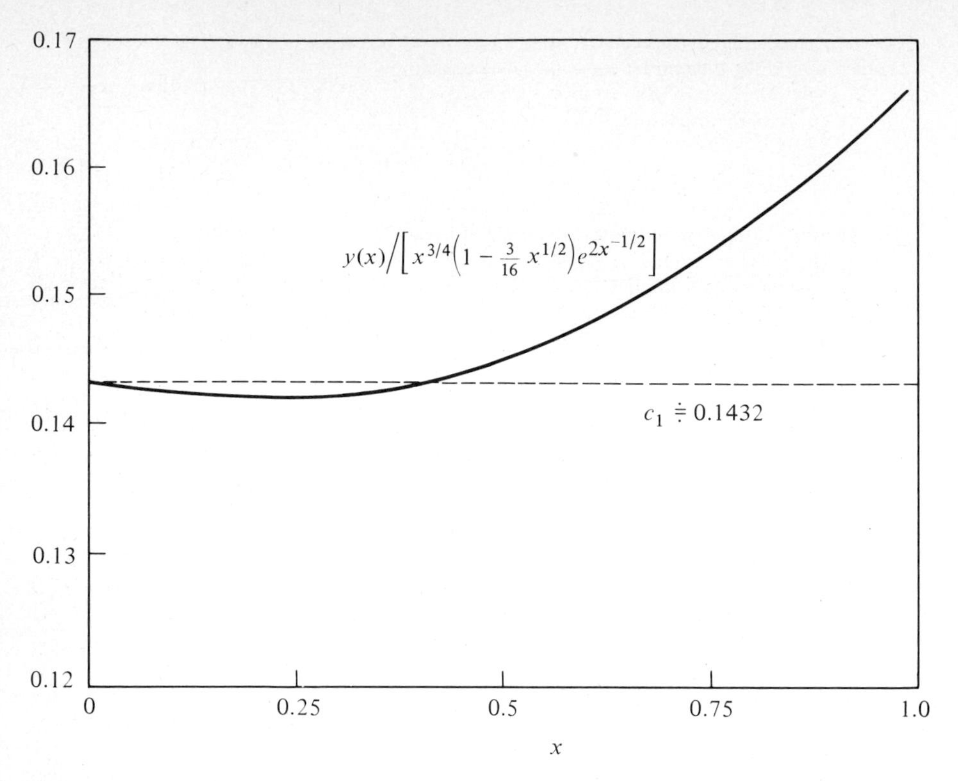

Figure 3.3 A plot of the ratio $y(x)$, the exact solution to (3.4.1) with $y(1) = 1$ and $y'(1) = 0$, to $x^{3/4}(1 - \frac{3}{16}x^{1/2}) \exp(2x^{-1/2})$ for $0 \le x \le 1$. A comparison of the results plotted in this figure with those of Fig. 3.2 shows that including the term $-\frac{3}{16}x^{1/2}$ in $w(x)$ improves the estimate of the numerical solution $y(x)$ as $x \to 0+$.

Equating coefficients of powers of $x^{1/2}$ gives the recursion relation

$$a_{n+1} = \frac{(2n-1)(2n+3)}{16(n+1)} a_n, \qquad n = 0, 1, 2, \ldots. \tag{3.4.21}$$

Thus, $a_0 = 1$, $a_1 = -\frac{3}{16}$, $a_2 = -\frac{15}{512}$, and so on. The general formula for a_n is

$$a_n = -[\Gamma(n - \tfrac{1}{2})\Gamma(n + \tfrac{3}{2})]/\pi 4^n n!, \qquad n = 0, 1, 2, \ldots,$$

where we have used $\Gamma(\frac{1}{2}) = \pi^{1/2}$ (see Prob. 2.6). This completes the local analysis of the solution to (3.4.1) whose leading behavior is (3.4.4*a*):

$$y(x) \sim -c_1 x^{3/4} e^{2x^{-1/2}} \sum_{n=0}^{\infty} \frac{\Gamma(n - \frac{1}{2})\Gamma(n + \frac{3}{2})}{\pi 4^n n!} x^{n/2}, \qquad x \to 0+. \tag{3.4.22a}$$

A simple application of the ratio test using (3.4.21) shows that the radius of convergence R of the series in (3.4.22*a*) is 0:

$$R = \lim_{n\to\infty} \left| \frac{a_n}{a_{n+1}} \right| = \lim_{n\to\infty} \left| \frac{16(n+1)}{(2n-1)(2n+3)} \right| = 0.$$

The series in (3.4.22*a*) diverges for all values of x! Nevertheless, this series is an asymptotic series and it will provide an accurate approximation to $y(x)$ as $x \to 0+$.

A similar analysis of the solutions to (3.4.1) whose leading behavior is (3.4.4*b*) gives the asymptotic series representation

$$y(x) \sim -c_2 x^{3/4} e^{-2x^{-1/2}} \sum_{n=0}^{\infty} (-1)^n \frac{\Gamma(n - \frac{1}{2})\Gamma(n + \frac{3}{2})}{\pi 4^n n!} x^{n/2}, \qquad x \to 0+. \tag{3.4.22b}$$

Now let us summarize the approach we have followed in this example. First, by means of the exponential substitution $y = e^S$ we determined the behavior of $S(x)$ up to terms that vanish as $x \to x_0$. This gave the leading behavior of $y(x)$. Next we refined this approximation to $y(x)$ by peeling or factoring off the leading behavior and expanding what remains as a series of fractional powers of $(x - x_0)$. This approach is very general and works for a wide class of differential equations.

Now that we have explained the technique of local analysis at an irregular singular point, let us review it by finding the local behavior of the solution to (3.4.2) in Example 3 whose leading behavior is given in (3.4.5).

Example 3 (revisited) *Determination of local behavior near an irregular singular point at 0.* The first step in determining the local behavior is finding the leading behavior. We do this by making the exponential substitution $y = e^S$ in (3.4.2); the result is

$$x^2 S'' + x^2(S')^2 + (1 + 3x)S' + 1 = 0. \tag{3.4.23}$$

Again, we neglect $x^2 S''$ relative to $x^2(S')^2$ and $3xS'$ relative to S' as $x \to 0+$, so we obtain the asymptotic differential equation $S' + 1 \sim -x^2(S')^2$ $(x \to 0+)$. The two roots of this quadratic equation for S' are

$$S' \sim [-1 \pm (1 - 4x^2)^{1/2}]/2x^2, \qquad x \to 0+.$$

Since x is small, the two solutions are

$$S' \sim -1, \qquad x \to 0+, \tag{3.4.24a}$$

$$S' \sim -x^{-2}, \qquad x \to 0+, \tag{3.4.24b}$$

Integrating (3.4.24*a*) gives $S \sim d$ $(x \to 0+)$, where d is some constant. This shows that the leading behavior of this solution is $y(x) \sim c_1$ $(x \to 0+)$, where $c_1 = e^d$. We have already obtained the complete asymptotic expansion of this solution in Example 3; the result is given in (3.4.3).

Integrating (3.4.24*b*), we obtain $S \sim x^{-1}$ $(x \to 0+)$. This gives the controlling factor $e^{x^{-1}}$ in the leading behavior (3.4.5). To find the full leading behavior, we substitute $S(x) = x^{-1} + C(x)$, where $C(x) \ll x^{-1}$ as $x \to 0+$, into (3.4.23) and obtain

$$x^2 C'' + x^2(C')^2 - (1 - 3x)C' - x^{-1} + 1 = 0. \tag{3.4.25}$$

Although there are six terms in this equation, it may be approximated as $x \to 0+$ by the asymptotic differential equation

$$C'(x) \sim -x^{-1}, \qquad x \to 0+ \tag{3.4.26}$$

(see Prob. 3.31). The solution to (3.4.26) is $C(x) \sim -\ln x$ $(x \to 0+)$. As in the previous example we have restricted the limit to be one sided to avoid complications resulting from imaginary numbers. We have now fully accounted for the x-dependence of the leading behavior in (3.4.5).

To find the full local behavior of this solution, we peel off the leading behavior by substituting $y(x) = c_2 x^{-1} e^{x^{-1}} w(x)$ into (3.4.2) with the intention of expanding $w(x)$ in a series of powers of x^α $(\alpha > 0)$. However, this analysis is unnecessary because, as is easily verified, the leading behavior (3.4.5) is already an exact solution to the differential equation!

Example 5 *Local behavior of solutions near an irregular singular point of a general nth-order Schrödinger equation.* In this example we derive an extremely simple and important formula for the leading behavior of solutions to the nth-order Schrödinger equation

$$\frac{d^n y}{dx^n} = Q(x)y \tag{3.4.27}$$

near an irregular singular point at x_0.

The exponential substitution $y = e^S$ and the asymptotic approximations $d^kS/dx^k \ll (S')^k$ as $x \to x_0$ for $k = 2, 3, \ldots, n$ give the asymptotic differential equation $(S')^n \sim Q(x)$ $(x \to x_0)$. Thus, $S(x) \sim \omega \int^x [Q(t)]^{1/n}\, dt$ $(x \to x_0)$, where ω is an nth root of unity. This result determines the n possible controlling factors of $y(x)$.

The leading behavior of $y(x)$ is found in the usual way (see Prob. 3.27) to be

$$\boxed{y(x) \sim c[Q(x)]^{(1-n)/2n} \exp\left\{\omega \int^x [Q(t)]^{1/n}\, dt\right\}, \qquad x \to x_0.} \tag{3.4.28}$$

If $x_0 \neq \infty$, (3.4.28) is valid if $|(x - x_0)^n Q(x)| \to \infty$ as $x \to x_0$. If $x_0 = \infty$, then (3.4.28) is valid if $|x^n Q(x)| \to \infty$ as $x \to \infty$. This important formula forms the basis of WKB theory and will be rederived perturbatively and in much greater detail in Sec. 10.2. If $Q(x) < 0$, solutions to (3.4.27) oscillate as $x \to \infty$; the nature of asymptotic relations between oscillatory functions is discussed in Sec. 3.7.

Here are some examples of the application of (3.4.28):

(*a*) For $y'' = y/x^5$, $y(x) \sim cx^{5/4}e^{\pm 2x^{-3/2}/3}$ $(x \to 0+)$.
(*b*) For $y''' = xy$, $y(x) \sim cx^{-1/3}e^{3\omega x^{4/3}/4}$ $(x \to +\infty)$, where $\omega^3 = 1$.
(*c*) For $d^4y/dy^4 = (x^4 + \sin x)y$, $y(x) \sim cx^{-3/2}e^{\omega x^2/2}$ $(x \to +\infty)$, where $\omega = \pm 1, \pm i$.

Concluding Remarks

We conclude this section with some philosophical commentary. You may recall our promise that the style of mathematics would change, beginning with this section. Let us now clarify what we meant. When we derive a local series about an ordinary point or a regular singular point, we use equal signs because such a series actually *converges* to a solution of the differential equation. And, if we truncate this series, we obtain an accurate approximation to the solution. We have learned in this section that the local behavior of a solution near an irregular singular point may be relatively, but *not* approximately, equal to the exact solution because the solution changes so rapidly near such a point. We have had to exchange equality and approximate equality for relative (asymptotic) equality. However, we do not begrudge this exchange. On the contrary, when a solution is changing rapidly, an asymptotic relation may be far more informative than an approximate equality. For example, discovering that $f(x)$ is approximately equal to 0 when $x \to 3$ is not nearly as dramatic as discovering that $f(x) \sim \exp[-(x-3)^{-2}]$ as $x \to 3$. Most of the approximations that will be derived in this book are asymptotic.

(E) 3.5 IRREGULAR SINGULAR POINT AT INFINITY

Many famous differential equations like the Bessel, Airy, and parabolic cylinder equations, which are discussed below, have an irregular singular point at ∞. In most applications a knowledge of the local behavior of the solutions to these equations for large $|x|$ is crucial. The development of the previous sections suggests an indirect way to perform a local analysis of such equations at ∞. We could map the singular point into the origin using the inversion transformation $x = 1/t$ and then evoke the methods of Sec. 3.4 to locally analyze the new differential equation at the origin. However, it is more efficient to derive and more convenient to represent the local behavior of solutions to the original equation directly in terms of asymptotic series valid at ∞. This section is mostly a collection of examples which will help to illustrate the relevant methods, but before tackling these examples we begin as we did in Sec. 3.4 with a small dose of formal theory, this time on the topic of asymptotic power series.

Asymptotic Power Series

We have derived several asymptotic power series in the previous section and you have probably arrived at a reasonably correct intuitive picture of what they are. Since we are about to derive several more asymptotic power series, this is the optimum moment to define them in precise terms and to summarize their mathematical properties. A more complete discussion is given in Sec. 3.8.

Definition The power series $\sum_{n=0}^{\infty} a_n(x - x_0)^n$ is said to be *asymptotic to the function* $y(x)$ *as* $x \to x_0$ and we write $y(x) \sim \sum_{n=0}^{\infty} a_n(x - x_0)^n$ $(x \to x_0)$ if $y(x) - \sum_{n=0}^{N} a_n(x - x_0)^n \ll (x - x_0)^N$ $(x \to x_0)$ for every N.

Thus, a power series is asymptotic to a function if the remainder after N terms is much smaller than the last retained term as $x \to x_0$. By this definition a series may be asymptotic to a function without being convergent.

An equivalent definition is $y(x) - \sum_{n=0}^{N} a_n(x - x_0)^n \sim a_M(x - x_0)^M$ $(x \to x_0)$, where a_M is the first nonzero coefficient after a_N. Thus, if there are an infinite number of nonzero a_n, an asymptotic power series is equivalent to an infinite sequence of asymptotic relations, one for each nonzero coefficient.

We also encounter asymptotic series in nonintegral powers of $x - x_0$. The series $\sum_{n=0}^{\infty} a_n(x - x_0)^{\alpha n}$ $(\alpha > 0)$ is asymptotic to the function $y(x)$ if $y(x) - \sum_{n=0}^{N} a_n(x - x_0)^{\alpha n} \ll (x - x_0)^{\alpha N}$ $(x \to x_0)$, for every N.

If $x_0 = \infty$ the corresponding definition is $y(x) \sim \sum_{n=0}^{\infty} a_n x^{-\alpha n}$ $(x \to \infty)$ if $y(x) - \sum_{n=0}^{N} a_n x^{-\alpha n} \ll x^{-\alpha N}$ $(x \to \infty)$ for every N.

Not all functions can be expanded in asymptotic power series. The function $y(x) = x^{-1}$ grows as $x \to 0$, so it cannot be asymptotic to a series of the form $\sum_{n=0}^{\infty} a_n x^{\alpha n}$ with $\alpha > 0$. Similarly, $y(x) = e^x$ has no asymptotic power series expansion of the form $\sum_{n=0}^{\infty} a_n x^{-\alpha n}$ as $x \to +\infty$ because e^x grows as $x \to +\infty$.

If $y(x)$ can be expanded in the asymptotic power series $y(x) \sim \sum_{n=0}^{\infty} a_n(x - x_0)^{\alpha n}$ $(x \to x_0)$, then the coefficients of this expansion are *unique*. This is because the above definition provides a way to determine the coefficients a_n

uniquely:

$$a_0 = \lim_{x \to x_0} y(x),$$

$$a_1 = \lim_{x \to x_0} \frac{y(x) - a_0}{(x - x_0)^\alpha},$$

and, in general,

$$a_N = \lim_{x \to x_0} \frac{y(x) - \sum_{n=0}^{N-1} a_n(x - x_0)^{\alpha n}}{(x - x_0)^{\alpha N}}.$$

The condition for a function to have an asymptotic power series expansion is that all these limits exist.

Most operations (addition, multiplication, division, differentiation, integration, and so on) can be performed on asymptotic power series term by term just as if they were convergent series, with important exceptions to be explained in Sec. 3.8.

Let us now return to the derivation of asymptotic series.

Example 1 *Behavior of a Taylor series for large x.* Consider the function $y(x)$ defined by the Taylor series

$$y(x) = \sum_{n=0}^{\infty} \frac{x^n}{(n!)^2}. \tag{3.5.1}$$

This series converges for all x, so $y(x)$ can, in principle, be calculated from it. However, while the series converges very rapidly for x small, it converges slowly for x large. If $x = 10$ only 16 terms are necessary to calculate $y(x)$ to 10 significant digits, while if $x = 10{,}000$ about 150 terms are required. (Why?) It would be nice to have a simple analytical formula for the sum of this series when x is large which does not require the addition of hundreds of numbers.

A similar Taylor series which also converges slowly when x is large is $\sum_{n=0}^{\infty} x^n/n!$. The sum of this series is exactly e^x.

Unfortunately, it is not possible to find an exact formula in terms of elementary functions for the sum of the series (3.5.1). However, it is possible to express the *local behavior* of the function $y(x)$ as $x \to +\infty$ using elementary functions.

To obtain the large-x behavior of $y(x)$ using the methods developed in this chapter, we construct a second-order differential equation satisfied by $y(x)$. Observe that $y' = \sum_{n=1}^{\infty} x^{n-1}/[n!\,(n-1)!]$ and that $(xy')' = \sum_{n=1}^{\infty} x^{n-1}/[(n-1)!]^2 = y$. Thus $y(x)$ is a solution to

$$xy'' + y' = y. \tag{3.5.2}$$

This equation has an irregular singular point at ∞.

As usual, we determine the leading behavior of $y(x)$ as $x \to +\infty$ by substituting $y = e^S$. The resulting equation

$$xS'' + x(S')^2 + S' = 1 \tag{3.5.3}$$

may be solved approximately if we first reduce it to the asymptotic differential equation $x(S')^2 \sim 1 - S'$ $(x \to +\infty)$, where we assume that $xS'' \ll x(S')^2$ as $x \to +\infty$. Solving for S' gives

$$S'(x) \sim [-1 \pm (1 + 4x)^{1/2}]/2x, \qquad x \to +\infty.$$

Thus, since x is large,

$$S'(x) \sim \pm x^{-1/2}, \qquad x \to +\infty.$$

Integrating this asymptotic relation gives

$$S(x) \sim \pm 2x^{1/2}, \qquad x \to +\infty,$$

or

$$S(x) = \pm 2x^{1/2} + C(x), \tag{3.5.4}$$

where $C(x) \ll 2x^{1/2}$ as $x \to +\infty$. [In general, if $S'(x) \sim x^\alpha$ as $x \to +\infty$, then: (a) $S(x) \sim x^{\alpha+1}/(\alpha + 1)$ as $x \to +\infty$ when $\alpha > -1$; (b) $S(x) \sim c$ as $x \to +\infty$ when $\alpha < -1$, where c is a constant; (c) $S(x) \sim \ln x$ as $x \to +\infty$ when $\alpha = -1$. The reader should compare these results with the corresponding ones for the integration of $S'(x) \sim (x - x_0)^{-b} (x \to x_0)$ in Prob. 3.28.]

Since the coefficients in the Taylor series (3.5.1) are all positive, $y(x)$ is an increasing function of x. Thus, its leading behavior must be governed by the positive sign in (3.5.4): $S(x) = 2x^{1/2} + C(x)$. Substituting this equation into (3.5.3) and combining terms gives an equation for $C(x)$:

$$xC'' + x(C')^2 + (2x^{1/2} + 1)C' + \tfrac{1}{2}x^{-1/2} = 0.$$

This equation may be approximated for large positive x using $1 \ll 2x^{1/2}$, $xC'' \ll \frac{1}{2}x^{-1/2}$, and $x(C')^2 \ll 2x^{1/2}C'$ as $x \to +\infty$, where the last two relations follow from $C' \ll x^{-1/2}$ and $C'' \ll x^{-3/2}$ $(x \to +\infty)$. We obtain the asymptotic differential equation

$$C' \sim -\tfrac{1}{4}x^{-1}, \qquad x \to +\infty,$$

whose solution is

$$C(x) \sim -\tfrac{1}{4} \ln x, \qquad x \to +\infty.$$

Hence the leading behavior of $y(x)$ is given by

$$y(x) \sim cx^{-1/4}e^{2x^{1/2}}, \qquad x \to +\infty. \tag{3.5.5}$$

The constant c cannot be determined from a local analysis of (3.5.2) because it is homogeneous. However, c can be determined by noticing that the Taylor series (3.5.1) for $y(x)$ is the same as that in (3.3.14) for the modified Bessel function I_0. Thus $y(x) = I_0(2x^{1/2})$. In Example 8 of Sec. 6.4, we derive the behavior of $I_0(x)$ as $x \to +\infty$ from an integral representation and show that $c = \frac{1}{2}\pi^{-1/2}$. This value for c may also be derived in another way using the Stirling formula for the large-n behavior of $n!$ (see Example 4 of Sec. 6.7).

How well does the leading asymptotic behavior (3.5.5) with $c = \frac{1}{2}\pi^{-1/2}$ approximate $y(x)$? We have measured the accuracy of (3.5.5) by plotting the ratio

$$\tfrac{1}{2}\pi^{-1/2}x^{-1/4}e^{2x^{1/2}}/y(x) \tag{3.5.6a}$$

as a function of x (see Fig. 3.4). We know that this ratio must approach 1 as $x \to +\infty$, but, in fact, this ratio is also near 1 for small values of x. Thus, the leading behavior is particularly useful because it is a good approximation over a large interval. We have also plotted in Fig. 3.4 the ratio

$$\left[\sum_{n=0}^{10} \frac{x^n}{(n!)^2}\right] \Bigg/ y(x). \tag{3.5.6b}$$

Observe that the first 11 terms in the Taylor series are a good polynomial approximation to $y(x)$ for $0 \le x < 50$ while just the first term in the asymptotic series is a good approximation to $y(x)$ for $x > 1$. The truncated Taylor series and the leading asymptotic behavior give good approximations in overlapping regions; together, they constitute a uniformly valid approximation to $y(x)$ over the entire region $0 \le x < \infty$.

We could drastically improve the accuracy and the range of validity of the asymptotic approximation by including more than just the leading term in the

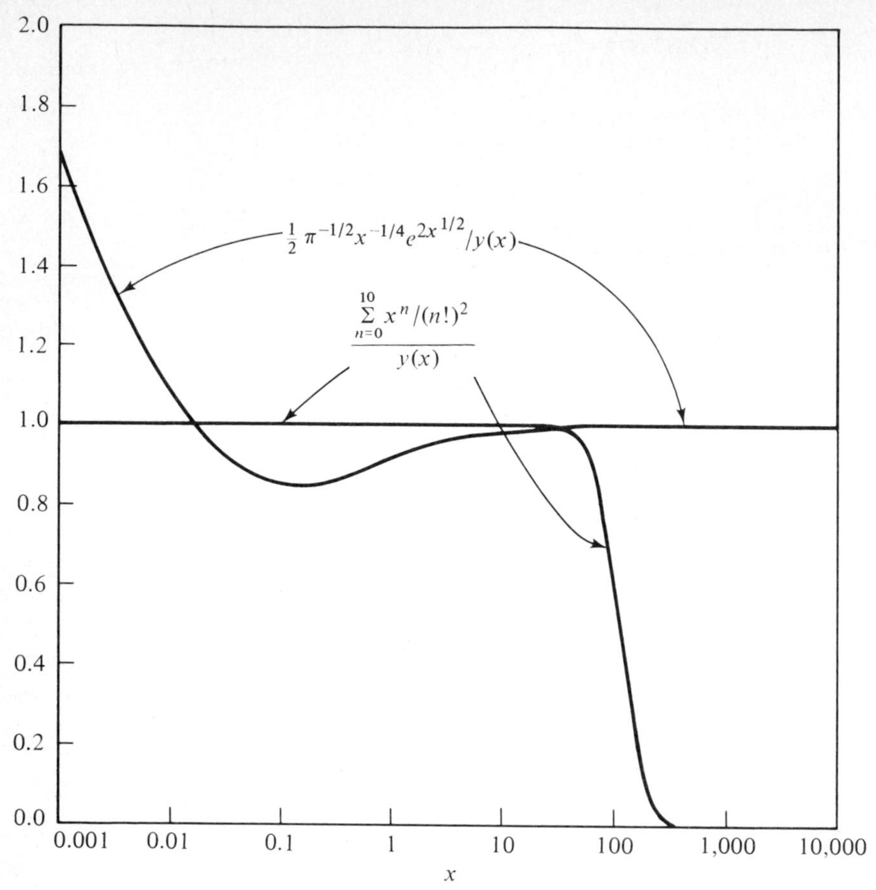

Figure 3.4 A comparison of the leading asymptotic approximation and an 11-term Taylor series approximation to the function $y(x)$ defined by the sum (3.5.1). The ratios of these approximations to $y(x)$ are plotted for $0.001 \le x \le 10{,}000$.

asymptotic series. In the next example we show how to *compute* the full asymptotic series for the modified Bessel function of arbitrary order ν. We do not *prove* here that the series that result are actually asymptotic to modified Bessel functions.

Example 2 *Behavior of modified Bessel functions for large x.* The modified Bessel equation $x^2y'' + xy' - (x^2 + \nu^2)y = 0$ has an irregular singular point at ∞. The leading behaviors of the solutions are [see Prob. 3.38(*a*)]

$$y(x) \sim c_1 x^{-1/2} e^x, \qquad x \to +\infty, \tag{3.5.7a}$$

$$y(x) \sim c_2 x^{-1/2} e^{-x}, \qquad x \to +\infty. \tag{3.5.7b}$$

Notice that (3.5.7*a*) is consistent with (3.5.5) for $I_0(2x^{1/2})$. Notice, also, that while the leading behaviors (3.5.7) do not depend on ν, ν will appear in the coefficients of the full asymptotic expansion of the solutions [see (3.5.9)].

We now derive the asymptotic expansion for the solution whose leading behavior is (3.5.7*a*). To do this we peel off the leading behavior by substituting

$$y(x) = c_1 x^{-1/2} e^x w(x) \tag{3.5.8a}$$

into the modified Bessel equation. The equation satisfied by $w(x)$ is

$$x^2 w'' + 2x^2 w' + (\tfrac{1}{4} - \nu^2)w = 0.$$

We seek a solution to this equation of the form $w(x) = 1 + \varepsilon(x)$ with $\varepsilon(x) \ll 1$ $(x \to +\infty)$.

What is the leading behavior of $\varepsilon(x)$? $\varepsilon(x)$ satisfies the equation

$$x^2\varepsilon'' + 2x^2\varepsilon' + (\tfrac{1}{4} - \nu^2)\varepsilon + \tfrac{1}{4} - \nu^2 = 0,$$

which may be simplified by the approximations

$$(\tfrac{1}{4} - \nu^2)\varepsilon \ll \tfrac{1}{4} - \nu^2, \qquad x^2\varepsilon'' \ll x^2\varepsilon', \qquad x \to +\infty.$$

We make the second of these approximations because we anticipate that ε decays like a power of x as $x \to +\infty$. The resulting asymptotic differential equation is

$$2x^2\varepsilon'(x) \sim \nu^2 - \tfrac{1}{4}, \qquad x \to +\infty.$$

Ordinarily the solution to this equation would be $\varepsilon(x) \sim c$ as $x \to +\infty$, where c is an integration constant. However, since $\varepsilon(x) \ll 1$ as $x \to +\infty$ we must set $c = 0$. The leading behavior of $\varepsilon(x)$ is then given by

$$\varepsilon(x) \sim (\tfrac{1}{8} - \tfrac{1}{2}\nu^2)x^{-1}, \qquad x \to +\infty.$$

This kind of analysis can be repeated to obtain all the terms in the asymptotic expansion of $w(x)$ as $x \to +\infty$. However, the leading behavior of $\varepsilon(x)$ suggests that $w(x)$ has a series expansion in inverse powers of x. Thus, to simplify the analysis, we assume at the outset that $w(x) \sim \sum_{n=0}^{\infty} a_n x^{-n}$ $(x \to +\infty;\, a_0 = 1)$. Substituting this expansion into the differential equation for w gives

$$0 \sim \sum_{n=0}^{\infty} n(n+1)a_n x^{-n} - 2\sum_{n=0}^{\infty} n a_n x^{1-n} + (\tfrac{1}{4} - \nu^2)\sum_{n=0}^{\infty} a_n x^{-n}, \qquad x \to +\infty.$$

You might think that this asymptotic relation is formulated poorly because in Sec. 3.4 we warned that a function could not be asymptotic to 0. However, by the definition of an asymptotic power series, the function 0 *does* have an asymptotic power series expansion whose coefficients are all 0. Therefore, since the coefficients of any asymptotic power series are unique, we may equate to 0 the coefficients of all powers of $1/x$ in the above relation:

$$x^{-n}: \qquad [(n + \tfrac{1}{2})^2 - \nu^2]a_n - 2(n+1)a_{n+1} = 0, \qquad n = 0, 1, 2, \ldots.$$

Solving this recursion relation and using $a_0 = 1$ we obtain

$$w(x) \sim 1 - \frac{(4\nu^2 - 1^2)}{1!\,8x} + \frac{(4\nu^2 - 1^2)(4\nu^2 - 3^2)}{2!\,(8x)^2} - \cdots, \qquad x \to +\infty. \tag{3.5.9a}$$

From the ratio test we see that the radius of convergence R of (3.5.9*a*) is

$$R = \lim_{n\to\infty}\left|\frac{a_n}{a_{n+1}}\right| = \lim_{n\to\infty}\frac{2(n+1)}{(n+\tfrac{1}{2})^2 - \nu^2} = 0$$

unless the series (3.5.9*a*) terminates, which it does when $\nu = \pm\tfrac{1}{2}, \pm\tfrac{3}{2}, \pm\tfrac{5}{2}, \ldots$. When this happens, the finite series (3.5.9*a*) when multiplied by $e^{-x}/\sqrt{x}$ gives an *exact* solution to the modified Bessel equation.

Similarly, the complete asymptotic series for the function whose leading behavior is given in (3.5.7*b*) is

$$y(x) = c_2 x^{-1/2} e^{-x} w(x), \tag{3.5.8b}$$

where

$$w(x) \sim 1 + \frac{(4\nu^2 - 1^2)}{1!\,8x} + \frac{(4\nu^2 - 1^2)(4\nu^2 - 3^2)}{2!\,(8x)^2} + \cdots, \qquad x \to +\infty. \tag{3.5.9b}$$

The modified Bessel functions $I_\nu(x)$ and $K_\nu(x)$ are special solutions of the modified Bessel equation which were introduced in Sec. 3.4. The function $I_\nu(x)$ grows exponentially as $x \to +\infty$; its asymptotic series is given in (3.5.8*a*) and (3.5.9*a*) with $c_1 = (2\pi)^{-1/2}$. The function $K_\nu(x)$ decays exponentially as $x \to +\infty$; its asymptotic series is given in (3.5.8*b*) and (3.5.9*b*) with $c_2 = (\pi/2)^{1/2}$. These values of the constants c_1 and c_2 will be derived in Sec. 6.4.

Until now, we have treated the derivation of an asymptotic series as a formal exercise. However, in the next example we show how a divergent asymptotic series can actually be used to approximate a function numerically.

Example 3 *Numerical evaluation of the asymptotic series for* $I_\nu(x)$. In the last example we obtained an asymptotic expansion for $I_\nu(x)$ valid as $x \to +\infty$. Let us use this expansion to compute $I_5(x)$ for various values of x. Setting $c_1 = (2\pi)^{-1/2}$ and $\nu = 5$ in (3.5.8*a*) and (3.5.9*a*), we obtain the expansion for $I_5(x)$:

$$I_5(x) \sim (2\pi)^{-1/2} e^x x^{-1/2} \left[1 - \frac{(100 - 1)}{1!\,8x} + \frac{(100 - 1)(100 - 9)}{2!\,(8x)^2} - \cdots\right], \qquad x \to +\infty. \tag{3.5.10}$$

Of course we have not proved it, but the series in (3.5.10) *is* asymptotic. From the definition of an asymptotic power series given at the beginning of this section, we know that if we terminate the series (3.5.10) after the x^{-N} term, the remainder, which is the difference between the value of the function and the sum of these $(N + 1)$ terms, is asymptotically the $(N + 2)$th term in the series as $x \to +\infty$. The first neglected term is thus a measure of the error as $x \to +\infty$. If x is large and held fixed (i.e., not allowed to tend to ∞), the first neglected term is only an estimate of the error. With this in mind we can formulate a simple rule for obtaining good numerical results from asymptotic series. We look over the individual terms in the asymptotic series; typically the terms get successively smaller for a while, but eventually, because the series is known to diverge, they get larger and larger and tend to infinity. For every given value of x we locate the smallest term. We then add all the preceding terms in the asymptotic series up to but *not* including the smallest term. This finite sum of terms usually gives the best estimate of the function because the next term, which approximates the error, is the smallest term in the series. The approximation obtained in this way is called the *optimal asymptotic approximation.*

We can compare this rule with the way we would evaluate the sum of the *convergent* power series $\sum_{n=0}^{\infty} a_n(x - x_0)^n$ for a fixed value of x. For this series, there is no limit to the accuracy; we can always improve the accuracy by taking more terms in the partial sum. However, for a divergent asymptotic series, for each given value of x there is an upper limit to the accuracy and if we take either more or less than the optimal number of terms in the partial sum according to our rule, we usually decrease the accuracy. If we are not satisfied with this maximal accuracy, then to improve it we must take x closer to x_0, or in the case of the series in (3.5.10) we must take x closer to $+\infty$.

Table 3.1 demonstrates how this rule works. We have used the series in (3.5.10) to evaluate $e^{-x}I_5(x)$ for $x = 3.0, 4.0, 5.0, 6.0, 7.0$. The entries in the columns are the partial sums truncated after the x^{-N} term. The underlined partial sum is the optimal approximation to $e^{-x}I_5(x)$ according to the rule. Observe that for each value of x the partial sums get closer for a while to the exact value of $e^{-x}I_5(x)$ and then rapidly veer off after reaching the optimal number of terms.

Table 3.1 Asymptotic approximations to $e^{-x}I_5(x)$ for five values of x using the series in (3.5.10)

Entries in the columns are the partial sums truncated after the x^{-N} term. Underlined partial sums are optimal asymptotic approximations. Notice that even when $x = 7$ the leading term in the asymptotic expansion gives a very poor approximation while the optimal asymptotic truncation is very accurate. The number in parentheses is the power of 10 multiplying the entry.

	x				
N	3.0	4.0	5.0	6.0	7.0
0	2.30324 (−1)	1.99471 (−1)	1.78412 (−1)	1.62868 (−1)	1.50786 (−1)
2	1.08147 (0)	4.59816 (−1)	2.39128 (−1)	1.45372 (−1)	1.00804 (−1)
4	2.01953 (−1)	4.74361 (−2)	2.52641 (−2)	2.35810 (−2)	2.61284 (−2)
6	2.11127 (−2)	1.14538 (−2)	1.49262 (−2)	1.98392 (−2)	2.45412 (−2)
7	1.16597 (−2)	1.03611 (−2)	1.47212 (−2)	1.97870 (−2)	2.45248 (−2)
8	5.50542 (−3)	9.82749 (−3)	1.46411 (−2)	1.97700 (−2)	2.45202 (−2)
9	1.20401 (−4)	9.47732 (−3)	1.45991 (−2)	1.97626 (−2)	2.45184 (−2)
10	−5.73580 (−3)	9.19172 (−3)	1.45717 (−2)	1.97585 (−2)	2.45176 (−2)
11	−1.33001 (−2)	8.91505 (−3)	1.45504 (−2)	1.97559 (−2)	2.45172 (−2)
12	−2.45677 (−2)	8.60595 (−3)	1.45314 (−2)	1.97540 (−2)	2.45169 (−2)
13	−4.35276 (−2)	8.21586 (−3)	1.45122 (−2)	1.97523 (−2)	2.45167 (−2)
14	−7.90210 (−2)	7.66817 (−3)	1.44907 (−2)	1.97508 (−2)	2.45166 (−2)
15	−1.52078 (−1)	6.82267 (−3)	1.44641 (−2)	1.97492 (−2)	2.45164 (−2)
20	−1.31437 (1)	−3.61663 (−2)	1.39178 (−2)	1.97329 (−2)	2.45155 (−2)
35	−3.12759 (10)	1.24079 (6)	−4.90286 (2)	−8.13340 (−1)	2.06197 (−2)
	Exact value of $e^{-x}I_5(x)$				
	4.54090 (−3)	9.24435 (−3)	1.45403 (−2)	1.97519 (−2)	2.45164 (−2)
	Relative error in optimal asymptotic approximation, %				
	21.0	0.57	0.069	0.0024	0.000071

Observe that as x increases the optimal number of terms increases and so does the accuracy of the corresponding partial sum. When $x = 3$ the relative error is 21 percent, when $x = 5$ it has improved to 0.07 percent, and when $x = 7$ it is a whopping 7×10^{-5} percent. It is nice to know that the asymptotic series which was derived by considering a small neighborhood of infinity is dependable for values of x so far from ∞.

In Fig. 3.5 we plot the optimal asymptotic approximation [obtained by truncating the asymptotic series (3.5.10) according to our rule] to $e^{-x}I_5(x)$ for x between 2.0 and 10.0. The graph shows that although $e^{-x}I_5(x)$ is a continuous function the optimal approximation has discontinuities at the points x where the optimal number of terms in the truncated series changes by one. For $x > 2.0$, these points occur when two successive terms in the series are equal; these values of x are $x = 2.63, 3.26, 3.88, 4.47, 5.05, 5.62, \ldots$. When $2 \le x \le 2.63$, we truncate after x^{-7}; when $2.63 \le x \le 3.26$, we truncate after x^{-8}; when $3.26 \le x \le 3.88$, we truncate after x^{-9}; and so on. These crossover points are given explicitly by the formula $x = [(2k+1)^2 - 100]/8(k+1)$ $(k = 8, 9, 10, \ldots)$.

The optimal truncation of the asymptotic series gives a good numerical approximation to $I_\nu(x)$ for all ν. However, Fig. 3.6 shows that the smallest value of x at which the optimal approximation is accurate increases approximately linearly with ν.

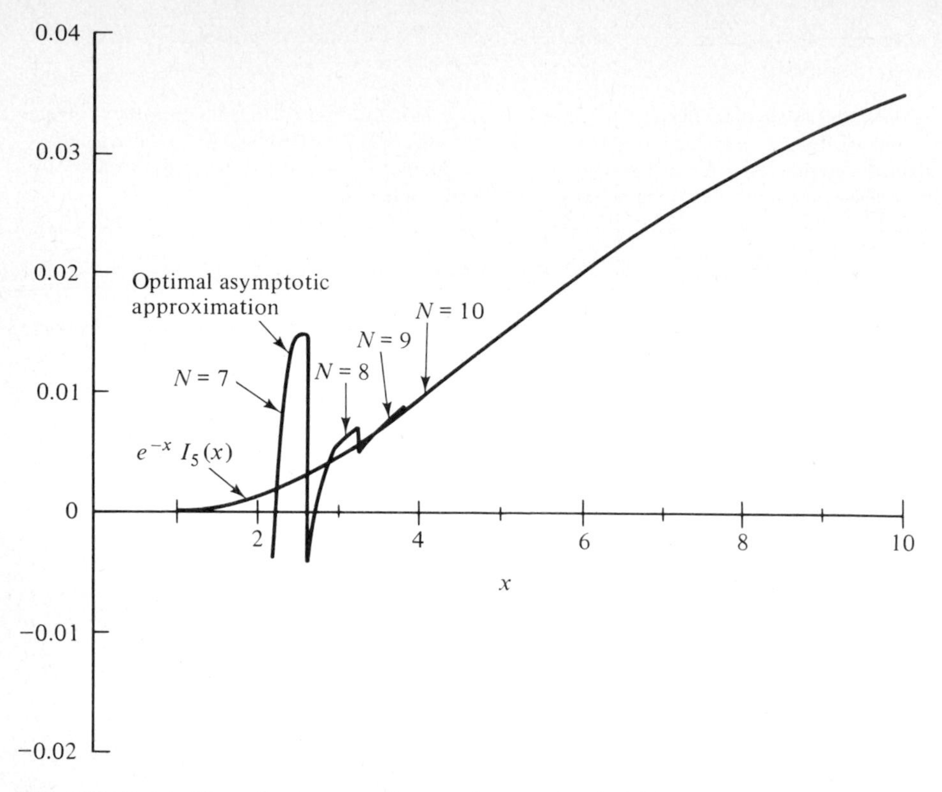

Figure 3.5 A plot of the optimal asymptotic approximation to $e^{-x}I_5(x)$ for $2 \le x \le 10$. For comparison, the exact numerical value of $e^{-x}I_5(x)$ is also shown for $0 \le x \le 10$. These two curves are indistinguishable when $x > 4$. The discontinuities in the optimal asymptotic approximation occur when the optimal number of terms increases by one. Each segment of the optimal asymptotic approximation is labeled by a number N which is the highest power of $1/x$ in the optimal truncation. [Note that we have chosen to plot $e^{-x}I_5(x)$ instead of $I_5(x)$ itself because $I_5(x)$ rapidly runs off scale as x increases.]

Despite these wonderful results in Table 3.1 and Figs. 3.5 and 3.6, you may be distressed about a rule for "summing" a divergent series which yields a maximal accuracy that cannot be surpassed. Maybe you are disappointed that the rest of the terms in the series must stand idle, unable to improve the optimal but relatively poor result for $x = 3$ obtained by adding up the first nine terms. Why bother to compute the full asymptotic series if only nine terms are usable? In fact, there are sophisticated rules for "summing" divergent series which make use of *all* the information contained in the terms of the asymptotic series. In many cases these rules surpass the limited accuracy of the optimal truncations and give *arbitrarily* accurate approximations provided sufficiently many terms in the series are used. The existence of these more powerful rules, which are discussed in Chap. 8, vastly increases the value of asymptotic series.

Example 4 *Behavior of parabolic cylinder functions for large* x. Let us examine the behavior of the solutions $y(x)$ to the parabolic cylinder equation

$$y'' + (\nu + \tfrac{1}{2} - \tfrac{1}{4}x^2)y = 0 \tag{3.5.11}$$

as $x \to +\infty$. In this equation ν is a parameter.

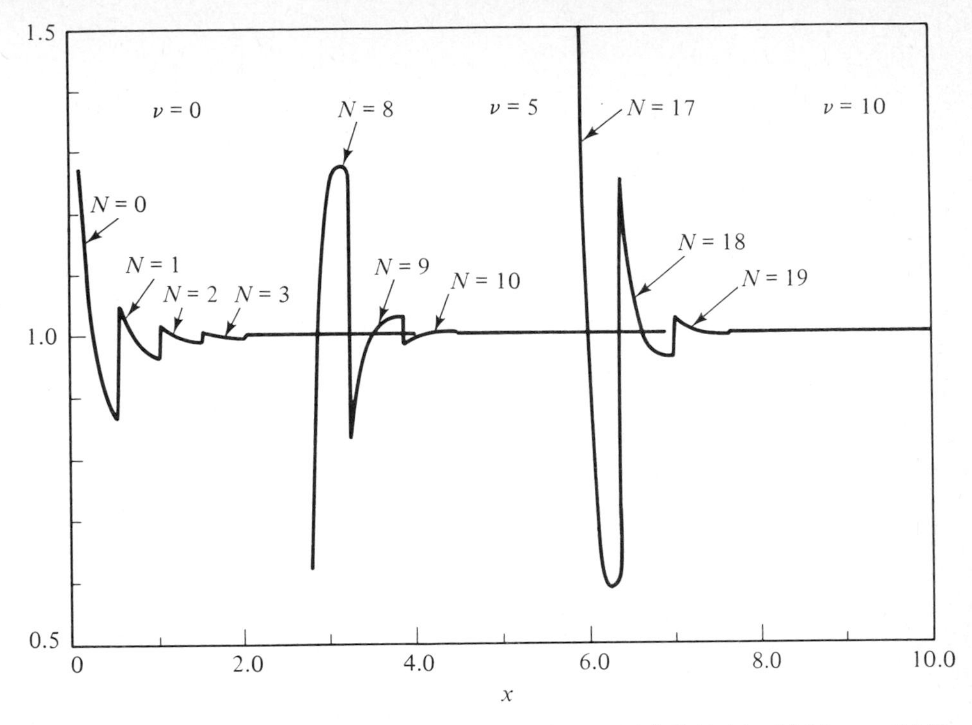

Figure 3.6 A plot of the optimal asymptotic approximation to $I_\nu(x)$ divided by $I_\nu(x)$ for $\nu = 0,5,10$. Observe that as ν increases, the smallest value of x at which the optimal asymptotic approximation gives a good estimate of $I_\nu(x)$ also increases (approximately linearly with ν). The number N is the highest power of x^{-1} in the optimal asymptotic approximation.

Since (3.5.11) has an irregular singular point at ∞, we make the exponential substitution $y = e^S$ as in Example 4 of Sec. 3.4 and obtain

$$S'' + (S')^2 + \nu + \tfrac{1}{2} - \tfrac{1}{4}x^2 = 0.$$

Making the approximations

$$S'' \ll (S')^2, \qquad \nu + \tfrac{1}{2} \ll \tfrac{1}{4}x^2, \qquad x \to +\infty,$$

gives the asymptotic differential equation $(S')^2 \sim \frac{1}{4}x^2$ $(x \to +\infty)$ whose solutions are

$$S(x) \sim \pm\tfrac{1}{4}x^2, \qquad x \to +\infty.$$

We have now determined that the possible controlling factors of the leading behavior of $y(x)$ are $e^{\pm x^2/4}$.

We leave as an exercise [Prob. 3.38(b)] the verification that the possible leading behaviors of $y(x)$ are

$$y(x) \sim c_1 x^{-\nu-1} e^{x^2/4}, \qquad x \to +\infty, \tag{3.5.12a}$$

or

$$y(x) \sim c_2 x^\nu e^{-x^2/4}, \qquad x \to +\infty. \tag{3.5.12b}$$

It is conventional to define the *parabolic cylinder function* $D_\nu(x)$ as that solution of the parabolic cylinder equation (3.5.11) whose asymptotic behavior is given by (3.5.12b) with $c_2 = 1$. This condition determines the function $D_\nu(x)$ uniquely because c_1 must be 0. (Why?) Like the

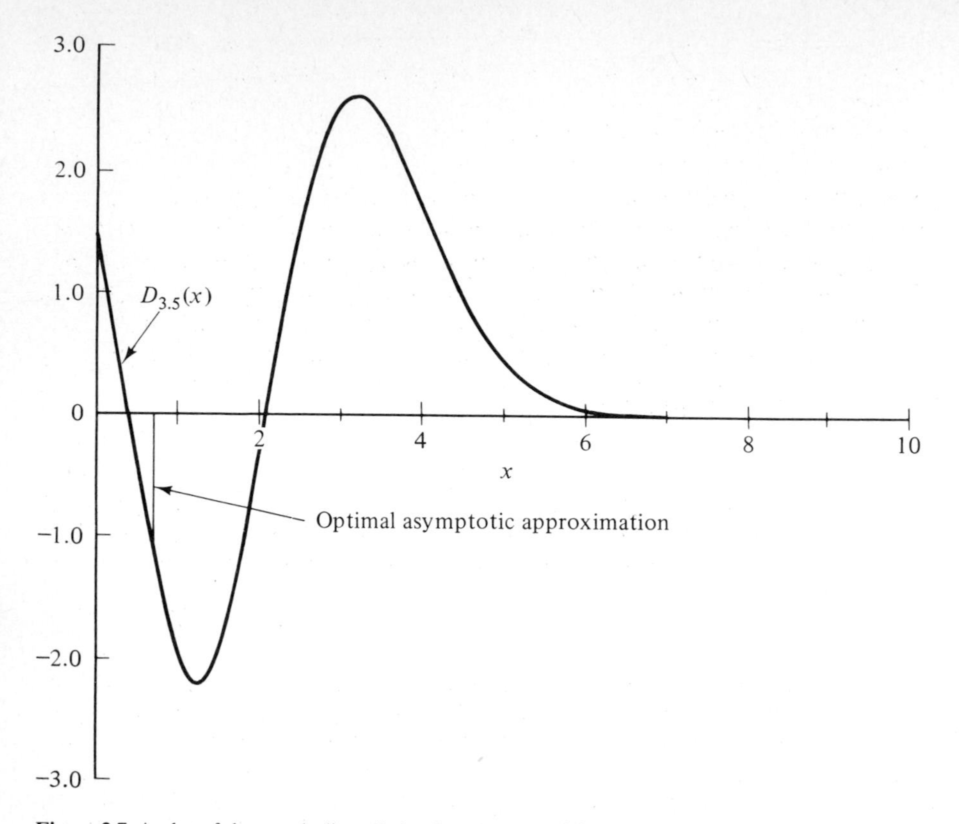

Figure 3.7 A plot of the parabolic cylinder function $D_{3.5}(x)$ and the optimal asymptotic approximation to $D_{3.5}(x)$ for $0 \le x \le 10$. The optimal asymptotic approximation is 0 for $0 \le x \le 0.68$. Notice how good the optimal asymptotic approximation is for $x \ge 0.69$.

Bessel and Airy functions, $D_\nu(x)$ is an important mathematical function that appears frequently in asymptotic analysis.

For further applications it will be necessary to know the complete asymptotic expansion of $D_\nu(x)$ valid as $x \to +\infty$. The expansion is obtained by substituting

$$D_\nu(x) = x^\nu e^{-x^2/4} w(x) \tag{3.5.13}$$

into (3.5.11). We obtain

$$x^2 w'' + (2\nu x - x^3)w' + \nu(\nu - 1)w = 0.$$

It may be shown in the usual way that the solution w, which tends to 1 as $x \to +\infty$, has an asymptotic series in powers of $1/x$:

$$w(x) \sim \sum_{n=0}^{\infty} a_n x^{-n}, \qquad x \to +\infty,$$

where $a_0 = 1$.

Substituting this expansion into the differential equation for $w(x)$ and equating coefficients of powers of x gives the following recursion relations:

$$x^{-1}: \qquad a_1 = 0,$$

$$x^{-n}: \qquad (n + 2)a_{n+2} = -(n - \nu)(n - \nu + 1)a_n, \qquad n = 0, 1, \ldots.$$

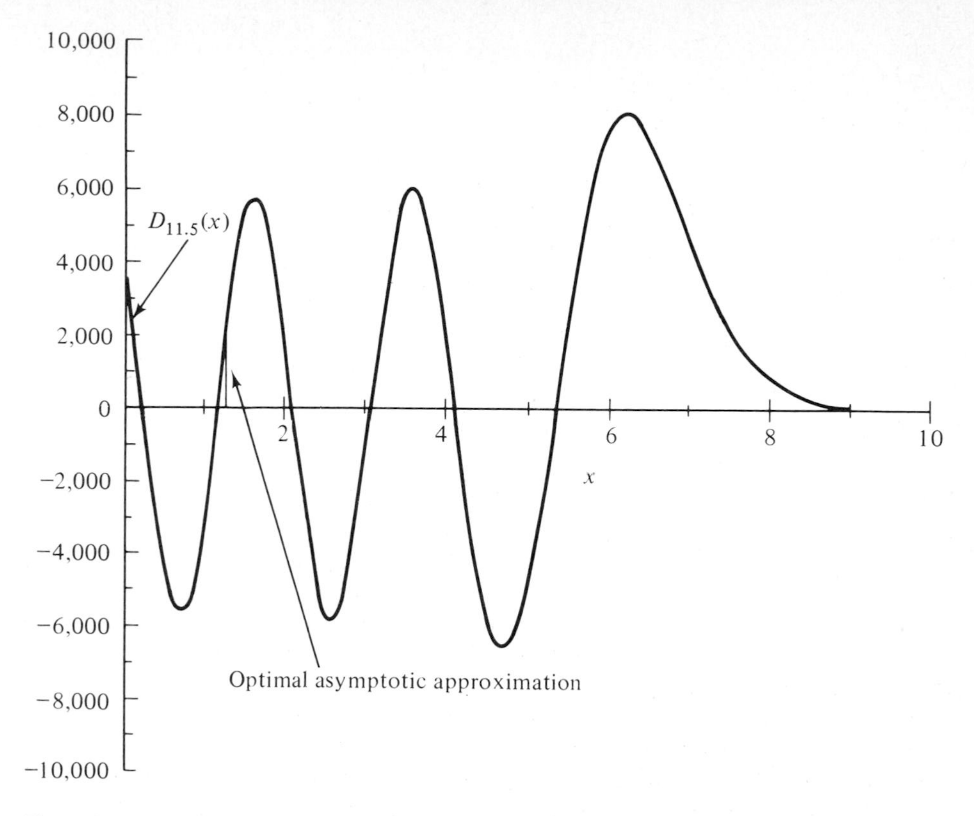

Figure 3.8 A plot of the parabolic cylinder function $D_{11.5}(x)$ and the optimal asymptotic approximation to $D_{11.5}(x)$ for $0 \le x \le 10$. The optimal asymptotic approximation is 0 for $x \le 1.37$. The error in the optimal approximation is less than the thickness of the curve for $x \ge 1.38$. [The cutoff point below which the optimal asymptotic approximation is zero and above which the optimal approximation is very accurate grows roughly as $\frac{1}{2}v^{1/2}$ as $v \to \infty$ (see Prob. 3.43).]

We conclude that $a_n = 0$ for n odd and that

$$w(x) \sim 1 - \frac{v(v-1)}{2^1 \cdot 1!\, x^2} + \frac{v(v-1)(v-2)(v-3)}{2^2 \cdot 2!\, x^4} - \cdots, \qquad x \to +\infty. \tag{3.5.14}$$

If v is not 0, 1, 2, 3, ..., then the ratio test applied to successive nonzero terms implies that R, the radius of convergence of (3.5.14), is 0:

$$R^2 = \lim_{n\to\infty} \left| \frac{a_n}{a_{n+2}} \right| = \lim_{n\to\infty} \left| \frac{n+2}{(n-v)(n-v+1)} \right| = 0.$$

However, if v is a nonnegative integer $v = N = 0, 1, 2, 3, \ldots$, then the series (3.5.14) terminates after $(\frac{1}{2}N + 1)$ terms if N is even and after $(\frac{1}{2}N + \frac{1}{2})$ terms if N is odd. When the series terminates, the parabolic cylinder function $D_N(x)$ takes the form of an Nth-degree polynomial $\mathrm{He}_N(x)$, called a Hermite polynomial, multiplied by a decaying exponential: $D_N(x) = e^{-x^2/4}x^N w(x) = e^{-x^2/4}\,\mathrm{He}_N(x)$. The first few Hermite polynomials determined from (3.5.14) are $\mathrm{He}_0(x) = 1$, $\mathrm{He}_1(x) = x$, $\mathrm{He}_2(x) = x^2 - 1$, $\mathrm{He}_3(x) = x^3 - 3x$, $\mathrm{He}_4(x) = x^4 - 6x^2 + 3$.

For values of v other than 0, 1, 2, 3, ..., (3.5.13) and the divergent series (3.5.14) give an excellent approximation to $D_v(x)$. In Figs. 3.7 and 3.8 we plot $D_v(x)$ for $v = 3.5$ and 11.5 along

Table 3.2 Relative error in optimal asymptotic approximation to the parabolic cylinder function $D_\nu(x)$ for $\nu = 3.5,\ 11.5$

	$\nu = 3.5$		$\nu = 11.5$	
x	$D_\nu(x)$	Relative error in optimal approximation, %	$D_\nu(x)$	Relative error in optimal approximation, %
1.0	-2.0368	2.3	-2.5726×10^3	100
1.5	-1.8583	3.0×10^{-1}	5.3550×10^3	2.7×10^{-2}
2.0	-1.8226×10^{-1}	3.3×10^{-1}	1.3327×10^3	5.5×10^{-3}
2.5	1.6604	2.6×10^{-3}	-5.7309×10^3	6.1×10^{-5}
3.0	2.5823	1.1×10^{-4}	-8.6290×10^2	1.3×10^{-5}
3.5	2.4320	5.2×10^{-6}	5.9345×10^3	4.8×10^{-8}

with the optimal asymptotic approximations to these functions. Observe how closely the asymptotic approximations hug the true functions. The relative errors in the optimal approximations to $D_{3.5}(x)$ and $D_{11.5}(x)$ are given in Table 3.2. It is necessary to augment the figures by this table because when x is large the magnitude of the error is less than the thickness of the plotted curves.

Example 5 *Behavior of Airy functions for large x.* The Airy equation

$$y'' = xy \tag{3.5.15}$$

has an irregular singular point at ∞. The leading behaviors of solutions for large x are determined by (3.4.28) with $n = 2$ and $Q(x) = x$ to be

$$y(x) \sim c_1 x^{-1/4} e^{-2x^{3/2}/3}, \qquad x \to +\infty, \tag{3.5.16a}$$

or

$$y(x) \sim c_2 x^{-1/4} e^{2x^{3/2}/3}, \qquad x \to +\infty. \tag{3.5.16b}$$

The Airy function Ai (x) is the unique solution to (3.5.15) that satisfies (3.5.16a) with $c_1 = \frac{1}{2}\pi^{-1/2}$:

$$\text{Ai}\,(x) \sim \tfrac{1}{2}\pi^{-1/2} x^{-1/4} e^{-2x^{3/2}/3}, \qquad x \to +\infty. \tag{3.5.17a}$$

Why does this one condition define Ai (x) uniquely?

The leading behavior of the other Airy function Bi (x) for large x is

$$\text{Bi}\,(x) \sim \pi^{-1/2} x^{-1/4} e^{2x^{3/2}/3}, \qquad x \to +\infty. \tag{3.5.17b}$$

This equation does *not* uniquely define Bi (x). Why?

The connection between the Taylor series for Ai (x) and Bi (x) in (3.2.1) and (3.2.2) and their leading behaviors must await our discussion of integrals (see Probs. 6.75 and 6.76).

In Figs. 3.9 and 3.10 we compare the Airy functions Ai (x) and Bi (x) with their leading asymptotic behaviors given in (3.5.17).

The full asymptotic expansion of Ai (x) or Bi (x) is found by peeling off the leading behavior and seeking an expansion of the remaining factor as a series of powers of $1/x$. To carry through this procedure for the Airy function Ai (x) we first peel off the leading behavior by substituting

$$y(x) = x^{-1/4} e^{-2x^{3/2}/3} w(x). \tag{3.5.18}$$

It follows that $w(x)$ is that solution to the differential equation

$$x^2 w'' - (2x^{5/2} + \tfrac{1}{2}x)w' + \tfrac{5}{16}w = 0 \tag{3.5.19}$$

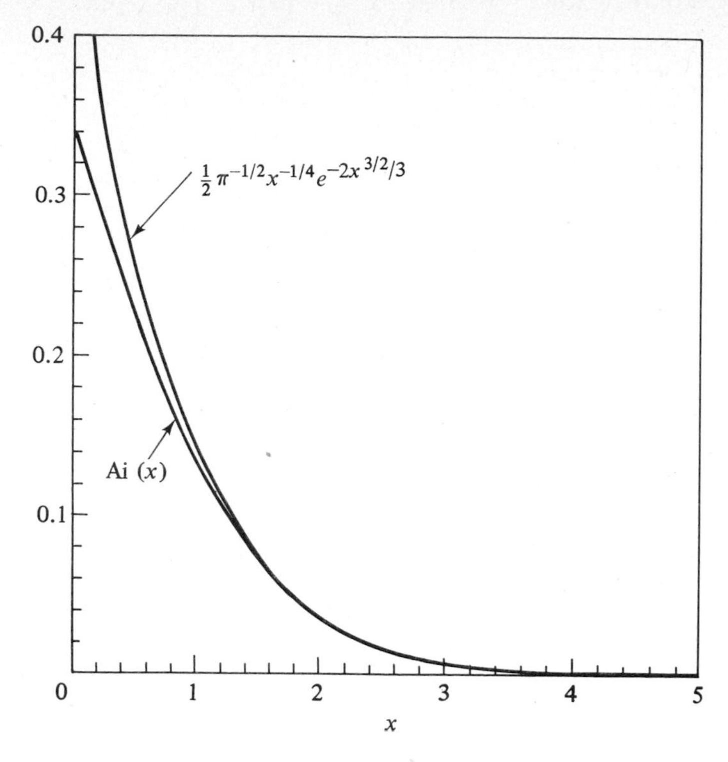

Figure 3.9 A plot of the Airy function Ai (x) and its leading behavior (3.5.17*a*) for $0 \le x \le 5$.

which approaches 1 as $x \to +\infty$. One way to analyze this solution is to: (*a*) write $w(x) = 1 + \varepsilon(x)$ with $\varepsilon(x) \ll 1$ as $x \to +\infty$; (*b*) determine the leading behavior of $\varepsilon(x)$ as $x \to +\infty$; and (*c*) find the full asymptotic series for $\varepsilon(x)$. Our past experience with this kind of analysis suggests that $\varepsilon(x)$ behaves like $x^{-\alpha}$ with $\alpha > 0$ as $x \to +\infty$ and that $\varepsilon(x)$ has an asymptotic expansion in powers of $x^{-\alpha}$.

Instead of repeating this kind of analysis once more, let us use a slightly different but equivalent approach to find the asymptotic expansion of $w(x)$. We assume at the outset that $w(x) \sim \sum_{n=0}^{\infty} a_n x^{\alpha n}$ $(x \to +\infty)$ with $\alpha < 0$ and $a_0 = 1$. Substituting this expansion into (3.5.19), it follows that

$$\tfrac{5}{16}a_0 - 2\alpha a_1 x^{\alpha+3/2} + (\alpha^2 - \tfrac{3}{2}\alpha + \tfrac{5}{16})a_1 x^{\alpha} - 4\alpha a_2 x^{2\alpha+3/2} + \cdots \sim 0, \qquad x \to +\infty. \tag{3.5.20}$$

In order that (3.5.20) be valid, it is necessary that the coefficient of each distinct power of x be zero. Since $a_0 = 1$ and $\alpha < 0$, it follows that $\alpha = -\frac{3}{2}$ and $a_1 = -\frac{5}{48}$. More generally, equating coefficients of $(x^{-3/2})^n$ in (3.5.20) to zero gives the recursion relation

$$a_{n+1} = -\frac{3}{4}\frac{(n+\frac{5}{6})(n+\frac{1}{6})}{n+1}a_n, \qquad n = 0, 1, \ldots.$$

The solution of this recursion relation is

$$a_n = \frac{1}{2\pi}\left(-\frac{3}{4}\right)^n \frac{\Gamma(n+\frac{5}{6})\Gamma(n+\frac{1}{6})}{n!},$$

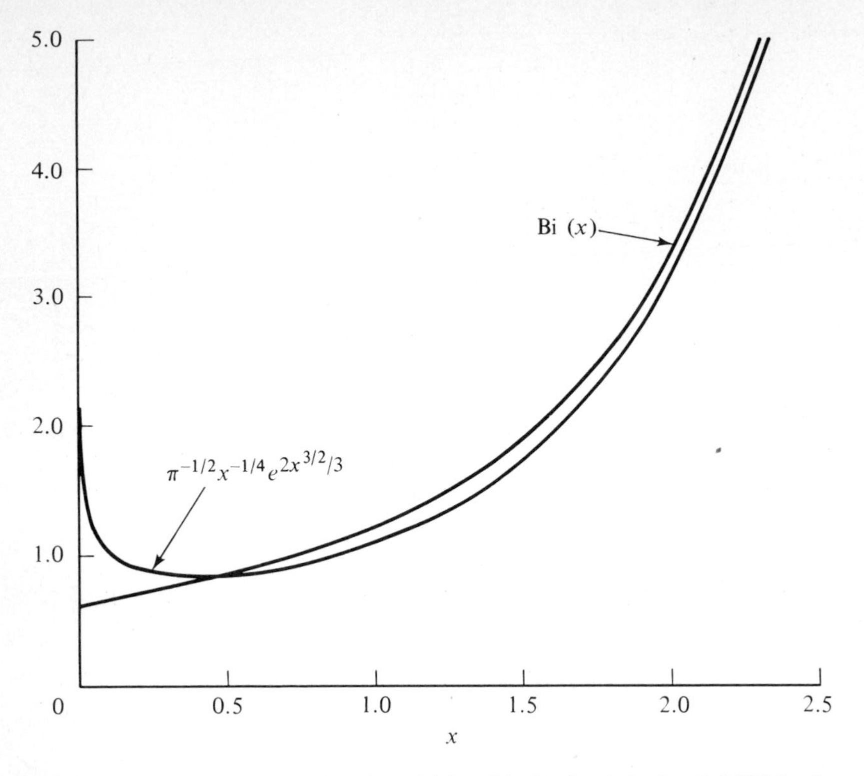

Figure 3.10 A plot of the Airy function Bi (x) and its leading behavior (3.5.17b) for $0 \le x \le 2.5$.

where we have used $a_0 = 1$ and $\Gamma(\frac{1}{6})\Gamma(\frac{5}{6}) = 2\pi$ (see Prob. 2.6). Thus,

$$\text{Ai }(x) \sim \tfrac{1}{2}\pi^{-1/2}x^{-1/4}e^{-2x^{3/2}/3}[1 - \tfrac{5}{48}x^{-3/2} + \tfrac{385}{4608}(x^{-3/2})^2 - \cdots]. \tag{3.5.21}$$

As usual, the ratio test applied to (3.5.20) gives a vanishing radius of convergence

$$R = \lim_{n\to\infty} \left| \frac{a_n}{a_{n+1}} \right| = \lim_{n\to\infty} \frac{4(n+1)}{3(n+\frac{1}{6})(n+\frac{5}{6})} = 0.$$

Asymptotic methods also apply to higher-order equations, as we see in the next example.

Example 6 "*Hyperairy*" *equation.* The hyperairy equation,

$$\frac{d^4y}{dy^4} = xy, \tag{3.5.22}$$

which is the fourth-order generalization of the Airy equation, has an irregular singular point at ∞. The four possible leading behaviors of its solution as $x \to +\infty$ are determined by (3.4.28) to be

$$y(x) \sim cx^{-3/8}e^{4\omega x^{5/4}/5}, \qquad x \to +\infty, \tag{3.5.23}$$

where c is a constant and $\omega = \pm 1, \pm i$.

By computer we have obtained the numerical solution $y(x)$ of the hyperairy equation which satisfies the initial conditions $y(0) = 2.0$, $y'(0) = -1.0$, $y''(0) = 0.0$, $y'''(0) = 1.0$. In Fig. 3.11 we

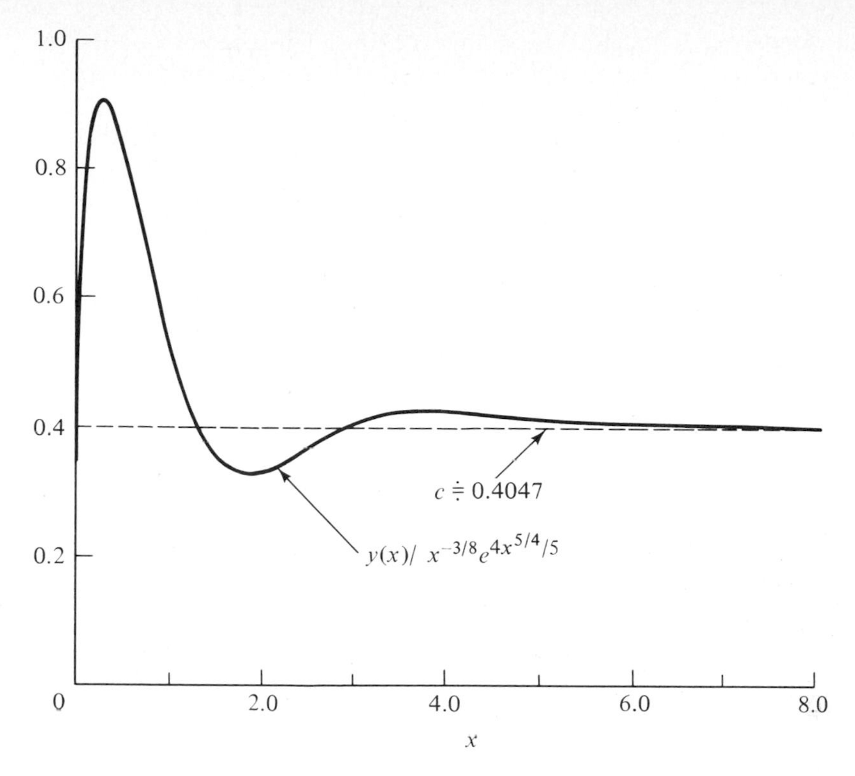

Figure 3.11 A plot of the ratio of the numerical solution of $d^4y/dx^4 = xy$ with $y(0) = 2$, $y'(0) = -1$, $y''(0) = 0$, $y'''(0) = 1$ to its leading asymptotic behavior in (3.5.23) with $\omega = +1$. As $x \to +\infty$, the ratio approaches the constant $c \doteqdot 0.4047$. The approach to this limit is oscillatory because the numerical solution is an admixture of four components, which oscillate or grow and decay exponentially. The exponentially growing solution dominates the leading behavior as $x \to +\infty$.

graph the ratio of this solution to the growing leading behavior [the solution for which $\omega = +1$ in (3.5.23)]. Note that the oscillating solutions (those for which $\omega = \pm i$) for small values of x are gradually overwhelmed by the growing solution as the ratio approaches the constant $c \doteqdot 0.4047$.

(E) 3.6 LOCAL ANALYSIS OF INHOMOGENEOUS LINEAR EQUATIONS

In this section we explore methods for determining the local behaviors of solutions to an inhomogeneous differential equation. We emphasize that it suffices to find the local behavior of *one* particular solution. The general local behavior can then be produced by adding in the behaviors of solutions to the associated homogeneous equation which we may assume are already known.

There are many ways to find the local behavior of a particular solution. One approach, which is rather indirect and can be complicated, relies on the method of variation of parameters. Variation of parameters (see Sec. 1.5) allows us to express a particular solution as an integral over solutions to the homogeneous equation. A local expansion of this integral, in which the integrand is approximated by replacing the exact solutions to the homogeneous equation with their local behaviors, is often sufficient to determine the behavior of a particular solution. We do not pursue this approach here because we have postponed all discussion of the local expansion of integrals until Chap. 6.

We prefer the more direct approach of applying the method of dominant balance to the differential equation. There are several cases to consider. First, we suppose that x_0 is an ordinary point of the associated homogeneous differential equation and a point of analyticity of the inhomogeneity. In this case the general solution has a Taylor series representation. (See Prob. 3.52.)

Example 1 *Taylor series representation of the general solution.* What is the local behavior of the general solution to

$$y' + xy = x^3 \tag{3.6.1}$$

at $x = 0$? Since $x = 0$ is an ordinary point of the homogeneous equation $y' + xy = 0$ and a point at which x^3 is analytic, we may assume a Taylor series representation for $y(x)$:

$$y(x) = \sum_{n=0}^{\infty} a_n x^n. \tag{3.6.2}$$

Substituting (3.6.2) into (3.6.1) and equating coefficients of like powers of x gives a recursion relation for a_n:

$$a_1 = 0,$$

$$na_n + a_{n-2} = \begin{cases} 0, & n \geq 2, n \neq 4, \\ 1, & n = 4. \end{cases}$$

This recursion relation determines a_n $(n \geq 1)$ in terms of a_0, which remains an arbitrary constant. This completes our determination of the local behavior of $y(x)$.

The Taylor series (3.6.2) has an infinite radius of convergence because neither the coefficient x nor the inhomogeneity x^3 has any singularities in the finite x plane (see Prob. 3.52).

If x_0 is an ordinary point of the differential equation but x_0 is not a point of analyticity of the inhomogeneity, then, although all solutions to the homogeneous equation can be expanded in Taylor series, a particular solution to the inhomogeneous equation does not have a Taylor series expansion. In this case we use the method of dominant balance to find the behavior of a particular solution.

Example 2 *Local behavior of solutions at an ordinary point of the associated homogeneous equation where the inhomogeneity is not analytic.* Let us find the leading behavior of a particular solution to

$$y' + xy = 1/x^4$$

near $x = 0$. We use the method of dominant balance, which was introduced in Sec. 3.4, to decide which terms in this equation are most important as $x \to 0$. There are three dominant balances to consider:

(*a*) $y' \sim -xy$, $x^{-4} \ll xy$ $(x \to 0)$. The solution of this asymptotic differential equation is $y \sim ae^{-x^2/2} \sim a$ $(x \to 0)$, which is not consistent with the condition that $x^{-4} \ll xy$ $(x \to 0)$.
(*b*) $xy \sim x^{-4}$, $y' \ll x^{-4}$ $(x \to 0)$. This asymptotic relation implies that $y \sim x^{-5}$ $(x \to 0)$, which violates the condition that $y' \ll x^{-4}$ $(x \to 0)$. This case is also inconsistent.

(c) $y' \sim x^{-4}$, $xy \ll x^{-4}$ $(x \to 0)$. The solution to this asymptotic differential equation is $y \sim -1/3x^3$ $(x \to 0)$, which *is* consistent with the condition that $xy \ll x^{-4}$ $(x \to 0)$.

Hence, the *only* consistent leading behavior of $y(x)$ as $x \to 0$ is $y(x) \sim -1/3x^3$ $(x \to 0)$. Note that this leading behavior contains no arbitrary constant and is thus independent of any initial condition [such as $y(5) = 3$].

The next step is to find the corrections to this leading behavior. If we set $y(x) = -1/3x^3 + C(x)$, where $C(x) \ll x^{-3}$ $(x \to 0)$, we find that $C(x)$ satisfies $C' + xC = 1/3x^2$. If we proceed as above using the method of dominant balance, we find that the leading behavior of $C(x)$ as $x \to 0$ is $C(x) \sim -1/3x$ $(x \to 0)$. Continuing in this fashion, we obtain the full local behavior of $y(x)$:

$$y(x) \sim -\frac{1}{3x^3} - \frac{1}{3x} + a_0 + \frac{1}{3}x - \frac{1}{2}a_0 x^2 + \cdots, \qquad x \to 0,$$

where a_0 is arbitrary. The parameter a_0 is determined by the initial conditions on the solution $y(x)$.

The method of dominant balance also works at singular points of the associated homogeneous equation.

Example 3 *Behavior of solutions to an inhomogeneous Airy equation at* ∞. Let us consider all possible solutions to

$$y'' = xy - 1 \tag{3.6.3}$$

which satisfy $y(+\infty) = 0$. These solutions can be found by variation of parameters; the result is

$$y(x) = \pi\left[\mathrm{Ai}\,(x)\int_0^x \mathrm{Bi}\,(t)\,dt + \mathrm{Bi}\,(x)\int_x^\infty \mathrm{Ai}\,(t)\,dt\right] + c\,\mathrm{Ai}\,(x),$$

where c is an arbitrary constant (see Prob. 3.51). One way to find the leading behavior of $y(x)$ as $x \to +\infty$ is to study the asymptotic behavior of these integrals using the techniques of Chap. 6 (see Prob. 3.51). Another way to find the leading behavior of $y(x)$ as $x \to +\infty$ is to use the method of dominant balance as shown below.

The balance $y'' \sim xy$, $1 \ll xy$ $(x \to +\infty)$ gives the two solutions of the homogeneous Airy equation Ai (x) and Bi (x). But we must exclude Ai (x) which violates $1 \ll xy$ $(x \to +\infty)$ and Bi (x) which violates $y(+\infty) = 0$. This balance is inconsistent.

The balance $y'' \sim -1$, $xy \ll 1$ $(x \to +\infty)$ is also inconsistent. (Why?)

The only consistent balance is $xy \sim 1$, $y'' \ll 1$ $(x \to +\infty)$, which gives

$$y \sim \frac{1}{x}, \qquad x \to +\infty. \tag{3.6.4}$$

Note that this leading behavior is unique and independent of an initial condition given, for example, at $x = 0$. For any value of $y(0)$ there is a unique solution of the differential equation (3.6.3) satisfying $y(+\infty) = 0$; solutions for different $y(0)$ differ by a multiple of Ai (x). The leading behavior (3.6.4) does not depend on $y(0)$ because the Airy function Ai (x) decays exponentially as $x \to +\infty$.

To determine the corrections to the leading behavior (3.6.4), we let $y(x) = 1/x + C(x)$, $C(x) \ll 1/x$ $(x \to +\infty)$. $C(x)$ satisfies the differential equation $2/x^3 + C'' = xC$. The method of dominant balance gives $C(x) \sim 2/x^4$ $(x \to +\infty)$. By continuing in this manner we find the full asymptotic power series expansion of $y(x)$ (see Prob. 3.51):

$$y(x) \sim \frac{1}{x} + \frac{2}{x^4} + \cdots + \frac{(3n)!}{3^n n!\, x^{3n+1}} + \cdots, \qquad x \to +\infty. \tag{3.6.5}$$

In Fig. 3.12 we compare the numerical solution to (3.6.3) which satisfies $y(0) = 1$ with the leading term $1/x$ and also the first two terms $1/x + 2/x^4$ in the asymptotic series (3.6.3).

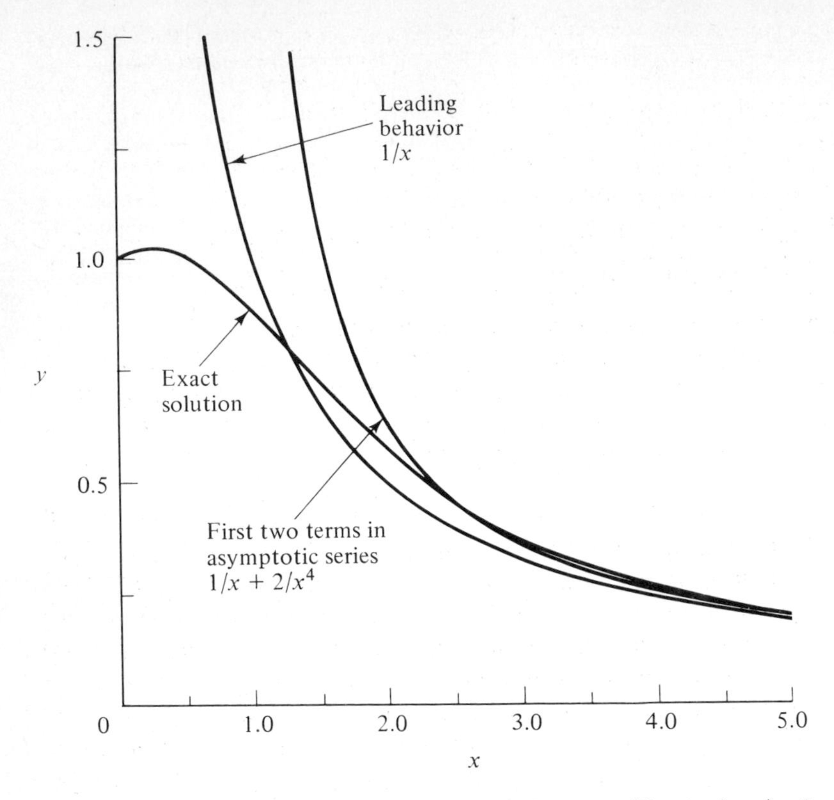

Figure 3.12 Comparison of the exact solution to $y'' = xy - 1$, $y(0) = 1$, $y(+\infty) = 0$ with the leading term and first two terms in the asymptotic series for $y(x)$ valid as $x \to +\infty$.

There is a mildly interesting subtlety that one may encounter when using the method of dominant balance. It may happen that *all* two-term balances are inconsistent. When this occurs, we must consider possible dominant balances among three or more terms.

Example 4 *Three-term dominant balance.* A local analysis of the equation $y' - y/x = (\cos x)/x^2$ near $x = 0$ shows that all three pair-wise dominant balances are inconsistent:

(*a*) If $x^{-2} \cos x \ll x^{-1}y$ $(x \to 0)$, then $y \sim cx$ $(x \to 0)$, where c is an integration constant. This result is inconsistent with the assumption that $x^{-2} \cos x \ll x^{-1}y$ $(x \to 0)$.
(*b*) If $x^{-1}y \ll x^{-2} \cos x$ $(x \to 0)$, then $y \sim -x^{-1}$ $(x \to 0)$. (Why?) This result is inconsistent with the assumption that $x^{-1}y \ll x^{-2} \cos x$ $(x \to 0)$.
(*c*) If $y' \ll x^{-2} \cos x$ $(x \to 0)$, then $y \sim -x^{-1} \cos x \sim -x^{-1}$ $(x \to 0)$. This result is inconsistent with $y' \ll x^{-2} \cos x$ $(x \to 0)$.

We must therefore seek a three-term balance in the asymptotic differential equation

$$y' - \frac{y}{x} \sim \frac{1}{x^2}, \qquad x \to 0. \tag{3.6.6}$$

No further approximations to this equation can be made, but an exact solution to (3.6.6) is easy to find. A simple dimensional analysis suggests that (3.6.6) has a solution of the form $y \sim cx^{-1}$ $(x \to 0)$; substituting this into (3.6.6) gives $c = -\frac{1}{2}$. Thus, the leading behavior of $y(x)$ is $y(x) \sim -1/2x$ $(x \to 0)$, independent of any initial condition [like $y(1) = a$].

For further practice in finding the behavior of solutions to inhomogeneous equations see Probs. 3.48 to 3.50.

(TI) 3.7 ASYMPTOTIC RELATIONS

This section presents a detailed study of asymptotic relations. We discuss asymptotic relations for oscillatory functions, asymptotic relations for complex functions, subdominance, Stokes lines, and the Stokes phenomenon.

Asymptotic Relations for Oscillatory Functions

To illustrate one of the subtle aspects of asymptotic relations, let us examine in detail the leading behavior of the solutions to the Airy equation $y'' = xy$ for large $|x|$. In Sec. 3.5 we saw that for large *positive* x, the leading behaviors of the Airy functions Ai (x) and Bi (x), whose Taylor series are given in (3.2.1) and (3.2.2), are

$$\text{Ai}(x) \sim \tfrac{1}{2}\pi^{-1/2}x^{-1/4}e^{-2x^{3/2}/3}, \qquad x \to +\infty, \tag{3.7.1}$$

$$\text{Bi}(x) \sim \pi^{-1/2}x^{-1/4}e^{2x^{3/2}/3}, \qquad x \to +\infty. \tag{3.7.2}$$

To determine the leading behavior of the solutions to the Airy equation for large *negative* x requires a bit of ingenuity. Let us see what happens if we proceed naively. Using our technique of substituting $y = e^S$ we obtain $S'' + (S')^2 = x$. Neglecting S'' as usual, we get the asymptotic differential equation $(S')^2 \sim x$ $(x \to -\infty)$. It follows that $S \sim \pm\frac{2}{3}i(-x)^{3/2}$ $(x \to -\infty)$. This gives the controlling factor of the leading behavior of y as $x \to -\infty$, namely $\exp[\pm\frac{2}{3}i(-x)^{3/2}]$. Proceeding in the usual way, we obtain the leading behavior

$$y \sim c(-x)^{-1/4}\exp[\pm\tfrac{2}{3}i(-x)^{3/2}], \qquad x \to -\infty. \tag{3.7.3}$$

No single one of the behaviors (3.7.3) can describe the leading behaviors of Ai (x) and Bi (x) as $x \to -\infty$ because the Airy functions are real when x is real; the Taylor series (3.2.1) and (3.2.2) for Ai (x) and Bi (x) have real coefficients. In order to represent the leading behaviors of Ai (x) and Bi (x) as $x \to -\infty$, we must form a real linear combination of the two leading behaviors (3.7.3):

$$y(x) \sim c_1(-x)^{-1/4}\sin[\tfrac{2}{3}(-x)^{3/2}] + c_2(-x)^{-1/4}\cos[\tfrac{2}{3}(-x)^{3/2}], \qquad x \to -\infty.$$

In fact, the graph of Ai (x) for large negative x (Fig. 3.13) very closely resembles the graph of

$$\pi^{-1/2}(-x)^{-1/4}\sin[\tfrac{2}{3}(-x)^{3/2} + \tfrac{1}{4}\pi],$$

and the graph of Bi (x) (Fig. 3.14) very closely resembles the graph of

$$\pi^{-1/2}(-x)^{-1/4}\cos[\tfrac{2}{3}(-x)^{3/2} + \tfrac{1}{4}\pi].$$

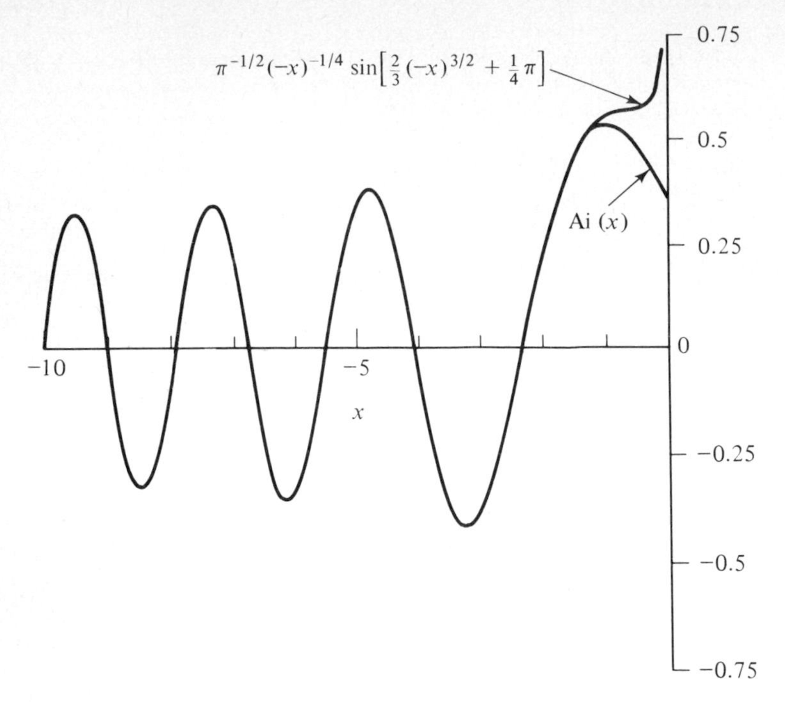

Figure 3.13 A plot of the Airy function Ai (x) and the function $\pi^{-1/2}(-x)^{-1/4}\sin[\frac{2}{3}(-x)^{3/2}+\frac{1}{4}\pi]$ for $-10 \le x \le 0$. Notice the good agreement between the graphs of these two functions for $x < -1$.

However, it is wrong to write

$$\text{Ai}\,(x) \sim \pi^{-1/2}(-x)^{-1/4}\sin\left[\tfrac{2}{3}(-x)^{3/2}+\tfrac{1}{4}\pi\right], \qquad x \to -\infty, \tag{3.7.4}$$

and

$$\text{Bi}\,(x) \sim \pi^{-1/2}(-x)^{-1/4}\cos\left[\tfrac{2}{3}(-x)^{3/2}+\tfrac{1}{4}\pi\right], \qquad x \to -\infty, \tag{3.7.5}$$

because the zeros of the right and left sides of these relations do not quite coincide for large negative x. Therefore, their ratios cannot approach the limit 1 as $x \to -\infty$, and this violates the definition of an asymptotic relation in Sec. 3.4. Nevertheless, it is clear from Figs. 3.13 and 3.14 that (3.7.4) and (3.7.5) want to be valid asymptotic relations. After all the successful asymptotic analysis of the previous three sections, it is surprising to encounter such a silly flaw in the definition of an asymptotic relation.

The next two examples illustrate in an elementary way how to resolve this problem with noncoincident zeros.

Example 1 *Functions having noncoincident zeros.* Consider the two functions $\sin x$ and $\sin(x + x^{-1})$. We would like to say that these two functions are asymptotic as $x \to \infty$ because their graphs become identical in this limit. However, these functions are not asymptotic according to the definition of an asymptotic relation because their zeros do not coincide for large x. This is similar to the difficulty encountered above with Airy functions.

The problem is to express the close similarity in the behavior of $\sin x$ and $\sin(x + x^{-1})$ without dividing by zero. One way to do this is to state that the arguments of the sine functions are asymptotic as $x \to \infty$: $x + x^{-1} \sim x$ $(x \to \infty)$. This avoids the trouble with noncoinciding zeros.

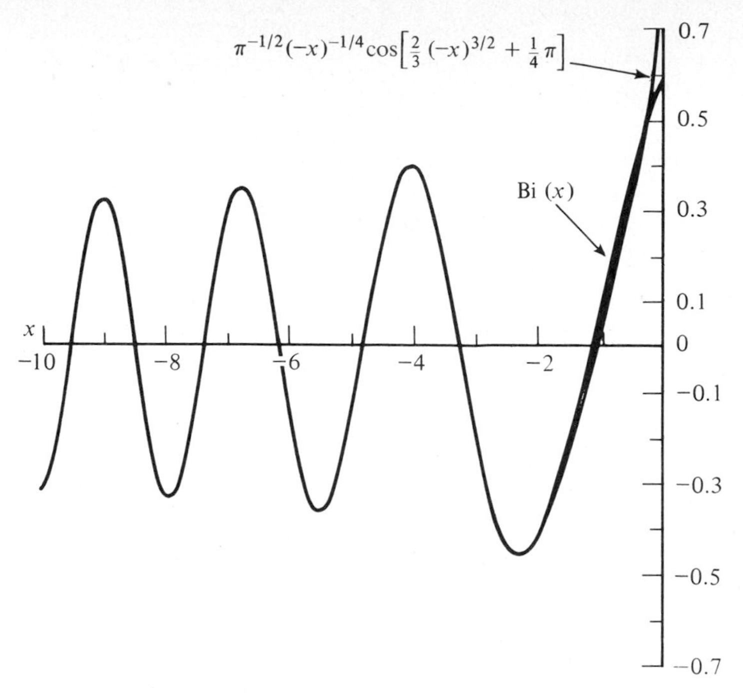

Figure 3.14 A plot of the Airy function Bi (x) and the function $\pi^{-1/2}(-x)^{-1/4}\cos[\frac{2}{3}(-x)^{3/2}+\frac{1}{4}\pi]$ for $-10 \le x \le 0$. Note the good agreement between the graphs of these two functions for $x < -0.5$.

Example 2 *Functions having noncoincident zeros.* A slightly different idea must be used to represent the similar behaviors of $\sin x$ and $(1+x^{-1})\sin(x+x^{-1})$ as $x \to \infty$. To do so, we write

$$(1+x^{-1})\sin(x+x^{-1}) = w_1(x)\sin x + w_2(x)\cos x,$$

where

$$w_1(x) = (1+x^{-1})\cos x^{-1} \quad \text{and} \quad w_2(x) = (1+x^{-1})\sin x^{-1}.$$

The close relation to the function $\sin x$ is expressed by the conditions $w_1(x) \sim 1$ and $w_2(x) \ll 1$ as $x \to \infty$.

In general, two functions cannot be asymptotic as $x \to x_0$ if they have noncoinciding zeros arbitrarily close to but not at x_0. Nevertheless, it is often necessary to represent the local behavior of very complicated, rapidly oscillatory functions in terms of simpler oscillatory functions, such as sines and cosines.

Let us now return to the problem of expressing the asymptotic behavior of Airy functions as $x \to -\infty$. Example 2 suggests that we first represent solutions of the Airy equation as

$$y(x) = w_1(x)(-x)^{-1/4}\sin\left[\tfrac{2}{3}(-x)^{3/2}+\tfrac{1}{4}\pi\right] + w_2(x)(-x)^{-1/4}\cos\left[\tfrac{2}{3}(-x)^{3/2}+\tfrac{1}{4}\pi\right] \tag{3.7.6}$$

and then seek the asymptotic behaviors of $w_1(x)$ and $w_2(x)$ as $x \to -\infty$. The notation $w_1(x)$ and $w_2(x)$ generalizes the notation $w(x)$ used in the previous two sections to represent what is left over after peeling off the leading behavior. Assuming that there is some truth in (3.7.4) and (3.7.5), we anticipate that the Airy function Ai (x) can be represented by (3.7.6) with $w_1(x) \sim \pi^{-1/2}$, $w_2(x) \ll w_1(x)$ $(x \to -\infty)$, and the Airy function Bi (x) can be represented by (3.7.6) with $w_2(x) \sim \pi^{-1/2}$, $w_1(x) \ll w_2(x)$ $(x \to -\infty)$. Of course, the constants $\pi^{-1/2}$ in these behaviors cannot be predicted by local analysis of the Airy equation.

Substituting (3.7.6) into the Airy equation $y'' = xy$ gives, after some simplification,

$$[w_1'' - \tfrac{1}{2}x^{-1}w_1' + 2(-x)^{1/2}w_2' + \tfrac{5}{16}x^{-2}w_1](-x)^{-1/4} \sin [\tfrac{2}{3}(-x)^{3/2} + \tfrac{1}{4}\pi]$$
$$+ [w_2'' - \tfrac{1}{2}x^{-1}w_2' - 2(-x)^{1/2}w_1' + \tfrac{5}{16}x^{-2}w_2](-x)^{1/4} \cos [\tfrac{2}{3}(-x)^{3/2} + \tfrac{1}{4}\pi] = 0. \tag{3.7.7}$$

The definition of $w_1(x)$ and $w_2(x)$ in (3.7.6) provides only one constraint on these two functions. Thus, $w_1(x)$ and $w_2(x)$ are not yet completely determined and we are free to impose a second constraint in addition to (3.7.7). This freedom is very similar to that encountered in the technique of variation of parameters. If we choose the second constraint to be

$$w_1'' - \tfrac{1}{2}x^{-1}w_1' + 2(-x)^{1/2}w_2' + \tfrac{5}{16}x^{-2}w_1 = 0, \tag{3.7.8}$$

then (3.7.7) implies that

$$w_2'' - \tfrac{1}{2}x^{-1}w_2' - 2(-x)^{1/2}w_1' + \tfrac{5}{16}x^{-2}w_2 = 0. \tag{3.7.9}$$

Thus, we obtain a pair of coupled differential equations for w_1 and w_2 which, as you can see, are explicitly free of the oscillatory terms in (3.7.6). In fact, a straightforward analysis of the leading behavior of solutions to (3.7.8) and (3.7.9) as $x \to -\infty$ shows that there are solutions that approach constants as $x \to -\infty$ (Prob. 3.59).

The full asymptotic expansions of $w_1(x)$ and $w_2(x)$ are series in powers of $(-x)^{-3/2}$ (see Prob. 3.59):

$$w_1(x) \sim \sum_{n=0}^{\infty} a_n(-x)^{-3n/2}, \qquad x \to -\infty, \tag{3.7.10}$$

$$w_2(x) \sim \sum_{n=0}^{\infty} b_n(-x)^{-3n/2}, \qquad x \to -\infty. \tag{3.7.11}$$

Substituting these series into (3.7.8) and (3.7.9), equating coefficients of powers of $(-x)^{3/2}$, and solving the resulting recurrence relations gives

$$a_{2n} = a_0(-1)^n c_{2n}, \qquad n = 0, 1, \ldots,$$
$$a_{2n+1} = b_0(-1)^n c_{2n+1}, \qquad n = 0, 1, \ldots,$$
$$b_{2n} = b_0(-1)^n c_{2n}, \qquad n = 0, 1, \ldots,$$
$$b_{2n+1} = a_0(-1)^{n+1} c_{2n+1}, \qquad n = 0, 1, \ldots,$$

where

$$c_n = \frac{(2n+1)(2n+3)\cdots(6n-1)}{144^n n!} = \frac{1}{2\pi}\left(\frac{3}{4}\right)^n \frac{\Gamma(n+\frac{5}{6})\Gamma(n+\frac{1}{6})}{n!}.$$

The asymptotic behavior of the Airy function Ai (x) as $x \to -\infty$ is represented in this way if $a_0 = \pi^{-1/2}$ and $b_0 = 0$. Then, (3.7.10) and (3.7.11) imply $w_1(x) \sim \pi^{-1/2}$ and $w_2(x) \ll w_1(x)$ as $x \to -\infty$. Thus, Ai (x) is given by (3.7.6) with

$$w_1(x) \sim \pi^{-1/2} \sum_{n=0}^{\infty} c_{2n} x^{-3n}, \qquad x \to -\infty, \tag{3.7.12a}$$

$$w_2(x) \sim -\pi^{-1/2}(-x)^{-3/2} \sum_{n=0}^{\infty} c_{2n+1} x^{-3n}, \qquad x \to -\infty. \tag{3.7.12b}$$

On the other hand, Bi (x) is represented by (3.7.6) and (3.7.10) and (3.7.11) if $a_0 = 0$ and $b_0 = \pi^{-1/2}$, so

$$w_1(x) \sim \pi^{-1/2}(-x)^{-3/2} \sum_{n=0}^{\infty} c_{2n+1} x^{-3n}, \qquad x \to -\infty, \tag{3.7.13a}$$

$$w_2(x) \sim \pi^{-1/2} \sum_{n=0}^{\infty} c_{2n} x^{-3n}, \qquad x \to -\infty. \tag{3.7.13b}$$

There are many other situations in which this kind of analysis must be used to find asymptotic representations.

Example 3 *Bessel functions for large positive x.* The Bessel equation of order ν,

$$x^2 y'' + xy' + (x^2 - \nu^2)y = 0,$$

has solutions which oscillate rapidly as $x \to +\infty$. The same kind of asymptotic analysis that we used to study the Airy equation for large negative x applies here. This analysis suggests a convenient representation for the large-x behavior of $y(x)$:

$$y(x) = w_1(x)x^{-1/2} \cos\left(x - \tfrac{1}{2}\nu\pi - \tfrac{1}{4}\pi\right) + w_2(x)x^{-1/2} \sin\left(x - \tfrac{1}{2}\nu\pi - \tfrac{1}{4}\pi\right). \tag{3.7.14}$$

The phase angle $-\frac{1}{2}\nu\pi - \frac{1}{4}\pi$ has been introduced into (3.7.14) so that the asymptotic expansion that results is a simple linear combination of the standard asymptotic expansions of the Bessel functions $J_\nu(x)$ and $Y_\nu(x)$ (see Appendix). In general, $w_1(x)$ and $w_2(x)$ have the asymptotic series representation (see Prob. 3.60):

$$w_1(x) \sim a_0 \sum_{n=0}^{\infty} (-1)^n c_{2n} x^{-2n} + b_0 \sum_{n=0}^{\infty} (-1)^n c_{2n+1} x^{-2n-1}, \qquad x \to +\infty, \tag{3.7.15a}$$

$$w_2(x) \sim b_0 \sum_{n=0}^{\infty} (-1)^n c_{2n} x^{-2n} - a_0 \sum_{n=0}^{\infty} (-1)^n c_{2n+1} x^{-2n-1}, \qquad x \to +\infty, \tag{3.7.15b}$$

where a_0 and b_0 are arbitrary constants, $c_0 = 1$, and

$$c_n = \frac{(4\nu^2 - 1^2)(4\nu^2 - 3^2)\cdots[4\nu^2 - (2n-1)^2]}{n!\, 8^n}, \qquad n = 1, 2, \ldots.$$

Observe that both series in (3.7.15) terminate when $\nu = n + \frac{1}{2}$, where n is an integer. In this case the asymptotic series converges for $x \neq 0$ and $\sim$ signs may be replaced by $=$ signs.

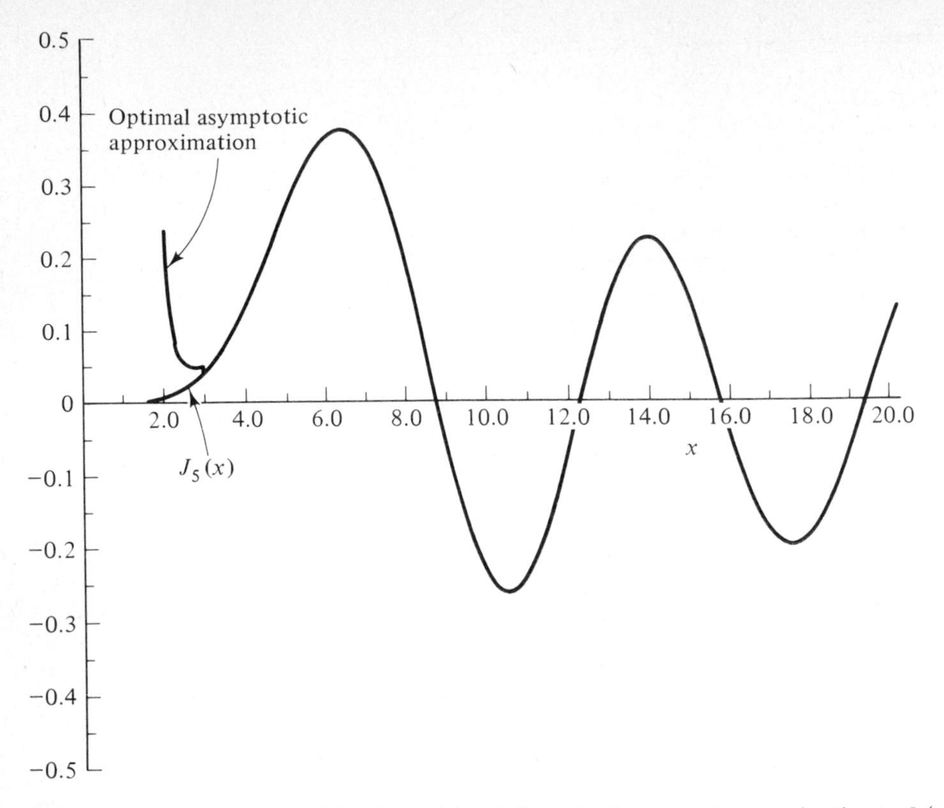

Figure 3.15 A plot of the Bessel function $J_5(x)$ and the optimal asymptotic approximation to $J_5(x)$ for $0 \le x \le 20$.

The Bessel function $J_\nu(x)$ is represented in this way if $a_0 = (2/\pi)^{1/2}$, $b_0 = 0$, while the Bessel function $Y_\nu(x)$ is so represented if $a_0 = 0$ and $b_0 = (2/\pi)^{1/2}$. The computation of these constants a_0 and b_0 must await our analysis of Bessel functions to be given in Sec. 6.4.

In Fig. 3.15 we plot $J_5(x)$ and the optimal asymptotic approximation to $J_5(x)$ obtained by truncating both series in (3.7.15) just prior to their smallest terms and substituting the result into (3.7.14). Like the optimal asymptotic approximation to $I_5(x)$, this approximation becomes very accurate for $x > 3$. When x is larger than 3, it is no longer possible to distinguish the two curves in Fig. 3.15; the numerical values are listed in Table 3.3.

This completes our generalization of asymptotic analysis to functions which oscillate rapidly as $x \to x_0$.

Asymptotic Relations in the Complex Plane

Until now, we have discussed asymptotic relations only on the real axis, but to fully appreciate the structure of asymptotic relations, one must examine them in the complex plane. It is not trivial to generalize the definition of asymptotic relations to complex functions. We encounter serious difficulties if we naively try to define $f(z) \sim g(z)$ $(z \to z_0)$ by $\lim_{z \to z_0} f(z)/g(z) = 1$, where the limit $z \to z_0$ is

Table 3.3 Relative error in optimal asymptotic approximation to the Bessel function $J_5(x)$

x	$J_5(x)$	Relative error in optimal asymptotic approximation, %
3.0	4.3028(−2)	1.6
4.0	1.3209(−1)	0.18
5.0	2.6114(−1)	0.0034
6.0	3.6209(−1)	0.00057

taken along arbitrary paths approaching z_0 in the complex plane. We illustrate these difficulties in the next few examples.

Example 4 *Behavior of Airy functions for large* $|z|$. The relation between the Airy function Ai (x) and its leading behavior for large positive x is Ai $(x) \sim \frac{1}{2}\pi^{-1/2}x^{-1/4}e^{-2x^{3/2}/3}$ $(x \to +\infty)$. Replacing $x \to +\infty$ by the complex limit $z \to \infty$ in this formula gives an immediate contradiction. Ai (z) is an entire function; its Taylor series (3.2.1) converges for all finite $|z|$. However, the function $\frac{1}{2}\pi^{-1/2}z^{-1/4}e^{-2z^{3/2}/3}$ is a multivalued function having branch points at $z = 0$ and $z = \infty$. Since the limit $z \to \infty$ includes all complex paths approaching complex ∞, we can reach all branches of the leading behavior. It does not make sense to say that an entire (single-valued) function is asymptotic to all the branches of a multivalued function.

Example 5 *Behavior of* sinh (z^{-1}) *for small* $|z|$. In the previous three sections we have been careful to distinguish between the limits $x \to x_0+$ and $x \to x_0-$ because in some problems these two limits give different asymptotic behaviors. However, if we allow x to approach x_0 along arbitrary paths in the complex plane we can no longer maintain the distinction between these special one-sided limits. For example, the relation sinh $(x^{-1}) \sim \frac{1}{2}e^{x^{-1}}$ $(x \to 0+)$ is not valid if $x \to 0-$. The correct relation for $x \to 0-$ is sinh $(x^{-1}) \sim -\frac{1}{2}e^{-x^{-1}}$ $(x \to 0-)$. Since the complex limit $z \to 0$ includes *both* real paths $z \to 0+$ (Im $z = 0$) and $z \to 0-$ (Im $z = 0$), sinh (z^{-1}) does not have a unique asymptotic behavior as $z \to 0$.

Example 6 *Nonuniqueness of leading behavior in complex plane.* Another example of the one-sidedness of asymptotic limits is provided by the Airy function. As we have seen, Ai (x) behaves differently as $|x| \to \infty$ along the positive or negative real axis. Thus, if we replace this real limit by the complex limit $z \to \infty$, we must conclude that the asymptotic behavior of Ai (z) for large $|z|$ is nonunique and depends on the particular choice of the complex path to ∞.

A satisfactory definition of a complex asymptotic relation must be path independent and unique. But the previous three examples show that if we are to extend the concept of an asymptotic relation into the complex plane, we must in many cases be careful to exclude complex paths which rotate around z_0 as they approach z_0. Such paths may destroy the one-sided nature of asymptotic relations and give nonunique limits. The simplest way to eliminate such paths is to insist that all paths lie entirely inside a sector or wedge-shaped region of the complex plane (see Fig. 3.16).

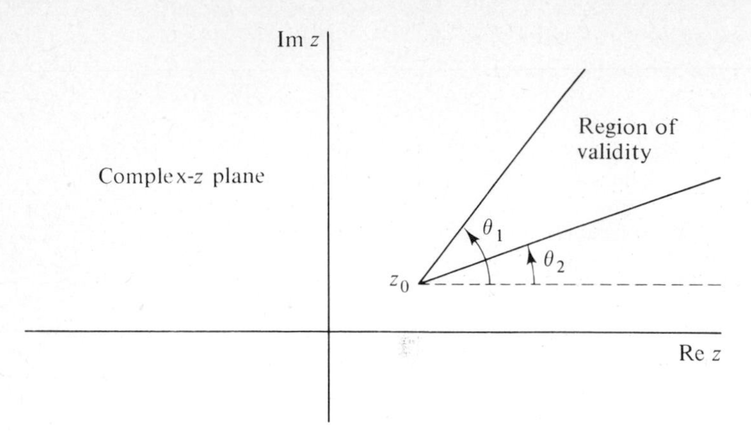

Figure 3.16 Sector of validity for an asymptotic relation.

If the boundaries of the sector are smooth curves in the neighborhood of z_0, they can be locally approximated by their tangent lines at z_0. The angle $\theta_2 - \theta_1$ between these tangent lines is called the *opening angle* of the sector (see Fig. 3.16). The statement that two functions are asymptotic as $z \to z_0$ in the complex plane must always be accompanied by the constraint that $z \to z_0$ along a path lying in the sector of validity where the size of the opening angle depends on the functions which are asymptotic. The necessity of introducing wedge-shaped regions is a wonderful subtlety that one could not have immediately predicted upon first reading the definition of an asymptotic relation in Sec. 3.4.

Here are some examples of complex asymptotic relations.

Example 7 *Asymptotic behavior of* $\sinh(z^{-1})$ *in the complex plane.*

$$\sinh(z^{-1}) \sim \tfrac{1}{2}e^{z^{-1}}, \qquad z \to 0;\ |\arg z| < \tfrac{1}{2}\pi,$$

$$\sinh(z^{-1}) \sim -\tfrac{1}{2}e^{-z^{-1}}, \qquad z \to 0;\ \tfrac{1}{2}\pi < \arg z < \tfrac{3}{2}\pi.$$

The sectors of validity are the half planes shown in Fig. 3.17.

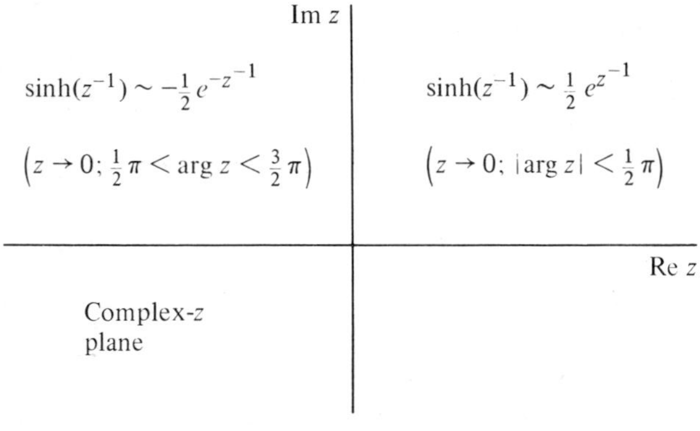

Im z

$\sinh(z^{-1}) \sim -\frac{1}{2}e^{-z^{-1}}$

$\left(z \to 0;\ \frac{1}{2}\pi < \arg z < \frac{3}{2}\pi\right)$

$\sinh(z^{-1}) \sim \frac{1}{2}e^{z^{-1}}$

$\left(z \to 0;\ |\arg z| < \frac{1}{2}\pi\right)$

Re z

Complex-z plane

Figure 3.17 Sectors of validity for the asymptotic behavior of $\sinh(z^{-1})$ as $z \to 0$.

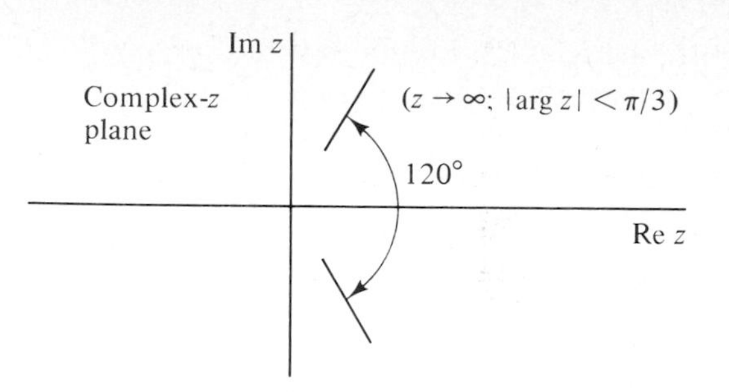

Figure 3.18 Sector of validity for the leading asymptotic behavior (3.7.16) of Bi (z).

Example 8 *Asymptotic behavior of* Bi (z) *in the complex plane.* Bi (z) is a solution of the Airy equation $y''(z) = zy(z)$ which grows exponentially as $z \to +\infty$ (Im $z = 0$). Its leading behavior in the complex plane is

$$\text{Bi}\,(z) \sim \pi^{-1/2} z^{-1/4} e^{2z^{3/2}/3}, \qquad z \to \infty;\ |\arg z| < \frac{\pi}{3}. \tag{3.7.16}$$

This relation is valid in the 120° wedge shown in Fig. 3.18. We compute the opening angle of this sector in Example 12.

Example 9 *Asymptotic relation valid for all directions in the complex plane.* The asymptotic relation $\sin z \sim z$ ($z \to 0$) is valid as $z \to 0$ along *any* path in the complex plane. Thus, the wedge is a full disk about $z = 0$.

Example 10 *Asymptotic relation valid in a cut plane.* The leading behavior (3.4.4*a*) of a solution to the differential equation $x^3 y'' = y$ that grows as $x \to 0+$ is valid in the cut plane $|\arg z| < \pi$: $y(z) \sim c_1 z^{3/4} e^{2z^{-1/2}}$ ($z \to 0$; $|\arg z| < \pi$). The opening angle of the sector of validity, as shown in Fig. 3.19, is 360°. The asymptotic relation is not valid when $|\arg z| = \pi$.

Stokes Phenomenon and Subdominance

Here is a brief explanation of how and why an asymptotic relation breaks down at the edge of the wedge of its validity. If two functions $f(z)$ and $g(z)$ are asymptotic as $z \to z_0$, the difference $f(z) - g(z)$ is small compared with $g(z)$ as $z \to z_0$: $\lim_{z \to z_0} [f(z) - g(z)]/g(z) = 0$. The difference $f(z) - g(z)$ is what we neglect to write down when we formulate an asymptotic relation; if we included this difference, we would have an equality: $f(z) = g(z) + [f(z) - g(z)]$. When z lies in the interior of the wedge, the difference $f(z) - g(z)$ is said to be *subdominant* or *recessive* as compared with $f(z)$ or $g(z)$ which are *dominant*.

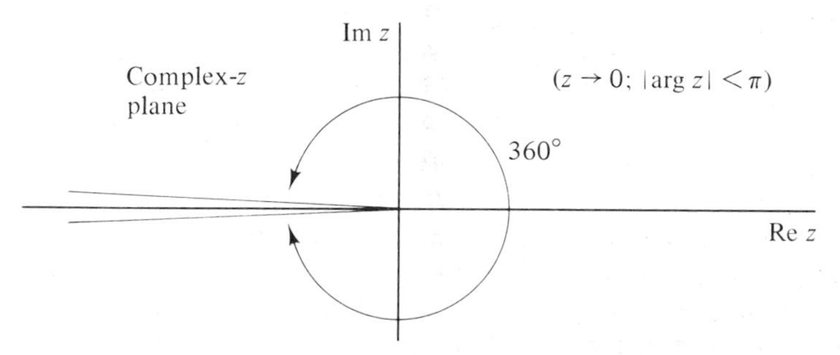

Figure 3.19 Sector of validity for the asymptotic relation (3.4.4*a*).

As z approaches the edge of the wedge of asymptoticity, the subdominant term $f - g$ grows in magnitude in comparison with the dominant term g. At the edge of the wedge, called a *Stokes line*, it is no longer valid to use the words dominant or subdominant because f and $f - g$ have the same order of magnitude. As z crosses the Stokes line, the dominant and subdominant terms exchange identities: on the other side of the Stokes line $f - g \gg g$ $(z \to z_0)$. The exchange of identities is called the *Stokes phenomenon.*

Example 11 *Stokes phenomenon for* $\sinh(z^{-1})$. Both $\sinh(z^{-1})$ and its asymptotic behavior $\frac{1}{2}e^{z^{-1}}$ grow exponentially as $z \to 0$ in the right half plane Re $z > 0$. The difference between $\sinh(z^{-1})$ and $\frac{1}{2}e^{z^{-1}}$ is $-\frac{1}{2}e^{-z^{-1}}$, which is exponentially small in the half plane Re $z > 0$. If we rotate z in the asymptotic relation $\sinh(z^{-1}) \sim \frac{1}{2}e^{z^{-1}}$ $(z \to 0)$ out of the right half plane, then this asymptotic relation is no longer valid because the subdominant error term $-\frac{1}{2}e^{-z^{-1}}$ is no longer small.

In general, the opening angle of the sector in which any asymptotic relation is valid is the largest angle for which the relative error between the left and right sides is small. The error term here is no longer exponentially small when z lies on the imaginary axis. This accounts for the wedge opening angle of 180°.

It is important to emphasize in the above example that the two possible leading asymptotic behaviors of $\sinh(z^{-1})$ as $z \to 0$, which are $\frac{1}{2}e^{z^{-1}}$ for Re $z > 0$ and $-\frac{1}{2}e^{-z^{-1}}$ for Re $z < 0$, are comparable in magnitude when they are purely oscillatory (on the Stokes line). The leading behaviors are most *unequal* in magnitude when they are purely real (exponentially growing or falling). The line along which the leading behaviors are most unequal is called an *anti-Stokes line.*

The Stokes phenomenon is a property of the exponential function. If the subdominant term is smaller than the dominant term by a power of $z - z_0$ as $z \to z_0$ in *all* directions in the complex plane, then there is no Stokes phenomenon. (Why?)

When Stokes lines do occur, their location is determined by the exponential contribution to the leading behavior. For example, if the leading behaviors of solutions to a second-order differential equation are $e^{S_1(z)}$ and $e^{S_2(z)}$ as $z \to z_0$, then the Stokes lines are the asymptotes as $z \to z_0$ of the curves

$$\text{Re}\,[S_1(z) - S_2(z)] = 0.$$

The anti-Stokes lines are the asymptotes as $z \to z_0$ of the curves

$$\text{Im}\,[S_1(z) - S_2(z)] = 0.$$

Observe that we have defined Stokes and anti-Stokes lines as asymptotes because these asymptotes determine the opening angles of the wedges of validity of asymptotic relations. Note that Stokes and anti-Stokes lines are a local property of functions; they are only meaningful in the immediate vicinity of z_0.

Example 12 *Stokes lines for the Airy equation.* The controlling factors of the large-z behaviors of solutions to the Airy equation $y'' = zy$ are given by $e^{\pm 2z^{3/2}/3}$. Hence, Stokes lines occur at Re $z^{3/2} = 0$ or arg $z = \pm\frac{1}{3}\pi$, π $(|z| \to \infty)$; anti-Stokes lines occur at Im $z^{3/2} = 0$ or arg $z = 0$, $\pm\frac{2}{3}\pi$ $(|z| \to \infty)$.

The leading asymptotic behavior of Bi (z) in (3.7.16) is valid between the Stokes lines at

$\arg z = \pm\frac{1}{3}\pi$. On the other hand, the leading asymptotic behavior of Ai (z) is valid in a much larger wedge:

$$\mathrm{Ai}\,(z) \sim \tfrac{1}{2}\pi^{-1/2}z^{-1/4}e^{-2z^{3/2}/3}, \qquad z \to \infty;\ |\arg z| < \pi. \tag{3.7.17}$$

Observe that the behavior (3.7.17) of Ai (z), which decays exponentially along the positive real axis, is valid in a much larger sector than the behavior (3.7.16) of Bi (z), which grows exponentially on the positive real axis. Despite the Stokes lines at $\arg z = \pm\frac{1}{3}\pi$, (3.7.17) is valid until the Stokes line at $\arg z = \pi$. This inequality of opening angles is a nontrivial property of Airy functions that we return to in Sec. 3.8.

Example 13 *Stokes lines for the parabolic cylinder equation.* The controlling factors of the leading behaviors of solutions to the parabolic cylinder equation $y'' + (\nu + \frac{1}{2} - \frac{1}{4}z^2)y = 0$ are $e^{\pm z^2/4}$ as $z \to \infty$. Thus, Stokes lines occur at $\arg z = \pm\frac{1}{4}\pi, \pm\frac{3}{4}\pi$ $(|z| \to \infty)$ and anti-Stokes lines occur at $\arg z = 0, \pm\pi/2, \pi$ $(|z| \to \infty)$.

The leading asymptotic behavior of the parabolic cylinder function $D_\nu(z)$ is

$$D_\nu(z) \sim z^\nu e^{-z^2/4}, \qquad z \to \infty;\ |\arg z| < \tfrac{3}{4}\pi. \tag{3.7.18}$$

Again observe that the behavior of $D_\nu(z)$ which decays exponentially along the positive real axis is valid beyond the Stokes lines at $\arg z = \pm\frac{1}{4}\pi$ and breaks down at the Stokes lines $\arg z = \pm\frac{3}{4}\pi$.

As a rule, that solution which decays most rapidly along the positive real axis [Ai (z) or $D_\nu(z)$] grows as z is rotated through the Stokes line nearest the positive real axis. For this kind of solution, it is correct to continue analytically the leading asymptotic behavior up to the Stokes lines beyond the ones nearest the real axis. It is incorrect to continue analytically the leading asymptotic behavior of solutions that grow along the positive real axis [Bi (z)] beyond the Stokes lines nearest the real axis. In Sec. 3.8 we show how to continue the full asymptotic series past Stokes lines, explain why some asymptotic relations are valid beyond Stokes lines, and discuss the connection between one-sided asymptotic relations that have different behaviors as $x \to x_0+$ and $x \to x_0-$.

We conclude with some comments on the origin of the Stokes phenomenon. As we have repeatedly said, the reason for using an asymptotic approximation is to replace a complicated transcendental function like Ai (z) by simpler expressions involving elementary functions like exponentials and powers of z. From a practical point of view, much is gained by such approximations. However, one pays for these advantages by having to deal with the complexities of the Stokes phenomenon. The Stokes phenomenon is not an intrinsic property of a function like Ai (z), but rather it is a property of the functions that are used to approximate it. The Stokes phenomenon reflects the presence of exponential functions in the asymptotic approximation. An asymptotic relation like Ai $(z) \sim$ Ai (z) $(z \to \infty)$ does not exhibit the Stokes phenomenon; this relation is valid as $z \to \infty$ in all directions. However, this relation contains no useful information. By contrast, the relation

$$\mathrm{Ai}\,(z) \sim \frac{1}{2\sqrt{\pi}}z^{-1/4}e^{-2z^{3/2}/3}, \qquad z \to \infty;\ |\arg z| < \pi,$$

represents progress, but at the same time it exhibits the Stokes phenomenon when $|\arg z| = \pi$.

(TD) 3.8 ASYMPTOTIC SERIES

In previous sections we have developed a formal procedure for finding asymptotic series representations of solutions to differential equations and have verified the validity of our results numerically. However, the approach has been intuitive. In this section, we outline the mathematical analysis necessary to justify the asymptotic methods we have used.

We begin by emphasizing the difference between convergent and asymptotic series. Then, we follow with several examples which illustrate what is involved in proving that a power series is asymptotic to a function and for these examples we show why the asymptotic series give good numerical approximations. Next, we review some of the mathematical properties of asymptotic series. We also show how to prove a formal power series is asymptotic to a solution of a differential equation. Finally, we consider asymptotic series in the complex plane and the Stokes phenomenon.

Convergent and Divergent Power Series

In Sec. 3.5 we defined $f(x) \sim \sum_{n=0}^{\infty} a_n(x - x_0)^n$ $(x \to x_0)$ to mean that for every N the remainder $\varepsilon_N(x)$ after $(N + 1)$ terms of the series is much smaller than the last retained term as $x \to x_0$: $\varepsilon_N(x) \equiv f(x) - \sum_{n=0}^{N} a_n(x - x_0)^n \ll (x - x_0)^N$ $(x \to x_0)$.

Example 1 *Taylor series as asymptotic series.* If the power series $\sum_{n=0}^{\infty} a_n(x - x_0)^n$ converges for $|x - x_0| < R$ to the function $f(x)$, then the series is also asymptotic to $f(x)$ as $x \to x_0$: $f(x) \sim \sum_{n=0}^{\infty} a_n(x - x_0)^n$ $(x \to x_0)$. Since $a_n = f^{(n)}(x_0)/n!$, repeated application of l'Hôpital's rule gives

$$\lim_{x \to x_0} \frac{f(x) - \sum_{n=0}^{N} a_n(x - x_0)^n}{(x - x_0)^{N+1}} = a_{N+1}.$$

Thus, $\varepsilon_N(x) = f(x) - \sum_{n=0}^{N} a_n(x - x_0)^n \ll (x - x_0)^N$ $(x \to x_0)$. We conclude that asymptotic series are generalizations of Taylor series because they include Taylor series as special cases.

A series need not be convergent to be asymptotic. Indeed, most asymptotic series are not convergent. Let us contrast convergent and asymptotic series. If $f(x) = \sum_{n=0}^{\infty} a_n(x - x_0)^n$ is a convergent series for $|x - x_0| < R$, then the remainder $\varepsilon_N(x)$ goes to zero as $N \to \infty$ for any fixed x, $|x - x_0| < R$:

$$\text{Convergent:} \quad \varepsilon_N(x) = \sum_{n=N+1}^{\infty} a_n(x - x_0)^n \to 0, \qquad N \to \infty;\ x \text{ fixed.}$$

On the other hand, if the series is asymptotic to $f(x)$, $f(x) \sim \sum_{n=0}^{\infty} a_n(x - x_0)^n$ $(x \to x_0)$, then the remainder $\varepsilon_N(x)$ goes to zero faster than $(x - x_0)^N$ as $x \to x_0$, but need not go to zero as $N \to \infty$ for fixed x:

Asymptotic: $\varepsilon_N(x) \ll (x - x_0)^N$, $\quad x \to x_0$; N fixed.

Convergence is an *absolute* concept; it is an intrinsic property of the expansion coefficients a_n. One can prove that a series converges without knowing the function to which it converges. However, asymptoticity is a *relative* property of the expansion coefficients *and* the function $f(x)$ to which the series is asymptotic. To prove that a power series is asymptotic to $f(x)$, one must consider *both* $f(x)$ and the expansion coefficients.

Let us clarify this distinction. Suppose you are given a power series and are asked to determine whether it is an asymptotic series as $x \to x_0$. The correct response is that you have been asked a stupid question! Why? Because every power series is asymptotic to some continuous function $f(x)$ as $x \to x_0$!

We present the construction of such a function as an example (see also Probs. 3.79 and 3.80).

Example 2 *Construction of a continuous function asymptotic to a given power series.* Given a formal power series $\sum_{n=0}^{\infty} a_n(x - x_0)^n$, we define the continuous function $\phi(x; \alpha)$, plotted in Fig. 3.20, as follows:

$$\phi(x; \alpha) = \begin{cases} 1, & |x| \le \frac{1}{2}\alpha, \\ 2\left(1 - \dfrac{|x|}{\alpha}\right), & \frac{1}{2}\alpha < |x| < \alpha, \\ 0, & \alpha \le |x|. \end{cases}$$

We also define a sequence of numbers $\alpha_n = \min\,(1/|a_n|,\ 2^{-n})$, where a_n are the arbitrary coefficients of the series $\sum_{n=0}^{\infty} a_n(x - x_0)^n$. The function $f(x)$ defined by

$$f(x) = \sum_{n=0}^{\infty} a_n \phi(x - x_0; \alpha_n)(x - x_0)^n$$

is finite, continuous, and satisfies

$$f(x) \sim \sum_{n=0}^{\infty} a_n(x - x_0)^n, \qquad x \to x_0.$$

It is finite and continuous for any $x \ne x_0$ because the series defining $f(x)$ truncates after at most N terms, where N is the smallest integer satisfying $2^{-N} \le |x - x_0|$. Also, if $|x - x_0| \le \frac{1}{2}\min\,(2^{-N}, 1/|a_0|, \ldots, 1/|a_N|) = R_N$, then $\phi(x - x_0; \alpha_n) = 1$ for $n = 0, \ldots, N$. Thus, if $|x - x_0| \le R_N$,

$$f(x) = a_0 + a_1(x - x_0) + \cdots + a_N(x - x_0)^N + \sum_{n=N+1}^{\infty} a_n \phi(x - x_0; \alpha_n)(x - x_0)^n.$$

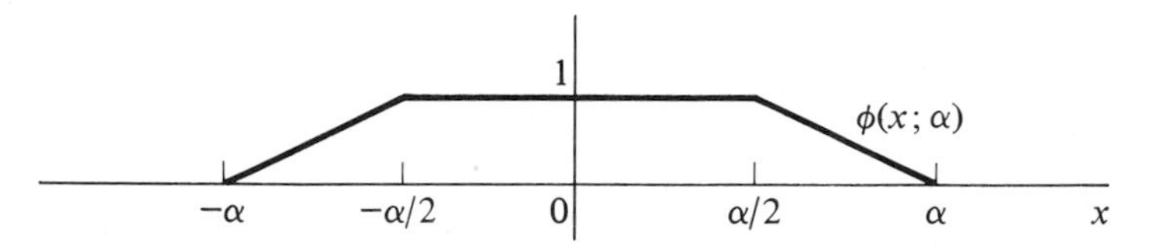

Figure 3.20 A plot of $\phi(x; \alpha)$ in Example 2.

Also, the definitions of α_n and $\phi(x;\alpha)$ imply that

$$|a_n\phi(x-x_0;\alpha_n)(x-x_0)| \le 1 \quad \text{and} \quad |\phi(x-x_0;\alpha_n)| \le 1$$

for all x and all n. Therefore, if $|x-x_0| \le R_N$, then

$$\begin{aligned}
|\varepsilon_N(x)| &= \left| f(x) - \sum_{n=0}^{N} a_n(x-x_0)^n \right| \\
&= \left| a_{N+1}\phi(x-x_0;\alpha_{N+1})(x-x_0)^{N+1} + \sum_{n=N+2}^{\infty} a_n\phi(x-x_0;\alpha_n)(x-x_0)^n \right| \\
&\le |a_{N+1}|\,|x-x_0|^{N+1} + \sum_{n=N+2}^{\infty} |x-x_0|^{n-1} \\
&= |x-x_0|^{N+1}[|a_{N+1}| + (1-|x-x_0|)^{-1}] \ll (x-x_0)^N, \qquad x \to x_0.
\end{aligned}$$

Thus, $\sum_{n=0}^{\infty} a_n(x-x_0)^n$ is asymptotic to $f(x)$ as $x \to x_0$.

Since *every* power series is an asymptotic power series, it is vacuous to ask whether a given series is asymptotic. However, it is meaningful to ask whether a power series is asymptotic to a given function $f(x)$ as $x \to x_0$. This is the reason why the definition given at the beginning of this section includes *both* a series and a function.

Examples of Asymptotic Series

Example 3 *Stieltjes series.* The prototype of an asymptotic series is the so-called Stieltjes series

$$\sum_{n=0}^{\infty} (-1)^n n!\, x^n. \tag{3.8.1}$$

We saw earlier that this series is a formal power series solution to the differential equation (3.4.2). We will now *prove* that this series is really asymptotic to a solution of (3.4.2) by "summing" the series and thereby reconstructing the exact solution to (3.4.2).

Of course, one cannot actually add up all the terms of a divergent series because the sum does not exist. By "summing" we mean finding a function to which the series is asymptotic. "Summation" is the inverse of expanding a function into an asymptotic series.

To sum the series (3.8.1) we invoke the integral identity $n! = \int_0^\infty e^{-t}t^n\,dt$:

$$\sum_{n=0}^{\infty} (-x)^n n! \to \sum_{n=0}^{\infty} (-x)^n \int_0^\infty e^{-t}t^n\,dt.$$

Next, we execute several sleazy maneuvers. We interchange the order of summation and integration,

$$\sum_{n=0}^{\infty} (-x)^n \int_0^\infty e^{-t}t^n\,dt \to \int_0^\infty dt\, e^{-t} \sum_{n=0}^{\infty} (-xt)^n,$$

and we sum the geometric series $\sum_{n=0}^{\infty} (-xt)^n \to 1/(1+xt)$, even though the sum diverges for those values of t such that $|xt| \ge 1$.

Despite these dubious manipulations, the resulting integral

$$y(x) = \int_0^\infty \frac{e^{-t}}{1+xt}\,dt, \tag{3.8.2}$$

which is called a *Stieltjes integral*, exists and defines an analytic function of x for all $x > 0$. Moreover, $y(x)$ exactly satisfies the differential equation (3.4.2):

$$x^2y'' + (1 + 3x)y' + y = \int_0^\infty \left[\frac{2x^2t^2}{(1 + xt)^3} - \frac{(1 + 3x)t}{(1 + xt)^2} + \frac{1}{1 + xt}\right] e^{-t}\, dt$$

$$= \int_0^\infty \frac{d}{dt}\left[\frac{t}{(1 + xt)^2} e^{-t}\right] dt = 0, \qquad x > 0.$$

Finally, we show that $y(x)$ in (3.8.2) has an asymptotic power series expansion valid as $x \to 0+$ which is precisely the Stieltjes series in (3.8.1).

Integrating by parts, we obtain the identity

$$\int_0^\infty (1 + xt)^{-n}e^{-t}\, dt = 1 - nx \int_0^\infty (1 + xt)^{-n-1}e^{-t}\, dt.$$

Repeated application of this formula gives

$$\begin{aligned} y(x) &= \int_0^\infty (1 + xt)^{-1}e^{-t}\, dt \\ &= 1 - x \int_0^\infty (1 + xt)^{-2}e^{-t}\, dt \\ &= 1 - x + 2x^2 \int_0^\infty (1 + xt)^{-3}e^{-t}\, dt \\ &\vdots \\ &= 1 - x + 2!\, x^2 - 3!\, x^3 + \cdots + (-1)^N N!\, x^N + \varepsilon_N(x), \end{aligned}$$

where

$$\varepsilon_N(x) = (-1)^{N+1}(N + 1)!\, x^{N+1} \int_0^\infty (1 + xt)^{-N-2}e^{-t}\, dt.$$

Finally, we use the inequality

$$\int_0^\infty (1 + xt)^{-N-2}e^{-t}\, dt \le \int_0^\infty e^{-t}\, dt = 1, \qquad x \to 0+,$$

which holds because $1 + xt > 1$ if $x > 0$ and $t > 0$, to show that

$$|\varepsilon_N(x)| \le (N + 1)!\, x^{N+1} \ll x^N, \qquad x \to 0+.$$

This completes the demonstration that the Stieltjes series (3.8.1) is asymptotic to the Stieltjes integral solution to the differential equation (3.4.2).

For the behavior of $y(x)$ as $x \to +\infty$, see Prob. 3.39(*i*).

Example 4 *General Stieltjes series and integrals.* A generalization of the Stieltjes integral (3.8.2) is given by

$$f(x) = \int_0^\infty \frac{\rho(t)}{1 + xt}\, dt, \tag{3.8.3}$$

where the *weight function* $\rho(t)$ is nonnegative for $t > 0$ and approaches zero so rapidly as $t \to \infty$ that the *moment integrals*

$$a_n = \int_0^\infty t^n \rho(t)\, dt \tag{3.8.4}$$

exist for all positive integers n.

Every Stieltjes integral has an asymptotic power series expansion whose coefficients are $(-1)^n a_n$:

$$f(x) \sim \sum_{n=0}^{\infty} (-1)^n a_n x^n, \qquad x \to 0+. \tag{3.8.5}$$

To prove this assertion, we note that

$$\begin{aligned} \varepsilon_N(x) &\equiv f(x) - \sum_{n=0}^{N} (-1)^n a_n x^n \\ &= \int_0^\infty \rho(t) \left[\frac{1}{1+xt} - \sum_{n=0}^{N} (-xt)^n \right] dt \\ &= \int_0^\infty \frac{\rho(t)}{1+xt} (-xt)^{N+1} \, dt \end{aligned} \tag{3.8.6}$$

for all N. Thus,

$$|\varepsilon_N(x)| \le x^{N+1} \int_0^\infty \rho(t) t^{N+1} \, dt = a_{N+1} x^{N+1} \ll x^N, \qquad x \to 0+, \tag{3.8.7}$$

where we use $1 + xt > 0$ and $\rho(t) > 0$ for $x > 0$ and $t > 0$. This completes the verification of (3.8.5).

Example 5 *Stieltjes series with weight function $K_0(t)$.* If $\rho(t) = K_0(t)$, the modified Bessel function of order 0, then (3.8.6) becomes

$$\int_0^\infty \frac{K_0(t)}{1+xt} \, dt \sim \frac{1}{2} \sum_{n=0}^{\infty} (-2x)^n \left[\Gamma\left(\frac{1}{2} n + \frac{1}{2}\right) \right]^2 \tag{3.8.8}$$

(see Prob. 3.76).

Numerical Approximations Using Asymptotic Series: The Optimal Truncation Rule

It is possible to improve our error estimates for Stieltjes series. Equation (3.8.7) shows that the error between the Stieltjes integral (3.8.2) and the first N terms of the Stieltjes series in (3.8.1) is smaller than the absolute value of the next term in the series $N!\, x^N$. Also, the sign of the error is the same as the sign of the next term. The same is true for the general Stieltjes series (3.8.5); the error between the Stieltjes integral (3.8.3) and N terms of the Stieltjes series (3.8.5) has the same sign and is less than the $(N + 1)$th term of the series.

These error bounds imply that for any fixed x, truncating the Stieltjes series (3.8.5) just before the smallest term will give a good numerical estimate of the Stieltjes integral (3.8.4). It is more difficult to justify this optimal truncation rule for asymptotic series that are not Stieltjes series. However, we have had remarkable success with this rule for truncating asymptotic series (see Sec. 3.5). In fact, even though the asymptotic series (3.5.8*b*) and (3.5.9*b*) for $K_\nu(x)$ and (3.7.14) and (3.7.15) for $J_\nu(x)$ are not Stieltjes series, it is still true that the error after N terms is less than the $(N + 1)$th term provided that N is larger than some number depending on ν but not on x (see Prob. 3.77).

Of course, not every asymptotic series has the property that the error after N terms is less than the $(N+1)$th term. For example, the error after N terms in the asymptotic series (3.5.8*a*) and (3.5.9*a*) for $I_\nu(x)$ is not similar in sign and smaller than the $(N+1)$th term. Nevertheless, our truncation procedure gave very good results (see Figs. 3.5 and 3.6 and Table 3.1).

How well do these optimal asymptotic approximations really work? For Stieltjes series, we can provide an accurate asymptotic estimate of the difference between the exact value of the Stieltjes integral and the optimal truncation of the Stieltjes series as $x \to 0+$.

Example 6 *Error estimate for an optimally truncated Stieltjes series.* According to (3.8.6), the error after N terms of the Stieltjes series (3.8.1) for which the weight function $\rho(t) = e^{-t}$ is

$$(-x)^N \int_0^\infty t^N \frac{e^{-t}}{1+xt}\, dt$$

for any N. The optimal truncation of (3.8.1) is obtained by choosing N equal to the largest integer less than or equal to $1/x$; this is true because the ratio of the $(n+1)$th term to the nth term of (3.8.1) is $-nx$. If we approximate this integral representation for the error (see Probs. 5.25 and 6.37), we find that the optimal error $\varepsilon_{\text{optimal}}(x)$ satisfies

$$|\varepsilon_{\text{optimal}}(x)| \sim \left(\frac{\pi}{2x}\right)^{1/2} e^{-1/x}, \qquad x \to 0+. \tag{3.8.9}$$

We have checked the validity of (3.8.9) numerically. In Fig. 3.21 we plot the ratio of $|\varepsilon_{\text{optimal}}(x)|$ determined numerically by optimally truncating the series (3.8.1) to its leading behavior given in (3.8.9). Observe that this ratio approaches 1 as $x \to 0+$.

Properties of Asymptotic Series

(a) Nonuniqueness We have given a successful prescription for obtaining good numerical results from divergent asymptotic series. Strangely enough, one must use this technique with caution because it produces a *unique* numerical answer! Actually, the "sum" of a divergent power series is not uniquely determined. For example, if $f(x) \sim \sum_{n=0}^\infty a_n(x-x_0)^n$ $(x \to x_0)$, then it is also true that $f(x) + e^{-(x-x_0)^{-2}} \sim \sum_{n=0}^\infty a_n(x-x_0)^n$ $(x \to x_0)$ because $e^{-(x-x_0)^{-2}} \ll (x-x_0)^n$ as $x \to x_0$ for all n. In fact, the series $\sum_{n=0}^\infty a_n(x-x_0)^n$ is asymptotic as $x \to x_0$ to any function which differs from $f(x)$ by a function $g(x)$ so long as $g(x) \to 0$ as $x \to x_0$ more rapidly than all powers of $x - x_0$. Such a function $g(x)$ is said to be *subdominant to the asymptotic power series*; the asymptotic expansion of $g(x)$ is

$$g(x) \sim \sum_{n=0}^\infty 0(x-x_0)^n, \qquad x \to x_0.$$

In short, an asymptotic series is asymptotic to a whole class of functions that differ from each other by subdominant functions. We do not change the asymptotic series by adding a subdominant function, even if the subdominant function is multiplied by a huge numerical coefficient.

For example, e^{-x^4} is subdominant with respect to the asymptotic expansion

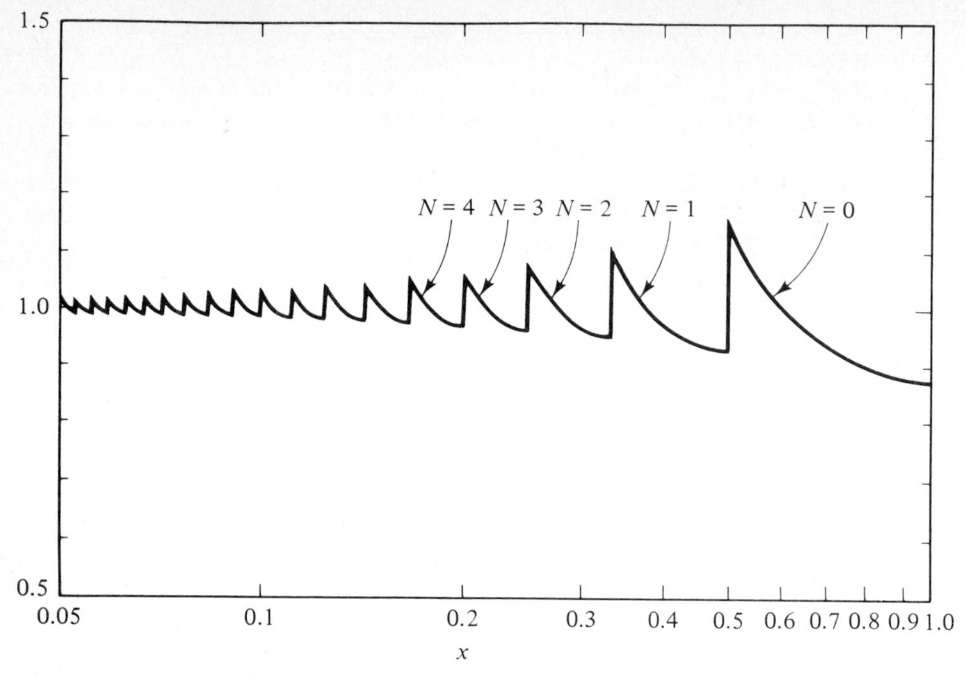

Figure 3.21 A computer plot of the ratio of $|\varepsilon_{\text{optimal}}(x)|/[(\pi/2x)^{1/2}\exp(-1/x)]$ for $0.05 \le x \le 1$. Here $\varepsilon_{\text{optimal}}(x)$ is the error in the optimal asymptotic approximation to the Stieltjes integral $\int_0^\infty e^{-t}(1+xt)^{-1}\,dt$ in (3.8.2); that is, it is the difference between the Stieltjes integral and the optimally truncated asymptotic series $\sum_{n=0}^{N}(-1)^n n!\, x^n$. Theoretically, the leading behavior of $|\varepsilon_{\text{optimal}}(x)|$ is $(\pi/2x)^{1/2}\exp(-1/x)$ as $x \to 0+$ [see (3.8.9)]. The graph clearly verifies this prediction.

(3.5.13) and (3.5.14) of $D_{3.5}(x)$ as $x \to +\infty$. Therefore, $f(x) = D_{3.5}(x) + 10^{10}e^{-x^4}$ has the same asymptotic expansion as $D_{3.5}(x)$ as $x \to +\infty$. What happens now if we compute the optimal asymptotic approximation to $f(x)$? We already know from Fig. 3.7 and Table 3.2 that the optimal asymptotic approximation is very close to $D_{3.5}(x)$ for $x > 1$. Therefore, since $10^{10}e^{-x^4} > |D_{3.5}(x)|$ for $0 \le x \le 2.1$, the optimal asymptotic approximation to $f(x)$ is not accurate for $1 \le x \le 2.1$. Nevertheless, when $x \ge 2.3$ the optimal asymptotic approximation is very close to $f(x)$.

The above discussion shows that the value of x for which the optimal asymptotic approximation becomes useful cannot be predicted from the asymptotic series itself. Rather, it depends on the admixture of subdominant functions. Thus, for any given problem we can never really know *a priori* whether or not asymptotic analysis will give good numerical results at a fixed value of x. However, experience has shown that asymptotic methods nearly always give spectacularly good results.

(b) Uniqueness Although there are many different functions asymptotic to a given power series, there is only *one* asymptotic power series for each function. Specifically, if a function $f(x)$ can be expanded as $f(x) \sim \sum_{n=0}^{\infty} a_n(x-x_0)^n$

$(x \to x_0)$, then the expansion coefficients are unique. The proof of uniqueness is given in Sec. 3.5.

(c) Equating coefficients in asymptotic series It is not strictly correct to write $\sum_{n=0}^{\infty} a_n(x - x_0)^n \sim \sum_{n=0}^{\infty} b_n(x - x_0)^n$ $(x \to x_0)$ because power series can only be asymptotic to functions, and not to other power series. However, we will occasionally use this notation; we define it to mean that the class of functions to which $\sum_{n=0}^{\infty} a_n(x - x_0)^n$ and $\sum_{n=0}^{\infty} b_n(x - x_0)^n$ are asymptotic as $x \to x_0$ are the same. It follows from the uniqueness of asymptotic expansions that two power series are asymptotic if and only if $a_n = b_n$ for all n. Thus, we may equate coefficients of like powers of $x - x_0$ in power series that are "asymptotic to each other."

(d) Arithmetical operations on asymptotic series Arithmetical operations may be performed term by term on asymptotic series. Specifically, suppose

$$f(x) \sim \sum_{n=0}^{\infty} a_n(x - x_0)^n, \qquad x \to x_0,$$

$$g(x) \sim \sum_{n=0}^{\infty} b_n(x - x_0)^n, \qquad x \to x_0.$$

Then

$$\alpha f(x) + \beta g(x) \sim \sum_{n=0}^{\infty} (\alpha a_n + \beta b_n)(x - x_0)^n, \qquad x \to x_0,$$

$$f(x)g(x) \sim \sum_{n=0}^{\infty} c_n(x - x_0)^n, \qquad x \to x_0,$$

$$\frac{f(x)}{g(x)} \sim \sum_{n=0}^{\infty} d_n(x - x_0)^n, \qquad x \to x_0,$$

where $c_n = \sum_{m=0}^{n} a_m b_{n-m}$, and if $b_0 \neq 0$, $d_0 = a_0/b_0$ and

$$d_n = \frac{a_n - \sum_{m=0}^{n-1} d_m b_{n-m}}{b_0}, \qquad n \geq 1.$$

The proofs of these results are elementary. For example, let us prove that asymptotic series can be multiplied term by term. Using the above expression for c_n, we obtain

$$\begin{aligned}
f(x)g(x) &- \sum_{n=0}^{N} c_n(x - x_0)^n \\
&= f(x)g(x) - \sum_{m=0}^{N} a_m(x - x_0)^m \sum_{n=m}^{N} b_{n-m}(x - x_0)^{n-m} \\
&= g(x)\left[f(x) - \sum_{m=0}^{N} a_m(x - x_0)^m\right] \\
&\quad + \sum_{m=0}^{N} a_m(x - x_0)^m \left[g(x) - \sum_{p=0}^{N-m} b_p(x - x_0)^p\right]
\end{aligned}$$

for all N. Since $\lim_{x\to x_0} g(x) = b_0$, $|g(x)| \le 2|b_0|$ for x sufficiently close to x_0, say $|x - x_0| \le R$. Hence, by the definition of asymptotic series,

$$|f(x)g(x) - \sum_{n=0}^{N} c_n(x - x_0)^n| \ll (2|b_0| + |a_0| + |a_1| + \cdots + |a_N|)|x - x_0|^N, \qquad x \to x_0,$$

for all N. Thus, asymptotic series can be multiplied term by term.

(e) Integration of asymptotic series Any asymptotic series $f(x) \sim \sum_{n=0}^{\infty} a_n(x - x_0)^n$ $(x \to x_0)$ can be integrated term by term if $f(x)$ is integrable near x_0:

$$\int_{x_0}^{x} f(t)\, dt \sim \sum_{n=0}^{\infty} \frac{a_n}{n+1}(x - x_0)^{n+1}, \qquad x \to x_0.$$

To prove this result we begin with the definition of an asymptotic power series: $|f(x) - \sum_{n=0}^{N} a_n(x - x_0)^n| \ll (x - x_0)^N$ $(x \to x_0)$. From this it follows that for any $\varepsilon > 0$ there exists an interval about x_0, say $|x - x_0| \le R$ (where R of course depends on ε), in which

$$|f(x) - \sum_{n=0}^{N} a_n(x - x_0)^n| \le \varepsilon |x - x_0|^N, \qquad |x - x_0| \le R.$$

Therefore,

$$\begin{aligned}\left|\int_{x_0}^{x} \left[f(t) - \sum_{n=0}^{N} a_n(t - x_0)^n\right] dt\right| &\le \int_{x_0}^{x} \left|f(t) - \sum_{n=0}^{N} a_n(t - x_0)^n\right| dt \\ &\le \varepsilon \int_{x_0}^{x} |t - x_0|^N\, dt \\ &= \frac{\varepsilon}{N+1}|x - x_0|^{N+1}, \qquad |x - x_0| \le R.\end{aligned}$$

Hence,

$$\left|\frac{\int_{x_0}^{x} f(t)\, dt - \sum_{n=0}^{N} [a_n/(n+1)](x - x_0)^{n+1}}{(x - x_0)^{N+1}}\right| \le \frac{\varepsilon}{N+1}, \qquad |x - x_0| \le R.$$

But $\varepsilon > 0$ is arbitrary, so

$$\int_{x_0}^{x} f(t)\, dt - \sum_{n=0}^{N} \frac{a_n}{n+1}(x - x_0)^{n+1} \ll (x - x_0)^{N+1}, \qquad x \to x_0,$$

for all N. Thus, asymptotic series can be integrated term by term.

If we wish to integrate an asymptotic series at infinity, there is a slight complication. The above argument can be extended to show that if $f(x) \sim \sum_{n=0}^{\infty} a_n x^{-n}$ $(x \to \infty)$, then

$$\int_{x}^{\infty} [f(t) - a_0 - a_1 t^{-1}]\, dt \sim \sum_{n=2}^{\infty} \frac{a_n}{n-1} x^{1-n}, \qquad x \to \infty.$$

(f) Differentiation of asymptotic series Asymptotic series cannot in general be differentiated term by term. For example, even if $f(x) \sim \sum_{n=0}^{\infty} a_n(x - x_0)^n$ $(x \to x_0)$, it does not necessarily follow that $f'(x) \sim \sum_{n=1}^{\infty} na_n(x - x_0)^{n-1}$ $(x \to x_0)$. The problem with differentiation is connected with subdominance: the functions $f(x)$ and

$$g(x) = f(x) + e^{-1/(x-x_0)^2} \sin\left(e^{1/(x-x_0)^2}\right)$$

differ by a subdominant function and thus have the same asymptotic series expansion as $x \to x_0$. However, it is not necessarily true that $f'(x)$ and

$$g'(x) = f'(x) - 2(x - x_0)^{-3} \cos\left(e^{1/(x-x_0)^2}\right) + 2(x - x_0)^{-3} e^{-1/(x-x_0)^2} \sin\left(e^{1/(x-x_0)^2}\right)$$

have the same asymptotic power series expansion as $x \to x_0$. Therefore, term-by-term differentiation of an asymptotic series may not be valid for both $f(x)$ and $g(x)$; asymptotic series cannot be differentiated termwise without additional restrictions.

Termwise *integration* of an asymptotic series, which we justified above, is an example of an *Abelian* theorem. In an Abelian theorem, asymptotic information about an average of a function (its integral) is deduced from asymptotic information about the function itself. Differentiation of asymptotic series relates to the converse process; namely, deducing asymptotic information about a derivative from asymptotic information about a function. Converses to Abelian theorems are called *Tauberian* theorems. Tauberian theorems require conditions supplementary to those of corresponding Abelian theorems to be valid. In the case of termwise differentiation of asymptotic series, there are several situations in which Tauberian-like theorems provide justification for termwise differentiation.

One such result is as follows. Suppose $f'(x)$ exists, is integrable, and $f(x) \sim \sum_{n=0}^{\infty} a_n(x - x_0)^n$ $(x \to x_0)$. Then it follows that

$$f'(x) \sim \sum_{n=1}^{\infty} na_n(x - x_0)^{n-1}, \qquad x \to x_0.$$

This result is an immediate consequence of the Abelian theorem for termwise integration of an asymptotic series proved above. To see this, suppose that $f'(x) \sim \sum_{n=0}^{\infty} b_n(x - x_0)^n$ $(x \to x_0)$. Then, integrating term by term gives

$$\begin{aligned} f(x) &= f(x_0) + \int_{x_0}^{x} f'(t)\, dt \\ &\sim f(x_0) + \sum_{n=0}^{\infty} \frac{b_n}{n+1}(x - x_0)^{n+1}, \qquad x \to x_0. \end{aligned}$$

But since $f(x) \sim \sum_{n=0}^{\infty} a_n(x - x_0)^n$ $(x \to x_0)$ and since asymptotic series are unique, we find that $a_0 = f(x_0)$, $a_{n+1} = b_n/(n + 1)$ $(n = 0, 1, \ldots)$. This proves the theorem.

There are other more technical Tauberian-like results for differentiation of asymptotic relations. A result of this kind is as follows. Suppose $f(x) \sim x^p$ $(x \to +\infty)$, where $p \geq 1$ and $f''(x)$ is positive. Then (see Prob. 3.84) $f'(x) \sim px^{p-1}$

$(x \to +\infty)$. This result concerns only the *leading* behavior of $f(x)$ and does *not* justify termwise differentiation of an asymptotic series (see Prob. 3.85).

Termwise differentiation of asymptotic series is much clearer in the complex domain. For example, suppose that $f(z)$ is analytic in the sector $\theta_1 \le \arg(z - z_0) \le \theta_2$, $0 < |z - z_0| < R$ and $f(z) \sim \sum_{n=0}^{\infty} a_n(z - z_0)^n$ $[z \to z_0; \theta_1 \le \arg(z - z_0) \le \theta_2]$. Then (see Prob. 3.72)

$$f'(z) \sim \sum_{n=1}^{\infty} na_n(z - z_0)^{n-1}, \qquad z \to z_0; \theta_1 < \arg(z - z_0) < \theta_2.$$

It should be clear enough from these three special cases that the variety and complexity of Tauberian theorems for differentiation of asymptotic series is bewildering. Fortunately, the whole situation is greatly simplified if a function is known to satisfy a linear differential equation whose *coefficients* can be expanded in asymptotic series.

Asymptotic Expansions of Solutions to Differential Equations

The formal procedures given in Secs. 3.4 to 3.7 for calculating asymptotic expansions of solutions to differential equations require justification. There are several possible difficulties. First, we have always assumed that after the leading-order behavior is factored off, an asymptotic series expansion of the solution remains. However, not all functions can be expanded in asymptotic series. For example, consider the function $y(t) = t^2 + e^{-t^2(1 - \sin t)}$ which has leading behavior t^2 as $t \to \infty$. As $t \to \infty$, there are narrow regions that occur periodically in which $\sin t$ is near 1 and the term $e^{-t^2(1 - \sin t)}$ is not negligible with respect to 1. The existence of these regions implies that there does not exist any asymptotic power series representation for $y(t)$ as $t \to +\infty$. We shall see that this difficulty does not afflict solutions of differential equations whose *coefficients* themselves have asymptotic power series expansions.

Second, asymptotic series cannot, in general, be differentiated termwise. Thus, the formal differentiation of asymptotic series, which allowed us to determine the coefficients in the expansions of solutions to differential equations, needs to be justified.

The proof that our formal methods are correct has two parts. First, we argue that if $y(x)$ is the solution of $y'' + py' + qy = 0$ where $p(x)$, $p'(x)$, $q(x)$ are expandable in asymptotic power series as $x \to x_0$ and if we *assume* that $y(x)$ is also expandable in an asymptotic power series as $x \to x_0$, then the derivatives of $y(x)$ are also expandable and their asymptotic power series are obtained by termwise differentiation of the asymptotic power series representing $y(x)$. The proof is elementary. Consider the special case of the differential equation $y''(x) + q(x)y(x) = 0$. If $q(x)$ and $y(x)$ possess asymptotic power series representations as $x \to x_0$, then the differential equation itself ensures that $y''(x)$ does also (because multiplication of asymptotic power series is permissible). Integrating the asymptotic power series representing $y''(x)$ shows that $y'(x)$ also has an asymptotic series, so termwise differentiation is justified. The argument for a general nth-order differential equation is left for an exercise (Prob. 3.81).

In summary, once it is known that $y(x)$ is expandable in an asymptotic power series, the differential equation itself ensures that derivatives of $y(x)$ can also be expanded; the uniqueness of asymptotic expansions ensures that formal calculation procedures yield correct answers. It remains only to prove that the solutions have asymptotic series representations. The method of proof will be illustrated by an example.

Example 7 *Existence of an asymptotic expansion for* $K_\nu(x)$. The quickest way to prove that (3.5.8*b*) and (3.5.9*b*) is the asymptotic expansion of $K_\nu(x)$ as $x \to +\infty$ is to expand an integral representation of $K_\nu(x)$ (see Prob. 3.77). However, this approach requires that one have an integral representation at hand. In fact, it is relatively rare to have an integral representation for a function and relatively common to have the differential equation. The method we are about to use proceeds directly from the differential equation and, as such, is very general.

Recall that $K_\nu(x)$ is defined as that solution of the modified Bessel equation

$$x^2y'' + xy' - (x^2 + \nu^2)y = 0 \tag{3.8.10}$$

which behaves as

$$K_\nu(x) \sim \left(\frac{\pi}{2x}\right)^{1/2} e^{-x}, \qquad x \to +\infty. \tag{3.8.11}$$

If we substitute $(\pi/2x)^{1/2}e^{-x}w(x)$ for $y(x)$ in (3.8.10), we obtain the differential equation for $w(x)$:

$$w'' - 2w' + \frac{\lambda}{x^2}w = 0, \tag{3.8.12}$$

where $\lambda = \frac{1}{4} - \nu^2$. We want to prove that $w(x)$ has an asymptotic expansion of the form $w(x) \sim 1 + a_1/x + a_2/x^2 + \cdots$, valid as $x \to +\infty$.

If there exists such a solution to (3.8.12), it is easy to verify that it must satisfy the integral equation

$$w(x) = 1 + \frac{1}{2}\lambda \int_x^\infty (e^{2(x-t)} - 1)\frac{w(t)}{t^2}\,dt. \tag{3.8.13}$$

In Prob. 3.82 it is shown that the integral equation (3.8.13) has a solution which is *bounded* for $x > a$: $|w(x)| \le B_0$ $(x > a)$, where B_0 and a are positive constants. This result is crucial. Once this bound is proved, the inequality $|e^{2(x-t)} - 1| < 1$ for $t \ge x$ gives

$$\left|\frac{1}{2}\lambda \int_x^\infty (e^{2(x-t)} - 1)\frac{w(t)}{t^2}\right| \le \frac{1}{2}|\lambda|\frac{B_0}{x}, \qquad x > a.$$

This and (3.8.13) imply that $w(x) \to 1$ as $x \to +\infty$ and that $w(x) = 1 + w_1(x)$, $|w_1(x)| \le B_1/x$ $(x > a)$, where $B_1 = \frac{1}{2}|\lambda B_0|$ is a positive constant. Substituting this result into (3.8.13) gives

$$w_1(x) = \frac{1}{2}\lambda \int_x^\infty (e^{2(x-t)} - 1)\frac{1}{t^2}\,dt + \frac{1}{2}\lambda \int_x^\infty (e^{2(x-t)} - 1)\frac{w_1(t)}{t^2}\,dt.$$

Integrating the first integral by parts gives

$$\frac{1}{2}\lambda \int_x^\infty (e^{2(x-t)} - 1)\frac{1}{t^2}\,dt = -\frac{1}{2}\frac{\lambda}{x} + \frac{\lambda}{4x^2} - \frac{\lambda}{2}\int_x^\infty \frac{e^{2x-2t}}{t^3}\,dt.$$

Hence, there is a positive constant B_2 such that $w(x) = 1 - \lambda/2x + w_2(x)$, $|w_2(x)| < B_2/x^2$ $(x > a)$, which is equivalent to $w(x) - 1 \sim -\lambda/2x$ $(x \to +\infty)$.

If we continue the process of substituting into the integral equation, integrating by parts, and bounding the remaining terms, we obtain all the terms in the asymptotic series (3.5.9*b*) for $w(x)$. We have proved the existence of the asymptotic series for $w(x)$ and constructed the series as well.

Asymptotic Power Series in the Complex Plane: Connection Between Various Wedges of Validity

In Sec. 3.7 we saw that asymptotic relations are valid in wedge-shaped regions in the complex plane. Asymptotic power series are also valid in sectors: if a power series is asymptotic to a single-valued function as $z \to z_0$ in a full 360° disk about z_0 and if z_0 is at worst an isolated singularity, then one can prove that the series is a Taylor series (see Prob. 3.83). Thus, if an asymptotic power series is *divergent*, it can only be valid in a sector whose opening angle is less than 360°.

Outside the sector in which a series is asymptotic to a function there are usually other series asymptotic to the function having their own sectors of validity. For example, we saw in Secs. 3.5 and 3.7 that the functions Ai (x) and Bi (x) have different asymptotic expansions for large $|x|$ depending on the sign of x. Let us consider the problem of representing the functions Ai (z) and Bi (z) as $z \to \infty$ in terms of asymptotic series. All the local asymptotic analysis we used when z was real still applies when z is complex. Therefore, any solution to the Airy equation $y'' = zy$ has the general asymptotic representation

$$y(z) \sim a_1 z^{-1/4} e^{-2z^{3/2}/3} \sum_{n=0}^{\infty} (-1)^n c_n z^{-3n/2} + a_2 z^{-1/4} e^{2z^{3/2}/3} \sum_{n=0}^{\infty} c_n z^{-3n/2}, \qquad z \to \infty, \tag{3.8.14}$$

where a_1 and a_2 are constants. The asymptotic expansions of special solutions to the Airy equation like Ai (z) and Bi (z) are determined by specifying the values of a_1 and a_2. The Stokes phenomenon is simply that the values of a_1 and a_2 for Ai (z) or Bi (z) may be different in different sectors of the complex plane. We summarize the values of a_1 and a_2 below. For Ai (z),

$$a_1 = \frac{1}{2\sqrt{\pi}}, \qquad a_2 = 0, \qquad |\arg z| < \pi; \tag{3.8.15}$$

$$a_1 = \frac{1}{2\sqrt{\pi}}, \qquad a_2 = \frac{i}{2\sqrt{\pi}}, \qquad \frac{\pi}{3} < \arg z < \frac{5\pi}{3}; \tag{3.8.16}$$

$$a_1 = \frac{1}{2\sqrt{\pi}}, \qquad a_2 = \frac{-i}{2\sqrt{\pi}}, \qquad -\frac{5\pi}{3} < \arg z < -\frac{\pi}{3}. \tag{3.8.17}$$

For Bi (z),

$$a_1 \text{ arbitrary}, \qquad a_2 = \frac{1}{\sqrt{\pi}}, \qquad |\arg z| < \frac{\pi}{3}; \tag{3.8.18}$$

$$a_1 = \frac{i}{2\sqrt{\pi}}, \qquad a_2 = \frac{1}{2\sqrt{\pi}}, \qquad \frac{\pi}{3} < \arg z < \frac{5\pi}{3}; \tag{3.8.19}$$

$$a_1 = -\frac{i}{2\sqrt{\pi}}, \qquad a_2 = \frac{1}{2\sqrt{\pi}}, \qquad -\frac{5\pi}{3} < \arg z < -\frac{\pi}{3}. \tag{3.8.20}$$

In (3.8.18) a_1 is arbitrary because it multiplies a function which is subdominant (exponentially small) for $|\arg z| < \pi/3$. Can you recover the asymptotic expansions (3.7.12) and (3.7.13) for Ai (x) and Bi (x) as $x \to -\infty$?

Even though the Airy functions are entire, one may not analytically continue their asymptotic expansions from one wedge of validity to another. What then is the connection between these different asymptotic expansions for large $|z|$?

To answer this question we will use two linear functional relations (see Prob. 3.78):

$$\mathrm{Ai}\,(z) = -\omega\,\mathrm{Ai}\,(\omega z) - \omega^2\,\mathrm{Ai}\,(\omega^2 z), \tag{3.8.21a}$$

$$\mathrm{Bi}\,(z) = i\omega\,\mathrm{Ai}\,(\omega z) - i\omega^2\,\mathrm{Ai}\,(\omega^2 z), \tag{3.8.21b}$$

where $\omega = e^{-2\pi i/3}$ is a cube root of unity.

Linear functional relations of this sort must exist because the four functions Ai (z), Bi (z), Ai (ωz), and Ai $(\omega^2 z)$ are all solutions of the Airy equation $y'' = zy$. Since only two of these functions can be linearly independent, it follows that there must be a linear relation between any three. To verify these relations one need only substitute the Taylor expansions for Ai (z) and Bi (z) from (3.2.1) and (3.2.2) into (3.8.21).

Let us use the functional relations (3.8.21) to derive the connection between (3.8.15) and the relations (3.8.16) and (3.8.19) valid for $\frac{1}{3}\pi < \arg z < \frac{5}{3}\pi$. If $\frac{1}{3}\pi < \arg z < \frac{5}{3}\pi$, then $-\pi/3 < \arg(\omega z) < \pi$ and $-\pi < \arg(\omega^2 z) < \frac{1}{3}\pi$. Therefore, (3.8.14) with a_1 and a_2 in (3.8.15) gives valid asymptotic expansions of Ai (ωz) and Ai $(\omega^2 z)$. Using (3.8.21), we obtain (3.8.16) and (3.8.19). Equations (3.8.21*a*,*b*) have allowed us to rotate the asymptotic expansions through an angle of 120°. The asymptotic expansion of Ai (z) is given for all arg z by (3.8.14) with (3.8.15) and (3.8.16). However, there is no overlap in the regions of validity of the asymptotic expansions for Bi (z) given by (3.8.14) with (3.8.18) to (3.8.20). Additional asymptotic expansions for Bi (z) valid for $-\pi < \arg z < \frac{1}{3}\pi$ and $-\frac{1}{3}\pi < \arg z < \pi$ may be similarly obtained (see Prob. 3.61). By deriving these different asymptotic expansions we explain the connection between the exponential behavior of Ai (x) and Bi (x) for large positive x and their oscillatory behavior for large negative x.

Example 8 *Parabolic cylinder function for large negative argument.* The technique we have just described can also be used to determine the behavior of the parabolic cylinder function $D_\nu(x)$ as $x \to -\infty$. We begin with the asymptotic expansion for $D_\nu(z)$ which was derived in Sec. 3.5:

$$D_\nu(z) \sim z^\nu e^{-z^2/4}\left[1 - \frac{\nu(\nu-1)}{2z^2} + \frac{\nu(\nu-1)(\nu-2)(\nu-3)}{2\cdot 4z^4} - \cdots\right],$$

$$z \to \infty;\ |\arg z| < \tfrac{3}{4}\pi. \tag{3.8.22}$$

It is not immediately evident why this asymptotic expansion is valid beyond the Stokes lines at $\arg z = \pm\pi/4$; we assume this expansion for now and postpone a discussion of it until the end of this section. However, as discussed earlier in this section, the rule is very simple: D_ν is the *subdominant* solution to the parabolic cylinder equation on the positive x axis, so its full asymptotic expansion is valid beyond the Stokes line nearest the positive x axis; the asymptotic expansion of D_ν only exhibits the Stokes phenomenon at subsequent Stokes lines.

We will now obtain the asymptotic expansion of $D_\nu(z)$ beyond the region of validity of

(3.8.22). To do this we will use a functional relation between $D_\nu(z)$, $D_\nu(-z)$, and $D_{-\nu-1}(-iz)$. Since each of these functions satisfies the parabolic cylinder equation $y'' + (\nu + \frac{1}{2} - \frac{1}{4}z^2)y = 0$, there must be a linear relation between them,

$$D_\nu(z) = aD_\nu(-z) + bD_{-\nu-1}(-iz),$$

valid for all z. We know that $D_\nu(-z)$ and $D_{-\nu-1}(-iz)$ are linearly independent solutions because (3.8.22) implies that their leading behaviors respectively are $(-z)^\nu e^{-z^2/4}$ and $(-iz)^{-\nu-1}e^{+z^2/4}$ for $z \to \infty$ along the negative real axis. To determine the constants a and b we require more information than that contained in the asymptotic expansion (3.8.22). For example, it may be shown (see Prob. 6.84) that

$$D_\nu(0) = \pi^{1/2}2^{\nu/2}/\Gamma(\tfrac{1}{2} - \tfrac{1}{2}\nu), \qquad D_\nu'(0) = -\pi^{1/2}2^{(\nu+1)/2}\Gamma(-\tfrac{1}{2}\nu).$$

These equations fix the values of a and b and we obtain the functional relation

$$D_\nu(z) = e^{i\nu\pi}D_\nu(-z) + \frac{(2\pi)^{1/2}}{\Gamma(-\nu)}e^{i(\nu+1)\pi/2}D_{-\nu-1}(-iz) \tag{3.8.23}$$

valid for all z, where we have used the identity (see Prob. 2.6) $\Gamma(-\nu) = -\pi^{1/2}2^{-\nu-1}\Gamma(\frac{1}{2} - \frac{1}{2}\nu)/[\Gamma(1 + \frac{1}{2}\nu)\sin\frac{1}{2}\pi\nu]$ to simplify the final expression. [Notice that $a = e^{i\nu\pi}$ can be determined by applying (3.8.22) to $D_\nu(ix) = aD_\nu(-ix) + bD_{-\nu-1}(x)$ for $x \to +\infty$ through positive real values. However, (3.8.22) does not determine the constant b.]

Let us now use the functional relation (3.8.23) to obtain an asymptotic expansion of $D_\nu(z)$ as $|z| \to \infty$ with $\frac{1}{4}\pi < \arg z < \frac{5}{4}\pi$. If $\frac{1}{4}\pi < \arg z < \frac{5}{4}\pi$, then $z_1 = -z = e^{-i\pi}z$ satisfies $-\frac{3}{4}\pi < \arg z_1 < \frac{1}{4}\pi$ while $z_2 = -iz = e^{-i\pi/2}z$ satisfies $-\frac{1}{4}\pi < \arg z_2 < \frac{3}{4}\pi$. Therefore, (3.8.22) gives valid asymptotic expansions of both $D_\nu(z_1)$ and $D_\nu(z_2)$ as $z \to \infty$. Thus, (3.8.23) implies

$$\begin{aligned} D_\nu(z) \sim{}& z^\nu e^{-z^2/4}\left[1 - \frac{\nu(\nu-1)}{2z^2} + \frac{\nu(\nu-1)(\nu-2)(\nu-3)}{2\cdot 4z^4} - \cdots\right] \\ &- \frac{(2\pi)^{1/2}}{\Gamma(-\nu)}e^{i\pi\nu}z^{-\nu-1}e^{z^2/4}\left[1 + \frac{(\nu+1)(\nu+2)}{2z^2} + \frac{(\nu+1)(\nu+2)(\nu+3)(\nu+4)}{2\cdot 4z^4} + \cdots\right], \end{aligned}$$
$$z \to \infty;\ \tfrac{1}{4}\pi < \arg z < \tfrac{5}{4}\pi. \tag{3.8.24}$$

Notice that the regions of validity of (3.8.22) and (3.8.24) overlap for $\frac{1}{4}\pi < \arg z < \frac{3}{4}\pi$. In this overlap region the two expansions differ by *subdominant* terms, so they are completely consistent with each other. For $\frac{3}{4}\pi < \arg z < \frac{5}{4}\pi$, (3.8.22) is not valid but (3.8.24) shows that

$$D_\nu(z) \sim -\frac{(2\pi)^{1/2}}{\Gamma(-\nu)}e^{i\nu\pi}z^{-\nu-1}e^{z^2/4}\left[1 + \frac{(\nu+1)(\nu+2)}{2z^2} + \frac{(\nu+1)(\nu+2)(\nu+3)(\nu+4)}{2\cdot 4z^4} + \cdots\right],$$
$$z \to \infty;\ \tfrac{3}{4}\pi < \arg z < \tfrac{5}{4}\pi. \tag{3.8.25}$$

The functional relation (3.8.23) can be applied again to find an asymptotic expansion of $D_\nu(z)$ valid for $\frac{3}{4}\pi < \arg z < \frac{7}{4}\pi$ because $z_1 = e^{-i\pi}z$ satisfies $-\frac{1}{4}\pi < \arg z_1 < \frac{3}{4}\pi$ so (3.8.22) applies, while $z_2 = e^{-i\pi/2}z$ satisfies $\frac{1}{4}\pi < \arg z_2 < \frac{5}{4}\pi$ so (3.8.24) applies. We find that

$$\begin{aligned} D_\nu(z) \sim{}& e^{-2i\nu\pi}z^\nu e^{-z^2/4}\left[1 - \frac{\nu(\nu-1)}{2z^2} + \frac{\nu(\nu-1)(\nu-2)(\nu-3)}{2\cdot 4z^4} - \cdots\right] \\ &- \frac{(2\pi)^{1/2}}{\Gamma(-\nu)}e^{i\nu\pi}z^{-\nu-1}e^{z^2/4}\left[1 + \frac{(\nu+1)(\nu+2)}{2z^2} + \frac{(\nu+1)(\nu+2)(\nu+3)(\nu+4)}{2\cdot 4z^4} + \cdots\right], \end{aligned}$$
$$z \to \infty;\ \tfrac{3}{4}\pi < \arg z < \tfrac{7}{4}\pi.$$

Repeating this argument once more gives

$$D_\nu(z) \sim e^{-2i\nu\pi}z^\nu e^{-z^2/4}\left[1 - \frac{\nu(\nu-1)}{2z^2} + \frac{\nu(\nu-1)(\nu-2)(\nu-3)}{2\cdot 4z^4} - \cdots\right], \qquad z \to \infty;\ \tfrac{5}{4}\pi < \arg z < \tfrac{9}{4}\pi.$$

Notice that the asymptotic expansions of $D_\nu(z)$ and $D_\nu(ze^{2\pi i})$ are identical, as they must be, because $D_\nu(z)$ is a single-valued function. The branch cut in the asymptotic expansion (3.8.22) when ν is nonintegral is only apparent and disappears when the expansions are properly continued through Stokes lines.

Example 9 *Eigenvalues of the parabolic cylinder equation.* The boundary-value problem $y'' + (\nu + \frac{1}{2} - \frac{1}{4}x^2)y = 0$ $(-\infty < x < \infty)$, $y \to 0$ as $|x| \to \infty$, is an eigenvalue problem for the parameter ν. (In physics this is the quantum harmonic oscillator problem.) In order for nonzero solutions to exist, it is necessary that ν assume special values. We can solve this eigenvalue problem using the asymptotic series (3.8.22) and (3.8.24).

First we note that the most general solution to the parabolic cylinder differential equation of index ν may be written as $y(x) = c_1 D_\nu(x) + c_2 D_{-\nu-1}(-ix)$ because $D_\nu(x)$ and $D_{-\nu-1}(-ix)$ are linearly independent. However, (3.8.22) shows that $D_{-\nu-1}(-ix)$ grows exponentially as $x \to +\infty$ for all values of ν. Thus, $c_2 = 0$. According to (3.8.22), $D_\nu(x) \to 0$ as $x \to +\infty$ so $y(x) = c_1 D_\nu(x)$ is the most general solution satisfying the boundary condition at $+\infty$.

Next, we observe that (3.8.24) implies that $D_\nu(x)$ grows exponentially as $x \to -\infty$ unless $1/\Gamma(-\nu)$, the coefficient of the growing component, is 0. However, $\Gamma(-\nu) = \infty$ only if ν is a nonnegative integer $\nu = 0, 1, 2, 3, \ldots$. Thus, the eigenvalue spectrum of the parabolic cylinder equation consists of the nonnegative integers.

An explicit demonstration that $\nu = n$ is an eigenvalue was given in Example 4 of Sec. 3.5, where we showed that the eigenfunctions have the form $D_n(x) = e^{-x^2/4}\,\mathrm{He}_n(x)$ where $\mathrm{He}_n(x)$ is the nth-degree Hermite polynomial. Thus, $D_n(x) \to 0$ as $|x| \to \infty$ as required. The argument of the present example shows that there are no other eigenvalues.

Until now we have assumed without proof that the functions Ai (z) and $D_\nu(z)$, which are the subdominant solutions to the Airy and parabolic cylinder equations for large positive z, have asymptotic expansions which are valid beyond the Stokes lines nearest the positive real axis. Specifically we have assumed that the expansion (3.8.14) with (3.8.15) for Ai (z) is valid for $|\arg z| < \pi$, and not just for $|\arg z| < \frac{1}{3}\pi$, and that the expansion (3.8.22) for $D_\nu(z)$ is valid for $|\arg z| < \frac{3}{4}\pi$, and not just for $|\arg z| < \frac{1}{4}\pi$. In the next two examples we argue that these assumptions are correct.

Our approach will be a generalization of that used earlier in this section to show that $K_\nu(x)$ has an asymptotic expansion as $x \to +\infty$. We will construct integral equations satisfied by Ai (z) and $D_\nu(z)$ and use these equations to establish the regions of validity of the asymptotic expansions. Some of the technical questions that arise are reserved for exercises, but our presentation will make clear the overall plan of attack.

Example 10 *Asymptotic expansion of* Ai (z) *for* $|\arg z| < \pi$. The most direct way to establish the Stokes behavior of Ai (z) is to utilize an integral representation (see Prob. 6.75). The phenomenon we are examining is also exhibited by functions for which integral representations are not known. The method of integral equations used here is more technical, but because it follows directly from the differential equations it is far more general.

The Airy function Ai (z) satisfies the integral equation (see Prob. 3.62)

$$y(z) = \frac{1}{2}\pi^{-1/2}z^{-1/4}e^{-2z^{3/2}/3}\left[1 + \int_\infty^z K(z, t)y(t)\,dt\right], \tag{3.8.26}$$

where $K(z, t) = \frac{5}{32}t^{-9/4}[\exp(\frac{2}{3}t^{3/2}) - \exp(\frac{4}{3}z^{3/2} - \frac{2}{3}t^{3/2})]$. The point at ∞ in (3.8.26) is real and positive, and restrictions on the complex path of integration in (3.8.26) connecting ∞ and z are given below.

The crucial step is demonstrating that there is a solution to (3.8.26) satisfying

$$|y(z)| < M|z|^{-1/4} \exp\left(-\tfrac{2}{3} \operatorname{Re} z^{3/2}\right), \qquad |\arg z| < \pi, \tag{3.8.27}$$

for all sufficiently large $|z|$ and some constant M (see Prob. 3.62). The key fact in this demonstration is that if $|\arg z| < \pi$, then there is a complex path of integration Γ from $t = +\infty$ to $t = z$ on which $\operatorname{Re} t^{3/2} \geq \operatorname{Re} z^{3/2}$ and on which t does not pass close to the origin (see Fig. 3.22). On such a path

$$|K(z, t)| \leq \tfrac{5}{16}|t|^{-9/4} \exp\left(\tfrac{2}{3} \operatorname{Re} t^{3/2}\right). \tag{3.8.28}$$

With these facts established it is straightforward to verify (3.8.15). Using (3.8.26) to (3.8.28), taking $|\arg z| < \pi$, and assuming $|z|$ to be so large that (3.8.27) holds on the path Γ, we have

$$\left|\frac{y(z)}{\frac{1}{2}\pi^{-1/2}z^{-1/4}e^{-2z^{3/2}/3}} - 1\right| \leq \int_\Gamma |K(z, t)|\,|y(t)|\,|dt|$$

$$\leq \tfrac{5}{16}M \int_\Gamma |t|^{-5/2}\,|dt| = \tfrac{5}{12}Mr^{-3/2},$$

where r is the smallest value of $|t|$ on the path Γ. It is not hard to show that there is an acceptable path Γ for which $r = |z|$ if $|\arg z| \leq 2\pi/3$ and $r = |\operatorname{Re} z^{3/2}|^{2/3}$ if $2\pi/3 < |\arg z| < \pi$ (see Prob. 3.62). This proves that

$$y(z) \sim (1/2\sqrt{\pi})z^{-1/4}e^{-2z^{3/2}/3}, \qquad z \to \infty;\ |\arg z| < \pi.$$

Continuing the argument as in Example 7 we generate the full asymptotic expansion (3.8.14) with (3.8.15) when $|\arg z| < \pi$ (see Prob. 3.62).

Example 11 *Asymptotic expansion of $D_\nu(z)$ for $|\arg z| < \frac{3}{4}\pi$.* The parabolic cylinder function $D_\nu(z)$ satisfies the integral equation (see Prob. 3.63)

$$y(z) = z^\nu e^{-z^2/4}\left[1 + \int_\infty^z K(z, t)y(t)\,dt\right], \tag{3.8.29}$$

where

$$K(z, t) = \nu(\nu - 1)t^{\nu-2}e^{-t^2/4}\int_z^t s^{-2\nu}e^{s^2/2}\,ds, \tag{3.8.30}$$

and the point at ∞ is real and positive. The complex paths in (3.8.29) and (3.8.30) are described below.

Again, the crucial step in the proof of (3.8.22) is showing that there is a solution to (3.8.29) satisfying

$$|y(z)| < M|z^\nu| \exp\left(-\tfrac{1}{4} \operatorname{Re} z^2\right), \qquad |\arg z| < \tfrac{3}{4}\pi, \tag{3.8.31}$$

for all sufficiently large $|z|$ and some constant M (see Prob. 3.63). The key facts for this proof are:

(*a*) If $|\arg z| < 3\pi/4$, then there is a complex path of integration Γ from $t = +\infty$ to $t = z$ on which $\operatorname{Re} t^2 \geq \operatorname{Re} z^2$ and on which t does not pass close to the origin (see Fig. 3.23).

(*b*) There is a constant C such that

$$|K(z, t)| < C\left[|t^{-\nu-3}| \exp\left(\frac{1}{4}\operatorname{Re} t^2\right) + \left|\frac{t^{\nu-2}}{z^{2\nu+1}}\right| \exp\left(\frac{1}{2}\operatorname{Re} z^2 - \frac{1}{4}\operatorname{Re} t^2\right)\right] \tag{3.8.32}$$

for all sufficiently large $|t|$ and $|z|$.

To complete the demonstration of (3.8.22) we proceed as in Examples 7 and 10. From (3.8.31) and (3.8.32), we obtain

$$\left|\frac{y(z)}{z^\nu e^{-z^2/4}} - 1\right| \leq \int_\Gamma |K(z, t)|\,|y(t)|\,|dt| \leq C_1 r^{-2},$$

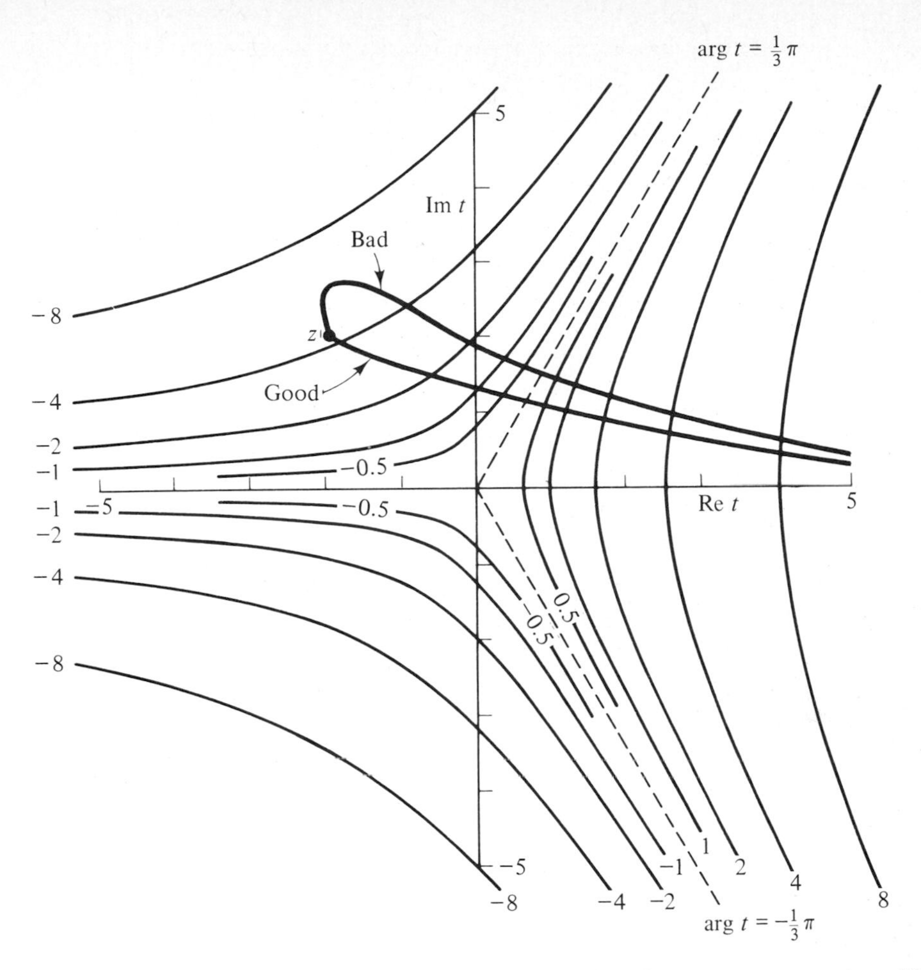

Figure 3.22 A graph of the complex-t plane for $|\text{Re } t| \le 5$, $|\text{Im } t| \le 5$. The contours of $\text{Re } t^{3/2} = c$ are shown for $c = \pm 0.5, \pm 1, \pm 2, \pm 4, \pm 8$ and are labeled by the value of c. This figure may be used as a guide for determining allowable contours of integration in (3.8.26); on an allowable contour $\text{Re } t^{3/2} \ge \text{Re } z^{3/2}$. Two contours are indicated: on the "good" contour $\text{Re } t^{3/2} \ge \text{Re } z^{3/2}$, while on the "bad" contour this condition is not satisfied.

where r is the smallest value of $|t|$ on the allowable path Γ and C_1 is a constant. It is not hard to show that there is an allowable path Γ for which $r = |z|$ if $|\arg z| \le \pi/2$ and $r = |\text{Re } z^2|^{1/2}$ if $\pi/2 < |\arg z| < 3\pi/4$ (see Prob. 3.63). This proves that

$$y(z) \sim z^{\nu} e^{-z^2/4}, \qquad z \to \infty; \ |\arg z| < \tfrac{3}{4}\pi.$$

Continuing in this way we may prove the validity of the full series (3.8.22) for $|\arg z| < 3\pi/4$ (see Prob. 3.63).

The techniques used in Examples 10 and 11 to establish the validity of asymptotic relations in the complex plane are admittedly complicated. It is easier to use integral representations of Ai (z) and $D_\nu(z)$ to get the same results (see Probs. 6.75

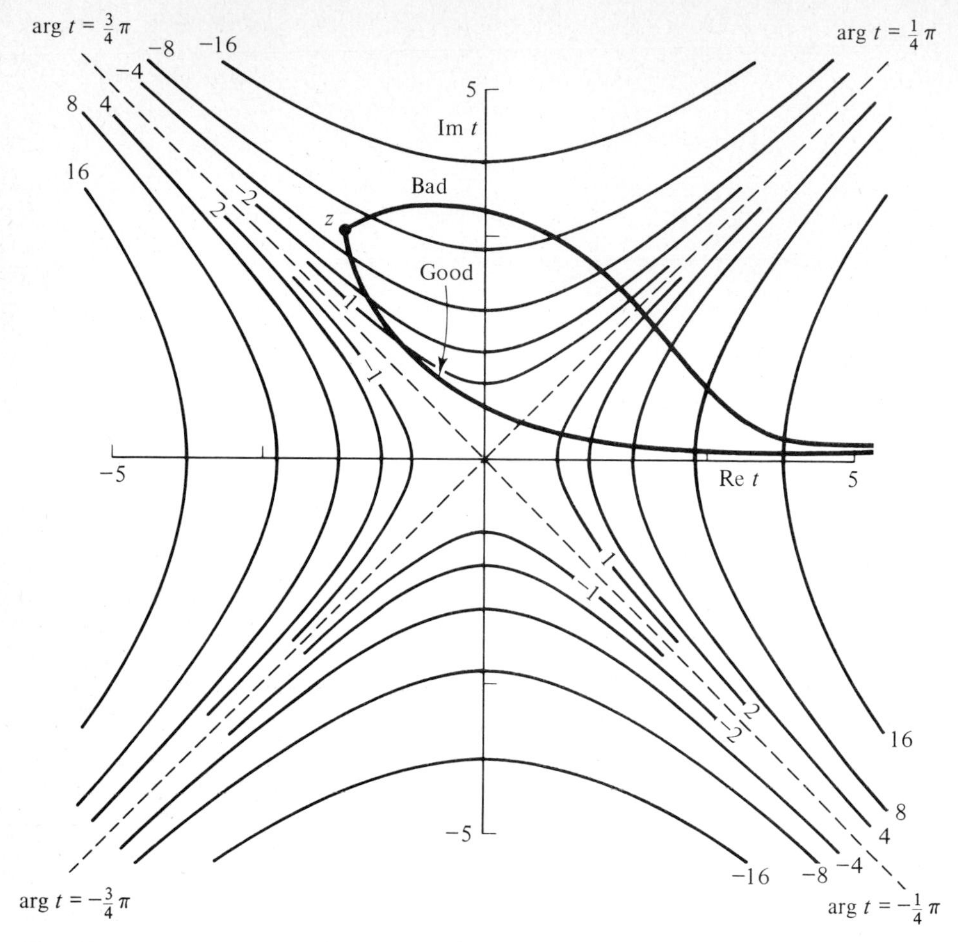

Figure 3.23 A graph of the complex-t plane for $|\text{Re } t| \leq 5$, $|\text{Im } t| \leq 5$. The contours of $\text{Re } t^2 = c$ are shown for $c = \pm 1, \pm 2, \pm 4, \pm 8, \pm 16$ and are labeled by the value of c. This figure may be used as a guide for determining allowable contours of integration in (3.8.29); on an allowable contour $\text{Re } t^2 \geq \text{Re } z^2$. "Good" and "bad" contours are shown which meet and do not meet these conditions.

and 6.84). However, we chose the present more-difficult route because the methods used here are very general. All linear differential equations can be converted to integral equations, while integral representations are only known for a small class of special functions.

PROBLEMS FOR CHAPTER 3

Section 3.1

(E) **3.1** We have used the transformation $x = 1/t$ to classify $x = \infty$ as an ordinary, regular singular, or irregular singular point. Would the transformations $x = t^{-n}$, $x = t^{-\alpha}$ $(\alpha > 0)$, $x = 1/\sinh t$ also work? Is there any advantage to $x = 1/t$?

(D) **3.2** It is not possible to solve the differential equation $(x^2 + 2)y'' + [(3x^2 - 4)/x]y' + (2 - 4x^2)y = 0$ in closed form. However, it is known that the general solution is analytic at $x = 0$. Verify this by finding an equivalent system of linear differential equations of the form (3.1.13) which is nonsingular at $x = 0$.

(E) **3.3** Classify all the singular points (finite and infinite) of the following differential equations:

(*a*) $y'' = xy$ (Airy equation);
(*b*) $x^2y'' + xy' + (x^2 - \nu^2)y = 0$ (Bessel equation);
(*c*) $x(1 - x)y'' + [c - (a + b + 1)x]y' - aby = 0$ (hypergeometric equation);
(*d*) $xy'' + (b - x)y' - ay = 0$ (Kummer's confluent hypergeometric equation);
(*e*) $x^2y'' + (\frac{1}{4} - \mu^2 + \kappa x - x^2/4)y = 0$ (Whittaker's confluent hypergeometric equation);
(*f*) $y'' + (\nu + \frac{1}{2} - \frac{1}{4}x^2)y = 0$ (Parabolic cylinder equation);
(*g*) $y'' + (h - 2\theta \cos 2x)y = 0$ (Mathieu equation);
(*h*) $(1 - x^2)y'' - 2xy' + [\lambda + 4\theta(1 - x^2) - \mu^2(1 - x^2)^{-1}]y = 0$ (spheroidal wave equation).

(E) **3.4** Classify the points at 0 and ∞ of the following differential equations:

(*a*) $x^7\, d^4y/dx^4 = y'$;
(*b*) $x^3y''' = y$;
(*c*) $y''' = x^3y$;
(*d*) $x^2y'' = e^{1/x}y$;
(*e*) $(\tan x)y' = y$;
(*f*) $y'' = (\ln x)y$.

Section 3.2

(E) **3.5** Find the Taylor expansion about $x = 0$ of the solution to the initial-value problem $(x - 1) \times (x - 2)y'' + (4x - 6)y' + 2y = 0$ $[y(0) = 2,\ y'(0) = 1]$. Where might one expect the series to converge? For which x does it actually converge?

(E) **3.6** Find the Taylor series about 0 of the solution to the initial-value problems:

(*a*) $y'' - 2xy' + 8y = 0$ $[y(0) = 4,\ y'(0) = 0]$;
(*b*) $y'' - 2xy' + 8y = 0$ $[y(0) = 0,\ y'(0) = 4]$;
(*c*) $(1 - x^2)y'' - 2xy' + 12y = 0$ $[y(0) = 0,\ y'(0) = 3]$;
(*d*) $y'' = (x - 1)y$ $[y(0) = 1,\ y'(0) = 0]$.

(E) **3.7** Estimate the number of terms in the Taylor series (3.2.1) and (3.2.2) that are necessary to compute Ai (x) and Bi (x) correct to three decimal places at $x = \pm 1, \pm 100, \pm 10{,}000$.

(E) **3.8** How many terms in the Taylor series solution to $y''' = x^3y$ $[y(0) = 1,\ y'(0) = y''(0) = 0]$ are needed to evaluate $\int_0^1 y(x)\, dx$ correct to three decimal places?

(TD) **3.9** Show that if $p_0(x), p_1(x), \ldots, p_{n-1}(x)$ are all analytic in the complex disk $|x| < R$, then any solution $y(x)$ to (3.1.1) is analytic for $|x| < R$.

Clues: (*a*) Use Cauchy's integral formula to show that the coefficients of the Taylor series expansions $p_i(x) = \sum_{m=0}^{\infty} p_{i,m} x^m$ $(i = 0, \ldots, n - 1)$ satisfy $|p_{im}| \le c$ for all i and m, $|p_{im}| \le R_1^{-m}$ for all i and $m \ge N$, for any $R_1 < R$; c and N are constants that depend on R_1.

(*b*) Show that the coefficients a_m of the Taylor series $y(x) = \sum_{m=0}^{\infty} a_m x^m$ satisfy the recursion relation $m^n a_m = -\sum_{k=0}^{m-1} a_k[k^{n-1}p_{n-1,m-k} + k^{n-2}p_{n-2,m-k} + \cdots + p_{0,m-k}]$.

(*c*) Use the above estimates to show by induction that $|a_m| \le A/R_1^m$ for all m, where A may depend on $R_1 < R$. Use this result to show that the Taylor series converges for all $|x| < R$.

(I) **3.10** Show that the power series representation for the series $\sum_{n=0}^{\infty} [x^n(x - 1)^{2n}]/n!$ cannot have three consecutive zero coefficients (Putnam Exam 1972).

(I) **3.11** If
$$u = 1 + \frac{x^3}{3!} + \frac{x^6}{6!} + \cdots,$$
$$v = \frac{x}{1!} + \frac{x^4}{4!} + \frac{x^7}{7!} + \cdots,$$
$$w = \frac{x^2}{2!} + \frac{x^5}{5!} + \frac{x^8}{8!} + \cdots,$$

prove that
$$u^3 + v^3 + w^3 - 3uvw = 1$$
(Putnam Exam 1939).

(I) **3.12** Show that
$$\frac{\dfrac{x}{1} + \dfrac{x^3}{1 \cdot 3} + \dfrac{x^5}{1 \cdot 3 \cdot 5} + \dfrac{x^7}{1 \cdot 3 \cdot 5 \cdot 7} + \cdots}{1 + \dfrac{x^2}{2} + \dfrac{x^4}{2 \cdot 4} + \dfrac{x^6}{2 \cdot 4 \cdot 6} + \cdots} = \int_0^x e^{-t^2/2}\, dt$$
(Putnam Exam 1950).

Section 3.3

(I) **3.13** Use the method of Sec. 3.3 to find series expansions of all solutions to the fourth-order differential equation $x^3\, d^4y/dx^4 = y$ about the regular singular point $x = 0$. (Wu and Pwu discovered that the general solution to this very difficult equation can be expressed as integrals of products of Bessel functions.)

(TE) **3.14** Show that if in the neighborhood of x_0 two linearly independent solutions to a second-order linear homogeneous differential equation are (*a*) two Frobenius series, or (*b*) a Frobenius series and a second solution of the form discovered under case II(*a*) or case II(*b*), then x_0 is a regular singular point of the differential equation.

Clue: Reconstruct the differential equation from the solutions using the method of Prob. 1.6.

(TD) **3.15** Develop the Fuchs-Frobenius theory for third-order differential equations in the neighborhood of a regular singular point x_0.

(*a*) Outline the form of the solutions for all values of the indices α_1, α_2, α_3. Enumerate cases according to whether $\alpha_1 - \alpha_2$ is a nonnegative integer, $\alpha_1 - \alpha_3$ is a nonnegative integer, and so on.

(*b*) Explain how to obtain series solutions for each of the above cases by differentiating, if necessary, with respect to the index α.

(*c*) Conversely, show that if all the solutions of a third-order equation have series expansions about x_0 as enumerated in the above cases, then x_0 is a regular singular point.

Clue: Use the method of the previous problem.

(TI) **3.16** (*a*) A singular point x_0 of a homogeneous linear differential equation is said to be *isolated* if the coefficient functions are singular at x_0 and are single-valued analytic functions in a punctured disk about x_0. Suppose it is known that (i) 0 is an isolated singular point of a second-order homogeneous linear differential equation and (ii) the linearly independent solutions have the form $y_1(x) = x^\alpha[A(x) + B(x) \ln x]$, $y_2(x) = x^\beta[C(x) + D(x) \ln x]$, where $A(x)$, $B(x)$, $C(x)$, and $D(x)$ are analytic at $x = 0$. Show that there is a nontrivial solution of the differential equation without logarithms: $y_3(x) = x^\alpha E(x)$, where $E(x)$ is analytic at 0.

Clue: Analytically continue the differential equation and its solutions $y_1(x)$ and $y_2(x)$ along a small closed curve encircling $x = 0$. Since the singularity at 0 is isolated, the differential equation recovers its original form after a full cycle. Therefore, $y_1(xe^{2\pi i})$ and $y_2(xe^{2\pi i})$ may be written as linear combinations of $y_1(x)$ and $y_2(x)$. Also show that if $B(x)$ or $D(x) \neq 0$, then $\alpha - \beta$ is an integer and $B(x)$ and $x^{\beta-\alpha}D(x)$ are constant multiples of $E(x)$.

(*b*) Explain how to generalize part (*a*) to *n*th-order equations.

(TI) **3.17** (*a*) Show that if the hypotheses (i) and (ii) of the preceding problem hold, then $x = 0$ is a regular singular point.

Clue: Suppose the differential equation is $y'' + p(x)y' + q(x)y = 0$. Use reduction of order. Set $y(x) = y_3(x)w(x)$, where $y_3 = x^\alpha E(x)$ is the solution found in Prob. 3.16. Show that w satisfies $p(x) = -2y_3'/y_3 - w''/w'$. Thus, show that $p(x)$ has at worst a first-order pole at $x = 0$. Finally, using $q(x) = -(1/y_3)[y_3'' + p(x)y_3']$, show that $q(x)$ has at worst a second-order pole at $x = 0$.

(*b*) Using induction on n, generalize the result of part (*a*) to *n*th-order homogeneous linear differential equations. Thus, if x_0 is an *isolated* singular point and there are n linearly independent solutions of the form $y = x^\alpha[A_1(x) + (\ln x)A_2(x) + \cdots + (\ln x)^{n-1}A_n(x)]$, where $A_1(x)$, $A_2(x)$, ..., $A_n(x)$ are analytic at $x = 0$, then $x = 0$ is a regular singular point of the differential equation.

(I) **3.18** Two solutions of a second-order homogeneous linear differential equation are 1 and $x \ln x$. What differential equation has these solutions? Show that $x = 0$ is not a regular singular point. Explain in what way these solutions are not of the form (3.1.2) to (3.1.5). Also, discuss why this problem is not a counterexample to Prob. 3.17(*a*).

(TE) **3.19** Can a homogeneous linear differential equation have only one singular point which is a regular singular point?

(TI) **3.20** What is the general form of an nth-order differential equation that is known to have just two regular singular points and no other singular points?

(TD) **3.21** Show that the general second-order equation with three regular singular points at 0, 1, and ∞ and no other singular points can be transformed into the hypergeometric equation $x(x-1)y'' + [(a+b+1)x - c]y' + aby = 0$, where a, b, and c are constants.

(I) **3.22** (*a*) One solution to $x^3y'' = y$ has a Taylor series at ∞. Find the series. If $y(\infty) = 1$, find $y(1)$ to three decimal places. Can you find the exact solution to this equation in terms of Bessel functions?

(*b*) The leading behavior of a particular solution to $x^3y'' = y$ is $y(x) \sim x$ $(x \to +\infty)$. What is the next largest term in the expansion of $y(x)$ for large positive x?

(TD) **3.23** Modify the argument used in Prob. 3.9 to show that the radius of convergence of the Frobenius series (3.3.1) is at least as large as the distance to the nearest complex singularity of the coefficient functions.

3.24 Find series expansions of all the solutions to the following differential equations about $x = 0$. Try to sum in closed form any infinite series that appear.

(I) (*a*) $x(x+2)y'' + 2(x+1)y' - 2y = 0$;
(I) (*b*) $xy'' + y = 0$;
(E) (*c*) $y'' + (e^x - 1)y = 0$;
(I) (*d*) $x(1-x)y'' - 3xy' - y = 0$;
(E) (*e*) $2xy'' - y' + x^2y = 0$;
(I) (*f*) $(\sin x)y'' - 2(\cos x)y' - (\sin x)y = 0$;
(E) (*g*) $y'' - x^2y = 0$;
(E) (*h*) $x(x+2)y'' + (x+1)y' - 4y = 0$;
(E) (*i*) $xy'' + (\frac{1}{2} - x)y' - y = 0$.

(E) **3.25** Show that all solutions of the modified Bessel equation (3.3.5) with $\nu = \frac{1}{2}, \frac{3}{2}, \frac{5}{2}, \ldots$ can be expanded in Frobenius series (without logarithmic terms).

(I) **3.26** Perform a local analysis of solutions to $(x-1)y'' - xy' + y = 0$ at $x = 1$. Use the results of this analysis to prove that a Taylor series expansion of any solution about $x = 0$ has an infinite radius of convergence.

Section 3.4

(I) **3.27** Derive (3.4.28).

(E) **3.28** (*a*) Show that if $f(x) \sim a(x - x_0)^{-b}$ as $x \to x_0+$, then

(i) $\int^x f\,dx \sim [a/(1-b)](x-x_0)^{1-b}$ $(x \to x_0+)$ if $b > 1$ and the path of integration does not pass through x_0;

(ii) $\int^x f\,dx \sim c$ $(x \to x_0+)$, where c is a constant if $b < 1$; if $c = 0$, then $\int^x f\,dx \sim [a/(1-b)] \times (x-x_0)^{1-b}$ $(x \to x_0+)$;

(iii) $\int_{x_0}^x f\,dx \sim [a/(1-b)](x-x_0)^{1-b}$ $(x \to x_0+)$ if $b < 1$;

(iv) $\int^x f\,dx \sim a \ln (x - x_0)$ $(x \to x_0+)$ if $b = 1$.

Clue: Use l'Hôpital's rule.

(*b*) Show that if $f(x) \sim g(x)$ as $x \to x_0$ and $g(x)$ is one-signed in a neighborhood of x_0, then $\int^x f(x)\,dx \sim \int^x g(x)\,dx + c$ $(x \to x_0)$, where c is some integration constant.

(E) **3.29** Repeat the analysis of Example 2 of Sec. 3.4 to derive the leading behavior of the solution to (3.4.1) whose controlling factor is $e^{-2x^{-1/2}}$ as $x \to 0+$.

(I) **3.30** Referring to (3.4.20), show that if $\varepsilon'' - 2x^{-3/2}\varepsilon' \sim 3x^{-2}/16$ $(x \to 0+)$, then it is not possible for all three terms to be of comparable magnitude as $x \to 0+$.

Clue: Express the solution in terms of integrals and use integration by parts (see Sec. 6.3 if necessary).

(E) **3.31** Show that (3.4.25) can be approximated by (3.4.26) as $x \to 0+$.

(D) **3.32** The differential equation $y'' + x^{-2}e^{1/x} \sin (e^{1/x})y = 0$ has an irregular singular point at $x = 0$. Show that if we make the exponential substitution $y = e^S$, it is *not* correct to assume $S'' \ll (S')^2$ as $x \to 0+$. What is the leading behavior of $y(x)$?

3.33 Find the leading behaviors as $x \to 0+$ of the following equations:

(E) (*a*) $x^4 y''' = y$;
(I) (*b*) $x^4 y''' - 3x^2 y' + 2y = 0$;
(I) (*c*) $y'' = \sqrt{x}\, y$;
(I) (*d*) $x^6 y'' = e^x y$;
(I) (*e*) $x^5 y''' - 2xy' + y = 0$;
(I) (*f*) $x^4 y'' - x^2 y' + \frac{1}{4} y = 0$;
(E) (*g*) $y'' = (\cot x)^4 y$.

(E) **3.34** (*a*) Give an example of an asymptotic relation $f(x) \sim g(x)$ $(x \to \infty)$ that cannot be exponentiated; that is, $e^{f(x)} \sim e^{g(x)}$ $(x \to \infty)$ is false.

(*b*) Show that if $f(x) - g(x) \ll 1$ $(x \to \infty)$, then $e^{f(x)} \sim e^{g(x)}$ $(x \to \infty)$.

(I) **3.35** Obtain the full asymptotic behaviors for small x of solutions to the equation $x^2 y'' + (2x + 1)y' - x^2[e^{2/x} + 1]y = 0$.

Section 3.5

(E) **3.36** Consider the function y defined by the differential equation $y'' = (x^3 + ax)y$ and the initial conditions $y(0) = 1$, $y'(0) = 0$. Show that the zeros of y are bounded above but unbounded below (Putnam Exam 1955).

(E) **3.37** Show that all solutions of the differential equation $y'' + e^x y = 0$ remain bounded as $x \to +\infty$ (Putnam Exam 1966).

(E) **3.38** (*a*) Show that the leading behaviors of the solutions to the modified Bessel equation of order ν are given by (3.5.7).

(*b*) Show that the leading behaviors of the solutions to the parabolic cylinder equation (3.5.11) are given by (3.5.12).

3.39 Find the leading asymptotic behaviors as $x \to +\infty$ of the following equations:

(E) (*a*) $x^3\, d^5y/dx^5 = y$;
(E) (*b*) $xy''' = y'$;
(I) (*c*) $d^6y/dx^6 = x\, d^4y/dx^4 + x^2 y'' + x^3 y$;
(E) (*d*) $y'' = \sqrt{x}\, y$;
(E) (*e*) $y'' = (\cosh x)y'$;
(I) (*f*) $y''' = x^{1/3} y + x^{2/3} y' + xy''$;
(E) (*g*) $y'' = (\ln x)^2 y$;
(E) (*h*) $y'' = e^{-3/x} y$;
(E) (*i*) $x^2 y'' + (1 + 3x)y' + y = 0$ [see (3.4.2)].

(E) **3.40** Show that the leading behaviors of solutions to $y'' = (x + x^\alpha)y$ as $x \to +\infty$ are the *same* as the leading behaviors of solutions to the Airy equation $y'' = xy$ as $x \to +\infty$ if Re $\alpha < -\frac{1}{2}$.

(D) **3.41** Find the leading asymptotic behaviors as $x \to +\infty$ of the solutions to the coupled pair of differential equations $x^2 y''(x) = (x + 1)y(x) + xz(x)$, $x^2 z''(x) = -(x + 1)z(x) - xy(x)$. For this particular system of equations it is incorrect to approximate $(x + 1)$ by x, even when x is large. Why?

3.42 Extend the investigation of Example 1 of Sec. 3.5 in two ways:

(I) (*a*) Obtain the next few corrections to the leading behavior (3.5.5). Then see how including these terms improves the numerical approximation to $y(x)$ in (3.5.1).

(D) (*b*) Find the leading behavior of $y(x) = \sum_{n=0}^{\infty} x^n/(n!)^j$ as $x \to +\infty$ for $j = 3, 4, 5, \ldots$.

(I) **3.43** Using the results of Prob. 5.26 that the optimal asymptotic approximation to $D_\nu(x)$ is accurate for x larger than roughly $\frac{1}{2}\nu^{1/2}$, show that the optimal asymptotic approximation to $D_\nu(x)$ gives an accurate approximation to an asymptotically finite fraction of the real zeros of $D_\nu(x)$ as $\nu \to +\infty$ (see Fig. 3.8). Using crude estimates show that roughly $1 - 1/\pi \doteq 0.682$ of the real positive zeros are given accurately as $\nu \to \infty$.

Clue: See Prob. 6.85.

(I) **3.44** The differential equation $x^3y'' - (2x^3 - x^2)y' + (x^3 - x^2 - 1)y = 0$ has an irregular singular point at ∞. Find the leading behaviors of the solutions as $x \to +\infty$. Explain the appearance of a logarithm in one of these leading behaviors.

3.45 A good way to ascertain the asymptotic behavior of certain integrals is to find differential equations that they satisfy and then to perform a local analysis of the differential equation. Use this technique to study the behavior of the following integrals:

(E) (*a*) $y(x) = \int_0^x e^{t^2}\, dt$ as $x \to +\infty$;

(D) (*b*) $y(x) = \int_0^\infty e^{-xt-1/t}\, dt$ as $x \to 0+$ and $x \to +\infty$;

(D) (*c*) $y(x) = \int_0^\infty e^{-xt-1/t^2}\, dt$ as $x \to 0+$ and $x \to +\infty$.

(I) **3.46** (*a*) What is the leading behavior of solutions to $y'' + x^{-3/2}y' - x^{-2}y = 0$ as $x \to +\infty$? Show that it is inconsistent to assume that $S'' \ll S'^2$ $(x \to +\infty)$. However, show that the approximate equation $S'' + (S')^2 \sim x^{-2}$ $(x \to +\infty)$ can be solved exactly by assuming a solution of the form $S' = c/x$.

(*b*) Analyze the leading behavior of solutions to $y'' + y'/(x \ln x) = y/x^2$ for large positive x. Show that it is not valid to assume that $S'' \ll (S')^2$ $(x \to +\infty)$. Does it help to make the substitution $t = \ln x$?

Section 3.6

(E) **3.47** Show that

$$x + \frac{2}{3}x^3 + \left(\frac{2}{3}\right)\left(\frac{4}{5}\right)x^5 + \left(\frac{2}{3}\right)\left(\frac{4}{5}\right)\left(\frac{6}{7}\right)x^7 + \cdots = \frac{\text{arc sin } x}{\sqrt{1 - x^2}}$$

(Putnam Exam 1948).

Clue: What differential equation does this Taylor series satisfy?

(I) **3.48** Find the first three terms in the local behavior as $x \to 0+$ of a particular solution $y(x)$ to

(*a*) $y' + xy = \cos x$;
(*b*) $y' + xy = x^{-3}$;
(*c*) $x^3y'' + y = x^{-4}$;
(*d*) $xy'' - 2y' + y = \cos x$;
(*e*) $x^3y''' - (\cosh x)y = x^{-1}$;
(*f*) $x^3y''' - 6(\cos x)y = x^3$.

(I) **3.49** Find the leading behavior as $x \to +\infty$ of the general solution to each of the following equations:

(*a*) $y'' + x^2y = \sin x$;
(*b*) $y'' + x^3y' + xy = 2x^4e^{-x^2}$;
(*c*) $y'' + xy = x^5$;
(*d*) $y'' + y/x^3 = x$;
(*e*) $d^4y/dx^4 + y/(1 + x^4) = x$;
(*f*) $y''' + x^2y = \sinh x$;
(*g*) $y'' - y/(1 + x) = x$.

(E) **3.50** Find the leading behavior of solutions to $y' - y/x = \cos x$ as $x \to 0+$. Show that the leading behavior is determined by a three-term dominant balance. Compare this leading behavior with the exact solution.

(I) **3.51** (*a*) Derive (3.6.5).

(*b*) Use variation of parameters to show that the most general solution to $y'' = xy - 1$, $y(+\infty) = 0$ is $y(x) = \pi[\text{Bi}(x) \int_x^\infty \text{Ai}(t)\, dt + \text{Ai}(x) \int_0^x \text{Bi}(t)\, dt] + c\, \text{Ai}(x)$, where c is an arbitrary constant.

(*c*) Rederive the leading behavior of $y(x)$ valid as $x \to +\infty$ in (3.6.4) by performing a local expansion of the above integral.

Clue: Replace Ai (t) and Bi (t) by their leading asymptotic behaviors as $x \to +\infty$ wherever possible and integrate by parts (see Sec. 6.3 if necessary).

(E) **3.52** (*a*) Show that the Taylor series solution to $y' + xy = x^3$ in (3.6.1) has an infinite radius of convergence.

(TD) (*b*) Show that, in general, if x_0 is an ordinary point of the homogeneous equation $Ly = 0$ and if x_0 is a point of analyticity of $h(x)$, then the general solution to $Ly = h$ is analytic at x_0 with a radius of convergence at least as large as the distance to the nearest singular point of the homogeneous equation $Ly = 0$ or the nearest singularity of $h(x)$ in the complex plane.

Section 3.7

(TE) **3.53** Suppose $f'(x) \to 0$ as $x \to +\infty$. Does it follow that $f \sim c$ as $x \to +\infty$, where c is a constant?

(E) **3.54** Show that although $[x/(1+x)] \cos x \sim \cos x \quad (x \to \infty)$, it does not follow that $\int_0^x [t/(1+t)] \cos t\, dt \sim \int_0^x \cos t\, dt \; (x \to \infty)$.

(I) **3.55** Find the locations of the Stokes lines as $z \to \infty$ for the following differential equations. Do not confuse branch cuts of the coefficient functions with Stokes lines. The differential equations have single-valued solutions on suitable Riemann surfaces determined by the coefficient functions. The problem is to find the locations of the Stokes lines on such Riemann surfaces.

(*a*) $y'' + (1 + 2z^{-1/2})y + \frac{1}{4}y = 0$;
(*b*) $y'' = z^{1/3}y$;
(*c*) $y''' = -zy$;
(*d*) $y'' = z^8 e^{1/z} y$;
(*e*) $d^4y/dz^4 = z^2 y$.

(D) **3.56** The asymptotic relation $\sinh \sinh z \sim \frac{1}{2} \exp\left(\frac{1}{2}e^z\right)$ $(|z| \to \infty)$ is certainly valid for $z \to \infty$ along the positive real axis. What is the opening angle of the wedge of validity of this relation?

(D) **3.57** Show that if $f(x) \ll 1$ and $f(x) - f(\frac{1}{2}x) \ll x$ as $x \to 0$, then $f(x) \ll x$ as $x \to 0$ (Putnam Exam 1954).

(E) **3.58** Suppose $f(x) \sim x^{-1+i\beta}$ $(x \to +\infty)$, where β is real. Show that the integrated relation $\int^x f(x)\, dx \sim x^{i\beta}/i\beta + c$ $(x \to +\infty)$ is not necessarily valid for any integration constant c.

Clue: Observe that $f(x) + (x \ln x)^{-1} \sim x^{-1+i\beta}$ $(x \to +\infty)$.

(I) **3.59** Show that there are solutions to (3.7.8) and (3.7.9) that approach constants as $x \to -\infty$.

(I) **3.60** Verify (3.7.15).

Section 3.8

(I) **3.61** Using the functional relations (3.8.21) derive asymptotic series expansions for Bi (z) for large $|z|$ valid in the following two sectors:

(*a*) $-\frac{1}{3}\pi < \arg z < \pi$;
(*b*) $-\pi < \arg z < \frac{1}{3}\pi$.

(TE) **3.62** (*a*) Prove that Ai (z) satisfies the integral equation in (3.8.26). The constant $1/2\sqrt{\pi}$ follows from a global analysis of an integral representation of Ai (z) and cannot be determined by the local analysis methods of this chapter.

(TD) (*b*) Verify that there is a solution of this equation satisfying the bound (3.8.27).

Clue: Generalize the approach of Prob. 3.82.

(TI) (*c*) Show that there is an integration contour Γ in the complex-t plane connecting z and $+\infty$ along which $\text{Re}\, t^{3/2} \geq \text{Re}\, z^{3/2}$ and $r = \min_\Gamma |t|$ satisfies $r = |z|$ if $|\arg z| \leq \frac{2}{3}\pi$ and $r = |\text{Re}\, z^{3/2}|^{2/3}$ if $\frac{2}{3}\pi \leq |\arg z| < \pi$.

Clue: If $\frac{2}{3}\pi \leq |\arg z| < \pi$, choose Γ to satisfy $\text{Re}\, t^{3/2} = \text{Re}\, z^{3/2}$ for $\frac{2}{3}\pi < |\arg t| < |\arg z|$.

(TD) (*d*) Using the method of Example 10 of Sec. 3.8, generate the first three terms of the asymptotic expansion of Ai (z) as $z \to \infty$ with $|\arg z| < \pi$ and bound the error.

(TE) **3.63** (*a*) Prove that $D_\nu(z)$ satisfies the integral equation in (3.8.29) and (3.8.30).

(TD) (*b*) Use integration by parts (see Sec. 6.3) to establish the bound (3.8.32) for $K(z, t)$.

(TD) (*c*) Verify that there is a solution of (3.8.29) and (3.8.30) satisfying the bound (3.8.31).

(TI) (*d*) Show that there is an integration contour Γ in the complex-t plane connecting z and $+\infty$

along which $\text{Re}\, t^2 \geq \text{Re}\, z^2$ and $r = \min_\Gamma |t|$ satisfies $r = |z|$ if $|\arg z| \leq \pi/2$ and $r = |\text{Re}\, z^2|^{1/2}$ if $\pi/2 \leq |\arg z| \leq 3\pi/4$.

Clue: If $\pi/2 \leq |\arg z| \leq 3\pi/4$, choose Γ to satisfy $\text{Re}\, t^2 = \text{Re}\, z^2$ for $\pi/2 < |\arg t| < |\arg z|$.

(TD) (*e*) Generate the first three terms of the asymptotic expansion $D_\nu(z)$ for $|\arg z| < 3\pi/4$ and bound the error.

(I) **3.64** Prove that

$$\int_0^\infty \frac{e^{-t}\, dt}{1 + xt^2} \sim \sum_{n=0}^\infty (-1)^n (2n)!\, x^n, \qquad x \to 0+.$$

Problems 3.65 to 3.68 are a sequence of problems that investigate the Stokes phenomenon for Bessel functions.

(D) **3.65** One solution to the modified Bessel equation (3.3.5) is $I_\nu(z)$, which is defined by the Frobenius series (3.3.8). We define a second solution $K_\nu(z)$ to the modified Bessel equation, which is linearly independent of $I_\nu(z)$, by $K_\nu(z) = \pi\{[I_{-\nu}(z) - I_\nu(z)]/(2 \sin \nu\pi)\}$, where the right side of this expression is defined by its limiting value if ν is an integer. Show that this definition of $K_\nu(z)$ is consistent with the series expansions of $K_0(z)$ in (3.3.15) and $K_1(z)$ in (3.3.21).

(TD) **3.66** Show that the asymptotic expansion (3.5.8*b*) and (3.5.9*b*) of the modified Bessel function $K_\nu(z)$, which is

$$K_\nu(z) \sim \left(\frac{\pi}{2z}\right)^{1/2} e^{-z} \left[1 + \frac{4\nu^2 - 1^2}{1!\, 8z} + \frac{(4\nu^2 - 1^2)(4\nu^2 - 3^2)}{2!\, (8z)^2} + \cdots\right], \qquad z \to \infty;\ |\arg z| < \tfrac{3}{2}\pi,$$

is valid beyond the Stokes lines at $|\arg z| = \frac{1}{2}\pi$ and only breaks down at the Stokes lines $|\arg z| = \frac{3}{2}\pi$. Note that $K_\nu(z)$ is the subdominant solution of the modified Bessel equation on the positive real axis. This is why one should expect that, like Ai (z) and $D_\nu(z)$, $K_\nu(z)$ has an asymptotic expansion valid past the Stokes lines nearest the positive real axis. It is interesting that the next Stokes lines are located at $\arg z = \pm\frac{3}{2}\pi$ which does not lie on the principal sheet of the Riemann surface of $K_\nu(z)$. Thus, the asymptotic expansion of $K_\nu(z)$ as $z \to \infty$ is valid on the entire principal sheet together with portions of the two adjacent sheets. Observe that these results do not contradict Prob. 3.83 because $K_\nu(z)$ is *not* a single-valued function of z.

(TI) **3.67** Obtain asymptotic expansions of $I_\nu(z)$ as $z \to \infty$ that are valid for $-\frac{3}{2}\pi < \arg z < \frac{1}{2}\pi$ and $-\frac{1}{2}\pi < \arg z < \frac{3}{2}\pi$. To do this, first use the Frobenius series (3.3.8) to show that $I_\nu(ze^{\pm i\pi}) = e^{\pm i\pi\nu} I_\nu(z)$, and then use Prob. 3.65 to show that

$$I_\nu(z) = \pm\frac{i}{\pi} K_\nu(ze^{\pm i\pi}) \mp \frac{ie^{\pm i\nu\pi}}{\pi} K_\nu(z).$$

(E) **3.68** (*a*) Show that $J_\nu(z)$ defined by

$$J_\nu(z) = \left(\frac{z}{2}\right)^\nu \sum_{n=0}^\infty \frac{(-z^2/4)^n}{n!\, \Gamma(\nu + n + 1)}$$

is a solution to the Bessel equation $z^2 y'' + zy' + (z^2 - \nu^2)y = 0$.

(E) (*b*) Show that $J_\nu(z) = e^{-i\nu\pi/2} I_\nu(ze^{i\pi/2})$.

(D) (*c*) Show that the asymptotic expansions (3.7.14) and (3.7.15) of $J_\nu(z)$ as $z \to \infty$ may be continued off the real-z axis and, in fact, are valid for $z \to \infty$, $|\arg z| < \pi$.

(D) (*d*) Obtain an asymptotic expansion of $J_\nu(z)$ valid for $z \to \infty$, $0 < \arg z < 2\pi$.

(I) **3.69** The subdominant solution as $x \to 0+$ of the differential equation $x^3 y'' = y$ has the asymptotic expansion (3.4.22*b*). What is the largest sector in the complex plane in which this expansion remains valid? Explain how the opening angle of this sector can be greater than 2π.

(E) **3.70** The subdominant solution as $x \to 0+$ of the differential equation $x^2 y'' + (1 + 3x)y' + y = 0$ is given exactly by $y(x) = c_1 x^{-1} e^{-x^{-1}}$. Discuss what happens at the Stokes lines at $\arg x = \pm\frac{1}{2}\pi, \pm\frac{3}{2}\pi$. Is there a Stokes phenomenon?

(D) **3.71** (*a*) Find the locations of the Stokes lines for the hyperairy equation $d^4y/dz^4 = zy$ as $z \to \infty$.
(*b*) The leading behavior of the solution to the hyperairy equation which is subdominant on the

positive real axis is $y \sim cx^{-3/8}e^{-4x^{5/4}/5}$ $(x \to +\infty)$. What is the opening angle of the largest sector in the complex plane in which this asymptotic relation is valid?

(TI) **3.72** Show that if $f(z)$ is analytic for $\alpha \le \arg z \le \beta$ and if $f(z) \sim \sum_{n=0}^{\infty} a_n z^n$ $(z \to 0; \alpha \le \arg z \le \beta)$ with $\alpha < \beta$, then $f'(z) \sim \sum_{n=0}^{\infty} n a_n z^{n-1}$ $(z \to 0; \alpha < \arg z < \beta)$.

Clue: Use the analyticity of $f(z)$ and the Cauchy integral formula to write

$$f(z) = \frac{1}{2\pi i} \oint_C \frac{f(t)}{t - z}\, dt,$$

where C is a circle of radius $K|z|$ which lies entirely within the wedge $\alpha \le \arg z \le \beta$ (K depends only on arg z but not on $|z|$). Differentiate this integral and use the assumed asymptotic expansion of $f(z)$ to uniformly approximate the resulting integral for $f'(z)$.

(TE) **3.73** Consider a function $f(z)$ which satisfies the criteria of Prob. 3.72 and suppose that the sector of validity of the asymptotic expansion $f(z) \sim \sum_{n=0}^{\infty} a_n z^n$ $(z \to 0)$ includes the positive real axis. Now construct a new function $g(z)$: $g(z) = f(z) + e^{-z^{-1}} \sin(e^{z^{-1}})$. It is no longer true that $g'(z) \sim \sum_{n=0}^{\infty} n a_n z^{n-1}$ $(z \to 0)$. Is this a counterexample to the theorem in Prob. 3.72?

(I) **3.74** Show that the integral $\int_0^\infty e^{-t}/(1 + xe^t)\, dt$ does not have an asymptotic expansion of the form $I(x) \sim \sum_{n=0}^{\infty} a_n x^n$ $(x \to 0+)$. What is the full asymptotic behavior of $I(x)$ as $x \to 0+$?

(I) **3.75** Find a function to which the series $\sum_{n=0}^{\infty} (-1)^n x^n \Gamma[(n + 1)/p]$ is asymptotic as $x \to 0+$, where $p > 0$.

(I) **3.76** Use the integral representation $K_0(x) = \int_1^\infty e^{-xt}(t^2 - 1)^{-1/2}\, dt$ to derive the asymptotic series (3.8.8).

Clue: Use the duplication formula and other properties of the gamma function $\Gamma(x)$ to show that $\int_1^\infty (t^2 - 1)^{-1/2} t^{-n-1}\, dt = 2^{n-1}\Gamma(\frac{1}{2}n + \frac{1}{2})^2/n!$.

(D) **3.77** The asymptotic series for the modified Bessel function $K_\nu(x)$ as $x \to +\infty$ is given by (3.5.8*b*) and (3.5.9*b*) with $c_2 = (\pi/2)^{1/2}$. Show that the remainder after the x^{-N} term of this series has the same sign and is smaller than the x^{-N-1} term, provided that $N \ge \nu - \frac{3}{2}$ and $\nu \ge 0$ and $x > 0$.

Clue: Use the integral representation

$$K_\nu(x) = \frac{\pi^{1/2}(x/2)^\nu}{\Gamma(\nu + \frac{1}{2})} \int_1^\infty e^{-xt}(t^2 - 1)^{\nu - 1/2}\, dt$$

and generate the asymptotic series using Taylor's theorem with an error term. (See Prob. 6.38.)

(E) **3.78** Verify the linear functional relations (3.8.21) between Ai (z), Bi (z), Ai (ωz), and Ai $(\omega^2 z)$.

Clue: Use the Taylor series in (3.2.1) and (3.2.2).

(TD) **3.79** Given the arbitrary power series $\sum_{n=0}^{\infty} a_n x^n$, show that there is a function analytic in a sector containing the positive real axis to which this series is asymptotic.

Clue: Modify the construction in Example 2 of Sec. 3.8 by redefining $\phi(x; a)$ as an analytic function.

3.80 We define a general asymptotic expansion as follows. Let $\phi_n(x)$ be a sequence of functions satisfying $\phi_{n+1} \ll \phi_n$ $(x \to x_0)$ for all n. Then we write $y(x) \sim \sum_{n=1}^{\infty} a_n \phi_n(x)$ $(x \to x_0)$ and say that the series is asymptotic to $y(x)$ as $x \to x_0$ if $y(x) - \sum_{n=1}^{N} a_n \phi_n(x) \ll \phi_N(x)$ $(x \to x_0)$ for all N. When $\phi_n(x) = (x - x_0)^{\alpha n}$, the definition of an asymptotic power series as given in Sec. 3.5 is recovered.

(TE) (*a*) Show that if $y \sim \sum a_n \phi_n$ and $z \sim \sum b_n \phi_n$ as $x \to x_0$, then $c_1 y + c_2 z \sim \sum (c_1 a_n + c_2 b_n)\phi_n$ $(x \to x_0)$.

(TI) (*b*) Show that if $\psi_n(x) = \int_{x_0}^x \phi_n(t)\, dt$ exists for each n and if all the functions $\phi_n(x)$ are positive for $x_0 < x < x_1$ for some x_1, then $y \sim \sum a_n \phi_n$ as $x \to x_0$ implies $\int_{x_0}^x y(t)\, dt \sim \sum_{n=1}^{\infty} a_n \psi_n(x)$ $(x \to x_0 +)$.

(TD) (*c*) Generalize the result of Example 2 of Sec. 3.8 by showing that every series $\sum a_n \phi_n(x)$ with finite a_n is asymptotic to some finite function as $x \to x_0$. Show that if each $\phi_n(x)$ is continuous, then there is a continuous function to which the series is asymptotic as $x \to x_0$.

(TI) **3.81** (*a*) Show that if $y(x)$, $p(x)$, $p'(x)$, $q(x)$ are expandable in asymptotic power series as $x \to x_0$ and if y satisfies $y'' + p(x)y' + q(x)y = 0$, then y' and y'' are also expandable in asymptotic series which may be obtained by differentiating the series for y termwise.

(TD) (*b*) Generalize this result to nth-order equations.

(TI) **3.82** Show that the integral equation (3.8.13) has a solution which is bounded for all x larger than some constant a.

Clue: Repeatedly iterate the integral equation to obtain a formal series expansion for the solution whose nth term $I_n(x)$ has the form of the n-fold integral

$$I_n(x) = \left(\frac{1}{2}\lambda\right)^n \int_x^\infty dt_1 \int_{t_1}^\infty dt_2 \cdots \int_{t_{n-1}}^\infty dt_n \frac{K(x, t_1)}{t_1^2} \frac{K(t_1, t_2)}{t_2^2} \cdots \frac{K(t_{n-1}, t_n)}{t_n^2},$$

where $K(x, t) = e^{2(x-t)} - 1$. Show that $|K(x, t)| < 1$ for $t > x$. From this deduce that the series $\sum I_n$ converges absolutely and that

$$\left|\sum_{n=0}^\infty I_n(x)\right| \le \exp\left(\frac{1}{2}|\lambda| x^{-1}\right)$$

for $x > 0$. Argue that since the series converges it is a solution of the integral equation.

(TI) **3.83** Show that if $f(z)$ is single valued and analytic for $0 < |z - z_0| < R$ for some $R > 0$ and if $f(z) \sim \sum_{n=0}^\infty a_n(z - z_0)^n$ $(z \to z_0)$ for all values of arg $(z - z_0)$, then the asymptotic series $\sum a_n(z - z_0)^n$ is *convergent* in a neighborhood of z_0.

Clue: Show that $f(z)$ is analytic at z_0.

(TI) **3.84** (*a*) Show that if $f(x) \sim x^p$ $(x \to +\infty)$ with $p \ge 1$ and $f''(x) > 0$ for sufficiently large x, then $f'(x) \sim px^{p-1}$ $(x \to +\infty)$.

Clue: First show that $f(x + h) - f(x) \ge hf'(x) \ge f(x) - f(x - h)$ for any $x > 0$, $h > 0$ provided that $x - h$ is sufficiently large. Then use the asymptotic property of $f(x)$ to prove the result.

(TD) (*b*) Show that if $p > 1$ the hypotheses $f'' > 0$ of part (*a*) can be weakened to $f''(x) > -cx^{p-2}$ for sufficiently large x, where c is an arbitrary constant.

(D) **3.85** Give an example of a function $f(x)$ satisfying $f(x) - x^p \sim ax^{p-1}$ $(x \to +\infty)$ for $p \ge 1$ and some constant a and $f''(x) > 0$ for all sufficiently large x but such that $f'(x) - px^{p-1} \sim a(p-1)x^{p-2}$ $(x \to +\infty)$ is not true.

CHAPTER

FOUR

APPROXIMATE SOLUTION OF NONLINEAR DIFFERENTIAL EQUATIONS

It is quite a three pipe problem, and I beg that you won't speak to me for fifty minutes.

—Sherlock Holmes, *The Red-Headed League*
Sir Arthur Conan Doyle

(E) 4.1 SPONTANEOUS SINGULARITIES

One cannot hope to obtain exact solutions to most nonlinear differential equations. As we saw in Chap. 1, there are only a limited number of systematic procedures for solving them, and these apply to a very restricted class of equations. Moreover, even when a closed-form solution is known, it may be so complicated that its qualitative properties are obscured. Thus, for most nonlinear equations it is necessary to have reliable techniques to determine the approximate behavior of the solutions.

The solutions of differential equations encountered in practice are regular at almost every point; in the neighborhood of ordinary points Taylor series provide an adequate description of the solution. However, the distinguishing features of the solution are its singularities. Determining the location and nature of these singularities, without actually solving the differential equation, requires the techniques of local analysis.

Solutions of nonlinear differential equations possess a richer spectrum of singular behaviors than solutions of linear differential equations. A solution of a linear equation can only be singular at points where the coefficient functions are singular, and at no other points. Since the locations of these singular points are independent of the choice of initial or boundary conditions, they are called *fixed singularities*. In contrast, solutions of nonlinear equations, in addition to having fixed singularities, may also exhibit new kinds of singularities which move around in the complex plane as the initial or boundary conditions vary. Such singularities are called *spontaneous* or *movable* singularities.

Example 1 *Appearance of a spontaneous singularity.* The linear differential equation $y' + y/(x-1) = 0$ $[y(0) = 1]$ has a regular singular point at $x = 1$. The solution $y(x) = 1/(1-x)$ has a pole at $x = 1$. The location of this pole does not change as the initial condition varies; if we replace $y(0) = 1$ with $y(0) = 2$, the new solution $y(x) = 2/(1-x)$ still has a pole at $x = 1$.

The nonlinear differential equation $y' = y^2$ $[y(0) = 1]$ also has the solution $y(x) = 1/(1-x)$. Even though the equation is not singular at $x = 1$, a pole has spontaneously appeared. If we replace the initial condition with $y(0) = 2$, the new solution is $y(x) = 2/(1-2x)$. The pole has moved.

In general, the solution $y(x)$ of the first-order differential equation $y'(x) = F(x, y)$ $[y(a) = b]$ is guaranteed to exist and be analytic in some neighborhood of $x = a$ as long as F is an analytic function of its two arguments at $x = a$, $y = b$ (see Prob. 4.4). However, when F is a nonlinear function of y, the extent of the region of analyticity and the radius of convergence of the Taylor series of $y(x)$ about $x = a$ may be difficult to predict because spontaneous singularities may appear.

Example 2 *Determination of the radius of convergence.* Let us perform a local analysis of

$$y' = \frac{y^2}{1 - xy}, \qquad y(0) = 1, \tag{4.1.1}$$

at $x = 0$. We seek a solution in the form of the Taylor series

$$y(x) = \sum_{n=0}^{\infty} a_n x^n, \qquad a_0 = 1. \tag{4.1.2}$$

As argued above, this series is guaranteed to have a nonzero radius of convergence.

Although it happens rarely with nonlinear equations, there is a lovely closed-form expression for the coefficients (Prob. 4.1):

$$a_n = \frac{(n+1)^{(n-1)}}{n!}. \tag{4.1.3}$$

The radius of convergence R of the series (4.1.2) is given by the ratio test as

$$R = \lim_{n \to \infty} \left| \frac{a_n}{a_{n+1}} \right| = \lim_{n \to \infty} \left(1 - \frac{1}{n+2}\right)^n = \frac{1}{e}.$$

Thus, $y(x)$ has a spontaneous singularity at a distance $1/e$ from the origin, a number which could not have been predicted by a cursory inspection of the differential equation. This singularity lies on the positive real-x axis because the coefficients a_n are all positive.

To verify that the radius of convergence does indeed have the value $1/e$, we note that the solution to the differential equation (4.1.1) satisfies the implicit functional relation

$$y = e^{xy}. \tag{4.1.4}$$

This is easily checked by substituting back into the differential equation. The solution to (4.1.4) is plotted in Fig. 4.1.

Equation (4.1.4) may be solved iteratively for y as a limit of nested exponentials:

$$y(x) = \lim_{n \to \infty} y_n(x), \tag{4.1.5}$$

where $y_{n+1} = \exp(xy_n)$; to wit, we choose $y_0 = y(0) = 1$ so $y_0 = 1$, $y_1 = \exp(x)$, $y_2 = \exp[x \exp(x)]$, $y_3 = \exp\{x \exp[x \exp(x)]\}$, $y_4 = \exp(x \exp\{x \exp[x \exp(x)]\})$, and so on. The limit exists when $-e \le x \le 1/e$ and ceases to exist when $x > 1/e$ (see Prob. 4.2).

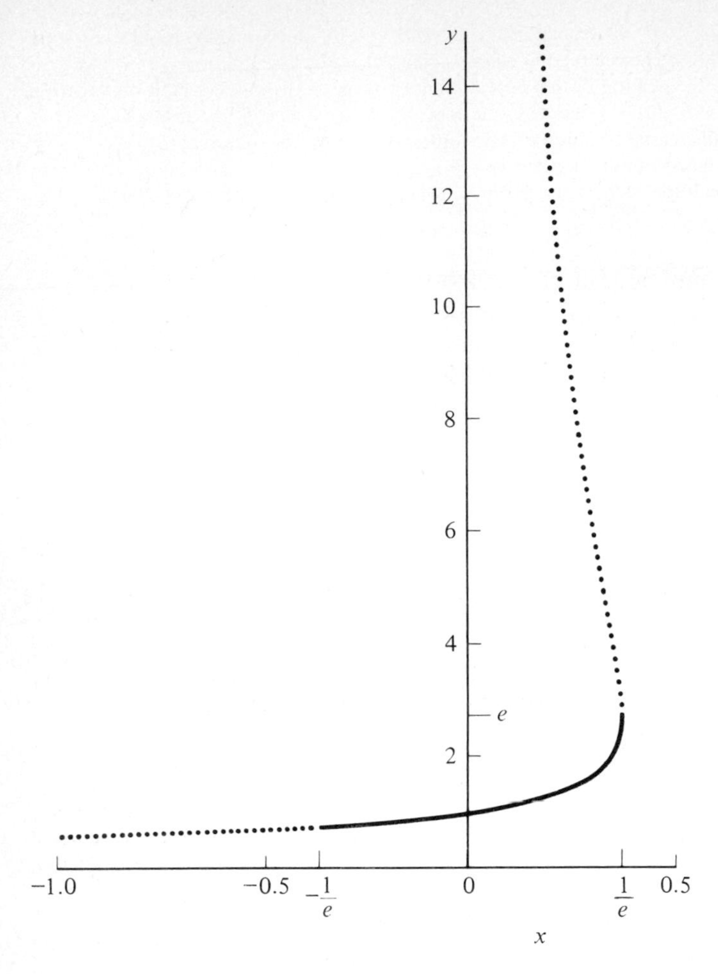

Figure 4.1 A graph of the solution to $y = \exp(xy)$. The series (4.1.2) converges to the solid portion of the curve. The spontaneous singularity at $x = 1/e$ appears as a point of infinite slope.

These examples illustrate how unpredictable the behavior of solutions to nonlinear differential equations may be. The rest of this chapter provides a collection of additional interesting examples which survey the range of techniques useful for local analysis.

(E) 4.2 APPROXIMATE SOLUTIONS OF FIRST-ORDER NONLINEAR DIFFERENTIAL EQUATIONS

Even simple-looking first-order nonlinear differential equations can have complicated solutions. This section contains two illustrative examples. The first is a numerical calculation and the second is a leading-order asymptotic analysis.

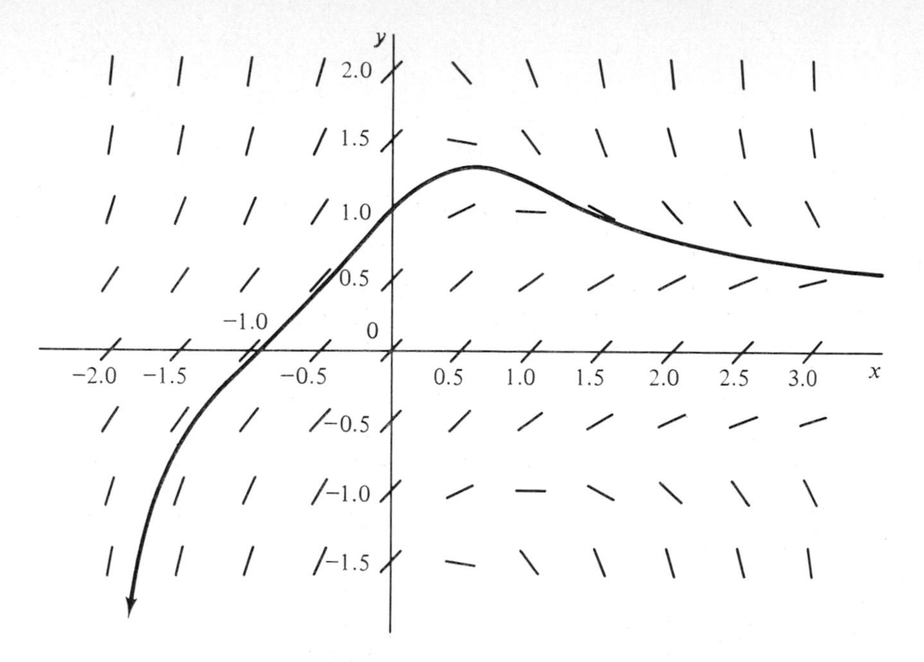

Figure 4.2 Tangent field for the equation $y' = 1 - xy^2$. The line segments indicate the slope of $y(x)$ for integer and half-integer values of x and y. The solution curve passing through $(x = 0,\ y = 1)$ is drawn; $y(x)$ reaches $-\infty$ when $x \doteq -2.12$.

Example 1 *Spontaneous singularities in the complex plane.* As shown in Sec. 4.1, nonlinear equations can generate spontaneous singularities. For example, the solution of the Riccati equation $y' = 1 - xy^2$ $[y(0) = 1]$ becomes singular at a *finite* negative value of x. The presence of this singularity can be understood from the graph of the tangent field given in Fig. 4.2. The tangent field indicates that the solution which satisfies the initial condition $y(0) = 1$ becomes large and negative for negative x. When y is sufficiently large and negative, 1 becomes negligible compared with $-xy^2$. The resulting approximate differential equation is

$$y' \sim -xy^2, \qquad y \to -\infty.$$

The solutions to $y' = -xy^2$, $y(x) = (x^2/2 + C)^{-1}$, that are negative somewhere have $C < 0$, so they become infinite for some finite negative x.

To find the location of this singularity numerically, we let $w(x) = 1/y(x)$. $w(x)$ satisfies the differential equation $w' = x - w^2$ $[w(0) = 1]$. Numerical integration of this differential equation gives a zero of w near $x = -2.12$. Thus, y becomes singular at $x \doteq -2.12$.

From this result one might expect the Taylor series solution about $x = 0$, $y(x) = \sum_{n=0}^{\infty} a_n x^n$, to have a radius of convergence of 2.12. However, a numerical evaluation of the Taylor coefficients a_n indicates that the true radius of convergence R is close to 1.228:

$$R = \lim_{n\to\infty} \left| \frac{a_n}{a_{n+1}} \right| \doteq 1.228.$$

R has this much smaller value because $y(x)$ also has *complex* spontaneous singularities. Further numerical integration shows that $w(x)$ has a zero in the complex plane at $x_0 \doteq 0.313 + 1.188i$. This is the zero of $w(x)$ which is nearest to the origin in the complex-x plane. Its distance to the origin is $|x_0| \doteq 1.228$. Therefore, it is this singularity and not the one at $x \doteq -2.12$ that determines the radius of convergence of the Taylor series for $y(x)$.

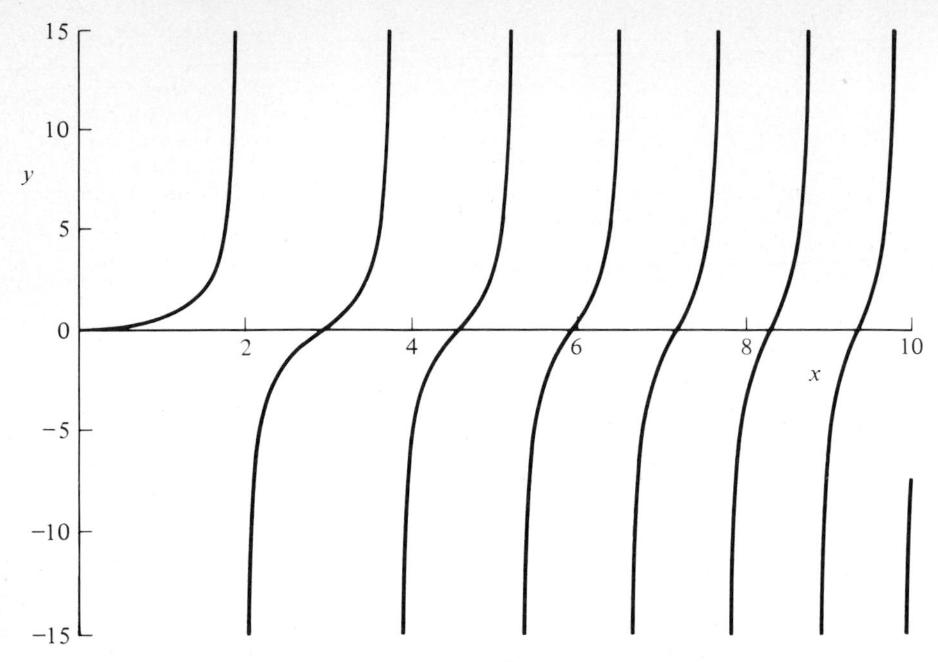

Figure 4.3 A computer plot of the solution to $y' = y^2 + x$ $[y(0) = 0]$ for $0 \le x \le 10$.

Example 2 *Infinite number of spontaneous singularities.* In this example we study the leading behavior of the solution to the Riccati equation

$$y' = y^2 + x \tag{4.2.1}$$

as $x \to +\infty$.

The previous example shows that the solution to a nonlinear differential equation may exhibit several spontaneous singularities. We will see that the solution to (4.2.1) has an *infinite* number of singularities along the positive real axis! Figure 4.3 is a computer plot of the solution to (4.2.1) satisfying the initial condition $y(0) = 0$. Note that the graph of $y(x)$ resembles that of the function $\tan x$.

The ultimate goal of our analysis is to construct a function which closely approximates $y(x)$ as $x \to +\infty$. However, we begin with a more modest investigation: let us try to determine the nature of the singularities of $y(x)$.

Can the singularities of $y(x)$ be poles? We know that in the neighborhood of a pole the leading behavior is given by $y(x) \sim A/(x-a)^b$ $(x \to a)$, where a is the location and b is the order of the pole. Substituting this asymptotic relation into the differential equation (4.2.1) and comparing leading terms gives $A = -1$, $b = 1$. Thus, solutions of the differential equation *probably* have simple poles. But to prove this conjecture we must show that in some neighborhood of $x = a$ there is a solution in the form of a (convergent) Laurent series

$$y(x) = -\frac{1}{x-a} + \sum_{n=0}^{\infty} a_n(x-a)^n. \tag{4.2.2}$$

It is left as an exercise (see Prob. 4.10) to compute the coefficients a_n directly from the differential equation and to verify that the series (4.2.2) converges in a neighborhood of $x = a$. Unfortunately, this series expression is only valid in a disk which does not contain any other singularity of y. It would be much more desirable to have a uniform description valid for large x which exhibits the *multiple* singularity structure of $y(x)$ (see Fig. 4.3).

To obtain such an expression it is necessary to approximate the differential equation by one that has an analytical solution. However, in this case an approximation which reveals the nature of the nonlinear differential equation is not easy to find! It would certainly be nice if one could neglect x in favor of y^2 or y^2 in favor of x in the differential equation. Unfortunately, a glance at Fig. 4.3 shows that as $x \to +\infty$, *sometimes* $y^2 > x$ and *sometimes* $x > y^2$; we need a more subtle approximation which is uniformly valid as x tends to $+\infty$.

An ingenious trick is to substitute

$$y(x) = x^{1/2}u(x).$$

The equation for $u(x)$ is then

$$u' = (1 + u^2)x^{1/2} - u/2x.$$

Now the term $u/2x$ is uniformly negligible for large x because

$$u \le 1 + u^2$$

for all u and

$$x^{-1} \ll x^{1/2}, \qquad x \to +\infty.$$

The resulting asymptotic differential equation

$$u' \sim (1 + u^2)x^{1/2}, \qquad x \to +\infty,$$

is easily solved because it is separable:

$$y(x) = x^{1/2}u(x) = x^{1/2} \tan \phi(x), \tag{4.2.3a}$$

where

$$\phi(x) \sim \tfrac{2}{3}x^{3/2}, \qquad x \to +\infty. \tag{4.2.3b}$$

This result suggests that for large x the solution of the Riccati equation (4.2.1) has an infinite sequence of first-order poles having an accumulation point at $x = \infty$.

Equation (4.2.3) could have been derived in another way. In Chap. 1 we showed that a Riccati equation is equivalent to a *linear* second-order equation. For equation (4.2.1) the appropriate transformation is $y(x) = -v'(x)/v(x)$ and the resulting second-order linear equation for $v(x)$ is $v'' = -xv$. Except for the minus sign, this is an Airy equation. The solutions to this equation are Ai $(-x)$ and Bi $(-x)$. From our analysis in Sec. 3.7 of the behavior of solutions to linear differential equations for large x, we know that

$$v(x) = ax^{-1/4} \cos \phi(x), \qquad \text{where } \phi(x) \sim \tfrac{2}{3}x^{3/2}, \qquad x \to +\infty.$$

Differentiating this result and retaining the largest term gives an approximate expression for $y(x)$:

$$y(x) = -v'(x)/v(x) = \sqrt{x} \tan \phi(x).$$

We have thus reproduced (4.2.3).

The accuracy of (4.2.3) may be tested in several ways. We could plot the function $\sqrt{x} \tan (\frac{2}{3}x^{3/2})$ for large x and compare the result with Fig. 4.3. However, a better test of (4.2.3) is to compute $y(x)$ numerically and to plot $[\arctan (x^{-1/2}y)]/\frac{2}{3}x^{3/2}$ and verify that this ratio approaches 1 as $x \to +\infty$. In Fig. 4.4 we plot this ratio for several choices of the initial value $y(0)$. Observe that our local analysis at $+\infty$ is independent of the initial conditions.

It is even more impressive to use (4.2.3) to compute the pole spacing for large x. Let Δ be the distance between consecutive poles and let x and $x + \Delta$ be the locations of two consecutive poles. Then the leading asymptotic approximation in (4.2.3) implies that $\frac{2}{3}(x + \Delta)^{3/2} - \frac{2}{3}x^{3/2} \sim \pi$ $(x \to +\infty)$. If we expand $(x + \Delta)^{3/2}$ using the binomial theorem, we find that the spacing of the poles is approximately

$$\Delta \sim \frac{\pi}{x^{1/2}}, \qquad x \to +\infty. \tag{4.2.4}$$

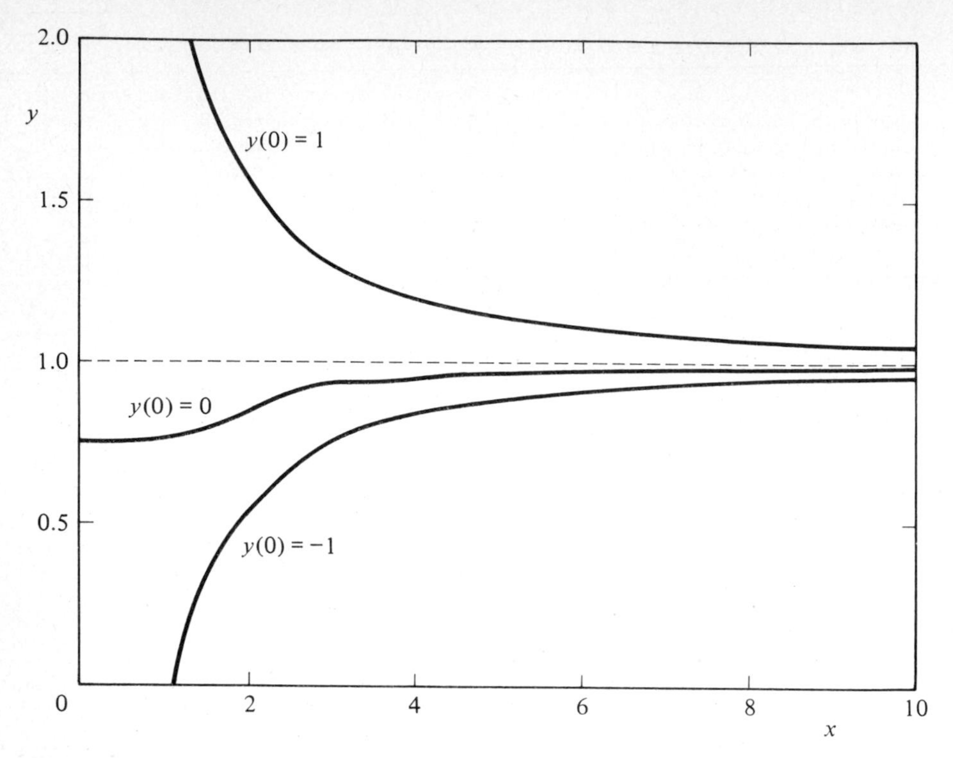

Figure 4.4 Three graphs of $[\arctan(x^{-1/2}y)]/\frac{2}{3}x^{3/2}$, where y is the solution of (4.2.1) and $y(0) = -1, 0,$ and 1. Observe that all three curves rapidly approach 1 as x gets large, thus verifying the formula for the leading behavior of $y(x)$ in (4.2.3).

To test this result we have computed the locations of the poles numerically. We denote the location of the nth pole on the positive real axis by x_n and define Δ_n to be $\Delta_n = x_n - x_{n-1}$. In Table 4.1 we list values of $\bar{x}_n^{1/2}\Delta_n$, where $\bar{x}_n = \frac{1}{2}(x_n + x_{n-1})$, the average location of two consecutive poles. We certainly would expect that as $n \to +\infty$, $\bar{x}_n^{1/2}\Delta_n$ approaches π, but the rapidity of this convergence is astonishing (see Prob. 4.5).

(I) 4.3 APPROXIMATE SOLUTIONS TO HIGHER-ORDER NONLINEAR DIFFERENTIAL EQUATIONS

This section is a collection of eight examples which illustrate techniques for determining the approximate behavior of solutions to higher-order nonlinear differential equations.

Asymptotic analysis of nonlinear differential equations is particularly difficult because it is hard to know when all possible behaviors have been found. There is no such difficulty with linear equations; we know that we have a complete description of the asymptotic behavior of an nth-order linear differential equation when n linearly independent behaviors have been found. Unfortunately, even if we have

Table 4.1 Locations of poles in the solution of $y' = y^2 + x$ $[y(0) = 0]$ **for** $0 \le x \le 20$

Asymptotic analysis predicts that $(x_n - x_{n-1})[(x_n + x_{n-1})/2]^{1/2} \to \pi$ as $n \to \infty$. Observe that the convergence to π is extremely rapid

Pole number n	Location of nth pole x_n	$(x_n - x_{n-1})\left(\frac{x_n + x_{n-1}}{2}\right)^{1/2}$
1	1.986 35	
2	3.825 34	3.134 84
3	5.295 62	3.139 83
4	6.584 31	3.140 80
5	7.757 33	3.141 15
6	8.847 53	3.141 31
7	9.874 28	3.141 40
8	10.850 2	3.141 45
9	11.784 0	3.141 49
10	12.682 2	3.141 51
11	13.549 6	3.141 53
12	14.390 2	3.141 55
13	15.206 8	3.141 56
14	16.002 1	3.141 56
15	16.778 1	3.141 57
16	17.536 5	3.141 58
17	18.278 9	3.141 58
18	19.006 5	3.141 58
19	19.720 4	3.141 58

found an n-parameter asymptotic approximation to the solution of an nth-order nonlinear differential equation, there may still be special solutions of the equation whose asymptotic behaviors are not obtained for any choice of the n parameters.

The origin of this difficulty does not lie in asymptotic analysis; rather, it is a characteristic property of nonlinear equations. It often happens that among the exact solutions to a nonlinear differential equation there are special solutions which cannot be obtained from the general solution for any choice of the constants of integration. For example, the general solution of the first-order nonlinear equation

$$(y')^2 + y^2 = 1 \tag{4.3.1}$$

is

$$y(x) = \sin(x + c_1), \tag{4.3.2}$$

where c_1 is a constant of integration. There are also two additional solutions, $y = \pm 1$; the general solution (4.3.2) does not reduce to these special solutions for any choice of c_1.

As another example, let us reexamine the differential equation in (1.1.9):

$$y'' = yy'/x. \tag{4.3.3}$$

The general solution to this equation is

$$y(x) = 2c_1 \tan(c_1 \ln x + c_2) - 1. \tag{4.3.4}$$

But there is also a special one-parameter family of solutions, $y = c_3$, which are not contained in (4.3.4) for any values of c_1 and c_2.

In the next two examples we determine the asymptotic behaviors of all solutions to two higher-order differential equations. In both examples we will see that, in addition to an n-parameter general asymptotic behavior, there are also special asymptotic behaviors. For these examples the special asymptotic behaviors are associated with singular limits of the parameters in the general asymptotic behavior.

Example 1 *Behavior of solutions to $y^3y'' = -1$ as $x \to +\infty$.* In this example we study the behavior of solutions to

$$y^3y'' = -1 \tag{4.3.5}$$

as $x \to +\infty$. Our objective is to find a complete description of the asymptotic behavior of y as $x \to +\infty$; that is, we will find the asymptotic behaviors of *all* solutions to (4.3.5).

There are no general procedures for finding the asymptotic behavior of solutions to a nonlinear equation. We have seen that in the case of homogeneous linear equations we are often led to an exponential behavior at an irregular singular point and to an algebraic and sometimes logarithmic behavior at an ordinary point or a regular singular point. The equation in (4.3.5), like most nonlinear equations, is not equidimensional in y and therefore does not admit an exponential asymptotic behavior. (An exponential behavior here would be inconsistent because the exponential factor does not cancel as it would in a linear equation. See Prob. 4.31.) But algebraic and logarithmic behaviors are not excluded, and it is a good rule to examine a nonlinear equation to see if it exhibits any of these behaviors before searching for something more complicated.

We quickly check that (4.3.5) does indeed admit an algebraic asymptotic behavior as $x \to +\infty$. Substituting $y(x) \sim Ax^\alpha$ $(x \to +\infty)$ into (4.3.5) gives $A^4\alpha(\alpha - 1)x^{4\alpha-2} \sim -1$ $(x \to +\infty)$. We conclude that $\alpha = \frac{1}{2}$ and that $A = \pm\sqrt{2}$. Notice that the multiplicative constant A is determined; in the case of linear equations the overall multiplicative constant of the leading behavior is arbitrary.

The behavior $y(x) \sim \pm\sqrt{2x}$ $(x \to +\infty)$ contains no free parameters. However, we note that (4.3.5) is translation invariant (autonomous); it is invariant under the translation $x \to x + a$. Therefore, we may generalize the above asymptotic behavior to the one-parameter family of behaviors

$$y(x) \sim \pm\sqrt{2(x + a)}, \qquad x \to +\infty. \tag{4.3.6}$$

The result in (4.3.6) is still not the most general asymptotic behavior of solutions to (4.3.5) because it does not contain two arbitrary constants. Moreover, it may not be generalized to include a second parameter (by arguing, for example, that it is the first term of a series in which there is another free parameter) because it is already an exact solution of (4.3.5)!

It is interesting that the missing asymptotic behaviors also behave algebraically and yet they were not found by substituting $y \sim Ax^\alpha$. The reason for this is that we did not consider leading behaviors which are linear in x and which therefore vanish after two differentiations. If such behaviors are possible, they must have the form

$$y(x) \sim bx + c + \varepsilon(x), \qquad x \to +\infty,$$

where $\varepsilon(x) \to 0$ as $x \to +\infty$. Substituting this relation into (4.3.5) gives an asymptotic differential equation for $\varepsilon(x)$,

$$b^3x^3\varepsilon''(x) \sim -1, \qquad x \to +\infty,$$

whose solution

$$\varepsilon(x) \sim -(2b^3x)^{-1}, \qquad x \to +\infty,$$

is consistent with the requirement that $\varepsilon(x) \to 0$ as $x \to +\infty$.

More accurate approximations to $\varepsilon(x)$ take the form of a series in inverse powers of x. Therefore, we attempt to represent the full asymptotic behavior of $y(x)$ as a formal Laurent series at ∞:

$$y(x) \sim bx + c + d/x + e/x^2 + \cdots, \qquad x \to +\infty.$$

Substituting this series into (4.3.5) gives

$$(2d/x^3 + 6e/x^4 + \cdots)(b^3x^3 + 3b^2cx^2 + \cdots) = 2db^3 + 6(b^3e + b^2cd)/x + \cdots \sim -1, \qquad x \to +\infty.$$

Comparing like powers of x gives an infinite sequence of equations for $d, e, \ldots$: $d = -1/2b^3$, $e = c/2b^4, \ldots$. Note that the two parameters b and c remain undetermined except for the requirement that $b \neq 0$. Thus, we have found the general two-parameter asymptotic behavior which does not reduce to (4.3.6) for any values of $b \neq 0$ and c:

$$y(x) \sim bx + c - \frac{1}{2b^3x} + \frac{c}{2b^4x^2} \cdots, \qquad x \to +\infty. \tag{4.3.7}$$

If we were to compute all of the coefficients in this series, we would find that (4.3.7) is just a binomial series whose sum has the form

$$y(x) = \pm\left[\frac{(c_1x + c_2)^2 - 1}{c_1}\right]^{1/2}, \qquad c_1 \neq 0, \tag{4.3.8}$$

which is an exact solution to (4.3.5) (see Prob. 4.35). The result in (4.3.6) may be regarded as the singular limit of (4.3.8) in which $c_1 \to 0$ and $c_2^2 \to 1$ in such a way that the ratio $(c_2^2 - 1)/2c_1 = a$ remains fixed.

The differential equation in (4.3.5) is a warm-up for the more difficult asymptotic analysis in later examples, but the equation is so simple that asymptotic analysis is not really necessary. Indeed, it would have been easier to solve the equation first and to study the behavior of the solutions instead of the differential equation. To demonstrate the power of asymptotic analysis, we consider next a differential equation whose exact solutions are not obtainable in closed form.

In this and the remaining examples of this section, we will often differentiate asymptotic relations without supplying any justification for doing so. If the reader feels so inclined, he will not find it difficult to prove rigorously that the final results we obtain are correct.

Example 2 *Behavior of solutions to* $y^2y''' = -\frac{1}{3}$ *as* $x \to +\infty$. The differential equation

$$y^2y''' = -\tfrac{1}{3} \tag{4.3.9}$$

affords a clear comparison of exact methods versus approximate methods. We will analyze this equation in two ways. First, in a glorious *tour de force*, we use the methods of Chap. 1 to reduce the differential equation to a soluble first-order equation! Unfortunately, we find that the exact solution is an implicit function of such complexity that it is of no value. Next, we use the methods of local analysis to obtain a complete and simple description of the behavior of $y(x)$ for large x.

The exact solution. Here we use the methods of Sec. 1.7 to solve (4.3.9) exactly. Noting that (4.3.9) is autonomous, we reduce its order by treating y as a new independent variable and introducing

$$A = A(y) = \frac{dy}{dx} \tag{4.3.10}$$

as a new dependent variable. The new *second*-order differential equation is

$$y^2A^2\frac{d^2A}{dy^2} + A\left(y\frac{dA}{dy}\right)^2 = -\frac{1}{3}. \tag{4.3.11}$$

This equation is equidimensional in y because it is invariant under the scale change $y \to ay$. Thus, the substitution

$$y = e^t, \qquad A(y) = B(t) \tag{4.3.12}$$

converts (4.3.11) to a second-order autonomous equation:

$$B^2\left(\frac{d^2B}{dt^2} - \frac{dB}{dt}\right) + B\left(\frac{dB}{dt}\right)^2 = -\frac{1}{3}. \tag{4.3.13}$$

We reduce this equation to one that is *first* order by letting

$$C = C(B) = \frac{dB}{dt}. \tag{4.3.14}$$

The result is

$$B^2\left(C\frac{dC}{dB} - C\right) + BC^2 = -\frac{1}{3}.$$

At first glance, this equation looks so complicated that one might expect no further progress. However, the substitution

$$C(B) = \frac{D(B)}{B} \tag{4.3.15}$$

gives a pleasing simplification:

$$D\frac{dD}{dB} - DB = -\frac{1}{3}. \tag{4.3.16}$$

The further substitution

$$D(B) = \tfrac{1}{2}B^2 + E(B), \tag{4.3.17}$$

which is motivated by first finding the large-B behavior of D in (4.3.16), gives a Riccati equation if we treat E as the independent variable and B as the dependent variable: $dB/dE + \frac{3}{2}B^2 + 3E = 0$.

The attainment of a Riccati equation represents major progress because all Riccati equations may be linearized. The appropriate substitution is

$$B(E) = \frac{2}{3F(E)}\frac{dF(E)}{dE}. \tag{4.3.18}$$

The result is

$$\frac{d^2F}{dE^2} + \frac{9}{2}EF = 0, \tag{4.3.19}$$

which we immediately recognize as an Airy equation! Its solution is a linear combination of $\text{Ai}\,[-(\frac{9}{2})^{1/3}E]$ and $\text{Bi}\,[-(\frac{9}{2})^{1/3}E]$.

Now we must work backward, undoing all the substitutions that led to the exact answer in terms of Airy functions. Substituting $F(E)$ into (4.3.18) gives

$$B(E) = -\left(\frac{4}{3}\right)^{1/3} \frac{c_1 \operatorname{Ai}'[-(\frac{9}{2})^{1/3}E] + \operatorname{Bi}'[-(\frac{9}{2})^{1/3}E]}{c_1 \operatorname{Ai}[-(\frac{9}{2})^{1/3}E] + \operatorname{Bi}[-(\frac{9}{2})^{1/3}E]}, \tag{4.3.20}$$

where c_1 is an integration constant.

Next we must find $D(B)$ in (4.3.17), but at this point we are thoroughly stymied; to find D requires that we solve (4.3.20) for E as a function of B, and this is hopeless. Although (4.3.20) is the exact solution to (4.3.9), we have achieved a Pyrrhic victory!

The asymptotic solution. The asymptotic solutions to (4.3.9) are similar in form to those for (4.3.5). The existence of a three-derivative term suggests that we might look for a solution of the form

$$y(x) \sim ax^2 + bx + c + \varepsilon(x), \qquad x \to +\infty,$$

where $\varepsilon(x) \to 0$ as $x \to +\infty$. Substituting this expression into (4.3.9) gives an asymptotic equation for $\varepsilon(x)$,

$$a^2x^4\varepsilon''' \sim -\tfrac{1}{3}, \qquad x \to +\infty,$$

whose solution

$$\varepsilon(x) \sim 1/18a^2x, \qquad x \to +\infty,$$

does indeed vanish as $x \to +\infty$.

Higher-order approximations to $\varepsilon(x)$ indicate that $\varepsilon(x)$ may be represented as a series in powers of $1/x$. Thus, the asymptotic behavior of $y(x)$ takes the form of a Laurent series at ∞:

$$y(x) \sim ax^2 + bx + c + \frac{d}{x} + \frac{e}{x^2} + \frac{f}{x^3} + \cdots. \tag{4.3.21}$$

If we substitute (4.3.21) into (4.3.9) and compare like powers of x, we obtain an infinite sequence of equations for $d, e, f, \ldots$ whose solutions are $d = 1/18a^2$, $e = -b/36a^3$, $f = (3b^2 - 2ac)/180a^4, \ldots$. Apart from the restriction that $a \neq 0$, the three parameters a, b, and c remain undetermined. Thus, we have found the general asymptotic behavior of $y(x)$ for large x:

$$y(x) \sim ax^2 + bx + c + \frac{1}{18a^2x} - \frac{b}{36a^3x^2} + \frac{3b^2 - 2ac}{180a^4x^3} + \cdots, \qquad x \to +\infty. \tag{4.3.22}$$

The convergence of this series is investigated in Prob. 4.36.

The asymptotic behavior in (4.3.22) becomes singular when $a = 0$. This suggests that in addition to (4.3.22) there is a special asymptotic behavior which is not contained in $y(x)$ for any values of a, b, and c. Is this behavior algebraic? If we substitute $y(x) \sim Ax^\alpha$ $(x \to +\infty)$ into (4.3.9), we get $A^3\alpha(\alpha-1)(\alpha-2)x^{3\alpha-3} \sim -\frac{1}{3}$ $(x \to +\infty)$. Comparing powers of x gives $3\alpha - 3 = 0$, or $\alpha = 1$; but this result is inconsistent because $\alpha = 1$ makes the left side of the equation vanish. The trouble is that the third derivative of x is 0 and not $1/x^2$. In any event we already know from (4.3.22) that $y(x)$ cannot grow like x for large x.

The remedy here is to introduce factors of $\ln x$. We try a solution whose leading behavior is

$$y(x) \sim Ax(\ln x)^\alpha, \qquad x \to +\infty.$$

Three derivatives of this expression are

$$y'''(x) \sim -A\alpha\frac{(\ln x)^{\alpha-1}}{x^2} + \frac{A\alpha(\alpha-1)(\alpha-2)(\ln x)^{\alpha-3}}{x^2}, \qquad x \to +\infty.$$

Neglecting $(\ln x)^{\alpha-3}$ compared with $(\ln x)^{\alpha-1}$ as $x \to +\infty$ and substituting into the differential equation (4.3.9) gives $-A^3\alpha(\ln x)^{3\alpha-1} \sim -\frac{1}{3}$ $(x \to +\infty)$ whose solution is $\alpha = \frac{1}{3}$ and $A = 1$. We have thus discovered a new possible leading behavior for solutions to (4.3.9):

$$y(x) \sim x(\ln x)^{1/3}, \qquad x \to +\infty. \tag{4.3.23}$$

What is the full asymptotic behavior of solutions whose leading behavior is given in (4.3.23)? The leading behavior is the first term in an infinite series which represents the asymptotic behavior. In this case the series is a power series in $(\ln x)^{-1}$:

$$y(x) \sim x(\ln x)^{1/3}[1 + A/(\ln x) + B/(\ln x)^2 + C/(\ln x)^3 + \cdots].$$

It is not easy to discover such a series; some trial and error and intelligent fiddling are required. But given the form of this series, it is easy to substitute it into the differential equation (4.3.9) and compute the coefficients. We find that A is arbitrary, $B = -\frac{10}{27} - A^2$, $C = \frac{50}{27}A + \frac{5}{3}A^3, \ldots$. We have therefore found a special one-parameter family of asymptotic behaviors of $y(x)$ for large x:

$$y(x) \sim x(\ln x)^{1/3}\left[1 + \frac{A}{\ln x} - \frac{A^2 + \frac{10}{27}}{(\ln x)^2} + \frac{\frac{5}{3}A^3 + \frac{50}{27}A}{(\ln x)^3} + \cdots\right], \qquad x \to +\infty. \tag{4.3.24}$$

One might think that since (4.3.9) is autonomous, we may replace x in (4.3.24) by $x + \alpha$, where α is a free parameter, thereby generalizing (4.3.24) to a two-parameter family of behaviors. However, if we reexpand the resulting expression as a series in inverse powers of $\ln x$, we learn that (see Prob. 4.37) terms containing the parameter α are not multiplied by x. In the limit $x \to +\infty$ such terms are subdominant with respect to all terms in the series (4.3.24) and should not be included in the formula for the asymptotic behavior of $y(x)$.

The series in (4.3.22) and (4.3.24) constitute a complete description of the asymptotic behavior of solutions to (4.3.9). In Figs. 4.5 and 4.6 we compare our predictions in (4.3.22) and (4.3.24) with numerical solutions to the differential equation.

Example 3 *Behavior of the first Painlevé transcendent as $x \to +\infty$.* What is the behavior of the solution to

$$y'' = y^2 + x \tag{4.3.25}$$

for large positive x?

This differential equation is the first of a set of six equations whose solutions are called the *Painlevé transcendents.* These equations were discovered by Painlevé in the course of classifying nonlinear differential equations. He considered all equations of the form

$$w'' = R(z, w)(w')^2 + S(z, w)w' + T(z, w)$$

having the properties (a) that R, S, and T are rational functions of w, but have arbitrary dependence on z, and (b) that the solutions may have various kinds of *fixed* singularities (poles, branch points, essential singularities), but may not have any movable singularities except for *poles.* There are 50 distinct types of equations having these properties. Of these, 44 types are soluble in terms of elementary transcendents (sines, cosines, exponentials), functions defined by *linear second-order* equations (Bessel functions, Legendre functions, and so on), or elliptic functions. The remaining six equations define the six Painlevé transcendents, one of which is (4.3.25).

Let us return now to the behavior of the differential equation (4.3.25). This differential equation is similar in form to the first-order equation in (4.2.1) and its asymptotic properties are also similar in some respects. However, because this is a second-order equation, a more sophisticated analysis is required.

We begin by arguing that $y(x)$ exhibits movable singularities. Since the curvature of $y(x)$ is positive ($y'' > x > 0$), it is likely that an arbitrary set of initial conditions will give rise to a solution which becomes singular at a finite value of x. To discover the leading behavior of such a singularity, we substitute

$$y(x) \sim A/(x - a)^b, \qquad x \to a,$$

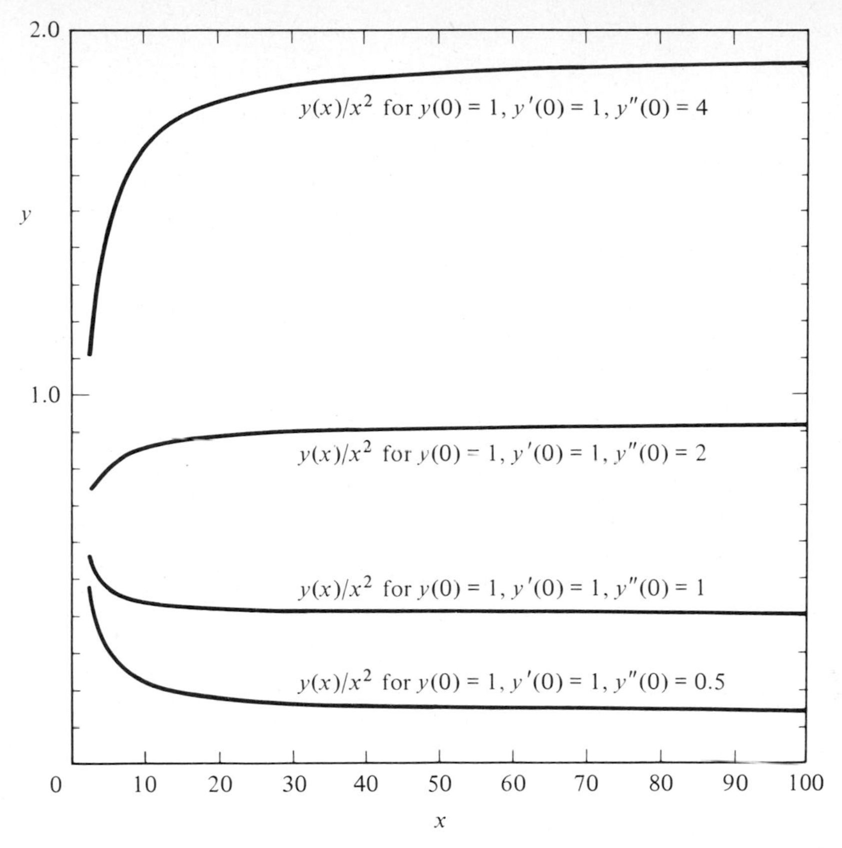

Figure 4.5 Four plots of $y(x)/x^2$, where $y(x)$ is the solution to $y'''y^2 = -\frac{1}{3}$, $y(0) = 1$, $y'(0) = 1$, and $y''(0)$ takes on the four values 0.5, 1, 2, 4. Note that all four curves approach a constant for large x. This verifies the prediction in (4.3.22) that the leading term in the general asymptotic behavior of $y(x)$ as $x \to +\infty$ is ax^2 (a is a constant which depends in a complicated way on the initial conditions).

into the differential equation (4.3.25). Comparing powers of $x - a$ gives $A = 6$ and $b = 2$. This suggests that $y(x)$ has movable second-order poles.

However, this does not prove that the movable singularities are poles (although they really are). To verify such a conjecture it is necessary to establish that a Laurent series solution of the form

$$y(x) = \frac{6}{(x-a)^2} + \sum_{n=-1}^{\infty} a_n(x-a)^n \tag{4.3.26}$$

exists in the neighborhood of $x = a$ (see Prob. 4.34).

Actually, $y(x)$ has an infinite number of second-order poles along the positive real axis and not just one! We have solved the differential equation (4.3.25) numerically, taking as initial conditions $y(0) = y'(0) = 0$, and have plotted the result in Fig. 4.7. Observe that there is a sequence of poles along the positive real axis (see Prob. 4.15).

A comparison of Figs. 4.3 and 4.7 shows that the solution to (4.2.1) is very similar to the solution of (4.3.5) with simple poles replaced by second-order poles. You may recall that we were

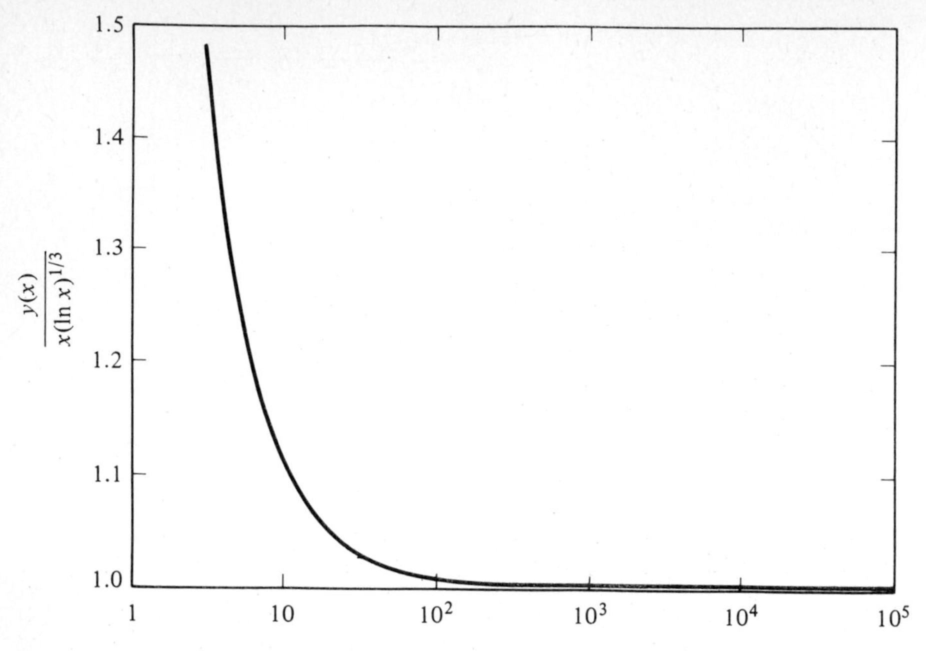

Figure 4.6 A plot of $y(x)/[x(\ln x)^{1/3}]$, where $y(x)$ is that solution to $y'''y^2 = -\frac{1}{3}$ for which $y(0) = 0$, $y'(0) = 1$, and $y''(0) \doteq 0.280\,376\,460\,87$. As $x \to +\infty$ the leading asymptotic behavior of most solutions to $y'''y^2 = -\frac{1}{3}$ is ax^2 (see Fig. 4.5), but these special initial conditions give a solution whose leading behavior is $x(\ln x)^{1/3}$ as $x \to +\infty$, as this graph clearly shows. The full asymptotic behavior of $y(x)$ is given in (4.3.24).

able to predict the spacing of the simple poles for (4.2.1) for large x from the leading asymptotic behavior of the solution. Can we predict the spacing of the second-order poles?

To answer this question we make the same transformation as in Example 2 of Sec. 4.2:

$$y = \sqrt{x}\, u(x).$$

We find that $u(x)$ satisfies

$$u'' = \sqrt{x}\,(u^2 + 1) - u'/x + u/4x^2.$$

Next we transform the independent variable x to remove the $\sqrt{x}$ that multiplies the $(u^2 + 1)$ term. The appropriate transformation is $x = s^{4/5}$. The new equation satisfied by $u(s)$ is

$$\frac{d^2u}{ds^2} = \frac{16}{25}(u^2 + 1) - \frac{1}{s}\frac{du}{ds} + \frac{4}{25}\frac{u}{s^2}. \tag{4.3.27}$$

The last two terms on the right side of this equation are negligible as $s \to +\infty$ $(x \to +\infty)$ (see Prob. 4.16). The resulting approximate differential equation

$$d^2u/ds^2 \sim 16(u^2 + 1)/25, \qquad s \to +\infty,$$

is autonomous and can now be solved in terms of elliptic functions. Multiplying the equation by u' and integrating gives

$$\tfrac{1}{2}(du/ds)^2 \sim 16(u^3/3 + u + c)/25, \qquad s \to +\infty.$$

Thus,

$$\int du(\tfrac{2}{3}u^3 + 2u + 2c)^{-1/2} \sim \pm\tfrac{4}{5}s, \qquad s \to +\infty.$$

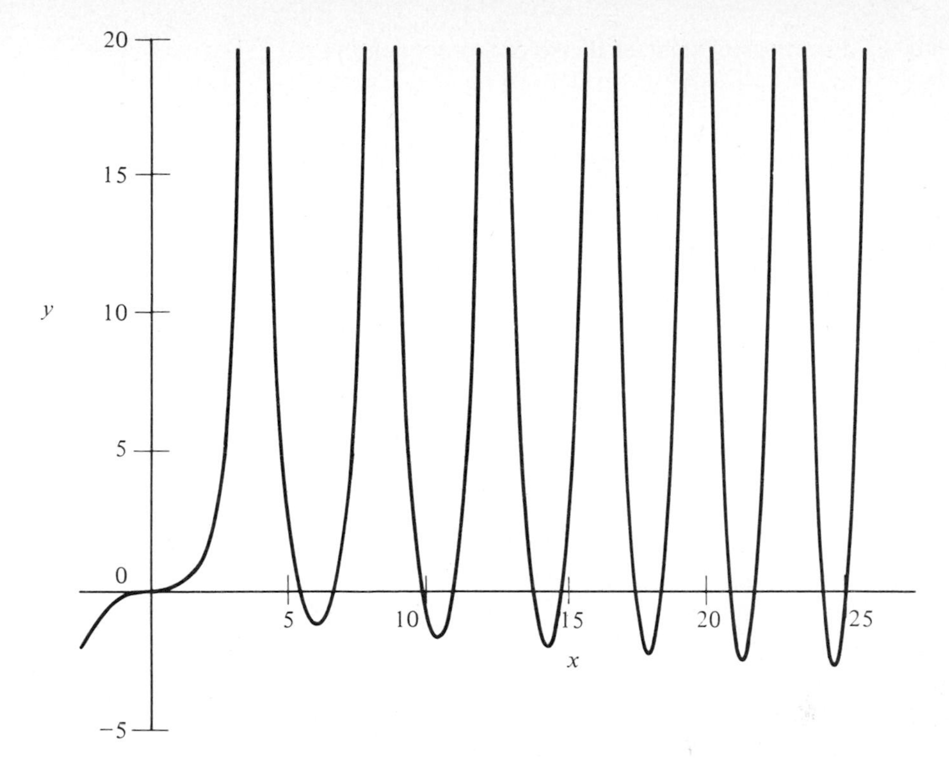

Figure 4.7 A computer plot of the solution to the initial-value problem $y'' = y^2 + x$ $[y(0) = y'(0) - 0]$. The solution has an infinite number of second-order poles on the positive real axis. (See Prob. 4.15.)

This integral defines $u = u(s)$ as an inverse function; $u(s)$ is an elliptic function (it is expressible in terms of the Weierstrass-$\mathscr{P}$ function). The only property of elliptic functions relevant to the discussion here is that they are periodic functions. In particular, the poles of $u(s)$ are separated by the constant period P. Therefore, in terms of the original independent variable x, the separation Δ of consecutive poles satisfies the following asymptotic relation: $(x + \Delta)^{5/4} - x^{5/4} \sim P$ $(x \to +\infty)$. Thus, using the binomial expansion,

$$\Delta \sim \tfrac{4}{5}Px^{-1/4}, \qquad x \to +\infty.$$

We conclude that $y(x)$ has an infinite sequence of second-order poles which bunch up as $x \to \infty$. We have now arrived at an impressively detailed approximate description of $y(x)$ for large positive x. We know both the nature and the separation of the poles. We have analyzed both the trees and the forest.

To test our predictions of the pole separation, we have determined the locations of the poles numerically and have listed our results in Table 4.2. As in Table 4.1, we denote the location of the nth pole by x_n and define Δ_n by $\Delta_n = x_{n+1} - x_n$. We then list values of $\bar{x}_n^{1/4}\Delta_n$, where $\bar{x}_n = \frac{1}{2}(x_n + x_{n+1})$, the average location of two consecutive poles. If our asymptotic analysis is valid, then the quantity $\bar{x}_n^{1/4}\Delta_n$ must approach the constant $4P/5$ as $n \to \infty$. In fact, the convergence is so rapid that, by the sixth pole, $\bar{x}_n^{1/4}\Delta_n$ is changing by only one part in 7×10^4!

Example 4 *Behavior of the first Painlevé transcendent as $x \to -\infty$.* The question of how solutions to (4.3.25) $y'' = y^2 + x$ behave as $x \to -\infty$ is particularly interesting because it involves a discussion of stability.

Table 4.2 Locations of some of the second-order poles in the solution of $y'' = y^2 + x$ [$y(0) = y'(0) = 0$] for $0 \le x \le 404$

Asymptotic analysis predicts that $(x_n - x_{n-1})[(x_n + x_{n-1})/2]^{1/4}$ should approach a constant K as $n \to \infty$; we observe rapid convergence to the value $K \doteqdot 7.276\,727$

Pole number n	Location of pole x_n	$(x_n - x_{n-1})\left(\frac{x_n + x_{n-1}}{2}\right)^{1/4}$
1	3.742 8	
2	8.375 8	7.268 83
3	12.426 5	7.274 58
4	16.168 2	7.275 75
5	19.703 9	7.276 17
6	20.087 1	7.276 37
7	26.350 5	7.276 47
8	29.515 6	7.276 54
9	32.598 0	7.276 58
10	35.609 1	7.276 61
40	110.893 5	7.276 720 7
41	113.130 3	7.276 721 0
42	115.356 0	7.276 721 2
43	117.571 1	7.276 721 5
44	119.775 8	7.276 721 7
45	121.970 4	7.276 722 0
46	124.155 1	7.276 722 2
47	126.330 3	7.276 722 4
48	128.496 2	7.276 722 6
49	130.653 0	7.276 722 7
100	232.035 3	7.276 725 8
101	233.897 9	7.276 725 9
102	235.756 7	7.276 725 9
103	237.612 0	7.276 725 9
104	239.463 6	7.276 725 9
105	241.311 6	7.276 725 9
106	243.156 1	7.276 725 9
107	244.997 1	7.276 726 0
108	246.834 6	7.276 726 0
109	248.668 8	7.276 726 0
190	388.399 8	7.276 726 5
191	390.038 1	7.276 726 5
192	391.674 7	7.276 726 5
193	393.309 5	7.276 726 5
194	394.942 7	7.276 726 5
195	396.574 2	7.276 726 5
196	398.204 0	7.276 726 5
197	399.832 1	7.276 726 5
198	401.458 6	7.276 726 5
199	403.083 4	7.276 726 5

When x is large and negative, it is reasonable to assume that there are solutions with the property that $y^2 \sim -x$ $(x \to -\infty)$. To demonstrate the consistency of this assumption, we solve for y,

$$y \sim \pm(-x)^{1/2}, \qquad x \to -\infty, \tag{4.3.28}$$

differentiate twice, $y'' \sim \mp\frac{1}{4}(-x)^{-3/2}$ $(x \to -\infty)$, and observe that y'' is indeed negligible compared with y^2 and x (see Prob. 4.27).

Of course, there are still solutions $y(x)$ having second-order poles on the negative real axis for which $y''(x)$ is *not* negligible compared with y^2 or x. However, when x is negative, it appears that a new kind of solution (4.3.28) may appear which does not have poles. When x is positive, (4.3.28) is complex, so it does not occur for real initial conditions.

Stability considerations arise in the study of the solutions (4.3.28) when we ask the following question. If we fix the initial conditions $y(0)$ and $y'(0)$ so that $y(x)$ does not have poles for negative x, which of the two functions, $+\sqrt{-x}$ or $-\sqrt{-x}$, does $y(x)$ approach as $x \to -\infty$?

To answer this question we examine the corrections to the leading behavior of $y(x)$ in (4.3.28). We set

$$y(x) = \pm\sqrt{-x} + \varepsilon(x), \tag{4.3.29}$$

where, by assumption,

$$\varepsilon(x) \ll \sqrt{-x}, \qquad x \to -\infty. \tag{4.3.30}$$

Substituting (4.3.29) into (4.3.25) gives the equation for $\varepsilon(x)$:

$$\mp\tfrac{1}{4}(-x)^{-3/2} + \varepsilon'' = \pm 2\varepsilon(-x)^{1/2} + \varepsilon^2.$$

This equation may be replaced by the asymptotic differential equation

$$\mp\tfrac{1}{4}(-x)^{-3/2} + \varepsilon'' \sim \pm 2\varepsilon(-x)^{1/2}, \qquad x \to -\infty, \tag{4.3.31}$$

because ε^2 is negligible compared with $\varepsilon(-x)^{1/2}$ according to (4.3.30).

Notice that *if* it is valid to differentiate (4.3.30) twice, we obtain $\varepsilon''(x) \ll (-x)^{-3/2}$ $(x \to -\infty)$, which allows a further simplification of (4.3.31):

$$\mp\tfrac{1}{4}(-x)^{-3/2} \sim \pm 2\varepsilon(-x)^{1/2}, \qquad x \to -\infty. \tag{4.3.32}$$

[As we will see, there are some solutions of (4.3.31) for which it is *not* valid to differentiate (4.3.30)! However, we differentiate anyway with the intention of checking that any solutions we find satisfy the constraint (4.3.30).] The solution to this equation is

$$\varepsilon(x) \sim -\tfrac{1}{8}(-x)^{-2}, \qquad x \to -\infty, \tag{4.3.33}$$

which satisfies the constraint (4.3.30).

The result in (4.3.33) is not a complete solution of (4.3.31) because there are no free parameters. Equation (4.3.31) is a linear inhomogeneous asymptotic equation so its general solution is a linear combination of solutions to the associated homogeneous asymptotic equation, plus a particular solution to the inhomogeneous asymptotic equation. The solution (4.3.33) is a particular solution.

We next examine the solutions of the homogeneous equation. For the lower choice of sign in (4.3.31) the techniques of Sec. 3.7 give (see Prob. 4.17)

$$\varepsilon(x) = c_1(-x)^{-1/8} \cos \phi(x) \tag{4.3.34a}$$

or

$$\varepsilon(x) = c_2(-x)^{-1/8} \sin \phi(x), \tag{4.3.34b}$$

where $\phi \sim \frac{4}{5}\sqrt{2}\,(-x)^{5/4}$ $(x \to -\infty)$. Note that both leading behaviors in (4.3.34) are acceptable because they satisfy the constraint (4.3.30). Yet, they are not solutions of (4.3.32) because their derivatives do not satisfy $\varepsilon'' \ll (-x)^{-3/2}$ $(x \to -\infty)$. For the upper choice of sign in (4.3.31) we have

$$\varepsilon(x) \sim c_1(-x)^{-1/8} \exp\left[\tfrac{4}{5}\sqrt{2}\,(-x)^{5/4}\right], \qquad x \to -\infty, \tag{4.3.34c}$$

or

$$\varepsilon(x) \sim c_2(-x)^{-1/8} \exp\left[-\tfrac{4}{5}\sqrt{2}\,(-x)^{5/4}\right], \qquad x \to -\infty. \tag{4.3.34d}$$

Now we can answer the question of which function, $+(-x)^{1/2}$ or $-(-x)^{1/2}$, is asymptotic to $y(x)$ if $y(x)$ has no poles as $x \to -\infty$. It is highly probable that $y(x)$ approaches $-(-x)^{1/2}$ because ε, the correction to this leading behavior, consists of two oscillatory parts with very slowly decaying amplitudes [see (4.3.34*a*,*b*)] and the negligible contribution in (4.3.33). Any combination of these three functions satisfies the condition in (4.3.30). We thus expect that any solution to (4.3.25) which approaches $-(-x)^{1/2}$ does so in an oscillatory fashion. Indeed, the solution to (4.3.25) in Fig. 4.7, if plotted for negative x, approaches $-(-x)^{1/2}$ as $x \to -\infty$ (see Fig. 4.8).

We argue that the approach to $-(-x)^{1/2}$ is *stable* because if the initial conditions $y(0)$ and $y'(0)$ are varied slightly, the relative amounts of (4.3.34*a*,*b*) in $\varepsilon(x)$ change accordingly, but the main qualitative feature of $y(x)$, namely, its oscillatory approach to $-(-x)^{1/2}$, does not change.

By contrast, it is highly improbable that $y(x)$ approaches $+(-x)^{1/2}$ because $\varepsilon(x)$ would almost certainly contain the exponentially growing component in (4.3.34*c*), which is inconsistent with the condition that $\varepsilon(x) \ll (-x)^{1/2}$ $(x \to -\infty)$. Nevertheless, it is possible to find special initial conditions, $y(0)$ and $y'(0)$, for which the growing exponential in (4.3.34*c*) does *not* appear (see Fig. 4.9); for such initial conditions $y(x)$ does approach $+(-x)^{1/2}$ and $\varepsilon(x) \sim -\frac{1}{8}x^{-2}$ as $x \to -\infty$. However, if these initial conditions are even minutely changed, the growing solution in (4.3.34*c*) appears and $y(x)$ rapidly veers away from $+(-x)^{1/2}$ (see Fig. 4.9). Thus, the approach to $+(-x)^{1/2}$ is *unstable*.

Example 5 *Beyond Painlevé transcendents.* What is the nature of the movable singularities of the solution to

$$y'' = y^2 + e^x? \tag{4.3.35}$$

This equation is a modification of the Painlevé equation (4.3.25) $y'' = y^2 + x$ whose movable singularities are second-order poles. Equation (4.3.35) is a relatively simple example of a differen-

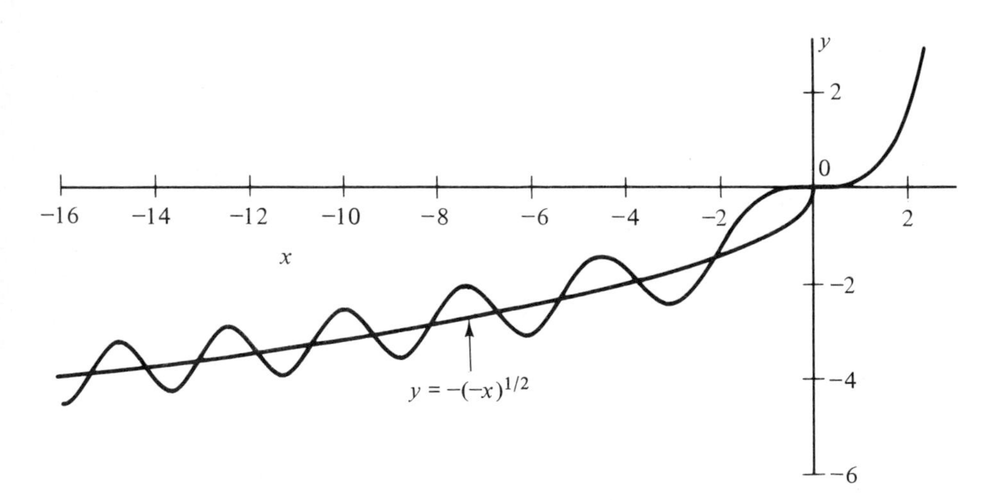

Figure 4.8 A computer plot of the solution to the initial-value problem $y'' = y^2 + x$ $[y(0) = y'(0) = 0]$ for $-16 \le x \le 2$. As $x \to -\infty$, the solution oscillates about $-(-x)^{1/2}$ with an amplitude that gradually decreases like $(-x)^{-1/8}$.

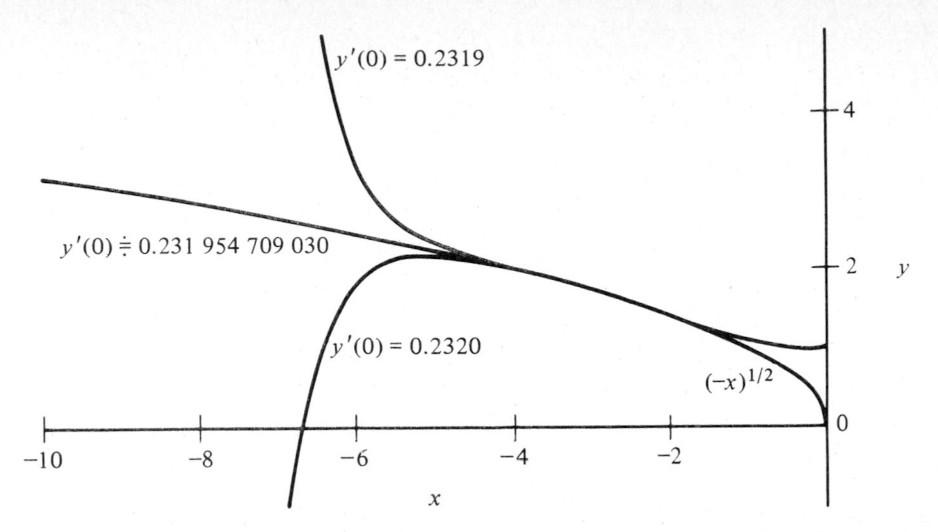

Figure 4.9 A computer plot of three solutions to the differential equation $y'' = y^2 + x$ $[y(0) = 1]$ and with slightly different values of $y'(0)$. The solution having $y'(0) \doteqdot 0.231\ 954\ 709\ 030$ is apparently asymptotic to the curve $y = (-x)^{1/2}$ as $x \to -\infty$. The other two solutions rapidly veer away from $y = (-x)^{1/2}$.

tial equation whose leading singular behavior is that of a second-order pole but whose singularity is *not* a pole. [Thus, (4.3.35) is not a Painlevé equation!]

Local analysis shows that the leading singular behavior of a solution $y(x)$ to (4.3.35) which becomes infinite at $x = x_0$ is $y(x) \sim 6/(x - x_0)^2$ $(x \to x_0)$. However, this leading behavior does not mean that $y(x)$ has a pole at $x = x_0$. If there is a pole at $x = x_0$, then $y = t^{-2}z(t)$, where $t = x - x_0$ and $z(t)$ is analytic in a neighborhood of $t = 0$. We will demonstrate that $z(t)$ is *not* analytic at $t = 0$ by showing that the assumption that

$$y(t) = (1/t^2)(6 + at + bt^2 + ct^3 + dt^4 + et^5 + ft^6 + \cdots)$$

leads to a contradiction. (It is remarkable that we need this many terms to see the contradiction!)

In terms of $t = x - x_0$, the differential equation (4.3.35) is

$$d^2y/dt^2 = y^2 + A(1 + t + t^2/2 + t^3/6 + \cdots),$$

where $A = e^{x_0}$. Substituting the series for $y(t)$, we find that

$$\begin{aligned}\frac{36}{t^4} + \frac{2a}{t^3} + 2d + 6et + 12ft^2 + \cdots = \frac{1}{t^4}[&36 + 12at + (12b + a^2)t^2 + (12c + 2ab)t^3 \\ &+ (b^2 + 2ac + 12d)t^4 + (12e + 2ad + 2bc)t^5 \\ &+ (c^2 + 2bd + 2ae + 12f)t^6 + \cdots] + A + At + \frac{At^2}{2!} + \cdots.\end{aligned}$$

Equating coefficients of t^{-4} gives $36 = 36$. This verifies that the leading behavior of $y(t)$ is indeed $y \sim 6t^{-2}$ $(t \to 0)$.

Next, we equate the coefficients of higher powers of t:

t^{-3}: $\quad 2a = 12a,\quad$ so $a = 0$;

t^{-2}: $\quad 12b + a^2 = 0,\quad$ so $b = 0$;

t^{-1}: $12c + 2ab = 0$, so $c = 0$;

t^0: $2d = b^2 + 2ac + 12d + A$, so $d = -A/10$;

t^1: $6e = 12e + 2ad + 2bc + A$, so $e = -A/6$;

t^2: $12f = c^2 + 2bd + 2ae + 12f + \frac{1}{2}A$, so $A = 0$. (!)

This last result is false; we have found a contradiction in sixth order in the series for $z(t)$, so $z(t)$ is *not* analytic near $t = 0$ and the singularity is not a pole.

What is the nature of the singularity at $x = x_0$? The impossible equation $\frac{1}{2}A = 0$ is reminiscent of the kind of equations one encounters in Frobenius theory when two indices differ by an integer. Thus, since the problem arises in sixth order when the ft^6 term formally cancels, we are led to try instead the combination of terms $ft^6 + f_1 t^6 \ln t$.

Now, separately equating coefficients of t^2 and $t^2 \ln t$ gives

$$12f + 7f_1 = c^2 + 2bd + 2ae + 12f + \tfrac{1}{2}A$$

for t^2 so $f_1 = A/14$; and

$$12f_1 = 12f_1$$

for $t^2 \ln t$ so f is an arbitrary constant. (There are now two arbitrary constants, x_0 and f.)

Now there is no contradiction, but the term $t^6 \log t$ induces a branch point at $x = x_0$. In fact, the correction term $f_1 t^6 \ln t$ is not all that is required. The presence of the nonlinear term y^2 in the differential equation causes terms like $(t^6 \ln t)^2$ to appear in twelfth order in the series for $z(t)$. Ultimately, all powers of $t^6 \ln t$ must be included in the series for $z(t)$. It is likely that the structure of the resulting singularity is extremely complicated. Singularities of this complexity can occur in differential equations of the type $y'' = y^2 + f(x)$ if $f(x)$ is not a linear function of x.

Example 6 *Approximate determination of radius of convergence.* Let us examine the equation $y'' = y^2 + e^x$ near $x = -\infty$. To do this we let $t = e^x$ and investigate the resulting differential equation

$$\left(t\frac{d}{dt}\right)\left(t\frac{d}{dt}\right)y = y^2 + t \tag{4.3.36}$$

near $t = 0$. What is the radius of convergence of that solution $y(t)$ which is analytic at $t = 0$ and which vanishes at $t = 0$?

First we must ascertain the leading local behavior of $y(t)$ near $t = 0$. There are three possible dominant balances. If $y^2 \ll t$ $(t \to 0)$, then

$$(t\, d/dt)^2 y \sim t, \qquad \text{so } y \sim t, \qquad t \to 0.$$

If $t \ll y^2$ $(t \to 0)$, then

$$(t\, d/dt)^2 y \sim y^2, \qquad \text{so } y \sim 6/(\ln t)^2, \qquad t \to 0,$$

which is not analytic at $t = 0$ and must be rejected. The third possibility is that

$$y^2 \sim -t, \qquad \text{so } y \sim (-t)^{1/2}, \qquad t \to 0.$$

But then $t \ll (t\, d/dt)^2 y$ $(t \to 0)$, so this possibility is inconsistent.

Hence, the Taylor series about $t = 0$ takes the form

$$y(t) = \sum_{n=1}^{\infty} a_n t^n, \qquad a_1 = 1. \tag{4.3.37}$$

Substituting this series in the differential equation and equating coefficients of powers of t gives the recurrence relation

$$a_n = \frac{1}{n^2} \sum_{j=1}^{n-1} a_j a_{n-j}, \qquad n \geq 2. \tag{4.3.38}$$

Thus, $a_2 = \frac{1}{4}$, $a_3 = \frac{1}{18}$, and so on. Since the coefficients a_n are all positive, the nearest singular point of $y(t)$ lies on the real axis.

Using (4.3.38) we can estimate that the radius of convergence R lies between 6 and $3\sqrt{6} \doteq 7.35$ (see Prob. 4.19). The upper bound may be sharpened somewhat to 7.1 (see Prob. 4.20). Numerical solution of the differential equation indicates that $R \doteq 6.8$.

Example 7 *Thomas-Fermi equation.* The Thomas-Fermi equation

$$y'' = y^{3/2}x^{-1/2}, \qquad y(0) = 1,\ y(+\infty) = 0, \tag{4.3.39}$$

provides a semiclassical description of the charge density in atoms of high atomic number. What is the leading behavior of the solution for large positive x?

Here, we are not trying to find a complete description of the asymptotic behaviors of all solutions to $y'' = y^{3/2}x^{-1/2}$ as $x \to +\infty$; we want to study only those solutions which approach 0 as $x \to +\infty$. Observe that it is consistent for $y(x)$ to fall off algebraically for large x. To check this, we substitute $y(x) \sim A/x^b$ $(x \to +\infty)$ into the differential equation:

$$\frac{Ab(b+1)}{x^{b+2}} \sim \frac{A^{3/2}}{x^{3b/2+1/2}}, \qquad x \to +\infty.$$

Comparing powers of x on each side gives $b = 3$ and $A = 144$. Thus, it is reasonable to expect that

$$y(x) \sim 144/x^3, \qquad x \to +\infty.$$

In fact, the function $y(x) = 144x^{-3}$ happens to be an exact solution of the differential equation. However, it does not satisfy the boundary condition at $x = 0$. If $144x^{-3}$ is also the leading behavior of the solution which *does* satisfy the boundary conditions, then we should be able to find the higher-order corrections to this behavior. Suppose that as $x \to +\infty$ the corrections take the form $y(x) = (144/x^3)[1 + \varepsilon(x)]$, where $\varepsilon(x) \ll 1$ $(x \to +\infty)$. When x is large and positive, $\varepsilon(x)$ satisfies an asymptotic Euler equation: $\varepsilon'' - 6\varepsilon'/x \sim 6\varepsilon/x^2$ $(x \to +\infty)$. The solutions to this equation have the form x^r, where $r = (7 + \sqrt{73})/2 \simeq 7.772$ and $r = (7 - \sqrt{73})/2 \simeq -0.772$; only the second is consistent with the assumption that $\varepsilon \to 0$ as $x \to +\infty$. Thus, we expect $y(x)$ to behave like

$$y(x) \sim \frac{144}{x^3}\left(1 + \frac{C}{x^{0.772}}\right), \qquad x \to +\infty, \tag{4.3.40}$$

where C is a constant which cannot be determined by the methods of local analysis. Unfortunately, the estimate $y(x) \sim 144x^{-3}$ $(x \to +\infty)$ has a relative error of $Cx^{-0.772}$ where C is unknown, so there is no way to predict how large x must be before this estimate becomes accurate.

To check the leading behavior $y \sim 144x^{-3}$ we have plotted in Fig. 4.10 the numerical solution to the Thomas-Fermi equation along with the predicted leading behavior. Observe that the agreement is terrible! When $x = 15$ the leading behavior is three times larger than the exact solution. In Table 4.3 we compare the solution to the Thomas-Fermi equation with its leading behavior for large values of x. Again we observe very poor agreement.

Apparently, the leading behavior is such a poor approximation to the solution of the Thomas-Fermi equation because by sheer perversity the value of C in (4.3.40) is not small. If we fit (4.3.40) to the numbers in Table 4.3, we find that $C \doteq -13.270\,973\,8$ (see Prob. 4.23). Thus, even when $x = 100$, the term $Cx^{-0.772}$ is not small compared with 1 and the leading asymptotic behavior is not very useful. The leading asymptotic behavior does not attain a relative error of 10 percent until x is about 550 and the first four terms in the asymptotic series do not attain a relative error of 10 percent until x is about 35.

Here, we have finally seen an example where local asymptotic analysis is not impressive. Nevertheless, it is easy to predict the qualitative features of the solution $y(x)$ for $x \geq 0$. First, we know that $y(x) \geq 0$; otherwise the solution is complex. Second, $y(x)$ has positive curvature

Table 4.3 A comparison of the numerical solution to the Thomas-Fermi equation $y'' = y^{3/2}x^{-1/2}$ $[y(0) = 1, y(\infty) = 0]$ with its leading behavior $144x^{-3}$ for various values of x

Here we see an example in which local analysis is unimpressive. The poor agreement is caused by the large magnitude of the error term $Cx^{-0.772}$ in (4.3.2). Also listed are the values of the first four terms in the expansion of $y(x)$ for large x. In this expansion $r = (7 - \sqrt{73})/2 \doteq -0.772\,001\,873$, $C \doteq -13.270\,973\,8$, $D \doteq 110.197\,059$, $E \doteq -732.467\,106$. This series does not attain an accuracy of 1 percent until $x > 70$ because the coefficients are so large

		Predicted leading behavior of $y(x)$ for large x		First four terms in asymptotic series for $y(x)$ for large x	
x	Exact solution $y(x)$ to Thomas-Fermi equation	$144x^{-3}$	Relative error, %	$(144/x^3)(1 + Cx^r + Dx^{2r} + Ex^{3r})$	Relative error, %
0	1.000 0	∞			
0.1	0.881 7	144,000			
0.2	0.793 1	18,000			
0.3	0.720 6	5,333			
0.4	0.659 5	2,250			
0.5	0.607 0	1,152			
0.6	0.561 2	666.7			
0.7	0.520 8	419.8			
0.8	0.484 9	281.3			
0.9	0.452 9	197.5			
1	0.424 0	144.0		−91,374	
2	0.243 0	18.00		−2,089	

3	0.156 6	5.333		−223.9	
4	0.108 4	2.250		−45.30	
5	0.788 1 (−1)	1.152		−12.98	
6	0.594 2 (−1)	0.666 7		−4.632	
7	0.461 0 (−1)	0.419 8		−1.921	
8	0.365 9 (−1)	0.281 3		−0.886 8	
9	0.295 9 (−1)	0.197 5		−0.443 3	
10	0.243 1 (−1)	0.144 0		−0.235 0	
20	0.578 5 (−2)	0.180 0 (−1)		0.100 1 (−2)	
30	0.225 6 (−2)	0.533 3 (−2)		0.180 8 (−2)	−20
40	0.111 4 (−2)	0.225 0 (−2)		0.103 1 (−2)	−7.4
50	0.632 3 (−3)	0.115 2 (−3)		0.610 2 (−3)	−3.5
60	0.393 9 (−3)	0.666 7 (−3)	69	0.386 4 (−3)	−1.9
70	0.262 3 (−3)	0.419 8 (−3)	60	0.259 3 (−3)	−1.1
80	0.183 5 (−3)	0.281 3 (−3)	53	0.182 1 (−3)	−0.73
90	0.133 5 (−3)	0.197 5 (−3)	48	0.132 9 (−3)	−0.50
100	0.100 2 (−3)	0.144 0 (−3)	44	0.998 9 (−4)	−0.35
200	0.145 018 0 (−4)	0.180 0 (−4)	24	0.144 962 8 (−4)	−0.38 (−1)
300	0.454 857 2 (−5)	0.533 3 (−5)	17	0.454 809 4 (−5)	−0.11 (−1)
400	0.197 973 3 (−5)	0.225 0 (−5)	14	0.197 964 9 (−5)	−0.42 (−2)
500	0.103 407 7 (−5)	0.115 2 (−5)	11	0.103 405 5 (−5)	−0.21 (−2)
600	0.606 868 8 (−6)	0.666 7 (−6)	9.9	0.606 861 6 (−6)	−0.12 (−2)
700	0.386 176 5 (−6)	0.419 8 (−6)	8.7	0.386 173 9 (−6)	−0.73 (−3)
800	0.260 813 7 (−6)	0.281 3 (−6)	7.8	0.260 812 5 (−6)	−0.48 (−3)
900	0.184 372 4 (−6)	0.197 5 (−6)	7.1	0.184 371 8 (−6)	−0.34 (−3)
1,000	0.135 127 5 (−6)	0.144 0 (−6)	6.6	0.135 127 2 (−6)	−0.24 (−3)

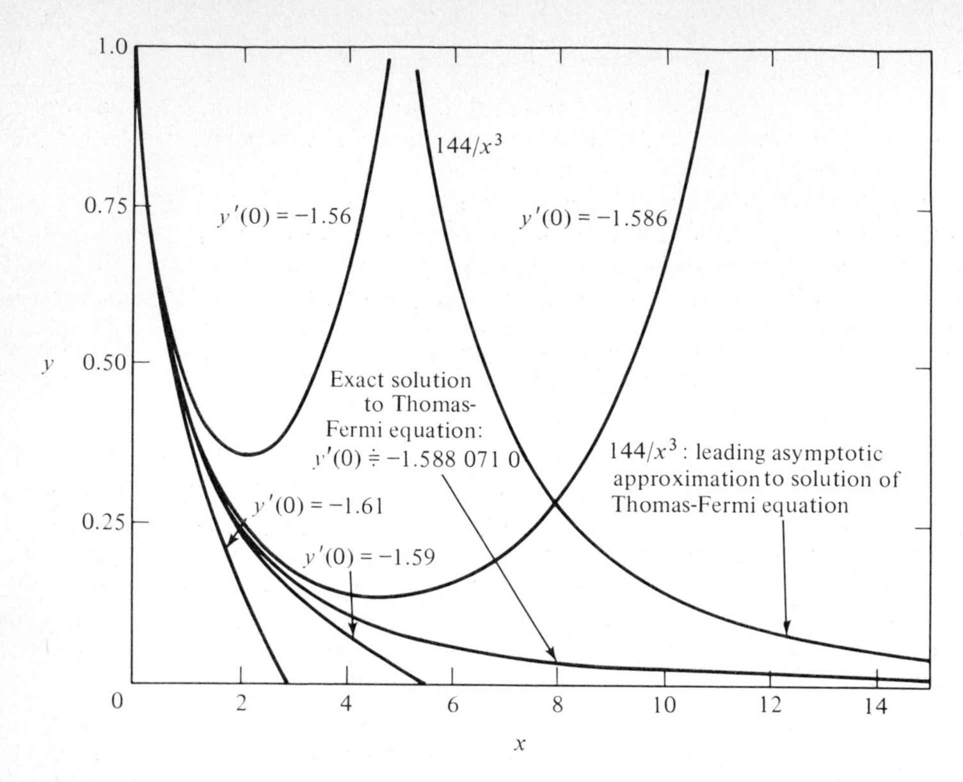

Figure 4.10 A computer plot of the solution to the Thomas-Fermi equation $y'' = y^{3/2}x^{-1/2}$ [$y(0) = 1$, $y(\infty) = 0$], along with its leading asymptotic behavior $144x^{-3}$. Observe that the leading behavior is a poor estimate of $y(x)$ even when $x = 15$. Also shown are four functions satisfying the differential equation $y'' = y^{3/2}x^{-1/2}$ and $y(0) = 1$, but having initial slopes which are slightly different from the initial slope needed to give $y(+\infty) = 0$. Two of these functions cross the x axis and become complex; the other two become infinite before $x = 15$.

($y'' > 0$), so the slope must *increase* with x. Consequently, $y'(x) < 0$; otherwise $y(x)$ could not satisfy the boundary condition $y(\infty) = 0$. Thus, $y(x)$ has no local minima or maxima; $y(x)$ smoothly decreases from $y = 1$ to $y = 0$ as $x \to \infty$. The initial slope $y'(0)$ must be chosen very precisely. If $y'(0)$ is too negative, then $y(x)$ crosses 0 at a finite value of x. if $y'(0)$ is not sufficiently negative, then $y(x)$ reaches a minimum, turns around, and rapidly becomes singular at some finite value of x (see Fig. 4.10). The leading behavior at this singularity is that of a fourth-order pole (see Probs. 4.24 and 4.25):

$$y(x) \sim \frac{400a}{(x-a)^4}, \qquad x \to a. \tag{4.3.41}$$

The graph of the solution to the Thomas-Fermi equation in Fig. 4.10 exhibits all of these qualitative characteristics.

Example 8 *Boundary-value problem on a finite interval.* In this final example we apply local analysis to study the properties of the solution to a nonlinear boundary-value problem near the boundaries. Let us examine the leading behavior of the solution to

$$yy'' = -1, \qquad y(0) = y(1) = 0, \tag{4.3.42}$$

as $x \to 0+$. (Note that this differential equation would not be consistent with the boundary conditions if -1 were replaced by 1. See Prob. 4.26.)

The simplest guess is that $y(x)$ approaches 0 algebraically as $x \to 0+$. If we assume that $y(x) \sim Ax^b$ $(x \to 0+)$ and substitute into the differential equation, we obtain

$$A^2b(b-1)x^{2b-2} \sim -1, \qquad x \to 0+.$$

It follows that $b = 1$. But this value is inconsistent because the left side of the equation vanishes. Unfortunately, the second derivative of x is zero and not $1/x$!

This argument shows that the behavior of $y(x)$ as $x \to 0+$ is *not* algebraic. Can it be logarithmic? We try a solution whose leading behavior is

$$y(x) \sim Ax(-\ln x)^b, \qquad x \to 0+.$$

Thus, $$y''(x) \sim -Ab(-\ln x)^{b-1}/x + Ab(b-1)(-\ln x)^{b-2}/x, \qquad x \to 0+.$$

Neglecting the second term on the right side of this equation because it is smaller by a factor of $(-\ln x)^{-1}$ as $x \to 0+$ and substituting the first term into the differential equation gives $-1 \sim -A^2b(-\ln x)^{2b-1}$ $(x \to 0+)$. It follows that $b = \frac{1}{2}$ and $A = \pm\sqrt{2}$. Hence the leading behavior of $y(x)$ as $x \to 0+$ is

$$y(x) \sim \pm x\sqrt{-2 \ln x}, \qquad x \to 0+. \tag{4.3.43}$$

Also, by a similar argument we have

$$y(x) \sim \pm(1-x)\sqrt{-2 \ln (1-x)}, \qquad x \to 1-. \tag{4.3.44}$$

The two asymptotic behaviors (4.3.43) and (4.3.44) may be verified by comparing them with the exact solution to (4.3.42). This comparison is given in Fig. 4.11.

(I) 4.4 NONLINEAR AUTONOMOUS SYSTEMS

Autonomous systems of equations, when they are interpreted as describing the motion of a point in phase space, are particularly susceptible to some very beautiful techniques of local analysis. By performing a local analysis of the system near what are known as critical points, one can make remarkably accurate predictions about the global properties of the solution.

Phase-Space Interpretation of Autonomous Equations

Differential equations which do not contain the independent variable explicitly are said to be *autonomous*. Any differential equation is equivalent to an autonomous equation of one higher order; to remove the explicit reference to the independent variable x one simply solves for x in terms of y and its derivatives and then differentiates the resulting equation once.

As we will see, it is convenient to study the approximate behavior of an autonomous equation of order n when it is in the form of a system of n coupled first-order equations. Also, by convention we will think of the independent vari-

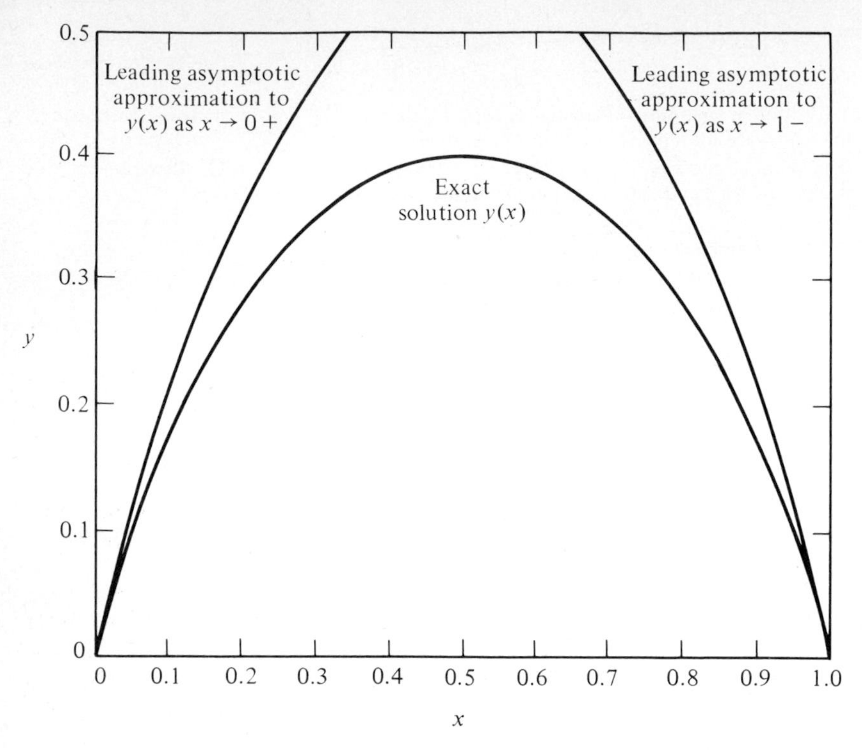

Figure 4.11 Comparison of an exact solution $y(x)$ to $yy'' = -1$ $[y(0) = y(1) = 0]$, with its leading asymptotic approximations $y(x) \sim x\sqrt{-2 \ln x}$ $(x \to 0+)$ and $y(x) \sim (1-x)\sqrt{-2 \ln (1-x)}$ $(x \to 1-)$ in (4.3.43) and (4.3.44). See Prob. 4.26(*b*).

able of the system as time t and the dependent variables $y_1, y_2, \ldots, y_n$ as position coordinates. The general form of such a system is

$$\begin{aligned} \dot{y}_1 &= f_1(y_1, y_2, \ldots, y_n), \\ \dot{y}_2 &= f_2(y_1, y_2, \ldots, y_n), \\ &\vdots \\ \dot{y}_n &= f_n(y_1, y_2, \ldots, y_n), \end{aligned} \tag{4.4.1}$$

where the dots indicate differentiation with respect to t.

The solution of the system (4.4.1) is a curve or *trajectory* in an n-dimensional space called *phase space*. The trajectory is parametrized in terms of t: $y_1 = y_1(t)$, $y_2 = y_2(t), \ldots, y_n = y_n(t)$.

We will assume that $f_1, f_2, \ldots, f_n$ are continuously differentiable with respect to each of their arguments. Thus, by the existence and uniqueness theorem of differential equations (see Sec. 1.2), any initial condition $y_1(0) = a_1$, $y_2(0) = a_2, \ldots, y_n(0) = a_n$ gives rise to a *unique* trajectory through the point $(a_1, a_2, \ldots, a_n)$.

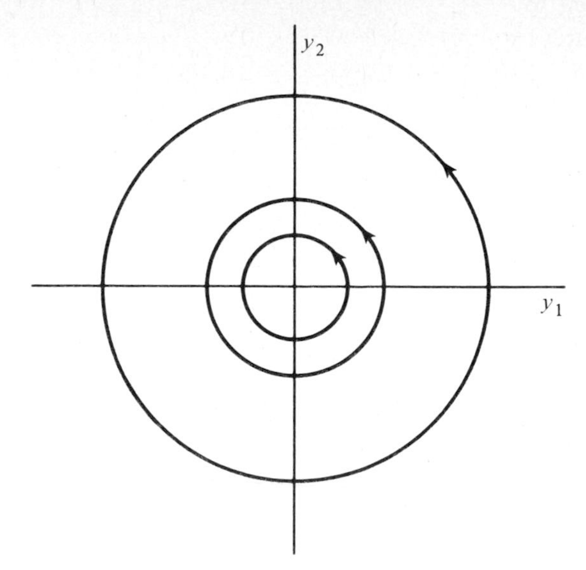

Figure 4.12 A plot of the trajectories in phase space for the two-dimensional system $\dot{y}_1 = y_2$, $\dot{y}_2 = -y_1$. Note that individual trajectories, which are circles, never intersect. The arrows indicate the direction of motion of the vector $[y_1(t), y_2(t)]$ for increasing t.

To understand this uniqueness property geometrically, note that at every point on the trajectory $[y_1(t), y_2(t), \ldots, y_n(t)]$ the system (4.4.1) assigns a unique velocity vector $[\dot{y}_1(t), \dot{y}_2(t), \ldots, \dot{y}_n(t)]$ which is tangent to the trajectory at that point. It immediately follows that *two trajectories cannot cross*; otherwise, the tangent vector at the crossing point would not be unique.

Example 1 *Phase plane for* $\ddot{y} + y = 0$. The equation $\ddot{y} + y = 0$ describes a particle undergoing harmonic motion (sinusoidal oscillation). To represent this motion in the phase plane we convert this equation into a first-order system; to wit, we let $y_1(t) = y(t)$, the position of the particle, and $y_2(t) = \dot{y}(t)$, the velocity of the particle. Then $\dot{y}_1 = y_2$, $\dot{y}_2 = -y_1$, is an equivalent first-order system. In phase space the trajectories are concentric circles (see Fig. 4.12).

Critical Points in Phase Space

If there are any solutions to the set of simultaneous algebraic equations

$$\begin{aligned} f_1(y_1, y_2, \ldots, y_n) &= 0, \\ f_2(y_1, y_2, \ldots, y_n) &= 0, \\ &\vdots \\ f_n(y_1, y_2, \ldots, y_n) &= 0, \end{aligned}$$

then there are special degenerate trajectories in phase space which are just points. (The velocity at these points is zero so the position vector does not move.) Such points are called *critical points.*

Note that while a trajectory can approach a critical point as $t \to \infty$, it cannot reach such a point in a finite time. The proof is simple. Suppose it were possible for a trajectory to reach a critical point in time T. Then the time-reversed system of equations (the equations obtained by replacing f_i with $-f_i$) would exhibit an impossible behavior: the position vector $[y_1(t), y_2(t), \ldots, y_n(t)]$ would rest motionless at the critical point and then suddenly begin to move at time T. (Recall that in an autonomous system $f_1, f_2, \ldots, f_n$, the components of the velocity vector, depend only on the position of the particle and not on the time.)

We will now see how a local analysis of solutions in the vicinity of the critical points actually enables one to deduce the *global* behavior of the solutions!

One-Dimensional Phase Space

One-dimensional phase space, which we call the *phase line*, is used to study solutions to the first-order autonomous system $\dot{y} = f(y)$. There are only three possibilities for the global behavior of a trajectory on a phase line:

1. The trajectory may approach a critical point as $t \to +\infty$.
2. The trajectory may approach $\pm\infty$ as $t \to +\infty$.
3. The trajectory may remain motionless at a critical point for all t.

In the neighborhood of a critical point there are three possibilities for the local behavior:

1. All trajectories may approach the critical point as $t \to +\infty$. We call such a critical point a *stable node* (see Fig. 4.13). The point $y = 0$ is a stable node for the equation $\dot{y} = -y$.
2. All trajectories may move away from the critical point as $t \to +\infty$. We call such a critical point an *unstable node* (see Fig. 4.13). The point $y = 0$ is an unstable node for the equation $\dot{y} = y^3$.
3. Trajectories on one side of the critical point may move toward it while trajectories on the other side of the critical point move away from it as $t \to +\infty$. We call such a point a *saddle point* (see Fig. 4.13). The point $y = 0$ is a saddle point for the equation $\dot{y} = y^2$.

Because there are only three kinds of global behaviors, it is easy to deduce the global time dependence of a one-dimensional system from a local analysis of the critical points.

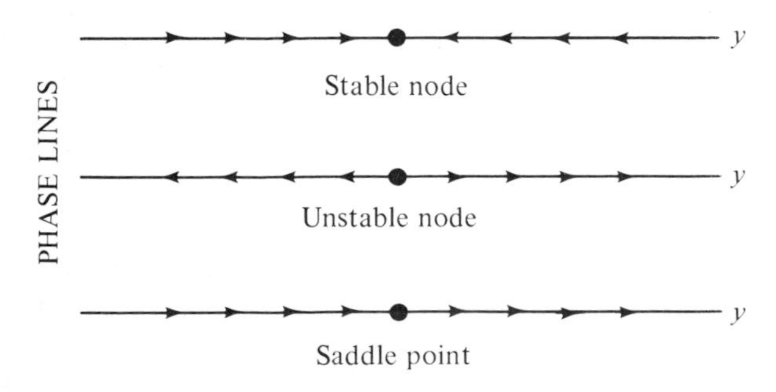

Figure 4.13 Phase line (one-dimensional phase space) showing the three kinds of critical points that may occur. Arrows indicate the motion of $y(t)$ as $t \to +\infty$. On a phase line the local behavior near critical points determines the global behavior of solutions.

Example 2 *Critical-point analysis of a one-dimensional system.* The equation $\dot{y} = y^2 - y$ represents a population y having a quadratic birthrate (y^2) and a linear deathrate ($-y$). It is easy to find the exact time dependence of $y(t)$ by solving the differential equation because the equation is separable (see Prob. 4.39). However, it is even easier to obtain a rough picture of the global behavior using critical-point analysis.

The critical points for $\dot{y} = y^2 - y$ solve the algebraic equation $y^2 - y = 0$. Thus, there are critical points at $y_1 = 0$ and at $y_2 = 1$. Next, we classify these critical points by finding the local behavior of $y(t)$ near y_1 and y_2. To do this we *linearize* the differential equation near these points. For example, when y is near 0 we let $y(t) = \varepsilon(t)$ and approximate the exact differential equation by $\dot{\varepsilon}(t) \sim -\varepsilon(t)$ ($\varepsilon \to 0$). The solutions to this equation decay exponentially, so $y = 0$ is a stable node. Similarly, when y is near 1 we let $y(t) = 1 + \varepsilon(t)$ and approximate the exact differential equation by $\dot{\varepsilon}(t) \sim \varepsilon(t)$ ($\varepsilon \to 0$). Since the solutions to this equation grow exponentially with time, we identify $y = 1$ as an unstable node.

Now we can infer the behavior of $y(t)$ for *any* initial value $y(0)$. Any trajectory on the phase line to the left of $y = 0$ must move to the right and approach $y = 0$ as $t \to +\infty$ because, by continuity, trajectories sufficiently near $y = 0$ move toward $y = 0$. If some trajectories were to move leftward toward $y = -\infty$ as $t \to +\infty$, then the dividing point between left-moving and right-moving trajectories would be a critical point. But there is no critical point to the left of $y = 0$. Similarly, all trajectories to the right of $y = 1$ move rightward toward $y = \infty$ as $t \to +\infty$ and all trajectories between $y = 0$ and $y = 1$ move leftward toward $y = 0$ as $t \to +\infty$. (Why?)

This completes our critical-point analysis of the differential equation and our conclusions about the structure of the phase line are depicted in Fig. 4.14. From this diagram we can predict the behavior of $y(t)$ as $t \to +\infty$ for any $y(0)$:

if $y(0) < 0$, then $y(t) \to 0$ as $t \to +\infty$,
if $y(0) = 0$, then $y(t) = 0$ for all t,
if $0 < y(0) < 1$, then $y(t) \to 0$ as $t \to +\infty$,
if $y(0) = 1$, then $y(t) = 1$ for all t,
if $y(0) > 1$, then $y(t) \to +\infty$ as $t \to +\infty$.

These conclusions are verified by the plots of $y(t)$ given in Fig. 4.14.

Observe that critical-point analysis does *not* predict how fast $y(t)$ approaches 0 or $+\infty$; that requires further *local* analysis. Rather, it answers the *global* question of how the initial condition $y(0)$ affects the behavior of $y(t)$ as $t \to +\infty$.

Two-Dimensional Phase Space

Two-dimensional phase space (the phase plane) is used to study a system of two coupled first-order equations. The phase plane is more complicated than the phase line, but, as we will see, it is still possible to make elegant global analyses of systems of two coupled differential equations. First, we enumerate the possible global behaviors of a trajectory in a two-dimensional system:

1. The trajectory may approach a critical point as $t \to +\infty$.
2. The trajectory may approach ∞ as $t \to +\infty$.
3. The trajectory may remain motionless at a critical point for all t.
4. The trajectory may describe a closed orbit or cycle (see Fig. 4.12).
5. The trajectory may approach a closed orbit (by spiraling inward or outward toward the orbit) as $t \to +\infty$.

Note that the first three possibilities also occur in one-dimensional systems, but the fourth and fifth are new configurations which cannot occur in a phase space of less than two dimensions.

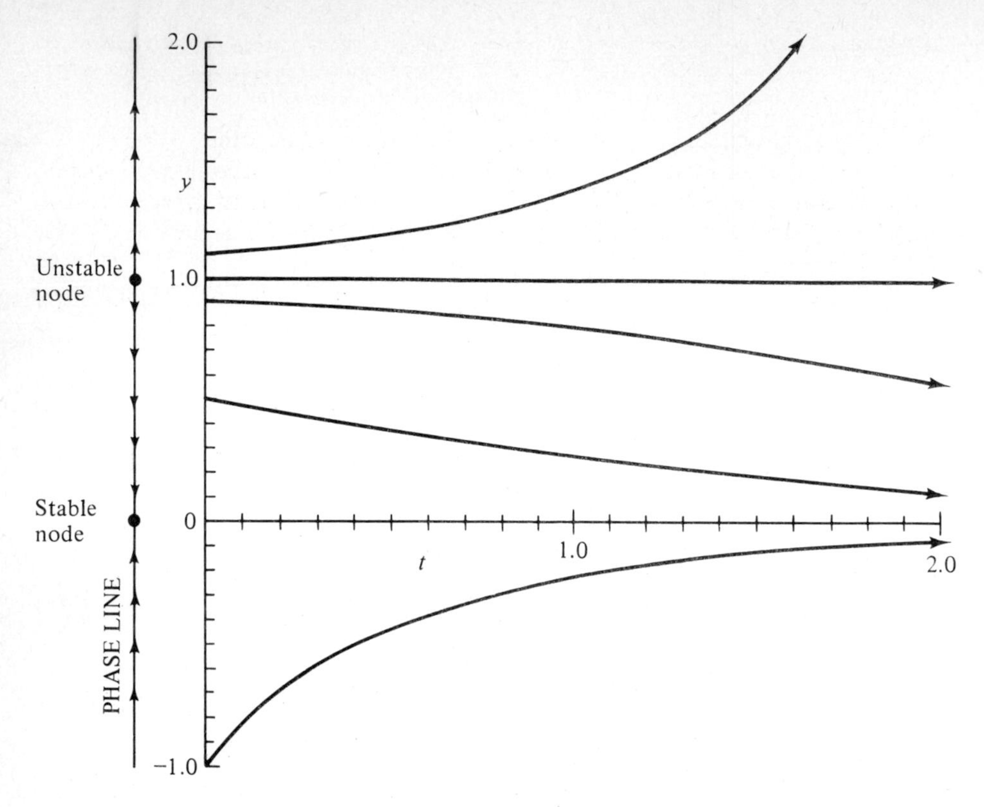

Figure 4.14 Phase-line analysis of the equation $\dot{y} = y^2 - y$ (see Example 2). To the left is the phase line for this equation on which the critical points at $y = 0$ and $y = 1$ are indicated. To the right is a plot of $y(t)$ for various initial values $y(0)$. Phase-line analysis enables us to predict the qualitative large-t behavior of $y(t)$ for any value of $y(0)$.

Next, we enumerate the possible local behaviors for trajectories near a critical point:

1. All trajectories may approach the critical point along curves which are asymptotically straight lines as $t \to +\infty$. We call such a critical point a *stable node.*
2. All trajectories may approach the critical point along spiral curves as $t \to +\infty$. Such a critical point is called a *stable spiral point.* [It is also possible for trajectories to approach the critical point along curves which are neither spirals nor asymptotic to straight lines (see Probs. 4.44 to 4.46).]
3. All time-reversed trajectories [that is, $y(t)$ with t decreasing] may move toward the critical point along paths which are asymptotically straight lines as $t \to -\infty$. Such a critical point is an *unstable node.* As *t increases*, all trajectories that start near an unstable node move *away* from the node along paths that are approximately straight lines, at least until the trajectory gets far from the node.
4. All time-reversed trajectories may move toward the critical point along spiral

curves as $t \to -\infty$. Such a critical point is called an *unstable spiral point*. As t *increases*, all trajectories move *away* from an unstable spiral point along trajectories that are, at least initially, spiral shaped.

5. Some trajectories may approach the critical point while others move away from the critical point as $t \to +\infty$. Such a critical point is called a *saddle point*.
6. All trajectories may form closed orbits about the critical point. Such a critical point is called a *center*. In Fig. 4.12 we see an example of a center.

Note that while nodes and saddle points occur in one-dimensional phase space, spiral points and centers cannot exist in less than two dimensions.

Linear Autonomous Systems

Since two-dimensional linear autonomous systems can exhibit any of the critical point behaviors that we have described above, it is appropriate to study linear systems before going on to nonlinear systems. With this in mind we introduce an easy method for solving linear autonomous systems.

The method uses elementary matrix algebra (see Prob. 4.41 for a review of the necessary theory). A two-dimensional linear autonomous system $\dot{y}_1 = ay_1 + by_2$, $\dot{y}_2 = cy_1 + dy_2$ may be rewritten in matrix form as

$$\dot{\mathbf{Y}} = M\mathbf{Y}, \tag{4.4.2}$$

where

$$\mathbf{Y} = \begin{pmatrix} y_1 \\ y_2 \end{pmatrix} \quad \text{and} \quad M = \begin{pmatrix} a & b \\ c & d \end{pmatrix}.$$

It is easy to verify that if the eigenvalues λ_1 and λ_2 of the matrix M are distinct and $\mathbf{V}_1$ and $\mathbf{V}_2$ are eigenvectors of M associated with the eigenvalues λ_1 and λ_2, then the general solution to (4.4.2) has the form

$$\mathbf{Y}(t) = c_1\mathbf{V}_1 e^{\lambda_1 t} + c_2\mathbf{V}_2 e^{\lambda_2 t}, \tag{4.4.3}$$

where c_1 and c_2 are constants of integration which are determined by the initial position $\mathbf{Y}(0)$. For the general solution to (4.4.2) when $\lambda_1 = \lambda_2$ see Prob. 4.42.

The linear system (4.4.2) has a critical point at the origin (0, 0). It is easy to classify this critical point once λ_1 and λ_2 are known. Note that λ_1 and λ_2 satisfy the eigenvalue condition

$$\det [M - I\lambda] = \det \begin{bmatrix} a - \lambda & b \\ c & d - \lambda \end{bmatrix}$$

$$= \lambda^2 - \lambda(a + d) + ad - bc = 0. \tag{4.4.4}$$

If λ_1 and λ_2 are real and negative, then all trajectories approach the origin as $t \to +\infty$ and (0, 0) is a stable node. Conversely, if λ_1 and λ_2 are real and positive, then all trajectories move away from (0, 0) as $t \to +\infty$ and (0, 0) is an unstable node. Also, if λ_1 and λ_2 are real but λ_1 is positive and λ_2 is negative, then (0, 0) is a saddle point; trajectories approach the origin in the direction $\mathbf{V}_2$ and move away from the origin in the direction $\mathbf{V}_1$.

Solutions λ_1 and λ_2 of (4.4.4) may be complex. However, when the matrix M is real, then λ_1 and λ_2 must be a complex conjugate pair. If λ_1 and λ_2 are pure imaginary, then the vector $\mathbf{Y}(t)$ represents a closed orbit for any c_1 and c_2 and the critical point at (0, 0) is a center. If λ_1 and λ_2 are complex with nonzero real part, then the critical point at (0, 0) is a spiral point. When Re $\lambda_{1,2} < 0$, then $\mathbf{Y}(t) \to 0$ as $t \to +\infty$ and (0, 0) is a stable spiral point; conversely, when Re $\lambda_{1,2} > 0$, (0, 0) is an unstable spiral point.

It is important to determine the directions of rotation of the trajectories in the vicinity of a spiral point or a center. To determine whether the rotation is clockwise or counterclockwise we simply let $y_2 = 0$, $y_1 > 0$, and see whether $\dot{y}_2$ is negative or positive.

Example 3 *Identification of critical points for linear autonomous systems.*

(*a*) The system $\dot{y}_1 = -y_1$, $\dot{y}_2 = -3y_2$ has a stable node at the origin because the eigenvalues of

$$M = \begin{pmatrix} -1 & 0 \\ 0 & -3 \end{pmatrix}$$

are -1 and -3 which are both negative.

(*b*) The system $\dot{y}_1 = y_2$, $\dot{y}_2 = -2y_1 - 2y_2$ has a stable spiral point at (0, 0) because the eigenvalues of

$$M = \begin{pmatrix} 0 & 1 \\ -2 & -2 \end{pmatrix}$$

are $\lambda_{1,2} = -1 \pm i$ which have negative real parts. All orbits approach (0, 0) following spirals in a clockwise direction. (Why?)

(*c*) The system $\dot{y}_1 = y_1$, $\dot{y}_2 = y_2$ has an unstable node at (0, 0). (Why?)

(*d*) The system $\dot{y}_1 = y_1 + 2y_2$, $\dot{y}_2 = -y_1 + y_2$ has an unstable spiral point at the origin. (Why?)

(*e*) The system $\dot{y}_1 = y_1 + 2y_2$, $\dot{y}_2 = y_1 + y_2$ has a saddle point at (0, 0). (Why?)

(*f*) The system $\dot{y}_1 = y_2$, $\dot{y}_2 = -y_1$ has a center at the origin because the eigenvalues of the matrix

$$M = \begin{pmatrix} 0 & 1 \\ -1 & 0 \end{pmatrix}$$

are $\lambda_{1,2} = \pm i$ which are pure imaginary.

Critical-Point Analysis of Two-Dimensional Nonlinear Systems

To illustrate the power of critical-point analysis, we use it to deduce the global features of some nonlinear systems of equations. The approach we take in the following examples is the same as in Example 2. We first identify the critical points. Then, we perform a local analysis of the system very near these critical points. As in Example 2, the exact system can usually be approximated by a linear autonomous system near a critical point. Using matrix methods we identify the nature of the critical points of the linear systems. Finally, we assemble the results of our local analysis and synthesize a qualitative global picture of the solution to the nonlinear system.

Example 4 *Nonlinear system having two critical points.* The nonlinear system

$$\begin{aligned} \dot{y}_1 &= y_1 - y_1 y_2, \\ \dot{y}_2 &= -y_2 + y_1 y_2, \end{aligned} \tag{4.4.5}$$

known as the Volterra equations, is a simple model of a predator-prey relation between two populations like that of rabbits and foxes: y_1 (the rabbit population) will grow out of bounds if y_2 (the fox population) is 0; however, if y_1 is 0 then y_2 will decay to 0 because of starvation. What happens if y_1 and y_2 are initially positive?

There are two critical points, (0, 0) and (1, 1). Near (0, 0) we let $(y_1, y_2) = (\varepsilon_1, \varepsilon_2)$ and approximate the exact differential equation by $\dot{\varepsilon}_1 \sim \varepsilon_1, \dot{\varepsilon}_2 \sim -\varepsilon_2$ $(\varepsilon_1, \varepsilon_2 \to 0)$. The solution to this system exhibits saddle-point behavior; trajectories near (0, 0) approach the origin vertically and move away from the origin horizontally as $t \to \infty$.

Near (1, 1) we let $(y_1, y_2) = (1 + \varepsilon_1, 1 + \varepsilon_2)$ and approximate the exact equation by $\dot{\varepsilon}_1 \sim -\varepsilon_2, \dot{\varepsilon}_2 \sim \varepsilon_1$ $(\varepsilon_1, \varepsilon_2 \to 0)$. The eigenvalues of the matrix M for this linear system are $\pm i$. Therefore, the critical point at (1, 1) is a center having closed orbits going counterclockwise as $t \to +\infty$. The counterclockwise rotation about (1, 1) is consistent with the directions of incoming and outgoing trajectories near the saddle point at (0, 0).

What can we infer about the global behavior? Suppose the initial condition were $y_1(0) = y_2(0) = a$ $(0 < a < 1)$. Then as t increases from 0, the vector $[y_1(t), y_2(t)]$ would move counterclockwise around the point (1, 1). [If it moved clockwise we would have discontinuous behavior because for a sufficiently near 1 we know that the rotation is counterclockwise. Check directly from (4.4.5) that the rotation really is counterclockwise.] As t increases, the vector must *continue* to rotate around (1, 1). It cannot cross the y_1 or y_2 axes because they are themselves trajectories. A deeper analysis (see Example 9 and Prob. 4.49) shows that this trajectory cannot approach ∞. Therefore, for some t, the vector must encircle the point (1, 1) and eventually recross the line connecting (0, 0) and (1, 1). Moreover, it must cross at the initial point (a, a) (see again Example 9 and Prob. 4.49).

In summary, all trajectories of (4.4.5) with $y_1(0) > 0$, $y_2(0) > 0$ are closed and encircle the point (1, 1) regardless of the initial condition at $t = 0$. Thus, the populations y_1 and y_2 oscillate with time [see Fig. 4.15 for a numerical solution to (4.4.5)]. However, it is important to note that while the conclusion that trajectories are exactly closed is correct, it cannot be justified by local analysis alone. Although it is often possible to infer global behavior from local analysis of stable and unstable critical points, in this example local analysis gives an incomplete description of the nature of phase-space trajectories.

Example 5 *Nonlinear system having three critical points.* Let us try to predict the global behavior of the system

$$\begin{aligned} \dot{y}_1 &= y_1(3 - y_1 - y_2), \\ \dot{y}_2 &= y_2(y_1 - 1), \end{aligned} \tag{4.4.6}$$

in the positive quadrant. We first locate the critical points by solving the simultaneous equations $y_1(3 - y_1 - y_2) = 0$, $y_2(y_1 - 1) = 0$. There are three solutions: (0, 0), (1, 2), (3, 0). Next, we investigate the local behavior of $y_1(t)$ and $y_2(t)$ in the vicinity of each of these critical points by linearizing (4.4.2) in the vicinity of each point as we did in Example 2.

Near the critical point at (0, 0) we let $(y_1, y_2) = (\varepsilon_1, \varepsilon_2)$ and approximate the exact differential equation by $\dot{\varepsilon}_1 \sim 3\varepsilon_1$, $\dot{\varepsilon}_2 \sim -\varepsilon_2$ $(\varepsilon_1, \varepsilon_2 \to 0)$. The solution to this system exhibits saddle-point behavior; trajectories near (0, 0) approach the origin vertically and move away from the origin horizontally as $t \to \infty$.

Near the critical point at (1, 2) we let $(y_1, y_2) = (1 + \varepsilon_1, 2 + \varepsilon_2)$ and approximate the differential equation by the linear system $\dot{\varepsilon}_1 \sim -\varepsilon_1 - \varepsilon_2, \dot{\varepsilon}_2 \sim 2\varepsilon_1$ $(\varepsilon_1, \varepsilon_2 \to 0)$. The solution to this linear system has a stable spiral point with counterclockwise rotation. (Why?)

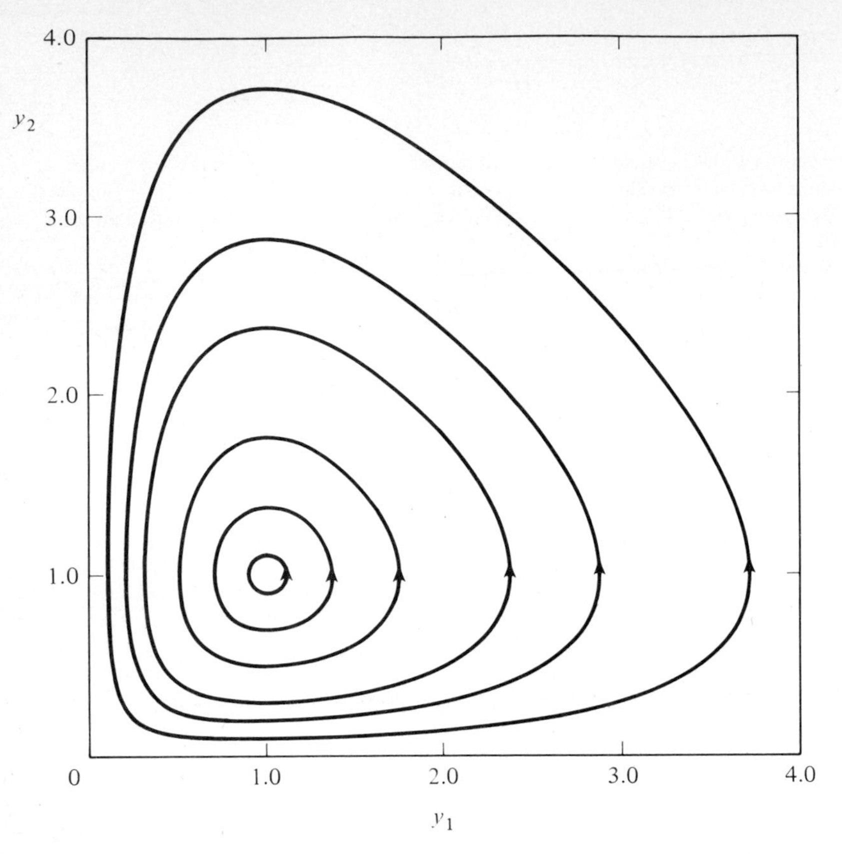

Figure 4.15 Numerical solution to the nonlinear system $\dot{y}_1 = y_1 - y_1 y_2$, $\dot{y}_2 = -y_2 + y_1 y_2$ in (4.4.5). The curves are trajectories in the phase plane and the arrows indicate the motion of the vector $[y_1(t), y_2(t)]$ for increasing t. There are critical points at (0,0) (a saddle point) and at (1,1) (a center).

Near the critical point at (3, 0) we let $(y_1, y_2) = (3 + \varepsilon_1, \varepsilon_2)$. The resulting approximate linear system, $\dot{\varepsilon}_1 \sim -3\varepsilon_1 - 3\varepsilon_2$, $\dot{\varepsilon}_2 \sim 2\varepsilon_2$ $(\varepsilon_1, \varepsilon_2 \to 0)$, is associated with the matrix

$$M = \begin{pmatrix} -3 & -3 \\ 0 & 2 \end{pmatrix}.$$

The eigenvalues of M are -3 and 2 and their eigenvectors are respectively (1, 0) and $(-3, 5)$. Thus the general solution to the linear system is

$$\begin{pmatrix} \varepsilon_1 \\ \varepsilon_2 \end{pmatrix} = c_1 \begin{pmatrix} 1 \\ 0 \end{pmatrix} e^{-3t} + c_2 \begin{pmatrix} -3 \\ 5 \end{pmatrix} e^{2t}.$$

We conclude that (3, 0) is a saddle point with ingoing trajectories lying parallel to the y_1 axis and outgoing trajectories having an asymptotic slope of $-\frac{5}{3}$ as $t \to +\infty$.

If we assemble the results of our local analysis, we deduce that as $t \to \infty$, *all* trajectories, regardless of the initial condition $[y_1(0), y_2(0)]$, spiral into the point (1, 2) in a counterclockwise direction. For example, we argue that (a) an initial condition for which y_1 is large and y_2 is small gives a trajectory which moves down along the y_1 axis until it reaches the saddle point at (3, 0),

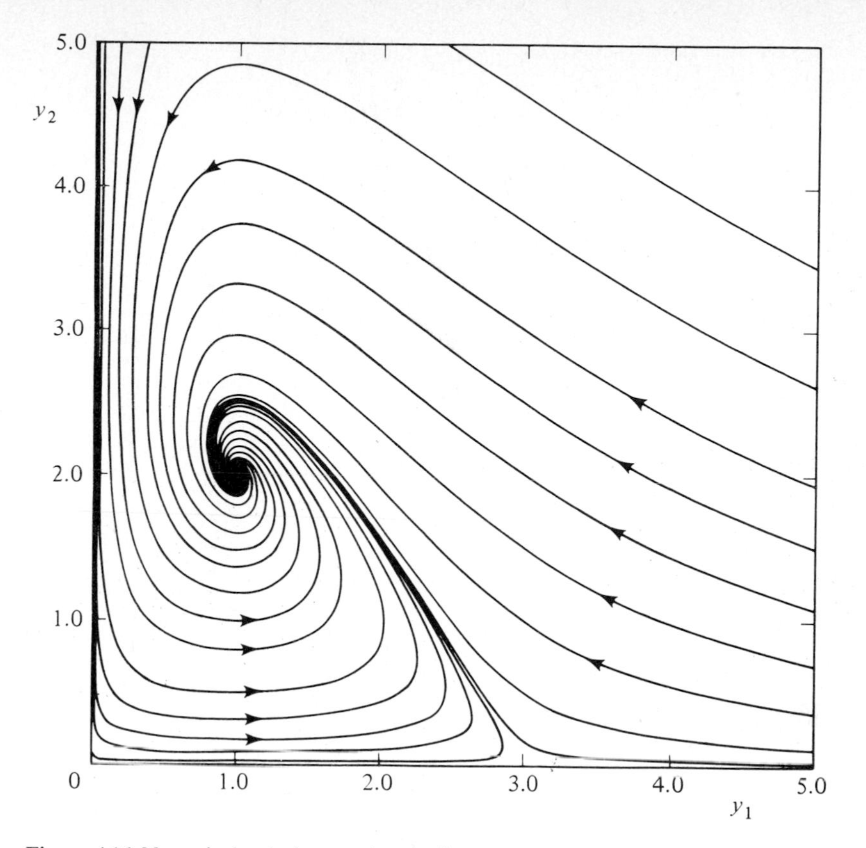

Figure 4.16 Numerical solution to the nonlinear system $\dot{y}_1 = y_1(3 - y_1 - y_2)$, $\dot{y}_2 = y_2(y_1 - 1)$ in (4.4.6). The curves are trajectories in the phase plane. Critical points lie at (0,0) and (3,0) (saddle points) and at (1,2) (a stable spiral point).

veers off at a slope of about $-\frac{5}{3}$, and gets trapped by the stable spiral point at (1, 2), and (*b*) an initial condition for which y_2 is large and y_1 is small gives a trajectory which moves down along the y_2 axis until it reaches the saddle point at (0, 0), makes a 90-degree turn and moves toward the saddle point at (3, 0), makes another turn, and moves off toward the spiral point at (1, 2). These qualitative arguments are verified by the plot of the exact trajectories in Fig. 4.16.

Example 6 *Nonlinear system having four critical points.* The system

$$
\begin{aligned}
\dot{y}_1 &= y_1^2 - y_1 y_2 - y_1, \\
\dot{y}_2 &= y_2^2 + y_1 y_2 - 2y_2,
\end{aligned}
\tag{4.4.7}
$$

is a more complicated version of the predator-prey equations (4.4.5) in Example 4 because it contains linear deathrate and quadratic birthrate terms. This system does *not* predict oscillating populations of rabbits y_1 and foxes y_2. Rather it predicts an unstable situation in which either the rabbits or the foxes grow out of bounds or else both species become extinct.

There are four critical points. It is good practice to verify that (0, 0) is a stable node; $(\frac{3}{2}, \frac{1}{2})$ is an unstable spiral point having a counterclockwise rotation; (1, 0) is a saddle point in whose neighborhood the general behavior of a solution has the form

$$\begin{pmatrix} y_1 \\ y_2 \end{pmatrix} \sim c_1 \begin{pmatrix} 1 \\ 0 \end{pmatrix} e^t + c_2 \begin{pmatrix} 1 \\ 2 \end{pmatrix} e^{-t},$$

so that trajectories of slope 2 move inward and trajectories of slope 0 move outward; (0, 2) is a saddle point in whose neighborhood the general solution has the form

$$\begin{pmatrix} y_1 \\ y_2 \end{pmatrix} \sim c_1 \begin{pmatrix} 5 \\ -2 \end{pmatrix} e^{-3t} + c_2 \begin{pmatrix} 0 \\ 1 \end{pmatrix} e^{2t},$$

so that trajectories of slope $-\frac{2}{5}$ move inward and vertical trajectories move outward.

From our local analysis we can predict that there is an approximate trapezoid of initial values bounded by the four critical points at (0, 0), (1, 0), $(\frac{3}{2}, \frac{1}{2})$, and (0, 2) for which trajectories all approach the origin and both populations $y_1(t)$ and $y_2(t)$ die out as $t \to +\infty$. Outside this trapezoid the initial conditions are unstable with either the fox or the rabbit populations becoming infinite as $t \to +\infty$. These predictions are consistent with the exact solution in Fig. 4.17. Is it possible for *both* $y_1(t)$ and $y_2(t)$ to become infinite as $t \to +\infty$ or must one of y_1 or y_2 always vanish? (See Prob. 4.50.)

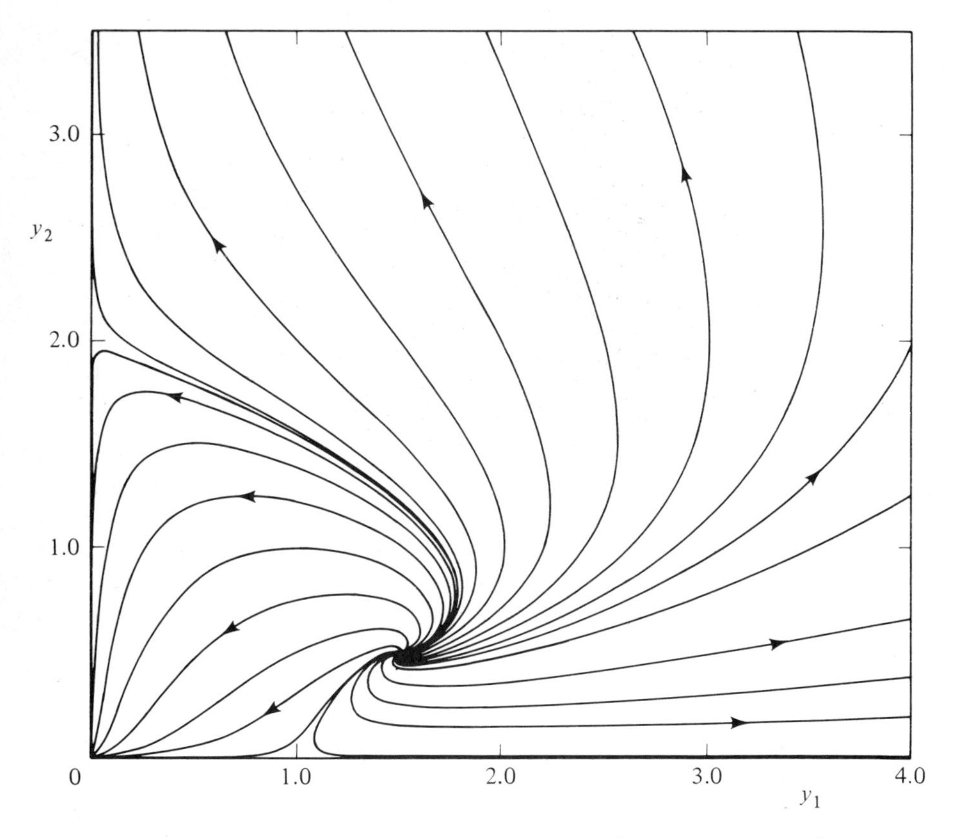

Figure 4.17 Numerical solution to the nonlinear system $\dot{y}_1 = y_1^2 - y_1 y_2 - y_1$, $\dot{y}_2 = y_2^2 + y_1 y_2 - 2y_2$ in (4.4.7). The curves are trajectories in the phase plane. There are four critical points: (0,0) is a stable node, (0,2) and (1,0) are saddle points, and $(\frac{3}{2}, \frac{1}{2})$ is an unstable spiral point.

Example 7 *Nonlinear system having a limit cycle.* The nonlinear system

$$\dot{y}_1 = y_1 + y_2 - y_1(y_1^2 + y_2^2),$$
$$\dot{y}_2 = -y_1 + y_2 - y_2(y_1^2 + y_2^2) \tag{4.4.8}$$

has a critical point at (0, 0). An analysis of the approximate linear system near (0, 0), $\dot{\varepsilon}_1 \sim \varepsilon_1 + \varepsilon_2$, $\dot{\varepsilon}_2 \sim -\varepsilon_1 + \varepsilon_2$ $(\varepsilon_1, \varepsilon_2 \to 0)$, shows that (0, 0) is an unstable spiral point having clockwise rotation.

Next, we examine (4.4.8) at ∞. If y_1 and y_2 are large, then $\dot{y}_1 \sim -y_1 z$, $\dot{y}_2 \sim -y_2 z$ $(y_1, y_2 \to \infty)$, where $z = y_1^2 + y_2^2$, which implies that distant trajectories move toward the origin as $t \to +\infty$.

How can it be that all trajectories near the origin spiral outward and that all distant trajectories move inward? By continuity, there must be at least one trajectory at a moderate distance from the origin that neither moves inward nor outward. Such a trajectory must be a closed orbit which encircles the origin. Trajectories outside or inside this orbit may approach but not cross it. We have thus inferred the existence of a limit cycle from pure local analysis.

By local analysis alone we cannot say how many limit cycles there are. (There might be several concentric closed orbits.) However, it is easy to solve (4.4.8) exactly and show that there is just one. We multiply the first and second equations of (4.4.8) by y_1 and y_2 and add the resulting equations. Setting $z = y_1^2 + y_2^2$ we obtain $\dot{z} = 2z - 2z^2$. This equation is similar to that analyzed in Example 2. It is thus clear that there is a circular limit cycle having a radius of $z = y_1^2 + y_2^2 = 1$. The exact solution to the system (4.4.8) is plotted in Fig. 4.18. It exhibits all of the qualitative features that we have deduced from local analysis.

Difficulties with Linear Critical-Point Analysis

The analysis in Examples 4 to 7 is called *linear* critical-point analysis because there is a linear approximation to the nonlinear system in the neighborhood of each critical point. Unfortunately, it is not always possible to find a linear approximation to a nonlinear system. For example, the system

$$\dot{y}_1 = y_1^2 + y_2^2 + y_1^6, \qquad \dot{y}_2 = \sin(y_1^4 + y_2^4)$$

has a critical point at (0, 0), but in the vicinity of this critical point the equations have the approximate form

$$\dot{\varepsilon}_1 \sim \varepsilon_1^2 + \varepsilon_2^2, \qquad \dot{\varepsilon}_2 \sim \varepsilon_1^4 + \varepsilon_2^4, \qquad \varepsilon_1, \varepsilon_2 \to 0,$$

which is still nonlinear.

The structure of nonlinear critical points can be much more complicated than that of linear critical points (see Probs. 4.44 to 4.46). There can be saddle points having many in and out directions and nodes for which the trajectories are not asymptotic to straight lines as $t \to \infty$. No simple matrix methods exist for identifying the structure of such nonlinear critical points. Extemporaneous analysis is often required.

There is a more subtle difficulty with linear critical-point analysis when matrix methods suggest that the critical point is a center. When the system is linear, then we can be sure that if the eigenvalues are imaginary the critical point is a center. However, if a linear approximation to a nonlinear system has a center, it is still *not correct* to conclude that the nonlinear system also has a center. A center is a very special kind of critical point for which the orbits *exactly* close. Any distortion or perturbation of a closed orbit, no matter how small, can give an open

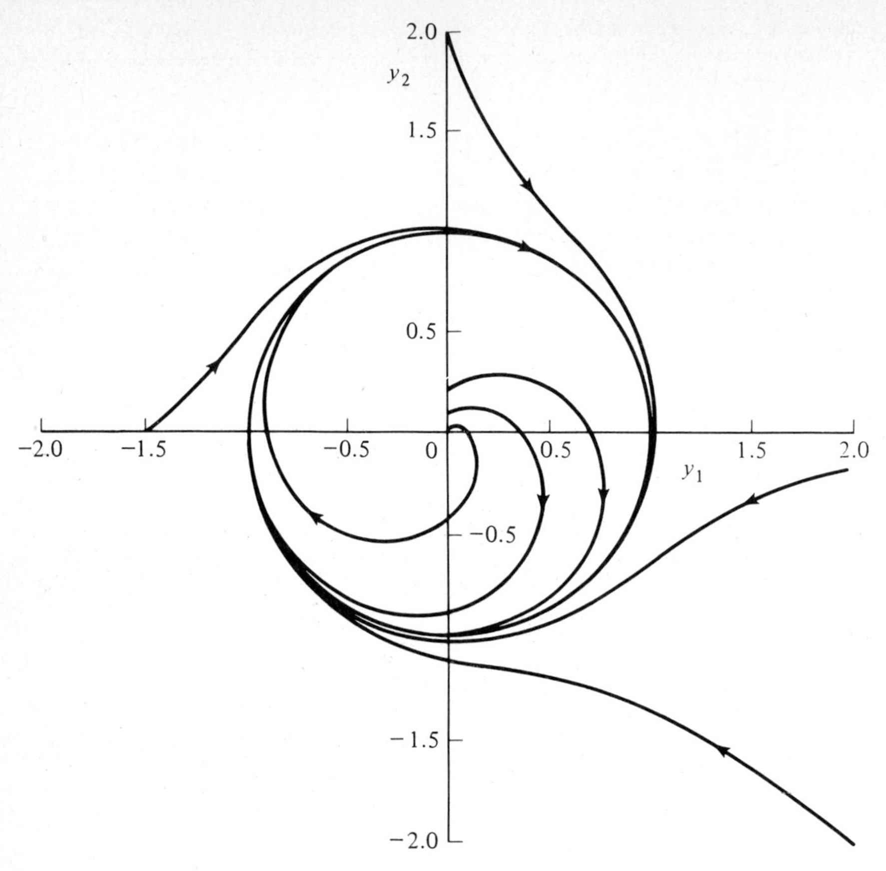

Figure 4.18 Numerical solution to the system $\dot{y}_1 = y_1 + y_2 - y_1(y_1^2 + y_2^2)$, $\dot{y}_2 = -y_1 + y_2 - y_2(y_1^2 + y_2^2)$ in (4.4.8). Curves are trajectories in the phase plane. The point (0,0) is an unstable spiral point and there is a circular limit cycle of radius 1. All trajectories approach the limit cycle as $t \to +\infty$.

orbit. (Small distortions of nodes, spiral points, and saddle points do not change the qualitative features of these critical points.) Therefore, even though a linear approximation to a nonlinear system may have a center, the nonlinear system may actually have a spiral point.

Example 8 *Nonlinear system having a spurious center.* The system

$$\begin{aligned} \dot{y}_1 &= -y_2 + y_1(y_1^2 + y_2^2), \\ \dot{y}_2 &= y_1 + y_2(y_1^2 + y_2^2) \end{aligned} \tag{4.4.9}$$

has a critical point at (0, 0). A linear approximation (4.4.9) in the vicinity of (0, 0) is $\dot{\varepsilon}_1 \sim -\varepsilon_2$, $\dot{\varepsilon}_2 \sim \varepsilon_1$ $(\varepsilon_1, \varepsilon_2 \to 0)$.

This linear system has a center. However, the exact nonlinear system has a *spiral point* and not a center! To see this we multiply the first and second equations of (4.4.9) by y_1 and y_2 and add the resulting equations. We obtain $y_1 \dot{y}_1 + y_2 \dot{y}_2 = (y_1^2 + y_2^2)^2$. In polar coordinates this becomes

$$\frac{1}{2}\frac{d}{dt}(r^2) = r^4. \tag{4.4.10}$$

Thus, the radius r *increases* with increasing t and we have an unstable spiral point. In fact, since the exact solution to (4.4.10) is $r(t) = r(0)/\sqrt{1 - 2r^2(0)t}$, we see that $r(t)$ reaches ∞ in the finite time $t = [2r^2(0)]^{-1}$.

To prove that a critical point is really a center one must demonstrate the existence of closed orbits. No approximate methods may be used in the proof. The usual technique consists of integrating the system of differential equations once to construct a time-independent quantity which is often called an energy integral. Of course, an energy integral for the system will be very difficult to find when the appropriate integrating factors are not obvious. In the following example we show how to construct energy integrals.

Example 9 *Proofs of the existence of closed orbits using energy integrals.*

(*a*) The system $\dot{y}_1 = -y_2, \dot{y}_2 = y_1$ is linear, so matrix methods are sufficient to prove that (0, 0) is a center. However, for illustrative purposes we use an energy integral. The trick consists of observing that (y_1, y_2) are integrating factors for the system. Multiplying the first equation by y_1 and the second by y_2 and adding the resulting equations gives an exact derivative: $\frac{1}{2}(d/dt)(y_1^2 + y_2^2) = 0$. Thus, $y_1^2 + y_2^2 = c$, where c is a constant. But this is the equation for a family of concentric circles about (0, 0). (See Fig. 4.12.) This proves that (0, 0) is a center.

(*b*) If we linearize the nonlinear system $\dot{y}_1 = -y_2 - y_2^3, \dot{y}_2 = y_1$ about the critical point at (0, 0), we find a center. To *prove* that (0, 0) is a center we note that this system has the same integrating factors as in (*a*). Multiplying by (y_1, y_2) and adding the equations gives

$$y_1\dot{y}_1 + y_2\dot{y}_2 = -y_1 y_2^3 = -y_2^3\dot{y}_2,$$

where we have used the second equation of the system. Again we have found an exact derivative which gives a time-independent quantity

$$y_1^2 + y_2^2 + \tfrac{1}{2}y_2^4 = c.$$

This expression represents a family of concentric closed curves about the point (0, 0). Hence (0, 0) is a center.

(*c*) The nonlinear system (4.4.5) in Example 4, $\dot{y}_1 = y_1 - y_1 y_2, \dot{y}_2 = -y_2 + y_1 y_2$, has a critical point at (1, 1) which we claimed (using matrix analysis) was a center. To prove this assertion we multiply the equations by the integrating factors $[(1 - y_1)/y_1, (1 - y_2)/y_2]$ and add the resulting equations together. The new equation is in the form of a total derivative. Integrating with respect to t gives

$$y_1 + y_2 - \ln(y_1 y_2) = c, \tag{4.4.11}$$

which represents a family of concentric closed curves containing the point (1, 1) (see Prob. 4.49). We have thus proved that (1, 1) is a center.

Further discussion of energy integrals is given in Sec. 11.1.

(I) 4.5 HIGHER-ORDER NONLINEAR AUTONOMOUS SYSTEMS

While the behavior of first- and second-order autonomous systems of differential equations is simple and easy to analyze in the phase plane, solutions to autonomous systems of order three or more can be very complicated. The properties of solutions to higher-order systems are the subject of much current research interest and our discussion of them here is limited to a brief nontechnical survey. The interested reader should consult the references.

Solutions to first- and second-order systems are simple because trajectories cannot cross in phase space. In three or more dimensions, the restriction that trajectories may not cross does not constrain the solutions to be simple; when orbits are not constrained to lie in a two-dimensional surface, they can twist, turn, and tangle themselves into fantastic knots as they develop in time t. Our intention here is to illustrate the richness of this very difficult subject.

Behavior Near a Stable Critical Point

A critical point is stable if the eigenvalues of the system of equations obtained by linearizing the nonlinear system in the neighborhood of the critical point have negative real parts. This is by far the simplest case. It can be proved that all trajectories of the full nonlinear equations that originate sufficiently close to a stable critical point always decay toward that critical point as $t \to +\infty$. The nonlinear effects do not change the qualitative (or even the quantitative) behavior of a system near a stable critical point.

Example 1 *Behavior of a third-order system near a stable critical point.* The point $x = y = z = 0$ is a stable critical point of the system

$$\frac{dx}{dt} = -x + x^3y^2 - x^2y^3z, \qquad \frac{dy}{dt} = -y + z^3, \qquad \frac{dz}{dt} = -z + x^4 - z^4$$

because the linearized system has three negative eigenvalues. The linearized system is just $\dot{x} = -x$, $\dot{y} = -y$, $\dot{z} = -z$, so the linearized behavior is just

$$x(t) = e^{-t}x(0), \qquad y(t) = e^{-t}y(0), \qquad z(t) = e^{-t}z(0).$$

This behavior persists in the nonlinear system. There, we can show that when t is large and $|x(0)|$, $|y(0)|$, and $|z(0)|$ are sufficiently small,

$$x(t) \sim \sum_{n=1}^{\infty} a_n e^{-nt}, \qquad y(t) \sim \sum_{n=1}^{\infty} b_n e^{-nt}, \qquad z(t) \sim \sum_{n=1}^{\infty} c_n e^{-nt}, \qquad t \to +\infty. \tag{4.5.1}$$

In fact, $a_2 = b_2 = c_2 = a_3 = c_3 = 0$, $b_3 = -\frac{1}{2}c_1^2$, and so on (see Prob. 4.52). To test these conclusions, we plot in Fig. 4.19 the ratios $x(t)/z(t)$ and $y(t)/z(t)$ for $x(0) = y(0) = z(0) = 1$. Equation (4.5.1) predicts that these ratios should approach the constants a_1/c_1 and b_1/c_1, respectively, as $t \to +\infty$. Figure 4.19 verifies this prediction.

Behavior Near a Center

A simple center is a critical point at which all the eigenvalues of the linearized system are pure imaginary and distinct. This case is perhaps the most difficult. To begin with, the solution to the linearized system need not be periodic because the eigenfrequencies may not be commensurate. For example, a real fourth-order linear system could have eigenvalues $\pm i$, $\pm i\sqrt{2}$. Can you see that there are solutions that are not periodic? However, all solutions to a linear system at a simple center are *almost periodic* in the sense that there exist arbitrarily large time periods T over which the solution repeats itself to any prespecified tolerance.

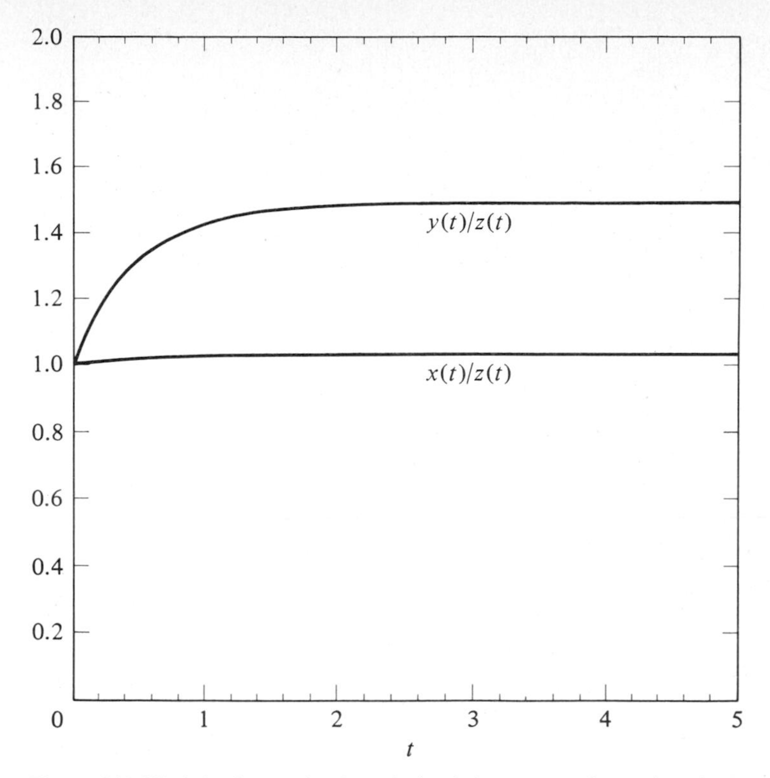

Figure 4.19 We infer from a local analysis of the system of equations in Example 1 that $x(t)/z(t)$ and $y(t)/z(t)$ approach constants as $t \to +\infty$. This graph verifies that prediction.

Mathematically speaking, for any $\varepsilon > 0$, there exists an unbounded sequence of time periods $T_1, T_2, T_3, \ldots$ such that, for each T_i, $|\mathbf{y}(t + T_i) - \mathbf{y}(t)| < \varepsilon$ for all t.

The behavior of a nonlinear system in the vicinity of a simple center can be even more complicated. As we have seen in Example 8 of Sec. 4.4, the existence of a nonlinear term may disrupt the orbits of the linear system entirely. The orbits may no longer be almost periodic; they may exhibit very complicated random behavior.

Example 2 *The Toda lattice: a simple center with almost periodic orbits.* The system

$$\frac{dq_j}{dt} = p_j, \tag{4.5.2}$$

$$\frac{dp_j}{dt} = \exp(q_{j-1} - q_j) - \exp(q_j - q_{j+1}), \tag{4.5.3}$$

where $j = 1, 2, \ldots, m$, $m > 1$, and $q_0 = q_m$, $q_{m+1} = q_1$ is known as the equations of the Toda lattice. The point $p_1 = p_2 = \cdots = p_m = 0$, $q_1 = q_2 = \cdots = q_m = 0$ is a simple center (see Prob. 4.53). Despite the nonlinearity of this system, it can be proved that the solutions to these equations are almost periodic for all m. To illustrate this almost periodic behavior, we have solved (4.5.2) and (4.5.3) numerically for the case $m = 3$. In Fig. 4.20 we have plotted $q_1(t)$ versus t. The repetitive structure of the curve suggests that the solution is almost periodic.

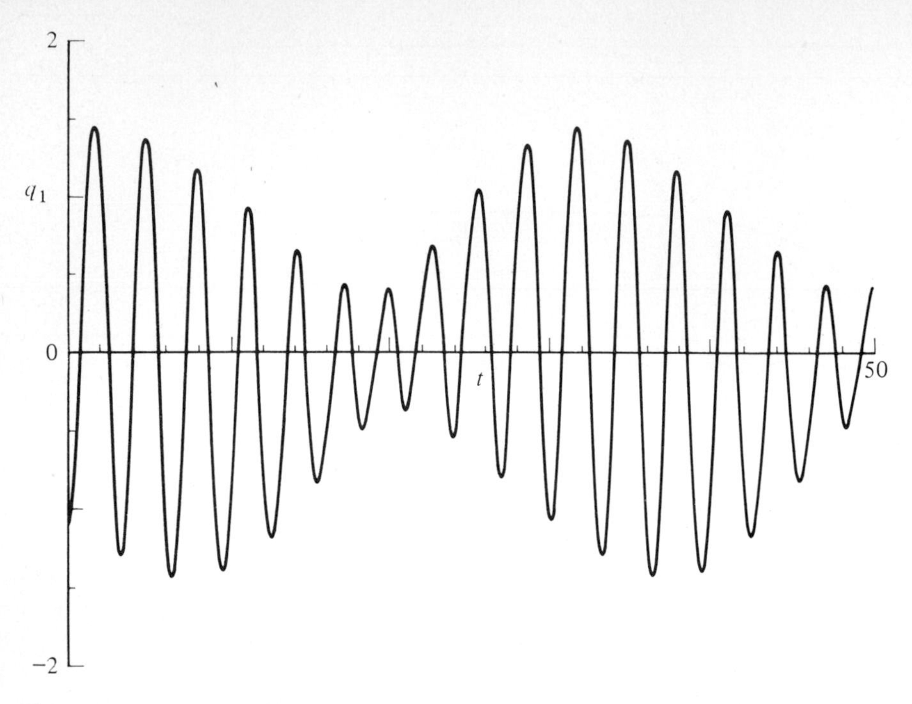

Figure 4.20 A plot of $q_1(t)$ versus t for the Toda lattice (4.5.2) and (4.5.3) with $m = 3$ and the initial conditions $p_1(0) = -1$, $p_2(0) = 0.7$, $p_3(0) = 0.3$, $q_1(0) = -1$, $q_2(0) = 0$, and $q_3(0) = 1$. Observe the almost periodic behavior of $q_1(t)$.

Example 3 *A simple center with complicated orbits.* The system (investigated by Hénon and Heiles)

$$\frac{dq_j}{dt} = p_j, \qquad j = 1, 2, \tag{4.5.4}$$

$$\frac{dp_1}{dt} = -q_1 - 2q_1 q_2, \tag{4.5.5}$$

$$\frac{dp_2}{dt} = -q_2 - q_1^2 + q_2^2 \tag{4.5.6}$$

has a simple center at $p_j = q_j = 0$ ($j = 1, 2$). (Why?) The orbits in the vicinity of this center are almost periodic; others exhibit random behavior. To illustrate the almost periodic and the random behaviors, we have constructed two Poincaré plots in Figs. 4.21 and 4.22. A Poincaré plot is a graph of $p_1(t)$ versus $q_1(t)$ at those discrete times t at which $q_2(t) = 0$ and $p_2(t) > 0$. That is, every time the trajectory passes through $q_2 = 0$ while q_2 is increasing, we plot a dot in the $p_1 - q_1$ plane. We observe two distinct behaviors. The initial condition I:

$$p_1(0) = \tfrac{1}{3}, \qquad p_2(0) = 0.129\,314\,4, \qquad q_1(0) = \tfrac{1}{4}, \qquad q_2(0) = \tfrac{1}{5}$$

gives rise to an almost periodic orbit while the initial condition II:

$$p_1(0) = 0.1, \qquad p_2(0) = 0.467\,618, \qquad q_1(0) = 0.1, \qquad q_2(0) = 0.1$$

gives rise to random behavior. In the Poincaré plot the initial condition I gives rise to a sequence of points that gradually fill out a number of closed curves which are called islands. The trajectory is almost periodic because the dots progress sequentially and regularly from island to island and nearly repeat themselves every so often. On the other hand, the initial

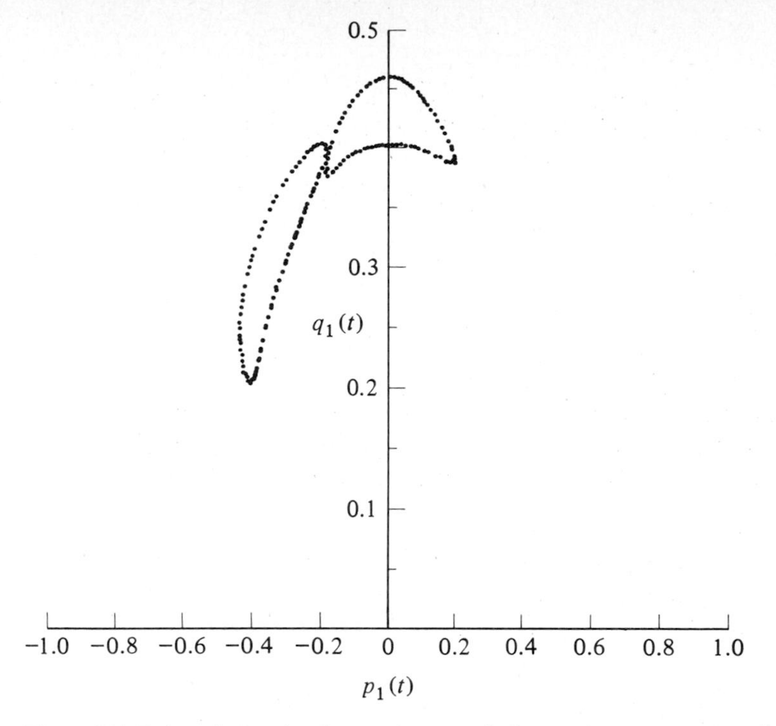

Figure 4.21 Poincaré plot showing an almost periodic solution to the system (4.5.4) to (4.5.6). The plot is constructed by graphing the points $[p_1(t), q_1(t)]$ for those discrete times t at which $q_2(t) = 0$ and $p_2(t) > 0$. For this plot we took $p_1(0) = \frac{1}{3}$, $p_2(0) = 0.129\,314\,4$, $q_1(0) = \frac{1}{4}$, $q_2(0) = \frac{1}{5}$, and plotted points for $0 \le t \le 1{,}000$. The trajectory is almost periodic because the dots alternate regularly between the islands, and the sequence of dots nearly repeats itself every so often.

condition II gives rise to a sequence of points in the Poincaré plot that jump around apparently at random in the region outside the islands. There is an interesting connection between the initial conditions I and II. (See Prob. 4.59.)

Notice from Fig. 4.22 that while the trajectory II is random it nevertheless does not move far away from the center at 0. The reason for this is that the system (4.5.4) to (4.5.6) satisfies the conditions of a profound theorem in analysis called the Arnold-Moser theorem. Loosely speaking, this theorem implies that trajectories of (4.5.4) to (4.5.6) that originate sufficiently close to the center remain close to the center for all time.

Hamiltonian Systems

The systems considered in Examples 2 and 3 are Hamiltonian systems. A Hamiltonian system has the form

$$\frac{dq_j}{dt} = \frac{\partial H}{\partial p_j}, \tag{4.5.7}$$

$$\frac{dp_j}{dt} = -\frac{\partial H}{\partial q_j}, \tag{4.5.8}$$

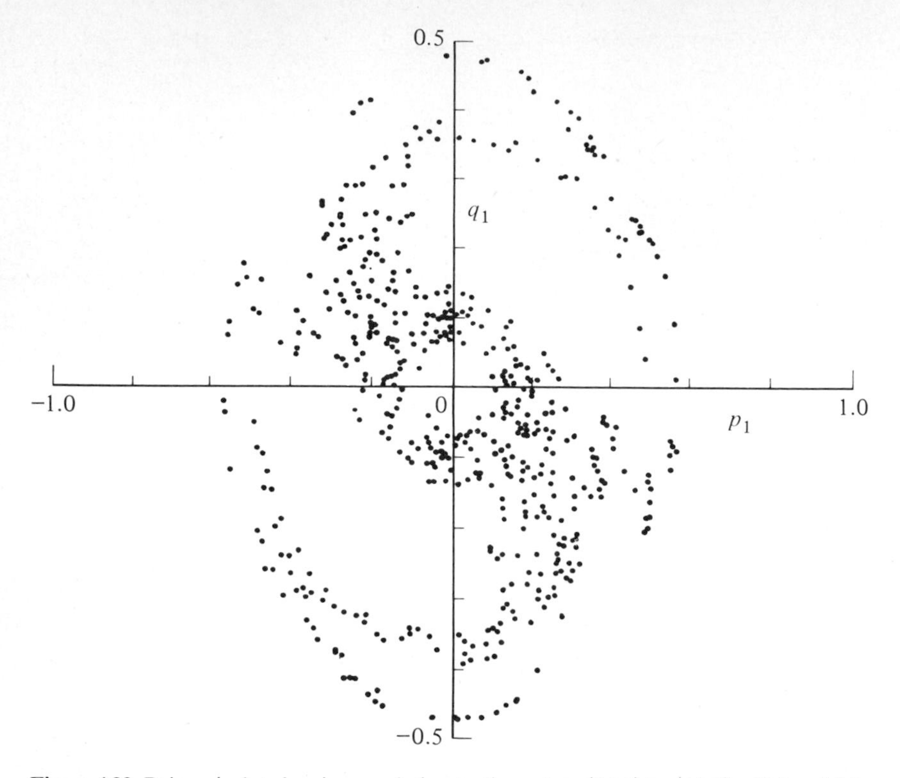

Figure 4.22 Poincaré plot showing a solution to the system (4.5.4) to (4.5.6) which exhibits random behavior. For this plot we took $p_1(0) = 0.1$, $p_2(0) = 0.467\,618$, $q_1(0) = 0.1$, $q_2(0) = 0.1$, and plotted points for $0 \le t \le 2{,}000$. It is remarkable that the plots in Figs. 4.21 and 4.22 represent solutions to the same system of equations.

where $j = 1, 2, \ldots, m$ and $H = H(p_1, p_2, \ldots, p_m, q_1, q_2, \ldots, q_m)$ is called the Hamiltonian. For the Toda lattice (4.5.2) and (4.5.3),

$$H = \tfrac{1}{2} \sum_{j=1}^{m} p_j^2 + \sum_{j=1}^{m} e^{q_{j-1} - q_j},$$

and for the system (4.5.4) to (4.5.6), $H = \frac{1}{2}(p_1^2 + p_2^2) + \frac{1}{2}(q_1^2 + q_2^2 + 2q_1^2 q_2 - \frac{2}{3}q_2^3)$.

Hamiltonian systems have two important properties. First, the Hamiltonian is an energy integral of the motion. This means that $H[\mathbf{p}(t), \mathbf{q}(t)] = H[\mathbf{p}(0), \mathbf{q}(0)]$:

$$\frac{dH}{dt}[\mathbf{p}(t), \mathbf{q}(t)] = \sum_j \left(\frac{\partial H}{\partial p_j} \frac{dp_j}{dt} + \frac{\partial H}{\partial q_j} \frac{dq_j}{dt} \right) = 0.$$

Second, Hamiltonian systems conserve volumes in phase space. This means that if we draw the trajectories that originate from all of the points inside a region of volume V in phase space at $t = 0$, then the endpoints of these trajectories at time t

fill a region with the same volume V for all t. Mathematically, this condition is that the Jacobian

$$J(t) = \frac{\partial[\mathbf{p}(t), \mathbf{q}(t)]}{\partial[\mathbf{p}(0), \mathbf{q}(0)]}$$

satisfies $J(t) = 1$ for all t (see Prob. 4.55).

Because Hamiltonian systems preserve volumes in phase space, stable critical points must be centers. A Hamiltonian system cannot exhibit the kind of stable critical point illustrated in Example 1. (Why?)

The Arnold-Moser theorem that we cited in Example 3 applies to Hamiltonian systems whose Hamiltonians are analytic functions of $\mathbf{p}$ and $\mathbf{q}$.

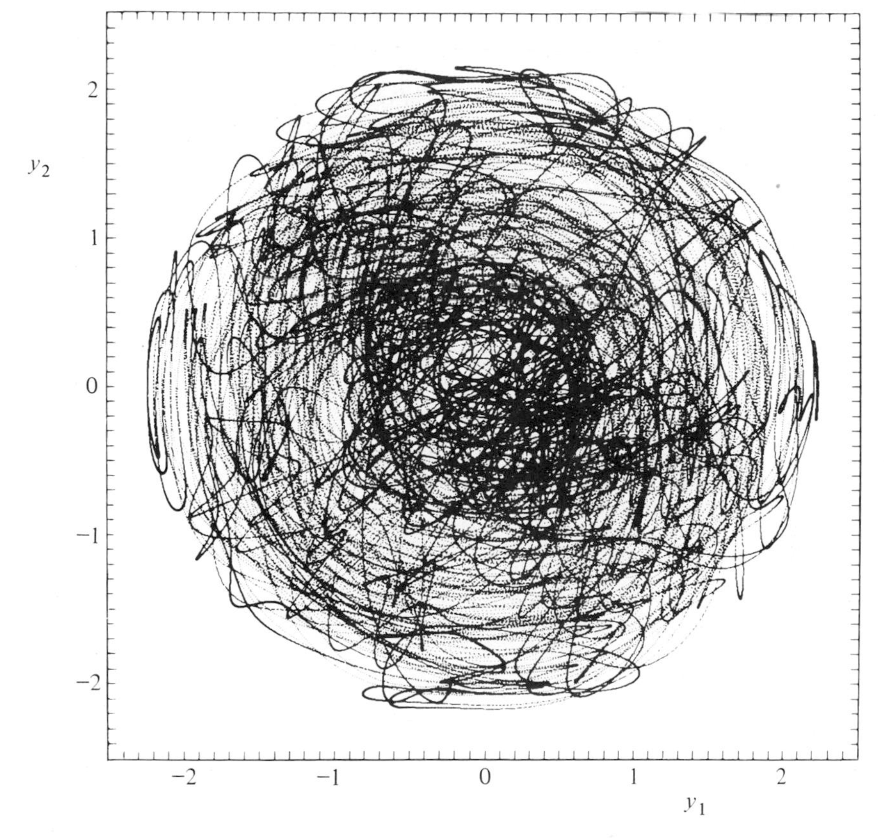

Figure 4.23 Plot of $y_1(t)$ versus $y_2(t)$ for a single trajectory of (4.5.9) for $0 \le t \le 3{,}000$. The initial conditions, $y_1(0) = 0.540\ 323$, $y_2(0) = -1.543\ 569$, $y_3(0) = -0.680\ 421$, $y_4(0) = -1.185\ 361$, $y_5(0) = -0.676\ 307$, satisfy $[y_1(0)]^2 + \cdots + [y_5(0)]^2 = 5$. Thus, the projection of the trajectory onto the (y_1, y_2) plane lies inside a circle of radius $\sqrt{5}$. The disordered wandering of the curve in the plot suggests that the solution does not have a simple asymptotic behavior as $t \to \infty$, but rather that the behavior is random.

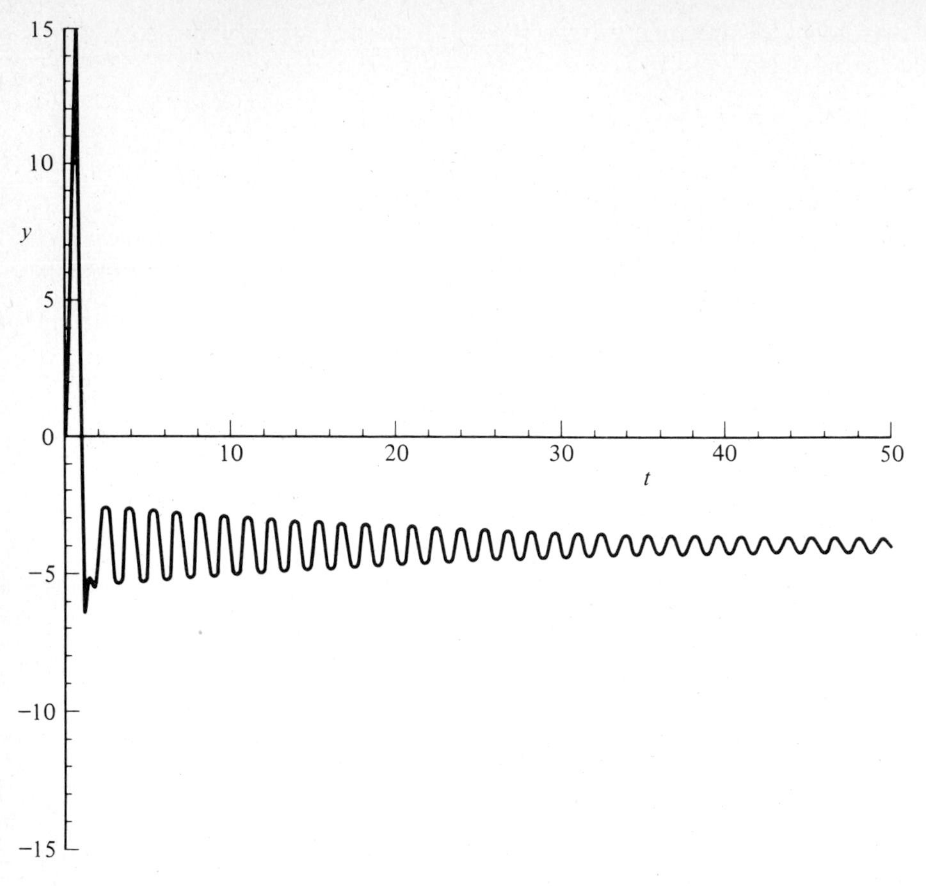

Figure 4.24 A plot of $y(t)$ versus t in the Lorenz model (4.5.10) to (4.5.12) for $r = 17$ and the initial conditions $x(0) = z(0) = 0$, $y(0) = 1$. Note the slow and regular oscillatory approach to the stable critical point at $x = -4$, $y = -4$, $z = 16$.

Systems That Exhibit Random Behavior

One class of equations known as C-systems exhibits random behavior. Roughly speaking, a C-system is one for which:

1. There exists an energy integral which confines the trajectory to a finite volume in phase space.
2. There are *no* critical points.
3. For each trajectory Γ in phase space there exist nearby trajectories, some of which move away from and some of which approach the trajectory Γ with increasing t.

There are no C-systems in less than three dimensions (see Prob. 4.56).

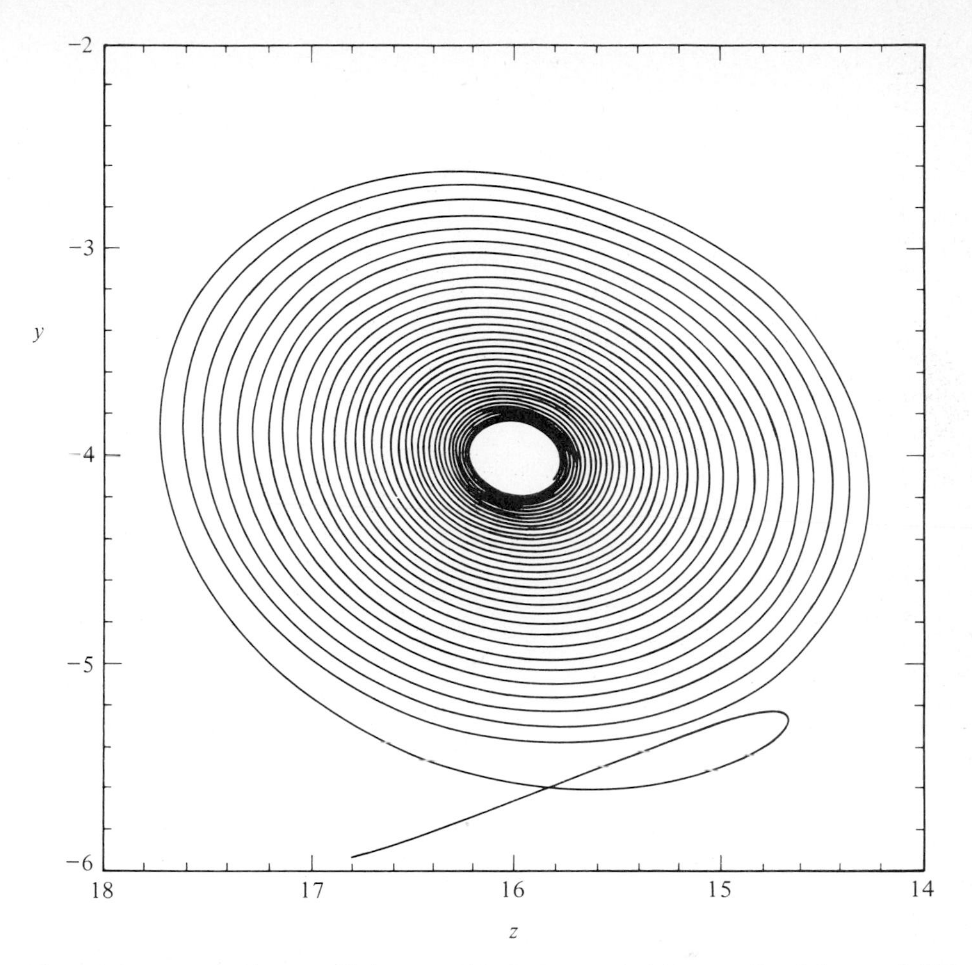

Figure 4.25 A plot of $y(t)$ versus $z(t)$ $(1 \le t \le 50)$ in the Lorenz model for $r = 17$ and the same initial conditions as in Fig. 4.24. Note the slow spiral approach to the critical point at $x = -4$, $y = -4$, $z = 16$.

There are systems of differential equations which are not C-systems (because they have critical points) which also exhibit random behavior. We give two examples.

Example 4 *A fifth-order system with an energy integral.* The fifth-order system

$$\frac{dy_i}{dt} = y_{i+1}y_{i+2} + y_{i-1}y_{i-2} - 2y_{i+1}y_{i-1}, \qquad i = 1, 2, \ldots, 5, \tag{4.5.9}$$

$y_i \equiv y_{i+5}$ for all i, has the energy integral $(d/dt) \sum_{i=1}^{5} y_i^2 = 0$, so orbits are confined to the surface of a ball in five-dimensional phase space. This system has critical points and they are all unstable (see Prob. 4.57). In Fig. 4.23 we plot $y_1(t)$ versus $y_2(t)$ for a single trajectory for $0 \le t \le$ 3,000. A glance at this figure suggests two conclusions. First, it is unlikely that there is a closed-form analytical expression for the trajectory. Second, the behavior is probably random.

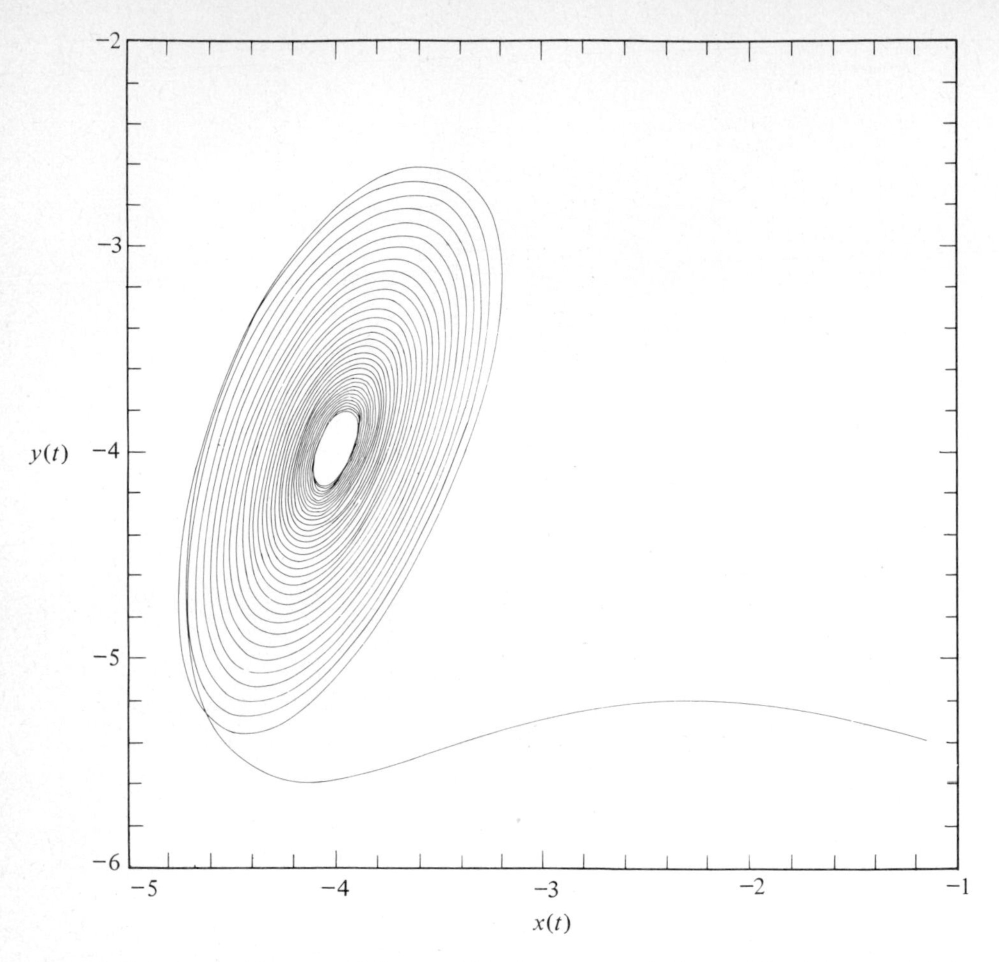

Figure 4.26 A plot of $x(t)$ versus $y(t)$ $(1 \le t \le 50)$ in the Lorenz model for $r = 17$ and the same initial conditions as in Fig. 4.24.

Example 5 *Lorenz model.* The Lorenz model is the third-order system

$$\frac{dx}{dt} = -3(x - y), \tag{4.5.10}$$

$$\frac{dy}{dt} = -xz + rx - y, \tag{4.5.11}$$

$$\frac{dz}{dt} = xy - z, \tag{4.5.12}$$

where r is a constant parameter.

If $r < 1$, the only critical point is at $x = y = z = 0$ and this point is stable. Thus, if $r < 1$, the

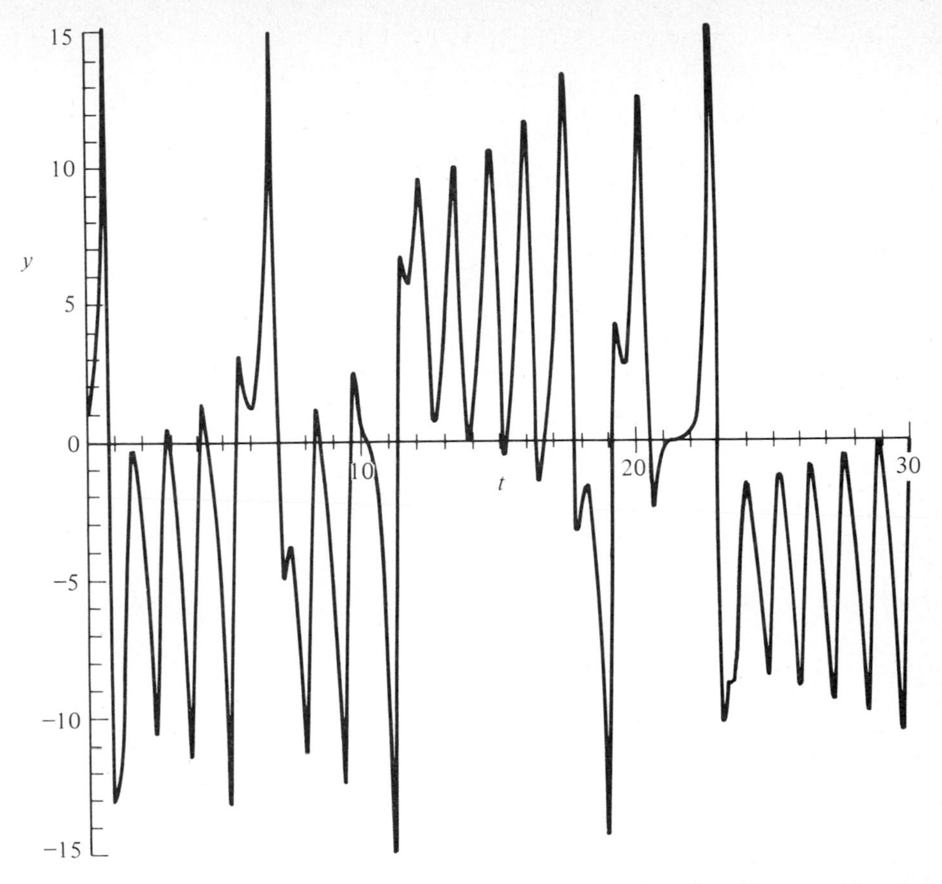

Figure 4.27 A plot of $y(t)$ versus t in the Lorenz model (4.5.10) to (4.5.12) for $r = 26$ and the initial conditions $x(0) = z(0) = 0$, $y(0) = 1$. This system exhibits random behavior which presents itself here as intermittent and irregular oscillation in contrast with the regular oscillation in Fig. 4.24.

system cannot exhibit random behavior if $x(0)$, $y(0)$, $z(0)$ are small (see Prob. 4.58). If $r > 1$, the origin is an unstable critical point.

However, if $r > 1$, there appear critical points at $x = y = \pm\sqrt{r-1}$, $z = r - 1$. These critical points are stable if $1 < r < 21$ (see Prob. 4.58). To illustrate the behavior of trajectories near these stable critical points, we take $r = 17$ and $x(0) = z(0) = 0$, $y(0) = 1$, and plot $y(t)$ versus t in Fig. 4.24. Observe that $y(t) \to -4$ as $t \to \infty$. (What is special about $y = -4$?) In Fig. 4.25 we plot $y(t)$ versus $z(t)$ and in Fig. 4.26 we plot $x(t)$ versus $y(t)$. There is a lovely oscillatory approach to the stable critical point at $(-4, -4, 16)$.

If $r > 21$, the critical points at $x = y = \pm\sqrt{r-1}$, $z = r - 1$ and at $x = y = z = 0$ are all unstable. In Fig. 4.27 we plot $y(t)$ versus t for the trajectory starting at $x(0) = z(0) = 0$, $y(0) = 1$ with $r = 26$. It has been proved that this system exhibits random behavior which is illustrated by the intermittent and irregular oscillation shown in Fig. 4.28. In Figs. 4.28 and 4.29 we plot $y(t)$ versus $z(t)$ and $x(t)$ versus $y(t)$, respectively, for the same conditions as in Fig. 4.27. The randomness of the trajectory consists of haphazard jumping back and forth from neighborhoods of the critical points at $x = y = 5$, $z = 25$ and $x = y = -5$, $z = 25$.

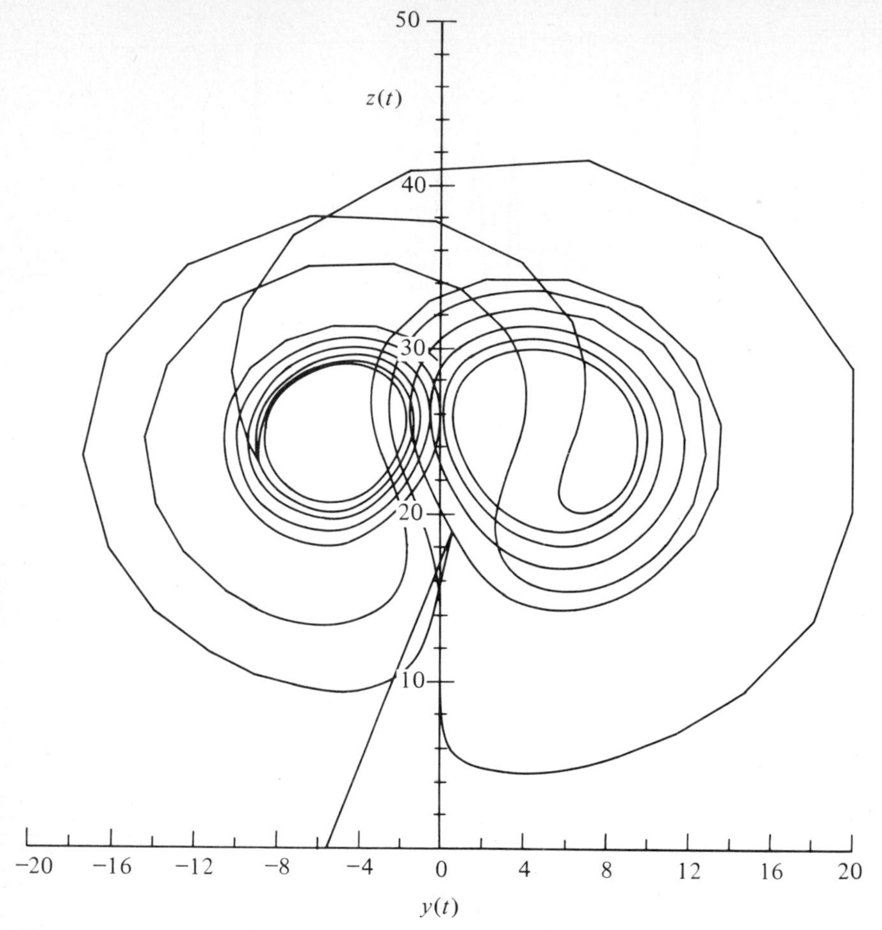

Figure 4.28 A plot of $z(t)$ versus $y(t)$ $(2.5 \le t \le 30)$ in the Lorenz model for $r = 26$ and the same initial conditions as in Fig. 4.27. The randomness of the trajectory presents itself as haphazard jumping back and forth from neighborhoods of the unstable critical points at $x = y = 5$, $z = 25$ and $x = y = -5$, $z = 25$. The segmented character of the outer portion of this plot and that of Fig. 4.29 reflect the discrete nature of the computer output.

PROBLEMS FOR CHAPTER 4

Section 4.1

(D) **4.1** Verify equation (4.1.3).

(D) **4.2** Prove that when $-e \le x \le 1/e$, the sequence $\{y_n(x)\}$ in (4.1.5) converges to the solution of the initial-value problem (4.1.1). Show that when $x < -e$, the sequence has *two* limiting functions, neither of which is the solution to $y = e^{xy}$. Which equation do these two limiting functions satisfy?

(E) **4.3** Perform a local analysis of the *algebraic* equation $y = e^{xy}$ near $x = 1/e$ by substituting $y = e + \delta(x)$, where $\delta \to 0$ as $x \to 1/e$. Solve approximately for $\delta(x)$ to show that near $x = 1/e$, $y(x)$ has a square-root singularity.

(TI) **4.4** Extend the argument given in Prob. 1.3 of Chap. 1 to show that if $F(x, y)$ is an analytic function of both its arguments in some neighborhood of $x = a$, $y = b$, then the solution $y(x)$ of the differential equation $y' = F(x, y)$, $y(a) = b$, exists and is analytic in some neighborhood of $x = a$.

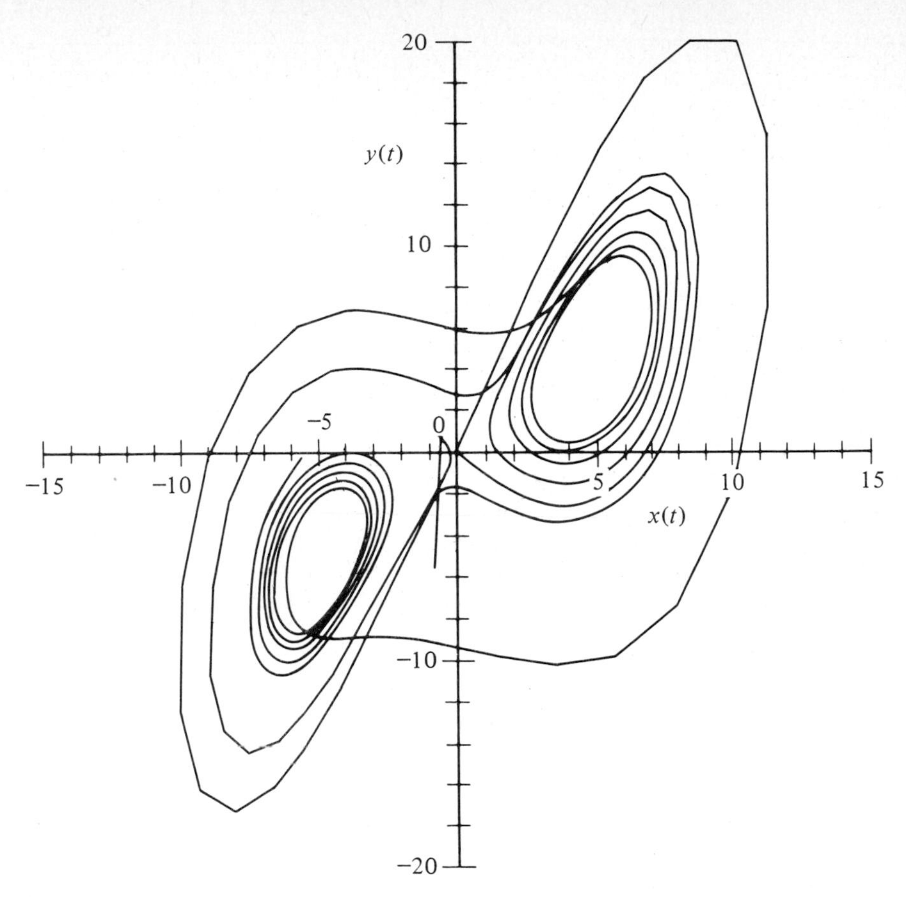

Figure 4.29 A plot of $y(t)$ versus $x(t)$ $(2.5 \leq t \leq 30)$ in the Lorenz model for $r = 26$ and the same initial conditions as in Fig. 4.27.

Section 4.2

(I) **4.5** Explain why $\bar{x}_n^{1/2}\Delta_n \to \pi$ very rapidly as $n \to \infty$ (see Table 4.1). Specifically, show that the approach is not so rapid if $\bar{x}_n$ is defined as x_n or x_{n+1}. Why does the averaging decrease the error so much?

(I) **4.6** Does the solution to the initial-value problem $y'(x) = \sqrt{x^2 + y^2}$ $[y(0) = a]$ remain finite for all x?

(I) **4.7** Can the solution to $y' = y^2 + 1/x^2$ $[y(-10) = a]$ become infinite at $x = 0$ if a is real? Solve the equation exactly.

(I) **4.8** Let $y(x)$ be a function such that $y(1) = 1$ and for $x \geq 1$, $y'(x) = [x^2 + y^2(x)]^{-1}$. Prove that $\lim_{x\to\infty} y(x)$ exists and is less than $1 + \pi/4$ (Putnam Exam, 1947).

(E) **4.9** Find several terms in the asymptotic series for $y(x)$ valid as $x \to +\infty$, where $y(x)$ satisfies $dy/dx = x - 1/3y$, $y \sim 1/3x$ $(x \to +\infty)$.

(I) **4.10** Compute the coefficients a_n in (4.2.2) and prove that the series converges for sufficiently small $|x - a| \neq 0$.

(E) **4.11** (*a*) Investigate the behavior of solutions to $y' = y^2 + x$ for large negative x.

(*b*) Find the leading behaviors of solutions to $y' = 1 - xy^2$ as $x \to +\infty$. Find the first corrections to the leading behaviors.

(TI) **4.12** Prove that if the general solution to the differential equation $dy/dx = N(x, y)/D(x, y)$, where N

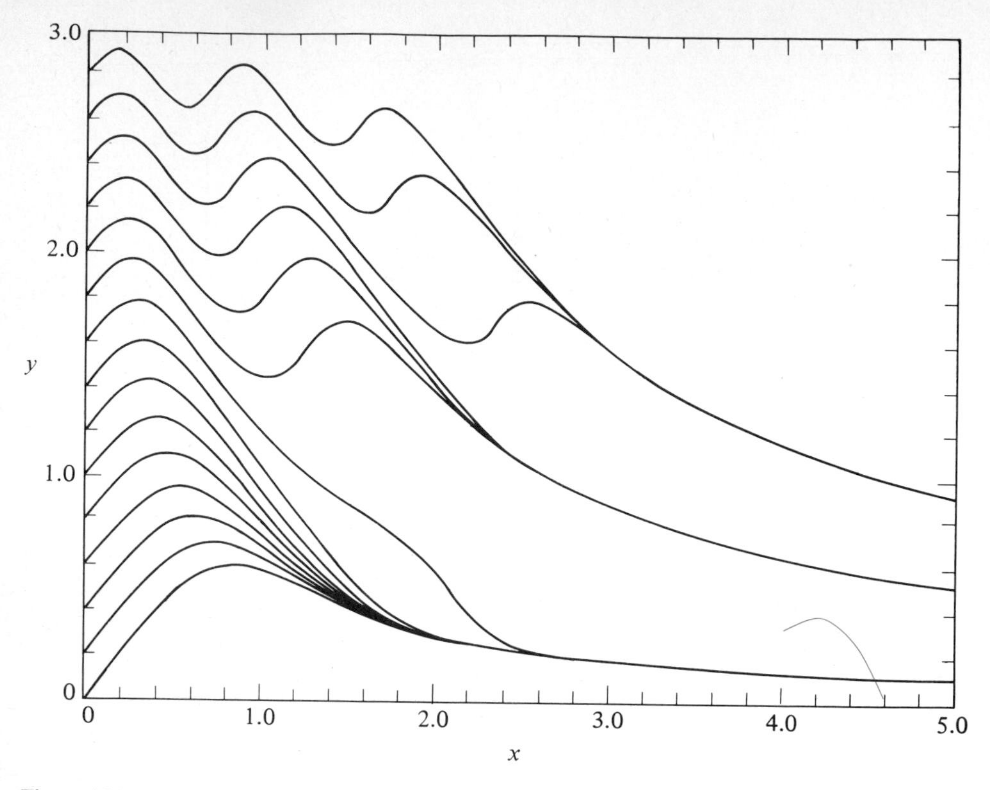

Figure 4.30 Solutions to $y' = \cos(\pi xy)$ for the initial values $y(0) = 0, 0.2, 0.4, 0.6, \ldots, 2.8$. Observe that as x increases, solutions bunch together. In Prob. 4.13 you are asked to explain this bunching phenomenon using asymptotic analysis.

and D are polynomials in y but arbitrary analytic functions of x, has no movable branch points, then the differential equation is a Riccati equation.

Clues:

(a) First, show that D is a function of x alone: $D = D(x)$. To do this argue that, by the fundamental theorem of algebra, if D depends on y, then it has a root $y = y_0(x)$. Perform a local analysis near $[x_0, y_0(x_0)]$ for any x_0 and show that $y(x)$ has a branch point at x_0 which contradicts the statement of the problem.

(b) Make the transformation $z = 1/y$ and repeat step (a) to show that $N(x, y)$ is at most quadratic in y.

(I) **4.13** The differential equation $y' = \cos(\pi xy)$ is too difficult to solve analytically. However, in Fig. 4.30 we have plotted the solutions to this equation for various values of $y(0)$. Note that the solutions bunch together as x increases. Could you have predicted this bunching phenomenon using asymptotic analysis? Find the possible leading behaviors of solutions as $x \to +\infty$. What are the corrections to these leading behaviors?

(E) **4.14** Show that if $(y')^2 + y^3 \to 0$ as $x \to +\infty$, then y and $y' \to 0$ as $x \to +\infty$ (Putnam Exam, 1974).

Section 4.3

(I) **4.15** There is a transformation which greatly facilitates the numerical solution of $y'' = y^2 + x$ plotted in Fig. 4.7. It is difficult to integrate numerically through the second-order poles of $y(x)$. However, as we approach a pole we let

$$y(x) = v^{-2}(x),$$

$$y'(x) = -\frac{2}{\sqrt{6}\, v^3(x)} - \frac{1}{2}\sqrt{6}\, xv(x) - 3v^2(x) + v^3(x)u(x).$$

Show that this gives the new system of equations

$$v' = \frac{1}{\sqrt{6}} + \frac{\sqrt{6}}{4}xv^4 + \frac{3}{2}v^5 - \frac{1}{2}uv^6,$$

$$u' = \frac{3}{4}x^2 v + \frac{9}{4}\sqrt{6}\, xv^2 + (9 - \sqrt{6}\, ux)v^3 - \frac{15}{2}uv^4 + \frac{3}{2}u^2v^5,$$

with initial conditions

$$v = -\frac{1}{\sqrt{y}}, \qquad u = \frac{3}{v} + \frac{\sqrt{6x}}{2v^2} + \frac{y'}{v^3} + \frac{2}{\sqrt{6}\, v^6},$$

and that these new equations may be used to integrate past the location of a pole of $y(x)$.

(I) **4.16** Show that to leading order the last two terms in (4.3.27) may be neglected as $s \to +\infty$.

(E) **4.17** Verify (4.3.34).

(I) **4.18** Consider the initial-value problem $y^2 y''' = -\frac{1}{3}$, $y(0) = a > 0$, $y'(0) = b$, $y''(0) = 0$. Show that $y(x)$ reaches $y = 0$ for some finite $x > 0$.

(E) **4.19** Prove by induction on (4.3.38) that a_n satisfies $a_n \le 6n/6^n$ (all n) and $a_n \ge 27n/4(3\sqrt{6})^n$ ($n \ge 3$). This shows that the radius of convergence of the series (4.3.37) lies between 6 and $3\sqrt{6} \doteqdot 7.35$.

(I) **4.20** Obtain a sharper upper bound on the radius of convergence of (4.3.37) than $3\sqrt{6}$ as follows:

(*a*) When $t > 0$ replace the exact differential equation $[t(d/dt)]^2 y = y^2 + t$ by $[t(d/dt)][t(d/dt)]y > y^2$.

(*b*) $y(t)$ is an increasing function because its Taylor series has positive coefficients so $y' \ge 0$. Multiply by $t(dy/dt)$, integrate from 0 to t, and obtain $y' \ge \sqrt{\frac{2}{3}}\, y^{3/2}/t$.

(*c*) Separate the dependences on y and t in this differential relation, integrate from a ($0 < a < t$) to t, and obtain $a \exp \sqrt{6/y(a)} \ge t \exp \sqrt{6/y(t)}$.

(*d*) Let $t \to t_0$, the radius of convergence of the series [note that since $a_n > 0$, $y(t_0) = +\infty$], and obtain $t_0 \le a \exp \sqrt{6/y(a)}$. Now, the goal is to find the value of a which minimizes the right side of the above inequality on t_0.

(*e*) First, use a one-term Taylor approximation to $y(a)$: $y(a) \sim a$ ($a \to 0$). Show that $a \exp \sqrt{6/a}$ has a minimum at $a = \frac{3}{2}$ and the value of the upper bound is $\frac{3}{2}e^2 \doteqdot 11.1$. This bound is terrible compared with the bound $t_0 \le 7.35$ of Prob. 4.19.

(*f*) Next, use a four-term Taylor approximation to $y(a)$ and approximate the rest of the series with the $6n/6^n$ approximate series of Prob. 4.19. Show that this procedure gives $t_0 \le 7.1$.

(I) **4.21** For the equation $y'' = y^2 + e^x$, make the substitutions $y = e^{x/2}u$, $s = e^{x/4}$ and obtain an equation whose solutions, asymptotically for large x, behave like elliptic functions of s. Deduce that the singularities of $y(x)$ are separated by a distance proportional to $e^{-x/4}$ as $x \to \infty$.

(TI) **4.22** The second Painlevé transcendent is defined by the differential equation $y'' = 2y^3 + xy + \alpha$.

(*a*) Discuss the behavior of $y(x)$ near a movable singularity and as $x \to \infty$.

(*b*) $y(x)$ has an infinite number of movable singularities along the real axis. Show that $\Delta(x)$, the separation between successive singularities, satisfies $\Delta(x) \sim c/\sqrt{x}$ ($x \to +\infty$), where c is a constant.

Clue: Let $z(t) = y(x)/\sqrt{x}$ where $t = x^{3/2}$. Approximate the resulting equation for $z(t)$ by an elliptic differential equation.

(I) **4.23** The next correction $\varepsilon_1(x)$ to the asymptotic relation in (4.3.40) takes the form $y(x) = 144[1 + Cx^{-0.772} + \varepsilon_1(x)]/x^3$. Determine the leading behavior of the function $\varepsilon_1(x)$ for large positive x. Then fit the resulting formula to the data in Table 4.3 and determine the value of C.

(E) **4.24** Show that the leading behavior of an explosive singularity of the Thomas-Fermi equation $y'' = y^{3/2}x^{-1/2}$ is correctly given in (4.3.41) as $y(x) \sim 400a/(x-a)^4$ $(x \to a)$.

(D) **4.25** Show that there exist solutions to the Thomas-Fermi equation which exhibit an infinite sequence of fourth-order poles along the real axis as $x \to +\infty$.

Clue: Let $y = xu$ and approximate the resulting equation by

$$\int \frac{du}{\sqrt{\frac{4}{5}u^{5/2} - C_1}} = x + C_2.$$

Argue that this integral defines u as a periodic function of x having fourth-order poles.

(I) **4.26** (*a*) Use integration by parts to show that $yy'' = 1$ $[y(0) = y(1) = 0]$ has no real solution.

(*b*) Show that the equation $yy'' = -1$ $[y(0) = y(1) = 0]$ has precisely two solutions that are nonsingular for $0 < x < 1$. (See Fig. 4.11.)

(I) **4.27** Find the first three terms in the asymptotic expansion of the two particular solutions to the Painlevé equation $y'' = y^2 + x$ whose leading behaviors are $y \sim \pm(-x)^{1/2}$ as $x \to -\infty$. That is, expand $\varepsilon(x)$ in (4.3.29).

(I) **4.28** Let $y\, d^4y/dx^4 = 1$ $[y(0) = y''(0) = y(1) = y''(1) = 0]$. Find the asymptotic behavior of $y(x)$ as $x \to 0+$. Try several terms involving combinations of logs and powers.

(D) **4.29** Find the general asymptotic behavior of solutions to $y''' + yy'' = 0$ (Blasius equation) as $x \to +\infty$.

(D) **4.30** Show that there exists a solution to $y'' = (2y^\alpha - x)y$ $(\alpha > 0)$ for which

$$y(x) \sim (x/2)^{1/\alpha}[1 + (1-\alpha)/\alpha^3x^3 + \cdots], \qquad x \to +\infty,$$

and $y(x) \sim k(\alpha)$ Ai $(-x)$ $(x \to -\infty)$ (de Boer and Ludford). (Note that this differential equation is a generalization of the second Painlevé transcendent in Prob. 4.22.)

(I) **4.31** Find the leading asymptotic behaviors of solutions to the following differential equations as $x \to +\infty$:

(*a*) $yy'' + x^2y'^2 + xyy' = 0$;

(*b*) $y'''y' = x^4yy''$;

(*c*) $yy'' = x^3y'^2$.

Note that these equations are equidimensional in y and therefore they admit exponential asymptotic behaviors. Also note that a constant is a solution to all these equations.

(D) **4.32** Find the general asymptotic behavior as $x \to +\infty$ of solutions to

(*a*) $yy'' + y' + xy = x^2$ (Levinson);

(*b*) $(y'')^2 = y' + y$ (Bellman);

(*c*) $y^2y'' = 1 + x$ (Strodt);

(*d*) $y'' = xy/(x^2 + y^2)$.

(I) **4.33** Find several terms in the asymptotic series for $y(x)$ valid as $x \to 0+$, where $y(x)$ satisfies $y'' - y'/x + y^2 = 1$ $[y'(0) = 0,\ y''(0) > 0]$.

(D) **4.34** Prove that the Laurent series (4.3.26) converges in a sufficiently small neighborhood of $x = a$.

(E) **4.35** Use the methods of Sec. 1.7 to solve $y^3y'' = -1$ in closed form. Derive the general two-parameter solution in (4.3.8) and the one-parameter special solution in (4.3.6).

(D) **4.36** Does the series in (4.3.22) converge for sufficiently large x?

(E) **4.37** Show that if we replace x by $x + \alpha$ in (4.3.24) and reexpand the result as a series in inverse powers of $\ln x$, the terms containing α are not multiplied by x. Conclude that in the limit $x \to +\infty$ terms which contain α are subdominant.

(E) **4.38** Find a complete description of the large-x behaviors of solutions to $y''y^3 = -1$. Show that there is a two-parameter general asymptotic behavior whose leading term is ax. What is the form of the rest of the series? There is also an algebraic behavior; find it.

Section 4.4

(E) **4.39** Solve $\dot{y} = y^2 - y$ analytically and use your result to verify the discussion in Example 2 of Sec. 4.4 and the plot in Fig. 4.14.

(E) **4.40** Describe the phase-line structure of the following equations:

(*a*) $\dot{y} = \sin y$;
(*b*) $\dot{y} = \sin^2 y$;
(*c*) $\dot{y} = \tan y$.

(E) **4.41** Find the eigenvalues and eigenvectors of the following matrices:

$$\begin{pmatrix} 1 & 2 \\ 3 & 4 \end{pmatrix}; \quad \begin{pmatrix} 1 & 2 \\ 0 & -1 \end{pmatrix}; \quad \begin{pmatrix} a & b \\ c & d \end{pmatrix}.$$

For all three matrices show that the product of the eigenvalues is the determinant of the matrix and that the sum of eigenvalues is the trace (sum of the diagonal entries) of the matrix.

(E) **4.42** In this problem we show how to solve the linear system

$$Y = M\mathbf{Y}, \tag{*}$$

where M is a constant matrix having repeated eigenvalues.

(*a*) Show that the matrices

$$M = \begin{pmatrix} 3 & -4 \\ 1 & -1 \end{pmatrix} \quad \text{and} \quad M = \begin{pmatrix} 1 & -1 \\ 1 & 3 \end{pmatrix}$$

have just one eigenvalue λ_1. Find λ_1. One solution to (*) is $\mathbf{Y} = \mathbf{V}e^{\lambda_1 t}$. Obtain a second linearly independent solution of the form $\mathbf{Y} = \mathbf{V}_1 e^{\lambda_1 t} + \mathbf{V}_2 te^{\lambda_1 t}$. Solve for $\mathbf{V}_1$ and $\mathbf{V}_2$ and show that the answer is not unique.

(*b*) Show that

$$M = \begin{pmatrix} 1 & 2 & -3 \\ 1 & 1 & 2 \\ 1 & -1 & 4 \end{pmatrix}$$

has one distinct eigenvalue λ_1. Find λ_1. One solution to (*) has the form $\mathbf{Y} = \mathbf{V}e^{\lambda_1 t}$. Obtain a second solution of the form $\mathbf{Y} = \mathbf{V}_1 e^{\lambda_1 t} + \mathbf{V}_2 te^{\lambda_1 t}$ and a third of the form $\mathbf{Y} = \mathbf{V}_3 e^{\lambda_1 t} + \mathbf{V}_4 te^{\lambda_1 t} + \mathbf{V}_5 t^2 e^{\lambda_1 t}$.

(I) **4.43** In this problem we show how to solve the inhomogeneous linear system

$$\dot{\mathbf{Y}} = M(t)\mathbf{Y} + \mathbf{H}(t), \tag{*}$$

when the solutions to the homogeneous system are known.

(*a*) Let $\mathbf{Y}_1(t), \mathbf{Y}_2(t), \ldots, \mathbf{Y}_n(t)$ be a complete set of linearly independent solutions to $\dot{\mathbf{Y}} = M\mathbf{Y}$. Construct the matrix $W(t)$ whose columns are the vectors $\mathbf{Y}_1(t), \ldots, \mathbf{Y}_n(t)$. Show that $\mathbf{Y} = W\mathbf{c}$, where

$$\mathbf{c} = \begin{pmatrix} c_1 \\ c_2 \\ \vdots \\ c_n \end{pmatrix}$$

is a constant vector, is the most general solution to $\dot{\mathbf{Y}} = M\mathbf{Y}$.

(*b*) Variation of parameters for matrix differential equations consists of looking for a general solution to (*) of the form $\mathbf{Y} = W\mathbf{c}(t)$. Substitute this expression for $\mathbf{Y}$ into (*) and obtain a formal expression for the vector $\mathbf{c}(t)$.

(*c*) Use variation of parameters to solve (*) in which

$$M = \begin{pmatrix} 2 & 3 \\ -1 & -2 \end{pmatrix} \quad \text{and} \quad H = \begin{pmatrix} t \\ e^t \end{pmatrix},$$

$$M = \begin{pmatrix} 2 & 1 \\ -5 & -2 \end{pmatrix} \quad \text{and} \quad H = \begin{pmatrix} t \\ t^2 \end{pmatrix},$$

$$M = \begin{pmatrix} 1 & 2 \\ 3 & 4 \end{pmatrix} \quad \text{and} \quad H = \begin{pmatrix} 1/t \\ t \end{pmatrix}.$$

(I) **4.44** Explain how to transform the system $\dot{r} = f(r, \theta)$, $\dot{\theta} = g(r, \theta)$ to Cartesian coordinates. Show that the system $\dot{r} = \sin \theta$, $\dot{\theta} = 0$ has a critical point at the origin with a critical line passing through it. What are the qualitative features of trajectories near the origin? Transform the system to Cartesian coordinates.

(I) **4.45** Find a system of equations in polar coordinates having a node-like critical point at the origin which trajectories approach in a wiggly fashion. (For example, $r = e^{-t}$, $\theta = 2 \sin^2 t$.) What is the form of the system in Cartesian coordinates?

(I) **4.46** The system $\dot{y}_1 = y_1$, $\dot{y}_2 = -y_2$ has a simple saddle point at $y_1 = y_2 = 0$ because there are only two "in" directions and two "out" directions for trajectories. Construct a system of equations which has a saddle point at $y_1 = y_2 = 0$ which exhibits three "in" and three "out" directions.

(I) **4.47** Examine the behavior of the trajectories in the phase plane for the following systems of equations:

(*a*) $\dot{y}_1 = y_1 - y_1^2 - y_1 y_2$, $\dot{y}_2 = \frac{1}{2}y_2 - \frac{1}{4}y_2^2 - \frac{3}{4}y_1 y_2$;

(*b*) $\dot{y}_1 = y_2$, $\dot{y}_2 = \pm a^2(1 - y_1^2)y_2 - y_1$ (Van der Pol's equation);

(*c*) $\dot{y}_1 = 2y_1 + y_2 + y_1 y_2^3$, $\dot{y}_2 = y_1 - 2y_2 - y_1 y_2$;

(*d*) $\dot{y}_1 = y_2$, $\dot{y}_2 = (\sin y_2) - y_1$ (show that this system has an infinite number of limit cycles which are alternately stable and unstable);

(*e*) $\dot{y}_1 = y_2 + 2y_1(1 - y_1^2 - y_2^2)$, $\dot{y}_2 = -y_1$ (Putnam Exam, 1960).

(I) **4.48** Discuss the nature and stability of the critical points at $y_1 = y_2 = 0$ for the following equations:

(*a*) $\dot{y}_1 = -y_1^3 + 2y_2^3$, $\dot{y}_2 = -2y_1 y_2^2$;

(*b*) $\dot{y}_1 = -y_1 - 3y_1^2 y_2$, $\dot{y}_2 = y_2 + y_2 \sin y_1$;

(*c*) $\dot{y}_1 = y_1^3 - 8y_2^3$, $\dot{y}_2 = 4y_1 y_2^3 + 4y_1^2 y_2 + 8y_2^3$;

(*d*) $\dot{y}_1 = y_1 \sinh y_2 - y_2 \cos y_1$, $\dot{y}_2 = y_2 \cosh y_1 + y_1 \sin y_2$.

(I) **4.49** Show that $y_1 + y_2 - \ln (y_1 y_2) = c$ represents a family of concentric closed curves about the point (1, 1). This proves that the critical point at (1, 1) in Example 4 of Sec. 4.4 is a center. [See (4.4.11).]

(I) **4.50** Perform a local analysis to determine whether there is a solution to (4.4.7) for which $y_1(t)$ and $y_2(t)$ both become infinite as $t \to +\infty$. See Fig. 4.17.

4.51 Prof. A. Toomre has suggested a problem on the autonomous system

$$\begin{aligned} \dot{x} &= yz, \\ \dot{y} &= -2xz, \\ \dot{z} &= xy, \end{aligned} \tag{*}$$

whose solution can be experimentally verified!

(E) (*a*) Show that for this system $x^2 + y^2 + z^2 =$ constant. If we choose this constant to be 1, then the autonomous system describes trajectories on the surface of a sphere of a radius 1.

(I) (*b*) Locate, classify, and examine the stability of the six critical points of the system at $(\pm 1, 0, 0)$, $(0, \pm 1, 0)$, $(0, 0, \pm 1)$.

(E) (*c*) Infer that the trajectories in Fig. 4.31 (by A. Toomre) are correctly drawn.

(E) (*d*) Show that the system (*) describes the rotational motion of a book tossed into the air. The angular-momentum vector for a freely rotating object is fixed in space because there is no external torque. This vector $\mathbf{L}(t)$, observed by someone sitting on the rotating object, appears to be spinning around in space. Its motion is governed by Euler's equation $0 = d\mathbf{L}/dt + \boldsymbol{\omega}\mathbf{L}$, where $\boldsymbol{\omega}$ is the angular velocity. Since $L_x = I_x \omega_x$, $L_y = I_y \omega_y$, $L_z = I_z \omega_z$, where I_x, I_y, I_z are the moments of inertia about the x, y, and z axes, we have

$$0 = \dot{L}_x + L_y L_z \left(\frac{1}{I_y} - \frac{1}{I_z} \right),$$

$$0 = \dot{L}_y + L_x L_z \left(\frac{1}{I_x} - \frac{1}{I_z} \right),$$

$$0 = \dot{L}_z + L_x L_y \left(\frac{1}{I_x} - \frac{1}{I_y} \right).$$

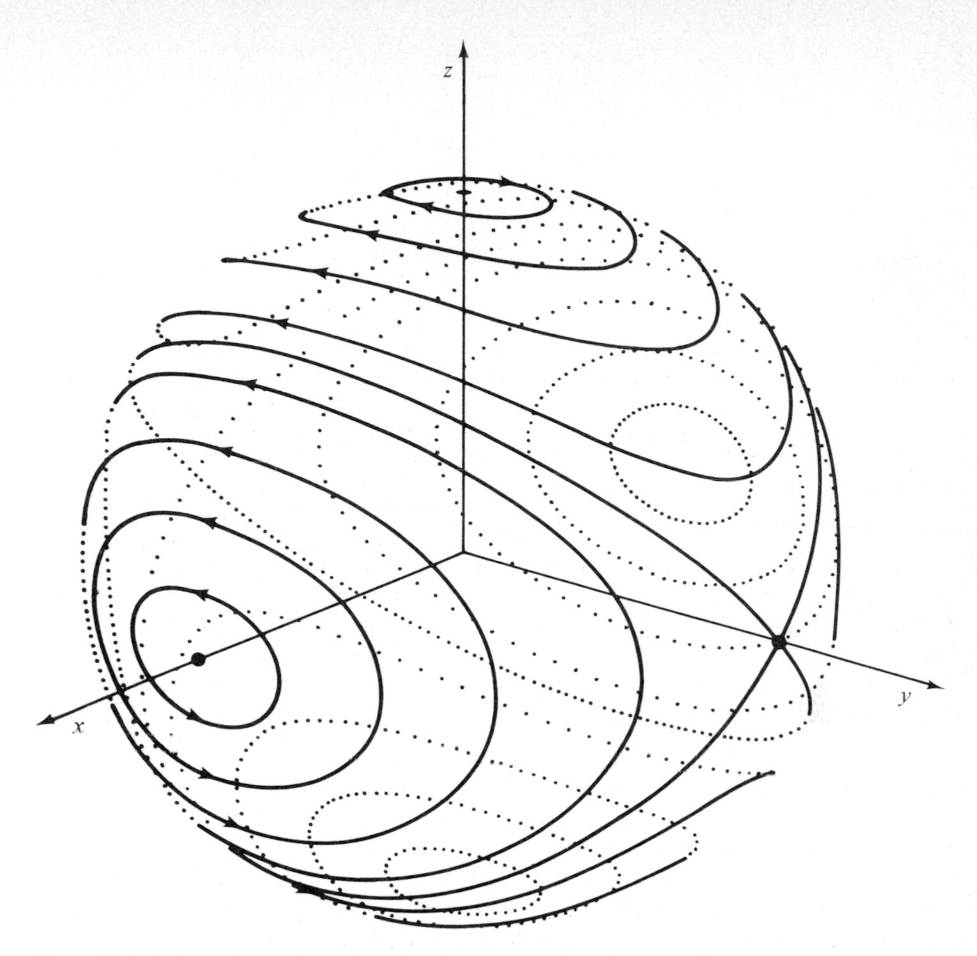

Figure 4.31 Trajectories on the surface of a phase sphere representing the free rotation of a rigid body. There are six critical points: $(\pm 1, 0, 0)$ and $(0, 0, \pm 1)$ are centers and $(0, \pm 1, 0)$ are saddle points (see Prob. 4.51).

If we now assume that the moments of inertia of a book about its x, y, and z axes are $I_x = 1$, $I_y = \frac{1}{2}$, $I_z = \frac{1}{3}$ (see Fig. 4.32), we recover the autonomous system (*) if we let $x = L_x$, $y = L_y$, $z = L_z$.

(E) (*e*) Experimentally verify the phase sphere diagram in Fig. 4.31 as follows. Argue from the diagram that a book tossed into the air, initially rotating about its x or z axes, will continue to do so in a stable fashion. (Try it!) Now argue that rotation about the y axis is *unstable* and predict that a book initially rotating about the y axis will somersault, no matter how carefully it is tossed into the air. (Try it!) This instability is called Eulerian wobble.

Section 4.5

(E) **4.52** Find the first three coefficients in each of the series in (4.5.1).

(I) **4.53** Show that the point $p_1 = p_2 = \cdots = p_m = 0$, $q_1 = q_2 = \cdots = q_m = 0$ is a simple center of the equations (4.5.2) and (4.5.3) which define the Toda lattice.

(D) **4.54** Show that when $m = 2$ the orbits of (4.5.2) and (4.5.3) are periodic. Are the orbits still periodic when $m = 3$?

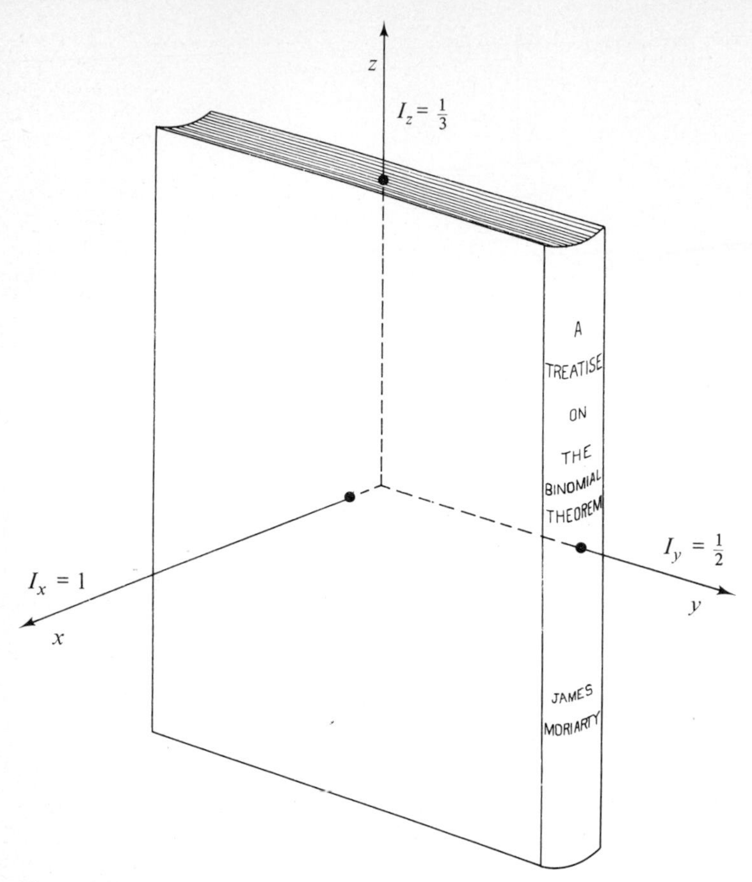

Figure 4.32 A book such as this, when tossed into the air, undergoes stable rotation about the x and z axes but wobbles and somersaults rather than rotating stably about the y axis. This motion is described by the trajectories on the phase sphere in Fig. 4.31 (see Prob. 4.51).

(E) **4.55** (*a*) Show that the Jacobian

$$J(t) = \frac{\partial[\mathbf{p}(t), \mathbf{q}(t)]}{\partial[\mathbf{p}(0), \mathbf{q}(0)]}$$

satisfies $J(t) = 1$ for all t if $\mathbf{p}$ and $\mathbf{q}$ satisfy the Hamiltonian system in (4.5.7) and (4.5.8).

(I) (*b*) Show that volumes are conserved in phase space if $J(t) \equiv 1$.

(I) **4.56** Show that there are no C-systems in less than three dimensions.

(I) **4.57** Show that the critical points of the system (4.5.9) are all unstable.

(I) **4.58** (*a*) Prove that when $r < 1$ the only critical point of the Lorenz model (4.5.10) to (4.5.12) is at $x = y = z = 0$ and it is stable. Show also that when $r > 1$ the origin becomes an unstable critical point. What happens when $r = 1$?

(*b*) Show that when $r > 1$ there are critical points at $x = y = \pm\sqrt{r-1}$, $z = r - 1$. Prove that these critical points are stable if $1 < r < 21$ and unstable if $r > 21$.

(I) **4.59** (*a*) Find the Hamiltonian for the system (4.5.4) to (4.5.6).

(*b*) Show that the trajectories in Figs. 4.21 and 4.22 have the same energy E; specifically, show that $E = \frac{1}{8}$.

CHAPTER

FIVE

APPROXIMATE SOLUTION OF DIFFERENCE EQUATIONS

What ineffable twaddle! I never read such rubbish in my life.

—Dr. Watson, *A Study in Scarlet*
Sir Arthur Conan Doyle

(E) 5.1 INTRODUCTORY COMMENTS

Difference equations (recursion relations) occur so frequently in applied mathematics that we allot a full chapter to a discussion of the behavior of their solutions. We will study the problem of determining the behavior of a_n, the solution to a difference equation, as $n \to \infty$. This is the most common kind of asymptotics problem involving difference equations.

Difference equations often arise in the course of a recursive or iterative solution to a problem. For example, the coefficients a_n of a Taylor series solution $y(x) = \sum a_n x^n$ to a differential equation satisfy a difference equation. In this example, although the large-n asymptotic behavior of a_n is only an approximate concept, the ratio

$$R = \lim_{n \to \infty} \left| \frac{a_n}{a_{n+1}} \right|$$

is *exactly* equal to the radius of convergence of the Taylor series. Thus, it is not necessary to solve the difference equation exactly to determine the region of analyticity of $y(x)$; a local analysis of the difference equation near $n = \infty$ is sufficient.

As we have already emphasized in Chap. 2, linear difference equations and linear differential equations have a very similar mathematical structure. For both kinds of equations the techniques for finding exact solutions, such as reduction of order and variation of parameters, are alike. This similarity extends to the techniques of local asymptotic analysis. For a linear difference equation, the point at $n = \infty$ may be classified as an ordinary, regular singular, or irregular singular point. The procedure for constructing the appropriate Taylor, Frobenius, or asymptotic series valid near the point $n = \infty$ is in close analogy with the corresponding procedure for differential equations.

The parallel between nonlinear difference and differential equations is not so strong. For example, the differential equation $y' = 1/y$ is easy to solve exactly while its simplest discrete analog $a_{n+1} - a_n = 1/a_n$ has not been solved in closed form. Unlike differential equations, even the simplest-looking nonlinear difference equations are rarely exactly solvable.

Nonlinear difference equations are distinctly different from nonlinear differential equations in another respect. They do not exhibit movable or spontaneous singularities. Consider the nonlinear differential equation $y' = y^2 - y$ $[y(0) = 2]$. The solution to this equation, $y(x) = 2/(2 - e^x)$, has an explosive singularity at a finite value of x; namely, at $x = \ln 2$. The solution a_n to the corresponding difference equation $a_{n+1} - a_n = a_n^2 - a_n$ $(a_0 = 2)$ is $a_n = 2^{2^n}$, which remains *finite* for all values of n. [As explained in Chap. 2, the solution to this difference equation is obtained by taking the logarithm of the equation: $b_n = \log_2 (a_n)$ satisfies $b_{n+1} = 2b_n$.] Observe that a_n becomes very strongly singular as $n \to \infty$, while $y(x)$ approaches 0 as $x \to \infty$. Thus, one must be cautious about forecasting the behavior of a difference equation from a study of the properties of the corresponding differential equation.

(I) 5.2 ORDINARY AND REGULAR SINGULAR POINTS OF LINEAR DIFFERENCE EQUATIONS

Techniques similar to those in Chap. 3 may be used to study the local behavior of solutions to linear difference equations as $n \to \infty$. Accordingly, in this section we show how to classify $n = \infty$ as an ordinary point, a regular singular point, or an irregular singular point of a linear difference equation. Then we investigate the leading behavior of solutions in the neighborhood of ordinary points and regular singular points. Finally, we develop expansions analogous to Taylor and Frobenius series to describe the full local behavior near ordinary points and regular singular points. The large-n behavior of solutions when ∞ is an irregular singular point is discussed in Secs. 5.3 to 5.5.

Classification of Linear Difference Equations at $n = \infty$

We consider the kth-order homogeneous linear difference equation

$$a_{n+k} + p_{k-1}(n)a_{n+k-1} + p_{k-2}(n)a_{n+k-2} + \cdots + p_0(n)a_n = 0, \qquad (5.2.1)$$

where $p_i(n)$ $(i = 0, \ldots, k-1)$ are given functions of the integer n. To classify this difference equation at $n = \infty$ we appeal to the differential-equation analog of

(5.2.1). Recall from Chap. 2 that in this formal analogy n plays the role of the independent variable x, a_n is the dependent variable $y(x)$, and we have the following correspondences:

$$\begin{aligned} n &\leftrightarrow x, \\ a_n &\leftrightarrow y(x), \\ Da_n &\equiv a_{n+1} - a_n \leftrightarrow y'(x), \\ D^2 a_n &\equiv D(Da_n) = a_{n+2} - 2a_{n+1} + a_n \leftrightarrow y''(x), \\ &\vdots \\ D^k a_n &= \sum_{j=0}^{k} (-1)^j \binom{k}{j} a_{n+k-j} \leftrightarrow y^{(k)}(x). \end{aligned}$$

These correspondences give a *unique* differential-equation analog of the difference equation (5.2.1) and the point $n = \infty$ is classified as an ordinary point, regular singular point, or irregular singular point of (5.2.1) if the point $x = \infty$ in the corresponding differential equation is so classified.

Example 1 *Classification of difference equations at $n = \infty$.*

(*a*) $a_{n+1} = a_n$. The differential-equation analog of this difference equation is just $y'(x) = 0$. Since $x = \infty$ is an ordinary point of the differential equation, $n = \infty$ is an ordinary point of the difference equation. Notice that the general solutions of the difference and differential equations are the same, $a_n = c$ and $y(x) = c$, where c is a constant.

(*b*) $a_{n+1} = -a_n$. The differential-equation analog is $y' = -2y$. The point $x = \infty$ is an irregular singular point, so $n = \infty$ is also an irregular singular point. Notice that the general solution of the difference equation is $a_n = (-1)^n c$ while the general solution of the differential equation is $y(x) = ce^{-2x}$. These solutions behave very differently as $n \to \infty$ and $x \to \infty$. The oscillation of the solution to the difference equation as $n \to \infty$ is exponentially rapid ($a_n = ce^{i\pi n}$); typically, one expects solutions to vary rapidly near an irregular singular point.

(*c*) $a_{n+1} = na_n$. Since $a_{n+1} - a_n = (n-1)a_n$, the corresponding differential equation is $y' = (x-1)y$ which has an irregular singular point at $x = \infty$. Thus, $n = \infty$ is an irregular singular point of the difference equation. Again, the solutions of the difference and differential equations, $a_n = c(n-1)!$ and $y(x) = ce^{(x^2-x)/2}$ behave very differently as $n = x \to \infty$.

(*d*) $a_{n+1} = a_n/n$. The corresponding differential equation is $y' = (x^{-1} - 1)y$, so $x = \infty$ and thus $n = \infty$ are irregular singular points. The solutions to the difference and differential equations $a_n = c/(n-1)!$ and $y(x) = cxe^{-x}$ behave in radically different ways as $n = x \to \infty$.

(*e*) $a_{n+1} = (1 + 1/n)a_n$. The corresponding differential equation $y' = x^{-1}y$ has a regular singular point at ∞, so $n = \infty$ is also a regular singular point. The solutions to the difference and differential equations, $a_n = cn$ and $y(x) = cx$, behave identically as $n = x \to \infty$.

(*f*) $a_{n+1} = (1 + 1/n^2)a_n$, $n = \infty$ is an ordinary point. Solutions to the difference equation and corresponding differential equation approach constants as $n = x \to \infty$.

(*g*) $a_{n+2} = a_n$. Since $a_{n+2} - 2a_{n+1} + a_n = -2(a_{n+1} - a_n)$, the corresponding differential equation is $y'' = -2y'$ and $x = \infty$ and $n = \infty$ are irregular singular points. The general solution of the difference equation is $a_n = c_1 + (-1)^n c_2$, while the general solution of the differential equation is $y(x) = c_1 + c_2 e^{-2x}$. The slowly varying solutions c_1 are identical, but the rapidly varying solutions $(-1)^n c_2$ and $c_2 e^{-2x}$ differ markedly as $x = n \to \infty$.

Leading Behavior as $n \to \infty$

In the preceding example we saw that when the point at ∞ is an ordinary point or a regular singular point, then the *leading behaviors* of solutions to the difference equation and its differential-equation analog are the same. In general, if $n = \infty$ is an ordinary or a regular singular point then, for any solution a_n of the difference equation, there is a solution $y(x)$ of the corresponding differential equation that satisfies

$$a_n \sim y(n), \qquad n \to \infty$$

(see Example 3). However, the higher-order behaviors of a_n and $y(n)$ as $n \to \infty$ are usually different. At an irregular singular point even the leading behaviors of a_n and $y(n)$ may be radically different, as we have illustrated. Here are some more examples.

Example 2 *Comparison between leading behaviors of difference and differential equations.*

(*a*) The differential equation which corresponds with the difference equation

$$a_{n+1} = \left(1 + \frac{\alpha}{n}\right) a_n \tag{5.2.2}$$

is $y' = \alpha x^{-1} y$, so ∞ is a regular singular point. Since the general solution to the differential equation is $y(x) = cx^\alpha$, we expect the leading behavior of solutions to the difference equation to be

$$a_n \sim cn^\alpha, \qquad n \to \infty, \tag{5.2.3}$$

for some constant c. This behavior may be checked from the exact solution to (5.2.2) which is $\ln (a_n/a_1) = \sum_{j=1}^{n-1} \ln (1 + \alpha/j)$. Exponentiating the relation

$$\sum_{j=1}^{n-1} \ln (1 + \alpha/j) - \alpha \ln n \sim k, \qquad n \to \infty,$$

where k is a constant (see Prob. 5.2), we obtain (5.2.3) with $c = a_1 e^k$.

The result (5.2.3) can also be verified from the exact solution of (5.2.2) written as

$$a_n = a_1 \frac{\Gamma(n + \alpha)}{\Gamma(n)\Gamma(\alpha + 1)},$$

where Γ is the gamma function (see Prob. 5.4).

We warn the reader that while the method of dominant balance is often very useful for difference equations, the leading behavior of solutions to (5.2.2) may *not* be obtained by neglecting α/n with respect to 1 as $n \to \infty$ because this gives the asymptotic difference equation

$$a_{n+1} \sim a_n, \qquad n \to \infty, \tag{5.2.4}$$

which is true but not useful. It is true because cn^α satisfies (5.2.4). It is not useful because cn^α is not the only function which satisfies this relation; $cn^\beta(\ln n)^\gamma e^{\delta n^\rho}$ satisfies (5.2.4) for any c, β, γ, δ and any $\rho < 1$! If we try to solve the asymptotic difference equation (5.2.4) by replacing the asymptotic sign by an equal sign, the solution, which is a constant, does not approximate the solution to (5.2.2) as $n \to \infty$.

It is often very helpful to rewrite difference equations in terms of discrete derivatives before terms are dropped. That is, (5.2.2) should be rewritten as $a_{n+1} - a_n = \alpha a_n/n$, and in this form it is not productive to neglect $\alpha a_n/n$ with respect to $a_{n+1} - a_n$ because it is wrong to neglect $\alpha x^{-1}y$ with respect to y' in the differential equation $y' = \alpha x^{-1}y$.

(*b*) The leading behavior of solutions to

$$a_{n+1} = \left(1 + \frac{\alpha}{n} + \frac{\beta}{n^2}\right) a_n \tag{5.2.5}$$

is also given by (5.2.3). Using discrete derivatives this difference equation takes the form $a_{n+1} - a_n = (\alpha/n + \beta/n^2)a_n$, whose differential-equation analog is $y' = (\alpha/x + \beta/x^2)y$. The general solution to this differential equation is $y(x) = cx^\alpha e^{-\beta/x}$ whose leading behavior is $y(x) \sim cx^\alpha$ $(x \to +\infty)$; the solutions to the difference and differential equations have the same leading behaviors as $n = x \to \infty$.

Note that the controlling factor n^α of the leading behavior cn^α can be found easily using a dominant-balance argument. Since $\beta n^{-2} \ll \alpha n^{-1}$ $(n \to \infty)$, it is valid to replace (5.2.5) by the asymptotic difference equation $a_{n+1} - a_n \sim \alpha a_n/n$ $(n \to \infty)$. The controlling factor of the solutions to this asymptotic difference equation is n^α. Correspondingly, neglecting the term $\beta x^{-2}y$ as $x \to \infty$ in the differential equation $y' = (\alpha x^{-1} + \beta x^{-2})y$ (assuming that $\alpha \neq 0$) and solving the resulting asymptotic differential equation gives the correct controlling factor x^α of $y(x)$ as $x \to +\infty$.

(*c*) The difference equation $a_{n+2} = (2 - 1/n)a_{n+1} - (1 - 1/n)a_n$ has the differential-equation analog $xy'' + y' = 0$ so ∞ is a regular singular point. Since the general solution of the differential equation is $y(x) = c_1 + c_2 \ln x$, we expect that

$$a_n \sim c_1 + c_2 \ln n, \qquad n \to \infty.$$

This prediction may be checked as follows. One solution of the difference equation is explicitly $a_n = c_1$. The other solution can be found by setting $b_n = a_{n+1} - a_n$ and thereby reducing the equation to first order. It follows that

$$b_{n+1} - b_n = -b_n/n.$$

The general solution of this difference equation is $b_n = c_2/(n-1)$. Thus,

$$a_n - a_2 = \sum_{k=2}^{n-1} b_k \sim c_2 \ln n, \qquad n \to \infty,$$

which verifies the predicted behavior of a_n as $n \to \infty$.

Equality of Leading Behaviors of Solutions to Difference and Differential Equations at Ordinary and Regular Singular Points

We asserted above that if the corresponding differential equation has an ordinary point or a regular singular point at ∞, then the leading behaviors of a_n and $y(x)$ as $n = x \to +\infty$ must agree. It is relatively easy to prove this assertion for the first-order difference equation

$$a_{n+1} = [1 + p(n)]a_n, \tag{5.2.6}$$

whose differential-equation analog is

$$y'(x) = p(x)y(x). \tag{5.2.7}$$

If $x = \infty$ is an ordinary or a regular singular point, then $p(x)$ may be expressed as the series

$$p(x) = \alpha/x + \beta/x^2 + \gamma/x^3 + \cdots,$$

which converges for sufficiently large $|x|$. The leading behaviors of the solutions to the differential equation (5.2.7) are

$$y(x) \sim cx^{\alpha}, \qquad x \to +\infty,$$

where c is a constant.

The simplest way to show that the solutions to the difference equation have the same leading behaviors is to solve the difference equation exactly and to examine the asymptotic behavior of the solution as $n \to \infty$. The exact solution to (5.2.6) is given by (see Chap. 2)

$$a_n = a_1 \exp\left\{\sum_{k=1}^{n-1} \ln\,[1 + p(k)]\right\}, \qquad n \geq 2.$$

Since

$$\ln\,[1 + p(k)] = \frac{\alpha}{k} + \frac{\beta - \frac{1}{2}\alpha^2}{k^2} + \cdots$$

for large k, we obtain

$$\sum_{k=1}^{n-1} \ln\,[1 + p(k)] - \alpha \ln n \sim C, \qquad n \to \infty,$$

for some constant C. Therefore,

$$a_n \sim a_1 e^C n^{\alpha}, \qquad n \to \infty,$$

which agrees with the asymptotic behavior of the solution to the differential equation (5.2.7) if $c = a_1 e^C$.

The proof of the equality of leading behaviors of solutions to higher-order difference and differential equations at ordinary and regular singular points is developed in Prob. 5.5.

Taylor and Frobenius-like Series Expansions

The higher-order behavior of a solution to a linear difference equation at an ordinary or regular singular point may be obtained by expanding the solution in a Taylor or Frobenius-like series about $n = \infty$. We present two different kinds of expansions which may be useful to approximate the solution of difference equations for large n. The first tends to be a bit clumsy when the difference equation has coefficients which are simple rational functions of n. [However, it may be much more appropriate to use this first version when the coefficients are not rational functions of n (see Prob. 5.7).]

Example 3 *Taylor series expansion (clumsy version).* The difference equation

$$a_{n+1} = \left(1 + \frac{1}{n^2}\right) a_n \tag{5.2.8}$$

has an ordinary point at ∞. Therefore, one should try to expand the solution as a Taylor series. One possible form for such a series,

$$a_n = \sum_{k=0}^{\infty} \frac{A_k}{n^k}, \qquad A_0 \neq 0, \tag{5.2.9}$$

turns out to be clumsy because, as we shall see, it leads to a complicated recursion relation for the expansion coefficients A_k. Substituting this expansion into (5.2.8) we obtain

$$\sum_{k=0}^{\infty} \frac{A_k}{n^k}\left(1+\frac{1}{n}\right)^{-k} = \sum_{k=0}^{\infty} \frac{A_k}{n^k} + \sum_{k=0}^{\infty} \frac{A_k}{n^{k+2}}.$$

Since

$$\left(1+\frac{1}{n}\right)^{-k} = \sum_{p=0}^{\infty} (-1)^p \frac{(k+p-1)!}{p!\,(k-1)!}\frac{1}{n^p},$$

it follows that

$$\sum_{k=1}^{\infty} \frac{A_k}{n^k}\left(1+\frac{1}{n}\right)^{-k} = \sum_{q=1}^{\infty} \frac{(q-1)!}{n^q} \sum_{\substack{p+k=q\\ 0\le p<q}} (-1)^p A_k \frac{1}{p!\,(k-1)!}.$$

Therefore, $A_1 = -A_0$ and

$$A_q = (q-1)! \sum_{p=1}^{q-1} (-1)^{p+1} A_{q-p} \frac{1}{(p+1)!\,(q-p-1)!} - \frac{A_{q-1}}{q}, \qquad q = 2, 3, \ldots. \tag{5.2.10}$$

This recurrence relation (5.2.10) is unnecessarily complicated.

The analysis of the behavior near $n = \infty$ can be simplified drastically if we recall from Example 3 of Sec. 2.1 that it is natural to replace n^{-k} by $\Gamma(n)/\Gamma(n+k)$ when analyzing difference equations. [Note that $\Gamma(n)/\Gamma(n+k) = [(n+k-1)\times(n+k-2)\cdots(n+1)(n)]^{-1} \sim n^{-k}$ $(n \to \infty)$.] Now the first discrete derivative of $\Gamma(n)/\Gamma(n+k)$ is

$$D\frac{\Gamma(n)}{\Gamma(n+k)} = \frac{\Gamma(n+1)}{\Gamma(n+k+1)} - \frac{\Gamma(n)}{\Gamma(n+k)}$$

$$= -k\frac{\Gamma(n)}{\Gamma(n+k+1)}. \tag{5.2.11}$$

This formula is analogous to $(d/dx)x^{-k} = -kx^{-k-1}$. Since $\Gamma(n)/\Gamma(n+k) \sim n^{-k}$ $(n \to \infty)$, these simple differentiation properties suggest that replacing the Taylor series expansion (5.2.9) by the modified expansion

$$a_n = \sum_{k=0}^{\infty} B_k \frac{\Gamma(n)}{\Gamma(n+k)} \tag{5.2.12}$$

will yield much simpler recurrence relations than those satisfied by A_k in (5.2.9) if the difference equation has polynomial coefficients.

Example 4 *Taylor series expansion (improved version).* To determine the coefficients B_k in (5.2.12) for the difference equation (5.2.8), we rewrite (5.2.8) as an equation in terms of discrete derivatives with polynomial coefficients:

$$n^2 Da_n = a_n. \tag{5.2.13}$$

Next, we substitute the general expansion (5.2.12) and use (5.2.11) to obtain

$$-\sum_{k=0}^{\infty} kn^2 B_k \frac{\Gamma(n)}{\Gamma(n+k+1)} = \sum_{k=0}^{\infty} B_k \frac{\Gamma(n)}{\Gamma(n+k)}. \tag{5.2.14}$$

Now we use the fact that the coefficients of (5.2.13) are polynomials in n. Specifically, any term of the form $n^p\Gamma(n)/\Gamma(n+q)$ where p, q are positive integers can be expressed as a finite sum of terms of the form $\Gamma(n)/\Gamma(n+r)$ with coefficients that do not depend on n. In the present example, we use the formula $n^2 = (n+k)(n+k-1) - (2k-1)(n+k) + k^2$ to obtain

$$n^2\frac{\Gamma(n)}{\Gamma(n+k+1)} = \frac{\Gamma(n)}{\Gamma(n+k-1)} - (2k-1)\frac{\Gamma(n)}{\Gamma(n+k)} + k^2\frac{\Gamma(n)}{\Gamma(n+k+1)}.$$

Substituting this result into (5.2.14) and shifting summation indices so that all sums contain the factor $\Gamma(n)/\Gamma(n+k)$ gives

$$-\sum_{k=0}^{\infty}[(k+1)B_{k+1} - k(2k-1)B_k + (k-1)^3B_{k-1}]\frac{\Gamma(n)}{\Gamma(n+k)} = \sum_{k=0}^{\infty}B_k\frac{\Gamma(n)}{\Gamma(n+k)}, \tag{5.2.15}$$

where we take $B_{-1} = 0$.

We find the recursion relation for B_k by equating coefficients of $\Gamma(n)/\Gamma(n+k)$ for $k = 0, 1, 2, \ldots$ in (5.2.15). The justification for this step is given in Prob. 5.10. The result is

$$-(k+1)B_{k+1} + k(2k-1)B_k - (k-1)^3B_{k-1} = B_k, \qquad k = 0, 1, 2, \ldots. \tag{5.2.16}$$

Thus, $B_1 = -B_0$, $B_2 = 0$, $B_3 = \frac{1}{3}B_0$, $B_4 = \frac{7}{6}B_0$, and so on.

Observe that the equation (5.2.16) for the coefficients B_k is significantly simpler than that in (5.2.10) for the coefficients A_k. While this equation for B_k does not have a simple analytic solution, it can be readily used to compute enough coefficients in the expansion (5.2.12) to obtain a very accurate description of a_n for large n.

Using the techniques of Sec. 5.3, the recursion relation (5.2.16) for B_k can be solved approximately for large k ($k = \infty$ is an irregular singular point):

$$B_k \sim c_{\pm}(k-1)!\,k^{-1\pm i}, \qquad k \to +\infty,$$

(see Prob. 5.38). This estimate shows that (5.2.9) converges for all n in the interval $0 < n < \infty$.

This series in (5.2.12) is not adequate to describe the behavior of solutions to difference equations if $n = \infty$ is a regular singular point because the leading behavior of the first nonzero term is $B_p n^{-p}$ as $n \to \infty$ for some $p = 0, 1, 2, 3, \ldots$. If $n = \infty$ is a regular singular point, at least one solution behaves like cn^α as $n \to \infty$ (and remaining solutions may behave like cn^α multiplied by factors of $\ln n$).

One way to modify the series in (5.2.12) is simply to multiply it by n^α to allow for algebraic behavior at $n = \infty$. However, we generally obtain a much simpler recursion relation for the expansion coefficients if we seek instead a series solution of the form

$$a_n = \sum_{k=0}^{\infty} A_k\frac{\Gamma(n)}{\Gamma(n+k-\alpha)}, \tag{5.2.17}$$

where α is determined by the requirement that $A_0 \neq 0$. Since

$$D[\Gamma(n)/\Gamma(n+k-\alpha)] = (\alpha - k)\Gamma(n)/\Gamma(n+k-\alpha+1)$$

and

$$\Gamma(n)/\Gamma(n+k-\alpha) \sim n^{\alpha-k}, \qquad n \to \infty$$

(see Sec. 5.4), we expect that $\Gamma(n)/\Gamma(n+k-\alpha)$ will be an appropriate choice as a discrete analog of the term $x^{\alpha-k}$ in the usual Frobenius series about $x = \infty$.

Example 5 *Frobenius series expansion.* Consider the difference equation

$$(n + Q)(a_{n+1} - a_n) = Pa_n, \tag{5.2.18}$$

where P and Q are noninteger constants such that $P + Q$ is not a negative integer. The leading behavior of a_n is easily shown to be cn^P. Therefore, we must choose $\alpha = P$ in (5.2.17). Substituting this value of α into the difference equation, we find that

$$(n + Q) \sum_{k=0}^{\infty} (P - k)A_k \frac{\Gamma(n)}{\Gamma(n + k - P + 1)} = \sum_{k=0}^{\infty} PA_k \frac{\Gamma(n)}{\Gamma(n + k - P)}.$$

Writing $n + Q = (n + k - P) + (Q + P - k)$, we obtain

$$\sum_{k=0}^{\infty} (P - k)A_k \frac{\Gamma(n)}{\Gamma(n + k - P)} + \sum_{k=0}^{\infty} (Q + P - k)(P - k)A_k \frac{\Gamma(n)}{\Gamma(n + k - P + 1)}$$

$$= \sum_{k=0}^{\infty} PA_k \frac{\Gamma(n)}{\Gamma(n + k - P)}.$$

Equating coefficients of $\Gamma(n)/\Gamma(n + k - P)$ for $k = 0, 1, 2, \ldots$ gives

$$kA_k = (k - P - Q - 1)(k - P - 1)A_{k-1}, \qquad k = 0, 1, 2, \ldots,$$

where $A_{-1} = 0$. Thus, A_0 is arbitrary and the solution to this difference equation for A_k is

$$A_k = A_0 \frac{\Gamma(k - P - Q)\Gamma(k - P)}{\Gamma(-P - Q)\Gamma(-P)k!}.$$

Thus, the solution to (5.2.18) is

$$a_n = A_0 \sum_{k=0}^{\infty} \frac{\Gamma(k - P - Q)\Gamma(k - P)\Gamma(n)}{\Gamma(-P - Q)\Gamma(-P)k!\, \Gamma(n + k - P)}. \tag{5.2.19}$$

Since the difference equation (5.2.18) also happens to be exactly soluble, we can calculate A_0 in terms of a_0 by rewriting (5.2.18) in the form $a_{n+1} = a_n(n + P + Q)/(n + Q)$. We then have

$$a_n = a_0 \frac{\Gamma(n + P + Q)\Gamma(Q)}{\Gamma(P + Q)\Gamma(n + Q)}. \tag{5.2.20}$$

Thus, comparing the asymptotic behavior as $n \to \infty$ of the expressions in (5.2.19) and (5.2.20) gives $A_0 = a_0\Gamma(Q)/\Gamma(P + Q)$. However, the formula for A_0 is a result of global *not* local analysis; we have used the exact solution (5.2.20) to obtain the connection between the initial value a_0 and the asymptotic behavior of a_n as $n \to \infty$.

Our discussion of Frobenius series must be further generalized when two indices are equal or differ by an integer to allow for the possibility of logarithmic terms. It is, of course, possible to multiply the generalized Frobenius series by $\ln n$ when these special cases occur. However, $\ln n$ has complicated properties with respect to the discrete difference operations:

$$D \ln n = \ln \left(1 + \frac{1}{n}\right) = \sum_{k=0}^{\infty} A_k \frac{\Gamma(n)}{\Gamma(n + k + 1)}.$$

In this expression $A_0 = 1$ but the higher coefficients A_k $(k = 1, 2, \ldots)$ are very complicated. Instead of $\ln n$, it is often simpler to use the *digamma* function

$\psi(n) = \Gamma'(n)/\Gamma(n)$ which satisfies

$$D\psi(n) = \frac{1}{n} \tag{5.2.21}$$

and $\psi(n) \sim \ln n$ $(n \to \infty)$. Since $(d/dx)(\ln x) = 1/x$, $\psi(n)$ is the appropriate discrete analog of $\ln x$ (see Prob. 5.8).

(E) 5.3 LOCAL BEHAVIOR NEAR AN IRREGULAR SINGULAR POINT AT INFINITY: DETERMINATION OF CONTROLLING FACTORS

We have argued in Sec. 5.2 that the first step in the analysis of the behavior of solutions to a linear difference equation as $n \to \infty$ should be to determine the nature of the point at ∞ of the corresponding differential equation. If the point at ∞ is an ordinary or a regular singular point, then, as we saw in Sec. 5.2, (1) the leading behaviors of solutions to the differential equation are the same as the leading behaviors of solutions to the difference equation, and (2) the higher corrections to this leading behavior of a_n take the form of a convergent series whose kth term is proportional to either $\Gamma(n)/\Gamma(n + k + \alpha)$ or $n^{-k-\alpha}$ (where α is an integer at an ordinary point).

The situation is more complicated if ∞ is an irregular singular point. In this case local analysis begins by determining the controlling factor (the most rapidly varying component) of the leading behavior. As in Chap. 3, the next step is to divide or peel off this controlling factor and to determine the next most rapidly varying component of the leading behavior. This process is repeated until the full leading behavior is found. The last stage of the local analysis consists of finding higher corrections to the leading behavior in the form of an infinite series.

In this section we discuss methods for finding the controlling factors of the behaviors of solutions as $n \to \infty$. In Secs. 5.4 and 5.5 we use the techniques developed here to obtain a complete description of the asymptotic behavior near the irregular singular point at $n = \infty$. There are three principal techniques for finding controlling factors:

METHOD I If the controlling factor of a solution $y(x)$ to the corresponding differential equation varies logarithmically or algebraically or even as rapidly as e^{ax^b} with $0 < b < 1$ as $x \to +\infty$, then there is a solution to the difference equation with the same controlling factor as $n \to \infty$.

METHOD II If the controlling factor of a_n varies rapidly (exponentially or faster) as $n \to \infty$, the substitution $a_n = e^{S_n}$, like the substitution $y(x) = e^{S(x)}$ for differential equations (see Sec. 3.4), usually reduces the problem of finding the controlling factor to that of solving a simple first-order difference equation.

METHOD III The method of dominant balance is often useful by itself (especially if methods I and II fail) or in conjunction with methods I and II.

Discussion of method I Method I works because if a function $y(x)$ varies logarithmically, algebraically, or no more rapidly than e^{ax^b} with $b < 1$ as $x \to +\infty$, then

$$Dy(x) = y(x+1) - y(x) \sim y'(x), \qquad x \to +\infty, \tag{5.3.1}$$

and

$$D^k y(x) \sim \frac{d^k y}{dx^k}, \qquad x \to +\infty, \tag{5.3.2}$$

for all k (see Prob. 5.11).

Example 1 *Functions satisfying $Dy \sim y'(x)$ $(x \to +\infty)$.*

(*a*) If $y(x) = \ln x$, then $Dy(x) = \ln(1 + 1/x) \sim 1/x = y'(x)$ $(x \to +\infty)$.
(*b*) If $y(x) = x^3$, then $Dy(x) = 3x^2 + 3x + 1 \sim 3x^2 = y'(x)$ $(x \to +\infty)$.
(*c*) If $y(x) = \exp(\sqrt{x})$, then

$$\begin{aligned} Dy(x) &= \exp(\sqrt{x+1}) - \exp(\sqrt{x}) \\ &= \exp(\sqrt{x})[\exp(\sqrt{x+1} - \sqrt{x}) - 1] \\ &= \exp(\sqrt{x})\left[\exp\left(\frac{1}{2\sqrt{x}} + \cdots\right) - 1\right] \\ &\sim \frac{\exp(\sqrt{x})}{2\sqrt{x}} = y'(x), \qquad x \to +\infty. \end{aligned}$$

(*d*) It is not true that $Dy \sim y'$ $(x \to +\infty)$ if $y(x) = e^x$ or e^{x^2}. (Why?)

If $y(x)$ satisfies (5.3.1) and (5.3.2) and solves a linear homogeneous differential equation, then there is a solution to the corresponding difference equation whose controlling factor for large n is the same as that of $y(x)$. This is because the controlling factor of $y(x)$ is determined by a dominant balance in the differential equation; when (5.3.1) and (5.3.2) are satisfied, the same dominant balance determines the controlling factor of a solution to the corresponding difference equation.

Example 2 *Use of method I to find controlling factors.*

(*a*) The controlling factor of the solution to $y' = y/\sqrt{x} + y/x$ is determined by the dominant balance $y' \sim y/\sqrt{x}$ $(x \to +\infty)$ to be $\exp(2\sqrt{x})$ as $x \to +\infty$. Similarly, the controlling factor of the solution to $Da_n \equiv a_{n+1} - a_n = a_n/\sqrt{n} + a_n/n$ is determined by the dominant balance $Da_n \sim a_n/\sqrt{n}$ $(n \to \infty)$ to be $\exp(2\sqrt{n})$ as $n \to \infty$.
(*b*) The controlling factors of the solutions to $xy'' = y$ are $\exp(\pm 2\sqrt{x})$ as $x \to +\infty$ [see (3.4.28)], so that the controlling factors for the corresponding difference equation $n(a_{n+2} - 2a_{n+1} + a_n) = a_n$ are $\exp(\pm 2\sqrt{n})$ as $n \to \infty$.
(*c*) One solution to the differential equation $2xy'' - (2x+1)y' + y = 0$ has controlling factor $\sqrt{x}$ as $x \to +\infty$. (Verify this!) The corresponding difference equation

$$2nD^2 a_n - (2n+1)Da_n + a_n = 0 \tag{5.3.3}$$

also has a solution whose controlling factor is $\sqrt{n}$ as $n \to \infty$. Note that the differential equation has a second solution whose controlling factor is e^x as $x \to +\infty$. (In fact, e^x is an exact solution.) However, the corresponding difference equation does not have any solution whose controlling factor is e^n. In fact, we use method II in Example 3 to show that there is a solution of the difference equation whose controlling factor is 2^n as $n \to \infty$.

Discussion of method II Method II is particularly useful for finding controlling factors which vary rapidly with n as $n \to \infty$. It begins with the substitution

$$a_n = e^{S_n}, \tag{5.3.4}$$

from which we have

$$Da_n = a_n(e^{DS_n} - 1), \tag{5.3.5a}$$

$$D^2 a_n = a_n[(e^{DS_n} - 1)^2 + e^{2DS_n}(e^{D^2 S_n} - 1)], \tag{5.3.5b}$$

and so on. If $D^2 S$ satisfies the two asymptotic relations

$$D^2 S_n \ll 1, \qquad D^2 S_n \ll (DS_n)^2, \qquad n \to \infty, \tag{5.3.6}$$

then we may disregard the second term in (5.3.5b) and obtain

$$D^2 a_n \sim a_n(e^{DS_n} - 1)^2, \qquad n \to \infty. \tag{5.3.7}$$

Using $a_{n+1} = a_n + Da_n$ and $a_{n+2} = a_n + 2Da_n + D^2 a_n$, we also have

$$a_{n+1} = a_n e^{DS_n}, \qquad n \to \infty, \tag{5.3.8a}$$

$$a_{n+2} \sim a_n e^{2DS_n}, \qquad n \to \infty. \tag{5.3.8b}$$

More generally, if we assume that

$$D^k S_n \ll 1, \qquad D^k S_n \ll (DS_n)^k, \qquad n \to \infty, \tag{5.3.9}$$

for $k = 2, 3, \ldots, p$, then

$$a_{n+p} \sim a_n e^{pDS_n}, \qquad n \to \infty. \tag{5.3.10}$$

Even if the asymptotic conditions (5.3.6) or (5.3.9) are violated, the asymptotic relations (5.3.8b) and (5.3.10) may still be valid (see Prob. 5.15).

Method II is a powerful technique because the above exponential substitutions can reduce an exact nth-order linear difference equation for a_n to an approximate first-order nonlinear equation for S_n.

Example 3 *Use of method II to find controlling factors.*

(a) The difference equation

$$a_{n+1} = na_n, \tag{5.3.11}$$

whose exact solution is $a_n = a_1(n-1)!$, has an irregular singular point at $n = \infty$; the solutions to the corresponding differential equation $y' = (x-1)y$ are $y = c \exp(\frac{1}{2}x^2 - x)$ which vary exponentially fast as $x \to +\infty$. If we make the substitution $a_n = e^{S_n}$, we obtain the exact (not just asymptotic) equation

$$DS_n = \ln n. \tag{5.3.12}$$

The solution to this equation is $S_n - S_1 = \sum_{j=1}^{n-1} \ln j$. Approximating the sum on the right by an integral, we obtain

$$S_n \sim \int^n \ln t \, dt \sim n \ln n, \qquad n \to \infty \tag{5.3.13}$$

(see Prob. 5.2). Thus, the controlling factor of $(n-1)!$ for large n is $e^{n \ln n} = n^n$.

(*b*) The difference equation

$$a_{n+2} = 3na_{n+1} - 2n^2 a_n \tag{5.3.14}$$

has an irregular singular point at $n = \infty$. The substitution $a_n = e^{S_n}$ together with (5.3.8) gives the asymptotic relation

$$(e^{DS_n})^2 \sim 3ne^{DS_n} - 2n^2, \qquad n \to \infty.$$

The solutions to this quadratic relation for e^{DS_n} are

$$e^{DS_n} \sim n, \qquad n \to \infty, \qquad \text{or} \qquad e^{DS_n} \sim 2n, \qquad n \to \infty.$$

Equivalently,

$$DS_n \sim \ln n, \qquad n \to \infty, \tag{5.3.15a}$$

or

$$DS_n \sim \ln (2n), \qquad n \to \infty. \tag{5.3.15b}$$

At this point, we observe that $D^2 S_n \sim 1/n$ $(n \to \infty)$ for both behaviors so that (5.3.6) is satisfied. Both solutions to (5.3.15) satisfy $S_n \sim \sum_{k=1}^{n} \ln k$ $(n \to \infty)$ (see Prob. 5.3). Approximating the sum on the right by an integral as in (5.3.13), we obtain $S_n \sim n \ln n$ $(n \to \infty)$. Thus, the controlling factor of all solutions to (5.3.14) is n^n as $n \to \infty$.

(*c*) The controlling factors e^{S_n} of solutions to $a_{n+3} = na_n$ satisfy $e^{3DS_n} \sim n$ $(n \to \infty)$. Therefore, $DS_n - \frac{1}{3} \ln n \sim \ln \omega$ $(n \to \infty)$, where $\omega^3 = 1$. Thus, $S_n \sim \frac{1}{3} n \ln n$ $(n \to \infty)$. Thus, the controlling factor of a_n as $n \to \infty$ is $n^{n/3}$. We can verify this result by examining the exact solution to $a_{n+3} = na_n$ which is a linear combination of the three expressions $3^n \Gamma(n/3)\omega^n$, where $\omega = 1$, $e^{\pm 2i\pi/3}$.

(*d*) The controlling factors e^{S_n} of solutions to $na_{n+2} - (4n - 2)a_{n+1} + (n - 1)a_n = 0$ satisfy

$$ne^{2DS_n} - (4n - 2)e^{DS_n} \sim 1 - n, \qquad n \to \infty.$$

This may be replaced by the even simpler asymptotic relation

$$e^{2DS_n} - 4e^{DS_n} \sim -1, \qquad n \to \infty,$$

by neglecting a constant compared with n in each term. Using the quadratic formula, the solutions for e^{DS_n} are

$$e^{DS_n} \sim 2 \pm \sqrt{3}, \qquad n \to \infty.$$

Therefore, the possible controlling factors of a_n are $(2 \pm \sqrt{3})^n$.

(*e*) Recall that method I fails to determine the controlling factor of the rapidly varying solution to the difference equation (5.3.3) in Example 2(*c*). However, the controlling factor is easily found using method II. Substituting (5.3.5*a*) and (5.3.7) into (5.3.3) gives

$$2n(e^{DS_n} - 1)^2 \sim (2n + 1)(e^{DS_n} - 1) - 1.$$

Solving this quadratic equation for $e^{DS_n} - 1$ gives

$$e^{DS_n} - 1 \sim 1, \qquad n \to \infty, \tag{5.3.16a}$$

and

$$e^{DS_n} - 1 \sim \frac{1}{2n}, \qquad n \to \infty. \tag{5.3.16b}$$

From (5.3.16*a*) we have $DS_n \sim \ln 2$ or $S_n \sim n \ln 2$ $(n \to \infty)$. Thus, a possible controlling factor is 2^n.

From (5.3.16*b*) we see that $DS_n \to 0$ as $n \to \infty$. Thus, expanding the exponent in (5.3.16*b*) gives $DS_n \sim 1/2n$ $(n \to \infty)$. If we integrate this relation with respect to n, we obtain $S_n \sim \frac{1}{2} \ln n$ $(n \to \infty)$ and thereby recover the controlling factor $\sqrt{n}$ which we found in Example 2(*c*). However, $S_n \sim \frac{1}{2} \ln n$ does not satisfy the conditions in (5.3.6) which justify using (5.3.7) in the first place. Why then did we obtain the correct answer? (See Prob. 5.15.)

Method III

Example 4 *Use of method III to find controlling factors.* Let us determine the controlling factors of solutions to

$$a_{n+2} - na_{n+1} + a_n = 0. \tag{5.3.17}$$

There are three possible dominant balances to consider: (*a*) $a_n \ll a_{n+2}$ $(n \to \infty)$; (*b*) $a_{n+2} \ll a_n$ $(n \to \infty)$; and (*c*) $na_{n+1} \ll a_n$ $(n \to \infty)$. Only the first two cases are discussed here, as the third may easily be shown to be inconsistent.

(*a*) $a_n \ll a_{n+2}$ $(n \to \infty)$. In this case (5.3.17) yields the asymptotic difference equation $a_{n+2} \sim na_{n+1}$ $(n \to \infty)$. Thus, the controlling factor of a_{n+1} is the same as that of $(n-1)!$; namely, n^n [see Example 3(*a*)].

(*b*) $a_{n+2} \ll a_n$ $(n \to \infty)$. In this case we obtain the asymptotic difference equation $a_{n+1} \sim a_n/n$ $(n \to \infty)$. Thus, the controlling factor of a_n is the same as that of $1/(n-1)!$; namely, n^{-n}.

Once the controlling factor of a solution a_n to a difference equation is found, we can then peel off this controlling factor and study the resulting difference equation. By applying the techniques of peeling off and asymptotic analysis of the controlling factor discussed above, we can find the full leading behavior of a_n. This procedure is developed further in Secs. 5.4 and 5.5.

(E) 5.4 ASYMPTOTIC BEHAVIOR OF $n!$ AS $n \to \infty$: THE STIRLING SERIES

The Stirling series is one of the oldest and most venerable of asymptotic series. It expresses the asymptotic behavior of the factorial function $n!$ for large values of n:

$$\Gamma(n) = (n-1)! \sim \left(\frac{2\pi}{n}\right)^{1/2} \left(\frac{n}{e}\right)^n \left(1 + \frac{A_1}{n} + \frac{A_2}{n^2} + \frac{A_3}{n^3} + \cdots\right), \quad n \to \infty, \tag{5.4.1}$$

where the first eight Stirling coefficients are

$$A_1 = \frac{1}{12} \doteqdot 0.083\,333\,33,$$

$$A_2 = \frac{1}{288} \doteqdot 0.003\,472\,22,$$

$$A_3 = -\frac{139}{51{,}840} \doteqdot -0.002\,681\,33,$$

$$A_4 = -\frac{571}{2{,}488{,}320} \doteqdot -0.000\,229\,47,$$

$$A_5 = \frac{163{,}879}{209{,}018{,}880} \doteqdot 0.000\,784\,04,$$

$$A_6 = \frac{5{,}246{,}819}{75{,}246{,}796{,}800} \doteqdot 0.000\,069\,73,$$

$$A_7 = -\frac{534{,}703{,}531}{902{,}961{,}561{,}600} \doteqdot -0.000\,592\,17,$$

$$A_8 = -\frac{4{,}483{,}131{,}259}{86{,}684{,}309{,}913{,}600} \doteqdot -0.000\,051\,72.$$

Although the coefficients A_j appear to be getting smaller as j increases, they eventually grow rapidly as $j \to \infty$. For example, $A_{15} \doteqdot -2.9 \times 10^{-2}$, $A_{25} \doteqdot 2.2 \times 10^3$, $A_{35} \doteqdot -1.1 \times 10^{10}$. In fact, it may be shown (see Probs. 5.18 to 5.20) that

$$A_{2j} \sim (-1)^{j+1} \frac{(2j-2)!}{6(2\pi)^{2j}}, \qquad j \to \infty, \tag{5.4.2a}$$

$$A_{2j+1} \sim (-1)^j \frac{2(2j)!}{(2\pi)^{2(j+1)}}, \qquad j \to \infty. \tag{5.4.2b}$$

Thus, the series (5.4.1) diverges for all values of $1/n$.

In this section we show how to derive the Stirling series directly from a local analysis of the difference equation $a_{n+1} = na_n$ satisfied by $a_n = (n-1)!$. It is more conventional to derive the Stirling series by performing an asymptotic expansion of the integral representation (2.2.2) of $\Gamma(n)$; we will show how to do that in Sec. 6.4. Our real purpose here is to strengthen our faith in the power of local analysis.

But before deriving the Stirling series, let us step back a moment and savor the properties of this marvelous old series. At first, it may be difficult to believe that a reasonable person would prefer the complexity of the Stirling series to the simplicity of the factorial function, even when n is large. It is often sensible to replace a complicated function by its asymptotic expansion because the latter usually consists of sums and products of *elementary functions* (exponentials, powers, logs, sines, cosines). Even though it is only approximate, an asymptotic series has a much more tractable form for further analysis; it is easy to evaluate, integrate, multiply, and so on. However, the factorial function is so simple that replacing the left side of (5.4.1) by its right side is not a sure sign of progress.

The reason why the Stirling series (5.4.1) is such an important result of applied mathematics is that if we replace the letter n by z, it becomes the full asymptotic expansion of the gamma function $\Gamma(z)$ for complex argument z. Specifically,

$$\Gamma(z) - \left(\frac{2\pi}{z}\right)^{1/2} \left(\frac{z}{e}\right)^z \left(1 + \sum_{j=1}^{N} A_j z^{-j}\right) \ll \left(\frac{z}{e}\right)^z z^{-1/2-N},$$

$$z \to \infty; \; |\arg z| < \pi. \tag{5.4.3}$$

Observe that the jth term in the Stirling series is proportional to $1/z^j$ rather than to $\Gamma(z)/\Gamma(z+j)$. This is contrary to our suggestion in Sec. 5.2 that series expansions of solutions to difference equations have simpler expansion coefficients in the latter form. There are two reasons for choosing the form (5.4.3) for the Stirling series. The first is obviously that an asymptotic expansion of $\Gamma(z)$ should not depend on $\Gamma(z)$ itself (see Prob. 5.21). More importantly, z is a continuous and not a discrete variable, so the Stirling series in the form (5.4.3) is easier to manipulate mathematically; it is much easier to integrate z^{-j} than $\Gamma(z)/\Gamma(z+j)$.

$\Gamma(z)$ has simple poles at $z = 0, -1, -2, -3, \ldots$ (see Fig. 2.1). These poles prevent the Stirling series from being valid on the negative real axes. This explains why the Stirling series is only valid in the sector $|\arg z| < \pi$.

$\Gamma(z)$ has no zeros, so $1/\Gamma(z)$ is an entire function; its Taylor series about $z = 0$ has an infinite radius of convergence. The Taylor series for $1/\Gamma(z)$ has the form

$$\frac{1}{\Gamma(z)} = \sum_{j=1}^{\infty} C_j z^j, \tag{5.4.4}$$

where the first 15 coefficients C_j are (see Prob. 6.30)

$$\begin{array}{lll}
C_1 = 1.000\,000\,00, & C_2 \doteqdot 0.577\,215\,66, & C_3 \doteqdot -0.655\,878\,07, \\
C_4 \doteqdot -0.042\,002\,64, & C_5 \doteqdot 0.166\,538\,61, & C_6 \doteqdot -0.042\,197\,73, \\
C_7 \doteqdot -0.009\,621\,97, & C_8 \doteqdot 0.007\,218\,94, & C_9 \doteqdot -0.001\,165\,17, \\
C_{10} \doteqdot -0.000\,215\,24, & C_{11} \doteqdot 0.000\,128\,05, & C_{12} \doteqdot -0.000\,020\,13, \\
C_{13} \doteqdot -0.000\,001\,25, & C_{14} \doteqdot 0.000\,001\,13, & C_{15} \doteqdot -0.000\,000\,21.
\end{array}$$

Since the radius of convergence of the Taylor series is infinite, we know that for any value of $|z|$, no matter how large, it must approximate $\Gamma(z)$ to within any prescribed error if we include enough terms. However, convergence alone does not make a series useful. A lovely way to illustrate this point is to hold a convergence race between the convergent Taylor and the divergent Stirling series. The Taylor series, like the tortoise, ultimately wins the race, but its rate of convergence is agonizingly slow. The Stirling series, like the hare, is doomed to lose the race, but it begins with an enormous burst of enthusiasm, and in just a few terms gives an extraordinarily accurate approximation to the actual value of $\Gamma(z)$. In Table 5.1 we evaluate both series at $z = 1$, $z = 2$, and $z = 4$; remember that $\Gamma(1) = 0! = 1$, $\Gamma(2) = 1! = 1$, and $\Gamma(4) = 3! = 6$. After eight terms, the Stirling series for $z = 1$, $z = 2$, and $z = 4$ are correct to within relative errors of 0.03, 1×10^{-4}, and 2×10^{-7} percent, respectively, while the Taylor series have hardly begun to converge. The Stirling series is asymptotic, so if we continue to increase n it will eventually reach optimal accuracy and then diverge. Meanwhile, the Taylor series must eventually converge. But how soon? Partial sums of more terms in the Taylor series for $z = 1, 2$, and 4 are given in Table 5.2. The Taylor series for $1/\Gamma(2)$ and $1/\Gamma(4)$ do eventually converge, but an accuracy of 1 percent requires 15 terms when $z = 2$ and 30 terms when $z = 4$.

Why is the Taylor series so sluggish? Since the Taylor series converges everywhere in the complex plane, it must approximate two very different kinds of behavior: $\Gamma(z)$ blows up very rapidly as $z \to +\infty$ while along the negative z axis $\Gamma(z)$ has poles at every integer. Apparently, when a Taylor series must represent two such different kinds of behavior and converge for all z, the series becomes cranky and takes its revenge in the form of slow convergence. The Stirling series,

Table 5.1 A term-by-term comparison of the rates of convergence of partial sums of the Taylor series for $1/\Gamma(z)$ and the Stirling series for $\Gamma(z)$ at $z = 1, 2, 4$

Although the Taylor series ultimately converges, the Stirling series is far more useful for numerical approximation

	$z = 1$		$z = 2$		$z = 4$	
	Taylor series	Stirling series	Taylor series	Stirling series	Taylor series	Stirling series
N	$\left(\sum_{k=0}^{N} C_k 1^k\right)^{-1}$	$\left(\frac{1}{e}\right)^1 \sqrt{2\pi}\left(\sum_{k=0}^{N} A_k 1^{-k}\right)$	$\left(\sum_{k=0}^{N} C_k 2^k\right)^{-1}$	$\left(\frac{2}{e}\right)^2 \sqrt{\pi}\left(\sum_{k=0}^{N} A_k 2^{-k}\right)$	$\left(\sum_{k=0}^{N} C_k 4^k\right)^{-1}$	$\left(\frac{4}{e}\right)^4 \sqrt{\frac{\pi}{2}}\left(\sum_{k=0}^{N} A_k 4^{-k}\right)$
0	—	0.922 137 01	—	0.959 502 18	—	5.876 543 80
1	1.000 000 00	0.998 981 76	0.500 000 00	0.999 481 43	0.250 000 00	5.998 971 78
2	0.634 028 70	1.002 183 62	0.230 798 04	1.000 314 33	0.075 554 66	6.000 247 07
3	1.085 378 48	0.999 711 07	−1.065 914 09	0.999 992 74	−0.034 793 81	6.000 000 87
4	1.137 223 07	0.999 499 47	−0.621 039 29	0.999 978 98	−0.025 320 67	5.999 995 60
5	0.956 138 51	1.000 222 46	0.268 887 21	1.000 002 49	0.007 611 34	6.000 000 10
6	0.996 337 63	1.000 286 76	0.981 955 12	1.000 003 54	−0.023 923 56	6.000 000 20
7	1.005 981 70	0.999 740 70	−4.689 642 38	0.999 999 09	−0.005 013 88	5.999 999 99
∞	1.000 000 00		1.000 000 00		6.000 000 00	

Table 5.2 A continuation of Table 5.1 showing the eventual convergence of the Taylor series for $1/\Gamma(z)$ for $z = 1$, 2, and 4

Because of severe roundoff error we have retained only three digits past the decimal point in the Taylor series for $1/\Gamma(4)$

N	$\left(\sum_{k=0}^{N} C_k 1^k\right)^{-1}$	$\left(\sum_{k=0}^{N} C_k 2^k\right)^{-1}$	$\left(\sum_{k=0}^{N} C_k 4^k\right)^{-1}$
8	0.998 722 81	0.611 690 53	0.004
10	1.000 107 61	1.222 732 58	−0.004
12	0.999 999 69	1.002 390 78	−0.017
14	0.999 999 80	0.944 100 61	0.006
16	1.000 000 00	1.000 403 15	−0.031
18	1.000 000 00	1.000 057 03	−0.036
20	1.000 000 00	0.999 994 16	0.105
22	1.000 000 00	0.999 999 77	0.455
24	1.000 000 00	1.000 000 03	−1.320
26	1.000 000 00	1.000 000 00	10.912
28	1.000 000 00	1.000 000 00	4.656
30	1.000 000 00	1.000 000 00	5.950
∞	$\Gamma(1) = 1$	$\Gamma(2) = 1$	$\Gamma(4) = 6$

on the other hand, only represents one of these two kinds of behavior, the divergence of $\Gamma(z)$ as $z \to \infty$. Moreover, we do not even require the Stirling series to converge. The Stirling series generously responds by giving very accurate approximations to $\Gamma(z)$, even for small values of z, though it is only expected to be valid in the limit $|z| \to \infty$.

The optimal asymptotic approximation to $\Gamma(z)$ is obtained as in Chap. 3 by truncating the Stirling series just before the smallest term $A_k z^{-k}$ in (5.4.3). In Fig. 5.1 we plot a graph of the optimal asymptotic approximation to $\Gamma(z)$ and $\Gamma(z)$ for $0 < z < 2$. Numerical values of the optimal asymptotic approximation are listed in Table 5.3. Notice that the optimal truncation of the Stirling series provides an approximation to $\Gamma(z)$ that is accurate to better than 1 percent for $z \geq 0.5$. The accuracy of the Stirling series for such small z is fortuitous and is due to the small size of the Stirling coefficients A_j for small j.

Example 1 *Positive minimum of* $\Gamma(z)$. $\Gamma(z)$ has a single minimum for $0 < z < \infty$. Suppose we approximate $\Gamma(z)$ by the first term in the Stirling series (5.4.1), $\Gamma(z) \sim (z/e)^z(2\pi/z)^{1/2}$. The positive minimum of this approximation occurs when $z \doteqdot 1.422$. If $\Gamma(z)$ is approximated by the two-term Stirling series $(z/e)^z(2\pi/z)^{1/2}(1 + 1/12z)$, the location of the minimum is determined to be $z \doteqdot 1.461$. The actual minimum of $\Gamma(z)$ is at $z \doteqdot 1.462$. The location of this minimum is determined accurately to five significant figures if $\Gamma(z)$ is represented by the optimal asymptotic approximation.

We presume that our demonstration of the marvels of the Stirling series has generated an appetite for its derivation. We propose to derive the Stirling series

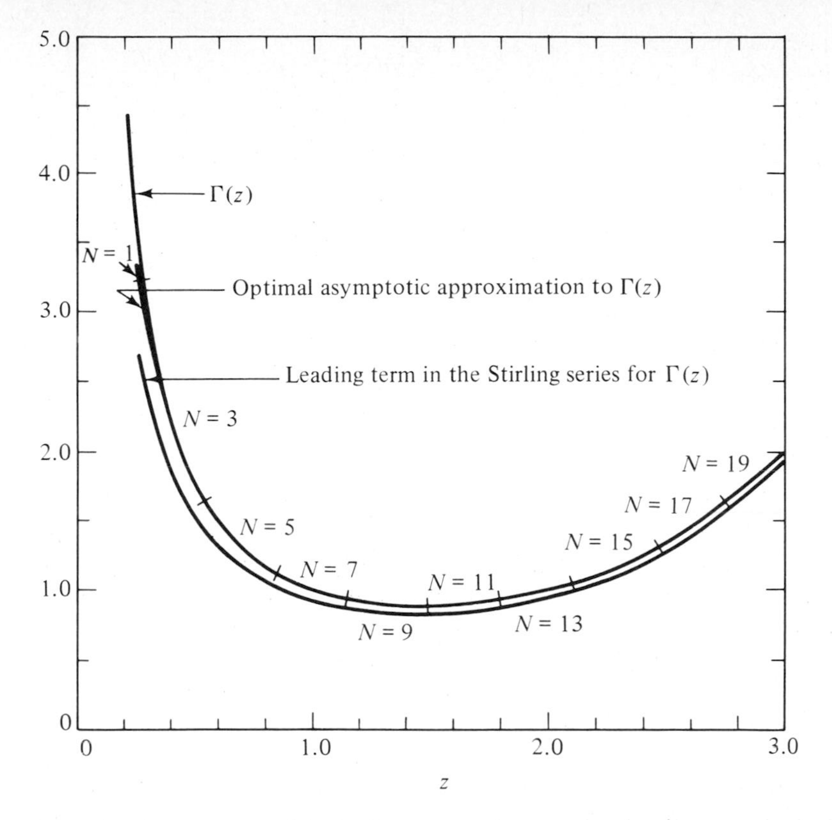

Figure 5.1 A comparison of the leading asymptotic approximation (first term in the Stirling series), the optimal asymptotic approximation to $\Gamma(z)$, and $\Gamma(z)$ for $0.2 \le z \le 3.0$. The relative error in the optimal truncation of the Stirling series is less than 1 percent when $z \ge 0.5$. The optimal asymptotic approximation is discontinuous at values of z for which the number of terms in the truncation changes. However, the discontinuities in the above graph are too small to be seen because the Stirling coefficients A_n are so small. The curve is divided into segments labeled by the number N which stands for the highest power of $1/z$ in the optimal asymptotic approximation.

(5.4.1) from a local analysis for large n of the difference equation

$$a_{n+1} = na_n, \tag{5.4.5}$$

whose solution is $a_n = (n-1)!$.

We begin by determining the controlling factor of the leading behavior of solutions to (5.4.5). As in Example 3(*a*) of Sec. 5.3, substituting $a_n = e^{S_n}$ into (5.4.5) gives $DS_n = \ln n$. Thus, the controlling factor of a_n is n^n.

The next step is to peel off the controlling factor of a_n by writing

$$a_n = b_n n^n \tag{5.4.6}$$

and then to compute the controlling factor of b_n (which should vary less rapidly

Table 5.3 The relative errors in the leading asymptotic behavior of $\Gamma(z)$ and in the optimal asymptotic approximation to $\Gamma(z)$ obtained by truncating the Stirling series just prior to the smallest term

z	$\Gamma(z)$	Leading behavior $(z/e)^z\sqrt{2\pi/z}$	Relative error of leading asymptotic behavior, %	$\Gamma_{\text{optimal}}(z)$	Relative error, %
0.5	1.772 453 850 906	1.520	14.0	1.762 241 600 912	0.57
1.0	1.0	0.922	7.7	0.999 740 700 425	2.6×10^{-2}
1.5	0.886 226 925 453	0.839	5.3	0.886 234 691 468	8.0×10^{-4}
2.0	1.0	0.960	4.0	1.000 000 358 004	3.6×10^{-5}
2.5	1.329 340 388 179	1.286	3.3	1.329 340 408 324	1.5×10^{-6}
3.0	2.0	1.945	2.7	1.999 999 998 888	5.6×10^{-8}
3.5	3.323 350 970 448	3.245	2.3	3.323 350 970 370	2.3×10^{-9}
4.0	6.0	5.877	2.1	6.000 000 000 005	9.1×10^{-11}
4.5	11.631 728 396 567	11.419	1.8	11.631 728 396 567	3.8×10^{-12}
5.0	24.0	23.604	1.7	24.000 000 000 000	1.5×10^{-13}
5.5	52.342 777 784 554	51.557	1.5	52.342 777 784 554	6.4×10^{-15}
6.0	120.0	118.346	1.4	120.000 000 000 000	2.6×10^{-16}

than n^n). If we substitute (5.4.6) into (5.4.5), we get the equation for b_n:

$$b_{n+1} = b_n \left(\frac{n+1}{n}\right)^{-n-1}. \tag{5.4.7}$$

For large n we may approximate the right side of (5.4.7) using the identity

$$\lim_{n \to \infty} \left(1 + \frac{1}{n}\right)^{-n-1} = \frac{1}{e}, \tag{5.4.8}$$

which we will examine carefully in the discussion following (5.4.11). Thus, when n is large, b_n satisfies the asymptotic difference equation $b_{n+1} \sim b_n/e$ $(n \to \infty)$. Thus, the controlling factor of b_n is e^{-n}. Note that e^{-n} varies less rapidly than n^n as $n \to \infty$.

The next step is to peel off the exponential behavior e^{-n} by setting $b_n = c_n e^{-n}$, or

$$a_n = (n/e)^n c_n. \tag{5.4.9}$$

We will find that the behavior of c_n for large n is algebraic: $c_n \sim kn^\alpha$ as $n \to \infty$, where k is a constant. The equation for c_n is found by substituting (5.4.9) into (5.4.5):

$$c_{n+1} = c_n e \left(\frac{n+1}{n}\right)^{-n-1}. \tag{5.4.10}$$

To determine the behavior of c_n we must carefully approximate the right side of this equation for large n. The simplest way to estimate

$$Q = e \left(\frac{n+1}{n}\right)^{n-1} \tag{5.4.11}$$

is to expand its logarithm:

$$\begin{aligned}
\ln Q &= \ln e - (n+1) \ln \left(1 + \frac{1}{n}\right) \\
&= 1 - (n+1)\left(\frac{1}{n} - \frac{1}{2n^2} + \frac{1}{3n^3} - \frac{1}{4n^4} + \cdots\right) \\
&= -\frac{1}{2n} + \frac{1}{6n^2} - \frac{1}{12n^3} + \cdots.
\end{aligned}$$

To solve for Q we exponentiate this series and continue to retain terms of order n^{-3}:

$$\begin{aligned}
Q &= \exp\left(-\frac{1}{2n} + \frac{1}{6n^2} - \frac{1}{12n^3} + \cdots\right) \\
&= \sum_{j=0}^{\infty} \frac{1}{j!}\left(-\frac{1}{2n} + \frac{1}{6n^2} - \frac{1}{12n^3} + \cdots\right)^j \\
&= 1 + \left(-\frac{1}{2n} + \frac{1}{6n^2} - \frac{1}{12n^3}\right) + \frac{1}{2}\left(\frac{1}{4n^2} - \frac{1}{6n^3}\right) + \frac{1}{6}\left(-\frac{1}{8n^3}\right) + \cdots \\
&= 1 - \frac{1}{2n} + \frac{7}{24n^2} - \frac{3}{16n^3} + \cdots.
\end{aligned}$$

Thus, c_n satisfies the equation

$$c_{n+1} = c_n \left(1 - \frac{1}{2n} + \frac{7}{24n^2} - \frac{3}{16n^3} + \cdots\right). \tag{5.4.12}$$

The leading behaviors of solutions to (5.4.12) may be found by using method I of Sec. 5.3. The corresponding differential equation is $y'(x) = y(x)(-1/2x + \cdots)$. Thus, the leading behavior of $y(x)$ is $y(x) \sim k/\sqrt{x}$ $(x \to +\infty)$, where k is a constant, and the leading behavior of c_n is therefore

$$c_n \sim k/\sqrt{n}, \qquad n \to \infty.$$

Hence, the leading behavior of a_n is

$$a_n \sim kn^n e^{-n} n^{-1/2}, \qquad n \to \infty,$$

and we have derived the first term of the Stirling series!

The rest of the Stirling series comes smoothly. Following the same kind of reasoning that we used for linear differential equations near irregular singular points, we treat $n^{-1/2}$ as the first term of a formal Frobenius series for large n:

$$c_n = kn^{-1/2} d_n, \tag{5.4.13}$$

where

$$d_n \sim 1 + A_1 n^{-1} + A_2 n^{-2} + A_3 n^{-3} + \cdots, \qquad n \to \infty. \tag{5.4.14}$$

The coefficients $A_1, A_2, \ldots$ will be determined by substituting the above expansion for c_n into the difference equation (5.4.12) for c_n. But first let us examine the equation for d_n:

$$\begin{aligned} d_{n+1} &= d_n \left(1 + \frac{1}{n}\right)^{1/2} \left(1 - \frac{1}{2} n^{-1} + \frac{7}{24} n^{-2} - \frac{3}{16} n^{-3} + \cdots\right) \\ &= d_n \left(1 - \frac{1}{12} n^{-2} + \frac{1}{12} n^{-3} + \cdots\right). \end{aligned}$$

Observe that the n^{-1} term has disappeared from the right side. We interpret this as a verification of the algebraic behavior $c_n \sim kn^{-1/2}$ $(n \to \infty)$. As we determine more and more of the coefficients A_i in the series for d_n, the $n^{-2}, n^{-3}, \ldots$ terms in the above difference equation will also disappear one by one.

For example, let us determine the first Stirling coefficient A_1 in the series for d_n. [A glance at (5.4.1) shows that A_1 has the value $\frac{1}{12}$.] We substitute $d_n = (1 + A_1 n^{-1}) e_n$ into the above difference equation and demand that the n^{-2} term disappear from the resulting equation satisfied by e_n:

$$\begin{aligned} e_{n+1} &= e_n \left(1 + \frac{A_1}{n+1}\right)^{-1} \left(1 + \frac{A_1}{n}\right) \left(1 - \frac{1}{12} n^{-2} + \frac{1}{12} n^{-3} + \cdots\right) \\ &= e_n \left[1 + \left(A_1 - \frac{1}{12}\right) n^{-2} + \left(\frac{1}{12} - A_1 - A_1^2\right) n^{-3} + \cdots\right]. \end{aligned}$$

Thus, if $A_1 = \frac{1}{12}$, the n^{-2} term vanishes.

Next, let us determine the first two Stirling coefficients A_1 and A_2 in the series for d_n. [A glance at (5.4.1) shows that A_2 has the value $\frac{1}{288}$.] We substitute $d_n = (1 + A_1 n^{-1} + A_2 n^{-2})e_n$ into the equation for d_n and demand that the n^{-2} and n^{-3} terms disappear from the resulting equation satisfied by e_n:

$$e_{n+1} = e_n \left[1 + \frac{A_1}{n+1} + \frac{A_2}{(n+1)^2}\right]^{-1} \left(1 + \frac{A_1}{n} + \frac{A_2}{n^2}\right)\left(1 - \frac{1}{12}n^{-2} + \frac{1}{12}n^{-3} + \cdots\right)$$

$$= e_n \left[1 + \left(A_1 - \frac{1}{12}\right)n^{-2} + \left(\frac{1}{12} - A_1 - A_1^2 + 2A_2\right)n^{-3} + \cdots\right].$$

Thus, if $A_1 = \frac{1}{12}$ and $A_2 = \frac{1}{288}$, the n^{-2} and n^{-3} terms vanish.

If the above process is continued forever, all of the Stirling coefficients will be determined. Each time a Stirling coefficient A_k is determined, the resulting difference equation for e_n predicts a gentler and more slowly varying behavior for e_n as $n \to \infty$. Ultimately, all the coefficients A_j in (5.4.14) may be determined. Summarizing the results of this analysis, we have shown that as $n \to \infty$ the solution a_n to the difference equation $a_{n+1} = na_n$ behaves like

$$a_n \sim kn^{n-1/2}e^{-n}\left(1 + \sum_{j=1}^{\infty} A_j n^{-j}\right), \qquad n \to \infty, \tag{5.4.15}$$

where $A_1 = \frac{1}{12}$ and $A_2 = \frac{1}{288}$. Thus, using local analysis, we have determined the Stirling series up to a multiplicative constant k.

Without further information k cannot be determined. The difference equation that we have solved is linear and homogeneous, so any arbitrary multiple of a solution is still a solution. However, the Stirling series (5.4.1) is the asymptotic expansion of the special solution $a_n = (n - 1)!$ of the difference equation satisfying the initial condition $a_1 = 1$. For this particular solution we know from (5.4.1) that $k = \sqrt{2\pi}$.

Unfortunately, the nature of local analysis precludes the use of information from values of n far from ∞ to determine the value of k. If we could analyze the difference equation simultaneously at $n = 1$ and $n = \infty$, we would be doing global and not local analysis. The advantage of the integral representation (2.2.2) is that it is equivalent to *both* the difference equation and the initial condition $a_1 = 1$. In Chap. 6, where we explain how to perform a local expansion of this integral for large $|z|$ and $|\arg z| < \pi$, we will be able to rederive the Stirling series with the correct multiplicative constant $\sqrt{2\pi}$.

(I) 5.5 LOCAL BEHAVIOR NEAR AN IRREGULAR SINGULAR POINT AT INFINITY: FULL ASYMPTOTIC SERIES

In the previous section we showed how to determine the leading behavior of the solution $\Gamma(n)$ to the difference equation $a_{n+1} = na_n$. In this section we show how to determine the leading behaviors of solutions to more complicated linear difference equations. We give three examples: the first two involve only local analysis while the third is a novel application of local analysis to an eigenvalue problem.

Example 1 *Behavior of Bessel functions for large order.* When the argument x is held fixed and the order n is allowed to tend to ∞, the Bessel functions $J_n(x)$ and $Y_n(x)$ exhibit the following asymptotic behaviors:

$$J_n(x) \sim \frac{(x/2)^n}{n!}, \qquad n \to \infty, \tag{5.5.1}$$

$$Y_n(x) \sim -\frac{1}{\pi}(n-1)!\,(2/x)^n, \qquad n \to \infty. \tag{5.5.2}$$

We will derive these behaviors from the recursion relation

$$a_{n+2} - \frac{2(n+1)}{x} a_{n+1} + a_n = 0, \tag{5.5.3}$$

which $J_n(x)$ and $Y_n(x)$ both satisfy for all n (see Prob. 5.29). Note that $n = \infty$ is an irregular singular point of (5.5.3).

To analyze (5.5.3), we use method III of Sec. 5.3, the method of dominant balance. There are three cases to consider:

(*a*) $a_n \ll a_{n+2}$ $(n \to \infty)$. In this case (5.5.3) yields the asymptotic difference equation $a_{n+2} \sim 2(n+1)a_{n+1}/x$ $(n \to \infty)$. The solution to this relation (with the $\sim$ sign replaced by an $=$ sign) is $C(n-1)!\,(2/x)^n$, where C is a constant. Since this expression satisfies $a_n \ll a_{n+2}$ $(n \to \infty)$, the dominant balance approximation is self-consistent. Therefore, we know that the controlling factor of a_n is the same as that of $(n-1)!$; namely, n^n. In fact, the dominant balance argument has, in this case, yielded the full leading behavior of a_n as $n \to \infty$. To verify this, we set

$$a_n = C\frac{(n-1)!\,2^n}{x^n} b_n \tag{5.5.4}$$

and substitute a_n into (5.5.3). The resulting equation for b_n is

$$b_{n+2} - b_{n+1} = -\frac{x^2}{4n(n+1)} b_n. \tag{5.5.5}$$

To show that the leading behavior of a_n as $n \to \infty$ is given correctly in (5.5.4) (with b_n replaced by 1), we must show that there is a solution to (5.5.5) satisfying $b_n \to 1$ as $n \to \infty$. In fact, we will show that b_n may be represented as an asymptotic series of the form

$$b_n \sim \sum_{k=0}^{\infty} A_k \frac{\Gamma(n)}{\Gamma(n+k)}, \qquad n \to \infty, \tag{5.5.6}$$

where $A_0 = 1$.

Substituting (5.5.6) into (5.5.5), multiplying by $n(n+1)$, noting the relations

$$D\Gamma(n+1)/\Gamma(n+k+1) = -kn\Gamma(n)/\Gamma(n+k+2)$$

and

$$n^2(n+1) = (n+k+1)(n+k)(n+k-1) - (3k-1)(n+k+1)(n+k) + k(3k+1)(n+k+1) - k(k+1)^2,$$

and equating the coefficients of $\Gamma(n)/\Gamma(n+k)$ (see Prob. 5.10), we have

$$-(k+1)A_{k+1} + k(3k-1)A_k - (k-1)^2(3k-2)A_{k-1} + (k-1)^2(k-2)^2 A_{k-2} = -\tfrac{1}{4}x^2 A_k \tag{5.5.7}$$

for $k = 0, 1, 2, \ldots$, in which $A_{-1} = A_{-2} = 0$ and $A_0 = 1$. Thus, $A_1 = \frac{1}{4}x^2$, $A_2 = \frac{1}{32}x^2(x^2 + 8)$, and so on. As $k \to \infty$, $A_k \sim A(x)(k-1)!$, where $A(x)$ is a function of x alone (see Prob. 5.30). Nevertheless, we have the surprising result that the series (5.5.6) converges for all $n > 1$ (see Prob. 5.30). Because we can consistently calculate all the coefficients A_k in the series (5.5.6), we conclude that the full leading behavior of a_n is given by the convergent series

$$a_n = C\frac{(n-1)!\,2^n}{x^n}\left[1 + \frac{x^2}{4n} + \frac{x^2(x^2+8)}{32n(n+1)} + \cdots\right]. \tag{5.5.8}$$

The Bessel function $Y_n(x)$ behaves in this way as $n \to \infty$. For this function, global analysis (see Prob. 6.25) shows that $C = -1/\pi$. The asymptotic expansion in (5.5.8) may be used to obtain accurate numerical approximations to $Y_n(5)$ for large n. Results of numerical calculations are given in Table 5.4.

(*b*) $a_{n+2} \ll a_n$ $(n \to \infty)$. In this case we obtain the asymptotic difference equation $a_{n+1} \sim \frac{1}{2}x(n+1)^{-1}a_n$ $(n \to \infty)$. Following the above procedure we find that $a_n \sim C(\frac{1}{2}x)^n/n!$ $(n \to \infty)$, where C is a constant. This behavior is consistent with the assumption that $a_{n+2} \ll a_n$ as $n \to \infty$. Higher corrections to this leading behavior in the form of an asymptotic series may be found as in case (*a*) above. Finally, we remark that the Bessel function $J_n(x)$ exhibits this behavior as $n \to \infty$ with fixed x; $J_n(x)$ is defined so that $C = 1$ [see (5.5.1) and Prob. 6.25].

(*c*) $2(n+1)a_{n+1}/x \ll a_n$ $(n \to \infty)$. This assumption is inconsistent. The only possible behaviors of a_n are those governed by cases (*a*) and (*b*) treated above.

Example 2 *Behavior of Legendre polynomials of large degree.* The Legendre polynomials $P_n(x)$ (n is the degree) satisfy the recursion relation

$$(n+2)a_{n+2} - (2n+3)xa_{n+1} + (n+1)a_n = 0, \qquad n = -1, 0, 1, \ldots, \tag{5.5.9}$$

for all x. To generate the Legendre polynomials we take as initial conditions $a_0 = P_0(x) = 1$, $a_1 = P_1(x) = x$. Thus, $P_2(x) = \frac{1}{2}(3x^2 - 1)$, $P_3(x) = \frac{1}{2}(5x^3 - 3x)$, and so on. The behavior of $P_n(x)$ as $n \to \infty$ for fixed x may be studied by considering the possible behaviors of solutions to (5.5.9) as $n \to \infty$.

As $n \to \infty$, (5.5.9) can be approximated by the constant-coefficient asymptotic difference equation

$$a_{n+2} \sim 2xa_{n+1} - a_n, \qquad n \to \infty, \tag{5.5.10}$$

obtained by approximating $n+2$ by n, $2n+3$ by $2n$, and $n+1$ by n. The difference equation (5.5.10) has solutions that behave exponentially at the irregular singular point $n = \infty$. Thus, the controlling factor of the behavior of a_n has the form r^n, where r satisfies $r^2 - 2xr + 1 = 0$. There are two solutions $r = r_+$ and $r = r_-$:

$$r_\pm = x \pm \sqrt{x^2 - 1}. \tag{5.5.11}$$

Note that if $|x| < 1$ then $r_\pm$ is complex.

To determine the full leading behavior of a_n for large n we substitute $a_n = b_n r_\pm^n$ into (5.5.9). For either r_+ or r_-, b_n satisfies the difference equation

$$(2xr - 1)(n+2)b_{n+2} - (2n+3)xrb_{n+1} + (n+1)b_n = 0. \tag{5.5.12}$$

Since we have already peeled off the exponential behavior of a_n, we seek a solution to (5.5.12) that behaves less singularly as $n \to \infty$; method I of Sec. 5.3 therefore applies. We find the controlling factor of b_n to be $n^{-1/2}$ as $n \to \infty$ (see Prob. 5.33).

The next step is to peel off the controlling factor $n^{-1/2}$ of b_n. However, this will introduce square roots into the resulting difference equation. Therefore, we resort to the trick of factoring off $\Gamma(n)/\Gamma(n+\frac{1}{2})$, whose large-$n$ behavior is also $n^{-1/2}$ (see Prob. 5.4). We write

$$b_n = c_n\Gamma(n)/\Gamma(n+\tfrac{1}{2})$$

Table 5.4 The exact values of $Y_n(5)$ for $5 \le n \le 100$ compared with the leading asymptotic approximation (5.5.2)

Also given are the next two corrections to (5.5.2) obtained by truncating the series in (5.5.8) after two and three terms

n	$Y_n(5)$	$-\frac{1}{\pi}\frac{(n-1)!\,2^n}{5^n}$	$-\frac{1}{\pi}\frac{(n-1)!\,2^n}{5^n}\left(1+\frac{5^2}{4n}\right)$	$-\frac{1}{\pi}\frac{(n-1)!\,2^n}{5^n}\left[1+\frac{5^2}{4n}+\frac{5^2(5^2+8)}{32n(n+1)}\right]$
5	-4.537×10^{-1}	-7.823×10^{-2}	-1.760×10^{-1}	-2.432×10^{-1}
10	-2.513×10^{1}	-1.211×10^{1}	-1.968×10^{1}	-2.252×10^{1}
15	-4.694×10^{4}	-2.980×10^{4}	-4.222×10^{4}	-4.542×10^{4}
20	-5.934×10^{8}	-4.257×10^{8}	-5.587×10^{8}	-5.849×10^{8}
30	-4.029×10^{18}	-3.245×10^{18}	-3.921×10^{18}	-4.011×10^{18}
40	-9.217×10^{29}	-7.849×10^{29}	-9.075×10^{29}	-9.199×10^{29}
50	-2.789×10^{42}	-2.454×10^{42}	-2.761×10^{42}	-2.786×10^{42}
100	-5.085×10^{115}	-4.774×10^{115}	-5.072×10^{115}	-5.084×10^{115}

and substitute into (5.5.12). We find that

$$(2xr-1)\left(1+\frac{1}{2n+3}\right)c_{n+2}-xr\left(2+\frac{1}{n+1}\right)c_{n+1}+\left(1+\frac{1}{2n}\right)c_n=0. \qquad (5.5.13)$$

Finally, we can use method I of Sec. 5.3 to verify that there is a solution of (5.5.13) satisfying $c_n \sim C(x)$ as $n \to \infty$, where $C(x)$ depends only on x (see Prob. 5.34). Thus, the leading behavior of a_n is

$$a_n \sim \frac{\Gamma(n)}{\Gamma(n+\frac{1}{2})}[C_+(x)r_+^n + C_-(x)r_-^n], \qquad n \to \infty. \qquad (5.5.14)$$

We can also develop the full asymptotic expansion of c_n for large n by substituting

$$c_n \sim \sum_{k=0}^{\infty} A_k \Gamma(n)/\Gamma(n+k), \qquad n \to \infty,$$

into (5.5.13) and deriving equations for the coefficients $A_k(x)$. In this way, we find $A_0(x) = C(x)$ and

$$A_1(x) = \left[-\frac{1}{4} + \frac{1}{8(xr-1)}\right] C(x) \qquad (5.5.15)$$

(see Prob. 5.34).

Using global analysis (see Prob. 6.29), it may be shown that the leading behavior of $P_n(x)$ for large n has the form (5.5.14) for all x. Specifically,

$$P_n(x) \sim (2\pi n)^{-1/2}(x^2-1)^{-1/4}(x+\sqrt{x^2-1})^{n+1/2}, \qquad n \to \infty;\ x > 1, \qquad (5.5.16a)$$

$$P_n(x) \sim (-1)^n(2\pi n)^{-1/2}(x^2-1)^{-1/4}(-x+\sqrt{x^2-1})^{n+1/2}, \qquad n \to \infty;\ x < -1, \qquad (5.5.16b)$$

while for $|x| < 1$,

$$P_n(\cos\theta) = (\pi n \sin\theta)^{-1/2}[C_n(\theta)\cos(n+\tfrac{1}{2})\theta + S_n(\theta)\sin(n+\tfrac{1}{2})\theta], \qquad (5.5.17)$$

where $C_n(\theta) \sim 1$ and $S_n(\theta) \sim 1$ as $n \to \infty$. Using (5.5.15) it may be shown that

$$C_n(\theta) - 1 \sim \frac{1}{n}\left[\frac{\sqrt{2}\cos(\theta-\frac{3}{4}\pi)}{8\sin\theta} - \frac{3}{8}\right], \qquad n \to \infty, \qquad (5.5.18a)$$

and

$$S_n(\theta) - 1 \sim -\frac{1}{n}\left[\frac{\sqrt{2}\sin(\theta-\frac{3}{4}\pi)}{8\sin\theta} + \frac{3}{8}\right], \qquad n \to \infty \qquad (5.5.18b)$$

(see Prob. 5.35).

The asymptotic formula in (5.5.17) with $C_n = S_n = 1$ provides an accurate numerical approximation to the zeros of $P_n(x)$ as $n \to \infty$. According to (5.5.17), the zeros of $P_n(x)$ (which all lie within $|x| < 1$) satisfy the asymptotic equation $\tan(n+\frac{1}{2})\theta \sim -1$ $(n \to \infty)$ or

$$x = \cos\theta, \qquad \theta \sim \frac{4m-1}{4n+2}\pi, \qquad n \to \infty;\ 1 \le m \le n. \qquad (5.5.19)$$

A more accurate determination of the zeros of $P_n(x)$ is obtained by using the more accurate expressions for C_n and S_n in (5.5.18).

In Table 5.5 we compare the exact zeros of $P_{20}(x)$ with the zeros determined approximately in the two ways described above. The relative error in the roots using (5.5.19) is roughly 1 part in 3,000. Using the first two terms in the expansions of C_n and S_n decreases the relative error to about 1 part in 50,000.

Example 3 *Difference-equation eigenvalue problem.* Eigenvalues are global properties of the solution to a boundary-value problem. Nevertheless, we will see in this example that local analysis

Table 5.5 Comparison of the exact zeros of the Legendre polynomial $P_{20}(x)$ of degree 20 with the zeros of the asymptotic approximation (5.5.17)

The second column lists the zeros (5.5.19) of the leading asymptotic approximation in (5.5.17) with $C_n = S_n = 1$. The third column lists the zeros of the two-term asymptotic approximation (5.5.17) obtained by using C_n and S_n as given in (5.5.18). Observe that the error in the final column is roughly 20 times smaller than the error in the second column. This is consistent with the fact that the asymptotic expansions of C_n and S_n are series in powers of $1/n$, the degree of the polynomial

Exact zeros of $P_{20}(x)$	Zeros of leading asymptotic approximation to $P_{20}(x)$	Zeros of two-term asymptotic approximation to $P_{20}(x)$
$\pm$0.076 527	$\pm$0.076 549	$\pm$0.076 526
$\pm$0.227 786	$\pm$0.227 854	$\pm$0.227 783
$\pm$0.373 706	$\pm$0.373 817	$\pm$0.373 702
$\pm$0.510 867	$\pm$0.511 019	$\pm$0.510 861
$\pm$0.636 054	$\pm$0.636 242	$\pm$0.636 046
$\pm$0.746 332	$\pm$0.746 553	$\pm$0.746 323
$\pm$0.839 117	$\pm$0.839 366	$\pm$0.839 106
$\pm$0.912 234	$\pm$0.912 504	$\pm$0.912 222
$\pm$0.963 972	$\pm$0.964 253	$\pm$0.963 957
$\pm$0.993 129	$\pm$0.993 402	$\pm$0.993 099

can greatly facilitate the solution of an eigenvalue problem. We are given the difference equation

$$a_{n+1} - Bna_n + na_{n-1} = Ea_n, \tag{5.5.20}$$

where E is the eigenvalue and $B > 1$ is a positive constant. a_n is required to exist (be finite) when $n = -1$ and to satisfy the boundary condition $\lim_{n\to\infty} a_n = 0$.

Setting $n = 0$ in (5.5.20) and noting that a_0 and a_{-1} are finite, we obtain $a_1 = Ea_0$. Once a_1 is so determined the difference equation (5.5.20) may be used to generate $a_2, a_3, a_4, \ldots$ in terms of a_0. For example, $a_2 = (E^2 + BE - 1)a_0$. Note that a_0 appears as an overall multiplicative factor in the expressions for a_n. Since the difference equation and boundary conditions are homogeneous, we are free to choose $a_0 = 1$.

There is a second, linearly independent solution of (5.5.20) for which $a_1 \neq Ea_0$. However, this solution must be rejected because it violates the finiteness of a_{-1}. Thus, we have uniquely determined a_n $(n = 0, 1, 2, \ldots)$ as functions of E and B.

Does this solution for a_n satisfy the boundary condition that $a_n \to 0$ as $n \to \infty$? To answer this question we must perform a local analysis of the irregular singular point of (5.5.20) at $n = \infty$. To find the possible controlling factors of a_n as $n \to \infty$, we replace (5.5.20) by the asymptotic difference equation

$$a_{n+1} + na_{n-1} \sim Bna_n, \qquad n \to \infty, \tag{5.5.21}$$

and use the method of dominant balance to study pairs of terms. There are three cases to consider:

(*a*) $a_{n+1} \sim Bna_n$ $(n \to \infty)$. The solution to this equation (with $\sim$ replaced by $=$) is proportional to $(n-1)!\, B^n$. This balance is consistent because for this solution $na_{n-1} \ll a_{n+1}$ $(n \to \infty)$. To find the full leading behavior of a_n we let $a_n = (n-1)!\, B^n b_n$. Then b_n satisfies

$$b_{n+1} - b_n + \frac{1}{B^2(n-1)} b_{n-1} = \frac{E}{Bn} b_n. \tag{5.5.22}$$

Using method I of Sec. 5.3 we find that the solution to (5.5.22) which varies less rapidly than B^n is $b_n \sim Cn^{(EB-1)/B^2}$ $(n \to \infty)$, where C is a constant, so the leading behavior of a_n is

$$a_n \sim C(n-1)!\, B^n n^{(EB-1)/B^2}, \qquad n \to \infty. \tag{5.5.23}$$

(*b*) $na_{n-1} \sim Bna_n$ $(n \to \infty)$. The solution to this equation (with $\sim$ replaced by $=$) is proportional to B^{-n}. This balance is consistent because for this solution $a_{n+1} \ll na_{n-1}$ $(n \to \infty)$. To find the full leading behavior of a_n we let $a_n = B^{-n}b_n$. Then b_n satisfies

$$B^{-1}b_{n+1} - Bnb_n + Bnb_{n-1} = Eb_n. \tag{5.5.24}$$

Using method I of Sec. 5.3, we find that the solution to (5.5.24) which varies less rapidly than B^{-n} is $b_n \sim Cn^{(1-EB)/B^2}$ $(n \to \infty)$, where C is a constant, so the leading behavior of a_n is

$$a_n \sim CB^{-n}n^{(1-EB)/B^2}, \qquad n \to \infty. \tag{5.5.25}$$

(*c*) $a_{n+1} \sim -na_{n-1}$ $(n \to \infty)$. This balance is inconsistent. (Why?)

This local analysis shows that in general a_n is a linear combination of growing and decaying solutions (5.5.23) and (5.5.25). An eigenvalue is a special value of E for which only the decaying solution occurs.

Although the results in (5.5.23) and (5.5.25) are only approximate, they may be translated into an exact condition for the existence of an eigenvalue. We define the function $f(z)$ by

$$f(z) = \sum_{n=0}^{\infty} \frac{a_n}{n!} z^n; \tag{5.5.26}$$

$f(z)$ is a generating function. If the boundary condition $a_n \to 0$ as $n \to \infty$ is satisfied, we know from (5.5.25) that $f(z)$ will be entire (it will converge for all complex z). If the boundary condition is violated, we know from (5.5.23) that $f(z)$ will have a singularity in the complex plane at $|z| = 1/B$. A leading local analysis is all that is required to determine the radius of convergence of $f(z)$.

Next, we multiply (5.5.20) by $z^n/n!$ and sum from $n = 0$ to $n = \infty$. This converts the difference equation for a_n into a differential equation for $f(z)$:

$$(1 - zB)f'(z) = (E - z)f(z). \tag{5.5.27}$$

This equation is easy to solve because it is separable:

$$f(z) = e^{z/B}(1 - Bz)^{(1-EB)/B^2}.$$

The solution is normalized so that $f(0) = a_0 = 1$.

As expected, $f(z)$ has a branch point or pole singularity at $|z| = 1/B$, except for special values of E for which the growing solution in (5.5.23) is absent; the eigenvalue condition is thus $(1 - EB)/B^2 = 0, 1, 2, 3, \ldots$, or

$$E = \frac{1}{B} - kB, \qquad k = 0, 1, 2, 3, \ldots, \tag{5.5.28}$$

which is the solution to the problem. When (5.5.28) is satisfied, $f(z)$ is entire because it is a polynomial multiplied by an exponential.

(E) 5.6 LOCAL BEHAVIOR OF NONLINEAR DIFFERENCE EQUATIONS

Nonlinear difference equations, like nonlinear differential equations, are so varied that it is not possible to formulate a general program for determining the local behavior of any given equation. Thus, in this section we parallel our approach to nonlinear differential equations by showing how to analyze a few examples. These examples have been selected because they allow us to explain and demonstrate the most broadly applicable difference-equation techniques. The first example, Newton's iteration method for finding the roots of an equation, illustrates techniques for examining the rate of convergence of a sequence $\{a_n\}$ as $n \to \infty$.

Example 1 *Rate of convergence of Newton's method.* In applied mathematics it is standard practice to solve problems by inventing an appropriate iteration procedure. In general, an iteration procedure is a repetitive calculation generating a sequence of approximations which rapidly approach a limit, the limit being the exact solution to the problem (see Example 2, Sec. 4.1). *Newton's method* is an efficient iteration procedure for finding the roots of the equation

$$f(x) = 0. \tag{5.6.1}$$

The appropriate sequence of approximations is defined by the difference equation

$$a_{n+1} = a_n - \frac{f(a_n)}{f'(a_n)}. \tag{5.6.2}$$

(For a justification of this equation see Prob. 5.42.) If one chooses an initial value a_0 which is reasonably close to the desired root x, one can show that the sequence $\{a_n\}$ does approach x (see Prob. 5.43). However, once this has been verified, the question that is of most concern for the applied mathematician is whether the solution to (5.6.2) converges *rapidly*. Usually, one estimates the rate of convergence by studying the magnitude of ε_n, the difference between the terms in the sequence and the limit x:

$$a_n = x + \varepsilon_n. \tag{5.6.3}$$

Substituting (5.6.3) into (5.6.2) gives an equation for ε_n:

$$\varepsilon_{n+1} = \varepsilon_n - f(x + \varepsilon_n)/f'(x + \varepsilon_n).$$

Assuming that ε_n is sufficiently small, we can Taylor expand the right side of this equation:

$$\frac{f(x+\varepsilon_n)}{f'(x+\varepsilon_n)} = \frac{\varepsilon_n f'(x) + \frac{1}{2}\varepsilon_n^2 f''(x) + \cdots}{f'(x) + \varepsilon_n f''(x) + \cdots}$$

$$= \varepsilon_n - \frac{1}{2}\varepsilon_n^2 \frac{f''(x)}{f'(x)} + \cdots.$$

Thus, ε_n satisfies the approximate equation

$$\varepsilon_{n+1} \sim \frac{1}{2}\frac{f''(x)}{f'(x)}\varepsilon_n^2, \qquad n \to \infty. \tag{5.6.4}$$

This equation shows that the convergence of a_n to x is extremely rapid because the error ε_{n+1} is proportional to the *square* of the error ε_n. This means that if one has already achieved three-decimal-place accuracy, then after the next iteration one will have about six-place accuracy, and then twelve, and so on. This is called *quadratic* convergence. Of course, the proof of quadratic convergence requires that $f'(x)$ in (5.6.4) not vanish. Thus, Newton's method is only useful for determining the simple zeros of $f(x)$; that is, values of x for which $f(x) = 0$ but $f'(x) \neq 0$ (see Prob. 5.48). If $f'(x) \neq 0$ but $f''(x) = 0$, then the convergence is at least cubic.

Example 2 *Newton's method for computing square roots.* To illustrate how fast the sequence $\{a_n\}$ does converge, let us use Newton's method to compute square roots. We may compute $\sqrt{Q}$ by finding the roots of the function $f(x) = x^2 - Q$. For this choice of $f(x)$ the iteration formula

(5.6.2) becomes

$$a_{n+1} = \frac{1}{2}\left(a_n + \frac{Q}{a_n}\right). \tag{5.6.5}$$

Let us use (5.6.5) to compute the square root of 4. We take $Q = 4$ and for fun we take as our first approximation $a_0 = 10$. The sequence of approximations is $a_0 = 10.000\,000\,000\,000$, $a_1 = 5.200\,000\,000\,000$, $a_2 = 2.984\,615\,384\,615$, $a_3 = 2.162\,410\,785\,091$, $a_4 = 2.006\,099\,040\,778$, $a_5 = 2.000\,009\,271\,302$, $a_6 = 2.000\,000\,000\,022$. The convergence is rapid even though the initial approximation a_0 is poor.

There are many other iteration schemes, such as

$$a_{n+1} = \frac{1}{3}\left(a_n + \frac{2Q}{a_n}\right), \tag{5.6.6}$$

which also generate sequences $\{a_n\}$ whose limits are $\sqrt{Q}$. However, (5.6.6) is inferior to Newton's method (5.6.5) because its convergence is linear rather than quadratic (Prob. 5.46). A recomputation of $\sqrt{4}$ using (5.6.6) with $Q = 4$ and $a_0 = 10$ gives $a_0 = 10.000\,000$, $a_1 = 3.600\,000$, $a_2 = 1.940\,741$, $a_3 = 2.020\,959$, $a_4 = 1.993\,158$, $a_5 = 2.002\,296$, $a_6 = 1.999\,236$, $a_7 = 2.000\,255$, $a_8 = 1.999\,915$, $a_9 = 2.000\,028$, $a_{10} = 1.999\,991$, $a_{11} = 2.000\,003$. Observe here that the accuracy is linearly proportional to the number of iterations. We are gaining about one decimal place of accuracy for every two iterations.

One way to demonstrate the convergence of Newton's method for computing square roots is to construct an exact analytical solution to (5.6.5). This difference equation is one of the few nonlinear difference equations whose solution may be expressed in closed form. To solve (5.6.5) we substitute

$$a_n = \sqrt{Q} \coth b_n.$$

The resulting equation for b_n is linear: $b_{n+1} = 2b_n$.

Thus, $b_n = 2^n K$, and

$$a_n = \sqrt{Q} \coth (2^n K), \tag{5.6.7}$$

where K is an arbitrary constant determined by the initial condition a_0. We can see that a_n approaches the roots $\pm\sqrt{Q}$ of $x^2 - Q = 0$ as $n \to \infty$ because $\coth (2^n K) \to \pm 1$ as $n \to \infty$, depending on the sign of K.

Let us now examine the behaviors of solutions of other nonlinear first-order difference equations as n tends to $+\infty$.

Example 3 *Failure of dominant balance.* If we modify Newton's iteration for computing square roots (5.6.5) only slightly by eliminating the factor of $\frac{1}{2}$, we obtain the equation

$$a_{n+1} = a_n + \frac{1}{a_n}, \tag{5.6.8}$$

which we can no longer solve in closed form. If the initial condition a_0 is chosen to be positive, then the solution a_n grows monotonically as $n \to \infty$. The solution cannot approach a finite limit as $n \to \infty$ because this limit would have to satisfy the equation

$$x = x + 1/x,$$

whose only solution is $x = \infty$.

How do we determine the asymptotic behavior of a_n as $n \to \infty$? In general, to perform a local analysis of a difference equation when it is known that $a_n \to \infty$ as $n \to \infty$, one should neglect those terms in the difference equation that become small in this limit. One hopes that the resulting dominant balance is then simple enough to solve in closed form. For the present equation it is surely true that $a_n \gg 1/a_n$ as $n \to \infty$. However, we cannot neglect the term $1/a_n$ because the solution of the difference equation $a_{n+1} = a_n$ is a constant, which violates the condition that $a_n \to \infty$ $(n \to \infty)$. Thus, naive application of the method of dominant balance is bootless.

To approximate the difference equation correctly when n is large we must first rewrite it in a more suitable form. The simplest approach is to square the equation (!):

$$a_{n+1}^2 = a_n^2 + 2 + a_n^{-2}. \tag{5.6.9}$$

Now, when n is large it *is* valid to neglect a_n^{-2} compared with 2. The resulting asymptotic difference equation becomes linear if we let $b_n = a_n^2$: $b_{n+1} - b_n \sim 2$ $(n \to \infty)$. The solution to this asymptotic equation is $b_n \sim 2n$ $(n \to \infty)$ [see Prob. 5.3(*c*)]. Thus,

$$a_n = \sqrt{b_n} \sim \sqrt{2n}, \qquad n \to \infty. \tag{5.6.10}$$

This is the leading asymptotic behavior of a_n as $n \to \infty$.

To determine the next-order correction to the leading behavior for large n, we let

$$a_n = \sqrt{2n + \varepsilon_n},$$

where we assume that ε_n is much smaller than $2n$ as $n \to \infty$. Substituting this expression into (5.6.9) gives

$$2n + 2 + \varepsilon_{n+1} = 2n + \varepsilon_n + 2 + 1/(2n + \varepsilon_n).$$

If we approximate the denominator of the fraction by $2n$ under the assumption that $\varepsilon_n \ll 2n$, we get

$$\varepsilon_{n+1} - \varepsilon_n \sim 1/2n, \qquad n \to \infty.$$

Thus, we find

$$\varepsilon_n \sim \tfrac{1}{2} \ln n, \qquad n \to \infty$$

[see Prob. 5.3(*c*)]. This result is consistent with our initial assumption that $\varepsilon_n \ll 2n$ as $n \to \infty$. Hence, for large n the higher-order corrections to the leading behavior (5.6.10) are

$$a_n - \sqrt{2n} \sim \sqrt{2n} + \varepsilon_n/2\sqrt{2n} - \sqrt{2n}$$

so that

$$a_n - \sqrt{2n} \sim \frac{\ln n}{4\sqrt{2n}}, \qquad n \to \infty. \tag{5.6.11}$$

This result may be somewhat surprising because one might have expected the higher-order corrections to take the form of a Frobenius series in powers of $1/n$: $a_n \sim \sqrt{2n} \sum \alpha_j n^{-j}$. One can conclude only that it is not easy to predict the formal structure of an approximate solution to a nonlinear equation.

In Table 5.6 we verify the predictions in (5.6.10) and (5.6.11) by comparing them with the exact values of a_n obtained on a computer by solving (5.6.8) with $a_1 = 1$. When $n = 10^6$, the prediction in (5.6.10) is accurate to 2 parts in 10^6 while the prediction in (5.6.11) is accurate to 1 part in 10^7.

In this example the leading large-n behavior of a_n could have been determined very quickly by solving the corresponding differential equation. The differential-equation analog of $a_{n+1} - a_n = 1/a_n$ is $y' = 1/y$. The solution to this differential equation is $y(x) = \sqrt{2x + C}$, where C is a constant. Note that when $x = n$ is large, the leading behavior of $y(x)$ agrees with that of a_n.

You will recall that dropping $1/a_n$ in (5.6.8) to give the asymptotic difference equation $a_{n+1} \sim a_n$ $(n \to \infty)$ was fruitless. In terms of the differential equation, this would be equivalent to dropping $1/y$ and obtaining $y' \sim 0$ $(x \to +\infty)$; it is wrong to neglect $1/y$ in comparison with 0.

This example is interesting because the naive use of dominant balance [discarding the term $1/a_n$ in (5.6.8)] is a dead end. However, it is important to point out that neglecting $1/a_n$ is *not* really

Table 5.6 Comparison of the exact solution to $a_{n+1} = a_n + 1/a_n$ $(a_1 = 1)$, with the asymptotic predictions in (5.6.10) and (5.6.11)

Observe that the error in (5.6.11) is only about 20 times smaller than that in (5.6.10) when $n = 10^6$ (see Prob. 5.52)

n	a_n	$\sqrt{2n}$	$\sqrt{2n} + \dfrac{\ln n}{4\sqrt{2n}}$
1	1.0	1.414 213 6	1.414 213 6
10	4.569 884 2	4.472 136 0	4.600 854 4
10^2	14.213 709	14.142 136	14.223 544
10^3	44.756 873	44.721 360	44.759 975
10^4	141.436 66	141.421 36	141.437 63
10^5	447.219 72	447.213 60	447.220 03
10^6	1,414.215 9	1,414.213 6	1,414.216 0

wrong; indeed, the resulting asymptotic relation $a_{n+1} \sim a_n$ $(n \to \infty)$ *is* satisfied by the leading behavior $\sqrt{2n}$ of a_n. However, many other functions, such as $n^{1/3}$, $\ln n$, $\exp(\sqrt{n})$, also satisfy $a_{n+1} \sim a_n$ $(n \to \infty)$ [see the discussion following (5.2.4)]. The solution $a_n = C$, a constant, of the equation obtained by replacing $\sim$ by $=$ is not consistent with the original approximation $1/a_n \ll a_n$ $(n \to \infty)$; this is why the method of dominant balance does not work here.

Example 4 *Use of differential equation to obtain leading behavior.* As in Example 3, dominant balance fails to yield the correct behavior as $n \to \infty$ of the solution to

$$a_{n+1} = a_n + \frac{1}{na_n}. \tag{5.6.12}$$

However, the failure is on a much more subtle level. If we were to neglect the term $1/na_n$ and solve $a_{n+1} = a_n$, we would conclude that $a_n = C$, a constant. This solution is *consistent* with dropping $1/na_n$ compared with a_n as $n \to \infty$ (in Example 3 dropping $1/a_n$ was *not* consistent). Nevertheless, a_n does *not* approach a constant as $n \to \infty$. To show this, we let $a_n = C + \varepsilon_n$, where $\varepsilon_n \to 0$ if $a_n \sim C$ $(n \to \infty)$. ε_n satisfies $\varepsilon_{n+1} = \varepsilon_n + 1/n(C + \varepsilon_n)$. If $\varepsilon_n \to 0$ $(n \to \infty)$ then this equation can be approximated by $\varepsilon_{n+1} - \varepsilon_n \sim 1/nC$ $(n \to \infty)$. The solution to this equation grows like $(\ln n)/C$ (see Prob. 5.3), which contradicts the assumption that $\varepsilon_n \to 0$ as $n \to \infty$. Therefore, $a_n \sim C$ $(n \to \infty)$ is wrong.

The behavior of solutions to (5.6.12) as $n \to \infty$ can be found by examining the corresponding differential equation $y' = 1/xy$. The exact solution to $y' = 1/xy$ is $y(x) = \sqrt{2 \ln x + c}$, where c is a constant. Thus, we expect that

$$a_n \sim \sqrt{2 \ln n}, \qquad n \to \infty.$$

This result can also be obtained by squaring the equation (5.6.12) and then using dominant balance.

Example 5 *Random behavior.* Sometimes the solution to a difference equation does not have a well-defined asymptotic behavior as $n \to \infty$, but rather jumps around in a random fashion. To construct such an equation we attempt to compute the square root of -3 using Newton's iteration:

$$a_{n+1} = \frac{1}{2}\left(a_n - \frac{3}{a_n}\right). \tag{5.6.13}$$

Of course, if we choose a_0 real then a_n cannot approach $\sqrt{-3}$ as $n \to \infty$ because a_n will be real for all n. Thus, what *is* the asymptotic behavior of a_n as $n \to \infty$?

To begin with, we observe that there are "cycles" such as $a_n = (-1)^n$. However, these cycles are unstable: if $a_n \sim 1 + \varepsilon$, then $a_{n+1} \sim -1 + 2\varepsilon$, $a_{n+2} \sim 1 + 4\varepsilon$, and so on. Thus, the sequence cannot approach such a cycle as $n \to \infty$. Second, one of the most powerful techniques for analyzing the structure of difference equations, namely, computing the first 50 terms numerically, seems to show nothing; the sequence $\{a_n\}$ jumps around apparently at random. These results indicate that a_n has no asymptotic behavior in the usual sense as $n \to \infty$. That is, a_n does not approach a limit or a simple function of n. We have already encountered random behavior of differential equations in Sec. 4.5.

Assuming that a_n does not have a well-defined asymptotic limit as $n \to \infty$, our approach will be to determine whether the sequence $\{a_n\}$ defines a *probability distribution function* $P(y)$ as $n \to \infty$. We interpret $P(y)$ as a probability distribution in the sense that $P(y)\,dy$ is the relative number of terms in the sequence $\{a_n\}$ that lie in the interval $[y, y + dy]$. If we can show that such a limiting distribution as $n \to \infty$ exists independently of the initial condition a_0 for almost all choices of a_0, then we will have performed a local analysis of the difference equation valid as $n \to \infty$.

If such a limiting distribution of the a_n exists independently of a_0, it must be stable under the action of the transformation

$$y = T(x) = \tfrac{1}{2}(x - 3/x),$$

because T just shifts the starting value from a_0 to a_1. It follows that the relative number of a_n in the interval $[y, y + dy]$ must equal the relative number of a_n in the interval $[x, x + dx]$, where $y = T(x)$. However, there are *two* intervals $[x, x + dx]$ that map into the one interval $[y, y + dy]$ because there are two solutions

$$x_\pm = y \pm \sqrt{y^2 + 3}$$

to the equation $y = T(x)$. Hence,

$$P(y)\,dy = P(x_+)\,dx_+ + P(x_-)\,dx_-, \tag{5.6.14}$$

where we assume that the intervals dx_+ and dx_- are sufficiently small that they do not overlap.

We compute

$$dx_\pm = \left(1 \pm \frac{y}{\sqrt{y^2 + 3}}\right) dy, \tag{5.6.15}$$

substitute (5.6.15) into (5.6.14), and divide out a common factor dy. We conclude that the distribution function $P(y)$, if it exists, must satisfy the following three-term functional equation:

$$P(y) = P(y - \sqrt{y^2 + 3})(1 - y/\sqrt{y^2 + 3}) + P(y + \sqrt{y^2 + 3})(1 + y/\sqrt{y^2 + 3}).$$

This monstrous-looking equation has a simple exact solution! If we multiply through by $y^2 + 3$ and set $f(y) = P(y)(y^2 + 3)$, then the functional equation for f simplifies to

$$f(y) = \tfrac{1}{2} f(y - \sqrt{y^2 + 3}) + \tfrac{1}{2} f(y + \sqrt{y^2 + 3}). \tag{5.6.16}$$

Observe that one solution to this equation is the linear function $f(y) = ay + b$. (We disregard the other more complicated solutions. See Probs. 5.53 and 5.54.) However, since $P(y)$ is a probability density, it must be positive for *all* y. Thus, $a = 0$ and $b > 0$. Hence, if the sequence generates an invariant probability distribution for arbitrary initial condition a_0, we predict that it may have the shape

$$P(y) \propto \frac{1}{y^2 + 3}. \tag{5.6.17}$$

This result was checked numerically. We used a computer to calculate the first 20,000 terms in the sequence $\{a_n\}$ and to sort these terms into four bins of equal area (equal probability) under $P(y)$. The results for four separate runs are given in Table 5.7. The fluctuations in the numerical results are within expected statistical limits.

Until now we have treated the difference equation (5.6.13) as an asymptotics problem; we have performed a local analysis at $n = \infty$. However, we have ignored the fact that (5.6.13), like (5.6.5), is exactly soluble. In particular, the substitution $a_n = \sqrt{3} \cot \theta_n$ converts the difference equation into $\cot (\theta_n) = \cot (2\theta_{n-1})$. Hence $\theta_n = 2^n \theta_0$ and

$$a_n = \sqrt{3} \cot (2^n \theta_0). \tag{5.6.18}$$

We may use this formula to verify that the difference equation does indeed generate the distribution function $P(y)$ in (5.6.17). The result in (5.6.18) shows that a small interval of θ_0 values in θ space (such an interval would exist if a_0 were defined, say, to 15 decimal places by the fixed bit length of computer words) gets expanded by a factor of 2^n into a much larger interval of θ_n values. When n is very large, this interval of θ_n is large compared with π, the periodicity of the cotangent function. Consequently, neglecting "end" effects, the distribution of a_n *must* approach that of uniform θ for $0 \le \theta \le \pi$; namely, $P(y)\, dy = d\theta/\pi$, where $y = \sqrt{3} \cot \theta$, so that

$$P(y) = \sqrt{3}/\pi(y^2 + 3).$$

This elementary proof of randomness depends on the measure-doubling character of the recurrence $\theta_n = 2\theta_{n-1}$. By contrast, problems of randomness in statistical mechanics involve measure-preserving transformations and are nontrivial.

The properties of some higher-order nonlinear difference equations are examined in the Problems.

Table 5.7 Results of a numerical experiment to test frequency distribution of the sequence a_n generated by the difference equation $a_{n+1} = \frac{1}{2}(a_n - 3/a_n)$

For each of four choices of a_0, a_n for $n = 1, 2, \ldots, 20{,}000$ was calculated and sorted into four bins $-\infty < y \le -\sqrt{3}$ (bin 1), $-\sqrt{3} < y \le 0$ (bin 2), $0 < y \le \sqrt{3}$ (bin 3), $\sqrt{3} < y < \infty$ (bin 4). These bins were chosen so that the area under the curve $P(y)$ given by (5.6.17) is the same for each bin. If our predictions are correct, the four bins should be equally populated. The results are consistent with equal population to within the expected statistical fluctuations

Run number	a_0	Number in bin 1	Number in bin 2	Number in bin 3	Number in bin 4	Total samples
1	1.234 567	4,855	4,988	4,988	5,169	20,000
2	10.0	4,961	4,961	4,962	5,116	20,000
3	100.0	5,025	4,990	4,991	4,994	20,000
4	$-\pi$	4,964	5,007	5,006	5,023	20,000
Total over all runs		19,805	19,946	19,947	20,302	

PROBLEMS FOR CHAPTER 5

Sections 5.1 and 5.2

(E) **5.1** Find the possible leading behaviors as $n \to \infty$ of solutions to the following difference equations:

(a) $a_{n+1} = a_n + n^{-3}a_n$;
(b) $a_{n+1} = ea_n$;
(c) $a_{n+1} = a_n + \ln n$;
(d) $a_{n+1} = (1 + 2/n)a_n + 1$;
(e) $a_{n+2} - 2a_{n+1} + a_n = 0$;
(f) $a_{n+2} - 2a_{n+1} + a_n = n$;
(g) $a_{n+2} + (3/n - 2)a_{n+1} + (2/n^2 - 3/n + 1)a_n = 0$;
(h) $a_{n+2} - 2(1 + 1/n)a_{n+1} + (1 + 2/n - 1/n^4)a_n = 0$;
(i) $a_{n+2} - 2a_{n+1} + (1 + 1/4n^2)a_n = 0$;
(j) $a_{n+2} + (1/n - 2)a_{n+1} + (1/n^2 - 1/n + 1)a_n = 0$.

(E) **5.2** Show that if p is an integer and $p > |\alpha|$, then $\sum_{j=p}^{n} \ln(1 + \alpha/j) - \alpha \ln n \sim k$ $(n \to \infty)$, where k is a constant.

(E) **5.3** (a) For which functions $f(n) > 0$ is it true that $a_{n+1} - a_n \sim f(n)$ $(n \to \infty)$ implies that $a_n \sim \sum_{k=1}^{n-1} f(k)$ $(n \to \infty)$?

(b) For which functions $f(n) > 0$ is it true that $a_{n+1} - a_n \sim f(n)$ $(n \to \infty)$ implies that $a_n \sim \int^n f(k)\,dk$ $(n \to \infty)$?

(c) For each of the following asymptotic relations determine whether the leading asymptotic behaviors of a_n as $n \to +\infty$ are determined by the given information and, if so, find the leading behavior of a_n: $a_{n+1} - a_n \sim 1$ $(n \to \infty)$; $a_{n+1} - a_n \sim 1/n$ $(n \to \infty)$; $a_{n+1} - a_n \sim n^\alpha$ $(n \to \infty)$.

(E) **5.4** Prove that the leading behavior of $a_n = \Gamma(n + \alpha)/\Gamma(n + \beta)$ is $Cn^{\alpha-\beta}$ as $n \to \infty$.

Clue: Show that a_n satisfies the difference equation $(n + \beta)a_{n+1} = (n + \alpha)a_n$. (See also Prob. 5.23.)

(I) **5.5** Show that if ∞ is an ordinary point or a regular singular point of a corresponding pair of difference and differential equations, then the leading behaviors of the corresponding solutions as $n = x \to \infty$ are the same.

Clue: If the solutions to the differential equation satisfy $y(x) \sim cx^\alpha$, $y' \sim c\alpha x^{\alpha-1}$ $(x \to +\infty)$, then $Dy \sim c\alpha x^{\alpha-1}$ $(x \to +\infty)$.

(I) **5.6** Find Taylor or Frobenius series solutions about the point at $n = \infty$ for:

(a) $$a_{n+1} = \left[1 + \frac{1}{n(n+1)(n+2)(n+3)}\right] a_n;$$

(b) $$a_{n+2} - 2a_{n+1} + \left[1 - \frac{1}{n(n+1)(n+2)(n+3)}\right] a_n = 0;$$

(c) $$a_{n+2} + \left[\frac{1}{n(n+1)} - 2\right] a_{n+1} + \left[\frac{1}{n(n+1)(n+2)(n+3)} - \frac{1}{n(n+1)} + 1\right] a_n = 0;$$

(d) $$a_{n+2} - \left(2 + \frac{1}{n}\right) a_{n+1} + \left[1 + \frac{2}{n} + \frac{2}{n(n+1)} + \frac{1}{n(n+1)(n+2)}\right] a_n = 0;$$

(e) $$a_{n+3} - 3a_{n+2} + 3a_{n+1} - \left[1 + \frac{1}{n(n+1)(n+2)(n+3)}\right] a_n = 0;$$

(f) verify (2.3.21).

(I) **5.7** Find the first three terms in the asymptotic expansion of a_n valid as $n \to +\infty$ for the following equations:

(a) $a_{n+2} - 2a_{n+1} + a_n e^{1/n^2} = 0$;
(b) $a_{n+1} - a_n[1 + \sin(1/n)/n] = 0$;
(c) $a_{n+2} + a_{n+1}e^{1/n}/n + a_n e^{1/n^2}/n^2 = 0$.

(I) **5.8** (*a*) Using the relation $\Gamma(z+1) = z\Gamma(z)$, verify the difference equation (5.2.21) satisfied by the digamma function $\psi(z)$: $\psi(z+1) = \psi(z) + 1/z$.

(*b*) Show that $\psi(n) \sim \ln n$ $(n \to \infty)$. (See Prob. 5.18.)

(*c*) Find series expansions valid as $n \to \infty$ for the solutions of

$$a_{n+2} + \left(\frac{1}{n} - 2\right) a_{n+1} + \left[\frac{1}{n(n+1)(n+2)} - \frac{1}{n} + 1\right] a_n = 0.$$

Clue: It is convenient to use the digamma function instead of $\ln n$ in one of these expansions.

(I) **5.9** (*a*) Find the leading behaviors of all solutions to

(i) $(4n^3 + 4)a_{n+2} - (8n^2 - 2)(n-1)a_{n+1} + (n-2)(4n^2-1)a_n = 0$;

(ii) $na_{n+2} - (2n+1)a_{n+1} + na_n = 0$.

(*b*) Find full series expansions of all solutions to the difference equations in part (*a*).

(TI) **5.10** (*a*) Show that if $\sum_{k=0}^{\infty} B_k \Gamma(n)/\Gamma(n+k) = 0$ for all sufficiently large n, then $B_k = 0$ for all k.

(*b*) Explain what is meant by the notation $a_n \sim \sum_{k=0}^{\infty} B_k \Gamma(n)/\Gamma(n+k)$ $(n \to \infty)$.

(*c*) Show that if $a_n \equiv 0$, then the definition introduced in part (*b*) implies $B_k = 0$ for all k.

Section 5.3

(E) **5.11** Verify (5.3.1) and (5.3.2).

(E) **5.12** (*a*) Show that if $\frac{1}{2} < \beta < 1$, then the leading behavior of solutions to $a_{n+1} = (1 + \alpha n^{-\beta})a_n$ is $a_n \sim c \exp[n^{1-\beta}\alpha/(1-\beta)]$ $(n \to \infty)$. Note that this behavior is the same as that of solutions to the corresponding differential equation.

(*b*) Show that if $0 < \beta < \frac{1}{2}$, then the leading behavior of a_n is different from the leading behaviors of the solutions to the corresponding differential equations.

(*c*) For which values of α and β does the general solution to $a_{n+2} - 2a_{n+1} + (1 + \alpha n^{-\beta})a_n = 0$ behave like $c_1 + c_2 n$ as $n \to \infty$?

(E) **5.13** Show that ∞ is an irregular singular point of the difference equations:

(*a*) $(n+3)a_{n+2} + (2-n)a_{n+1} + 6a_n = 0$;

(*b*) $a_{n+2} + (3-n)a_{n+1} + (3+n)a_n = 0$.

Then show that there exists a solution to each of these equations that has the same controlling factor for $n \to \infty$ as a solution to the corresponding differential equations.

(E) **5.14** Bernoulli's method for finding the roots of the polynomial

$$P(x) = x^k + p_{k-1}x^{k-1} + \cdots + p_1 x + p_0$$

consists of computing the sequence a_n determined by

$$a_{n+k} = -p_{k-1}a_{n+k-1} - p_{k-2}a_{n+k-2} - \cdots - p_1 a_{n+1} - p_0 a_n$$

for $n = 0, 1, \ldots$. Show that for typical initial values $a_0, a_1, \ldots, a_{k-1}$, $\lim_{n\to\infty} a_n/a_{n-1} = r_1$, where r_1 is that root of $P(x)$ of largest absolute value. (If there are two roots with absolute value r_1, then $\lim_{n\to\infty} a_n/a_{n-1}$ does not exist for typical starting values.)

Clue: Suppose first that the roots of $P(x)$ are $r_1, r_2, \ldots, r_k$ with $|r_1| > |r_j|$ for $j \neq 1$ and that these roots are all distinct. Show that $a_n = \sum_{p=1}^{k} c_p r_p^n$ for some constants c_p. Use this to show that $\lim_{n\to\infty} a_n/a_{n-1} = r_1$ if $c_1 \neq 0$.

(E) **5.15** Show that if $a_n = n^\alpha$ then (5.3.6) and (5.3.9) do not hold but that the asymptotic relations in (5.3.8*b*) and (5.3.10) are still valid.

(I) **5.16** Find the controlling factors of the leading behaviors to:

(*a*) $a_{n+2} - n^2 a_{n+1} + (2n^2 - 4)a_n = 0$;

(*b*) $(n+3)a_{n+2} - (n+3)a_{n+1} - 2na_n = 0$;

(*c*) $a_{n+2} - na_{n+1} - na_n = 0$;

(*d*) $a_{n+2} - (n+1)a_{n+1} - (n+1)a_n = 0$;

(*e*) $a_{n+2} - na_{n+1} - n^2 a_n = 0$;

(*f*) $a_{n+2} - a_{n+1} - n^2 a_n = 0$;

(g) $na_{n+2} - (n-2)a_{n+1} - a_n = 0$;
(h) $n^2D^2a_n - 2nDa_n + 2a_n = n^2$;
(i) $a_{n+2} - 2na_{n+1} + n(n-1)a_n = n!$;
(j) $a_{n+2} - 2a_{n+1} + n(n-1)a_n = n!$.

(TI) **5.17** (a) Use the methods of Sec. 5.3 to verify the Poincaré-Perron theorems: If $\lim_{n\to\infty} p(n) = p_0$, $\lim_{n\to\infty} q(n) = q_0$, there are solutions to the difference equation $a_{n+2} + p(n)a_{n+1} + q(n)a_n = 0$ that satisfy $\lim_{n\to\infty} a_{n+1}/a_n = \lambda_1$ and $\lim_{n\to\infty} a_{n+1}/a_n = \lambda_2$, where λ_1 and λ_2 are the roots of $\lambda^2 + p_0\lambda + q_0 = 0$.

(b) Generalize the above theorem to nth-order difference equations.

Section 5.4

(I) **5.18** (a) Using the difference equation $\psi(z+1) = \psi(z) + 1/z$ satisfied by the digamma function $\psi(z) = \Gamma'(z)/\Gamma(z)$, show that the asymptotic expansion of $\psi(z)$ as $z \to +\infty$ is $\psi(z) \sim \ln z + C + \sum_{n=1}^{\infty} b_n z^{-n}$ $(z \to +\infty)$. Global analysis gives $C = 0$. Show that the expansion coefficients b_n satisfy

$$b_n = (n-1)! \sum_{k=2}^{n} \frac{(-1)^k}{k!\,(n-k)!} b_{n+1-k} + \frac{(-1)^n}{n(n+1)}, \qquad n \geq 1.$$

Thus, $b_1 = -\frac{1}{2}$, $b_2 = -\frac{1}{12}$, $b_3 = 0$, $b_4 = +\frac{1}{120}$, $b_5 = 0$, $b_6 = -\frac{1}{252}$.

(b) We define the Bernoulli numbers B_n by $B_0 = 1$ and $B_n \equiv (-1)^{n+1} n b_n$ $(n \geq 1)$. Thus, B_n satisfies $\sum_{k=0}^{n} B_k/k!\,(n+1-k)! = 0$ $(n \geq 1)$. Show that $\sum_{n=0}^{\infty} B_n t^n/n! = t/(e^t - 1)$.

Clue: Multiply the recursion relation for B_k by t^n and sum from $n = 1$ to ∞.

(c) Show that $B_{2n+1} = 0$ for $n \geq 1$.

(E) **5.19** Using Prob. 5.18 show that $\ln \Gamma(z)$ has the asymptotic expansion $\ln \Gamma(z) \sim (z - \frac{1}{2}) \ln z - z + C' + \sum_{n=1}^{\infty} b'_n z^{-n}$ $(z \to +\infty)$. Stirling's formula implies that $C' = \frac{1}{2} \ln (2\pi)$. Find the relation between b'_n and b_n.

(I) **5.20** (a) What is the relation between the Stirling numbers A_n and the Bernoulli numbers B_n? Here the numbers A_n are defined by the asymptotic series (5.4.1) and the numbers B_n are defined in Prob. 5.18.

(b) In Prob. 6.79 it is established that $B_{2n} \sim 2(-1)^{n+1}(2n)!\,(2\pi)^{-2n}$. Show that this result is consistent with the series given in Prob. 5.18(b).

(c) Using the result of (b) and those of Probs. 5.18 and 5.19, verify (5.4.2).

(I) **5.21** Compute the coefficients A'_0, A'_1, and A'_2 for a modified Stirling series of the form $\Gamma(z) \sim \sqrt{2\pi/z}\,(z/e)^z \sum_{j=1}^{\infty} A'_j \Gamma(z)/\Gamma(z+j)$ $(z \to \infty)$.

(I) **5.22** (a) As you can observe from Fig. 5.1, the points at which the numbers of terms in the optimal asymptotic approximation to the Stirling series (5.4.3) for $\Gamma(z)$ increases by 2 appear to be evenly spaced. Demonstrate this fact by showing that the number of terms in the optimal asymptotic approximation to $\Gamma(z)$ is asymptotic to $2\pi z$ as $z \to +\infty$. (It follows that the spacing between points where the optimal number of terms changes approaches $1/\pi$ as $z \to +\infty$.)

(b) Use (a), (5.4.2), and Stirling's approximation to $n!$ to show that the error in the optimal asymptotic approximation to $\Gamma(x)$ can be approximated by $e^{-2\pi x}/2\pi\sqrt{x}$ as $x \to +\infty$.

(E) **5.23** Use the Stirling approximation to show that $\Gamma(n+\alpha)/\Gamma(n+\beta) \sim n^{\alpha-\beta}$ $(n \to \infty)$ (see Prob. 5.4).

(E) **5.24** Find the leading behaviors of:
(a) $(2n)!/(n!)^2$;
(b) $(kn)!/(n!)^k$;
(c) $(2n+1)!! = (2n+1)(2n-1)(2n-3)\cdots(5)(3)(1)$;
(d) $(3n-1)(3n-2)(3n-4)(3n-5)\cdots(5)(4)(2)(1)$;
(e) $\ln [(n!)!]$;
(f) $\ln \Gamma(\ln n)$.

(E) **5.25** The error in the optimal asymptotic truncation of the asymptotic expansion $\sum_{n=0}^{\infty} n!\,(-x)^n$ of the Stieltjes integral $\int_0^\infty e^{-t}/(1+xt)\,dt$ as $x \to 0+$ is asymptotic to half the smallest term in the series as

$x \to 0+$ (see Example 6 of Sec. 3.8 and Prob. 6.37). Use Stirling's approximation and this result to establish (3.8.9).

(I) **5.26** (*a*) Find an approximation to the error in the optimal asymptotic truncation of the asymptotic expansion (3.5.13) and (3.5.14) of the parabolic cylinder function $D_\nu(x)$ as $x \to +\infty$.

Clue: Even though this asymptotic series is not a series of Stieltjes (why?), assume that the optimal error is less than the smallest term in the series. As a difficult exercise, you may try justifying this assumption for sufficiently large x. Show that the optimal asymptotic approximation to $D_\nu(x)$ is accurate for x larger than roughly $\nu^{1/2}/2$.

(*b*) As in part (*a*), find an approximation to the error in the optimal truncation of the asymptotic series (3.5.21) for the Airy function Ai (x) as $x \to +\infty$.

(*c*) Do the same for the asymptotic series (3.5.8) and (3.5.9) of the modified Bessel functions $I_\nu(x)$ and $K_\nu(x)$ as $x \to +\infty$.

(I) **5.27** Test the following series for convergence:

(*a*) $\displaystyle\sum_{k=2}^{\infty} \frac{1}{\ln (k!)}$;

(*b*) $\displaystyle\sum_{k=1}^{\infty} k^{-(k+1)/k}$;

(*c*) $\displaystyle\sum_{k=1}^{\infty} \frac{1}{3^{1+1/2+\cdots+1/k}}$;

(*d*) $\displaystyle\sum_{k=1}^{\infty} \frac{1}{2^{1+1/2+\cdots+1/k}}$.

(E) **5.28** (*a*) The function $\sqrt{1+x}$ has a Taylor series expansion about $x = 0$: $\sqrt{1+x} = \sum_{n=0}^{\infty} a_n x^n$. Find the leading behavior of a_n as $n \to \infty$. Express your answer in terms of elementary functions.

(*b*) What is the radius of convergence of the Taylor series $\sum_{n=1}^{\infty} a_n n^{-2n} x^n$, where a_n satisfies $a_{n+1} = n^2 a_n$?

Section 5.5

(E) **5.29** Show that if $a_n(x)$ and $a_{n+1}(x)$ satisfy the Bessel differential equation $x^2 y'' + xy' + (x^2 - \nu^2)y = 0$ with $\nu = n$ and $\nu = n + 1$, respectively, then $a_{n+2}(x)$ defined by the recursion relation (5.5.3) satisfies this differential equation with $\nu = n + 2$.

(E) **5.30** (*a*) Show that the leading behavior A_k in (5.5.7) is given by $A_k \sim A(x)(k-1)!$ as $k \to \infty$.

(*b*) Show that (5.5.7) converges for all $n > 1$.

(D) **5.31** Relate the series for $Y_n(x)$ given by (5.5.8) with $C = -1/\pi$ (here $n \to \infty$ with x fixed) to the Frobenius series for $Y_n(x)$ about $x = 0$ given in the Appendix.

(E) **5.32** The Legendre polynomial $P_n(x)$ is the solution of the differential equation

$$\frac{d}{dx}(1-x^2)\frac{dP_n}{dx} + n(n+1)P_n = 0$$

satisfying $P_n(\pm 1) = (\pm 1)^n$. Show that $P_n(x)$ satisfies (5.5.9).

(E) **5.33** Using method I of Sec. 5.3 show that there is a solution b_n of (5.5.12) whose controlling factor is $n^{-1/2}$ as $n \to \infty$.

(I) **5.34** (*a*) Use method I of Sec. 5.3 to verify that there is a solution of (5.5.13) satisfying $c_n \sim C(x)$ as $n \to \infty$, where $C(x)$ is a function of x alone.

(*b*) Show that $c_n - C(x) \sim [-\frac{1}{4} + 1/8(xr - 1)]C(x)/n$ $(n \to \infty)$ as stated in (5.5.15).

(*c*) Show that the asymptotic expansion $C_n \sim \sum_{k=0}^{\infty} A_k \Gamma(n)/\Gamma(n+k)$ $(n \to \infty)$ diverges for all n.

(D) **5.35** Derive (5.5.17) and (5.5.18).

Clue: The Stirling series for $\Gamma(n)$ can be used to show that $\Gamma(n)/\Gamma(n+\frac{1}{2}) \sim n^{-1/2}(1 + 1/8n + \cdots)$ $(n \to \infty)$.

(I) **5.36** Find the possible leading behaviors as $n \to \infty$ of the solutions a_n to the difference equations:
(a) $a_{n+2} - 3na_{n+1} + 2n^2 a_n = 0$;
(b) $a_{n+3} = na_n$;
(c) $(n+1)a_{n+2} - (2n+1)a_{n+1} + (n+1)a_n = 0$;
(d) $a_{n+2} - 2e^n a_{n+1} + e^{2n} a_n = 0$;
(e) $a_{n+2} - 3e^{n^2} a_{n+1} + 2e^{2n^2} a_n = 0$;
(f) $a_{n+2} - 5e^{e^n} a_{n+1} + 4e^{2e^n} a_n = 0$;
(g) $a_{n+3} - (3 + 1/n)a_{n+2} + (3 - 1/n^2)a_{n+1} - (1 + 1/n^3)a_n = 0$;
(h) $a_{n+2} + na_{n+1} + n^2 a_n = 0$;
(i) $a_{n+2} - 4na_{n+1} + 4n^2 a_n = 0$;
(j) $a_{n+2} - a_{n+1} = na_n$;
(k) $a_{n+2} - a_{n+1} = na_n + n^2$;
(l) $a_{n+2} - 3na_{n+1} + 2n^2 a_n = n^3$;
(m) $a_{n+2} - a_{n+1} = na_n + n!$;
(n) $a_{n+2} - 3e^n a_{n+1} + 2e^{2n} a_n = n!$;
(o) $a_{n+2} - 2e^n a_{n+1} + e^{2n} a_n = 2^n$.

(D) **5.37** Find full asymptotic series descriptions as $n \to \infty$ of the solutions to each of the difference equations in Prob. 5.36.

(I) **5.38** Find the leading behaviors of the solutions to the recursion relation in (5.2.16) valid as $k \to +\infty$. Show that $B_k \sim C_\pm k!\, k^{-1\pm i}$ $(k \to +\infty)$. From this estimate, show that (5.2.12) converges for all n in the interval $1 < n < \infty$.

(I) **5.39** Let $a_{n+1} = f(n)a_n - g(n)$, where

$$f(n) = \frac{2n(n+1)(4n+7)}{(n+2)(4n+3)(2n+3)}$$

and

$$g(n) = \frac{\Gamma(n - \frac{1}{2})(4n+5)(4n+7)}{8\sqrt{\pi}(n+3)!\,(n+2)}.$$

Assume that a_0 is finite and therefore that $a_1 = -g(0)$. Show that $A_n \sim 2n^{-5/2}/\sqrt{\pi}$ $(n \to +\infty)$.

(I) **5.40** Laguerre polynomials $L_n(x)$ of degree n satisfy the difference equation

$$(n+1)L_{n+1}(x) - (2n+1-x)L_n(x) + nL_{n-1}(x) = 0.$$

Show that there is a solution to this difference equation satisfying $L_n(x) \sim [F(x)/n^{1/4}] \cos(2\sqrt{nx} - \frac{1}{4}\pi)$ $(n \to \infty)$, and thus infer the leading asymptotic behavior as $n \to \infty$ of the zeros of Laguerre polynomials. The constant $\pi/4$ is not determined by local analysis.

(I) **5.41** Hermite polynomials $\mathrm{He}_n(x)$ of degree n satisfy the difference equation

$$\mathrm{He}_{n+1}(x) - x\,\mathrm{He}_n(x) + n\,\mathrm{He}_{n-1}(x) = 0.$$

Use this difference equation to obtain the leading asymptotic behavior of $\mathrm{He}_n(x)$ as $n \to \infty$ and the leading asymptotic behavior of the zeros of $\mathrm{He}_n(x)$ as $n \to \infty$ (see Prob. 6.85).

Section 5.6

(TE) **5.42** Use a graphical argument to derive (5.6.2).

Clue: Define a_{n+1} as the point at which the tangent to the curve $y = f(x)$ at $x = a_n$ crosses the x axis.

(TI) **5.43** Show that if x is a simple zero of $f(x)$ and if a_0 is sufficiently close to x, then the sequence of approximations defined in (5.6.2) converges to x as $n \to \infty$.

Clue: Show that $|a_{n+1} - x| < K|a_n - x|$ for some $K < 1$ and all n if a_0 is sufficiently close to x. You may assume that f'' is continuous.

(I) **5.44** Find an iteration scheme that converges quadratically to a double root of $f(x) = 0$. How does the scheme behave near a simple root? a cubic root?

(I) **5.45** Can you invent an iteration scheme for finding simple zeros of $f(x)$ which exhibits cubic convergence?

(I) **5.46** Show that although the solution to $a_{n+1} = \frac{1}{3}(a_n + 2Q/a_n)$ may approach $\sqrt{Q}$, this iteration scheme is not as efficient as that in (5.5.5) because the convergence is linear.

(I) **5.47** Use Newton's method to find an iteration scheme which converges quadratically to the cube root of Q. Solve the difference equation exactly.

(TI) **5.48** Investigate the rate of convergence of Newton's method to double roots of $f(x) = 0$.

(I) **5.49** Assuming that $a_0 > 0$, find the leading large-n behaviors of solutions to the following difference equations:

(*a*) $a_{n+1} = a_n + n/a_n$;
(*b*) $a_{n+1} = (\sin a_n)^2$;
(*c*) $a_{n+1} = \exp(-a_n)$;
(*d*) $a_{n+1} = \ln(1 + a_n)$.

(I) **5.50** Show that the first two terms in the asymptotic expansion of the solution to $a_{n+1} = \sin a_n$ are $a_n \sim \sqrt{3/n} - 3^{3/2} \ln n/10n^{3/2}$ $(n \to +\infty)$.

(I) **5.51** Consider the difference equation $a_{n+1} = a_n^2 + \varepsilon$ $(a_0 = 0)$.

(*a*) Show that when $\varepsilon > \frac{1}{4}$, a_n does not approach a limit as $n \to +\infty$. What is the behavior of a_n for large n?

(*b*) Show that when $-\frac{3}{4} \le \varepsilon \le \frac{1}{4}$, a_n does approach a limit as $n \to +\infty$. What is the rate at which a_n approaches this limit?

(I) **5.52** Find the asymptotic behavior of the first correction to the result given in (5.6.11). Does this correction explain the behavior of the errors of the results given in Table 5.6?

(TI) **5.53** Show that the functional equation (5.6.16) has solutions that are not linear functions of y.
Clue: Choose $f(y)$ arbitrarily for $y < 0$ and use (5.6.16) to determine $f(y)$ for $y > 0$.

(TD) **5.54** Is $f(y) - ay + b$ the most general entire solution to the functional relation (5.5.16)?

(I) **5.55** (*a*) Let $a_{n+1} = a_n(2 - \alpha a_n)$ for $n = 0, 1, \ldots$ where α is a real positive number. For what range of real values of a_0 does $a_n \to 1/\alpha$ as $n \to \infty$ (Putnam Exam, 1957)?

(*b*) Let $0 < a_1 < 1$ and $a_{n+1} = a_n(1 - a_n)$. Show that $a_n \sim 1/n$ $(n \to \infty)$ (Putnam Exam, 1966).

(I) **5.56** Let a_0 be any real number. If $a_{n+1} = \cos a_n$ prove that $\lim_{n\to\infty} a_n$ exists and is independent of a_0 (Putnam Exam, 1952).

(I) **5.57** Show that there is a solution to the functional difference equation $f[x + f(x)] = f(x) + 1$ whose leading behavior is $f(x) \sim \sqrt{2x}$ $(x \to +\infty)$. What are the corrections to this leading behavior?

(I) **5.58** Show that the sequence $\sqrt{7}, \sqrt{7 - \sqrt{7}}, \sqrt{7 - \sqrt{7 + \sqrt{7}}}, \ldots$ converges and evaluate the limit (Putnam Exam, 1953).

(D) **5.59** Justify the statement that

$$3 = \sqrt{1 + 2\sqrt{1 + 3\sqrt{1 + 4\sqrt{1 + 5\sqrt{1 + \cdots}}}}}$$

(Putnam Exam, 1966).

(I) **5.60** For what range of values of a_0 does the solution to $a_{n+1} = a_n + \ln(k - a_n)$ approach $k - 1$ as $n \to \infty$?

(I) **5.61** If $a_{n+1}(2 - a_n) = 1$ for $n = 1, 2, \ldots$ prove that $\lim_{n\to\infty} a_n$ exists and is equal to 1 (Putnam Exam, 1947).

(E) **5.62** The sequence $x_0, x_1, x_2, \ldots$ is defined by $x_0 = a$, $x_1 = b$, $x_{n+1} = [x_{n-1} + (2n - 1)x_n]/2n$, where a and b are given numbers. Express $\lim_{n\to\infty} x_n$ concisely in terms of a and b (Putnam Exam, 1950).

(I) **5.63** a_1, b_1, c_1 are positive numbers whose sum is 1 and for $n = 1, 2, \ldots$ we define $a_{n+1} = a_n^2 + 2b_n c_n$, $b_{n+1} = b_n^2 + 2a_n c_n$, $c_{n+1} = c_n^2 + 2a_n b_n$. Show that a_n, b_n, c_n approach limits as $n \to \infty$ and find those limits (Putnam Exam, 1947).

(I) **5.64** The infinite sequence 1, 2, 2, 3, 3, 4, 4, 4, 5, 5, 5, 6, 6, 6, 6, 7, 7, 7, 7, 8, ... is characterized by the conditions that its nth term $a(n)$ is a positive integer, $a(n+1) \geq a(n)$, and $a(n)$ *is the number of times that n appears in the sequence.*

(*a*) Show that $a(n)$ satisfies the functional difference equation $a\{n + a[a(n)]\} = a(n) + 1$.

(*b*) Use this functional difference equation to find the leading behavior of $a(n)$ as $n \to \infty$.

(I) **5.65** Let $a_{n+1} = a_n + \sum_{j=0}^{n} a_j a_{n-j}$ $(a_0 = 1)$. Show that $a_n \sim c^n$ $(n \to +\infty)$ where $c \doteqdot 5.8$.
Clue: Introduce the generating function $f(x) = \sum_{n=0}^{\infty} a_n x^n$.

(I) **5.66** Let $d_n = (4n)!/[n!\,(2n)!\,4^n]$. Define c_n by $d_n - c_n = 1/n \sum_{j=1}^{n-1} jc_j d_{n-j}$ $(n = 2, 3, \ldots; c_1 = 3)$. Show that $c_n/d_n \sim 1 + a/n + b/n^2 + \cdots$ $(n \to \infty)$, where $a = -\frac{3}{16}$, $b = -\frac{183}{512}$, ...

(I) **5.67** Consider the nonlinear difference equation $a_{n+1} = (\alpha n + \beta) \sum_{j=0}^{n} a_j a_{n-j}$ $(n = 0, 1, 2, \ldots)$, with $a_0 = 1$ and $\alpha \neq 0$. Find the controlling factor of the large-n behavior of a_n.
Clue: The largest contribution to the convolution sum when $\alpha \neq 0$ comes from the endpoints $j = 0$ and $j = n$. What happens if the largest terms are kept? and the next largest also?

(D) **5.68** For the case $\alpha = \beta = 1$ in Prob. 5.67 show that the leading behavior of a_n is $a_n \sim (2n)!/2^n n!\,e$ $(n \to \infty)$. (It is easy to check this answer on a computer but the proof is difficult.)

(I) **5.69** Let a_n be a sequence of numbers having the property that $a_n \sim (-3)^n \Gamma(n + \frac{1}{2})$ $(n \to \infty)$. Define b_n by

$$b_n = \sum_{j=0}^{n} \frac{a_j 6^{n-j} \Gamma(n + j/2 - \frac{1}{2})}{\Gamma(3j/2 - \frac{1}{2})(n-j)!}.$$

Show that the ratio a_n/b_n approaches the constant e^3 as $n \to \infty$.
Clue: Use Stirling's formula.

(D) **5.70** Consider the difference equation $a_{n+1} = \lambda a_n(1 - a_n)$ $(0 < a_0 < 1)$. Show that:

(*a*) There is some number $4 > \lambda_1 > 0$ such that for $0 < \lambda < \lambda_1$, $a_n \to 0$ or $a_n \to 1 - 1/\lambda$ as $n \to \infty$.

(*b*) There is a number $4 > \lambda_2 > \lambda_1$ such that $a_{n+2}/a_n \to 1$ as $n \to \infty$, but that $a_{n+1}/a_n \nrightarrow 1$ as $n \to \infty$. (This indicates that a_n approaches a two-cycle. In an N-cycle, $a_{n+N} = a_n$ for all n.)

(*c*) Can there be four-cycles? eight-cycles?

(*d*) Investigate on a computer what happens when λ approaches 4 from below.

(*e*) What happens when $\lambda > 4$?

(I) **5.71** The function e^{e^x} can be expanded in the Taylor series $e^{e^x} = \sum_{n=0}^{\infty} a_n x^n$.

(*a*) What is the radius of convergence of this series?

(*b*) Find a difference equation satisfied by a_n.
Clue: Find a first-order differential equation satisfied by e^{e^x} and substitute the Taylor series into it. The resulting difference equation for a_n contains a convolution sum.

(*c*) Use the difference equation found in (*b*) to find the controlling factor of the leading behavior of a_n as $n \to \infty$.
Clue: The answer has the form $\exp[-n^{\alpha}(\ln n)^{\beta}(\ln \ln n)^{\gamma}(\ln \ln \ln n)^{\delta}]$. You are to find $\alpha, \beta, \gamma, \delta$. $a_n \to 0$ much slower than $1/n!$ as $n \to \infty$. Assume that $a_{n+k}/a_n \sim [f(n)]^k$ as $n \to \infty$ with k fixed, where $f(n)$ is some function of n. For an alternative derivation of the asymptotic behavior of a_n as $n \to \infty$, see Prob. 6.78.

(E) **5.72** There is an elegant algorithm discovered by E. Salamin for computing π which involves generating sequences of arithmetic and geometric means. Let $a_0 = 1$, $b_0 = 1/\sqrt{2}$ and define the sequences a_n, b_n by $a_{n+1} = \frac{1}{2}(a_n + b_n)$, $b_{n+1} = \sqrt{a_n b_n}$. If we set

$$c_n = \frac{4a_n b_n}{1 - \sum_{j=1}^{n} 2^{j+1}(a_j^2 - b_j^2)},$$

then $\lim_{n\to\infty} c_n = \pi$. (The derivation of this result is difficult and requires knowledge of elliptic integrals.)

(*a*) Show that $\lim_{n\to\infty} a_n$ and $\lim_{n\to\infty} b_n$ exist and are equal.

(*b*) Show that if $c_n \to \pi$ as $n \to \infty$, then the error $|c_{n+1} - \pi|$ is roughly proportional to $|c_n - \pi|^2$. Therefore, the number of decimal places of accuracy of c_{n+1} is roughly double that of c_n. In 1977, this algorithm was used to calculate π to 2×10^7 decimal places.

CHAPTER

SIX

ASYMPTOTIC EXPANSION OF INTEGRALS

In five minutes you will say that it is all so absurdly simple.

—Sherlock Holmes, *The Adventure of the Dancing Men*
Sir Arthur Conan Doyle

(E) 6.1 INTRODUCTION

The analysis of differential and difference equations in Chaps. 3 to 5 is pure local analysis; there we predict the behavior of solutions near one point, but we do not incorporate initial-value or boundary-value data at other points. As a result, our predictions of the local behavior usually contain unknown constants. However, when the differential or difference equation is soluble, we can use the boundary and initial data to make parameter-free predictions of local behavior.

Example 1 *Effect of initial data upon asymptotic behavior.* The solution to the initial-value problem $y'' = y$ $[y(0) = 1, y'(0) = 0]$ is $y(x) = \cosh x$. Local analysis of the differential equation $y'' = y$ at the irregular singular point at $x = \infty$ gives two kinds of possible behaviors for $y(x)$ as $x \to +\infty$: $y(x) \sim Ae^x$ $(x \to +\infty)$, where A is some nonzero constant, or, if $A = 0$, $y(x) \sim Be^{-x}$ $(x \to +\infty)$, where B is some constant. Since the equation is soluble in closed form, we know from the initial conditions that $A = \frac{1}{2}$. Similarly, if the initial conditions were $y(0) = 1$, $y'(0) = -1$, we would find that $A = 0$ and $B = 1$.

In very rare cases, it is possible to solve a differential equation in closed form. There, as we have seen in Example 1, it is easy to incorporate initial or boundary conditions. Among those equations that are not soluble in terms of elementary functions, it is sometimes possible (though still very rare) to find a representation for the solution of the equation as an integral in which the independent variable x appears as a parameter. Typically, this integral representation of the solution contains all of the initial-value or boundary-value data. In this chapter we will show how to perform a local analysis of integral representations containing x as a parameter. We will see that the local behavior of the solution $y(x)$ is completely determined by this analysis of the integral and contains no arbitrary constants.

Example 2 *Integral representation of a solution to an initial-value problem.* The solution to the initial-value problem $y' = xy + 1$ $[y(0) = 0]$ can be expressed as the integral $y(x) =$

$e^{x^2/2} \int_0^x e^{-t^2/2}\,dt$. This integral representation implies that $y(x) \sim \sqrt{\pi/2}\, e^{x^2/2}$ $(x \to +\infty)$ because $\lim_{x\to+\infty} \int_0^x e^{-t^2/2}\,dt = \int_0^\infty e^{-t^2/2}\,dt = \sqrt{\pi/2}$. Note that a direct local analysis of the differential equation $y' = xy + 1$ predicts only that $y(x) \sim Ae^{x^2/2}$ $(x \to +\infty)$ (where A is unknown) because it does not utilize the initial condition $y(0) = 0$.

Here is a more striking example of the usefulness of integral representations.

Example 3 *Integral representation of a solution to a boundary-value problem.* Consider the boundary-value problem

$$xy''' + 2y = 0, \qquad y(0) = 1, \qquad y(+\infty) = 0. \tag{6.1.1}$$

The problem is to find the behavior of $y(x)$ as $x \to +\infty$. Using the techniques of Chap. 3, local analysis gives three possible behaviors of $y(x)$ for large positive x: $y(x) \sim Ax^{1/3}e^{-3\omega(x/2)^{2/3}}$ $(x \to +\infty)$, where ω is one of the cube roots of unity: $1, (-1 \pm i\sqrt{3})/2$. The condition $y(+\infty) = 0$ implies that we must choose $\omega = 1$ to avoid the exponentially growing solutions. Therefore, the solution to (6.1.1) satisfies

$$y(x) \sim Ax^{1/3}e^{-3(x/2)^{2/3}}, \qquad x \to +\infty. \tag{6.1.2}$$

Here A is a constant that cannot be determined by the methods of Chap. 3.

However, this problem is rigged; there is a delightful integral representation for the solution of the boundary-value problem (6.1.1):

$$y(x) = \int_0^\infty \exp\left(-t - \frac{x}{\sqrt{t}}\right) dt. \tag{6.1.3}$$

Notice that this integral satisfies the boundary conditions $y(0) = 1$ (set $x = 0$ and evaluate the integral) and $y(+\infty) = 0$. (Why?) To prove that (6.1.3) satisfies the differential equation when $x > 0$, we differentiate three times under the integral sign and integrate the result by parts once (see Prob. 6.1).

The integral representation (6.1.3) can be used to evaluate the constant A in the asymptotic behavior (6.1.2) of the solution $y(x)$. Laplace's method (see Sec. 6.4 and Prob. 6.31) gives $A = \pi^{1/2}2^{2/3}3^{-1/2}$.

Example 4 *Integral representation for $n!$.* The factorial function $a_n = (n-1)!$ satisfies the first-order difference equation $a_{n+1} = na_n$ $(a_1 = 1)$. The direct local analysis of this difference equation in Sec. 5.4 gives the large-n behavior of a_n (the Stirling formula) apart from an overall multiplicative constant: $a_n \sim Cn^n e^{-n} n^{-1/2}$ $(n \to \infty)$. The constant C is determined by the initial condition $a_1 = 1$ which cannot be used in the analysis of the difference equation at $n = \infty$.

The integral representation (2.2.2) for a_n, $a_n = \int_0^\infty t^{n-1}e^{-t}\,dt$, is equivalent to *both* the difference equation and the initial condition. When n is large, the asymptotic behavior of this integral may be found using Laplace's method (see Example 10 in Sec. 6.4). The result is $C = \sqrt{2\pi}$.

Example 5 *Integral representation for the solution of a difference equation.* The solution of the initial-value problem

$$(n+1)a_{n+1} = 2na_n - na_{n-1}, \qquad a_0 = 1,\ a_1 = 0, \tag{6.1.4}$$

may be expressed as an integral which is equivalent to the difference equation together with the initial conditions:

$$a_n = \frac{1}{n!}\int_0^\infty e^{1-t}t^n J_0(2\sqrt{t})\,dt, \tag{6.1.5}$$

where J_0 is the Bessel function of order zero (see Prob. 6.2).

Local analysis of the difference equation gives two possible behaviors for large n, $a_n \sim A_\pm \exp(\pm 2i\sqrt{n})n^{-1/4}$ $(n \to +\infty)$, where the constants $A_\pm$ cannot be determined. However, an analysis of the integral representation (6.1.5) using Laplace's method (see Prob. 6.32) shows that

$$a_n \sim \sqrt{e/\pi}\, n^{-1/4} \cos(2\sqrt{n} - \tfrac{1}{4}\pi), \qquad n \to +\infty. \tag{6.1.6}$$

The asymptotic expansion of integral representations is an extremely important technique because all of the special (Bessel, Airy, gamma, parabolic cylinder, hypergeometric) functions commonly used in mathematical physics and applied mathematics have integral representations. The asymptotic properties of these special functions are derived from their integral representations. Many of these properties are used in Part IV to obtain the global behavior of general classes of differential equations whose solutions are not expressible as integrals.

Sometimes, an integral representation of the solution to a differential equation is derived by following the systematic procedure of taking an integral (Fourier, Laplace, Hankel) transform of the equation. However, many integral representations are the product of imaginative guesswork. Unfortunately, apart from a small number of equations, one cannot hope to find integral representations for the solutions. This chapter is concerned only with the local analysis of integrals and not with the construction of integral representations. After all, discovering an integral representation is the same as solving the equation in closed form in terms of known functions, and this book is primarily concerned with those equations which *cannot* be solved exactly. Thus, the construction and subsequent expansion of integral representations is not a general method of global analysis. That is why, although the examples in this chapter provide a first glimpse of how to obtain global information about the solutions to differential and difference equations, we include this chapter in Part II, Local Analysis, rather than in Part IV, Global Analysis. This chapter concerns the local analysis of integrals rather than the global analysis of differential and difference equations.

(E) 6.2 ELEMENTARY EXAMPLES

It is sometimes possible to determine the behavior of an integral without using any techniques beyond those introduced in our first exposure to asymptotic analysis in Chap. 3. Consider, for example, the integral

$$I(x) = \int_0^2 \cos\left[(xt^2 + x^2t)^{1/3}\right] dt.$$

It is hard to evaluate this integral in closed form when x is nonzero. However, to determine the leading behavior of $I(x)$ as $x \to 0$, we simply set $x = 0$ in the integral and do the trivial integration. The result is $I(x) \sim 2$ $(x \to 0)$.

More generally, suppose we are asked to find the leading behavior of the integral

$$I(x) = \int_a^b f(x, t)\, dt \qquad \text{as } x \to x_0.$$

If it is given that

$$f(t, x) \sim f_0(t), \qquad x \to x_0,$$

uniformly for $a \le t \le b$ [that is, $\lim_{x \to x_0} f(t, x)/f_0(t) = 1$ uniformly in t], then the leading behavior of $I(x)$ as $x \to x_0$ is just

$$I(x) = \int_a^b f(t, x)\, dt \sim \int_a^b f_0(t)\, dt, \qquad x \to x_0, \tag{6.2.1}$$

provided that the right side of this relation is finite and nonzero. (See Prob. 6.3 for the details of the argument.)

This simple idea may be easily extended to give the full asymptotic expansion of $I(x)$ as $x \to x_0$. If $f(t, x)$ possesses the asymptotic expansion

$$f(t, x) \sim \sum_{n=0}^{\infty} f_n(t)(x - x_0)^{\alpha n}, \qquad x \to x_0,$$

for some $\alpha > 0$, uniformly for $a \le t \le b$, then

$$\int_a^b f(t, x)\, dt \sim \sum_{n=0}^{\infty} (x - x_0)^{\alpha n} \int_a^b f_n(t)\, dt, \qquad x \to x_0, \tag{6.2.2}$$

provided that all the terms on the right are finite (see Prob. 6.3). In the following examples we illustrate the use of formulas (6.2.1) and (6.2.2) and introduce some new twists that extend the applicability of these elementary ideas.

Example 1 $\int_0^1 [(\sin tx)/t]\, dt$ as $x \to 0$. Since the Taylor expansion $(\sin tx)/t = x - x^3t^2/6 + x^5t^4/120 - \cdots$ converges uniformly for $0 \le t \le 1$, $|x| \le 1$, it follows that

$$\int_0^1 \frac{\sin tx}{t}\, dt \sim x - \frac{1}{18}x^3 + \frac{1}{600}x^5 - \cdots, \qquad x \to 0.$$

The series on the right converges for all x.

Example 2 $\int_0^x t^{-1/2}e^{-t}\, dt$ as $x \to 0+$. The expansion

$$t^{-1/2}e^{-t} = t^{-1/2} - t^{1/2} + \tfrac{1}{2}t^{3/2} - \tfrac{1}{6}t^{5/2} + \cdots$$

converges for all $t \ne 0$ but does not converge at $t = 0$. However, this series is asymptotic uniformly for $0 \le t \le x$ as $x \to 0+$. Thus, term-by-term integration gives

$$\int_0^x t^{-1/2}e^{-t}\, dt \sim 2x^{1/2} - \tfrac{2}{3}x^{3/2} + \tfrac{1}{5}x^{5/2} - \tfrac{1}{21}x^{7/2} + \cdots, \qquad x \to 0+. \tag{6.2.3}$$

This result can be rederived by substituting $s = t^{1/2}$ in the integral:

$$\int_0^x t^{-1/2}e^{-t}\, dt = 2\int_0^{\sqrt{x}} e^{-s^2}\, ds.$$

The Taylor expansion of e^{-s^2} does converge uniformly for $0 \le s \le 1$, so

$$\int_0^{\sqrt{x}} e^{-s^2}\, ds = \int_0^{\sqrt{x}} 1 - s^2 + \tfrac{1}{2}s^4 - \tfrac{1}{6}s^6 + \cdots\, ds$$

$$= [s - \tfrac{1}{3}s^3 + \tfrac{1}{10}s^5 - \tfrac{1}{42}s^7 + \cdots]_0^{\sqrt{x}},$$

which reproduces (6.2.3).

Example 3 $\int_x^\infty e^{-t^4}\,dt$ as $x \to 0$. Term-by-term integration of the convergent Taylor series $e^{-t^4} = 1 - t^4 + \frac{1}{2}t^8 - \frac{1}{6}t^{12} + \cdots$ gives a divergent result. The proper way to apply (6.2.2) is to write

$$\int_x^\infty e^{-t^4}\,dt = \int_0^\infty e^{-t^4}\,dt - \int_0^x e^{-t^4}\,dt$$

and then to substitute the Taylor series only in the second term on the right. The result is

$$\int_x^\infty e^{-t^4}\,dt = \Gamma(\tfrac{5}{4}) - x + \tfrac{1}{5}x^5 - \tfrac{1}{18}x^9 + \tfrac{1}{78}x^{13} - \cdots, \tag{6.2.4}$$

where we have used the substitution $s = t^4$ to obtain

$$\int_0^\infty e^{-t^4}\,dt = \tfrac{1}{4}\int_0^\infty s^{-3/4}e^{-s}\,ds = \tfrac{1}{4}\Gamma(\tfrac{1}{4}) = \Gamma(\tfrac{5}{4}).$$

The series (6.2.4) converges for all x, although it is not very useful if $|x|$ is large (see Example 2 of Sec. 6.3).

Example 4 *Incomplete gamma function* $\Gamma(a, x)$ *as* $x \to 0+$. The incomplete gamma function $\Gamma(a, x)$ is defined by $\Gamma(a, x) = \int_x^\infty t^{a-1}e^{-t}\,dt$ $(x > 0)$. To discuss the behavior of $\Gamma(a, x)$ as $x \to 0+$, we distinguish three cases: (*a*) $a > 0$; (*b*) $a < 0$ but nonintegral; and (*c*) $a = 0, -1, -2, -3, \ldots$. Only cases (*b*) and (*c*) present new difficulties.

(*a*) $a > 0$. As in Example 3, we find that

$$\begin{aligned}\Gamma(a, x) &= \int_0^\infty t^{a-1}e^{-t}\,dt - \int_0^x t^{a-1}(1 - t + \tfrac{1}{2}t^2 - \tfrac{1}{6}t^3 + \cdots)\,dt \\ &= \Gamma(a) - \sum_{n=0}^\infty (-1)^n \frac{x^{a+n}}{n!\,(a+n)},\end{aligned} \tag{6.2.5}$$

where the series converges for all x.

(*b*) $a < 0$ but nonintegral. In this case we write

$$\begin{aligned}\Gamma(a, x) &= \int_x^\infty t^{a-1}\left[1 - t + \cdots + (-1)^N \frac{t^N}{N!}\right] dt - \int_0^x t^{a-1}\left[e^{-t} - \sum_{n=0}^N \frac{(-t)^n}{n!}\right] dt \\ &\quad + \int_0^\infty t^{a-1}\left[e^{-t} - \sum_{n=0}^N \frac{(-t)^n}{n!}\right] dt,\end{aligned} \tag{6.2.6}$$

where N is the largest integer less than $-a$. Notice that all three integrals on the right side of (6.2.6) converge. The first two integrals may be performed term by term and the third may be done by repeated integration by parts (see Prob. 6.4):

$$\int_0^\infty t^{a-1}\left[e^{-t} - \sum_{n=0}^N \frac{(-t)^n}{n!}\right] dt = \Gamma(a). \tag{6.2.7}$$

We conclude that the series in (6.2.5) is still valid for this case.

There is a slightly simpler derivation of (6.2.5) for case (*b*) which uses the relation $\partial\Gamma(a, x)/\partial x = -x^{a-1}e^{-x}$. Integrating the series expansion of $x^{a-1}e^{-x}$ gives

$$\Gamma(a, x) \sim C - \sum_{n=0}^N (-1)^n \frac{x^{a-n}}{n!\,(a+n)}, \qquad x \to 0+,$$

for some constant C. The value $C = \Gamma(a)$ is determined by evaluating an integral like that in (6.2.7).

(c) $a = 0, -1, -2, \ldots$. For simplicity, we consider the case $a = 0$:

$$E_1(x) = \int_x^\infty \frac{e^{-t}}{t}\,dt,$$

which is the *exponential integral*. Since

$$dE_1(x)/dx = -e^{-x}/x = -1/x + 1 - \tfrac{1}{2}x + \tfrac{1}{6}x^2 - \cdots,$$

we have

$$E_1(x) \sim C - \ln x + x - \tfrac{1}{4}x^2 + \tfrac{1}{18}x^3 - \cdots, \qquad x \to 0+, \tag{6.2.8}$$

where C is a constant. Note the appearance of the function $\ln x$ in this series.

Next, we compute C. From (6.2.8) we see that $E_1(x) + \ln x \sim C$ $(x \to 0+)$, so

$$\begin{aligned} C &= \lim_{x \to 0+} \left(\int_x^\infty \frac{e^{-t}}{t}\,dt + \ln x \right) \\ &= -\gamma, \end{aligned} \tag{6.2.9}$$

where $\gamma \doteqdot 0.5772$ is Euler's constant [see Prob. 6.5(*a*)].

The expansion of $E_1(x)$ may be obtained more directly by writing

$$E_1(x) = \int_x^\infty \frac{1}{t(t+1)}\,dt + \int_0^\infty \left(e^{-t} - \frac{1}{t+1}\right)\frac{1}{t}\,dt - \int_0^x \left(e^{-t} - \frac{1}{t+1}\right)\frac{1}{t}\,dt. \tag{6.2.10}$$

The first integral on the right equals $-\ln x + \ln(1 + x) = -\ln x + x - \tfrac{1}{2}x^2 + \tfrac{1}{3}x^3 - \cdots$, the second integral equals $-\gamma$ [see Prob. 6.5(*b*)], and Taylor expanding the third integrand gives

$$\int_0^x \left(e^{-t} - \frac{1}{t+1}\right)\frac{1}{t}\,dt = -\frac{1}{4}x^2 + \frac{5}{18}x^3 - \cdots.$$

Combining these results recovers (6.2.8) with $C = -\gamma$. More generally, if $a = -N$ $(N = 0, 1, 2, \ldots)$, then a similar analysis (see Prob. 6.6) gives

$$\Gamma(-N, x) \sim C_N + \frac{(-1)^{N+1}}{N!}\ln x - \sum_{\substack{n=0 \\ n \neq N}}^{\infty} (-1)^n \frac{x^{n-N}}{n!\,(n-N)}, \qquad x \to 0+, \tag{6.2.11}$$

where

$$C_N = \frac{(-1)^{N+1}}{N!}\left(\gamma - \sum_{n=1}^{N} \frac{1}{n}\right), \qquad N > 0, \tag{6.2.12}$$

and $C_0 = -\gamma$. This series contains the function $\ln x$, a new feature not found in the series (6.2.5). The appearance of a logarithmic term when $a = 0, -1, -2, \ldots$ is reminiscent of the special cases of the Frobenius method for differential equations in which logarithms also appear.

(E) 6.3 INTEGRATION BY PARTS

Integration by parts is a particularly easy procedure for developing asymptotic approximations to many kinds of integrals. We explain the technique by applying it to a variety of problems.

Example 1 *Derivation of an asymptotic power series.* If $f(x)$ is differentiable near $x = 0$, then the local behavior of $f(x)$ near 0 may be studied using integration by parts. We merely represent $f(x)$

as the integral $f(x) = f(0) + \int_0^x f'(t)\,dt$. Integrating once by parts gives

$$f(x) = f(0) + (t - x)f'(t)\Big|_0^x + \int_0^x (x - t)f''(t)\,dt$$

$$= f(0) + xf'(0) + \int_0^x (x - t)f''(t)\,dt.$$

Repeating this process $(N - 1)$ times gives

$$f(x) = \sum_{n=0}^{N} \frac{x^n}{n!} f^{(n)}(0) + \frac{1}{N!}\int_0^x (x - t)^N f^{(N+1)}(t)\,dt.$$

If the remainder term (the integral on the right) exists for all N and sufficiently small positive x, then

$$f(x) \sim \sum_{n=0}^{\infty} \frac{x^n}{n!} f^{(n)}(0), \qquad x \to 0+.$$

Why? Moreover, if this series converges, then it is just the Taylor expansion of $f(x)$ about $x = 0$.

Example 2 *Behavior of* $\int_x^\infty e^{-t^4}\,dt$ *as* $x \to +\infty$. We have already found the behavior of the integral

$$I(x) = \int_x^\infty e^{-t^4}\,dt \tag{6.3.1}$$

for small values of the parameter x. In Example 3 of Sec. 6.2 we showed that

$$I(x) = \int_0^\infty e^{-t^4}\,dt - \int_0^x (1 - t^4 + \tfrac{1}{2}t^8 - \tfrac{1}{6}t^{12} + \cdots)\,dt$$

$$= \Gamma(\tfrac{5}{4}) - x + \tfrac{1}{5}x^5 - \tfrac{1}{18}x^9 + \tfrac{1}{78}x^{13} - \cdots. \tag{6.3.2}$$

Although the series (6.3.2) converges for all x, it is not very useful if x is large (see Fig. 6.1).

In order to study $I(x)$ for large x we must develop an asymptotic series for $I(x)$ in inverse powers of x; to wit, we rewrite (6.3.1) as

$$I(x) = -\frac{1}{4}\int_x^\infty \frac{1}{t^3}\frac{d}{dt}(e^{-t^4})\,dt$$

and integrate by parts:

$$I(x) = -\frac{1}{4t^3}e^{-t^4}\Big|_x^\infty - \frac{3}{4}\int_x^\infty \frac{1}{t^4}e^{-t^4}\,dt$$

$$= \frac{1}{4x^3}e^{-x^4} - \frac{3}{4}\int_x^\infty \frac{1}{t^4}e^{-t^4}\,dt. \tag{6.3.3}$$

But
$$\int_x^\infty \frac{1}{t^4}e^{-t^4}\,dt < \frac{1}{x^4}\int_x^\infty e^{-t^4}\,dt = \frac{1}{x^4}I(x) \ll I(x), \qquad x \to +\infty,$$

so the leading behavior of $I(x)$ is

$$I(x) \sim \frac{1}{4x^3}e^{-x^4}, \qquad x \to +\infty. \tag{6.3.4}$$

Repeated integration by parts gives the full asymptotic expansion of $I(x)$. To systematize the argument we define the integrals

$$I_n(x) = \int_x^\infty \frac{1}{t^{4n}}e^{-t^4}\,dt. \tag{6.3.5}$$

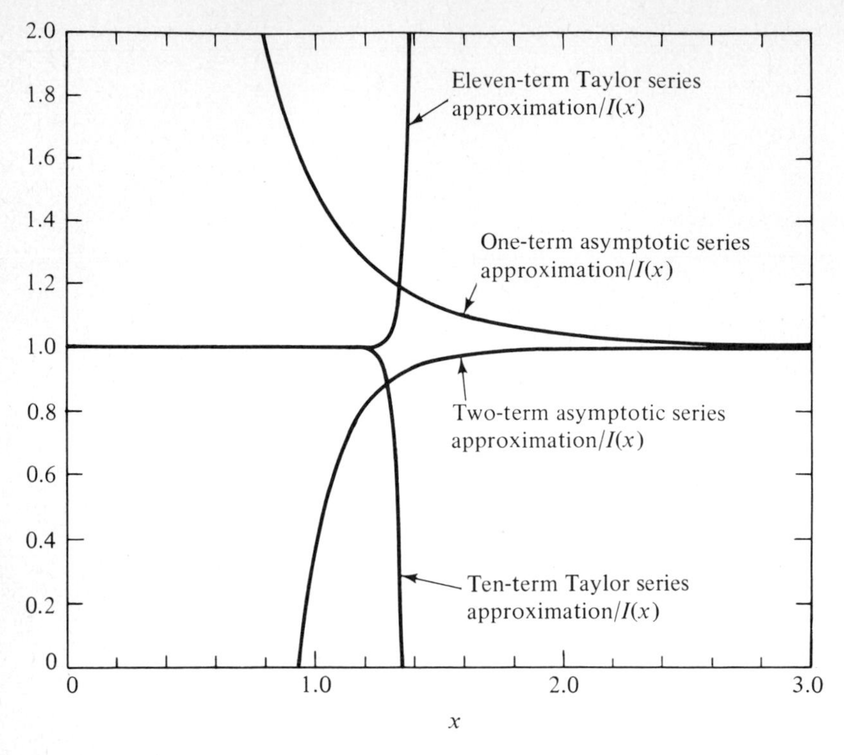

Figure 6.1 A comparison of small-x and large-x approximations to $I(x) = \int_x^\infty \exp(-t^4)\,dt$. The small-$x$ approximation to $I(x)$ is the Taylor series in (6.3.2) which converges for all x. The large-x approximation to $I(x)$ is the asymptotic series in (6.3.8) which is valid as $x \to +\infty$. All approximations to $I(x)$ are normalized by dividing by $I(x)$. It is clear that the truncated asymptotic series is useful in a much larger region than the truncated Taylor series.

Now, (6.3.3) becomes $I(x) = e^{-x^4}/4x^3 - \frac{3}{4}I_1(x)$. Integrating

$$I_n(x) = -\frac{1}{4}\int_x^\infty \frac{1}{t^{4n+3}}\frac{d}{dt}e^{-t^4}\,dt$$

by parts gives

$$\begin{aligned} I_n(x) &= -\frac{1}{t^{4n+3}}e^{-t^4}\Big|_x^\infty - \left(n+\frac{3}{4}\right)\int_x^\infty \frac{1}{t^{4n+4}}e^{-t^4}\,dt \\ &= \frac{1}{4x^{4n+3}}e^{-x^4} - \left(n+\frac{3}{4}\right)I_{n+1}(x). \end{aligned} \tag{6.3.6}$$

Therefore,

$$\begin{aligned} I(x) = \frac{1}{4x^3}e^{-x^4}&\left[1 - \frac{3}{4x^4} + \frac{(3)(7)}{(4x^4)^2} - \frac{(3)(7)(11)}{(4x^4)^3} + \cdots + (-1)^{n-1}\frac{(3)(7)(11)\cdots(4n-5)}{(4x^4)^{n-1}}\right] \\ &+ (-1)^n\frac{(3)(7)(11)\cdots(4n-1)}{4^n}I_n(x), \end{aligned} \tag{6.3.7}$$

which is an identity valid for all $x > 0$.

The integrand of (6.3.5) is positive, so $I_{n+1}(x) > 0$ for $x > 0$. Thus, (6.3.6) implies that

$$|I_n(x)| < \frac{1}{4x^{4n+3}} e^{-x^4} \ll \frac{1}{x^{4n-1}} e^{-x^4}, \qquad x \to +\infty,$$

so the term proportional to $I_n(x)$ in (6.3.7) is asymptotically much smaller than the last retained term in square brackets in (6.3.7). Therefore, the full asymptotic expansion of $I(x)$ is

$$I(x) \sim \frac{1}{4x^3} e^{-x^4} \left[1 + \sum_{n=1}^{\infty} (-1)^n \frac{(3)(7)(11)\cdots(4n-1)}{(4x^4)^n}\right], \qquad x \to +\infty. \tag{6.3.8}$$

The accuracy of this asymptotic approximation and its advantage over the convergent series in (6.3.2) when $x > 1$ is demonstrated in Fig. 6.1.

Example 3 *Behavior of* $\int_0^x t^{-1/2} e^{-t}\, dt$ *as* $x \to +\infty$. We may use integration by parts to find the behavior of $I(x) = \int_0^x t^{-1/2} e^{-t}\, dt$ for large x, but we must be careful. Immediate integration by parts gives an indeterminate expression which is the difference of two infinite terms:

$$I(x) = \int_0^x t^{-1/2} e^{-t}\, dt = -t^{-1/2} e^{-t}\Big|_0^x - \frac{1}{2}\int_0^x t^{-3/2} e^{-t}\, dt. \tag{6.3.9}$$

It is best to express $I(x)$ as the difference of two integrals:

$$\int_0^x t^{-1/2} e^{-t}\, dt = \int_0^\infty t^{-1/2} e^{-t}\, dt - \int_x^\infty t^{-1/2} e^{-t}\, dt.$$

The first integral on the right is finite and has the value $\Gamma(\frac{1}{2}) = \sqrt{\pi}$; the second may be integrated by parts successfully because the contribution from the endpoint at ∞ vanishes:

$$\int_0^x t^{-1/2} e^{-t}\, dt = \sqrt{\pi} + \int_x^\infty t^{-1/2} \frac{d}{dt}(e^{-t})\, dt$$

$$= \sqrt{\pi} - \frac{e^{-x}}{\sqrt{x}} + \frac{1}{2}\int_x^\infty t^{-3/2} e^{-t}\, dt.$$

Repeated use of integration by parts gives the full asymptotic expansion of $I(x)$ (see Prob. 6.11):

$$\int_0^x t^{-1/2} e^{-t}\, dt \sim \sqrt{\pi} - \frac{e^{-x}}{\sqrt{x}}\left[1 + \sum_{n=1}^{\infty} (-1)^n \frac{(1)(3)(5)\cdots(2n-1)}{(2x)^n}\right], \qquad x \to +\infty. \tag{6.3.10}$$

The general rule to be learned from this example is that integration by parts will not work if the contribution from one of the limits of integration is much larger than the size of the integral. In this example the integral $I(x)$ is finite for all $x > 0$, but the endpoint $t = 0$ contributes a spurious infinity in (6.3.9).

Example 4 *Behavior of* $\int_0^x e^{t^2}\, dt$ *as* $x \to +\infty$. Here, it is wrong to write $\int_0^x e^{t^2}\, dt = \int_0^\infty e^{t^2}\, dt - \int_x^\infty e^{t^2}\, dt$ because the right side has the form $\infty - \infty$. But it is also wrong to integrate directly by parts:

$$\int_0^x e^{t^2}\, dt = \frac{1}{2}\int_0^x \frac{1}{t}\frac{d}{dt}(e^{t^2})\, dt$$

$$= \frac{1}{2t} e^{t^2}\Big|_0^x + \frac{1}{2}\int_0^x t^{-2} e^{t^2}\, dt,$$

which also has the form $\infty - \infty$. To obtain a correct asymptotic expansion of this integral about $x = \infty$, we introduce a cutoff parameter a and write the integrals

$$\int_0^x e^{t^2}\, dt = \int_0^a e^{t^2}\, dt + \int_a^x e^{t^2}\, dt \tag{6.3.11}$$

for some fixed $0 < a < x$. We will see that for fixed a, the full asymptotic expansion of (6.3.11) is independent of the first integral on the right and is also independent of the cutoff a!

We begin our analysis by expanding the second integral on the right side of (6.3.11):

$$\int_a^x e^{t^2}\,dt = \frac{1}{2t}e^{t^2}\bigg|_a^x + \frac{1}{2}\int_a^x \frac{1}{t^2}e^{t^2}\,dt$$

$$= \frac{1}{2x}e^{x^2} - \frac{1}{2a}e^{a^2} + \frac{1}{2}\int_a^x \frac{1}{t^2}e^{t^2}\,dt. \tag{6.3.12}$$

Note that $\frac{1}{2}\int_a^x e^{t^2}\,dt/t^2 \ll e^{x^2}/2x$ $(x \to +\infty)$ because by l'Hôpital's rule

$$\lim_{x\to+\infty} \frac{\frac{1}{2}\int_a^x t^{-2}e^{t^2}\,dt}{\frac{1}{2}x^{-1}e^{x^2}} = \lim_{x\to+\infty} \frac{x^{-2}e^{x^2}}{(2 - x^{-2})e^{x^2}} = 0.$$

Also,

$$\int_0^a e^{t^2}\,dt \ll \int_0^x e^{t^2}\,dt, \qquad x \to +\infty,$$

$$\frac{1}{2a}e^{a^2} \ll \frac{1}{2x}e^{x^2}, \qquad x \to +\infty.$$

Hence, from (6.3.11) and (6.3.12) we have

$$\int_0^x e^{t^2}\,dt \sim \frac{1}{2x}e^{x^2}, \qquad x \to +\infty, \tag{6.3.13}$$

which is the leading asymptotic behavior of the integral. Observe that integration by parts works because the endpoint contribution in (6.3.12) from $t = a$ is negligible compared with that from $t = x$ when $0 < a < x$.

Repeated integration by parts in (6.3.12) establishes the full asymptotic expansion of $\int_0^x e^{t^2}\,dt$ as $x \to +\infty$ (see Prob. 6.11):

$$\int_0^x e^{t^2}\,dt \sim \frac{1}{2x}e^{x^2}\left[1 + \sum_{n=1}^{\infty} \frac{(1)(3)(5)\cdots(2n-1)}{(2x^2)^n}\right], \qquad x \to +\infty. \tag{6.3.14}$$

Notice that the arbitrary cutoff a does not appear in this asymptotic expansion. In fact, the endpoint contributions from $t = a$ are exponentially smaller than those from $t = x$ after any number of integrations by parts.

Example 5 *Behavior of integrals of Airy functions.* The asymptotic expansion of the integral $\int_0^x \text{Ai}(t)\,dt$ as $x \to +\infty$ is obtained by following the procedure of Example 3. First, we write $\int_0^x \text{Ai}(t)\,dt = \frac{1}{3} - \int_x^\infty \text{Ai}(t)\,dt$. [Here we have used the property of Ai (t) that $\int_0^\infty \text{Ai}(t)\,dt = \frac{1}{3}$; see Prob. 10.20.] This decomposition avoids endpoint contributions (from $t = 0$) that are larger than the integral itself. Next, we integrate by parts using $\text{Ai}''(t) = t\,\text{Ai}(t)$, the differential equation satisfied by the Airy function:

$$\int_x^\infty \text{Ai}(t)\,dt = \int_x^\infty \frac{1}{t}\text{Ai}''(t)\,dt$$

$$= \frac{1}{t}\text{Ai}'(t)\bigg|_x^\infty + \int_x^\infty \frac{1}{t^2}\text{Ai}'(t)\,dt.$$

But

$$\text{Ai}'(x) \sim -\frac{1}{2\sqrt{\pi}}x^{1/4}e^{-2x^{3/2}/3}, \qquad x \to +\infty,$$

and
$$\left|\int_x^\infty \frac{1}{t^2} \text{Ai}'(t)\,dt\right| < \frac{1}{x^2}\left|\int_x^\infty \text{Ai}'(t)\,dt\right| = \frac{1}{x^2}\text{Ai}(x)$$
$$\sim \frac{1}{2\sqrt{\pi}} x^{-9/4} e^{-2x^{3/2}/3}, \qquad x \to +\infty.$$

Thus,
$$\int_x^\infty \text{Ai}(t)\,dt \sim \frac{1}{2\sqrt{\pi}} x^{-3/4} e^{-2x^{3/2}/3}, \qquad x \to +\infty. \tag{6.3.15}$$

The full asymptotic expansion of $\int_x^\infty \text{Ai}(t)\,dt$ as $x \to +\infty$ is most easily obtained by using the differential equation $y''' = xy'$ that it satisfies to find suitable recursion relations (see Prob. 6.12).

Another Airy function integral is $\int_0^x \text{Bi}(t)\,dt$. The asymptotic behavior of this integral as $x \to +\infty$ may be found by following the procedure in Example 4. First, we write $\int_0^x \text{Bi}(t)\,dt = \int_0^a \text{Bi}(t)\,dt + \int_a^x \text{Bi}(t)\,dt$, where $0 < a < x$, in order to avoid infinite endpoint contributions. Then, integrating by parts in the second integral on the right using $\text{Bi}''(t) = t\,\text{Bi}(t)$ gives

$$\int_a^x \text{Bi}(t)\,dt = \int_a^x \frac{1}{t}\text{Bi}''(t)\,dt$$
$$= \frac{1}{x}\text{Bi}'(x) - \frac{1}{a}\text{Bi}'(a) + \int_a^x \frac{1}{t^2}\text{Bi}'(t)\,dt.$$

Next, we note a number of asymptotic relations:

$$\frac{1}{a}\text{Bi}'(a) \ll \frac{1}{x}\text{Bi}'(x), \qquad x \to +\infty,$$
$$\int_0^a \text{Bi}(t)\,dt \ll \int_0^x \text{Bi}(t)\,dt, \qquad x \to +\infty,$$
$$\int_a^x \frac{1}{t^2}\text{Bi}'(t)\,dt \ll \frac{1}{x}\text{Bi}'(x), \qquad x \to +\infty,$$

this last asymptotic relation following from l'Hôpital's rule, and

$$\text{Bi}(x) \sim \frac{1}{\sqrt{\pi}} x^{-1/4} e^{2x^{3/2}/3}, \qquad x \to +\infty,$$
$$\text{Bi}'(x) \sim \frac{1}{\sqrt{\pi}} x^{1/4} e^{2x^{3/2}/3}, \qquad x \to +\infty.$$

From this heap of asymptotic inequalities we deduce that

$$\int_0^x \text{Bi}(t)\,dt \sim \frac{1}{\sqrt{\pi}} x^{-3/4} e^{2x^{3/2}/3}, \qquad x \to +\infty \tag{6.3.1}$$

(see Prob. 6.13).

Since integration of one-signed asymptotic relations is permissible (see Prob. 3.28), (6.3.1 could have been obtained by integrating the asymptotic behavior of Bi (x):

$$\int_0^x \text{Bi}(t)\,dt \sim \frac{1}{\sqrt{\pi}} \int_0^x t^{-1/4} e^{2t^{3/2}/3}\,dt, \qquad x \to +\infty.$$

Integration by parts gives, for any $a > 0$,

$$\int_a^x t^{-1/4} e^{2t^{3/2}/3}\, dt = \int_a^x t^{-3/4} \frac{d}{dt} e^{2t^{3/2}/3}\, dt$$

$$= t^{-3/4} e^{2t^{3/2}/3} \Big|_a^x + \frac{3}{4} \int_a^x t^{-7/4} e^{2t^{3/2}/3}\, dt,$$

and (6.3.16) is easily recovered.

Integration by Parts for Laplace Integrals

Until now we have only considered integrals where the parameter x appears as a limit of integration. A *Laplace* integral has the form

$$I(x) = \int_a^b f(t) e^{x\phi(t)}\, dt \tag{6.3.17}$$

in which x appears as part of the integrand.

To obtain the asymptotic behavior of $I(x)$ as $x \to +\infty$, we try integrating by parts:

$$I(x) = \frac{1}{x} \int_a^b \frac{f(t)}{\phi'(t)} \frac{d}{dt} [e^{x\phi(t)}]\, dt$$

$$= \frac{1}{x} \frac{f(t)}{\phi'(t)} e^{x\phi(t)} \Big|_a^b - \frac{1}{x} \int_a^b \frac{d}{dt} \left[\frac{f(t)}{\phi'(t)}\right] e^{x\phi(t)}\, dt. \tag{6.3.18}$$

(We assume that the new integral on the right exists, of course.) The formula in (6.3.18) is useful if the integral on the right side is asymptotically smaller than the boundary terms as $x \to \infty$. If this is true, then the boundary terms in (6.3.18) are asymptotic to $I(x)$:

$$I(x) \sim \frac{1}{x} \frac{f(b)}{\phi'(b)} e^{x\phi(b)} - \frac{1}{x} \frac{f(a)}{\phi'(a)} e^{x\phi(a)}, \qquad x \to +\infty. \tag{6.3.19}$$

In general, (6.3.19) is a correct asymptotic relation if $\phi(t)$, $\phi'(t)$, and $f(t)$ are continuous (possibly complex) functions and one of the following three conditions is satisfied:

1. $\phi'(t) \neq 0$ $(a \le t \le b)$ and at least one of $f(a)$ and $f(b)$ are not zero. These conditions are sufficient to ensure that the remainder integral on the right side of (6.3.18) exists. Once we know that this integral exists we can prove (see Prob. 6.15) that it becomes negligible compared with the boundary term in (6.3.18) as $x \to \infty$ and therefore that (6.3.19) is valid.
2. Re $\phi(t) <$ Re $\phi(b)$ $(a \le t < b)$, Re $\phi'(b) \neq 0$, and $f(b) \neq 0$. These conditions are insufficient to imply that the integral on the right side of (6.3.18) exists. Nevertheless, they *are* strong enough to ensure that

$$I(x) \sim \frac{1}{x} \frac{f(b)}{\phi'(b)} e^{x\phi(b)}, \qquad x \to +\infty. \tag{6.3.20}$$

This result is explained using Laplace's method in Sec. 6.4 [see (6.4.19*b*)].

3. Re $\phi(t) <$ Re $\phi(a)$ $(a \le t \le b)$, Re $\phi'(a) \ne 0$, and $f(a) \ne 0$. As in condition 2, these conditions are again insufficient to imply that the integral on the right side of (6.3.18) exists, but they are strong enough to ensure that

$$I(x) \sim -\frac{1}{x}\frac{f(a)}{\phi'(a)} e^{x\phi(a)}, \qquad x \to +\infty \tag{6.3.21}$$

[see (6.4.19*a*)].

Example 6 *Leading behavior of simple Laplace integrals.* Using the formulas in (6.3.19) to (6.3.21) we have

$$(a) \int_1^2 e^{x \cosh t}\, dt \sim \frac{e^{x \cosh 2}}{x \sinh 2}, \qquad x \to +\infty;$$

$$(b) \int_{-1}^3 e^{x \cosh^2 t} \sim \frac{e^{x \cosh^2 3}}{2x \sinh 3 \cosh 3}, \qquad x \to +\infty.$$

If the integral on the right side of (6.3.18) meets one of the three conditions stated above, we may continue integrating by parts. Apparently, each integration by parts introduces a new factor of $1/x$; for example, if Re $\phi(b) >$ Re $\phi(a)$ the full asymptotic expansion of $I(x)$ in (6.3.17) has the form

$$I(x) \sim e^{x\phi(b)} \sum_{n=1}^{\infty} A_n x^{-n}, \qquad x \to +\infty. \tag{6.3.22}$$

Failure of Integration by Parts

The method of integration by parts is rather inflexible; it can only produce asymptotic series of the form in (6.3.22) which contain integral powers of $1/x$. However, Laplace integrals like $I(x)$ in (6.3.17) can have large-x asymptotic expansions which contain fractional powers of x. It is clear, therefore, that the method of integration by parts is inadequate to find the asymptotic expansion of all such integrals. If we have no prior knowledge of the correct expansion of an integral, how then do we know whether or not integration by parts will work? Generally, the symptoms that integration by parts is breaking down are easy to detect and interpret: when integration by parts produces an integral which does not exist it is not working. We know that integration by parts is about to fail when $\phi'(t)$ has a zero somewhere in $a \le t \le b$. Here are some examples.

Example 7 *Failure of integration by parts for* $\int_0^\infty e^{-xt^2}\, dt$. The integral $\int_0^\infty e^{-xt^2}\, dt$ has the exact value $\frac{1}{2}\sqrt{\pi/x}$. Since its asymptotic behavior as $x \to +\infty$ is not an asymptotic series of powers of $1/x$ like that in (6.3.22), we expect integration by parts to fail. Comparing this integral with the general form in (6.3.17) shows that $\phi(t) = -t^2$. Since $\phi'(t) = -2t$ vanishes at $t = 0$, integration by parts gives a nonexistent integral,

$$\int_0^\infty e^{-xt^2}\, dt = \int_0^\infty \left(\frac{1}{-2xt}\right)(-2xte^{-xt^2})\, dt = \frac{e^{-xt^2}}{-2xt}\bigg|_0^\infty - \int_0^\infty \frac{1}{2xt^2} e^{-xt^2}\, dt,$$

a sure sign that integration by parts is not applicable.

Example 8 *Failure of integration by parts for* $\int_0^\infty e^{-x\sinh^2 t}\,dt$. For this integral $\phi(t) = -\sinh^2 t$. Since $\phi'(t)$ vanishes at $t = 0$, we do not expect that integration by parts will be useful for finding the large-x behavior of this integral.

In Example 6(*b*) of Sec. 6.4 we will use Laplace's method to show that the leading behavior of this integral as $x \to +\infty$ is $\frac{1}{2}\sqrt{\pi/x}$.

Example 9 *Leading behavior of* $\int_0^\infty \ln(1+t)\, e^{-x\sinh^2 t}\,dt$. For some integrals, integration by parts yields several terms in the expansion of $I(x)$ and then breaks down. The integral $I(x) = \int_0^\infty \ln(1+t)\, e^{-x\sinh^2 t}\,dt$ has this property. For this integral $f(t) = \ln(1+t)$ and $\phi(t) = \sinh^2 t$. Thus, although $\phi'(t) = \sinh(2t)$ vanishes at $t = 0$, $f(t)$ also vanishes there. As a result, it is correct to integrate by parts once:

$$I(x) = -\frac{1}{x}\int_0^\infty \frac{\ln(1+t)}{\sinh(2t)}\frac{d}{dt}\left(e^{-x\sinh^2 t}\right)dt$$

$$= -\frac{1}{x}\frac{\ln(1+t)}{\sinh(2t)}e^{-x\sinh^2 t}\Bigg|_0^\infty + \frac{1}{x}\int_0^\infty \left[\frac{d}{dt}\frac{\ln(1+t)}{\sinh(2t)}\right]e^{-x\sinh^2 t}\,dt. \tag{6.3.23}$$

Using Laplace's method, one can show that the last integral on the right vanishes like $x^{-3/2}$ as $x \to +\infty$ (see Prob. 6.33). Hence, $I(x) \sim 1/2x$ $(x \to +\infty)$. Thus, integration by parts gives the leading behavior of $I(x)$ correctly. However, integration by parts cannot be used to find the next term in the asymptotic expansion of $I(x)$ for large x because the next term is proportional to $x^{-3/2}$, which is not an integer power of $1/x$.

Example 10 *Stieltjes integral.* Integration by parts is useful for finding the behavior of the Stieltjes integral (see Sec. 3.8):

$$I(x) = \int_0^\infty \frac{e^{-t}}{1+xt}\,dt$$

for small positive x but not for large x. (The Stieltjes integral is not a Laplace integral.)

To derive the small-x behavior of $I(x)$, we organize the integration by parts so that one new factor of x is introduced at each stage:

$$I(x) = -\int_0^\infty \frac{1}{1+xt}\frac{d}{dt}e^{-t}\,dt$$

$$= -\frac{1}{1+xt}e^{-t}\Bigg|_0^\infty - x\int_0^\infty \frac{1}{(1+xt)^2}e^{-t}\,dt$$

$$= 1 + x\frac{1}{(1+xt)^2}e^{-t}\Bigg|_0^\infty + 2x^2\int_0^\infty \frac{1}{(1+xt)^3}e^{-t}\,dt$$

$$\vdots$$

$$= 1 - x + 2!\,x^2 - \cdots + (-1)^{n-1}(n-1)!\,x^{n-1}$$

$$+ (-1)^n n!\int_0^\infty \frac{1}{(1+xt)^{n+1}}e^{-t}\,dt.$$

Since this procedure works for all n, we have successfully derived the full asymptotic behavior of $I(x)$ for small x: $I(x) \sim \sum_{n=0}^\infty (-1)^n n!\,x^n$ $(x \to 0+)$.

Now let us see how integration by parts fails to give the large-x behavior of $I(x)$. If integration by parts did work it would have to introduce an additional factor of $1/x$ at each stage. Thus, our best hope is to write $I(x) = (1/x)\int_0^\infty e^{-t}(d/dt)\ln(1+xt)\,dt$ and to use integration by parts to obtain $I(x) = (1/x) \int_0^\infty e^{-t} \ln(1 + xt)\, dt$. Integrating by parts once again gives

$$I(x) = \frac{1}{x^2} \int_0^\infty e^{-t} \frac{d}{dt}[(1 + xt) \ln(1 + xt) - (1 + xt)]$$

$$= \frac{1}{x^2} + \frac{1}{x^2} \int_0^\infty e^{-t}[(1 + xt) \ln(1 + xt) - (1 + xt)]\, dt.$$

Have we shown that the leading behavior of $I(x)$ is x^{-2} $(x \to +\infty)$? No, because the last term on the right is not small compared to x^{-2}. In fact

$$\lim_{x \to +\infty} \int_0^\infty e^{-t}[(1 + xt) \ln(1 + xt) - (1 + xt)]\, dt = \infty \tag{6.3.24}$$

(see Prob. 6.16).

To understand why integration by parts has failed we need only recall [see Example 3 of Sec. 3.8 and Prob. 3.39(i)] that $I(x) \sim (\ln x)/x$ $(x \to +\infty)$; the large-x behavior of $I(x)$ is not a power series in $1/x$.

(E) 6.4 LAPLACE'S METHOD AND WATSON'S LEMMA

Laplace's method is a very general technique for obtaining the asymptotic behavior as $x \to +\infty$ of integrals in which the large parameter x appears in an exponential:

$$I(x) = \int_a^b f(t) e^{x\phi(t)}\, dt. \tag{6.4.1}$$

Here, we assume that $f(t)$ and $\phi(t)$ are real continuous functions. Integrals of this form are called Laplace integrals and were introduced in Sec. 6.3.

Laplace's method rests on an important idea involved in many standard techniques of asymptotic analysis of integrals, such as the methods of stationary phase and steepest descents which are discussed in Secs. 6.5 and 6.6. The idea is this: if the real continuous function $\phi(t)$ has its maximum on the interval $a \le t \le b$ at $t = c$ and if $f(c) \ne 0$, then it is only the immediate neighborhood of $t = c$ that contributes to the full asymptotic expansion of $I(x)$ for large x. That is, we may approximate the integral $I(x)$ by $I(x; \varepsilon)$, where

$$I(x; \varepsilon) = \int_{c-\varepsilon}^{c+\varepsilon} f(t) e^{x\phi(t)}\, dt \tag{6.4.2a}$$

if $a < c < b$,

$$I(x; \varepsilon) = \int_a^{a+\varepsilon} f(t) e^{x\phi(t)}\, dt \tag{6.4.2b}$$

if the maximum of $\phi(t)$ is at $t = a$, and

$$I(x; \varepsilon) = \int_{b-\varepsilon}^{b} f(t) e^{x\phi(t)}\, dt \tag{6.4.2c}$$

if the maximum of $\phi(t)$ is at $t = b$. Here ε may be chosen to be an arbitrary positive number (such that the restricted integration range $c - \varepsilon \le t \le c + \varepsilon$ is a subinterval of $a \le t \le b$). It is crucial that the full asymptotic expansion of $I(x; \varepsilon)$ as $x \to +\infty$ (1) does not depend on ε and (2) is identical to the full asymptotic expansion of $I(x)$ as $x \to +\infty$. Both of these rather surprising results are true because (we assume here that $a < c < b$) $|\int_a^{c-\varepsilon} f(t)e^{x\phi(t)}\, dt| + |\int_{c+\varepsilon}^{b} f(t)e^{x\phi(t)}\, dt|$ is subdominant (exponentially small) with respect to $I(x)$ as $x \to +\infty$. This is so because for all t on the intervals $a \le t \le c - \varepsilon$ and $c + \varepsilon \le t \le b$, $e^{x\phi(t)}$ is exponentially smaller than $e^{x\phi(c)}$ as $x \to +\infty$. To show that $I(x) - I(x; \varepsilon)$ is subdominant as $x \to +\infty$, we use integration by parts (see Prob. 6.23). The result that $I(x) - I(x; \varepsilon)$ is subdominant as $x \to +\infty$ sometimes holds even if $f(c) = 0$; we discuss this point later.

It is helpful to approximate $I(x)$ by $I(x; \varepsilon)$ because $\varepsilon > 0$ may be chosen so small that it is valid to replace $f(t)$ and $\phi(t)$ by their Taylor or asymptotic series expansions about $t = c$.

Example 1 *Leading behavior of* $\int_0^{10} (1 + t)^{-1}e^{-xt}\, dt$ *as* $x \to +\infty$. We use Laplace's method to approximate this integral. Here $\phi(t) = -t$ has a maximum in the integration region $0 \le t \le 10$ at $t = 0$. Therefore, we may replace the integral by

$$I(x; \varepsilon) = \int_0^{\varepsilon} (1 + t)^{-1}e^{-xt}\, dt$$

for any $\varepsilon > 0$ at the cost of introducing errors which are exponentially small as $x \to +\infty$. Next we choose ε so small that we can replace $(1 + t)^{-1}$ by 1, the first term in its Taylor series about $t = 0$. This replacement makes the integral easy to evaluate. Thus,

$$\int_0^{10} (1 + t)^{-1}e^{-xt}\, dt \sim \int_0^{\varepsilon} e^{-xt}\, dt = (1 - e^{-\varepsilon x})/x, \qquad x \to +\infty.$$

Since $e^{-\varepsilon x} \ll 1$ as $x \to +\infty$ for any $\varepsilon > 0$, we obtain

$$\int_0^{10} (1 + t)^{-1}e^{-xt}\, dt \sim \frac{1}{x}, \qquad x \to +\infty. \tag{6.4.3}$$

Note that the final result in (6.4.3) does not depend on the arbitrary parameter ε; ε appears in a subdominant term only.

Example 1 (revisited) *Full asymptotic expansion of* $\int_0^{10} (1 + t)^{-1}e^{-xt}\, dt$ *as* $x \to +\infty$. Laplace's method gives the full asymptotic series expansion of this integral. As in Example 1, we can replace 10, the upper limit of integration, by $\varepsilon < 1$ and introduce only exponentially small errors as $x \to +\infty$. Instead of replacing $(1 + t)^{-1}$ by 1, as we did in Example 1, we use the full Taylor expansion $(1 + t)^{-1} = \sum (-t)^n$, which converges for $|t| < 1$:

$$I(x; \varepsilon) = \int_0^{\varepsilon} (1 + t)^{-1}e^{-xt}\, dt = \sum_{n=0}^{\infty} \int_0^{\varepsilon} (-t)^n e^{-xt}\, dt.$$

The easiest way to evaluate the integral $\int_0^{\varepsilon} (-t)^n e^{-xt}\, dt$ for any n is to replace it by $\int_0^{\infty} (-t)^n e^{-xt}\, dt$. It may seem surprising that it is valid to replace the small parameter ε by ∞! However, this replacement introduces only exponentially small errors as $x \to +\infty$ because the integral from ε to ∞ is subdominant with respect to $\int_0^{\infty} (-t)^n e^{-xt}\, dt = (-1)^n n!\, x^{-n-1}$ as

$x \to +\infty$. We verify this using integration by parts; integration by parts gives

$$\int_{\varepsilon}^{\infty} (-t)^n e^{-xt}\, dt \sim (-\varepsilon)^n e^{-\varepsilon x}/x, \qquad x \to +\infty,$$

which is indeed exponentially smaller than $\int_0^\infty (-t)^n e^{-xt}\, dt$ as $x \to +\infty$. Assembling these results, we obtain the full asymptotic expansion of $\int_0^{10} (1+t)^{-1} e^{-xt}\, dt$ as $x \to +\infty$:

$$\int_0^{10} (1+t)^{-1} e^{-xt}\, dt \sim \sum_{n=0}^{\infty} (-1)^n n!\, x^{-n-1}, \qquad x \to +\infty. \tag{6.4.4}$$

Let us pause a moment to review the procedure we have just used. There are three steps involved in Laplace's method applied to an integral $I(x)$. First, we approximate $I(x)$ by $I(x; \varepsilon)$ by restricting the original integration region to a narrow region surrounding the maximum of $\phi(t)$. Second, we expand the functions $f(t)$ and $\phi(t)$ in series which are valid near the location of the maximum of $\phi(t)$. This allows us to expand $I(x; \varepsilon)$ into a series of integrals. Finally, the most convenient way to evaluate the integrals in the series for $I(x; \varepsilon)$ is to extend the integration region in each integral to infinity. It is this third step that is hardest to grasp. It may seem foolish to first replace the finite number 10 in Example 1 (revisited) by ε and then to replace ε by ∞! However, we must choose ε to be small in order to expand the integrand of $I(x; \varepsilon)$ and thereby obtain a series. We then let the integration region become infinite in order to evaluate the terms in the series. Each time we change the limits of integration, we introduce only exponentially small errors. Note that had we not replaced the integration limit 10 by $\varepsilon < 1$, we could not have used the Taylor expansion for $(1+t)^{-1}$, which is only valid for $|t| < 1$.

Watson's Lemma

'Pon my word, Watson, you are coming along wonderfully.
You have really done very well indeed.

—Sherlock Holmes, *A Case of Identity*
Sir Arthur Conan Doyle

In Example 1 we obtained the asymptotic expansion valid as $x \to +\infty$ of an integral which belongs to a broad class of integrals of the form

$$I(x) = \int_0^b f(t) e^{-xt}\, dt, \qquad b > 0. \tag{6.4.5}$$

There is a general formula, usually referred to as Watson's lemma, which gives the full asymptotic expansion of any integral of this type provided that $f(t)$ is continuous on the interval $0 \le t \le b$ and that $f(t)$ has the asymptotic series expansion

$$f(t) \sim t^{\alpha} \sum_{n=0}^{\infty} a_n t^{\beta n}, \qquad t \to 0+. \tag{6.4.6}$$

Note that we must require that $\alpha > -1$ and $\beta > 0$ for the integral to converge at $t = 0$. Note also that if $b = +\infty$, it is necessary that $f(t) \ll e^{ct}$ $(t \to +\infty)$, for some positive constant c, for the integral (6.4.5) to converge.

Watson's lemma states that if the above conditions hold then

$$I(x) \sim \sum_{n=0}^{\infty} \frac{a_n \Gamma(\alpha + \beta n + 1)}{x^{\alpha + \beta n + 1}}, \qquad x \to +\infty. \tag{6.4.7}$$

We prove this result as follows. First, we replace $I(x)$ by $I(x; \varepsilon)$, where

$$I(x; \varepsilon) = \int_0^{\varepsilon} f(t) e^{-xt}\, dt. \tag{6.4.8}$$

This approximation introduces only exponentially small errors for any positive value of ε. In particular, we can choose ε so small that the first n terms in the asymptotic series for $f(t)$ are a good approximation to $f(t)$:

$$\left| f(t) - t^{\alpha} \sum_{n=0}^{N} a_n t^{\beta n} \right| \le K t^{\alpha + \beta(N+1)}, \qquad 0 \le t \le \varepsilon, \tag{6.4.9}$$

where K is a nonzero constant. Substituting the first N terms in the series for $f(t)$ into (6.4.8) and using (6.4.9) gives

$$\begin{aligned} \left| I(x; \varepsilon) - \sum_{n=0}^{N} a_n \int_0^{\varepsilon} t^{\alpha + \beta n} e^{-xt}\, dt \right| &\le K \int_0^{\varepsilon} t^{\alpha + \beta(N+1)} e^{-xt}\, dt \\ &\le K \int_0^{\infty} t^{\alpha + \beta(N+1)} e^{-xt}\, dt \\ &= K \frac{\Gamma(\alpha + \beta + \beta N + 1)}{x^{\alpha + \beta + \beta N + 1}}. \end{aligned}$$

Finally, we replace ε by ∞ and use the identity $\int_0^{\infty} t^{\alpha + \beta n} e^{-xt}\, dt = [\Gamma(\alpha + \beta n + 1)]/x^{\alpha + \beta n + 1}$ to obtain

$$I(x) - \sum_{n=0}^{N} a_n \Gamma(\alpha + \beta n + 1)/x^{\alpha + \beta n + 1} \ll x^{-\alpha - \beta N - 1}, \qquad x \to +\infty.$$

Since this asymptotic relation is valid for all N, we have established the validity of the asymptotic series in (6.4.7) and have proved Watson's lemma.

Example 2 *Application of Watson's lemma.* To expand the integral

$$I(x) = \int_0^5 \frac{e^{-xt}}{1 + t^2}\, dt$$

for large x, we replace $(1 + t^2)^{-1}$ by its Taylor series about $t = 0$:

$$\frac{1}{1 + t^2} = 1 - t^2 + t^4 - t^6 + \cdots. \tag{6.4.10}$$

Watson's lemma allows us to substitute (6.4.10) into the integral, interchange orders of integration and summation, and replace the upper limit of integration 5 by ∞. This gives $I(x) \sim 1/x - 2!/x^3 + 4!/x^5 - 6!/x^7 + \cdots$ $(x \to +\infty)$.

Example 3 *Asymptotic expansion of $K_0(x)$ using Watson's lemma.* A standard integral representation of the modified Bessel function $K_0(x)$ is $K_0(x) = \int_1^\infty (s^2 - 1)^{-1/2} e^{-xs}\, ds$. In order to apply Watson's lemma, we substitute $s = t + 1$. This shifts the lower endpoint of integration to $t = 0$:

$$K_0(x) = e^{-x} \int_0^\infty (t^2 + 2t)^{-1/2} e^{-xt}\, dt. \tag{6.4.11}$$

When $|t| < 2$, the binomial theorem gives

$$(t^2 + 2t)^{-1/2} = (2t)^{-1/2}(1 + t/2)^{-1/2}$$
$$= (2t)^{-1/2} \sum_{n=0}^{\infty} (-t/2)^n \frac{\Gamma(n + \frac{1}{2})}{n!\, \Gamma(\frac{1}{2})}.$$

Watson's lemma then gives

$$K_0(x) \sim e^{-x} \sum_{n=0}^{\infty} (-1)^n \frac{[\Gamma(n + \frac{1}{2})]^2}{2^{n+1/2} n!\, \Gamma(\frac{1}{2}) x^{n+1/2}}, \qquad x \to +\infty. \tag{6.4.12}$$

Example 4 *Asymptotic expansion of $D_\nu(x)$ using Watson's lemma.* An integral representation of the parabolic cylinder function $D_\nu(x)$ which is valid when Re $(\nu) < 0$ is

$$D_\nu(x) = \frac{e^{-x^2/4}}{\Gamma(-\nu)} \int_0^\infty t^{-\nu-1} e^{-t^2/2} e^{-xt}\, dt. \tag{6.4.13}$$

To obtain the behavior of $D_\nu(x)$ as $x \to +\infty$, we expand $e^{-t^2/2}$ as a power series in t: $e^{-t^2/2} = \sum_{n=0}^\infty (-1)^n t^{2n}/2^n n!$. Watson's lemma then gives

$$D_\nu(x) \sim x^\nu \frac{e^{-x^2/4}}{\Gamma(-\nu)} \sum_{n=0}^{\infty} (-1)^n \frac{\Gamma(2n - \nu)}{2^n n!\, x^{2n}}, \qquad x \to +\infty, \tag{6.4.14}$$

in agreement with (3.5.13) and (3.5.14). [The expansion in (6.4.14) is also valid when Re $\nu \geq 0$.]

Asymptotic Expansion of General Laplace Integrals

Watson's lemma only applies to Laplace integrals $I(x)$ of the form (6.4.1) in which $\phi(t) = -t$. For more general $\phi(t)$, there are two possible approaches. If $\phi(t)$ is sufficiently simple, it may be useful to make a change of variable by substituting

$$s = -\phi(t) \tag{6.4.15}$$

into (6.4.1) and to rewrite the integral in the form $\int_{-\phi(a)}^{-\phi(b)} F(s) e^{-xs}\, ds$, where $F(s) = -f(t)/\phi'(t)$. Watson's lemma applies to this transformed integral.

Example 5 *Indirect use of Watson's lemma.* Watson's lemma does not apply directly to the integral

$$I(x) = \int_0^{\pi/2} e^{-x \sin^2 t}\, dt \tag{6.4.16}$$

because $\phi(t) = -\sin^2 t$. However, if we let $s = \sin^2 t$ and rewrite the integral as $I(x) = \frac{1}{2} \int_0^1 [s(1 - s)]^{-1/2} e^{-xs}\, ds$, then Watson's lemma does apply. Since

$$[s(1 - s)]^{-1/2} = \sum_{n=0}^{\infty} \frac{\Gamma(n + \frac{1}{2}) s^{n-1/2}}{n!\, \Gamma(\frac{1}{2})}$$

for $|s| < 1$, Watson's lemma gives

$$I(x) \sim -\frac{1}{2} \sum_{n=0}^{\infty} \frac{[\Gamma(n + \frac{1}{2})]^2}{n!\, \Gamma(\frac{1}{2}) x^{n+1/2}}, \qquad x \to +\infty. \tag{6.4.17}$$

Sometimes the substitution (6.4.15) is unwieldy because the inverse function $t = \phi^{-1}(-s)$ is a complicated multivalued function. In this case, it may be simpler to use a more direct method than Watson's lemma for obtaining the first few terms in the asymptotic series for $I(x)$. We discuss the mechanical aspects of this calculation first and postpone a more theoretical discussion till Examples 7 and 8.

To obtain the leading behavior of $I(x)$ as $x \to +\infty$, we argue as follows. If $\phi(t)$ has a maximum at $t = c$, then we can approximate $I(x)$ by $I(x; \varepsilon)$, as in (6.4.2). In the narrow region $|t - c| \le \varepsilon$, we replace $\phi(t)$ by the first few terms in its Taylor series. There are several cases to consider. If c lies at one of the endpoints a or b and if $\phi'(c) \neq 0$, then we approximate $\phi(t)$ by

$$\phi(c) + (t - c)\phi'(c). \tag{6.4.18a}$$

If $\phi'(c) = 0$ (this must happen when c is interior to the interval $a \le t \le b$) but $\phi''(c) \neq 0$, then we approximate $\phi(t)$ by

$$\phi(c) + \tfrac{1}{2}(t - c)^2\phi''(c). \tag{6.4.18b}$$

More generally, if $\phi'(c) = \phi''(c) = \cdots = \phi^{(p-1)}(c) = 0$ and $\phi^{(p)}(c) \neq 0$, then we approximate $\phi(t)$ by

$$\phi(c) + \frac{1}{p!}(t - c)^p\phi^{(p)}(c). \tag{6.4.18c}$$

In each of these cases, we also expand $f(t)$ about $t = c$ and retain just the leading term. For simplicity, let us assume that $f(t)$ is continuous and that $f(c) \neq 0$ [that is, the leading behavior of $f(t)$ as $t \to c$ is $f(c)$]. We treat cases in which $f(c) = 0$ later in Example 6 and in Prob. 6.24.

Now we substitute these approximations in $I(x; \varepsilon)$ and evaluate the leading behavior of the resulting integral by extending the range of integration to infinity. When (6.4.18a) holds, c must be an endpoint, $c = a$ or $c = b$. If $c = a$, then $\phi'(a) < 0$ and

$$I(x; \varepsilon) \sim \int_a^{a+\varepsilon} f(a)e^{x[\phi(a) + (t-a)\phi'(a)]}\, dt, \qquad x \to +\infty,$$

$$\sim f(a)e^{x\phi(a)} \int_a^{\infty} e^{x(t-a)\phi'(a)}\, dt, \qquad x \to +\infty.$$

Thus,

$$I(x) \sim -\frac{f(a)e^{x\phi(a)}}{x\phi'(a)}, \qquad x \to +\infty. \tag{6.4.19a}$$

If $c = b$, then $\phi'(b) > 0$ and a similar computation gives

$$I(x) \sim \frac{f(b)e^{x\phi(b)}}{x\phi'(b)}, \qquad x \to +\infty. \tag{6.4.19b}$$

When (6.4.18b) holds and $a < c < b$, then $\phi''(c) < 0$ [because $\phi(t)$ has a maximum at $t = c$] and

$$I(x;\,\varepsilon) \sim \int_{c-\varepsilon}^{c+\varepsilon} f(c)e^{x[\phi(c)+(t-c)^2\phi''(c)/2]}\,dt, \qquad x \to +\infty,$$

$$\sim f(c)e^{x\phi(c)} \int_{-\infty}^{\infty} e^{x(t-c)^2\phi''(c)/2}\,dt, \qquad x \to +\infty,$$

$$= \frac{\sqrt{2}\,f(c)e^{x\phi(c)}}{\sqrt{-x\phi''(c)}} \int_{-\infty}^{\infty} e^{-s^2}\,ds,$$

where we have substituted $s^2 = -x(t-c)^2\phi''(c)$. Recall that $\int_{-\infty}^{\infty} e^{-s^2}\,ds = \int_0^\infty u^{-1/2}e^{-u}\,du = \Gamma(\frac{1}{2}) = \sqrt{\pi}$, so

$$I(x) \sim \frac{\sqrt{2\pi}\,f(c)e^{x\phi(c)}}{\sqrt{-x\phi''(c)}}, \qquad x \to +\infty. \tag{6.4.19c}$$

This result holds if c is interior to the interval (a, b); if $c = a$ or $c = b$, the result in (6.4.19c) must be multiplied by a factor $\frac{1}{2}$. (Why?)

When (6.4.18c) holds and $a < c < b$, then p must be even and $\phi^{(p)}(c) < 0$ [otherwise $\phi(c)$ would not be a maximum]. Then

$$I(x;\,\varepsilon) \sim \int_{c-\varepsilon}^{c+\varepsilon} f(c)e^{x[\phi(c)+(t-c)^p\phi^{(p)}(c)/p!]}\,dt, \qquad x \to +\infty,$$

$$\sim f(c)e^{x\phi(c)} \int_{-\infty}^{\infty} e^{x(t-c)^p\phi^{(p)}(c)/p!}\,dt, \qquad x \to +\infty,$$

$$= f(c)e^{x\phi(c)}[-x\phi^{(p)}(c)/p!]^{-1/p} \int_{-\infty}^{\infty} e^{-s^p}\,ds.$$

Now recall that $\int_{-\infty}^{\infty} e^{-s^p}\,ds = 2\Gamma(1/p)/p$, so

$$I(x) \sim \frac{2\Gamma(1/p)(p!)^{1/p}}{p[-x\phi^{(p)}(c)]^{1/p}} f(c)e^{x\phi(c)}, \qquad x \to +\infty. \tag{6.4.19d}$$

Example 6 *Use of Laplace's method to determine leading behavior.*

(*a*) $\int_0^{\pi/2} e^{-x\tan t}\,dt \sim 1/x$ $(x \to +\infty)$ because (6.4.19a) applies.
(*b*) $\int_0^\infty e^{-x\sinh^2 t}\,dt \sim \frac{1}{2}\sqrt{\pi/x}$ $(x \to +\infty)$. Here (6.4.19c) with the result multiplied by $\frac{1}{2}$ applies because $c = 0$ is an endpoint.
(*c*) $\int_{-1}^{1} e^{-x\sin^4 t}\,dt \sim [\Gamma(\frac{1}{4})]/2x^{1/4}$ $(x \to +\infty)$ because (6.4.19b) applies with $p = 4$.
(*d*) $\int_{-\pi/2}^{\pi/2} (t+2)e^{-x\cos t}\,dt \sim 4/x$ $(x \to +\infty)$, where we have added together contributions obtained using both (6.4.19a) and (6.4.19b) because there are maxima at both $t = -\pi/2$ and $t = \pi/2$.

$$(e)\quad \int_0^1 \sin t e^{-x\sinh^4 t}\,dt \sim \int_0^\varepsilon t e^{-xt^4}\,dt, \qquad x \to +\infty,$$

$$\sim \int_0^\infty t e^{-xt^4}\,dt, \qquad x \to +\infty,$$

$$= \frac{1}{4x^{1/2}} \int_0^\infty s^{-1/2} e^{-s}\,ds = \frac{\Gamma(\frac{1}{2})}{4x^{1/2}} = \frac{1}{4}\sqrt{\frac{\pi}{x}}.$$

This example is interesting because the maximum of ϕ occurs at a point where $f(t)$ vanishes. Laplace's method works because the contribution to the integral from outside the interval $0 \le t \le \varepsilon$ is subdominant for any $\varepsilon > 0$. In general, if $f(t)$ vanishes algebraically at the maximum of ϕ, Laplace's method works as explained above.

$$(f)\quad \int_0^\infty \frac{e^{-x\cosh t}}{\sqrt{\sinh t}}\,dt \sim \int_0^\varepsilon \frac{e^{-x(1+t^2/2)}}{\sqrt{t}}\,dt, \qquad x \to +\infty,$$

$$\sim e^{-x} \int_0^\infty \frac{e^{-xt^2/2}}{\sqrt{t}}\,dt, \qquad x \to +\infty,$$

$$= (8x)^{-1/4} e^{-x} \int_0^\infty s^{-3/4} e^{-s}\,ds$$

$$= \Gamma\left(\frac{1}{4}\right)(8x)^{-1/4} e^{-x}.$$

This example is interesting because $f(t)$ is infinite at the maximum of ϕ. Again, Laplace's method works because the contribution to the integral from outside the interval $0 \le t \le \varepsilon$ is subdominant.

(*g*) The modified Bessel function $K_\nu(x)$ of order ν has the integral representation

$$K_\nu(x) = \int_0^\infty e^{-x\cosh t} \cosh(\nu t)\,dt, \tag{6.4.20}$$

which is valid for $x > 0$ (see Prob. 6.36). Therefore, (6.4.19*c*) with an extra factor $\frac{1}{2}$ gives

$$K_\nu(x) \sim \sqrt{\frac{\pi}{2x}}\, e^{-x}, \qquad x \to +\infty, \tag{6.4.21}$$

in agreement with (3.8.11).

(*h*) The integral representation (6.4.20) can also be used to find the asymptotic behavior of $K_\nu(p\nu)$ as $\nu \to +\infty$ with p fixed: $K_\nu(p\nu) = \frac{1}{2}\int_0^\infty e^{-\nu(p\cosh t + t)}\,dt + \frac{1}{2}\int_0^\infty e^{-\nu(p\cosh t - t)}\,dt$, where we have substituted $\cosh(\nu t) = (e^{\nu t} + e^{-\nu t})/2$. Laplace's method applied to the first integral on the right gives its leading behavior as $e^{-p\nu}/2\nu$. Laplace's method also applies to the second integral on the right. Here $\phi(t) = t - p\cosh t$ has a maximum at $t = c$ where $p \sinh c = 1$. Therefore, (6.4.19*c*) gives

$$K_\nu(p\nu) \sim \sqrt{\frac{\pi}{2\nu}} \frac{e^{-\nu q}}{(1+p^2)^{1/4}}, \qquad \nu \to +\infty;\ p > 0, \tag{6.4.22}$$

where $q = p\cosh c - c = \sqrt{1+p^2} - \ln[(1+\sqrt{1+p^2})/p]$. The derivation of (6.4.22) is facilitated by the formulas $p\cosh c = \sqrt{1+p^2}$ and $e^c = \cosh c + \sinh c = (\sqrt{1+p^2}+1)/p$.

The procedures used to derive the results in (6.4.19) are correct but not fully justified. In the next two examples, we apply Laplace's method to an integral taking care to justify and explain our approximations more carefully.

Example 7 *Careful application of Laplace's method to* $\int_0^{\pi/2} e^{-x \sin^2 t}\, dt$. Here $\phi(t) = -\sin^2 t$ has a maximum at $t = 0$, so (6.4.2) shows that for any ε $(0 < \varepsilon < \pi/2)$, $I(x) = \int_0^{\pi/2} e^{-x \sin^2 t}\, dt \sim \int_0^{\varepsilon} e^{-x \sin^2 t}\, dt$ $(x \to +\infty)$ with only exponentially small errors. Recall that this step is justified in Prob. 6.23. If ε is small, $\sin t$ can be approximated by t for all t $(0 \le t \le \varepsilon)$. Thus, we expect that it is valid to approximate $I(x)$ by $I(x) \sim \int_0^{\varepsilon} e^{-xt^2}\, dt$ $(x \to +\infty)$. For finite ε, the Gaussian integral on the right cannot be evaluated in terms of elementary functions (it is an error function). However, the evaluation of the integral on the right is easy if we extend the integration region to ∞:

$$\int_0^{\infty} e^{-xt^2}\, dt = \frac{1}{2\sqrt{x}} \int_0^{\infty} s^{-1/2} e^{-s}\, ds = \frac{1}{2\sqrt{x}} \Gamma\left(\frac{1}{2}\right) = \frac{1}{2}\sqrt{\frac{\pi}{x}},$$

where we have used the substitution $s = xt^2$. Thus,

$$I(x) \sim \tfrac{1}{2}\sqrt{\pi/x}, \qquad x \to +\infty. \tag{6.4.23}$$

Note that this result agrees with the first term in (6.4.17).

To justify the asymptotic relation (6.4.23) we must verify that

$$\int_0^{\varepsilon} e^{-x \sin^2 t}\, dt \sim \int_0^{\varepsilon} e^{-xt^2}\, dt, \qquad x \to +\infty, \tag{6.4.24}$$

when $0 < \varepsilon < 2\pi$.

The proof of (6.4.24) is delicate and illustrates one of the more subtle aspects of Laplace's method, so we go through it in some detail. The idea of the proof is that over the range of t that contributes substantially to either of the integrals in (6.4.24), the integrands of both integrals are nearly identical. Outside of this range the two integrands are quite different, but both are exponentially small.

The proof of (6.4.24) begins by breaking up the range of integration $0 \le t \le \varepsilon$ into the two ranges $0 \le t \le x^{-\alpha}$ and $x^{-\alpha} < t \le \varepsilon$, where $\frac{1}{4} < \alpha < \frac{1}{2}$. This restriction on α is made so that when $t = x^{-\alpha}$, $xt^2 \to +\infty$ but $xt^4 \to 0$ as $x \to +\infty$. When $t \le x^{-\alpha}$ with $\frac{1}{4} < \alpha < \frac{1}{2}$,

$$x \sin^2 t - xt^2 \ll 1, \qquad x \to +\infty. \tag{6.4.25}$$

To prove (6.4.25) we use the inequalities $t - t^3/6 \le \sin t \le t$ which hold for all $t > 0$ (see Prob. 6.40) to obtain

$$|x \sin^2 t - xt^2| = x|\sin t + t|\,|\sin t - t| < x(2t)(t^3/6) = xt^4/3.$$

But $xt^4/3 \to 0$ as $x \to +\infty$ when $t \le x^{-\alpha}$, with $\alpha > \frac{1}{4}$, which proves (6.4.25). Exponentiating (6.4.25) gives

$$e^{-x \sin^2 t} \sim e^{-xt^2}, \qquad t \le x^{-\alpha},\ x \to +\infty. \tag{6.4.26}$$

Integrating this asymptotic relation gives

$$\int_0^{x^{-\alpha}} e^{-x \sin^2 t}\, dt \sim \int_0^{x^{-\alpha}} e^{-xt^2}\, dt, \qquad x \to +\infty.$$

To complete the proof of (6.4.24), we must estimate the contribution to each integral from the interval $x^{-\alpha} \le t \le \varepsilon$. Note that (6.4.26) is not true for all t in this range. In fact, when t is of order ε and $x \to +\infty$, $e^{-x \sin^2 t}$ is exponentially larger in magnitude than e^{-xt^2}. Here lies the subtlety in the proof of (6.4.24); (6.4.24) remains valid despite the discrepancy between the magnitudes of the two integrands $e^{-x \sin^2 t}$ and e^{-xt^2} when t is of the order ε. The point is simply that the contribution to each integral in (6.4.24) from the interval $x^{-\alpha} < t \le \varepsilon$ is subdominant as $x \to +\infty$ with respect to the contribution from $0 \le t \le x^{-\alpha}$ when $\alpha < \frac{1}{2}$. In fact, if $t > x^{-\alpha}$ with $\alpha < \frac{1}{2}$, then $e^{-x \sin^2 t}$ and e^{-xt^2} are both smaller than $e^{-x \sin^2 (x^{-\alpha})} \sim e^{-x^{1-2\alpha}}$ $(x \to +\infty)$ which is exponentially small. In Fig. 6.2, we plot the integrands on the left and right sides of (6.4.24) for $x = 100$. Observe that the integrands are nearly the same for those values of t that contribute substantially to the integrals and that the integrands differ substantially only for those values of t that make a negligible contribution to the integrals.

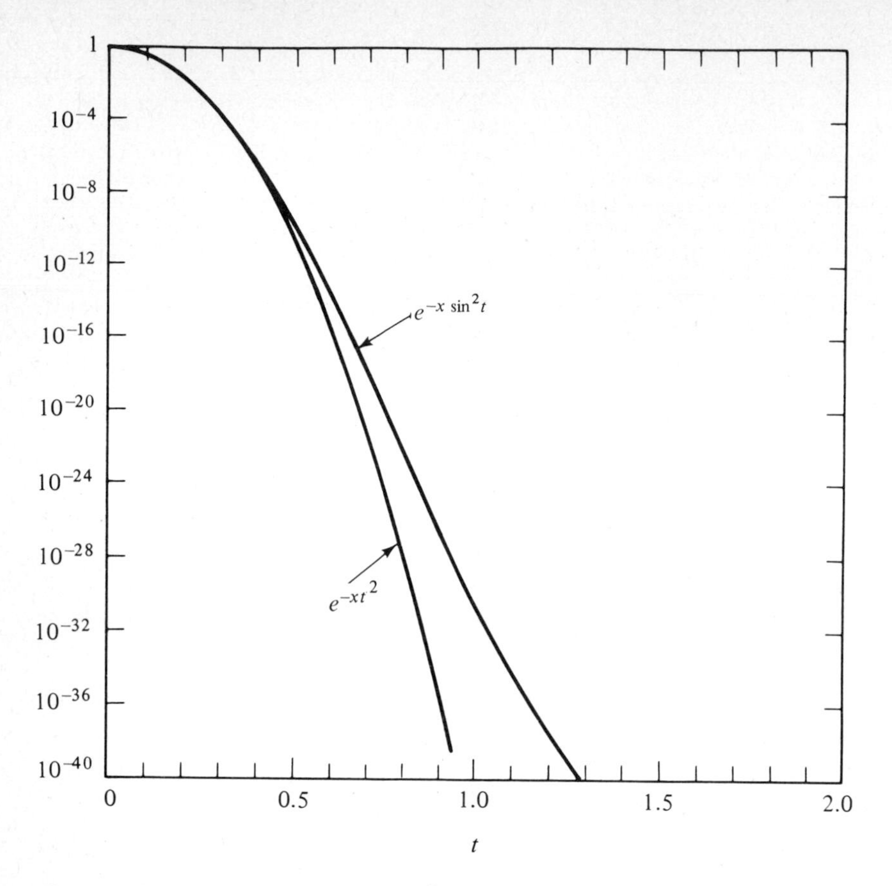

Figure 6.2 A graphical demonstration that $\int_0^\varepsilon \exp(-x \sin^2 t)\, dt \sim \int_0^\varepsilon \exp(-xt^2)\, dt$ $(x \to +\infty)$ for any positive $\varepsilon < 2\pi$ [see (6.4.24)]. The graph compares the two integrands for $x = 100$. Observe that over the range of t that contributes significantly to either of the integrands, both integrands are nearly equal. Outside of this range the integrands differ substantially but are negligibly small.

Example 8 *Leading behavior of $I_n(x)$ as $x \to +\infty$.* It may be shown (see Prob. 6.41) that

$$I_n(x) = \frac{1}{\pi} \int_0^\pi e^{x \cos t} \cos(nt)\, dt, \tag{6.4.27}$$

where $I_n(x)$ is the modified Bessel function of order n. Local analysis of the modified Bessel equation (see Example 2 of Sec. 3.5) establishes only that

$$I_n(x) \sim \frac{1}{\sqrt{x}}(c_1 e^x + c_2 e^{-x}), \qquad x \to +\infty, \tag{6.4.28}$$

for some constants c_1 and c_2, but it does not establish the values of c_1 and c_2. However, asymptotic analysis of the integral representation (6.4.27) shows that $c_1 = 1/\sqrt{2\pi}$; the value of c_2 is *not* determined by asymptotic analysis of (6.4.27) because e^{-x} is subdominant with respect to e^x as $x \to +\infty$. Observe that integration by parts cannot be used to derive (6.4.28) because the leading behavior of $I_n(x)$ involves $1/\sqrt{x}$ and not $1/x$.

To use Laplace's method we note that $\phi(t) = \cos t$ has a maximum at $t = 0$. Thus,

$$I_n(x) \sim \int_0^\varepsilon e^{x \cos t} \cos(nt)\, dt, \qquad n \to +\infty. \tag{6.4.29}$$

According to (6.4.18*b*), we must replace $\cos t$ by the first two terms in its Taylor expansion about 0. What happens if, instead, we approximate $\cos t$ by 1 and $\cos(nt)$ by 1 for $0 \le t \le \varepsilon$? The resulting approximation to $I_n(x)$, $\int_0^\varepsilon e^x(1)\, dt = \varepsilon e^x$, is *not* correct because the dependence on the arbitrary constant ε has not dropped out. The trouble here is that e^x is not a good approximation to $e^{x \cos t}$ over the subinterval of $0 \le t \le \varepsilon$ that gives the dominant contribution to the integral (see Fig. 6.3).

A correct application of Laplace's method to (6.4.29) is obtained by approximating $\cos t$ by the first *two* terms of its Taylor series $1 - t^2/2$ and by approximating $\cos(nt)$ by 1:

$$I_n(x) \sim \frac{1}{\pi} \int_0^\varepsilon e^{x(1 - t^2/2)}\, dt, \qquad x \to +\infty, \tag{6.4.30}$$

$$\sim \frac{1}{\sqrt{2\pi x}} e^x, \qquad x \to +\infty. \tag{6.4.31}$$

The dependence on ε has disappeared, as it should.

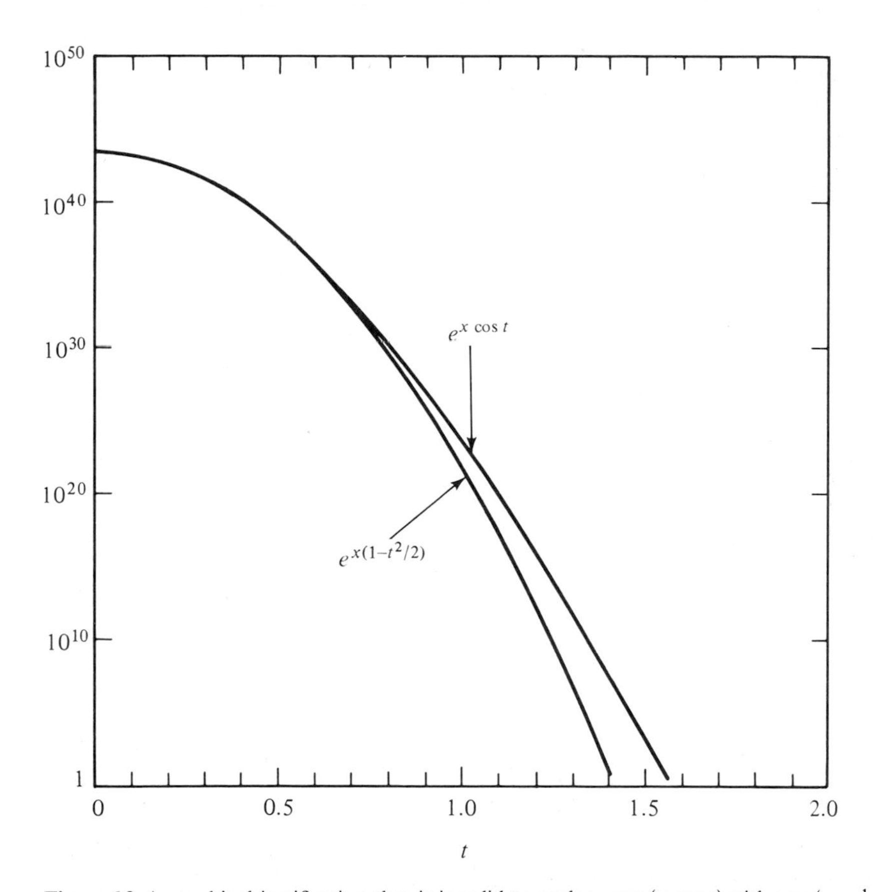

Figure 6.3 A graphical justification that it is valid to replace $\exp(x \cos t)$ with $\exp(x - \frac{1}{2}xt^2)$ in the integrand in (6.4.29) when $x \to \infty$. The graph compares these two expressions when $x = 100$. Note that the two curves are nearly identical over many orders of magnitude.

The leading behavior of $I_n(x)$ as $x \to +\infty$ does not depend on n; however, higher-order terms in the asymptotic expansion of $I_n(x)$ do depend on n. In fact, the complete asymptotic expansion of $I_n(x)$ as $x \to +\infty$ is given in (3.5.8) and (3.5.9) with $c_1 = 1/\sqrt{2\pi}$.

The only step that really needs justification is (6.4.30). The argument is nearly the same as that used to justify (6.4.24). The range of integration from $t = 0$ to $t = \varepsilon$ is broken up into the two ranges $0 \le t \le x^{-\alpha}$ and $x^{-\alpha} < t \le \varepsilon$, where $\frac{1}{4} < \alpha < \frac{1}{2}$. Now for fixed n, $\cos(nt)\, e^{x \cos t} \sim e^{x(1-t^2/2)}$ $(x \to +\infty)$ uniformly for all t satisfying $0 \le t \le x^{-\alpha}$ because $1 - t^2/2 \le \cos t \le 1 - t^2/2 + t^4/24$. Therefore, $\int_0^{x^{-\alpha}} \cos(nt)\, e^{x\cos t}\, dt \sim \int_0^{x^{-\alpha}} e^{x(1-t^2/2)}\, dt$ $(x \to +\infty)$. Also, when $x^{-\alpha} < t \le \varepsilon$, the integrands on both sides of (6.4.30) are subdominant with respect to e^x, so the contribution to (6.4.30) from the integration range $x^{-\alpha} < t \le \varepsilon$ is exponentially small compared to the contribution from the range $0 \le t \le x^{-\alpha}$.

Laplace's Method—Determination of Higher-Order Terms

The approach we have used to obtain the leading asymptotic behavior of integrals by Laplace's method can be extended to give the higher-order terms in the asymptotic expansion of the integral. To do this one would naturally expect to have to retain more terms in the expansions of $\phi(t)$ and $f(t)$ than those used to obtain (6.4.19). We illustrate the mechanics of this procedure for the case in which $\phi'(c) = 0$, $\phi''(c) < 0$, $f(c) \ne 0$, and $a < c < b$, where c is the location of the maximum of $\phi(t)$.

By (6.4.2), $I(x) \sim \int_{c-\varepsilon}^{c+\varepsilon} f(t) e^{x\phi(t)}\, dt$ $(x \to +\infty)$ with exponentially small errors. The leading behavior of $I(x)$ given by (6.4.19*c*) is obtained by replacing $f(t)$ by $f(c)$ and $\phi(t)$ by $\phi(c) + \frac{1}{2}(t-c)^2\phi''(c)$. To compute the first correction to (6.4.19*c*) we must approximate $f(t)$ and $\phi(t)$ by two *more* terms in their Taylor series:

$$I(x) \sim \int_{c-\varepsilon}^{c+\varepsilon} [f(c) + f'(c)(t-c) + \tfrac{1}{2}f''(c)(t-c)^2] \times \exp\{x[\phi(c) + \tfrac{1}{2}(t-c)^2\phi''(c) + \tfrac{1}{6}(t-c)^3\phi'''(c) + \tfrac{1}{24}(t-c)^4(d^4\phi/dt^4)(c)]\}\, dt, \qquad x \to +\infty. \tag{6.4.32}$$

It is somewhat surprising that *two* additional terms in the series for $\phi(t)$ and $f(t)$ are required to compute just the next term in (6.4.19*c*). We will see shortly why this is so.

Because ε may be chosen small, we Taylor expand the integrand in (6.4.32) as follows:

$$\exp\{x[\tfrac{1}{6}(t-c)^3\phi'''(c) + \tfrac{1}{24}(t-c)^4(d^4\phi/dt^4)(c)]\} = 1 + x[\tfrac{1}{6}(t-c)^3\phi'''(c) + \tfrac{1}{24}(t-c)^4(d^4\phi/dt^4)(c)] + \tfrac{1}{72}x^2(t-c)^6[\phi'''(c)]^2 + \cdots.$$

Substituting this expansion into (6.4.32) and collecting powers of $t - c$ gives

$$I(x) \sim \int_{c-\varepsilon}^{c+\varepsilon} e^{x\phi(c) + x(t-c)^2\phi''(c)/2} \times \Big\{ f(c) + \frac{(t-c)^2}{2} f''(c) + (t-c)^4[\tfrac{1}{24}xf(c)(d^4\phi/dt^4)(c) + \tfrac{1}{6}xf'(c)\phi'''(c)] + \tfrac{1}{72}(t-c)^6 x^2 f(c)[\phi'''(c)]^2 + \cdots \Big\}\, dt, \qquad x \to +\infty, \tag{6.4.33}$$

where we have excluded odd powers of $t-c$ because they vanish upon integration. Only the displayed terms in (6.4.33) contribute to the next term in (6.4.19*c*). Notice that we do *not* Taylor expand $\exp\left[\frac{1}{2}x(t-c)^2\phi''(c)\right]$; we return to this point shortly.

Next we extend the range of integration in (6.4.33) to $(-\infty, \infty)$ and substitute $s = \sqrt{x}(t-c)$:

$$I(x) \sim \frac{1}{\sqrt{x}} e^{x\phi(c)} \int_{-\infty}^{\infty} e^{s^2\phi''(c)/2} \times \left\{ f(c) + \frac{1}{x}\left[\tfrac{1}{2}s^2 f''(c) + \tfrac{1}{24}s^4 f(c)(d^4\phi/dt^4)(c) + \tfrac{1}{6}s^4 f'(c)\phi'''(c) + \tfrac{1}{72}s^6[\phi'''(c)]^2 f(c)\right]\right\} ds, \qquad x \to +\infty. \tag{6.4.34}$$

Observe that all the displayed terms in (6.4.33) contribute to the coefficient of $1/x$ in (6.4.34); the additional terms that we have neglected in going from (6.4.32) to (6.4.34) contribute to the coefficients of $1/x^2$, $1/x^3$, and so on.

To evaluate the integrals in (6.4.34) we use integration by parts to derive the general formula $\int_{-\infty}^{\infty} e^{-s^2/2}s^{2n}\, ds = \sqrt{2\pi}(2n-1)(2n-3)(2n-5)\cdots(5)(3)(1)$. Thus, we have

$$I(x) \sim \sqrt{\frac{2\pi}{-x\phi''(c)}}\, e^{x\phi(c)} \left\{ f(c) + \frac{1}{x}\left[-\frac{f''(c)}{2\phi''(c)} + \frac{f(c)(d^4\phi/dt^4)(c)}{8[\phi''(c)]^2} + \frac{f'(c)\phi'''(c)}{2[\phi''(c)]^2} - \frac{5[\phi'''(c)]^2 f(c)}{24[\phi''(c)]^3} \right]\right\}, \qquad x \to +\infty. \tag{6.4.35}$$

One aspect of the derivation of (6.4.35) requires explanation. In proceeding from (6.4.32) to (6.4.34) we did not Taylor expand $\exp\left[\frac{1}{2}x(t-c)^2\phi''(c)\right]$, but we did Taylor expand the cubic and quartic terms in the exponential. If we had Taylor expanded $\exp\left[\frac{1}{2}x(t-c)^2\phi''(c)\right]$ and retained only a finite number of terms, the resulting approximation to $I(x)$ would depend on ε (see Example 8). If we had not expanded the cubic and quartic terms and if $(d^4\phi/dt^4)(c)$ were nonnegative, then extending the range of integration from $(c-\varepsilon, c+\varepsilon)$ to $(-\infty, \infty)$ would yield a divergent integral which would be a poor approximation to $I(x)$ indeed! If we had not expanded the cubic and quartic terms and if $(d^4\phi/dt^4)(c) < 0$, then extending the range of integration from $(c-\varepsilon, c+\varepsilon)$ to $(-\infty, \infty)$ would yield a convergent integral. However, this convergent integral might not be asymptotic to $I(x)$ because replacing $\phi(t)$ by the four-term Taylor series in (6.4.32) can introduce new relative maxima which lie outside $(c-\varepsilon, c+\varepsilon)$ which would dominate the integral on the right side of (6.4.32). In summary, there are three reasons why we must Taylor expand the cubic and quartic terms in the exponential before we extend the range of integration to $(-\infty, \infty)$:

1. The resulting integrals are always convergent and depend on ε only through subdominant terms.
2. It is easy to evaluate the resulting Gaussian integrals.
3. We avoid introducing any spurious maxima into the integrand.

To illustrate the above discussion we consider the integral

$$I(x) = \int_0^{\pi/2} e^{-x \sin^2 t}\, dt. \tag{6.4.36}$$

To obtain a higher-order approximation to this integral than that in (6.4.23), we Taylor expand $\sin^2 t$ through t^4:

$$I(x) \sim \int_0^{\varepsilon} e^{-x(t^2 - t^4/3)}\, dt, \qquad x \to +\infty. \tag{6.4.37}$$

Taylor expanding the quartic term, we obtain

$$\begin{aligned} I(x) &\sim \int_0^{\varepsilon} e^{-xt^2}\left(1 + \frac{1}{3}xt^4\right) dt, \qquad x \to +\infty, \\ &\sim \int_0^{\infty} e^{-xt^2}\left(1 + \frac{1}{3}xt^4\right) dt \\ &= \frac{1}{2}\sqrt{\frac{\pi}{x}}\left(1 + \frac{1}{4x}\right), \qquad x \to +\infty, \end{aligned} \tag{6.4.38}$$

in agreement with (6.4.17). In Fig. 6.4 we plot the integrands of (6.4.36) to (6.4.38) for $x = 100$. Observe that all three integrands are nearly identical for small t, but that the integrand in (6.4.37) blows up as $t \to +\infty$. The integrands of (6.4.36) and (6.4.38) do differ when t is large, but large t makes a negligible contribution to both (6.4.36) and (6.4.38).

Laplace's Method for Integrals with Movable Maxima

There are two kinds of problems where Laplace's method is useful but does not apply directly. First, we know what to do when $f(t)$ vanishes algebraically at $t = c$, the maximum of $\phi(t)$. But what if $f(t)$ vanishes exponentially fast at c? Second, it can happen that the Laplace integral (6.4.1) converges but that max $\phi(t) = \infty$. What do we do then? We consider each of these cases in the following two examples.

Example 9 *Leading behavior of* $\int_0^\infty e^{-xt-1/t}\, dt$. Here $f(t) = e^{-1/t}$ vanishes exponentially fast at $t = 0$, the maximum of $\phi(t) = -t$. If we apply Watson's lemma (6.4.7), we obtain the asymptotic series expansion $0 + 0/x + 0/x^2 + \cdots$ $(x \to +\infty)$ because the coefficients of the asymptotic power series of $e^{-1/t}$ as $t \to 0+$ are all zero. Watson's lemma does not determine the behavior of $I(x)$ because Watson's lemma can only produce a series of inverse powers of x. Here, $I(x)$ is smaller than any power of x; it decreases exponentially fast as $x \to +\infty$.

In order to determine the correct behavior of $I(x)$, let us determine the location of the true maximum of the full integrand $e^{-xt-1/t}$. This maximum occurs when $(d/dt)(-xt - 1/t) = 0$ or $t = 1/\sqrt{x}$. We call such a maximum a movable maximum because its location depends on x.

For this kind of movable maximum problem, Laplace's method can be applied if we first transform the movable maximum to a fixed maximum. This is done by making the change of variables $t = s/\sqrt{x}$:

$$I(x) = \frac{1}{\sqrt{x}} \int_0^\infty e^{-\sqrt{x}(s+1/s)}\, ds.$$

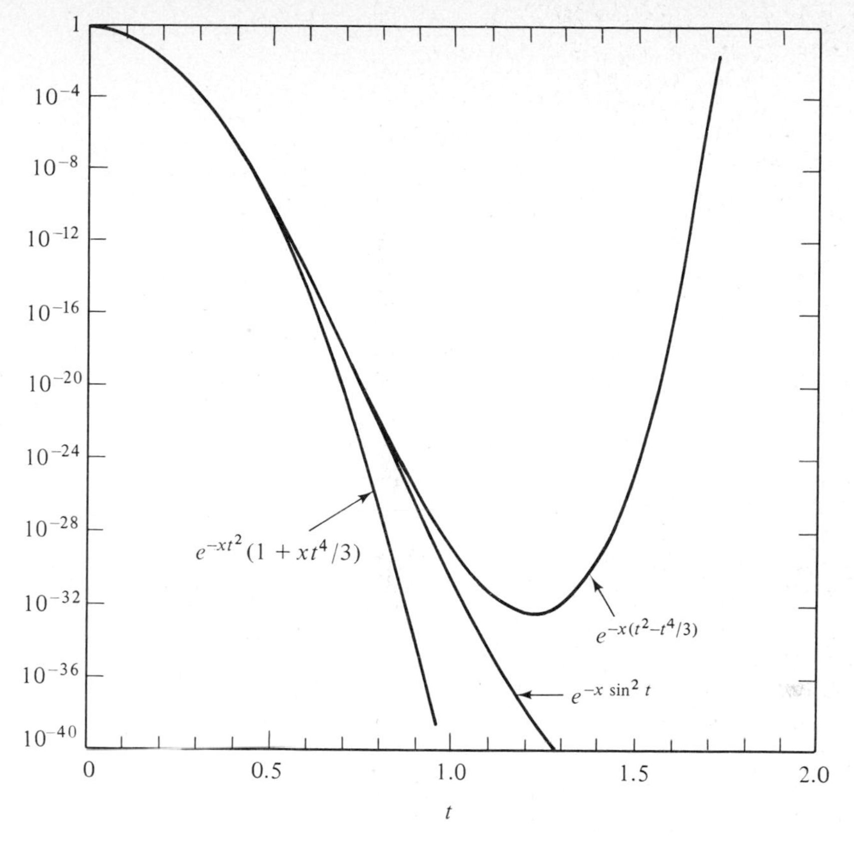

Figure 6.4 A comparison of the three integrands in (6.4.36) to (6.4.38) for $x = 100$. All three differ when t is large, but in the restricted range near $t = 0$ all three contribute equally to the integral as $x \to \infty$.

In this form, $f(s) = 1$ and $\phi(s) = s + 1/s$ and Laplace's method applies directly. The maximum of the new function $\phi(s)$ occurs at $s = 1$, so (6.4.19c) gives

$$I(x) \sim \sqrt{\pi}\, e^{-2\sqrt{x}}/x^{3/4}, \qquad x \to +\infty.$$

Example 10 *Derivation of Stirling's formula for* $\Gamma(x)$. A convergent integral representation for $\Gamma(x)$ is $\Gamma(x) = \int_0^\infty e^{-t}t^{x-1}\,dt$ $(x > 0)$ (see Sec. 2.2). Here $f(t) = e^{-t}/t$ and $\phi(t) = \ln t$. Note that $\max \phi(t) = \infty$ for $0 \le t < \infty$, so Laplace's method is not immediately applicable. This example is very similar to the previous example because the maximum of $\phi(t)$ occurs as $t \to \infty$ where $f(t)$ is exponentially small. As in Example 9, we find the location of the maximum of $e^{-t}t^x$, neglecting the factor $1/t$ which vanishes algebraically at ∞. This maximum occurs where $(d/dt)e^{-t}t^x = 0$ or $t = x$. Again, we encounter a movable maximum.

If we make the change of variables $t = sx$, we obtain

$$\Gamma(x) = x^x \int_0^\infty e^{-x(s - \ln s)}\,\frac{ds}{s}.$$

Now $f(s) = 1/s$ and $\phi(s) = -s + \ln s$. Laplace's method applies directly to this transformed integral. The maximum of $\phi(s)$ occurs at $s = 1$ so (6.4.19c) gives

$$\Gamma(x) \sim x^x e^{-x} \sqrt{2\pi/x}, \qquad x \to +\infty, \tag{6.4.39}$$

in agreement with (5.4.1). To obtain the next term in the Stirling series we note that $\phi(1) = -1$, $\phi'(1) = 0$, $\phi''(1) = -1$, $\phi'''(1) = 2$, $(d^4\phi/ds^4)(1) = -6$, $f(1) = 1$, $f'(1) = -1$, $f''(1) = 2$. Substituting these coefficients into the formula (6.4.35), we obtain

$$\Gamma(x) \sim x^x e^{-x} \sqrt{\frac{2\pi}{x}} \left(1 + \frac{1}{12x}\right), \qquad x \to +\infty, \tag{6.4.40}$$

in agreement with (5.4.1).

The distinction between ordinary and movable maxima is examined in Probs. 6.45 to 6.47.

(I) 6.5 METHOD OF STATIONARY PHASE

There is an immediate generalization of the Laplace integrals studied in Sec. 6.4 which we obtain by allowing the function $\phi(t)$ in (6.4.1) to be complex. Note that, if we wish, we may assume that $f(t)$ is real; if it were complex, $f(t)$ could be decomposed into a sum of its real and imaginary parts. However, allowing $\phi(t)$ to be complex poses new and nontrivial problems. In this section we consider the special case in which $\phi(t)$ is pure imaginary: $\phi(t) = i\psi(t)$, where $\psi(t)$ is real. The resulting integral

$$I(x) = \int_a^b f(t) e^{ix\psi(t)}\, dt \tag{6.5.1}$$

with $f(t)$, $\psi(t)$, a, b, x all real is called a generalized Fourier integral. When $\psi(t) = t$, $I(x)$ is an ordinary Fourier integral. The general case in which $\phi(t)$ is complex is considered in Sec. 6.6.

To study the behavior of $I(x)$ in (6.5.1) as $x \to +\infty$, we can use integration by parts to develop an asymptotic expansion in inverse powers of x so long as the boundary terms are finite and the resulting integrals exist.

Example 1 *Asymptotic expansion of a Fourier integral as $x \to +\infty$.* We use integration by parts to find an asymptotic approximation to the Fourier integral

$$I(x) = \int_0^1 \frac{e^{ixt}}{1+t}\, dt.$$

After one integration by parts we obtain

$$I(x) = -\frac{i}{2x} e^{ix} + \frac{i}{x} - \frac{i}{x} \int_0^1 \frac{e^{ixt}}{(1+t)^2}\, dt. \tag{6.5.2}$$

The integral on the right side of (6.5.2) is negligible compared with the boundary terms as $x \to +\infty$; in fact, it vanishes like $1/x^2$ as $x \to +\infty$. To see this, we integrate by parts again:

$$-\frac{i}{x} \int_0^1 \frac{e^{ixt}}{(1+t)^2}\, dt = -\frac{1}{4x^2} e^{ix} + \frac{1}{x^2} - \frac{2}{x^2} \int_0^1 \frac{e^{ixt}}{(1+t)^3}\, dt.$$

The integral on the right is bounded because

$$\left|\int_0^1 \frac{e^{ixt}}{(1+t)^3}\,dt\right| \le \int_0^1 (1+t)^{-3}\,dt = \frac{3}{8}.$$

Since the integral on the right in (6.5.2) does vanish like $1/x^2$ as $x \to +\infty$, $I(x)$ is asymptotic to the boundary terms: $I(x) \sim -(i/2x)e^{ix} + i/x$ $(x \to +\infty)$.

Repeated application of integration by parts gives the complete asymptotic expansion of $I(x)$ as $x \to +\infty$: $I(x) = e^{ix}u(x) + v(x)$ where

$$u(x) \sim -\frac{i}{2x} - \frac{1}{4x^2} + \cdots + \frac{(-i)^n(n-1)!}{(2x)^n} + \cdots, \qquad x \to +\infty,$$

$$v(x) \sim \frac{i}{x} + \frac{1}{x^2} + \cdots - \frac{(-i)^n(n-1)!}{x^n} + \cdots, \qquad x \to +\infty.$$

Example 2 *Integration by parts applied to* $\int_0^1 \sqrt{t}\, e^{ixt}\,dt$. Integration by parts can be used just once for the Fourier integral $I(x) = \int_0^1 \sqrt{t}\, e^{ixt}\,dt$. One integration by parts gives

$$I(x) = -\frac{i}{x}e^{ix} + \frac{i}{2x}\int_0^1 \frac{e^{ixt}}{\sqrt{t}}\,dt. \tag{6.5.3}$$

The integral on the right side of (6.5.3) vanishes more rapidly than the boundary term as $x \to +\infty$. We cannot use integration by parts to verify this because the resulting integral does not exist. (Why?) However, we can use the following simple scaling argument. We let $s = xt$ and obtain

$$\frac{i}{2x}\int_0^1 \frac{e^{ixt}}{\sqrt{t}}\,dt = \frac{i}{2x^{3/2}}\int_0^x \frac{e^{is}}{\sqrt{s}}\,ds \sim \frac{i}{2x^{3/2}}\int_0^\infty \frac{e^{is}}{\sqrt{s}}\,ds, \qquad x \to +\infty.$$

To evaluate the last integral we rotate the contour of integration from the real-s axis to the positive imaginary-s axis in the complex-s plane and obtain

$$\int_0^\infty \frac{e^{is}}{\sqrt{s}}\,ds = \sqrt{\pi}\, e^{i\pi/4}. \tag{6.5.4}$$

(See Prob. 6.49 for the details of this calculation.) Therefore,

$$I(x) + \frac{i}{x}e^{ix} \sim \frac{i}{2x^{3/2}}\sqrt{\pi}\, e^{i\pi/4}, \qquad x \to +\infty. \tag{6.5.5}$$

Clearly, this result cannot be found by direct integration by parts of the integral on the right side of (6.5.3) because a fractional power of x has appeared. However, it is possible to find the full asymptotic expansion of $I(x)$ as $x \to +\infty$ by an indirect application of integration by parts (see Prob. 6.50).

In Example 1 we used integration by parts to argue that the integral on the right side of (6.5.2) vanishes more rapidly than the boundary terms as $x \to +\infty$. In Example 2 we used a scaling argument to show that the integral on the right side of (6.5.3) vanishes more rapidly than the boundary terms as $x \to +\infty$. There is, in fact, a very general result called the Riemann-Lebesgue lemma that guarantees that

$$\int_a^b f(t)e^{ixt}\,dt \to 0, \qquad x \to +\infty, \tag{6.5 6}$$

provided that $\int_a^b |f(t)|\, dt$ exists. This result is valid even when $f(t)$ is not differentiable and integration by parts or scaling do not work. We will cite the Riemann-Lebesgue lemma repeatedly throughout this section; we could have used it to justify neglecting the integrals on the right sides of (6.5.2) and (6.5.3).

We reserve a proof of the Riemann-Lebesgue lemma for Prob. 6.51. Although the proof of (6.5.6) is messy, it is easy to understand the result heuristically. When x becomes large, the integrand $f(t)e^{ixt}$ oscillates rapidly and contributions from adjacent subintervals nearly cancel.

The Riemann-Lebesgue lemma can be extended to cover generalized Fourier integrals of the form (6.5.1). It states that $I(x) \to 0$ as $x \to +\infty$ so long as $|f(t)|$ is integrable, $\psi(t)$ is continuously differentiable for $a \le t \le b$, and $\psi(t)$ is not constant on any subinterval of $a \le t \le b$ (see Prob. 6.52). The lemma implies that $\int_0^{10} t^3 e^{ix \sin^2 t}\, dt \to 0$ $(x \to +\infty)$, but it does not apply to $\int_0^{10} t^3 e^{2ix}\, dt$.

Integration by parts gives the leading asymptotic behavior as $x \to +\infty$ of generalized Fourier integrals of the form (6.5.1), provided that $f(t)/\psi'(t)$ is smooth for $a \le t \le b$ and nonvanishing at one of the endpoints a or b. Explicitly,

$$I(x) = \frac{f(t)}{ix\psi'(t)} e^{ix\psi(t)} \Bigg|_{t=a}^{t=b} - \frac{1}{ix} \int_a^b \frac{d}{dt} \frac{f(t)}{\psi'(t)} e^{ix\psi(t)}\, dt.$$

The Riemann-Lebesgue lemma shows that the integral on the right vanishes more rapidly than $1/x$ as $x \to +\infty$. Therefore, $I(x)$ is asymptotic to the boundary terms (assuming that they do not vanish):

$$I(x) \sim \frac{f(t)}{ix\psi'(t)} e^{ix\psi(t)} \Bigg|_{t=a}^{t=b}, \qquad x \to +\infty. \tag{6.5.7}$$

Observe that when integration by parts applies, $I(x)$ vanishes like $1/x$ as $x \to +\infty$.

Integration by parts may not work if $\psi'(t) = 0$ for some t in the interval $a \le t \le b$. Such a point is called a *stationary* point of ψ. When there are stationary points in the interval $a \le t \le b$, $I(x)$ must still vanish as $x \to +\infty$ by the Riemann-Lebesgue lemma, but $I(x)$ usually vanishes less rapidly than $1/x$ because the integrand $f(t)e^{ix\psi(t)}$ oscillates less rapidly near a stationary point than it does near a point where $\psi'(t) \ne 0$. Consequently, there is less cancellation between adjacent subintervals near the stationary point.

The method of stationary phase gives the *leading* asymptotic behavior of generalized Fourier integrals having stationary points. This method is very similar to Laplace's method in that the leading contribution to $I(x)$ comes from a small interval of width ε surrounding the stationary points of $\psi(t)$. We will show that if c is a stationary point and if $f(c) \ne 0$, then $I(x)$ goes to zero like $x^{-1/2}$ as $x \to +\infty$ if $\psi''(c) \ne 0$, like $x^{-1/3}$ if $\psi''(c) = 0$ but $\psi'''(c) \ne 0$, and so on; as $\psi(t)$ becomes flatter at $t = c$, $I(x)$ vanishes less rapidly as $x \to +\infty$.

Since any generalized Fourier integral can be written as a sum of integrals in which $\psi'(t)$ vanishes only at an endpoint, we can explain the method of stationary phase for the special integral (6.5.1) in which $\psi'(a) = 0$ and $\psi'(t) \ne 0$ for $a < t \le b$.

We decompose $I(x)$ into two terms:

$$I(x) = \int_a^{a+\varepsilon} f(t)e^{ix\psi(t)}\,dt + \int_{a+\varepsilon}^{b} f(t)e^{ix\psi(t)}\,dt, \tag{6.5.8}$$

where ε is a small positive number to be chosen later. The second integral on the right side of (6.5.8) vanishes like $1/x$ as $x \to +\infty$ because there are no stationary points in the interval $a + \varepsilon \le t \le b$.

To obtain the leading behavior of the first integral on the right side of (6.5.8), we replace $f(t)$ by $f(a)$ and $\psi(t)$ by $\psi(a) + \psi^{(p)}(a)(t-a)^p/p!$ where $\psi^{(p)}(a) \neq 0$ but $\psi'(a) = \cdots = \psi^{(p-1)}(a) = 0$:

$$I(x) \sim \int_a^{a+\varepsilon} f(a) \exp\left\{ix\left[\psi(a) + \frac{1}{p!}\psi^{(p)}(a)(t-a)^p\right]\right\} dt, \qquad x \to +\infty. \tag{6.5.9}$$

Next, we replace ε by ∞, which introduces error terms that vanish like $1/x$ as $x \to +\infty$ and thus may be disregarded, and let $s = (t - a)$:

$$I(x) \sim f(a)e^{ix\psi(a)} \int_0^\infty \exp\left[\frac{ix}{p!}\psi^{(p)}(a)s^p\right] ds, \qquad x \to +\infty. \tag{6.5.10}$$

To evaluate the integral on the right, we rotate the contour of integration from the real-s axis by an angle $\pi/2p$ if $\psi^{(p)}(a) > 0$ and make the substitution

$$s = e^{i\pi/2p}\left[\frac{p!\,u}{x\psi^{(p)}(a)}\right]^{1/p} \tag{6.5.11a}$$

with u real or rotate the contour by an angle $-\pi/2p$ if $\psi^{(p)}(a) < 0$ and make the substitution

$$s = e^{-i\pi/2p}\left[\frac{p!\,u}{x\,|\psi^{(p)}(a)|}\right]^{1/p}. \tag{6.5.11b}$$

Thus,

$$I(x) \sim f(a)e^{ix\psi(a) \pm i\pi/2p}\left[\frac{p!}{x\,|\psi^{(p)}(a)|}\right]^{1/p}\frac{\Gamma(1/p)}{p}, \qquad x \to +\infty, \tag{6.5.12}$$

where we use the factor $e^{i\pi/2p}$ if $\psi^{(p)}(a) > 0$ and the factor $e^{-i\pi/2p}$ if $\psi^{(p)}(a) < 0$.

The formula in (6.5.12) gives the leading behavior of $I(x)$ if $f(a) \neq 0$ but $\psi'(a) = 0$. If $f(a)$ vanishes, it is necessary to decide whether the contribution from the stationary point still dominates the leading behavior. When it does, the behavior is slightly more complicated than (6.5.12) (see Prob. 6.53).

Example 3 *Leading behavior of* $\int_0^{\pi/2} e^{ix\cos t}\,dt$ *as* $x \to +\infty$. The function $\psi(t) = \cos t$ has a stationary point at $t = 0$. Since $\psi''(0) = -1$, (6.5.12) with $p = 2$ gives $I(x) \sim \sqrt{\pi/2x}\, e^{i(x-\pi/4)}$ $(x \to +\infty)$.

Example 4 *Leading behavior of* $\int_0^\infty \cos(xt^2 - t)\,dt$ *as* $x \to +\infty$. To use the method of stationary phase, we write this integral as $\int_0^\infty \cos(xt^2 - t)\,dt = \text{Re}\int_0^\infty e^{i(xt^2-t)}\,dt$. The function $\psi(t) = t^2$ has a stationary point at $t = 0$. Since $\psi''(0) = 2$, (6.5.12) with $p = 2$ gives $\int_0^\infty \cos(xt^2 - t)\,dt \sim \text{Re}\,\frac{1}{2}\sqrt{\pi/x}\,e^{i\pi/4} = \frac{1}{2}\sqrt{\pi/2x}$ $(x \to +\infty)$.

Example 5 *Leading behavior of* $J_n(n)$ *as* $n \to \infty$. When n is an integer, the Bessel function $J_n(x)$ has the integral representation

$$J_n(x) = \frac{1}{\pi}\int_0^\pi \cos(x \sin t - nt)\,dt \tag{6.5.13}$$

(see Prob. 6.54). Therefore, $J_n(n) = \text{Re}\int_0^\pi e^{in(\sin t - t)}\,dt/\pi$. The function $\psi(t) = \sin t - t$ has a stationary point at $t = 0$. Since $\psi''(0) = 0$, $\psi'''(0) = -1$, (6.5.12) with $p = 3$ gives

$$\begin{aligned} J_n(n) &\sim \frac{1}{\pi}\text{Re}\left[\frac{1}{3}e^{-i\pi/6}\left(\frac{6}{n}\right)^{1/3}\Gamma\left(\frac{1}{3}\right)\right], & x \to +\infty, \\ &= \frac{1}{\pi}2^{-2/3}3^{-1/6}\Gamma\left(\frac{1}{3}\right)n^{-1/3}, & n \to \infty. \end{aligned} \tag{6.5.14}$$

Observe that because $\psi''(0) = 0$, $J_n(n)$ vanishes less rapidly than $n^{-1/2}$ as $n \to \infty$.

If n is not an integer, (6.5.14) still holds (see Prob. 6.55).

In this section we have obtained only the leading behavior of generalized Fourier integrals. Higher-order approximations can be complicated because non-stationary points may also contribute to the large-x behavior of the integral. Specifically, the second integral on the right in (6.5.8) must be taken into account when computing higher-order terms because the error incurred in neglecting this integral is usually algebraically small. By contrast, recall that the approximation in (6.4.2) for Laplace's method is valid to all orders because the errors are exponentially, rather than algebraically, small. To obtain the higher-order corrections to (6.5.12), one can either use the method of asymptotic matching (see Sec. 7.4) or the method of steepest descents (see Sec. 6.6).

(I) 6.6 METHOD OF STEEPEST DESCENTS

The method of steepest descents is a technique for finding the asymptotic behavior of integrals of the form

$$I(x) = \int_C h(t)e^{x\rho(t)}\,dt \tag{6.6.1}$$

as $x \to +\infty$, where C is an integration contour in the complex-t plane and $h(t)$ and $\rho(t)$ are analytic functions of t. The idea of the method is to use the analyticity of the integrand to justify deforming the contour C to a new contour C' on which $\rho(t)$ has a constant imaginary part. Once this has been done, $I(x)$ may be evaluated asymptotically as $x \to +\infty$ using Laplace's method. To see why, observe that on the contour C' we may write $\rho(t) = \phi(t) + i\psi$, where ψ is a real constant and $\phi(t)$ is a real function. Thus, $I(x)$ in (6.6.1) takes the form

$$I(x) = e^{ix\psi}\int_{C'} f(t)e^{x\phi(t)}\,dt. \tag{6.6.2}$$

Although t is complex, (6.6.2) can be treated by Laplace's method as $x \to +\infty$ because $\phi(t)$ is real.

Our motivation for deforming C into a path C' on which Im $\rho(t)$ is a constant is to eliminate rapid oscillations of the integrand when x is large. Of course, one could also deform C into a path on which Re $\rho(t)$ is a constant and then apply the method of stationary phase. However, we have seen that Laplace's method is a much better approximation scheme than the method of stationary phase because the full asymptotic expansion of a generalized Laplace integral is determined by the integrand in an arbitrarily small neighborhood of the point where Re $\rho(t)$ is a maximum on the contour. By contrast, the full asymptotic expansion of a generalized Fourier integral typically depends on the behavior of the integrand along the entire contour. As a consequence, it is usually easier to obtain the full asymptotic expansion of a generalized Laplace integral than of a generalized Fourier integral.

Before giving a formal exposition of the method of steepest descents, we consider three preliminary examples which illustrate how shifting complex contours can greatly simplify asymptotic analysis. In the first example we consider a Fourier integral whose asymptotic expansion is difficult to find by the methods used in Sec. 6.5. However, deforming the contour reduces the integral to a pair of integrals that are easy to evaluate by Laplace's method.

Example 1 *Conversion of a Fourier integral into a Laplace integral by deforming the contour.* The behavior of the integral

$$I(x) = \int_0^1 \ln t \, e^{ixt} \, dt \tag{6.6.3}$$

as $x \to +\infty$ cannot be found directly by the methods of Sec. 6.5 because there is no stationary point. Also, integration by parts is useless because $\ln 0 = -\infty$. Integration by parts is doomed to fail because, as we will see, the leading asymptotic behavior of $I(x)$ contains the factor $\ln x$ which is not a power of $1/x$.

To approximate $I(x)$ we deform the integration contour C, which runs from 0 to 1 along the real-t axis, to one which consists of three line segments: C_1, which runs up the imaginary-t axis from 0 to iT; C_2, which runs parallel to the real-t axis from iT to $1 + iT$; and C_3, which runs down from $1 + iT$ to 1 along a straight line parallel to the imaginary-t axis (see Fig. 6.5). By Cauchy's theorem, $I(x) = \int_{C_1+C_2+C_3} \ln t \, e^{ixt} \, dt$. Next we let $T \to +\infty$. In this limit the contribution from C_2 approaches 0. (Why?) In the integral along C_1 we set $t = is$, and in the integral along C_3 we set $t = 1 + is$, where s is real in both integrals. This gives

$$I(x) = i \int_0^\infty \ln (is) \, e^{-xs} \, ds - i \int_0^\infty \ln (1 + is) \, e^{ix(1+is)} \, ds. \tag{6.6.4}$$

The sign of the second integral on the right is negative because C_3 is traversed downward.

Observe that both integrals in (6.6.4) are Laplace integrals. The first integral can be done exactly. We substitute $u = xs$ and use $\ln (is) = \ln s + i\pi/2$ and the identity $\int_0^\infty e^{-u} \ln u \, du = -\gamma$, where $\gamma = 0.5772\ldots$ is Euler's constant, and obtain

$$i \int_0^\infty \ln (is) \, e^{-xs} \, ds = -i(\ln x)/x - (i\gamma + \pi/2)/x.$$

We apply Watson's lemma to the second integral on the right in (6.6.4) using the Taylor expansion $\ln (1 + is) = -\sum_{n=1}^\infty (-is)^n/n$, and obtain

$$-i \int_0^\infty \ln (1 + is) \, e^{ix(1+is)} \, ds \sim ie^{ix} \sum_{n=1}^\infty \frac{(-i)^n (n-1)!}{x^{n+1}}, \qquad x \to +\infty.$$

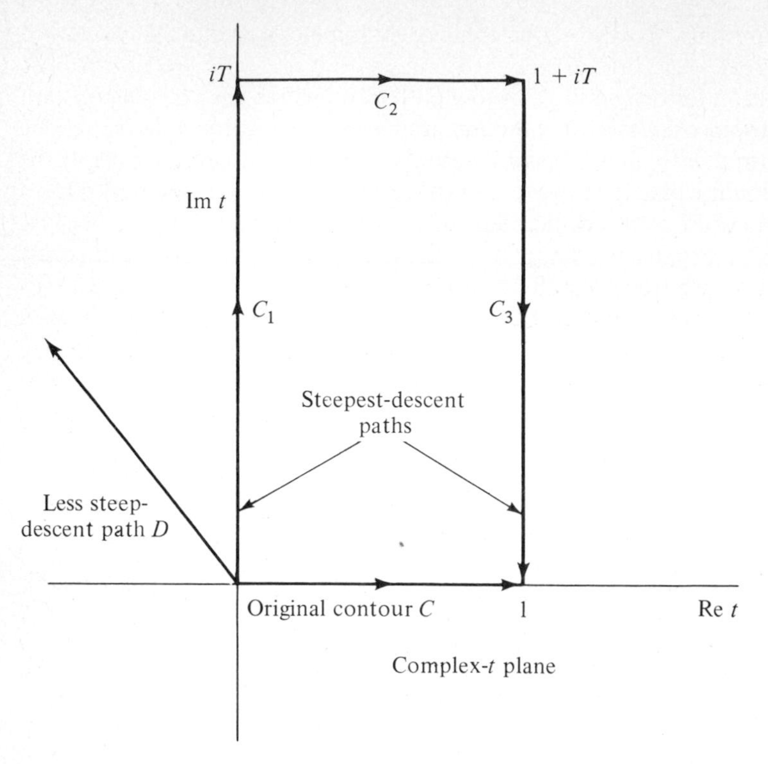

Figure 6.5 It is possible to convert the Fourier integral $I(x)$ in (6.6.3) into a Laplace integral merely by deforming the original contour C into $C_1 + C_2 + C_3$ as shown above and then allowing $T \to \infty$. C_1 and C_3 are called steepest-descent paths because $|\exp[x\rho(t)]|$ decreases most rapidly along these paths as t moves up from the real-t axis; $|\exp[x\rho(t)]|$ also decreases along D, but less rapidly per unit length than along C_1.

Combining the above two expansions gives the final result:

$$I(x) \sim -\frac{i \ln x}{x} - \frac{i\gamma + \pi/2}{x} + ie^{ix} \sum_{n=1}^{\infty} \frac{(-i)^n (n-1)!}{x^{n+1}}, \qquad x \to +\infty.$$

Let us review the calculation in the preceding example. For the integral (6.6.3), $\rho(t) = it$. For this function, paths of constant Im $\rho(t)$ are straight lines parallel to the imaginary-t axis. On the particular contours C_1 and C_3, Im $\rho(t) = 0$ and 1, respectively. Note that Im $\rho(t)$ is not the same constant on C_1 and C_3, but this does not matter; we have applied Laplace's method separately to each of the integrals on the right side of (6.6.4). Since Im $\rho(t = 0) \neq$ Im $\rho(t = 1)$, it is clear that there is no continuous contour joining $t = 0$ and $t = 1$ on which Im $\rho(t)$ is constant. This is why it is necessary to deform the original contour C into C_1 and C_3 which are joined at ∞ by C_2 along which the integrand vanishes. In general, we expect that if Im $\rho(t)$ is not the same at the endpoints of the original integration contour C, then we cannot deform C into a continuous contour on which Im $\rho(t)$ is constant; the best one can hope for is to be able to deform C into distinct constant-phase contours which are joined by a contour on which the integrand vanishes.

Now we can explain why the procedure used in Example 1 is called the method of steepest descents. The contours C_1 and C_3 are called contours of constant phase because the phase of the complex number $e^{x\rho(t)}$ is constant. At the same time, C_1 and C_3 are also called steepest-descent paths because $|e^{x\rho(t)}|$ decreases most rapidly along these paths as t ranges from the endpoints 0 and 1 toward ∞. Any path originating at the endpoints 0 and 1 and moving upward in the complex-t plane is a path on which $|e^{x\rho(t)}|$ decreases (see Fig. 6.5). However, after traversing any given length of arc, $|e^{x\rho(t)}|$ decreases more along the vertical paths C_1 and C_3 than along any other path leaving the endpoints 0 and 1, respectively. We will explain this feature of steepest-descent paths later in this section.

Example 2 *Full asymptotic behavior of* $\int_0^1 e^{ixt^2}\,dt$ *as* $x \to +\infty$. The method of stationary phase can be used to find the leading behavior of the integral $I(x) = \int_0^1 e^{ixt^2}\,dt$. Here $\psi(t) = t^2$, so the stationary point lies at $t = 0$ and, using (6.5.12), $I(x) \sim \frac{1}{2}\sqrt{\pi/x}\, e^{i\pi/4}$ $(x \to +\infty)$. The method of steepest descents gives an easy way to determine the full asymptotic behavior of $I(x)$. [The method of integration by parts also works (see Prob. 6.57).]

As in Example 1, we try to deform the contour C: $0 \le t \le 1$ into contours along which Im $\rho(t)$ is constant, where $\rho(t) = it^2$. We begin by finding a contour which passes through $t = 0$ and on which Im $\rho(t)$ is constant. Writing $t = u + iv$ with u and v real, we obtain Im $\rho(t) = u^2 - v^2$. At $t = 0$, Im $\rho = 0$. Therefore, constant-phase contours passing through $t = 0$ must satisfy $u = v$ or $u = -v$ everywhere along the contour (see Fig. 6.6). On the contour $u = -v$, Re $\rho(t) = 2v^2$, so $|e^{x\rho(t)}| = e^{2xv^2}$ increases as $t = (i - 1)v \to \infty$. This is called a steepest-ascent contour; since there is no maximum of $|e^{x\rho(t)}|$ on this contour, Laplace's method cannot be applied. On the other hand, the contour $u = v$ is a steepest-descent contour because Re $\rho(t) = -2v^2$, so $|e^{x\rho(t)}| = e^{-2xv^2}$ decreases as $t = (1 + i)v \to \infty$. The contour C_1: $t = (1 + i)v$ $(0 \le v < \infty)$ is comparable to the contour C_1 employed in Example 1.

Next, we must find a steepest-descent contour passing through $t = 1$ along which Im $\rho(t)$ is constant. At $t = 1$, the value of Im $\rho(t)$ is 1. Therefore, the constant-phase contour passing through $u = 1$, $v = 0$ is given by $u = \sqrt{v^2 + 1}$. Since Re $\rho(t) = -2uv$ decreases as $t = u + iv \to \infty$ along the portion of this constant-phase contour with $0 \le v < \infty$, the steepest-descent contour passing through $t = 1$ is given by C_3: $t = \sqrt{v^2 + 1} + iv$, $0 \le v < \infty$. Note that C_1 and C_3 become tangent as $v \to +\infty$ (see Fig. 6.6).

The next step is to deform the original contour C: $0 \le t \le 1$ into $C_1 + C_3$, in which C_3 is traversed from $t = \infty$ to $t = 1$. Along C_1, Im $\rho(t) = 0$, while along C_3, Im $\rho(t) = 1$. Since the value of Im $\rho(t)$ is different on C_1 and C_3, it is clear that the original contour cannot be continuously deformed into $C_1 + C_3$. Rather, we must include a third contour C_2 which bridges the gap between C_1 and C_3. We take C_2 to be the straight line connecting the points $(1 + i)V$ on C_1 and $\sqrt{V^2 + 1} + iV$ on C_3 (see Fig. 6.6). C can be continuously deformed into C_2 together with the portions of C_1 and C_3 satisfying $0 \le v \le V$. Now, as $V \to \infty$, the contribution from the contour C_2 vanishes. (Why?) Thus,

$$I(x) = \int_{C_1} e^{ixt^2}\,dt - \int_{C_3} e^{ixt^2}\,dt. \tag{6.6.5}$$

The integral along C_1 can be evaluated exactly. Setting $t = (1 + i)v$, we obtain

$$\int_{C_1} e^{ixt^2}\,dt = (1 + i)\int_0^\infty e^{-2xv^2}\,dv = \frac{1}{2}\sqrt{\frac{\pi}{x}}\,e^{i\pi/4}. \tag{6.6.6}$$

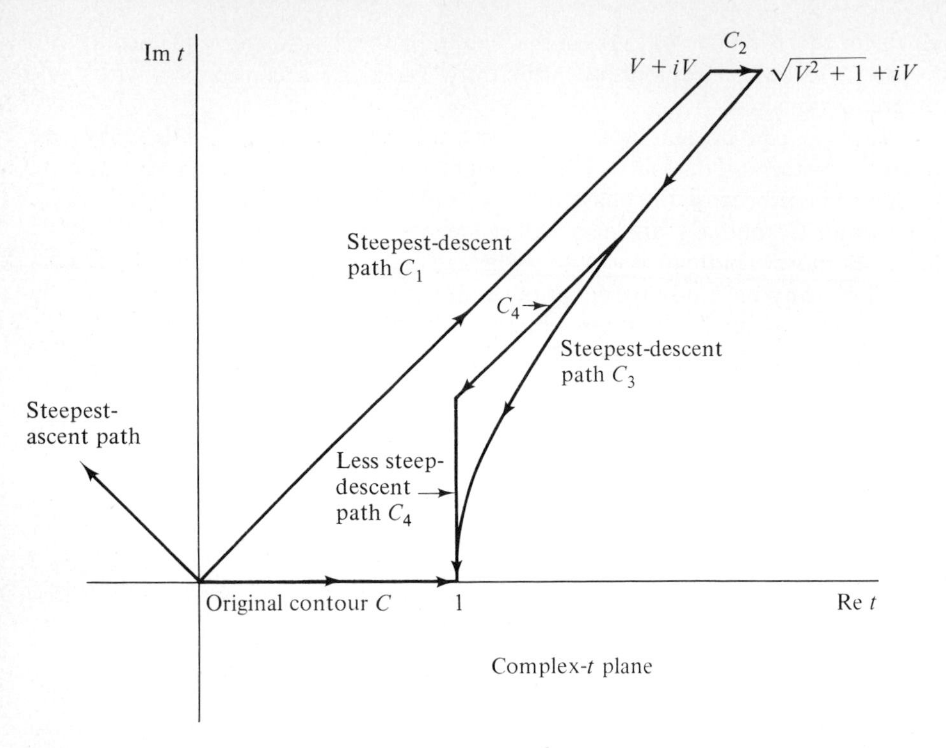

Figure 6.6 The Fourier integral $I(x)$ in Example 2 becomes a pair of Laplace integrals if the original contour C is distorted into $C_1 + C_2 + C_3$ and V is allowed to approach ∞. To simplify the evaluation of the integral along C_3, we can replace the lower part of the contour C_3 by C_4.

This contribution is precisely the leading behavior of $I(x)$ as $x \to +\infty$ that we found using the method of stationary phase.

Now we evaluate the contribution to $I(x)$ from the integral on C_3. Note that if we substitute $t = \sqrt{v^2 + 1} + iv$, $0 \le v < \infty$, then $\rho(t) = it^2 = i - 2v\sqrt{v^2 + 1}$. This verifies that C_3 is a curve of constant phase; it is also a curve of steepest descent. An easy way to obtain the full asymptotic expansion of the integral over C_3 is to use Watson's lemma. To do this, the integral must be expressed in the form $\int_0^\infty f(s)e^{-xs}\,ds$. This motivates the change of variables from t to s where s is defined by

$$\rho(t) = it^2 = i - s; \tag{6.6.7}$$

observe that $s = 2v\sqrt{v^2 + 1}$ is real and satisfies $0 \le s < \infty$ along C_3. Since $t = (1 + is)^{1/2}$, $dt/ds = \frac{1}{2}i(1 + is)^{-1/2}$, so

$$\int_{C_3} e^{ixt^2}\,dt = \tfrac{1}{2}ie^{ix} \int_0^\infty \frac{e^{-xs}}{\sqrt{1 + is}}\,ds.$$

To apply Watson's lemma, we use the Taylor expansion

$$(1 + is)^{-1/2} = \sum_{n=0}^{\infty} (-is)^n \Gamma(n + \tfrac{1}{2})/n!\,\Gamma(\tfrac{1}{2}).$$

We obtain

$$\int_{C_3} e^{ixt^2}\, dt \sim \frac{1}{2} ie^{ix} \sum_{n=0}^{\infty} (-i)^n \frac{\Gamma(n+\frac{1}{2})}{\Gamma(\frac{1}{2})x^{n+1}}, \qquad x \to +\infty. \tag{6.6.8}$$

Combining this result with that in (6.6.6) gives the full asymptotic expansion of $I(x)$ as $x \to +\infty$:

$$I(x) \sim \frac{1}{2}\sqrt{\frac{\pi}{x}}\, e^{i\pi/4} - \frac{1}{2} ie^{ix} \sum_{n=0}^{\infty} (-i)^n \frac{\Gamma(n+\frac{1}{2})}{\Gamma(\frac{1}{2})x^{n+1}}, \qquad x \to +\infty. \tag{6.6.9}$$

Finally, we mention an alternative way to obtain the result in (6.6.8) for the integral on C_3. The substitution in (6.6.7) is an exact parametrization of the curve C_3 in terms of the real parameter s. However, as we know from our discussion of Laplace's method in Sec. 6.4, it is only the immediate neighborhood of the maximum at $t = 1$ that contributes to the full asymptotic expansion of the integral on C_3. Therefore, it is not necessary to follow the curve C_3 exactly. It is correct to shift the integration path C_3 to one which still passes through the maximum at $t = 1$ and which is a descent contour in the sense that $|e^{x\rho(t)}|$ decreases and rejoins C_3 for large $|t|$. Any deformation of C_3 of this kind does not change the value of the integral because the integrand is analytic. For the present example, a convenient alternative to C_3 is a contour C_4 which originates at $t = 1$, goes vertically upward parallel to the imaginary-t axis, and then rejoins the contour C_3 at any point in the upper half plane (see Fig. 6.6). Only the vertical straight-line portion of C_4 in the immediate vicinity of $t = 1$ contributes to the full asymptotic expansion of the integral. We can parametrize the straight-line portion of C_4 near $t = 1$ by $t = 1 + iv$, where v is real and $0 \le v \le \varepsilon$ with ε small. Thus,

$$\int_{C_3} e^{ixt^2}\, dt = \int_{C_4} e^{ixt^2}\, dt \sim i \int_0^{\varepsilon} e^{ix(1+iv)^2}\, dv$$

$$= ie^{ix} \int_0^{\varepsilon} e^{-2xv} e^{-ixv^2}\, dv, \qquad x \to +\infty.$$

Using Laplace's method

$$\int_0^{\varepsilon} e^{-2xv} e^{-ixv^2}\, dv \sim \int_0^{\varepsilon} e^{-2xv} \sum_{n=0}^{\infty} \frac{(-ix)^n v^{2n}}{n!}\, dv$$

$$\sim \sum_{n=0}^{\infty} \frac{(-i)^n (2n)!}{2^{2n+1} n!\, x^{n+1}}, \qquad x \to +\infty.$$

Since $(2n)!/(2^{2n}n!) = \Gamma(n+\frac{1}{2})/\Gamma(\frac{1}{2})$, we have reproduced (6.6.8) exactly.

This alternative calculation, in which we have replaced the curved path C_3 by a path C_4 which begins as a straight line, is an important computational device that is frequently helpful in the method of steepest descents. Note that C_4 is neither a curve of constant phase nor a curve of steepest descent, although it is a curve of descent of $|e^{x\rho(t)}|$. Other descent curves could be used instead of C_4 (see Prob. 6.58).

Example 3 *Sophisticated example of the method of steepest descents.* What is the leading behavior of the generalized Fourier integral

$$I(x) = \int_0^1 \exp\,(ixe^{-1/s})\, ds \tag{6.6.10}$$

as $x \to +\infty$? This is a sophisticated example because $s = 0$ is an infinite-order stationary point; i.e., all derivatives of $e^{-1/s}$ vanish as $s \to 0+$. We know from our discussion of the method of stationary phase that if the first nonvanishing derivative of ψ in (6.5.1) at a stationary point is $\psi^{(p)}$, then $I(x)$ must vanish like $x^{-1/p}$ as $x \to +\infty$. Therefore, we expect that if the integrand has an infinite-order stationary point, $I(x)$ vanishes less rapidly than any power of $1/x$ as $x \to +\infty$. However, the Riemann-Lebesgue lemma guarantees that $I(x)$ does indeed vanish as $x \to +\infty$.

How fast does $I(x)$ in (6.6.10) vanish? It is hard to apply the method of stationary phase to $I(x)$ directly. (Try it!) However, the method of steepest descents provides a relatively easy approach. We begin by making the substitution $t = e^{-1/s}$:

$$I(x) = \int_0^{1/e} \frac{e^{ixt}}{t(\ln t)^2}\,dt.$$

The form of this integral is similar to that of the integral (6.6.3) considered in Example 1. Therefore, as in Example 1, we shift the contour $C: 0 \le t \le 1/e$ to two vertical lines parallel to the imaginary-t axis:

$$I(x) = \int_{C_1} \frac{e^{ixt}}{t(\ln t)^2}\,dt - \int_{C_3} \frac{e^{ixt}}{t(\ln t)^2}\,dt, \tag{6.6.11}$$

where C_1 is the path $t = iv$ $(0 \le v < \infty)$ and C_3 is the path $t = 1/e + iv$ $(0 \le v < \infty)$. Now we find the leading behavior of each of the integrals on the right in (6.6.11).

The integral on the path C_3 requires only a straightforward application of Laplace's method. We substitute $t = 1/e + iv$ $(0 \le v < \infty)$ and obtain

$$\int_{C_3} \frac{e^{ixt}}{t(\ln t)^2}\,dt = ie^{ix/e} \int_0^\infty \frac{e^{-xv}}{(1/e + iv)[\ln(1/e + iv)]^2}\,dv \sim ie^{ix/e}e/x, \qquad x \to +\infty. \tag{6.6.12}$$

The integral on C_1 is more difficult. We simplify the integral by substituting $t = iv$ $(0 \le v < \infty)$ and perform one integration by parts:

$$I_1(x) = \int_{C_1} \frac{e^{ixt}}{t(\ln t)^2}\,dt = \int_0^\infty \frac{e^{-xv}}{v[\ln(iv)]^2}\,dv = -x \int_0^\infty \frac{e^{-xv}}{\ln(iv)}\,dv. \tag{6.6.13}$$

The integral on the right side of (6.6.13) is a Laplace integral; we can restrict the range of integration to the vicinity of $v = 0$ without altering its asymptotic expansion as $x \to +\infty$. Thus,

$$I_1(x) \sim -x \int_0^\varepsilon \frac{e^{-xv}}{\ln(iv)}\,dv, \qquad x \to +\infty.$$

This integral does not yield to a straightforward application of Laplace's method because the integrand vanishes at $v = 0$. Moreover, the conventional treatment of a moving maximum [see the derivation of the Stirling series for $\Gamma(x)$ given in Example 10 of Sec. 6.4] does not work because the moving maximum of the integrand is too broad (see Prob. 6.47). A good way to proceed is to substitute $r = xv$ and thus obtain

$$I_1(x) \sim -\int_0^{\varepsilon x} \frac{e^{-r}}{\ln r - \ln x + i\pi/2}\,dr, \qquad x \to +\infty,$$

where we have used the relation $\ln(iv) = \ln v + i\pi/2$. Next, we argue that the immediate vicinity of the origin, say $0 \le r \le 1/x^{1/2}$, does not contribute to the asymptotic expansion of the integral as $x \to +\infty$. To prove this we bound the contribution to $I_1(x)$ from $0 \le r \le 1/x^{1/2}$:

$$\left| \int_0^{1/x^{1/2}} \frac{e^{-r}}{\ln r - \ln x + i\pi/2}\,dr \right| \le \frac{2}{\pi x^{1/2}},$$

because $|\ln r - \ln x + i\pi/2| \ge \pi/2$ and $|e^{-r}| \le 1$. This contribution to $I_1(x)$ is negligible because, as we shall see, the full asymptotic expansion of $I_1(x)$ is a series in inverse powers of $\ln x$. Thus,

$$I_1(x) \sim -\int_{1/x^{1/2}}^{\varepsilon x} \frac{e^{-r}}{\ln r - \ln x + i\pi/2}\,dr, \qquad x \to +\infty. \tag{6.6.14}$$

To expand the integral in (6.6.14), we Taylor expand the integrand in powers of $1/\ln x$:

$$\frac{1}{\ln r - \ln x + i\pi/2} = -\frac{1}{\ln x} \sum_{n=0}^\infty \left(\frac{i\pi/2 + \ln r}{\ln x}\right)^n, \qquad x^{-1/2} \le r \le \varepsilon x,\ x \to +\infty.$$

Thus,
$$I_1(x) \sim \frac{1}{\ln x} \sum_{n=0}^{\infty} \int_{1/x^{1/2}}^{\varepsilon x} e^{-r} \left(\frac{i\pi/2 + \ln r}{\ln x} \right)^n dr, \qquad x \to +\infty. \tag{6.6.15}$$

The range of each of the integrals in (6.6.15) can be extended to $0 \le r < \infty$ with an error smaller than any inverse power of $\ln x$. (Why?) Evaluating the first two integrals, we obtain

$$I_1(x) \sim \frac{1}{\ln x} + \frac{i\pi/2 - \gamma}{(\ln x)^2} + \cdots, \tag{6.6.16}$$

where we have used $\int_0^\infty \ln r e^{-r} \, dr = -\gamma$. The coefficient of the general term in (6.6.16) may be expressed in terms of derivatives of $\Gamma(t)$ at $t = 1$ (see Prob. 6.59).

Combining the results (6.6.12) and (6.6.16) with (6.6.11), we obtain the final result

$$I(x) \sim \frac{1}{\ln x} + \frac{i\pi/2 - \gamma}{(\ln x)^2} + \cdots, \qquad x \to +\infty. \tag{6.6.17}$$

One could not have guessed this result from a cursory inspection of the original integral in (6.6.10)! Does this asymptotic series diverge? (See Prob. 6.60.) In Table 6.1 we compare numerical values of $I(x)$ with the asymptotic results for $I(x)$ given in (6.6.17).

Formal Discussion of Steepest-Descent Paths in the Complex Plane

In the previous three introductory examples, we have shown that deforming contours of integration in the complex-t plane can facilitate the asymptotic evaluation of integrals. It is now appropriate to give a more general discussion of steepest-descent (constant-phase) contours.

We begin by recalling the role of the gradient in elementary calculus. If $f(u, v)$ is a differentiable function of two variables, then the gradient of f is the vector $\nabla f = (\partial f/\partial u, \partial f/\partial v)$. This vector points in the direction of the most rapid change of f at the point (u, v). In terms of the gradient, the directional derivative df/ds in the direction of the unit vector $\mathbf{n}$ is $df/ds = \mathbf{n} \cdot \nabla f$. This directional derivative is the rate of change of f in the direction $\mathbf{n}$. Thus, the largest directional derivative is in the direction $\mathbf{n} = \nabla f / |\nabla f|$ and has magnitude $|\nabla f|$. On a two-dimensional contour plot of $f(u, v)$, the vector ∇f is perpendicular to the contours of constant f

Table 6.1 Comparison between the exact value of the integral $I(x)$ in (6.6.10) and one-term and two-term asymptotic approximations to $I(x)$ in (6.6.17) obtained using the method of steepest descents

$\ln x$	Exact value of $I(x)$	One-term asymptotic approximation	Two-term asymptotic approximation
0	$0.9814 + 0.1467i$	∞	∞
2	$0.3077 + 0.5419i$	0.5000	$0.3557 + 0.3927i$
4	$0.2499 + 0.0643i$	0.2500	$0.2139 + 0.0982i$
6	$0.1428 + 0.0423i$	0.1667	$0.1506 + 0.0436i$
8	$0.1146 + 0.0227i$	0.1250	$0.1160 + 0.0245i$
10	$0.0935 + 0.0143i$	0.1000	$0.0942 + 0.0157i$
12	$0.0790 + 0.0100i$	0.0833	$0.0793 + 0.0109i$

(level curves). Note that the directional derivative in the direction of the tangents to a level curve is 0.

We will now give a formal proof that constant-phase contours are also steepest contours. Let $\rho(t) = \phi(t) + i\psi(t)$ be an analytic function of the complex variable $t = u + iv$. Also, for the moment, we restrict ourselves to regions of the complex-t plane in which $\rho'(t) \neq 0$.

We define a constant-phase contour of $e^{x\rho(t)}$ where $x > 0$ as a contour on which $\psi(t)$ is constant. A steepest contour is defined as a contour whose tangent is parallel to $\nabla |e^{x\rho(t)}| = \nabla e^{x\phi(t)}$, which is parallel to $\nabla\phi$. That is, a steepest contour is one on which the magnitude of $e^{x\rho(t)}$ is changing most rapidly with t.

Now we will show that if $\rho(t)$ is analytic, then constant-phase contours are steepest contours. If $\rho(t)$ is analytic, then it satisfies the Cauchy-Riemann equations

$$\partial\phi/\partial u = \partial\psi/\partial v, \qquad \partial\phi/\partial v = -\partial\psi/\partial u.$$

Therefore,

$$(\partial\phi/\partial u)(\partial\psi/\partial u) + (\partial\phi/\partial v)(\partial\psi/\partial v) = 0.$$

However, this equation can be written in vector form as $\nabla\phi \cdot \nabla\psi = 0$, so $\nabla\phi$ is perpendicular to $\nabla\psi$ and the directional derivative in the direction of $\nabla\phi$ satisfies $d\psi/ds = 0$. Thus, ψ is constant on contours whose tangents are parallel to $\nabla\phi$, showing that constant-phase contours are also steepest contours.

There is a slightly more sophisticated way to establish that constant-phase contours are steepest contours. It is well known that an analytic function $\rho(t)$ is a conformal (angle-preserving) mapping from the complex-t plane (u, v) to the complex-ρ plane (ϕ, ψ) if $\rho'(t) \neq 0$. Therefore, since lines of constant u are perpendicular to lines of constant v, lines of constant ϕ are perpendicular to lines of constant ψ. But lines of constant ϕ are also perpendicular to steepest curves of ϕ. This reestablishes the identity of steepest and constant-phase contours.

In the above two arguments, it was necessary to assume that $\rho'(t) \neq 0$. In the second argument, this condition was necessary because a map is not conformal at a point where $\rho'(t) = 0$. Where was this condition used in the first argument?

Saddle Points

When the contour of integration in (6.6.1) is deformed into constant-phase contours, the asymptotic behavior of the integral is determined by the behavior of the integrand near the local maxima of $\phi(t)$ along the contour. These local maxima of $\phi(t)$ may occur at endpoints of constant-phase contours (see Examples 1 to 3) or at an interior point of a constant-phase contour. If $\phi(t)$ has an interior maximum then the directional derivative along the constant-phase contour $d\phi/ds = |\nabla\phi|$ vanishes. The Cauchy-Riemann equations imply that $\nabla\phi = \nabla\psi = 0$ so $\rho'(t) = 0$ at an interior maximum of ϕ on a constant-phase contour.

A point at which $\rho'(t) = 0$ is called a *saddle point*. Saddle points are special because it is only at such a point that two distinct steepest curves can intersect. When $\rho'(t_0) \neq 0$, there is only one steepest curve passing through t and its tangent is parallel to $\nabla\phi$. In the direction of $\nabla\phi$, $|e^{x\rho}|$ is increasing so this portion of the curve is a steepest-*ascent* curve; in the direction of $-\nabla\phi$, $|e^{x\rho}|$ is decreasing so this portion of the curve is a steepest-*descent* curve. On the other hand, when $\rho'(t_0) = 0$ there are two or more steepest-ascent curves and two or more steepest-descent curves emerging from the point t_0.

To study the nature of the steepest curves emerging from a saddle point, let us study the region of the complex-t plane near t_0.

Example 4 *Steepest curves of e^{xt^2} near the saddle point $t = 0$.* Here $\rho(t) = t^2$. Observe that $\rho'(t) = 2t$ vanishes at $t = 0$, which verifies that 0 is a saddle point. We substitute $t = u + iv$ and identify the real and imaginary parts of $\rho(t)$:

$$\rho(t) = u^2 - v^2 + 2iuv, \qquad \phi(t) = u^2 - v^2, \qquad \psi(t) = 2uv.$$

Since $\rho(0) = 0$, the constant-phase contours that pass through $t = 0$ must satisfy $\psi(t) = 0$ everywhere. The constant-phase contours $u = 0$ (the imaginary axis) and $v = 0$ (the real axis) cross at the saddle point $t = 0$.

All four curves that emerge from $t = 0$, (*a*) $u = 0$ with v positive, (*b*) $u = 0$ with v negative, (*c*) $v = 0$ with u positive, and (*d*) $v = 0$ with u negative, are steepest curves because $\rho'(t) \neq 0$ except at $t = 0$. Which of these four curves are steepest-ascent curves and which are steepest-descent curves? On curves (*a*) and (*b*), $\phi(t) = -v^2$, so ϕ is decreasing away from $t = 0$; these curves are steepest-descent curves. On curves (*c*) and (*d*), $\phi(t) = u^2$, so ϕ is increasing away from $t = 0$; these curves are steepest-ascent curves. A plot showing these steepest-ascent and -descent curves as well as the level curves of ϕ away from $t = 0$ is given in Fig. 6.7.

Example 5 *Steepest curves of $e^{ix \cosh t}$ near the saddle point $t = 0$.* Here $\rho(t) = i \cosh t$, so $\rho'(t) = i \sinh t$ vanishes at $t = 0$. If we substitute $t = u + iv$ and use the identity

$$\cosh (u + iv) = \cosh u \cos v + i \sinh u \sin v,$$

we obtain the real and imaginary parts of $\rho(t)$:

$$\phi(t) = -\sinh u \sin v, \qquad \psi(t) = \cosh u \cos v.$$

Since $\rho(0) = i$, the constant-phase contours passing through $t = 0$ must satisfy $\psi(t) = \text{Im}\, \rho(t) = 1$. Thus, the constant-phase contours through $t = 0$ are given by

$$\cosh u \cos v = 1.$$

Other constant-phase contours (steepest-descent and -ascent curves) are given by $\cosh u \cos v = c$, where c is a constant. On Fig. 6.8 we plot the constant-phase contours for various values of c. Observe that steepest curves never cross except at saddle points.

Example 6 *Steepest curves of $e^{x(\sinh t - t)}$ near the saddle point at $t = 0$.* Here $\rho(t) = \sinh t - t$, so $\rho'(t) = \cosh t - 1$ vanishes at $t = 0$. Note that $\rho''(t) = \sinh t$ also vanishes at 0 and that the lowest nonvanishing derivative of ρ at $t = 0$ is $\rho'''(t)$. We call such a saddle point a third-order saddle point. At $t = 0$ six constant-phase contours meet. To find these contours we substitute $t = u + iv$ and identify the real and imaginary parts of ρ:

$$\rho = \phi + i\psi = (\sinh u \cos v - u) + i(\cosh u \sin v - v).$$

But $\rho(0) = 0$. Thus, constant-phase contours passing through $t = 0$ satisfy $\cosh u \sin v - v = 0$. Solutions to this equation are $v = 0$ (the u axis) and $u = \text{arc cosh } (v/\sin v)$.

In Prob. 6.61 you are asked to verify that (*a*) a total of six steepest paths emerge from $t = 0$; (*b*) paths emerge at 60° angles from adjacent paths; (*c*) as t moves away from 0, the paths alternate between steepest-ascent and steepest-descent paths; (*d*) the paths approach $\pm\infty$, $\pm\infty + i\pi$, $\pm\infty - i\pi$. All these results are shown on Fig. 6.9.

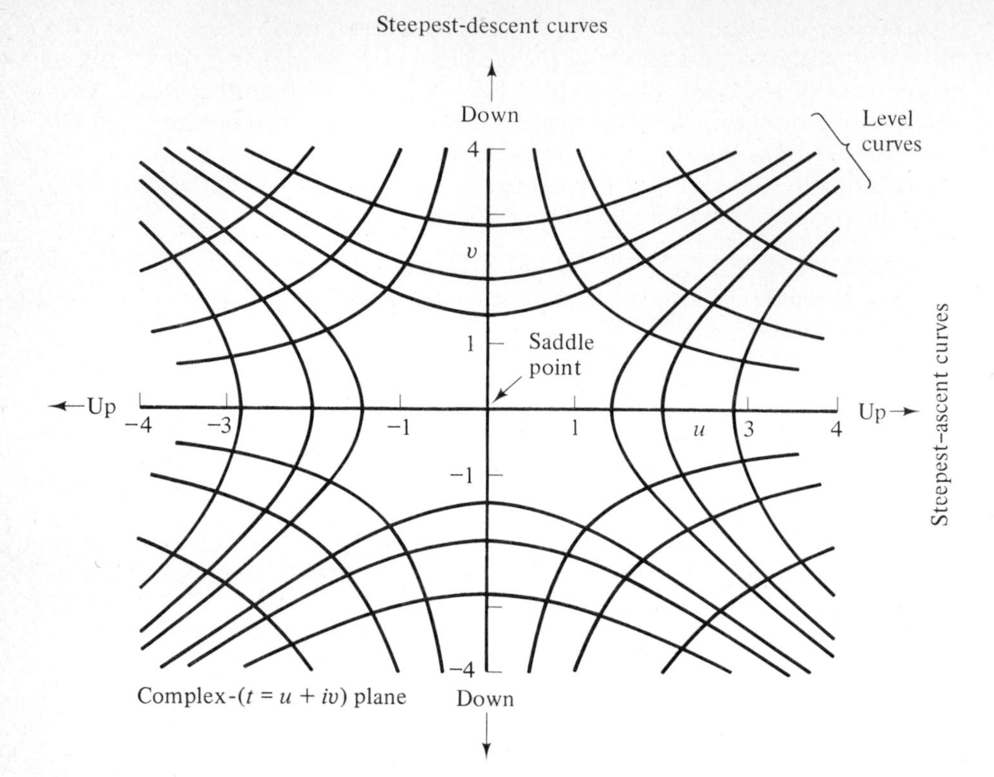

Figure 6.7 Steepest curves of e^{xt^2} near the saddle point at $t = 0$ in the complex-t plane. The steepest curves satisfy uv = constant. The level curves of ϕ satisfy $u^2 - v^2$ = constant and are orthogonal to the steepest curves.

Example 7 *Steepest curves of* $e^{x(\cosh t - t^2/2)}$ *near the saddle point at* $t = 0$. Here $\rho(t) = \cosh t - t^2/2$. Note that $\rho'(t)$, $\rho''(t)$, and $\rho'''(t)$ all vanish at $t = 0$. The first nonvanishing derivative of $\rho(t)$ at $t = 0$ is $d^4\rho/dt^4$, so we call $t = 0$ a fourth-order saddle point. Eight constant-phase curves meet at $t = 0$. Note that

$$\rho(t) = \cosh u \cos v + (v^2 - u^2)/2 + i(\sinh u \sin v - uv).$$

Thus, constant-phase contours emerging from $t = 0$ satisfy $\psi = \sinh u \sin v - uv = 0$. Solutions to this equation are $u = 0$ (the imaginary axis), $v = 0$ (the real axis), and $(\sinh u)/u = v/\sin v$.

In Prob. 6.62 you are asked to verify the results on Fig. 6.10. Namely, that (*a*) eight steepest paths emerge from $t = 0$, all equally spaced at 45° from each other; (*b*) as t moves away from 0, the paths alternate between steepest-ascent and steepest-descent paths; (*c*) the four steepest-ascent paths lie on the u and v axes; (*d*) the four steepest-descent paths approach $\pm\infty + i\pi$, $\pm\infty - i\pi$.

Steepest-Descent Approximation to Integrals with Saddle Points

We have seen that by shifting the integration contour so that it follows a path of constant phase we can treat an integral of the form in (6.6.1) by Laplace's method.

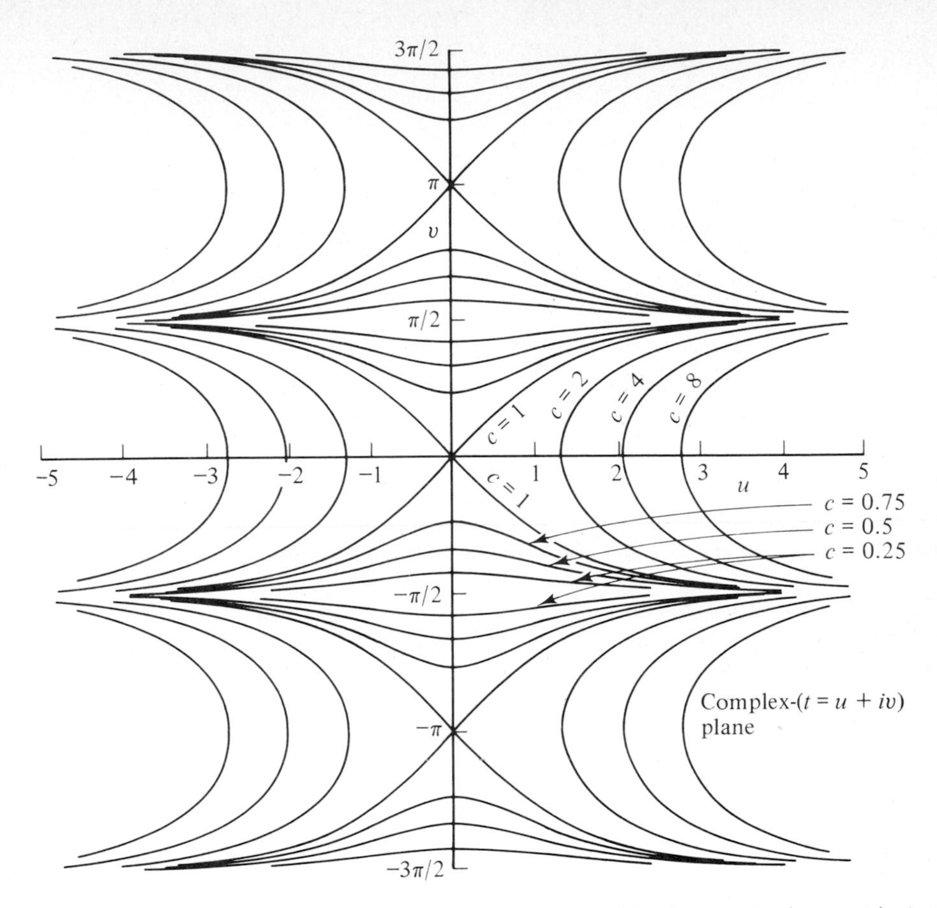

Figure 6.8 Constant-phase (steepest) contours of exp $(ix \cosh t)$ in the complex-$(t = u + iv)$ plane. Constant-phase contours satisfy $(\cosh u)(\cos v) = c$, where c is a constant. Saddle points lying at $t = 0$ and $t = \pm i\pi$ are shown.

What happens when the constant-phase contour passes through a saddle point? In the following examples we encounter this situation.

Example 8 *Asymptotic expansion of* $J_0(x)$ *as* $x \to +\infty$. A standard integral representation for $J_0(x)$ [see (6.5.13)] is $J_0(x) = \int_{-\pi/2}^{\pi/2} \cos(x \cos\theta)\, d\theta/\pi$, which can be transformed into

$$J_0(x) = \operatorname{Re} \frac{1}{i\pi} \int_{-i\pi/2}^{i\pi/2} dt\, e^{ix \cosh t} \tag{6.6.18a}$$

by substituting $t = i\theta$.

We can certainly use the method of stationary phase to find the leading behavior of this integral as $x \to +\infty$ (see Prob. 6.54). However, it is better to use the method of steepest descents to find the higher-order corrections to the leading behavior. (Why?)

To apply the method of steepest descents we extend the contour to infinity. Note that the integrals

$$\frac{1}{i\pi} \int_{-\infty - i\pi/2}^{-i\pi/2} dt\, e^{ix \cosh t}, \tag{6.6.18b}$$

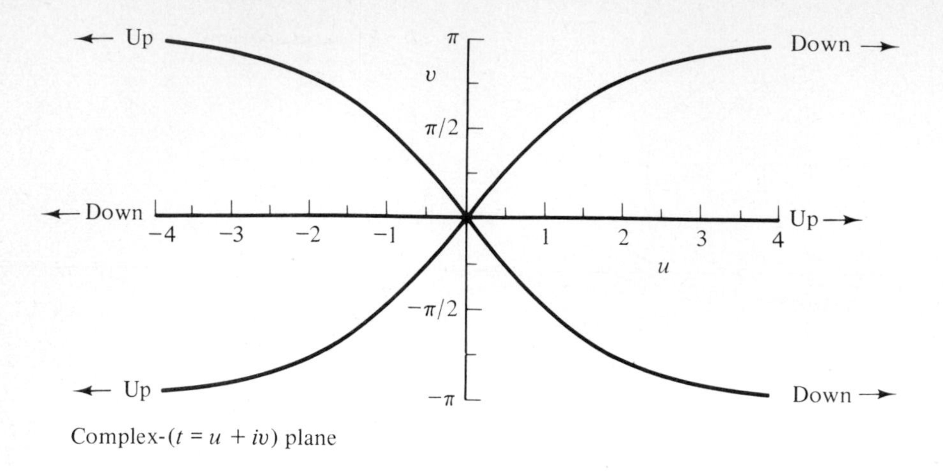

Figure 6.9 Steepest curves of $\exp\,[x(-t + \sinh t)]$ near the third-order saddle point at $t = 0$. The plot indicates that three steepest-descent curves and three steepest-ascent curves meet at $t = 0$.

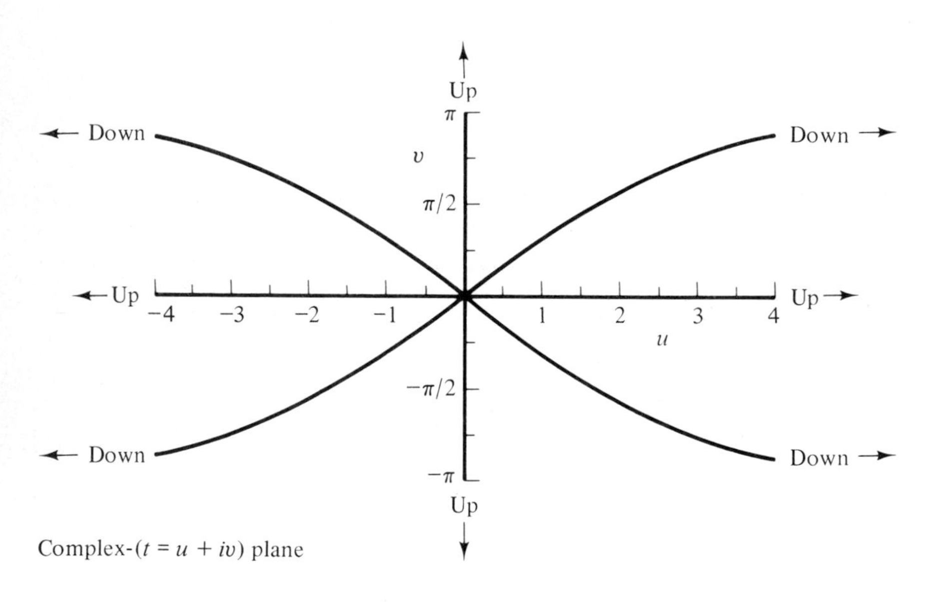

Figure 6.10 Steepest curves of $\exp\,[x(-\frac{1}{2}t^2 + \cosh t)]$ near the fourth-order saddle point at $t = 0$. The graph shows that four steepest-descent curves and four steepest-ascent curves meet at $t = 0$. In Example 12 the structure of the saddle point is the same as the one in this graph shifted by $i\pi$; the steepest-descent curve used in Example 12 consists of the curves in the third and fourth quadrants of this figure.

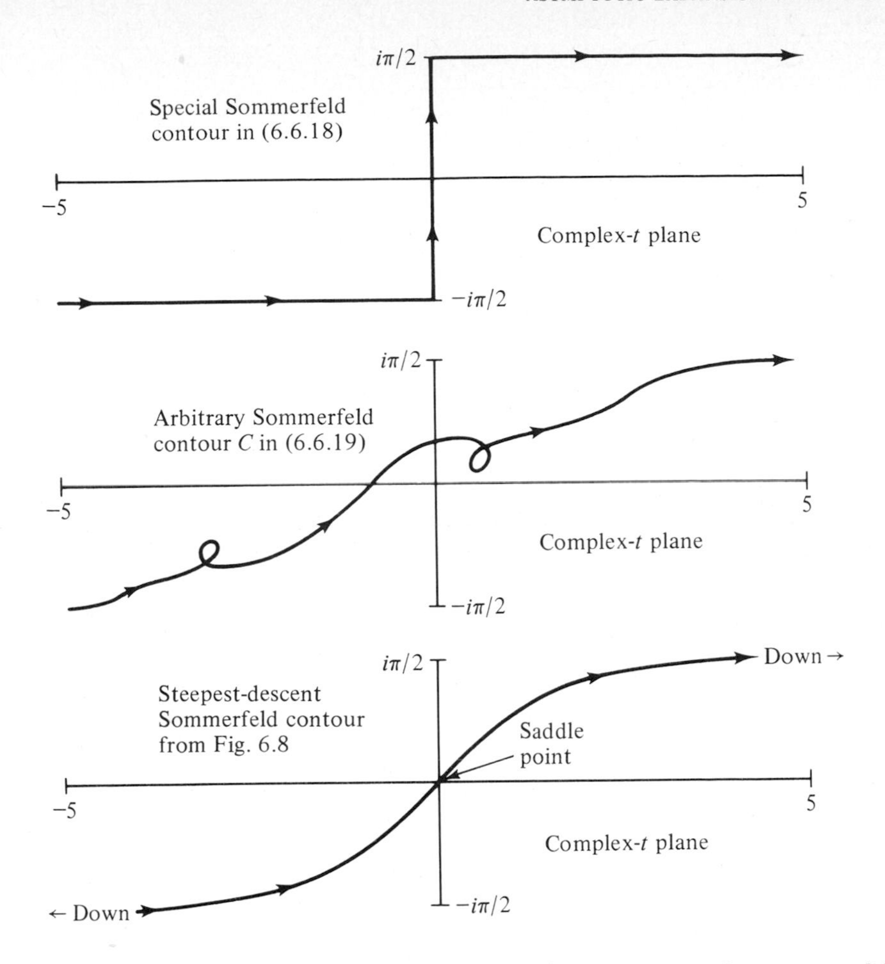

Figure 6.11 To find the asymptotic behavior of $J_0(x)$ as $x \to +\infty$ we first represent $J_0(x)$ as an integral along the special contour in (6.6.18*a*, *b*, *c*). Second, we observe that any Sommerfeld contour C from $-\infty - i\pi/2$ to $+\infty + i\pi/2$ is equally good. Third, to approximate the integral in (6.6.19) we choose that Sommerfeld contour which is also a path of steepest descent through the saddle point at $t = 0$.

where the contour extends along a line parallel to and below the real axis, and

$$\frac{1}{i\pi}\int_{i\pi/2}^{\infty + i\pi/2} dt\, e^{ix\cosh t}, \tag{6.6.18c}$$

where the contour extends along a line parallel to and above the real axis, are convergent and pure imaginary (see Prob. 6.63). Thus, we have constructed the rather fancy representation

$$J_0(x) = \text{Re}\, \frac{1}{i\pi}\int_C dt\, e^{ix\cosh t}, \tag{6.6.19}$$

where C is *any* contour which ranges from $-\infty - i\pi/2$ to $+\infty + i\pi/2$ (see Fig. 6.11). Such a contour is called a Sommerfeld contour.

From here on the steepest-descent analysis is easy because there is a curve of constant phase which ranges from $-\infty - i\pi/2$ to $+\infty + i\pi/2$ (see Fig. 6.8)! We have seen in Example 5 that this curve passes through a saddle point at $t = 0$. The equation for this curve is $\cosh u \cos v = 1$. Note that $|e^{ix \cosh t}|$ *attains its maximum value on the contour at the saddle point at* $t = 0$. Thus, we know from our study of Laplace's method that as $x \to +\infty$ the entire asymptotic expansion is determined by a small neighborhood about $t = 0$.

To find the leading behavior of $J_0(x)$ as $x \to \infty$, we approximate the steepest-descent path in a small neighborhood of $t = 0$ by the straight line $t = (1 + i)s$ (s real) and approximate $\cosh t$ near $s = 0$ by $\cosh t \sim 1 + is^2$ ($s \to 0$). Thus,

$$J_0(x) \sim \text{Re}\,[(1+i)/i\pi] \int_{s=-\varepsilon}^{\varepsilon} e^{ix - xs^2}\, ds, \qquad x \to +\infty.$$

Extending the limits of integration to ∞ and evaluating the integral gives

$$J_0(x) \sim \text{Re}\,[(1+i)/i\pi] e^{ix}\sqrt{\pi/x} = \sqrt{2/\pi x}\,\cos(x - \pi/4), \qquad x \to +\infty.$$

To find the full asymptotic expansion of $J_0(x)$ as $x \to +\infty$, we use Watson's lemma. It is simplest to parametrize the integration path in terms of $\phi = \text{Re}\,\rho(t)$. We know that along the steepest-descent contour $\rho(t) = i + \phi(t)$, where $\phi(t)$ is real and ranges from $\phi = 0$ at $t = 0$ to $\phi = -\infty$ at $t = \pm(\infty \pm i\pi/2)$. Also, we have $\phi(t) = i \cosh t - i$, so $d\phi = i \sinh t\, dt$. Thus, $dt = d\phi/i\sqrt{-\phi^2 - 2i\phi}$. Substituting this result into (6.6.19) and replacing ϕ by $-\phi$ gives

$$J_0(x) = \text{Re}\, \frac{e^{ix - i\pi/4}}{\pi} \sqrt{2} \int_0^\infty \frac{d\phi}{\sqrt{\phi}} e^{-\phi x} \left(1 - \frac{i\phi}{2}\right)^{-1/2}.$$

To apply Watson's lemma, we expand the square root:

$$\left(1 - \frac{i\phi}{2}\right)^{-1/2} = \sum_{n=0}^{\infty} \frac{(i\phi/2)^n \Gamma(n + \frac{1}{2})}{n!\, \Gamma(\frac{1}{2})}$$

and integrate term by term:

$$J_0(x) \sim \text{Re}\, \frac{e^{ix - i\pi/4}}{\pi^{3/2}} \sqrt{2} \sum_{n=0}^{\infty} \frac{[\Gamma(n + \frac{1}{2})]^2}{n!\, \sqrt{x}} \left(\frac{i}{2x}\right)^n, \qquad x \to +\infty.$$

Thus, the full asymptotic expansion of $J_0(x)$ is given by

$$J_0(x) = \sqrt{\frac{2}{x\pi}}\,[\alpha(x)\cos(x - \pi/4) + \beta(x)\sin(x - \pi/4)], \tag{6.6.20}$$

where

$$\alpha(x) \sim \sum_{k=0}^{\infty} \frac{[\Gamma(2k + \frac{1}{2})]^2(-1)^k}{\pi(2k)!\,(2x)^{2k}}, \qquad x \to +\infty,$$

and

$$\beta(x) \sim \sum_{k=0}^{\infty} \frac{[\Gamma(2k + \frac{3}{2})]^2(-1)^{k+1}}{\pi(2k+1)!\,(2x)^{2k+1}}, \qquad x \to +\infty.$$

The trick of adding the contour integrals (6.6.18*b*,*c*) to (6.6.18*a*) to derive (6.6.19) could have been avoided by deforming the contour from $-i\pi/2$ to $i\pi/2$ into three constant-phase contours: C_1: $t = -i\pi/2 + u$ ($-\infty < u \le 0$); C_2: $\cosh u \cos v = 1$; and C_3: $t = i\pi/2 + u$ ($0 \le u < \infty$). The contributions from C_1 and C_3 cancel exactly in this problem.

Example 9 *Asymptotic expansion of* $\Gamma(x)$ *as* $x \to +\infty$. In Example 10 of Sec. 6.4 we used Laplace's method to show that

$$\Gamma(x) \sim x^x e^{-x}\sqrt{2\pi/x} \tag{6.6.21}$$

[see (6.4.39)]. In this example we use the method of steepest descents to rederive this result from a complex-contour integral representation of $\Gamma(x)$ [see Prob. 2.6(f)]:

$$\frac{1}{\Gamma(x)} = \frac{1}{2\pi i}\int_C e^t t^{-x}\, dt, \tag{6.6.22}$$

where C is a contour that begins at $t = -\infty - ia$ ($a > 0$), encircles the branch cut that lies along the negative real axis, and ends up at $-\infty + ib$ ($b > 0$) (see Fig. 6.12). The branch cut is present when x is nonintegral because t^{-x} is a multivalued function. The advantage of (6.6.22) over the integral representation used in Example 10 is that it converges for *all* complex values of x and not just those x for which Re $x > 0$. Nevertheless, in this example we will only investigate the behavior of $\Gamma(x)$ in the limit $x \to +\infty$.

We begin our analysis by making the same substitution that was made in Example 10 of Sec. 6.4; namely, $t = xs$. This substitution converts the integrand from one that has a movable saddle point to one that has a fixed saddle point. (Why?) The resulting integral representation is

$$\frac{1}{\Gamma(x)} = \frac{1}{2\pi i x^{x-1}}\int_C ds\, e^{x(s-\ln s)}. \tag{6.6.23}$$

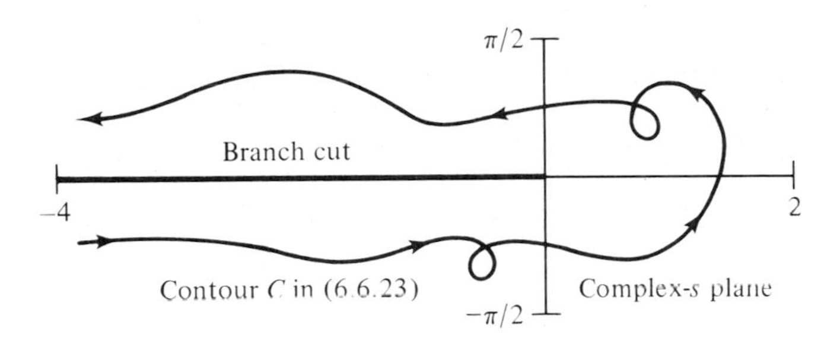

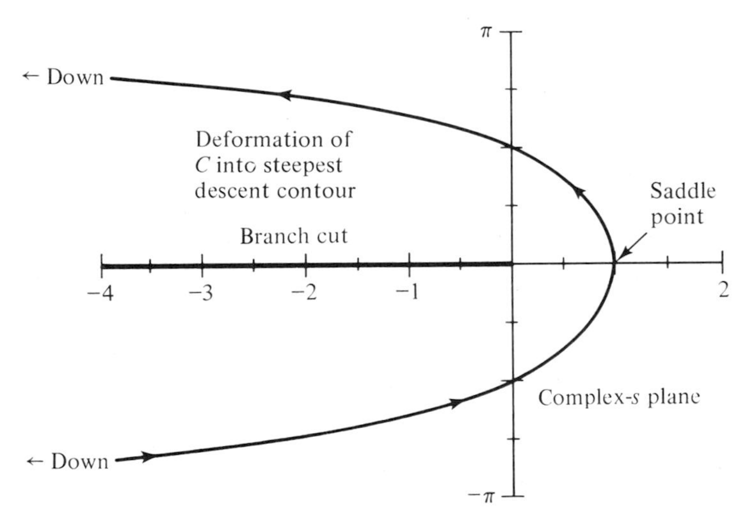

Figure 6.12 To find the asymptotic behavior of $\Gamma(x)$ as $x \to +\infty$, we represent $\Gamma(x)$ as the integral in (6.6.23) along a contour C in the complex-s plane which goes around the branch cut on the negative real axis. Then we distort C into a steepest-descent contour which passes through the saddle point at $s = 1$.

For this integral $\rho(s) = s - \ln s$. Thus, $\rho'(s) = 1 - 1/s$ and $\rho' = 0$ when $s = 1$. So there is a simple (second-order) saddle point at $s = 1$.

To ascertain the structure of the saddle point we let $s = u + iv$ and identify the real and imaginary parts of ρ: $\rho(s) = u - \ln \sqrt{u^2 + v^2} + i(v - \arctan v/u)$. At $s = 1$, $\rho = 1$. Therefore, paths of constant phase (steepest curves) emerging from $s = 1$ must satisfy

$$v - \arctan v/u = 0.$$

There are two solutions to this equation: $v = 0$ and $u = v \cot v$. These two curves are shown on Fig. 6.12. In Prob. 6.64 you are asked to verify that (*a*) the steepest-descent curves are correctly shown on Fig. 6.12; (*b*) as s moves away from $s = 1$, steepest-descent curves emerge from $s = 1$ initially parallel to the Im $s = v$ axis; (*c*) the steepest-descent curves cross the v axis at $\pm i\pi/2$ and approach $s = -\infty \pm i\pi$.

To use the method of steepest descents, we simply shift the contour C so that it is just the steepest-descent contour on Fig. 6.12 which passes through the saddle point at $s = 1$. Let us review why we choose such a contour. In general, we always choose a steepest-descent contour because on such a contour we can apply the techniques of Laplace's method directly to complex integrals. If the steepest-descent contour is finite and does not pass through a saddle point, then the maximum value of $|e^{x\rho}|$ must occur at an endpoint of the contour and we need only perform a local analysis of the integral at this endpoint. However, in the present example the contour has no endpoint and is infinitely long. It is crucial that it pass through a saddle point because $|e^{x\rho}|$ reaches its maximum at the saddle point and decays exponentially as $s \to \infty$ along *both* of the steepest-descent curves. If there were no saddle point, then, although $|e^{x\rho}|$ would decrease in one direction along the contour, it would increase in the other direction and the integral would not even converge!

Now we proceed with the asymptotic expansion of the integral in (6.6.23). We can approximate the steepest-descent contour in the neighborhood of $s = 1$ by the straight line $s = 1 + iv$. This gives the Laplace integral

$$\frac{1}{\Gamma(x)} \sim \frac{1}{2\pi x^{x-1}} \int_{-\varepsilon}^{\varepsilon} dv\, e^{x(1-v^2/2)}, \qquad x \to +\infty,$$

which we evaluate by letting $\varepsilon \to \infty$:

$$\frac{1}{\Gamma(x)} \sim \frac{1}{2\pi x^{x-1}} \frac{e^x}{\sqrt{x}} \sqrt{2\pi}, \qquad x \to +\infty.$$

We thereby recover the result in (6.6.21).

Example 10 *Steepest-descents approximation of a real integral where Laplace's method fails.* In this example we consider the real integral

$$I(x) = \int_0^1 dt\, e^{-4xt^2} \cos(5xt - xt^3) \tag{6.6.24}$$

in the limit $x \to +\infty$. This integral is *not* a Laplace integral because the argument of the cosine contains x. Nonetheless, one might think that one could use the ideas of Laplace's method to approximate the integral. To wit, one would argue that as $x \to +\infty$, the contribution to the integral is localized about $x = 0$. Thus, a very naive approach is simply to replace the argument of the cosine by 0. If this reasoning were correct, then we would conclude that

$$I(x) \sim \int_0^1 dt\, e^{-4xt^2} \sim \sqrt{\frac{\pi}{16x}}, \qquad x \to +\infty. \qquad \text{(WRONG)}$$

This result is clearly incorrect because e^{-xt^2} does not become exponentially small until t is larger than $1/\sqrt{x}$. Thus, when $t \sim 1/\sqrt{x}$ $(x \to +\infty)$, the argument of the cosine is *not* small. In particular, the term $5xt$ is large and the cosine oscillates rapidly. This suggests that there is destructive interference and that $I(x)$ decays much more rapidly than $\sqrt{\pi/16x}$ as $x \to +\infty$.

Can we correct this approach by including the $5xt$ term but neglecting the xt^3 term? After all, when t lies in the range from 0 to $1/\sqrt{x}$, the term $xt^3 \to 0$ as $x \to +\infty$. Thus, xt^3 does not even shift the phase of the cosine more than a fraction of a cycle. If we were to include just the $5xt$ term, we would obtain

$$I(x) \sim \int_0^1 dt\, e^{-4xt^2} \cos(5xt), \qquad x \to +\infty,$$

$$\sim \int_0^\infty dt\, e^{-4xt^2} \cos(5xt), \qquad x \to +\infty,$$

$$= \tfrac{1}{2} \int_{-\infty}^\infty dt\, e^{-4xt^2 + 5ixt}$$

$$= \tfrac{1}{2} \int_{-\infty}^\infty dt\, e^{-x(2t-5i/4)^2 - 25x/16}$$

$$= \tfrac{1}{4}\sqrt{\pi/x}\, e^{-25x/16}, \qquad x \to +\infty. \qquad \text{(WRONG)}$$

Although this result is exponentially smaller than the previous wrong result, it is also wrong! It is incorrect to neglect the xt^3 term (see Prob. 6.65).

But if we cannot neglect even the xt^3 term, then how can we make any approximation at all? It should not be necessary to do the integral exactly to find its asymptotic behavior!

The correct approach is to use the method of steepest descents to approximate the integral at a saddle point in the complex plane. To prepare for this analysis we rewrite the integral in the following convenient form:

$$I(x) = \tfrac{1}{2} \int_{-1}^1 dt\, e^{-4xt^2 + 5ixt - ixt^3}$$

$$= \tfrac{1}{2} e^{-2x} \int_{-1}^1 dt\, e^{x\rho(t)}, \tag{6.6.25}$$

where

$$\rho(t) = -(t-i)^2 - i(t-i)^3. \tag{6.6.26}$$

Our objective now is to find steepest-descent (constant-phase) contours that emerge from $t = 1$ and $t = -1$, to distort the original contour of integration t: $-1 \to 1$ into these contours, and then to use Laplace's method. To find these contours we substitute $t = u + iv$ and identify the real and imaginary parts of ρ:

$$\rho(t) = \phi + i\psi$$

$$= -v^3 + 4v^2 - 5v + 3u^2v - 4u^2 + 2 + i(3uv^2 - 8uv + 5u - u^3). \tag{6.6.27}$$

Note that the phase of $\psi = \text{Im}\, \rho$ at $t = 1$ and at $t = -1$ is different: $\text{Im}\, \rho(-1) = -4$, $\text{Im}\, \rho(1) = 4$. Thus, there is no single constant-phase contour which connects $t = -1$ to $t = 1$.

Our method is similar to that used in Examples 1 and 2. We follow steepest-descent contours C_1 and C_2 from $t = -1$ and from $t = 1$ out to ∞. Next, we join these two contours at ∞ by a third contour C_3 which is also a path of constant phase. C_3 must pass through a saddle point because its endpoints lie at ∞; otherwise, the integral along C_3 will not converge (see the discussion in Example 9).

There are two saddle points in the complex plane because $\rho'(t) = -2(t-i) - 3i(t-i)^2 = 0$ has two roots, $t = i$ and $t = 5i/3$. The contour C_3 happens to pass through the saddle point at i. On Fig. 6.13 we plot the three constant-phase contours C_1, C_2, and C_3. It is clear that the original contour C can be deformed into $C_1 + C_2 + C_3$. (In Prob. 6.66 you are to verify the results on Fig. 6.13.)

The asymptotic behavior of $I(x)$ as $x \to +\infty$ is determined by just three points on the contour $C_1 + C_2 + C_3$: the endpoints of C_1 and C_2 at $t = -1$ and at $t = +1$ and the saddle point at i. However, the contributions to $I(x)$ at $t = \pm 1$ are exponentially small compared with

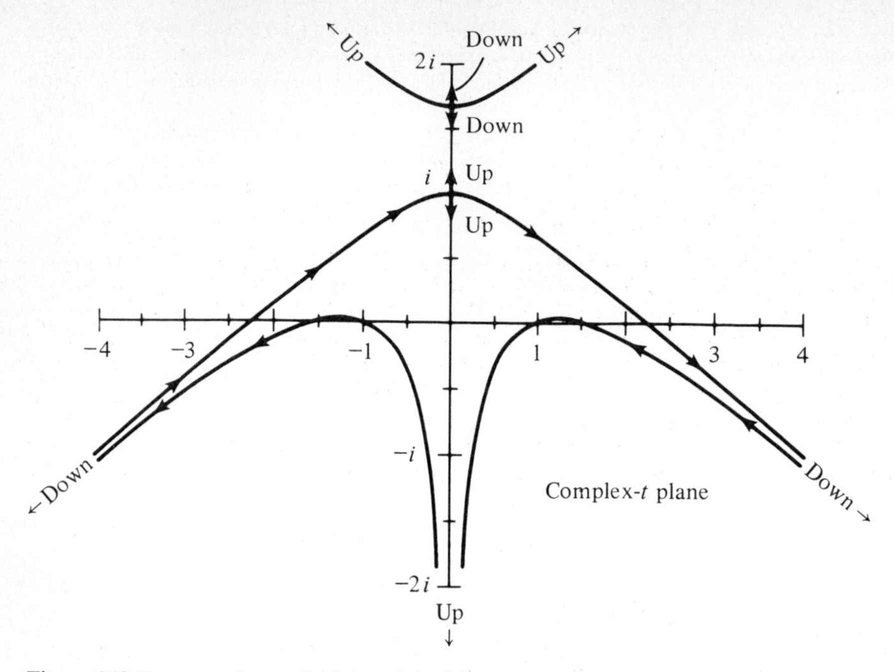

Figure 6.13 To approximate the integral in (6.6.25) by the method of steepest descents we deform the original contour connecting the points $t = -1$ to $t = 1$ along the real axis into the three distinct steepest-descent contours above, one of which passes through a saddle point at $t = i$. Steepest-ascent and -descent curves near a second saddle point at $t = 5i/3$ and steepest-ascent curves going from 1 and -1 to $-i\infty$ are also shown, but these curves play no role in the calculation.

that at $t = i$ (see Prob. 6.67). Near $t = i$ we can approximate the contour C_3 by the straight line $t = i + u$ and $\rho(t)$ by $\rho(t) \sim -u^2$ $(u \to 0)$. Thus,

$$I(x) \sim \tfrac{1}{2}e^{-2x} \int_{-\varepsilon}^{\varepsilon} e^{-xu^2}\, du, \qquad x \to +\infty,$$

$$\sim \tfrac{1}{2}e^{-2x}\sqrt{\pi/x}, \qquad x \to +\infty. \tag{6.6.28}$$

This, finally, is the correct asymptotic behavior of $I(x)$! This splendid example certainly shows the subtlety of asymptotic analysis and the power of the method of steepest descents.

Example 11 *Steepest-descents analysis with a third-order saddle point.* In Example 5 of Sec. 6.5 and Prob. 6.55 we showed that

$$J_x(x) \sim \frac{1}{\pi} 2^{-2/3} 3^{-1/6} \Gamma\left(\frac{1}{3}\right) x^{-1/3}, \qquad x \to +\infty. \tag{6.6.29}$$

Here we rederive (6.6.29) using the method of steepest descents.

We begin with the complex-contour integral representation for $J_\nu(x)$:

$$J_\nu(x) = \frac{1}{2\pi i} \int_C dt\, e^{x \sinh t - \nu t}, \tag{6.6.30}$$

where C is a Sommerfeld contour that begins at $+\infty - i\pi$ and ends at $+\infty + i\pi$. Setting $\nu = x$ gives

$$J_x(x) = \frac{1}{2\pi i} \int_C dt\, e^{x(\sinh t - t)}. \tag{6.6.31}$$

For this integral $\rho(t) = \sinh t - t$ has a third-order saddle point at $t = 0$.

We have already analyzed the steepest curves of this $\rho(t)$ in Example 6 (see Fig. 6.9). Note that we can deform the contour C so that it follows steepest-descent paths to and from the saddle point at $t = 0$.

The contribution to $J_x(x)$ as $x \to +\infty$ comes entirely from the neighborhood of the saddle point. In the vicinity of the saddle point we can approximate the contours approaching and leaving $t = 0$ by the straight lines $t = re^{-i\pi/3}$ and $t = re^{i\pi/3}$. Substituting into (6.6.30) gives

$$J_x(x) \sim \frac{1}{2\pi i} \int_{r=\varepsilon}^{0} dr\, e^{-i\pi/3} e^{-xr^3/6} + \frac{1}{2\pi i} \int_{r=0}^{\varepsilon} dr\, e^{i\pi/3} e^{-xr^3/6}, \qquad x \to +\infty.$$

To evaluate these integrals, we replace ε by ∞:

$$J_x(x) \sim \frac{e^{i\pi/3} - e^{-i\pi/3}}{2\pi i} \int_{r=0}^{\infty} dr\, e^{-xr^3/6}, \qquad x \to +\infty,$$

$$= \frac{\sin(\pi/3)}{\pi} (6/x)^{1/3} \int_0^{\infty} e^{-r^3}\, dr, \qquad x \to +\infty.$$

But $\int_0^\infty e^{-r^3}\, dr = \frac{1}{3} \int_0^\infty e^{-s} s^{-2/3}\, ds = \frac{1}{3}\Gamma(\frac{1}{3})$. Thus,

$$J_x(x) \sim \frac{\sin(\pi/3)}{3\pi} \left(\frac{6}{x}\right)^{1/3} \Gamma\left(\frac{1}{3}\right),$$

which reproduces the result in (6.6.29).

Example 12 *Steepest-descents analysis with a fourth-order saddle point.* What is the leading asymptotic behavior of the real integral

$$I(x) = \int_0^{\infty} dt\, \cos(x\pi t)\, e^{-x(\cosh t + t^2/2)} \tag{6.6.32}$$

as $x \to +\infty$? To analyze $I(x)$ we first rewrite the integral as

$$I(x) = \frac{1}{2} \int_{-\infty}^{\infty} e^{x(i\pi t - \cosh t - t^2/2)}$$

$$= \frac{1}{2} e^{-x\pi^2/2} \int_{-\infty}^{\infty} dt\, e^{x[\cosh(t - i\pi) - (t - i\pi)^2/2]}. \tag{6.6.33}$$

For this integral $\rho(t) = \cosh(t - i\pi) - (t - i\pi)^2/2$ has a fourth-order saddle point at $t = i\pi$ (see Example 7). The steepest-descent contours from this saddle point are drawn in Fig. 6.10 (the saddle point in Fig. 6.10 is shifted downward by π).

To approximate $I(x)$ we shift the original integration path t: $-\infty \to +\infty$ from the real axis into the complex plane so that it follows a steepest-descent curve passing through the saddle point. The asymptotic behavior of $I(x)$ is completely determined by the contribution from the saddle point. In the neighborhood of the saddle point at $i\pi$, we can approximate the steepest-descent contour by the straight lines $t = i\pi + re^{i\pi/4}$ to the left of the saddle point and $t = i\pi + re^{-i\pi/4}$ to the right of the saddle point. In terms of r, (6.6.33) becomes

$$I(x) \sim \frac{1}{2} e^{-x\pi^2/2} \left[\int_{-\varepsilon}^{0} e^{i\pi/4}\, dr\, e^{x(1 - r^4/24)} + \int_0^{\varepsilon} e^{-i\pi/4}\, dr\, e^{x(1 - r^4/24)} \right], \qquad x \to +\infty,$$

$$= e^{-x\pi^2/2 + x} \cos(\pi/4) \int_0^{\infty} dr\, e^{-xr^4/24}, \qquad x \to +\infty.$$

But $\int_0^\infty dr\, e^{-r^4} = \frac{1}{4} \int_0^\infty dr\, r^{-3/4} e^{-r} = \frac{1}{4}\Gamma(\frac{1}{4})$. Thus, we obtain the final result that

$$I(x) \sim \tfrac{1}{4} e^{x(1 - \pi^2/2)} (6/x)^{1/4} \Gamma(\tfrac{1}{4}). \tag{6.6.34}$$

This result could not have been obtained by performing a Laplace-like analysis of the real integral in (6.6.32). Suppose, for example, we argue that as $x \to +\infty$ the contribution to (6.6.32) comes entirely from the neighborhood of the origin $t = 0$. Then it would seem valid to replace $\cosh t$ by $1 + t^2/2$, the first two terms in its Taylor series. If we do this, we obtain an integral which we can evaluate exactly:

$$\int_0^\infty dt \cos(x\pi t)\, e^{-x(1+t^2)} = \tfrac{1}{2}\int_{-\infty}^{\infty} dt\, e^{ix\pi t - x(1+t^2)}$$

$$= \tfrac{1}{2}\int_{-\infty}^{\infty} dt\, e^{-x(t-i\pi/2)^2} e^{-x(1+\pi^2/4)}$$

$$= \tfrac{1}{2}\sqrt{\pi/x}\, e^{-x(1+\pi^2/4)}.$$

But this does not agree with (6.6.34) and is therefore *not* the asymptotic behavior of $I(x)$ as $x \to +\infty$! What is wrong with this argument? (See Prob. 6.68.)

Steepest Descents for Complex x and the Stokes Phenomenon

Until now, x in (6.6.1) has been treated as a large *real* parameter. However, the method of steepest descents can be used to treat problems where x is complex. As we have already seen in Secs. 3.7 and 3.8, an asymptotic relation is valid as $x \to \infty$ in a wedge-shaped region of the complex-x plane. At the edge of the wedge, the asymptotic relation ceases to be valid and must be replaced by another asymptotic relation. This change from one asymptotic relation to another is called the Stokes phenomenon.

The Stokes phenomenon usually surfaces in the method of steepest descents in a relatively simple way. For example, as x rotates in the complex plane, the structure of steepest-descent paths can change abruptly. When this happens, the asymptotic behavior of the integral changes accordingly. The integral representation of Ai (x) behaves in this manner (see Prob. 6.75). The Stokes phenomenon can also appear when the contribution from an endpoint of the contour suddenly becomes subdominant relative to the contribution from a saddle point (or vice versa). We consider this case in the next example.

Example 13 *Reexamination of Example 10 for complex x.* In this example we explain how the Stokes phenomenon arises in the integral (6.6.25). It is essential that the reader master Example 10 before reading further.

The integral $I(x)$ in (6.6.25) exhibits the Stokes phenomenon at $\arg x = \pm \arctan \tfrac{1}{2} \doteq 26.57°$ and at $\pm\pi$. When $|\arg x| < \arctan \tfrac{1}{2}$, the contribution to $I(x)$ from the saddle point at $t = i$ dominates the endpoint contributions. As in (6.6.28), this gives

$$I(x) \sim \tfrac{1}{2}e^{-2x}\sqrt{\pi/x}, \qquad x \to \infty,\ |\arg x| < \arctan \tfrac{1}{2}. \tag{6.6.35}$$

When $\arctan \tfrac{1}{2} < \arg x < \pi$, the endpoint contribution from $t = -1$ dominates. We obtain (see Prob. 6.69)

$$I(x) \sim \frac{i-4}{68x} e^{-4(i+1)x}, \qquad x \to \infty,\ \arctan \tfrac{1}{2} < \arg x < \pi. \tag{6.6.36}$$

When $-\pi < \arg x < -\arctan \tfrac{1}{2}$, the endpoint contribution from $t = 1$ dominates, giving

$$I(x) \sim -\frac{i+4}{68x} e^{4(i-1)x}, \qquad x \to \infty,\ -\pi < \arg x < -\arctan \tfrac{1}{2}. \tag{6.6.37}$$

It is interesting to see what happens to the steepest-descent contours as x is rotated into the complex-x plane. We have plotted the steepest-descent contours for $I(x)$ for $\arg x = 0°, 30°, 75°$, and $135°$ in Figs. 6.13 to 6.16. Observe that as $\arg x$ increases from $0°$ to $75°$, the contours through the endpoints at $t = \pm 1$ and the saddle point at $t = i$ tilt and distort slightly. Note that the asymptotes of these contours at ∞ rotate by $-(\arg x)/3$ as $\arg x$ increases. This is so because

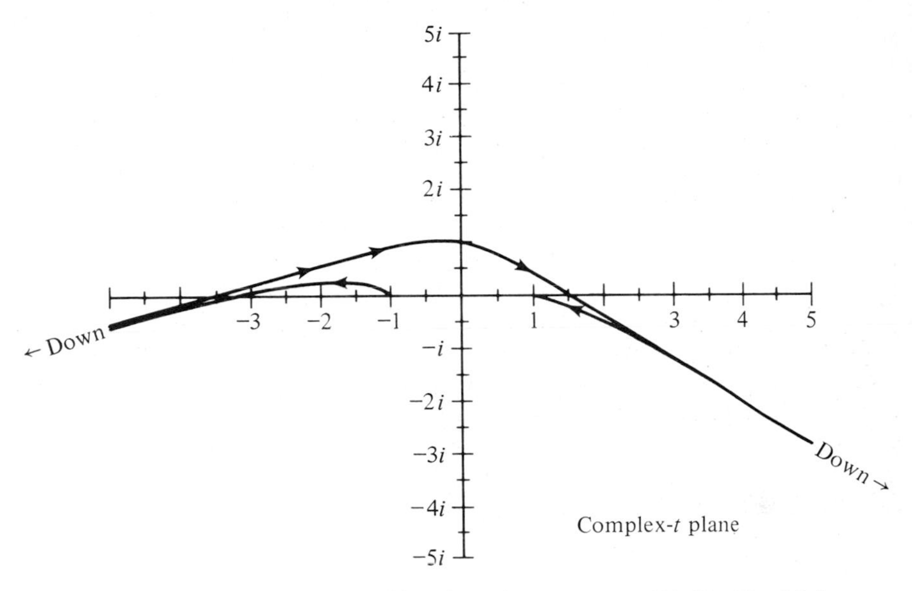

Figure 6.14 Steepest-descent path for $I(x)$ in (6.6.25) when $\arg x = 30°$. (See Fig. 6.13.)

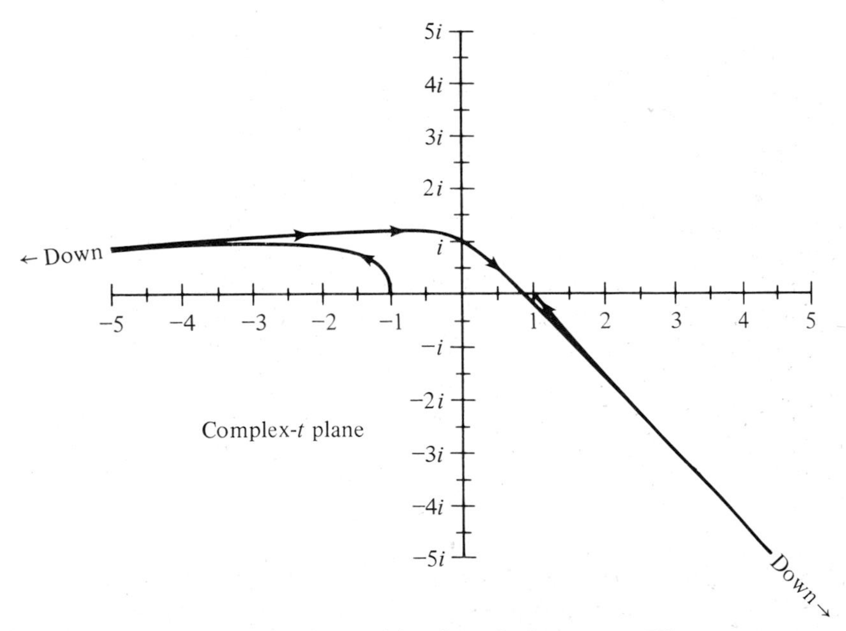

Figure 6.15 Steepest-descent path for $I(x)$ in (6.6.25) when $\arg x = 75°$.

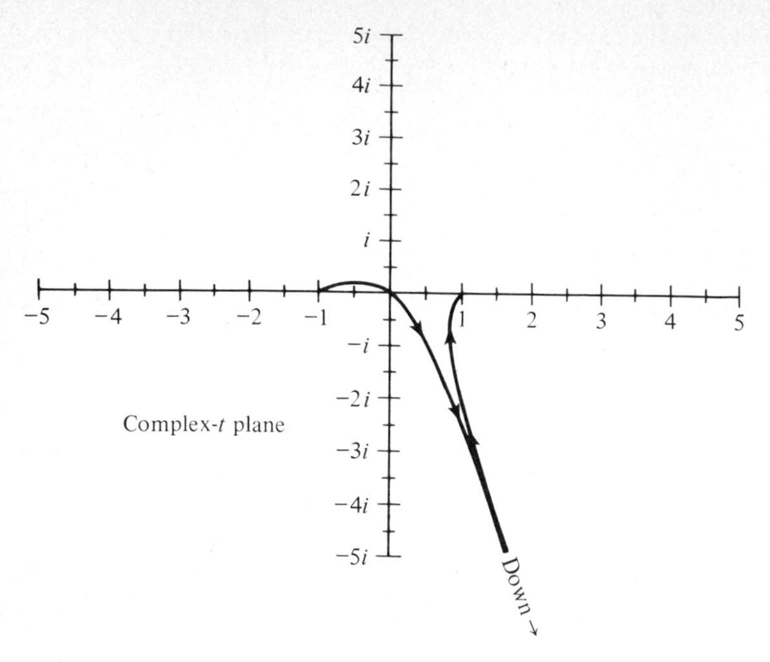

Figure 6.16 Steepest-descent path for $I(x)$ in (6.6.25) when $\arg x = 135°$. Note that the steepest-descent path no longer passes through the saddle point at $t = i$ as it does in Figs. 6.13 to 6.15.

$\text{Im}\,[x\rho(t)]$ must be constant at $t = \infty$. The constancy of $\text{Im}\,[x\rho(t)]$ on the steepest-descent contours implies that the endpoint contours passing through $t = \pm 1$ rotate by $-\arg x$ near $t = \pm 1$ and that the contour through $t = i$ rotates by $-(\arg x)/2$ near $t = i$. There is *no* abrupt or discontinuous change in the configuration of the steepest-descent contours as $\arg x$ increases past arc tan $\frac{1}{2}$. In this example the Stokes phenomenon is not associated with any discontinuity in the structure of the steepest-descent path. It occurs because the contribution from the saddle point becomes subdominant with respect to the contribution from the endpoint as $\arg x$ increases past arc tan $\frac{1}{2}$.

When $\arg x$ reaches $\pi - \text{arc tan } 2 \doteqdot 116.57°$, there is a discontinuous change in the steepest-descent path for $I(x)$ (see Prob. 6.69). As illustrated in Fig. 6.16, when $\arg x = 135°$, the steepest-descent contour no longer passes through the saddle point at $t = i$. When $\arg x > 116.57°$, the steepest-descent contours from $t = \pm 1$ meet at ∞, so it is no longer necessary to join them by a constant-phase contour passing through the saddle point at i. The abrupt disappearance of the saddle-point contour from the steepest-descent path when $\arg x$ increases beyond $116.57°$ does not affect the asymptotic behavior of $I(x)$ because the saddle-point contribution from $t = i$ is subdominant when $\text{arc tan } \frac{1}{2} < |\arg x| < \pi$.

(I) 6.7 ASYMPTOTIC EVALUATION OF SUMS

In this section we discuss methods for finding the asymptotic behavior of sums which depend on a large parameter x. We consider four methods in all: truncating the sum after a finite number of terms, approximating the sum by a Riemann integral, Laplace's method for sums, and the Euler-Maclaurin sum formula. The first method is very elementary and rarely applicable, but for completeness we illustrate it in the following brief example:

Example 1 *Behavior of* $\sum_{n=0}^{\infty} e^{-n^2x^2}$ *as* $x \to +\infty$. The behavior of this convergent series is easy to find because the sum of the terms with $n = 1, 2, 3, \ldots$ is clearly subdominant with respect to the first term as $x \to +\infty$. Thus, $1 + e^{-x^2} + e^{-4x^2} + \cdots \sim 1$ $(x \to +\infty)$.

Approximation of Sums by Riemann Integrals

The Riemann integral $I = \int_a^b f(t)\, dt$ is defined as the limit of the Riemann sum,

$$I = \lim_{N \to \infty} \sum_{n=0}^{N-1} f(\bar{t}_n)(t_{n+1} - t_n),$$

where $f(t)$ is continuous, $\bar{t}_n$ is any point in the interval $t_n \le \bar{t}_n \le t_{n+1}$, and $t_n = a + n(b - a)/N$. In the next two examples we show how to use this formula to find the leading behavior of sums.

Example 2 *Behavior of* $S(x) = \sum_{n=0}^{\infty} 1/(n^2 + x^2)$ *as* $x \to +\infty$. Note that each term in this series decays to 0 like x^{-2} as $x \to +\infty$ and that this series converges. One might therefore be tempted to conclude that $S(x)$ decays to 0 like x^{-2}. However, the correct behavior is

$$S(x) \sim \frac{\pi}{2x}, \qquad x \to +\infty. \tag{6.7.1}$$

The origin of this surprising result is that as x increases, more terms contribute significantly to the leading behavior of $S(x)$. Roughly speaking, x terms, each of size x^{-2}, contribute to $S(x)$ causing $S(x)$ to decay like x^{-1} as $x \to +\infty$.

We can establish (6.7.1) by converting the series to a Riemann sum. We simply multiply by x, rewrite the series as

$$xS(x) = \sum_{n=0}^{\infty} \frac{1}{1 + (n/x)^2} \frac{1}{x},$$

and observe that as $x \to +\infty$ the series becomes a Riemann sum for the convergent integral

$$\int_0^{\infty} \frac{1}{1 + t^2}\, dt = \frac{\pi}{2}. \tag{6.7.2}$$

Specifically, a Riemann sum for this integral is obtained by choosing the discrete points $t_n = \bar{t}_n = n/x$ $(n = 0, 1, 2, \ldots)$. This gives

$$\sum_{n=0}^{\infty} \frac{1}{1 + t_n^2} (t_{n+1} - t_n).$$

This Riemann sum converges to the integral in (6.7.2) as the interval $t_{n+1} - t_n = x^{-1} \to 0$. Thus, we obtain $\lim_{x \to \infty} xS(x) = \pi/2$, which is just (6.7.1). For a higher-order approximation to $S(x)$ see Prob. 6.91.

Example 3 *Behavior of* $\sum_{1 \le n < x} n^{\alpha}$ *as* $x \to +\infty$. We will show that the leading behavior of

$$S(x) = \sum_{1 \le n < x} n^{\alpha} \tag{6.7.3}$$

for large x is

$$S(x) \sim \begin{cases} \dfrac{x^{\alpha+1}}{\alpha+1}, & x \to +\infty;\ \alpha > -1, & (6.7.4a) \\ \ln x, & x \to +\infty;\ \alpha = -1, & (6.7.4b) \\ \zeta(-\alpha), & x \to +\infty;\ \alpha < -1, & (6.7.4c) \end{cases}$$

where $\zeta(p)$ is the Riemann zeta function defined by the convergent series

$$\zeta(p) = \sum_{n=1}^{\infty} n^{-p}, \qquad \text{Re } p > 1. \tag{6.7.5}$$

To verify (6.7.4a) we reason as in Example 2. We observe that as $x \to \infty$, $x^{-\alpha-1}S(x) = \sum_{1 \le n < x} (n/x)^{\alpha}/x$ is a Riemann sum for the integral $\int_0^1 t^{\alpha}\, dt = 1/(\alpha+1)$, where we have approximated the integral as a Riemann sum on the discrete points $t_n = n/x$ for $n = 1, 2, \ldots, [x]$ ($[x]$ stands for the largest integer less than x). This justifies (6.7.4a). The proof of (6.7.4b) was given in Prob. 5.8. Finally, when $\alpha < -1$, the sum $S(x)$ converges as $x \to +\infty$ to $S(+\infty) = \zeta(-\alpha)$. This proves (6.7.4$c$).

Laplace's Method for Sums

In the next example we show how to obtain the leading behavior of a sum by following the philosophy of Laplace's method which was introduced in Sec. 6.4.

Example 4 *Leading behavior of* $\sum_{n=0}^{\infty} x^n (n!)^{-k}$ *as* $x \to +\infty$. We have already encountered this problem in Example 1 of Sec. 3.5 and Prob. 3.42. There we solved the problem by finding a differential equation satisfied by the sum and then determining the asymptotic behavior of its solutions. However, you will recall that because the differential equation was linear, local analysis could not determine the overall multiplicative constant in the asymptotic behavior of the sum. In this example we find the leading behavior of the sum directly and encounter no such ambiguities.

To find the leading behavior of this sum by Laplace's method, we must identify the largest term in the series. Let us examine the ratio of the $(n-1)$th term to the nth term in the series:

$$x^{n-1}[(n-1)!]^{-k}/x^n(n!)^{-k} = n^k/x.$$

Note that this ratio is less than 1 if $n < x^{1/k}$ and greater than 1 if $n > x^{1/k}$. Therefore, the terms in the series increase as n increases until n reaches $[x^{1/k}]$. This is the largest term in the series. The remaining terms decrease with increasing n.

As $x \to +\infty$, the terms in the series peak sharply near the $n = [x^{1/k}]$ term. Therefore, using the principles of Laplace's method, we expect that for any $\varepsilon > 0$

$$S(x) = \sum_{n=0}^{\infty} x^n (n!)^{-k} \sim \sum_{n=[x^{1/k}(1-\varepsilon)]}^{[x^{1/k}(1+\varepsilon)]} \frac{x^n}{(n!)^k}, \tag{6.7.6}$$

with errors that are subdominant with respect to every power of $1/x$ as $x \to +\infty$.

Next we use the Stirling formula (see Sec. 5.4) to approximate each of the terms retained in (6.7.6). If $n = x^{1/k} + t$, where t is small compared with $x^{1/k}$, then by Stirling's formula we have

$$\begin{aligned} n! &\sim (n/e)^n \sqrt{2\pi n}, && n \to +\infty, \\ &= (x^{1/k} + t)^n e^{-n} \sqrt{2\pi n} \\ &= x^{n/k} e^{n \ln(1 + tx^{-1/k})} e^{-n} \sqrt{2\pi n} \\ &\sim x^{n/k} e^{-x^{1/k}} e^{t^2 x^{-1/k}/2} \sqrt{2\pi}\, x^{1/2k}, && x \to +\infty;\ t^3 \ll x^{2/k}, \end{aligned}$$

where we have truncated the Taylor expansion of $\ln(1 + tx^{-1/k})$ after quadratic terms in t. Therefore,

$$x^n(n!)^{-k} \sim e^{kx^{1/k}} e^{-t^2 kx^{-1/k}/2} (2\pi)^{-k/2} x^{-1/2}, \qquad x \to +\infty;\ n = x^{1/k} + t,\ t^3 \ll x^{2/k}.$$

Finally, we substitute this result into the sum on the right side of (6.7.6) and extend the region of summation to ∞, just as we do with integrals. We obtain

$$S(x) \sim \sum_{t=-\infty}^{\infty} e^{kx^{1/k}} e^{-t^2 kx^{-1/k}/2} (2\pi)^{-k/2} x^{-1/2}, \qquad x \to +\infty. \tag{6.7.7}$$

The leading behavior of the sum on the right side of (6.7.7) can be obtained as in Examples 2 and 3 by observing that the sum is a Riemann sum for the integral $\int_{-\infty}^{\infty} e^{kx^{1/k}} e^{-t^2 kx^{-1/k}/2} (2\pi)^{-k/2} x^{-1/2}\, dt$. Hence, we evaluate this integral and find that the leading behavior of $S(x)$ as $x \to +\infty$ is

$$S(x) \sim (2\pi)^{(1-k)/2} k^{-1/2} x^{(1-k)/2k} e^{kx^{1/k}}, \qquad x \to +\infty. \tag{6.7.8}$$

Can you find the next-higher-order correction to this result? (See Prob. 6.89.)

Euler-Maclaurin Sum Formula

The Euler-Maclaurin sum formula is an elegant and general expression for the asymptotic expansion of sums of the form

$$F(n) = \sum_{k=0}^{n} f(k) \tag{6.7.9}$$

as $n \to \infty$. For example, this formula can be used to find the full asymptotic expansion of sums like $\sum_{k=1}^{n} k^\alpha$, $\sum_{k=1}^{n} \ln k$ as $n \to \infty$ and even sums like $\sum_{k=0}^{\infty} 1/(k^2 + x^2)$, $\sum_{k=0}^{\infty} (k + x)^{-\alpha}$ as $x \to +\infty$.

The Euler-Maclaurin sum formula involves Bernoulli polynomials $B_n(x)$ which are defined as the nth derivative of $te^{xt}/(e^t - 1)$ evaluated at $t = 0$. The first few Bernoulli polynomials are $B_0(x) = 1$, $B_1(x) = x - \frac{1}{2}$, $B_2(x) = x^2 - x + \frac{1}{6}$. $B_n(x)$ is a polynomial of degree n. Some of the properties of these polynomials are studied in Prob. 6.88. The Bernoulli numbers B_n are defined in terms of Bernoulli polynomials as $B_n = B_n(0)$. The first few Bernoulli numbers are $B_0 = 1$, $B_1 = -\frac{1}{2}$, $B_2 = \frac{1}{6}$, $B_3 = 0$, $B_4 = -\frac{1}{30}$. Some of the properties of Bernoulli numbers are examined in Prob. 5.18. There it is shown that $B_{2n+1} = 0$ for $n \geq 1$.

In terms of B_n and $B_n(x)$, the full asymptotic expansion of $f(n)$ in (6.7.9) is

$$F(n) \sim \frac{1}{2} f(n) + \int_0^n f(t)\, dt + C + \sum_{j=1}^{\infty} (-1)^{j+1} \frac{B_{j+1}}{(j+1)!} f^{(j)}(n), \qquad n \to \infty, \tag{6.7.10}$$

where C is a constant given by the rather messy formula

$$C = \lim_{m \to \infty} \left[\sum_{j=1}^{m} \frac{(-1)^j B_{j+1}}{(j+1)!} f^{(j)}(0) + \frac{1}{2} f(0) + \frac{(-1)^m}{(m+1)!} \int_0^\infty B_{m+1}(t - [t]) f^{(m+1)}(t)\, dt \right] \tag{6.7.11}$$

Table 6.2 Comparison between the exact sum of the first n reciprocal integers and the one-, two-, and three-term asymptotic approximations to this sum given by the Euler-Maclaurin sum formula in (6.7.12)

n	$\sum_{k=1}^{n} \frac{1}{k}$	$\ln n$	$\ln n + \gamma$	$\ln n + \gamma + \frac{1}{2n}$
10	2.928 968 25	2.302 585 09	2.879 800 76	2.929 800 76
50	4.499 205 34	3.912 023 01	4.489 238 67	4.499 238 67
100	5.187 377 52	4.605 170 19	5.182 385 85	5.187 385 85
500	6.792 823 43	6.214 608 10	6.791 823 76	6.792 823 76
1,000	7.485 470 86	6.907 755 28	7.484 970 94	7.485 470 94
5,000	9.094 508 85	8.517 193 19	9.094 408 86	9.094 508 86

and $[t]$ is the largest integer less than t. The proof of these formulas is left to Prob. 6.88.

Example 5 *Full asymptotic behavior of* $\sum_{k=1}^{n} 1/k$ *as* $n \to \infty$. By (6.7.10) with $f(t) = 1/(t+1)$ and n replaced by $n - 1$,

$$\sum_{k=1}^{n} \frac{1}{k} \sim \ln n + C + \frac{1}{2n} - \frac{B_2}{2n^2} - \frac{B_4}{4n^4} - \frac{B_6}{6n^6} - \cdots, \qquad n \to \infty. \tag{6.7.12}$$

The constant C, as given in (6.7.11), is evaluated in Prob. 6.90; the result is $C = \gamma \doteq 0.5772$. In Table 6.2 we test the accuracy of the expansion in (6.7.12).

Example 6 *Full asymptotic behavior of* $\ln (n!)$ *as* $n \to \infty$. Using $f(t) = \ln (1 + t)$ and n replaced by $n - 1$, (6.7.10) gives

$$\ln (n!) \sim \left(n + \frac{1}{2}\right) \ln n - n + C + \frac{B_2}{1 \cdot 2n} + \frac{B_4}{3 \cdot 4n^3} + \frac{B_6}{5 \cdot 6n^5} + \cdots, \qquad n \to \infty. \tag{6.7.13}$$

Here $C = \lim_{n\to\infty} [\ln (n!) - (n + \frac{1}{2}) \ln n + n] = \frac{1}{2} \ln 2\pi$, by Stirling's formula.

For more examples, see the problems for Sec. 6.7.

PROBLEMS FOR CHAPTER 6

Sections 6.1 and 6.2

(E) **6.1** Show that the integral in (6.1.3) satisfies the differential equation (6.1.1).
Clue: Differentiate three times under the integral sign and integrate by parts once.

(E) **6.2** Show that a_n in (6.1.5) satisfies the difference equation (6.1.4) and the initial conditions $a_0 = 1$, $a_1 = 0$.

(TE) **6.3** (*a*) Prove (6.2.1).
Clue: Assuming that the asymptotic relation $f(t, x) \sim f_0(t)$ $(x \to x_0)$ is uniform in t, show that for any $\varepsilon > 0$, $\left|\int_a^b f(t, x)\, dt - \int_a^b f_0(t)\, dt\right| < \varepsilon \left|\int_a^b f_0(t)\, dt\right|$ for x sufficiently close to x_0.
(*b*) Prove (6.2.2).

(I) **6.4** Verify (6.2.7). (Note that N is the largest integer less than $-a$ and $a < 0$.)

(D) **6.5** Euler's constant γ is defined by $\gamma \equiv \lim_{n\to\infty} (1 + \frac{1}{2} + \frac{1}{3} + \cdots + 1/n - \ln n) \doteq 0.5772$. Show that γ can also be represented as

$$(a)\ \gamma = -\lim_{x\to 0+}\left(\int_x^\infty t^{-1}e^{-t}\,dt + \ln x\right);$$

$$(b)\ \gamma = \int_0^\infty (dt/t)[1/(1+t) - e^{-t}].$$

(I) **6.6** Verify (6.2.11) and (6.2.12).

(E) **6.7** Find the leading behavior as $x \to 0+$ of the following integrals:

$$(a)\ \int_x^1 \cos(xt)\,dt;$$

$$(b)\ \int_0^1 \sqrt{\sinh(xt)}\,dt;$$

$$(c)\ \int_0^1 e^{-x/t}\,dt;$$

$$(d)\ \int_x^1 e^{-1/t}\,dt;$$

$$(e)\ \int_x^1 \sin(xt)\,dt;$$

$$(f)\ \int_0^1 [dt/(1-t)](e^x - e^{xt});$$

$$(g)\ \int_0^{1/x} e^{-t^2}\,dt;$$

$$(h)\ \int_0^1 [e^{-xt}/(1+t^2)]\,dt;$$

$$(i)\ \int_0^\infty J_0(xt)t^{-1/2}\,dt.$$

(E) **6.8** Find the full asymptotic behavior as $x \to 0+$ of the following integrals:

$$(a)\ \int_0^1 [e^{-t}/(1+x^2t^3)]\,dt;$$

$$(b)\ \int_0^1 [dt/(1-t)](e^x - e^{xt});$$

$$(c)\ \int_0^x [e^{-t}/(t+a)]\,dt.$$

(I) **6.9** Show that $\int_0^1 [dt/(1-t)](e^x - e^{xt}) \sim e^x \ln x + e^x\gamma + \cdots\ (x \to +\infty)$.

(D) **6.10** Let $I(x) = \int_0^\infty [e^{-t}/(1 + xe^{t^2})]\,dt$.
(a) Show that $I(x) - 1 \sim -\exp(-\sqrt{\ln x})\ (x \to 0+)$.
(b) Find the full asymptotic expansion of $I(x)$ as $x \to 0+$.

Section 6.3

(I) **6.11** (a) Verify (6.3.10).
(b) Verify (6.3.14).

(I) **6.12** Find the full asymptotic expansion of $\int_0^x \text{Ai}(t)\,dt$ as $x \to +\infty$ and compare three methods for obtaining this result:

(*a*) Use direct integration by parts.

(*b*) Use the asymptotic expansion of Ai (x) valid for large x.

(*c*) Show that $\int_0^x \text{Ai}(t)\,dt$ satisfies the differential equation $y''' = xy'$ and use this equation to generate the series.

Which method is easiest?

(I) **6.13** Verify (6.3.16).

(I) **6.14** Find the full asymptotic expansion of $\int_0^x \text{Bi}(t)\,dt$ as $x \to +\infty$.

(TD) **6.15** Show that the integral term on the right side of (6.3.18) becomes negligible compared with the boundary term as $x \to +\infty$, assuming that the integral exists and that $\phi'(t) \neq 0$ for $a \le t \le b$.

Clue: Divide the integration region into small subintervals and bound the integral on each subinterval.

(E) **6.16** Verify (6.3.24).

Clue: $\int_0^\infty e^{-t}(1 + xt) \ln (1 + xt)\,dt > x \int_0^\infty te^{-t} \ln (xt)\,dt$.

(I) **6.17** (*a*) The Fresnel integrals are $\int_x^\infty \cos(t^2)\,dt$, $\int_x^\infty \sin(t^2)\,dt$. Find the full asymptotic expansions as $x \to 0$ and as $x \to +\infty$.

(*b*) The generalized Fresnel integral is defined by $F(x, a) = \int_x^\infty t^{-a}e^{it}\,dt$ for $a > 0$. Find the full asymptotic expansion of $F(x, a)$ as $x \to +\infty$.

(E) **6.18** Find the leading behaviors of:

(*a*) $\displaystyle\int_x^\infty e^{-at^b}\,dt$ as $x \to +\infty$, where $a > 0$ and $b > 0$;

(*b*) $\displaystyle\int_1^\infty \cos(xt)t^{-1}\,dt$ as $x \to 0+$.

(I) **6.19** (*a*) How many terms of the asymptotic expansion of $\int_0^{\pi/4} \cos(xt^2) \tan^2 t\,dt$ as $x \to +\infty$ can be computed using integration by parts? Compute them.

(*b*) Do the same for $\int_0^\pi \cos(xt) \sin(t^2)\,dt$ as $x \to +\infty$.

(I) **6.20** Find the leading behavior as $x \to +\infty$ of the following integrals:

(*a*) $\int_x^\infty K_0(t)\,dt$, where K_0 is a modified Bessel function of order 0;

(*b*) $\int_0^x I_0(t)\,dt$, where I_0 is a modified Bessel function of order 0;

(*c*) $\int_x^\infty D_\nu(t)\,dt$, where D_ν is a parabolic cylinder function of order ν.

(I) **6.21** (*a*) Show that if f is infinitely differentiable, then $A_n = \int_0^\pi f(\cos\theta) \cos(n\theta)\,d\theta$ defined for integer n vanishes more rapidly than any finite power of $1/n$ as $n \to \infty$. (This proves that the Fourier expansion of any even, 2π-periodic, infinitely differentiable function is uniformly convergent and can be differentiated termwise an arbitrary number of times.)

(*b*) What is the leading behavior of A_n as $n \to \infty$ through nonintegral values?

(I) **6.22** Find the leading behavior as $n \to \infty$ (through integer values) of the integral $A_n = \int_0^1 t \sin(\frac{1}{2}\pi t)J_0(\lambda_n t)\,dt$, where J_0 is the Bessel function of order 0 and λ_n is its nth zero.

Clue: Use the asymptotic expansion of $J_0(x)$ as $x \to +\infty$ to show that $\lambda_n \sim (n - \frac{1}{4})\pi$ $(n \to \infty)$. Also, note that $J_0(\lambda_n t)$ satisfies the differential equation $(ty')' + t\lambda_n^2 y = 0$. Note that the Fourier-Bessel expansion of $\sin(\frac{1}{2}\pi x)$ is $2 \sum_{n=1}^\infty A_n J_0(\lambda_n x)/J_0'(\lambda_n)^2$.

Section 6.4

(TI) **6.23** Use integration by parts to show that the difference between $I(x)$ in (6.4.1) and $I(x; \varepsilon)$ in (6.4.2) is subdominant with respect to $I(x)$.

(TI) **6.24** (*a*) Find the leading behavior as $x \to +\infty$ of Laplace integrals of the form $\int_a^b (t - a)^\alpha g(t)e^{x\phi(t)}\,dt$, where $\phi(t)$ has a maximum at $t = a$ and $g(a) = 1$. Here $\alpha > -1$ and $\phi'(a) < 0$.

(*b*) Repeat the analysis of part (*a*) when $\alpha > -1$ and $\phi'(a) = \phi''(a) = \cdots = \phi^{(p-1)}(a) = 0$ and $\phi^{(p)}(a) < 0$.

(I) **6.25** (*a*) Use the integral representation

$$J_\nu(x) = \frac{(x/2)^\nu}{\sqrt{\pi}\,\Gamma(\nu + \frac{1}{2})} \int_0^\pi \cos(x \cos\theta) \sin^{2\nu}\theta \, d\theta,$$

which is valid for $\nu > -\frac{1}{2}$, to show that the Bessel function $J_\nu(x)$ satisfies $J_\nu(x) \sim (x/2)^\nu/\Gamma(\nu + 1)$ $(\nu \to \infty)$.

(*b*) Use the integral representation

$$Y_\nu(x) = \frac{1}{\pi} \int_0^\pi \sin(x \sin\theta - \nu\theta)\, d\theta - \frac{1}{\pi} \int_0^\infty [e^{\nu t} + e^{-\nu t} \cos(\nu\pi)] e^{-x \sinh t}\, dt$$

to show that the Bessel function $Y_\nu(x)$ satisfies $Y_\nu(x) \sim -\Gamma(\nu)(2/x)^\nu/\pi$.

(E) **6.26** (*a*) Obtain three terms of the asymptotic expansion of $\int_0^{\pi/2} e^{-x \tan^2\theta}\, d\theta$ as $x \to \infty$.

(*b*) Find the leading behavior of $\int_0^{2\pi} (1 + t^2) e^{x \cos t}\, dt$ as $x \to +\infty$. Note that two maxima contribute to this leading behavior.

(I) **6.27** Show that $\int_0^\infty \mathrm{Ai}(xt)/(1 + t^2)\, dt \sim 1/(3x)$ $(x \to +\infty)$, where $\mathrm{Ai}(s)$ is the Airy function. Can you find the full asymptotic behavior of this integral as $x \to +\infty$?

(I) **6.28** Find the leading behaviors of

(*a*) $\displaystyle\int_0^{\pi/2} \sqrt{\sin t}\, e^{-x \sin^4 t}\, dt$ as $x \to \infty$;

(*b*) $\displaystyle\int_0^1 \sqrt{t(1 - t)}\, (t + a)^{-x}\, dt$ as $x \to +\infty$ with $a > 0$;

(*c*) $\displaystyle\int_0^{\pi/4} \sqrt{\tan t}\, e^{-xt^2}\, dt$ as $x \to +\infty$;

(*d*) $\displaystyle\int_0^{\pi^2/2} ds \int_0^{\pi^2/2} dt\, e^{x \cos\sqrt{s+t}}$ as $x \to +\infty$.

(I) **6.29** Define $P_n(z) = (1/\pi) \int_0^\pi [z + (z^2 - 1)^{1/2} \cos\theta]^n\, d\theta$ $(z > 1)$ where the positive square root is taken. $P_n(z)$ is the nth-degree Legendre polynomial. Show that for large n,

$$P_n(z) \sim \frac{1}{\sqrt{2\pi n}} \frac{[z + (z^2 - 1)^{1/2}]^{n + 1/2}}{(z^2 - 1)^{1/4}}.$$

(I) **6.30** (*a*) The digamma function $\psi(z) = \Gamma'(z)/\Gamma(z)$ has the integral representation

$$\psi(z) = \ln z - 1/2z - \int_0^\infty [(e^t - 1)^{-1} - t^{-1} + \tfrac{1}{2}] e^{-tz}\, dt.$$

Use this integral representation to generate the first three terms of the asymptotic expansion of $\psi(z) - \ln z + 1/2z$ as $z \to \infty$.

(*b*) Show that

$$\psi(z) + \gamma = \int_0^\infty \frac{e^{-t} - e^{-zt}}{1 - e^{-t}}\, dt.$$

(*c*) Show that $\psi(z) + 1/z + \gamma = \sum_{k=2}^\infty (-1)^k \zeta(k) z^{k-1}$ $(|z| < 1)$, where the Riemann ζ function is defined by $\zeta(k) = \sum_{n=1}^\infty n^{-k}$.

(*d*) Use the series in part (*c*) to derive a recursion relation for the coefficients C_j in the Taylor series (5.4.4) for $1/\Gamma(z)$.

(I) **6.31** Use Laplace's method for a moving maximum on (6.1.3) to show that A in (6.1.2) is given by $\pi^{1/2} 2^{2/3} 3^{-1/2}$.

(I) **6.32** Use Laplace's method for a moving maximum on (6.1.5) to verify (6.1.6).

(I) **6.33** Show that the last integral on the right side of (6.3.23) vanishes like $x^{-3/2}$ as $x \to +\infty$.

(D) **6.34** Calculate two terms in the asymptotic expansion of $\int_0^\infty e^{-t-x/t^2}\,dt$ and $\int_0^\infty e^{-xt}(1+t)^{-1/2}\,dt$ as $x \to 0+$ and as $x \to +\infty$.

(I) **6.35** Find the leading behavior of $\int_0^\infty e^{-t-x/t^\alpha}\,dt$ for $\alpha > 0$ as $x \to 0+$ and as $x \to +\infty$.

(I) **6.36** (*a*) Verify that the integral representation in (6.4.20) satisfies the differential equation for $K_\nu(x)$.
(*b*) Use the integral representation (6.4.20) to show that $K_\nu(x) \sim \sqrt{\pi/2\nu}\,(2\nu/ex)^\nu$ $(\nu \to +\infty)$.

(I) **6.37** (*a*) Show that

$$\int_0^\infty \frac{t^{x-1}e^{-t}}{t+x}\,dt \sim \frac{1}{2x}\Gamma(x), \qquad x \to +\infty.$$

(*b*) Find the leading behavior as $x \to +\infty$ of

$$\frac{\displaystyle\int_0^\infty [t^{x-1}/(x+t)]\,e^{-t^x}\,dt}{\displaystyle\int_0^\infty t^{x-1}e^{-t^x}\,dt}.$$

(I) **6.38** Solve Prob. 3.77 using Watson's lemma.

(I) **6.39** The logarithmic integral function li (x) is defined as li $(x) = P\int_0^x dt/\ln t$, where P indicates that the Cauchy principal part of the integral is taken when $x > 1$. Show that li $(e^a) \sim e^a \sum_{n=0}^\infty n!/a^{n+1}$ $(a \to +\infty)$.

(I) **6.40** Prove that

$$\sum_{n=0}^{2N+1} (-1)^n \frac{t^{2n+1}}{(2n+1)!} < \sin t < \sum_{n=0}^{2N} (-1)^n \frac{t^{2n+1}}{(2n+1)!}$$

for all $t > 0$ and all integers N.

Clue: Prove that $\sin t < t$ by integrating $\cos t < 1$. In the same way, use repeated integration to establish the general result.

(I) **6.41** Show that (6.4.27) is an integral representation of the modified Bessel function $I_n(x)$. In other words, show that the integral satisfies the differential equation $x^2y'' + xy' - (x^2 + \nu^2)y = 0$ and the relation $I_n(x) \sim (x/2)^n/n!$ $(x \to 0+)$.

(I) **6.42** Use Laplace's method for a movable maximum to find the next correction to (6.4.40). In particular, show that

$$\Gamma(x) \sim x^{x-1/2}e^{-x}\sqrt{2\pi}\left(1 + \frac{1}{12x} + \frac{1}{288x^2} + \cdots\right), \qquad x \to +\infty.$$

(I) **6.43** (*a*) Show that Laplace's method for expanding integrals consists of approximating the integrand by a δ function. In particular, show how the representation $\delta(t) = \lim_{x\to\infty} \sqrt{x/\pi}\,e^{-xt^2}$ reproduces the leading behavior of a Laplace integral for which $\phi'(c) = 0$ but $\phi''(c) < 0$. [See (6.4.19*c*) and (1.5.10*c*).]

(*b*) What is the appropriate δ-function representation for the case in which $\phi(t) < \phi(a)$ for $a < t < b$ and $\phi'(a) < 0$? [See (6.4.19*a*).]

(*c*) What is the appropriate δ-function representation for the case in which $\phi'(c) = \phi''(c) = \cdots = \phi^{(p-1)}(c) = 0$, $\phi^{(p)}(c) < 0$ with p even? [See (6.4.19*d*).]

(D) (*d*) Extend the δ-function analysis of parts (*a*) to (*c*) to give the higher-order corrections to the leading behavior.

Clue: The answer is given in (6.4.35).

(D) **6.44** Find the leading behavior of the double integral $\int_0^\infty ds \int_0^\pi dt\, e^{-x[\nu s + (\cos t)(\sin s)]}$ as $x \to +\infty$ for $0 < \nu < 1$, $\nu = 1$, and $\nu > 1$. Sketch the function for large x.

Clue: Show that when $0 < \nu < 1$, the exponent has four stationary points. As $\nu \to 1-$, these stationary points merge into two. When $\nu > 1$, there are no stationary points.

(I) **6.45** What happens if we try to treat an ordinary Laplace integral $\int_a^b f(t)e^{x\phi(t)}\,dt$ using the methods appropriate for a moving maximum? Suppose we rewrite the integral as $\int_a^b e^{x\phi(t)+\ln f(t)}\,dt$ and expand about the maximum of the integrand. Show that now an interior maximum is shifted slightly from $t = c$ where $\phi'(c) = 0$, but that this does not affect the result given by Laplace's method in (6.4.19*c* to *d*).

(I) **6.46** Show that naive application of Laplace's method for a moving maximum to the integral $I(x, \alpha) = \int_0^\infty t^\alpha e^{-xt}\,dt = x^{-\alpha-1}\Gamma(\alpha+1)$ gives the wrong answer! Show that the maximum of the integrand occurs at $t = \alpha/x$ and retaining only quadratic terms gives

$$I(x, \alpha) \sim e^{-\alpha}\alpha^{\alpha+1/2}x^{-\alpha-1}\sqrt{2}\int_{-\sqrt{\alpha/2}}^{\infty} e^{-u^2}\,du, \qquad x \to +\infty.$$

Explain why we have obtained the wrong answer.

(I) **6.47** (*a*) Show that

$$\int_0^{1/e} \frac{e^{-xt}}{\ln t}\,dt \sim -\frac{1}{x \ln x}, \qquad x \to +\infty.$$

Clue: See Example 3 of Sec. 6.6.

(*b*) Show that

$$\int_0^{1/e} \frac{e^{-xt}}{\ln t}\,dt \sim -\frac{1}{x \ln x}\sum_{n=0}^{\infty} (\ln x)^{-n}\int_0^\infty (\ln s)^n e^{-s}\,ds, \qquad x \to +\infty.$$

(*c*) Explain why naive use of Laplace's method for a moving maximum fails to give the results (*a*) and (*b*) above.

(D) **6.48** Find the leading behaviors of

(*a*) $\int_0^\infty e^{-xt}e^{-a(\ln t)^2}t^{-1}\,dt$ as $x \to \infty$;

(*b*) $(d^n/dx^n)\Gamma(x)|_{x=1}$ as $n \to \infty$.

Section 6.5

(I) **6.49** Show that $\int_0^\infty e^{is}s^{\alpha-1}\,ds = e^{i\pi\alpha/2}\Gamma(\alpha)$ for $0 < \text{Re}\,\alpha < 1$.

Clue: Substitute $s = it$ and rotate the contour of integration from the negative imaginary-t axis to the positive real-t axis.

(I) **6.50** Use integration by parts to show that the full asymptotic expansion $I(x)$ in (6.5.3) is

$$I(x) \sim \frac{i\sqrt{\pi}}{2x^{3/2}}e^{i\pi/4} - \frac{i}{x\sqrt{\pi}}e^{ix}\left[1 + \frac{1}{2}\sum_{n=0}^{\infty}\left(\frac{-i}{x}\right)^{n+1}\Gamma\left(n+\frac{1}{2}\right)\right], \qquad x \to +\infty.$$

Clue: Write

$$\int_0^1 \frac{e^{ixt}}{\sqrt{t}}\,dt = \int_0^\infty \frac{e^{ixt}}{\sqrt{t}}\,dt - \int_1^\infty \frac{e^{ixt}}{\sqrt{t}}\,dt$$

and use integration by parts on the second integral on the right.

(TD) **6.51** Prove the Riemann-Lebesgue lemma by showing that $\int_a^b f(t)e^{ixt}\,dt \to 0$ $(x \to +\infty)$ provided that $|f(t)|$ is integrable.

Clue: Break up the region of integration into small subintervals and bound the integral on each subinterval.

(TI) **6.52** Show that $\int_a^b f(t)e^{ix\psi(t)}\,dt \to 0$ $(x \to +\infty)$ provided that $|f(t)|$ is integrable, $\psi(t)$ is continuously differentiable, and $\psi(t)$ is not constant on any subinterval of $a \le t \le b$.

Clue: Use the Riemann-Lebesgue lemma.

(D) **6.53** Find the leading behavior of $\int_a^b f(t)e^{ix\psi(t)}\,dt$ under the following assumptions: $\psi'(a) = \cdots = \psi^{(p-1)}(a) = 0$; $\psi^{(p)}(a) \ne 0$; $f(t) \sim A(t-a)^\alpha$ $(t \to a+)$ with $\alpha > -1$.

(*a*) What is the leading contribution to the behavior of $I(x)$ from the neighborhood of the stationary point at $t = a$? This result is a generalization of the formula in (6.5.12).

(b) For which values of α and p does the contribution found in (a) equal the leading behavior of $I(x)$ as $x \to \infty$. Assume that $\psi'(t) \neq 0$ for $a < t \leq b$ and that $f(t)$ is continuous and nonvanishing for $a < t \leq b$.

(I) **6.54** (a) Show that (6.5.13) is an integral representation of $J_n(x)$.

Clue: Show that the integral satisfies the Bessel equation $x^2y'' + xy' + (x^2 - n^2)y = 0$ and behaves like $(x/2)^n/n!$ as $x \to 0$.

(b) Use the integral representation (6.5.13) to find the leading behavior of $J_n(x)$ as $x \to +\infty$.

(I) **6.55** When ν is not an integer, (6.5.13) generalizes for positive x to

$$J_\nu(x) = \frac{1}{\pi}\int_0^\pi \cos(x \sin t - \nu t)\, dt - \frac{\sin(\nu\pi)}{\pi}\int_0^\infty e^{-x \sinh t - \nu t}\, dt.$$

(a) Verify the validity of this integral representation of $J_\nu(x)$.

(b) Use this integral representation to find the leading behavior of $J_\nu(\nu)$ as $\nu \to +\infty$. The result shows that (6.5.14) remains valid when n is not an integer.

(E) **6.56** Use the method of stationary phase to find the leading behavior of the following integrals as $x \to +\infty$:

(a) $\int_0^1 e^{ixt^2} \cosh t^2\, dt$;

(b) $\int_0^1 \cos(xt^4) \tan t\, dt$;

(c) $\int_0^1 e^{ix(t - \sin t)}\, dt$;

(d) $\int_0^1 \sin[x(t + \frac{1}{6}t^3 - \sinh t)] \cos t\, dt$;

(e) $\int_{-1}^1 \sin[x(t - \sin t)] \sinh t\, dt$.

Section 6.6

(I) **6.57** Evaluate the full asymptotic behavior of $\int_0^1 e^{ixt^2}\, dt$ using integration by parts (see Example 2 of Sec. 6.6).

(I) **6.58** Explain why the contour C_3 can be replaced by the contour C_4 in Fig. 6.6. Show that this replacement introduces errors that are smaller than any term in the asymptotic expansion (6.6.8).

(E) **6.59** Show that the coefficient of the general term in (6.6.16) can be expressed in terms of derivatives of $\Gamma(t)$ at $t = 1$.

(I) **6.60** Find the radius of convergence of the asymptotic series (6.6.17).

Clue: See Probs. 6.48(b) and (6.59).

(E) **6.61** Verify the features of Fig. 6.9. In particular, show that six steepest paths emerge from $t = 0$ with $60°$ angular separations, that the six paths are alternately ascent and descent curves, and that the paths approach $\pm\infty$, $\pm\infty + i\pi$, $\pm\infty - i\pi$.

(E) **6.62** Verify the features of Fig. 6.10. In particular, show that eight steepest paths emerge from $t = 0$ with $45°$ angular separations, that the eight paths are alternately ascent and descent curves, and that the paths approach $\pm\infty$, $\pm i\infty$, $\pm\infty + i\pi$, $\pm\infty - i\pi$.

(E) **6.63** Verify (6.6.19) by showing that the integrals (6.6.18) are pure imaginary.

(E) **6.64** Verify that the steepest-descent curve through the saddle point of (6.6.23) is correctly drawn in Fig. 6.12. Show that this curve is vertical at $s = 1$, crosses the imaginary-s axis at $s = \pm i\pi/2$, and approaches $-\infty \pm i\pi$.

(I) **6.65** Formulate a convincing explanation that if we neglect the term xt^3 in (6.6.24), we will obtain the wrong asymptotic behavior of the integral as $x \to +\infty$.

Clue: On the assumption that the term is small, expand the cosine and estimate the error terms. Show that the error is not smaller than the exponentially small integral obtained by neglecting xt^3.

(E) **6.66** Verify the results in Fig. 6.13. In particular, show that C_1, C_2, and C_3 are steepest-descent contours.

(E) **6.67** Show that the contributions to $I(x)$ in (6.6.25) from C_1 and C_2 in Fig. 6.13 are exponentially small compared with the contribution from the saddle point on C_3.

(I) **6.68** Formulate a convincing explanation that if we replace $\cosh t$ by $1 + t^2/2$ in (6.6.32), we will obtain the wrong asymptotic behavior of (6.6.32) as $x \to +\infty$.

Clue: See Prob. 6.65.

(D) **6.69** (*a*) Verify (6.6.35) to (6.6.37).

(*b*) Show that as $\arg x$ in (6.6.25) increases past $\pi - \arctan 2$ the steepest descent path through the endpoints at $t = \pm 1$ no longer passes through the saddle point at $t = i$. (See Figs. 6.13 to 6.16.)

(I) **6.70** The integrand in (6.6.19) has saddle points at $t = in\pi$ $(n = 0, \pm 1, \pm 2, \ldots)$. Why don't we distort the contour C to pass through all or some of these points instead of just through $t = 0$ as in Example 8 of Sec. 6.6?

(D) **6.71** An integral representation of the modified Bessel function $K_\nu(x)$ is $K_\nu(x) = \frac{1}{2}\int_{-\infty}^{\infty} e^{-x\cosh t + \nu t}\, dt$. Show that $K_{ip}(x) = \sqrt{2\pi}\,(p^2 - x^2)^{-1/4} e^{-p\pi/2} \sin \phi(x)$, where $\phi(x) - p\cosh^{-1}(p/x) + \sqrt{p^2 - x^2} \sim \pi/4$ $(x \to +\infty,\ p/x \to +\infty)$.

Clue: The contribution comes from the neighborhood of two saddle points satisfying $\sinh t = ip/x$. Explain why it is that although there are an infinite number of saddle points, only two contribute to the leading behavior.

(D) **6.72** Investigate the Stokes phenomenon for the integral $I(x) = \int_0^1 e^{xt^3}\, dt$ as $x \to \infty$. Specifically, evaluate the asymptotic behavior of $I(x)$ as $x \to \infty$ with $\arg x$ fixed for several values of $\arg x$ and show that there are Stokes lines at $|\arg x| = \pi/2$.

(I) **6.73** Find the first two terms in the asymptotic behavior of $\int_0^{\pi/4} \cos(xt^2) \tan^2 t\, dt$ as $x \to +\infty$.

(I) **6.74** Find three terms in the asymptotic behavior of $\int_0^1 \ln(1 + t)\, e^{ix\sin^2 t}\, dt$ as $x \to +\infty$.

(D) **6.75** (*a*) Show that an integral representation of the Airy function $\text{Ai}(x)$ is given by

$$\text{Ai}(x) = \frac{1}{2\pi i}\int_C e^{xt - t^3/3}\, dt,$$

where C is a contour which originates at $\infty e^{-2\pi i/3}$ and terminates at $\infty e^{2\pi i/3}$.

(*b*) Use this integral representation to show that the Taylor series expansion of $\text{Ai}(x)$ about $x = 0$ is as given in (3.2.1).

(*c*) Using the method of steepest descents, find the asymptotic behavior of $\text{Ai}(x)$ as $x \to +\infty$.

(*d*) Extend the steepest-descent argument used in part (*c*) to show that the same asymptotic behavior is valid for $x \to \infty$ with $|\arg x| < \pi$ and that there is no Stokes phenomenon at $|\arg x| = \pi/3$.

(*e*) Show that there is no Stokes phenomenon at $|\arg x| = \pi/3$ in a different way. Transform the integral in (*a*) to

$$\text{Ai}(x) = \frac{1}{\pi} e^{-2x^{3/2}/3} \int_0^\infty e^{-x^{1/2}t^2} e^{it^3/3}\, dt$$

and then use Laplace's method.

Clue: For real positive x, deform C in (*a*) into the straight-line contour connecting $-x - i\infty$ to $-x + i\infty$ and then allow x to be complex with $|\arg x^{1/2}| < \pi/2$.

(*f*) Find the leading behavior of $\text{Ai}(x)$ as $x \to -\infty$.

Clue: Show that the steepest-descent contour connecting $\infty e^{-2\pi i/3}$ to $\infty e^{2\pi i/3}$ consists of two pieces, one passing through the saddle point at $t = -i\sqrt{-x}$ and one passing through the saddle point at $t = +i\sqrt{-x}$.

(D) **6.76** (*a*) Show that an integral representation of the Airy function Bi (x) is given by

$$\text{Bi}(x) = \frac{1}{2\pi}\int_{C_+} e^{xt - t^3/3}\, dt + \frac{1}{2\pi}\int_{C_-} e^{xt - t^3/3}\, dt,$$

where $C_\pm$ is a contour which originates at $\infty e^{\pm 2\pi i/3}$ and terminates at $+\infty$.

(*b*) Use this integral representation to show that the Taylor series expansion of Bi (x) about $x = 0$ is given in (3.2.2).

(*c*) Using the method of steepest descents, find the asymptotic behavior of Bi (x) as $x \to +\infty$.

(*d*) Extend the steepest-descent argument used in (*c*) to show that this leading asymptotic behavior of Bi (x) is still valid for $x \to \infty$ with $|\arg x| < \pi/3$.

(*e*) Use the integral representation of (*a*) to find the asymptotic behavior of Bi (x) as $x \to \infty$ with $\pi/3 < |\arg x| < \pi$. Show that as $|\arg x|$ increases beyond $\pi/3$, the contribution from the saddle point at $t = -\sqrt{x}$ overwhelms the contribution from the saddle point at $t = +\sqrt{x}$.

(D) **6.77** Consider the integral $I(n) = \int_0^\pi \cos(nt)\, e^{ia \cos t}\, dt$ where n is an integer.

(*a*) Use integration by parts to show that $I(n)$ for n integral decays to 0 faster than any power of $1/n$ as $n \to \infty$. What if n is nonintegral?

(*b*) Use the method of steepest descents in the complex-t plane to show that $I(n) \sim \sqrt{\pi/2n}\,(\tfrac{1}{2}iea/n)^n$ $(n \to \infty)$.

(*c*) By setting $z = \cos t$ show that

$$I(n) = \frac{1}{\pi}\int_C \frac{e^{iaz}}{(z + \sqrt{z^2 - 1})^n}\frac{dz}{\sqrt{1 - z^2}}$$

where the branch cut of $\sqrt{z^2 - 1}$ is chosen to lie along the real axis from $z = -1$ to $z = +1$, the branch is chosen so that $\sqrt{z^2 - 1} \to +z$ as $z \to \infty$, and C is any contour that loops around the branch cut just once in the counterclockwise sense.

(*d*) Use the integral representation in (*c*) and the method of steepest descents to reproduce the results in (*b*). Notice that there are no endpoint contours required with this alternative derivation.

(D) **6.78** Let $f(z)$ be an entire (everywhere analytic) function of z whose Taylor series about $z = 0$ is $f(z) = \sum_{n=0}^{\infty} a_n z^n$. Cauchy's theorem gives a contour integral representation for a_n:

$$a_n = \frac{1}{2\pi i}\int_C \frac{f(t)}{t^{n+1}}\, dt,$$

where C loops the origin once in the counterclockwise sense. Using the method of steepest descents, find the leading asymptotic behavior of a_n as $n \to \infty$ for the following functions:

(*a*) $f(z) = e^z$ (the result is a rederivation of the Stirling formula for $n!$);

(*b*) $f(z) = e^{z^m}$;

(*c*) $f(z) = \exp(e^z)$ (see Prob. 5.71).

(I) **6.79** Using the approach of Prob. 6.78, verify the result stated in Prob. 5.20 for the asymptotic behavior of the Bernoulli numbers $B_n = n!\, f^{(n)}(0)$ where $f(z) = z/(e^z - 1)$.

(D) **6.80** Find two terms in the asymptotic expansion of $\int_0^\pi e^{ix \cos t}\, dt$ as $x \to +\infty$. Where are the Stokes lines in the complex plane?

(D) **6.81** Use the method of steepest descents to find the full asymptotic behaviors of

(*a*) $\int_0^1 e^{ixt^3}\, dt \quad (x \to +\infty)$;

(*b*) $\int_0^1 t^\alpha e^{ixt^3}\, dt \quad (x \to +\infty)$.

(D) **6.82** Find the leading behavior of the integral $\int_{-\infty+ia}^{\infty+ia} t^{-1/2}e^{ix(t^3/3-t)}\,dt$ as $x \to +\infty$, where $a > 0$ and the contour of integration is an infinite straight line parallel to the real-t axis lying in the upper half of the complex-t plane. What is the effect of the branch point at $t = 0$?

(D) **6.83** (*a*) Show that $y(x) = \int_C e^{xt-t^5/5}\,dt$, where C is any contour connecting the points ∞, $\infty e^{\pm 2\pi i/5}$, $\infty e^{\pm 4\pi i/5}$ at which the integrand vanishes, satisfies the hyperairy equation (3.5.22).

(*b*) Use the integral representation given in (*a*) to find the possible leading behaviors of solutions to the hyperairy equation as $x \to +\infty$.

(*c*) Investigate fully the Stokes phenomenon for that solution of the hyperairy equation obtained by choosing C to be a contour connecting $\infty e^{-4\pi i/5}$ to $\infty e^{-2\pi i/5}$.

(D) **6.84** (*a*) Show that $D_\nu(x) = (e^{x^2/4}/i\sqrt{2\pi}) \int_C t^\nu e^{-xt+t^2/2}\,dt$, where C is a contour connecting $-i\infty$ to $+i\infty$ on which $\text{Re}\, t > 0$, satisfies the parabolic cylinder equation (3.5.11) and the initial conditions $D_\nu(0) = \pi^{1/2}2^{\nu/2}/\Gamma(\frac{1}{2} - \nu/2)$, $D_\nu'(0) = -\pi^{1/2}2^{(\nu+1)/2}\Gamma(-\nu/2)$.

(*b*) Use the method of steepest descents to show that $D_\nu(x) \sim x^\nu e^{-x^2/4}$ $(x \to +\infty)$.

(*c*) Extend the steepest descent argument of (*b*) to show that $D_\nu(x) \sim x^\nu e^{-x^2/4}$ is still valid if $x \to \infty$ with $|\arg x| < \pi/2$.

Clue: Show that the branch point of t^ν at $t = 0$ does not affect the steepest descent calculation if $|\arg x| < \pi/2$.

(*d*) Show that the asymptotic behavior found in (*b*) and (*c*) is valid beyond $|\arg x| = \pi/2$ and only breaks down at $|\arg x| = 3\pi/4$ when ν is not a nonnegative integer.

Clue: Show that the contribution from the branch point at $t = 0$ found in part (*c*) becomes significant as $|\arg x|$ increases beyond $3\pi/4$.

(*e*) Find the asymptotic behavior of $D_\nu(x)$ as $x \to \infty$ with $3\pi/4 < |\arg x| < 5\pi/4$.

(D) **6.85** (*a*) Using the integral representation of Prob. 6.84(*a*), show that if $\text{Re}\,(\nu) > -1$, $D_\nu(x) = \sqrt{2/\pi}\, e^{x^2/4} \int_0^\infty e^{-t^2/2}t^\nu \cos(xt - \nu\pi/2)\,dt$.

(*b*) Show that $D_\nu(x) = \sqrt{2}\,(\nu/e)^{\nu/2} \cos\theta$, where $\theta \sim x\nu^{1/2} - \nu\pi/2$ $(\nu \to \infty;\ x \text{ fixed})$.

(D) **6.86** Find the leading behavior as $x \to +\infty$ of the integral

$$\int_C \frac{e^{ix(t^3/3-t)}}{t-a}\,dt,$$

where C is a contour connecting $-\infty$ to $+\infty$ in the upper-half t plane and a is a real constant. Investigate separately the cases $|a| < 1$, $|a| > 1$, $a = \pm 1$.

(D) **6.87** (*a*) Find the leading behavior of $I_n(x) = \int_0^\infty e^{xt-t^n/n}\,dt$ as $x \to +\infty$. (See Prob. 6.83.)

(*b*) Investigate the Stokes phenomenon for $I_n(x)$ as $x \to \infty$ in the complex-x plane.

Section 6.7

(I) **6.88** Derive the Euler-Maclaurin sum formula (6.7.10) as follows:

(*a*) Verify that

$$\tfrac{1}{2}[f(k) + f(k+1)] - \int_k^{k+1} f(t)\,dt = \int_k^{k+1} (t - k - \tfrac{1}{2})f'(t)\,dt.$$

(*b*) Sum the identity in (*a*) to show that $F(n)$ in (4.7.9) satisfies $F(n) = \frac{1}{2}[f(0) + f(n)] + \int_0^n f(t)\,dt + \int_0^n B_1(t - [t])f'(t)\,dt$, where $B_1(s)$ is the Bernoulli polynomial of degree 1, $B_1(s) = s - \frac{1}{2}$.

(*c*) Show that the Bernoulli polynomials $B_n(x)$, which are defined by the formula $te^{xt}/(e^t - 1) = \sum_{n=0}^\infty B_n(x)t^n/n!$, satisfy $B_n'(x) = nB_{n-1}(x)$ and $B_n(0) = B_n(1)$ for $n \geq 2$.

(*d*) Integrate the result of (*b*) repeatedly by parts to show that

$$F(n) = \tfrac{1}{2}[f(0) + f(n)] + \int_0^n f(t)\,dt + \sum_{j=1}^m (-1)^{j+1} \frac{B_{j+1}}{(j+1)!}$$

$$\times [f^{(j)}(n) - f^{(j)}(0)] + \frac{(-1)^m}{(m+1)!} \int_0^n B_{m+1}(t) f^{(m+1)}(t)\,dt.$$

(*e*) Let $m \to \infty$ and thus derive (6.7.10) and (6.7.11).

(D) **6.89** Find the first correction to the result in (6.7.8).

(D) **6.90** Euler's constant γ is defined as $\gamma = \lim_{n\to\infty} (\sum_{k=1}^{n} 1/k - \ln n)$. Use the formula (6.7.11) to show that the constant C in (6.7.12) is Euler's constant γ.

Clue: Reverse the process of integration by parts used in Prob. 6.88.

(D) **6.91** (*a*) Show that $\sum_{k=0}^{\infty} 1/(k^2 + x^2) - \pi/2x \sim 1/2x^2$ $(x \to +\infty)$.

Clue: Follow the derivation given in Prob. 6.88 to derive an analog of the Euler-Maclaurin sum formula for $x \to +\infty$.

(*b*) Evaluate $\sum_{k=0}^{\infty} (k^2 + x^2)^{-1}$ exactly by representing the sum as the contour integral

$$\frac{1}{2\pi i}\int_C \frac{\cot t}{t^2 + x^2}\,dt$$

over an appropriate contour.

(*c*) By comparing the results of (*a*) and (*b*), show that the error in (*a*) is exponentially small as $x \to +\infty$.

(D) **6.92** Find two terms in the asymptotic behavior as $x \to +\infty$ of the following sums:

(*a*) $\sum_{k=0}^{\infty} (k + x)^{-\alpha}$ $(\alpha > 1)$;

(*b*) $\sum_{k=0}^{\infty} (k^2 + x^2)^{-2}$.

(D) **6.93** Find three terms in the asymptotic behavior as $n \to \infty$ of the following sums:

(*a*) $\sum_{k=1}^{n} (-1)^k/k$;

(*b*) $\sum_{k=1}^{n} B_{2k}$;

(*c*) $\sum_{k=1}^{n} \sin k/k$.

(D) **6.94** Show that

$$\sum_{k=1}^{\infty} \frac{1}{k(k^2 + x^2)} \sim \frac{\ln x}{x^2} + \frac{\gamma}{x^2} - \sum_{n=1}^{\infty} \frac{(-1)^n B_{2n}}{2nx^{2n+2}}, \qquad x \to +\infty.$$

(I) **6.95** Show that

(*a*) $\sum_{1 \le n \le x} 2^{-(1+1/2+1/3+\cdots+1/n)} \sim x^{1-\ln 2}/(1 - \ln 2)$ $(x \to +\infty)$;

(*b*) $\sum_{1 \le n \le x} n^{1/2} e^{n^{1/2}} \sim 2xe^{x^{1/2}}$ $(x \to +\infty)$;

(*c*) $\sum_{n=0}^{\infty} (n + 1)^3/[(n + \frac{1}{2})^2 + x^2]^{5/2} \sim 2/3x$ $(x \to +\infty)$;

(*d*) $\sum_{j=0}^{n} j^k(n - j)^k \sim (n/2)^{2k+1}\sqrt{\pi/k}$ $(n \to \infty)$.

PART THREE

PERTURBATION METHODS

When I hear you give your reasons, the thing always appears to me to be so ridiculously simple that I could easily do it myself, though at each successive instance of your reasoning I am baffled until you explain your process.

—Dr. Watson, *A Scandal in Bohemia*
Sir Arthur Conan Doyle

The local analysis methods of Part II are powerful tools, but they cannot provide global information on the behavior of solutions at two distantly separated points. They cannot predict how a change in initial conditions at $x = 0$ will affect the asymptotic behavior as $x \to +\infty$. To answer such questions we must apply the methods of global analysis which will be developed in Part IV. Since global methods are perturbative in character, in this part we will first introduce the requisite mathematical concepts: perturbation theory in Chap. 7 and summation theory in Chap. 8.

Perturbation theory is a collection of methods for the systematic analysis of the global behavior of solutions to differential and difference equations. The general procedure of perturbation theory is to identify a small parameter, usually denoted by ε, such that when $\varepsilon = 0$ the problem becomes soluble. The global solution to the given problem can then be studied by a local analysis about $\varepsilon = 0$. For example, the differential equation $y'' = [1 + \varepsilon/(1 + x^2)]y$ can only be solved in terms of elementary functions when $\varepsilon = 0$. A perturbative solution is constructed by local analysis about $\varepsilon = 0$ as a series of powers of ε:

$$y(x) = y_0(x) + \varepsilon y_1(x) + \varepsilon^2 y_2(x) + \cdots.$$

This series is called a perturbation series. It has the attractive feature that $y_n(x)$ can be computed in terms of $y_0(x), \ldots, y_{n-1}(x)$ as long as the problem obtained by setting $\varepsilon = 0$, $y'' = y$, is soluble, which it is in this case. Notice that the perturbation series for $y(x)$ is *local* in ε but that it is *global* in x. If ε is very small, we expect that $y(x)$ will be well approximated by only a few terms of the perturbation series.

The local analysis methods of Part II are other examples of perturbation theory. There the expansion parameter is $\varepsilon = x - x_0$ or $\varepsilon = 1/x$ if $x_0 = \infty$.

Perturbation series, like asymptotic expansions, often diverge for all $\varepsilon \neq 0$. However, since ε is not necessarily a small parameter, the optimal asymptotic approximation may give very poor numerical results. Thus, to extract maximal information from perturbation theory, it is necessary to develop sophisticated techniques to "sum" divergent series and to accelerate the convergence of slowly converging series. Methods to achieve these goals are presented in Chap. 8. Summation methods also apply to the local series expansions derived in Part II.

In perturbation theory it is convenient to have an asymptotic order relation that expresses the relative magnitudes of two functions more precisely than $\ll$ but less precisely than $\sim$. We define

$$f(x) = \mathrm{O}[g(x)], \qquad x \to x_0,$$

and say "$f(x)$ is at most of order $g(x)$ as $x \to x_0$" or "$f(x)$ is 'O' of $g(x)$ as $x \to x_0$" if $f(x)/g(x)$ is bounded for x near x_0; that is, $|f(x)/g(x)| < M$, for some constant M if x is sufficiently close to x_0. Observe that if $f(x) \sim g(x)$ or if $f(x) \ll g(x)$ as $x \to x_0$, then $f(x) = \mathrm{O}[g(x)]$ as $x \to x_0$. If $f \ll g$ as $x \to x_0$, then any $M > 0$ satisfies the definition, while if $f \sim g$ $(x \to x_0)$, only $M > 1$ can work.

In perturbation theory one may calculate just a few terms in a perturbation series. Whether or not this series is convergent, the notation "O" is very useful for expressing the order of magnitude of the first neglected term when that term has not been calculated explicitly.

Examples

1. $x \sin x = \mathrm{O}(x)$ $(x \to 0$ or $x \to \infty)$;
2. $e^{-1/x} = \mathrm{O}(x^n)$ $(x \to 0+)$ for all n;
3. $x^5 = \mathrm{O}(x^2)$ $(x \to 0+)$;
4. $e^x = 1 + x + (x^2/2) + \mathrm{O}(x^3)$ $(x \to 0)$;
5. $\mathrm{Ai}\,(x) = \frac{1}{2}\pi^{-1/2}x^{-1/4}e^{-2x^{3/2}/3}[1 - \frac{5}{48}x^{-3/2} + \mathrm{O}(x^{-3})]$ $(x \to +\infty)$.

CHAPTER

SEVEN

PERTURBATION SERIES

You have erred perhaps in attempting to put colour and life into each of your statements instead of confining yourself to the task of placing upon record that severe reasoning from cause to effect which is really the only notable feature about the thing. You have degraded what should have been a course of lectures into a series of tales.

—Sherlock Holmes, *The Adventure of the Copper Beeches*
Sir Arthur Conan Doyle

(E) 7.1 PERTURBATION THEORY

Perturbation theory is a large collection of iterative methods for obtaining approximate solutions to problems involving a small parameter ε. These methods are so powerful that sometimes it is actually advisable to introduce a parameter ε temporarily into a difficult problem having no small parameter, and then finally to set $\varepsilon = 1$ to recover the original problem. This apparently artificial conversion to a perturbation problem may be the only way to make progress.

The thematic approach of perturbation theory is to decompose a tough problem into an infinite number of relatively easy ones. Hence, perturbation theory is most useful when the first few steps reveal the important features of the solution and the remaining ones give small corrections.

Here is an elementary example to introduce the ideas of perturbation theory.

Example 1 *Roots of a cubic polynomial.* Let us find approximations to the roots of

$$x^3 - 4.001x + 0.002 = 0. \tag{7.1.1}$$

As it stands, this problem is not a perturbation problem because there is no small parameter ε. It may not be easy to convert a particular problem into a tractable perturbation problem, but in the present case the necessary trick is almost obvious. Instead of the single equation (7.1.1) we consider the *one-parameter family* of polynomial equations

$$x^3 - (4 + \varepsilon)x + 2\varepsilon = 0. \tag{7.1.2}$$

When $\varepsilon = 0.001$, the original equation (7.1.1) is reproduced.

It may seem a bit surprising at first, but it is easier to compute the approximate roots of the family of polynomials (7.1.2) than it is to solve just the one equation with $\varepsilon = 0.001$. The reason for this is that if we consider the roots to be functions of ε, then we may further assume a perturbation series in powers of ε:

$$x(\varepsilon) = \sum_{n=0}^{\infty} a_n \varepsilon^n. \tag{7.1.3}$$

To obtain the first term in this series, we set $\varepsilon = 0$ in (7.1.2) and solve

$$x^3 - 4x = 0. \tag{7.1.4}$$

This expression is easy to factor and we obtain in *zeroth-order* perturbation theory $x(0) = a_0 = -2, 0, 2$.

A *second-order* perturbation approximation to the first of these roots consists of writing (7.1.3) as $x_1 = -2 + a_1 \varepsilon + a_2 \varepsilon^2 + O(\varepsilon^3)$ $(\varepsilon \to 0)$, substituting this expression into (7.1.2), and neglecting powers of ε beyond ε^2. The result is

$$(-8 + 8) + (12a_1 - 4a_1 + 2 + 2)\varepsilon + (12a_2 - a_1 - 6a_1^2 - 4a_2)\varepsilon^2 = O(\varepsilon^3), \qquad \varepsilon \to 0. \tag{7.1.5}$$

It is at this step that we realize the power of generalizing the original problem to a family of problems (7.1.2) with variable ε. It is because ε is *variable* that we can conclude that the coefficient of each power of ε in (7.1.5) is *separately* equal to zero. This gives a sequence of equations for the expansion coefficients $a_1, a_2, \ldots$:

$$\varepsilon^1: \quad 8a_1 + 4 = 0; \qquad\qquad \varepsilon^2: \quad 8a_2 - a_1 - 6a_1^2 = 0;$$

and so on. The solutions to the equations are $a_1 = -\frac{1}{2}$, $a_2 = \frac{1}{8}$, ... Therefore, the perturbation expansion for the root x_1 is

$$x_1 = -2 - \tfrac{1}{2}\varepsilon + \tfrac{1}{8}\varepsilon^2 + \cdots. \tag{7.1.6}$$

If we now set $\varepsilon = 0.001$, we obtain x_1 from (7.1.6) accurate to better than one part in 10^9. The same procedure gives

$$x_2 = 0 + \tfrac{1}{2}\varepsilon - \tfrac{1}{8}\varepsilon^2 + O(\varepsilon^3), \qquad x_3 = 2 + 0 \cdot \varepsilon + 0 \cdot \varepsilon^2 + O(\varepsilon^3), \qquad \varepsilon \to 0.$$

(Successive coefficients in the perturbation series for x_3 all vanish because $x_3 = 2$ is the exact solution for all ε.) All three perturbation series for the roots converge for $\varepsilon = 0.001$. Can you prove that they converge for $|\varepsilon| < 1$? (See Prob. 7.6.)

This example illustrates the three steps of perturbative analysis:

1. *Convert the original problem into a perturbation problem by introducing the small parameter* ε.
2. *Assume an expression for the answer in the form of a perturbation series and compute the coefficients of that series.*
3. *Recover the answer to the original problem by summing the perturbation series for the appropriate value of* ε.

Step (1) is sometimes ambiguous because there may be many ways to introduce an ε. However, it is preferable to introduce ε in such a way that the *zeroth-order* solution (the leading term in the perturbation series) is obtainable as a closed-form analytic expression. Perturbation problems generally take the form of a soluble equation [such as (7.1.4)] whose solution is altered slightly by a perturbing term [such as $(2 - x)\varepsilon$]. Of course, step (1) may be omitted when the original problem already has a small parameter if a perturbation series can be developed in powers of that parameter.

Step (2) is frequently a routine iterative procedure for determining successive coefficients in the perturbation series. A *zeroth-order* solution consists of finding the leading term in the perturbation series. In Example 1 this involves solving the *unperturbed problem*, the problem obtained by setting $\varepsilon = 0$ in the perturbation problem. A *first-order* solution consists of finding the first two terms in the perturbation series, and so on. In Example 1 each of the coefficients in the perturbation series is determined in terms of the previous coefficients by a simple linear equation, even though the original problem was a nonlinear (cubic) equation.

Generally it is the existence of a closed-form zeroth-order solution which ensures that the higher-order terms may also be determined as closed-form analytical expressions.

Step (3) may or may not be easy. If the perturbation series converges, its sum is the desired answer. If there are several ways to reduce a problem to a perturbation problem, one chooses the way that is the best compromise between difficulty of calculation of the perturbation series coefficients and rapidity of convergence of the series itself. However, many series converge so slowly that their utility is impaired. Also, we will shortly see that perturbation series are frequently divergent. This is not necessarily bad because many of these divergent perturbation series are asymptotic. In such cases, one obtains a good approximation to the answer when ε is very small by summing the first few terms according to the optimal truncation rule (see Sec. 3.5). When ε is not small, it may still be possible to obtain a good approximation to the answer from a slowly converging or divergent series using the summation methods discussed in Chap. 8.

Let us now apply these three rules of perturbation theory to a slightly more sophisticated example.

Example 2 *Approximate solution of an initial-value problem.* Consider the initial-value problem

$$y'' = f(x)y, \qquad y(0) = 1,\ y'(0) = 1, \tag{7.1.7}$$

where $f(x)$ is continuous. This problem has no closed-form solution except for very special choices for $f(x)$. Nevertheless, it can be solved perturbatively.

First, we introduce an ε in such a way that the unperturbed problem is solvable:

$$y'' = \varepsilon f(x)y, \qquad y(0) = 1,\ y'(0) = 1. \tag{7.1.8}$$

Second, we assume a perturbation expansion for $y(x)$ of the form

$$y(x) = \sum_{n=0}^{\infty} \varepsilon^n y_n(x), \tag{7.1.9}$$

where $y_0(0) = 1$, $y_0'(0) = 1$, and $y_n(0) = 0$, $y_n'(0) = 0$ $(n \geq 1)$.

The zeroth-order problem $y'' = 0$ is obtained by setting $\varepsilon = 0$, and the solution which satisfies the initial conditions is $y_0 = 1 + x$. The nth-order problem $(n \geq 1)$ is obtained by substituting (7.1.9) into (7.1.8) and setting the coefficient of ε^n $(n \geq 1)$ equal to 0. The result is

$$y_n'' = y_{n-1} f(x), \qquad y_n(0) = y_n'(0) = 0. \tag{7.1.10}$$

Observe that perturbation theory has replaced the intractable differential equation (7.1.7) with a sequence of inhomogeneous equations (7.1.10). In general, any inhomogeneous equation may be solved routinely by the method of variation of parameters whenever the solution of the associated homogeneous equation is known (Sec. 1.5). Here the homogeneous equation is precisely the unperturbed equation. Thus, it is clear why it is so crucial that the unperturbed equation be soluble.

The solution to (7.1.10) is

$$y_n = \int_0^x dt \int_0^t ds\, f(s) y_{n-1}(s), \qquad n \geq 1. \tag{7.1.11}$$

Equation (7.1.11) gives a simple iterative procedure for calculating successive terms in the perturbation series (7.1.9):

$$\begin{aligned} y(x) = 1 + x + \varepsilon \int_0^x dt \int_0^t ds(1+s) f(s) \\ + \varepsilon^2 \int_0^x dt \int_0^t ds\, f(s) \int_0^s dv \int_0^v du\,(1+u) f(u) + \cdots. \end{aligned} \tag{7.1.12}$$

Third, we must sum this series. It is easy to show that when N is large, the Nth term in this series is bounded in absolute value by $\varepsilon^N x^{2N} K^N (1 + |x|)/(2N)!$, where K is an upper bound for $|f(t)|$ in the interval $0 \leq |t| \leq |x|$. Thus, the series (7.1.12) is convergent for all x. We also conclude that if $x^2 K$ is small, then the perturbation series is rapidly convergent for $\varepsilon = 1$ and an accurate solution to the original problem may be achieved by taking only a few terms.

How do these perturbation methods for differential equations compare with the series methods that were introduced in Chap. 3? Suppose $f(x)$ in (7.1.7) has a convergent Taylor expansion about $x = 0$ of the form

$$f(x) = \sum_{n=0}^{\infty} f_n x^n. \tag{7.1.13}$$

Then another way to solve for $y(x)$ is to perform a local analysis of the differential equation near $x = 0$ by substituting the series solution

$$y(x) = \sum_{n=0}^{\infty} a_n x^n, \qquad a_0 = a_1 = 1, \tag{7.1.14}$$

and computing the coefficients a_n. As shown in Chap. 3, the series in (7.1.14) is guaranteed to have a radius of convergence at least as large as that in (7.1.13).

By contrast, the perturbation series (7.1.9) converges for all finite values of x, and not just those inside the radius of convergence of $f(x)$. Moreover, the perturbation series converges even if $f(x)$ has no Taylor series expansion at all.

Example 3 *Comparison of Taylor and perturbation series.* The differential equation

$$y'' = -e^{-x} y, \qquad y(0) = 1,\ y'(0) = 1, \tag{7.1.15}$$

may be solved in terms of Bessel functions as

$$y(x) = \frac{[Y_0(2) + Y_0'(2)] J_0(2e^{-x/2}) - [J_0(2) + J_0'(2)] Y_0(2e^{-x/2})}{J_0(2) Y_0'(2) - J_0'(2) Y_0(2)}.$$

The local expansion (7.1.14) converges everywhere because e^{-x} has no finite singularities. Nevertheless, a fixed number of terms of the perturbation series (7.1.9) (see Prob. 7.11) gives a much better approximation than the same number of terms of the Taylor series (7.1.14) if x is large and positive (see Fig. 7.1).

In addition, the perturbation methods of Example 2 are immediately applicable to problems where local analysis cannot be used. For example, an approximate solution of the formidable-looking nonlinear two-point boundary-value problem

$$y'' + y = \frac{\cos x}{3 + y^2}, \qquad y(0) = y\left(\frac{\pi}{2}\right) = 2, \tag{7.1.16}$$

may be readily obtained using perturbation theory (see Prob. 7.14).

Thus, the ideas of perturbation theory apply equally well to problems requiring local or global analysis.

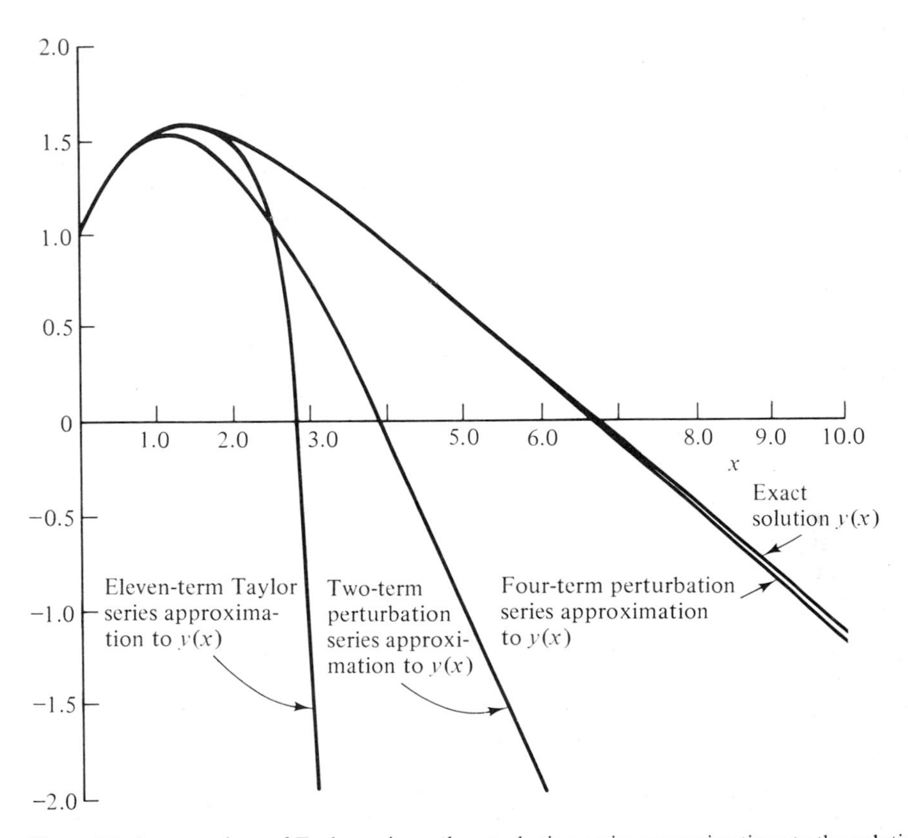

Figure 7.1 A comparison of Taylor series and perturbation series approximations to the solution of the initial-value problem $y'' = -e^{-x}y$ [$y(0) = 1$, $y'(0) = 1$] in (7.1.15). The exact solution to the problem is plotted. Also plotted are an 11-term Taylor series approximation of the form in (7.1.14) and 2- and 4-term perturbation series approximations of the form in (7.1.3) with $\varepsilon = 1$. The global perturbative approximation is clearly far superior to the local Taylor series.

(E) 7.2 REGULAR AND SINGULAR PERTURBATION THEORY

The formal techniques of perturbation theory are a natural generalization of the ideas of local analysis of differential equations in Chap. 3. Local analysis involves approximating the solution to a differential equation near the point $x = a$ by developing a series solution about a in powers of a small parameter, either $x - a$ for finite a or $1/x$ for $a = \infty$. Once the leading behavior of the solution near $x = a$ (which we would now refer to as the zeroth-order solution!) is known, the remaining coefficients in the series can be computed recursively.

The strong analogy between local analysis of differential equations and formal perturbation theory may be used to classify perturbation problems. Recall that there are two different types of series solutions to differential equations. A series solution about an *ordinary* point of a differential equation is always a Taylor series having a nonvanishing radius of convergence. A series solution about a *singular* point does not have this form (except in rare cases). Instead, it may either be a convergent series not in Taylor series form (such as a Frobenius series) or it may be a divergent series. Series solutions about singular points often have the remarkable property of being meaningful near a singular point yet not existing at the singular point. [The Frobenius series for $K_0(x)$ does not exist at $x = 0$ and the asymptotic series for Bi (x) does not exist at $x = \infty$.]

Perturbation series also occur in two varieties. We define a *regular* perturbation problem as one whose perturbation series is a power series in ε having a nonvanishing radius of convergence. A basic feature of all regular perturbation problems (which we will use to identify such problems) is that the exact solution for small but nonzero $|\varepsilon|$ smoothly approaches the unperturbed or zeroth-order solution as $\varepsilon \to 0$.

We define a *singular* perturbation problem as one whose perturbation series either does not take the form of a power series or, if it does, the power series has a vanishing radius of convergence. In singular perturbation theory there is sometimes no solution to the unperturbed problem (the exact solution as a function of ε may cease to exist when $\varepsilon = 0$); when a solution to the unperturbed problem does exist, its qualitative features are distinctly different from those of the exact solution for arbitrarily small but nonzero ε. In either case, the exact solution for $\varepsilon = 0$ is *fundamentally different in character* from the "neighboring" solutions obtained in the limit $\varepsilon \to 0$. If there is no such abrupt change in character, then we would have to classify the problem as a regular perturbation problem.

When dealing with a singular perturbation problem, one must take care to distinguish between the *zeroth-order* solution (the leading term in the perturbation series) and the solution of the unperturbed problem, since the latter may not even exist. There is no difference between these two in a regular perturbation theory, but in a singular perturbation theory the zeroth-order solution may depend on ε and may exist only for nonzero ε.

The examples of the previous section are all regular perturbation problems. Here are some examples of singular perturbation problems:

Example 1 *Roots of a polynomial.* How does one determine the approximate roots of

$$\varepsilon^2 x^6 - \varepsilon x^4 - x^3 + 8 = 0? \tag{7.2.1}$$

We may begin by setting $\varepsilon = 0$ to obtain the unperturbed problem $-x^3 + 8 = 0$, which is easily solved:

$$x = 2,\ 2\omega,\ 2\omega^2, \tag{7.2.2}$$

where $\omega = e^{2\pi i/3}$ is a complex root of unity. Note that the unperturbed equation has only three roots while the original equation has six roots. This abrupt change in the character of the solution, namely the disappearance of three roots when $\varepsilon = 0$, implies that (7.2.1) is a singular perturbation problem. Part of the exact solution ceases to exist when $\varepsilon = 0$.

The explanation for this behavior is that the three missing roots tend to ∞ as $\varepsilon \to 0$. Thus, for those roots it is no longer valid to neglect $\varepsilon^2 x^6 - \varepsilon x^4$ compared with $-x^3 + 8$ in the limit $\varepsilon \to 0$. Of course, for the three roots near 2, 2ω, and $2\omega^2$, the terms $\varepsilon^2 x^6$ and εx^4 are indeed small as $\varepsilon \to 0$ and we may assume a regular perturbation expansion for these roots of the form

$$x_k(\varepsilon) = 2e^{2\pi i k/3} + \sum_{n=1}^{\infty} a_{n,k}\varepsilon^n, \qquad k = 1, 2, 3. \tag{7.2.3}$$

Substituting (7.2.3) into (7.2.1) and comparing powers of ε, as in Example 1 of Sec. 7.1, gives a sequence of equations which determine the coefficients $a_{n,k}$.

To track down the three missing roots we first estimate their orders of magnitude as $\varepsilon \to 0$. We do this by considering all possible dominant balances between pairs of terms in (7.2.1). There are four terms in (7.2.1) so there are six pairs to consider:

(*a*) Suppose $\varepsilon^2 x^6 \sim \varepsilon x^4$ $(\varepsilon \to 0)$ is the dominant balance. Then $x = O(\varepsilon^{-1/2})$ $(\varepsilon \to 0)$. It follows that the terms $\varepsilon^2 x^6$ and εx^4 are both $O(\varepsilon^{-1})$. But $\varepsilon x^4 \ll x^3 = O(\varepsilon^{-3/2})$ as $\varepsilon \to 0$, so x^3 is the biggest term in the equation and is not balanced by any other term. Thus, the assumption that $\varepsilon^2 x^6$ and εx^4 are the dominant terms as $\varepsilon \to 0$ is inconsistent.

(*b*) Suppose $\varepsilon x^4 \sim x^3$ as $\varepsilon \to 0$. Then $x = O(\varepsilon^{-1})$. It follows that $\varepsilon x^4 \sim x^3 = O(\varepsilon^{-3})$. But $x^3 \ll \varepsilon^2 x^6 = O(\varepsilon^{-4})$ as $\varepsilon \to 0$. Thus, $\varepsilon^2 x^6$ is the largest term in the equation. Hence, the original assumption is again inconsistent.

(*c*) Suppose $\varepsilon^2 x^6 \sim 8$ so that $x = O(\varepsilon^{-1/3})$ $(\varepsilon \to 0)$. Hence, $x^3 = O(\varepsilon^{-1})$ is the largest term, which is again inconsistent.

(*d*) Suppose $\varepsilon x^4 \sim 8$ so that $x = O(\varepsilon^{-1/4})$ $(\varepsilon \to 0)$. Then $x^3 = O(\varepsilon^{-3/4})$ is the biggest term, which is also inconsistent.

(*e*) Suppose $x^3 \sim 8$. Then $x = O(1)$. This is a consistent assumption because the other two terms in the equation, $\varepsilon^2 x^6$ and εx^4, are negligible compared with x^3 and 8, and we recover the three roots of the unperturbed equation $x = 2$, 2ω, and $2\omega^2$.

(*f*) Suppose $\varepsilon^2 x^6 \sim x^3$ $(\varepsilon \to 0)$. Then $x = O(\varepsilon^{-2/3})$. This is consistent because $\varepsilon^2 x^6 \sim x^3 = O(\varepsilon^{-2})$ is bigger than $\varepsilon x^4 = O(\varepsilon^{-5/3})$ and $8 = O(1)$ as $\varepsilon \to 0$.

Thus, the magnitudes of the three missing roots are $O(\varepsilon^{-2/3})$ as $\varepsilon \to 0$. This result is a clue to the structure of the perturbation series for the missing roots. In particular, it suggests a *scale transformation* for the variable x:

$$x = \varepsilon^{-2/3} y. \tag{7.2.4}$$

Substituting (7.2.4) into (7.2.1) gives

$$y^6 - y^3 + 8\varepsilon^2 - \varepsilon^{1/3} y^4 = 0. \tag{7.2.5}$$

This is now a *regular* perturbation problem for y in the parameter $\varepsilon^{1/3}$ because the unperturbed problem $y^6 - y^3 = 0$ has six roots $y = 1, \omega, \omega^2, 0, 0, 0$. Now, no roots disappear in the limit $\varepsilon^{1/3} \to 0$.

The perturbative corrections to these roots may be found by assuming a regular perturbation expansion in powers of $\varepsilon^{1/3}$ (it would not be possible to match powers in an expansion having only integral powers of ε):

$$y = \sum_{n=0}^{\infty} y_n(\varepsilon^{1/3})^n. \tag{7.2.6}$$

Having established that we are dealing with a singular perturbation problem, it is no surprise that the perturbation series for the roots x is not a series in integral powers of ε.

Nevertheless, when $y_0 = 0$ we find that $y_1 = 0$ and $y_2 = 2, 2\omega$, and $2\omega^2$. Thus, since the first two terms in this series vanish, $x = \varepsilon^{-2/3}y$ is not really $O(\varepsilon^{-2/3})$ but rather $O(1)$ and we have reproduced the three finite roots near $x = 2, 2\omega, 2\omega^2$. Moreover, only every third coefficient in (7.2.6), $y_2, y_5, y_8, \ldots$, is nonvanishing, so we have also reproduced the regular perturbation series in (7.2.3)!

Example 2 *Appearance of a boundary layer.* The boundary-value problem

$$\varepsilon y'' - y' = 0, \qquad y(0) = 0,\ y(1) = 1, \tag{7.2.7}$$

is a singular perturbation problem because the associated unperturbed problem

$$-y' = 0, \qquad y(0) = 0,\ y(1) = 1, \tag{7.2.8}$$

has no solution. (The solution to this first-order differential equation, $y =$ constant, cannot satisfy both boundary conditions.) The solution to (7.2.7) cannot have a regular perturbation expansion of the form $y = \sum_{n=0}^{\infty} y_n(x)\varepsilon^n$ because y_0 does not exist.

There is a close parallel between this example and the previous one. Here, the highest derivative is multiplied by ε and in the limit $\varepsilon \to 0$ the unperturbed solution loses its ability to satisfy the boundary conditions because a solution is lost. In the previous example the highest power of x is multiplied by ε and in the limit $\varepsilon \to 0$ some roots are lost.

The exact solution to (7.2.7) is easy to find:

$$y(x) = \frac{e^{x/\varepsilon} - 1}{e^{1/\varepsilon} - 1}. \tag{7.2.9}$$

This function is plotted in Fig. 7.2 for several small positive values of ε. For very small but nonzero ε it is clear from Fig. 7.2 that y is almost constant except in a very narrow interval of thickness $O(\varepsilon)$ at $x = 1$, which is called a *boundary layer.* Thus, outside the boundary layer the exact solution satisfies the left boundary condition $y(0) = 0$ and almost but not quite satisfies the unperturbed equation $y' = 0$.

It is not obvious how to construct a perturbative approximation to a differential equation whose highest derivative is multiplied by ε until it is known how to construct an analytical expression for the zeroth-order approximation. A new technique called asymptotic matching must be introduced (see Sec. 7.4 and Chap. 9) to solve this problem.

Example 3 *Appearance of rapid variation on a global scale.* In the previous example we saw that the exact solution varies rapidly in the neighborhood of $x = 1$ for small ε and develops a discontinuity there in the limit $\varepsilon \to 0+$. A solution to a boundary-value problem may also develop discontinuities throughout a large region as well as in the neighborhood of a point.

The boundary-value problem $\varepsilon y'' + y = 0$ [$y(0) = 0$, $y(1) = 1$] is a singular perturbation problem because when $\varepsilon = 0$, the solution to the unperturbed problem, $y = 0$, does not satisfy the boundary condition $y(1) = 1$. The exact solution, when ε is not of the form $(n\pi)^{-2}$ ($n = 0, 1, 2, \ldots$), is $y(x) = \sin(x/\sqrt{\varepsilon})/\sin(1/\sqrt{\varepsilon})$. Observe that $y(x)$ becomes discontinuous throughout the inter-

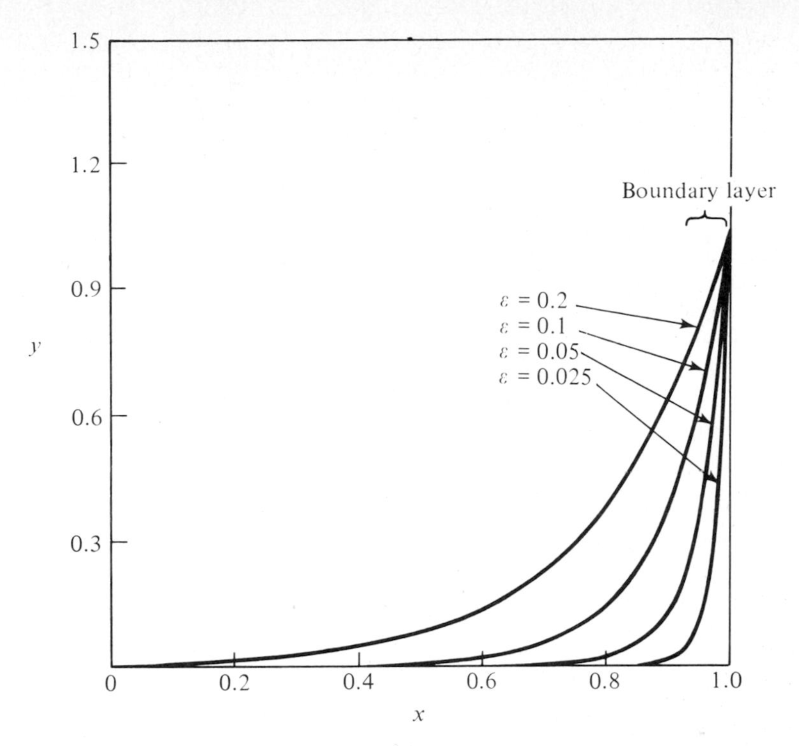

Figure 7.2 A plot of $y(x) = (e^{x/\varepsilon} - 1)/(e^{1/\varepsilon} - 1)$ $(0 \le x \le 1)$ for $\varepsilon = 0.2, 0.1, 0.05, 0.025$. When ε is small $y(x)$ varies rapidly near $x = 1$; this localized region of rapid variation is called a boundary layer. When ε is negative the boundary layer is at $x = 0$ instead of $x = 1$. This abrupt jump in the location of the boundary layer as ε changes sign reflects the singular nature of the perturbation problem.

val $0 \le x \le 1$ in the limit $\varepsilon \to 0+$ (see Fig. 7.3). When $\varepsilon = (n\pi)^{-2}$, there is no solution to the boundary-value problem.

When the solution to a differential-equation perturbation problem varies rapidly on a global scale for small ε, it is not obvious how to construct a leading-order perturbative approximation to the exact solution. The best procedure that has evolved is called WKB theory (see Chap. 10).

Example 4 *Perturbation theory on an infinite interval.* The initial-value problem

$$y'' + (1 - \varepsilon x)y = 0, \qquad y(0) = 1,\ y'(0) = 0, \tag{7.2.10}$$

is a regular perturbation problem in ε over the finite interval $0 \le x \le L$. In fact, the perturbation solution is just

$$y(x) = \cos x + \varepsilon(\tfrac{1}{4}x^2 \sin x + \tfrac{1}{4}x \cos x - \tfrac{1}{4} \sin x)$$
$$+ \varepsilon^2(-\tfrac{1}{32}x^4 \cos x + \tfrac{5}{48}x^3 \sin x + \tfrac{7}{16}x^2 \cos x - \tfrac{7}{16}x \sin x) + \cdots, \tag{7.2.11}$$

which converges for all x and ε, with increasing rapidity as $\varepsilon \to 0+$ for fixed x.

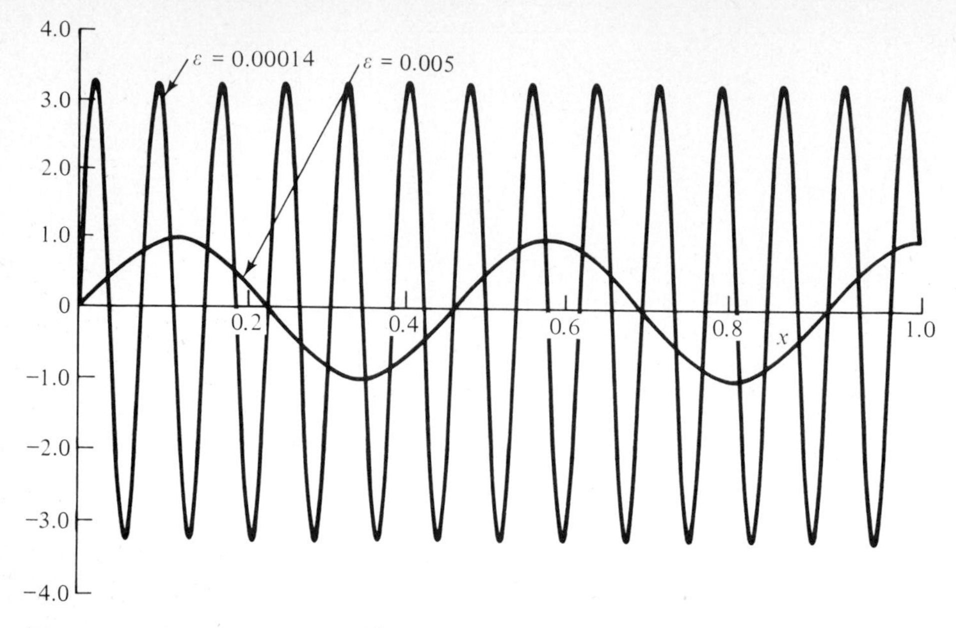

Figure 7.3 A plot of $y(x) = [\sin(x\varepsilon^{-1/2})]/[\sin(\varepsilon^{-1/2})]$ $(0 \le x \le 1)$ for $\varepsilon = 0.005$ and 0.00014. As ε gets smaller the oscillations become more violent; as $\varepsilon \to 0+$, $y(x)$ becomes discontinuous over the entire interval. The WKB approximation is a perturbative method commonly used to describe functions like $y(x)$ which exhibit rapid variation on a global scale.

However, this same initial-value problem must be reclassified as a singular perturbation problem over the semi-infinite interval $0 \le x < \infty$. While the exact solution does approach the solution to the unperturbed problem as $\varepsilon \to 0+$ for fixed x, it does not do so uniformly for all x (see Fig. 7.4). The zeroth-order solution is bounded and oscillatory for all x. But when $\varepsilon > 0$, local analysis of the exact solution for large x shows that it is a linear combination of exponentially increasing and decreasing functions (Prob. 7.20). This change in character of the solution occurs because it is certainly wrong to neglect εx compared with 1 when x is bigger than $1/\varepsilon$. In fact, a more careful argument shows that the term εx is not a small perturbation unless $x \ll \varepsilon^{-1/2}$ (Prob. 7.20).

Example 4 shows that the interval itself can determine whether a perturbation problem is regular or singular. We examine more examples having this property in the next section on eigenvalue problems. The feature that is common to all such examples is that an nth-order perturbative approximation bears less and less resemblance to the exact solution as x increases.

For these sorts of problems Chap. 11 introduces new perturbative procedures called multiple-scale methods which substantially improve the rather poor predictions of ordinary perturbation theory. The particular problem in Example 4 is reconsidered in Prob. 11.13.

Example 5 *Roots of a high-degree polynomial.* When a perturbation problem is regular, the perturbation series is convergent and the exact solution is a smooth analytic function of ε for

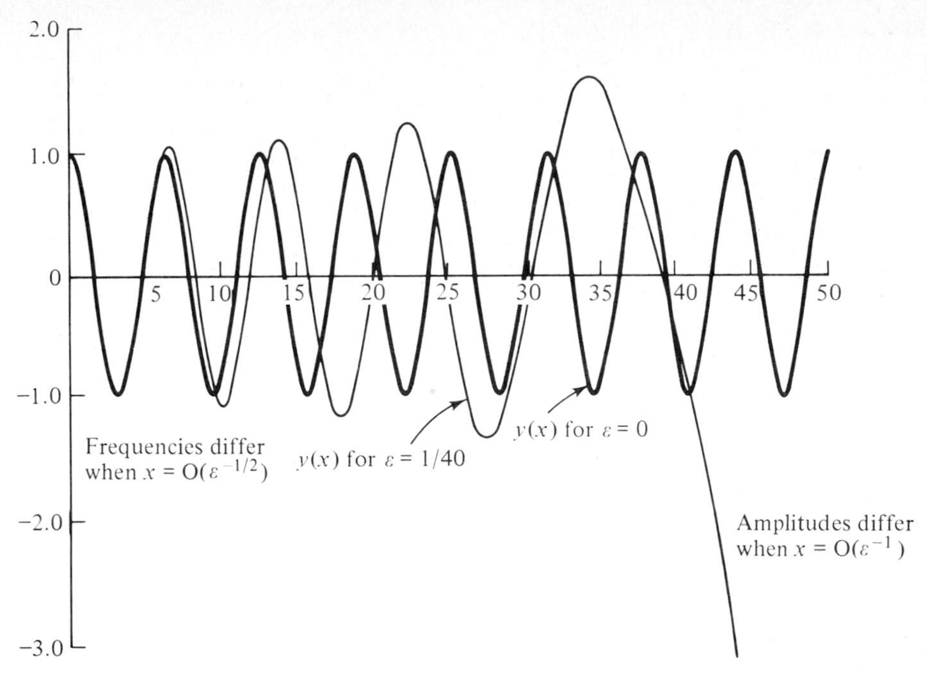

Figure 7.4 Exact solutions to the initial-value problem $y'' + (1 - \varepsilon x)y = 0$ $[y(0) = 1,\ y'(0) = 0]$ in (7.2.10) for $\varepsilon = 0$ and $\varepsilon = \frac{1}{40}$. Although this is a regular perturbation problem on the finite interval $0 \le x \le L$, it is a singular perturbation problem on the infinite interval $0 \le x \le \infty$ because the perturbed solution ($\varepsilon > 0$) is not close to the unperturbed solution ($\varepsilon = 0$), no matter how small ε is. When $x = O(\varepsilon^{-1/2})$ the frequencies begin to differ (the curves become phase shifted) and when $x = O(\varepsilon^{-1})$ the amplitudes differ (one curve remains finite while the other grows exponentially).

sufficiently small ε. However, just what is "sufficiently small" may vary enormously from problem to problem. A striking example by Wilkinson concerns the roots of the polynomial

$$\prod_{k=1}^{20} (x - k) + \varepsilon x^{19} = x^{20} - (210 - \varepsilon)x^{19} + \cdots + 20! \tag{7.2.12}$$

The perturbation εx^{19} is regular, since no roots are lost in the limit $\varepsilon \to 0$; the roots of the unperturbed polynomial lie at 1, 2, 3, ..., 20.

Let us now take $\varepsilon = 10^{-9}$ so that the perturbation in the coefficient of x^{19} is of relative magnitude $10^{-9}/210$, or roughly 10^{-11}. For such a small regular perturbation one might expect the 20 roots to be only very slightly displaced from their $\varepsilon = 0$ values. The actual displaced roots are given in Table 7.1. One is surprised to find that while some roots are relatively unchanged by the perturbation, others have paired into complex conjugates. The qualitative effect on the roots of varying ε is shown in Figs. 7.5 and 7.6. In these plots the paths of the roots are traced as a function of ε. As $|\varepsilon|$ increases, the roots coalesce into pairs of complex conjugate roots. Evidently, a "small" perturbation is one for which $|\varepsilon| < 10^{-11}$, while $|\varepsilon| \gtrsim 10^{-10}$ is a "large" perturbation for at least some of the roots. Low-order regular perturbation theory may be used to understand this behavior (Probs. 7.22 and 7.23).

Table 7.1 Roots of the Wilkinson polynomial (7.2.12) with $\varepsilon = 10^{-9}$

The first column lists the unperturbed ($\varepsilon = 0$) roots 1, 2, ..., 20; the second column gives the results of first-order perturbation theory (see Prob. 7.22); the third column gives the exact roots. The unperturbed roots at 13 and 14, 15 and 16, and 17 and 18 are perturbed into complex-conjugate pairs. Observe that while first-order perturbation theory is moderately accurate for the real perturbed roots near 1, 2, ..., 12, 19, 20, it cannot predict the locations of the complex roots (but see Prob. 7.23)

Unperturbed root	First-order perturbation theory	Exact root
1	1.000 000 000 0	1.000 000 000 0
2	2.000 000 000 0	2.000 000 000 0
3	3.000 000 000 0	3.000 000 000 0
4	4.000 000 000 0	4.000 000 000 0
5	5.000 000 000 0	5.000 000 000 0
6	5.999 999 941 8	5.999 999 941 8
7	7.000 002 542 4	7.000 002 542 4
8	7.999 994 030 4	7.999 994 031 5
9	9.000 839 327 5	9.000 841 033 5
10	9.992 405 941 6	9.992 518 124 0
11	11.046 444 571	11.050 622 592
12	11.801 496 835	11.832 935 987
13	13.605 558 629	$13.349\,018\,036 \pm 0.532\,765\,750\,0i$
14	12.667 031 557	
15	17.119 065 220	$15.457\,790\,724 \pm 0.899\,341\,526\,2i$
16	13.592 486 027	
17	18.904 402 150	$17.662\,434\,477 \pm 0.704\,285\,236\,9i$
18	17.004 413 300	
19	19.309 013 459	19.233 703 334
20	19.956 900 195	19.950 949 654

This example shows that the roots of high-degree polynomials may be extraordinarily sensitive to changes in the coefficients of the polynomial, even though the perturbation problem so obtained is regular. It should serve as ample warning to a "number cruncher" not to trust computer output without sufficient understanding of the nature of the problem being solved.

(I) 7.3 PERTURBATION METHODS FOR LINEAR EIGENVALUE PROBLEMS

In this section we show how perturbation theory can be used to approximate the eigenvalues and eigenfunctions of the Schrödinger equation

$$\left[-\frac{d^2}{dx^2} + V(x) + W(x) - E\right] y(x) = 0, \tag{7.3.1}$$

subject to the boundary condition

$$\lim_{|x| \to \infty} y(x) = 0. \tag{7.3.2}$$

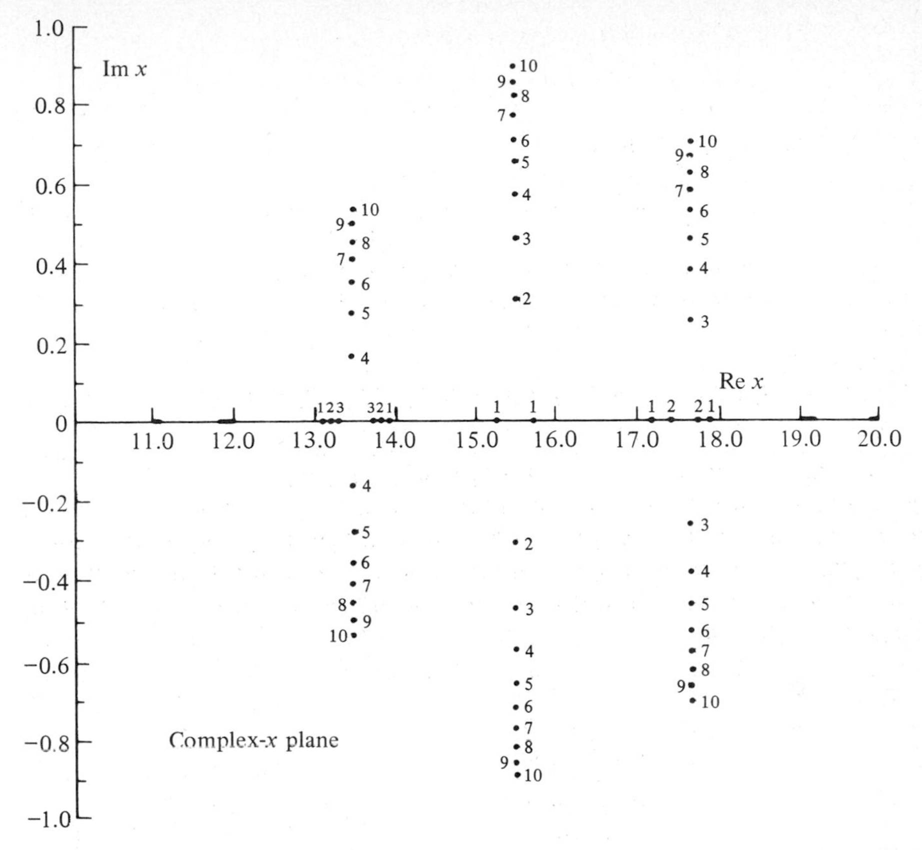

Figure 7.5 Roots of the Wilkinson polynomial $(x-1)(x-2)(x-3)\cdots(x-20)+\varepsilon x^{19}$ in (7.2.12) for 11 values of ε. When $\varepsilon = 0$ the roots shown are 10, 11, . . . , 20. As ε is allowed to increase very slowly, the roots move toward each other in pairs along the real-x axis and then veer off in opposite directions into the complex-x plane. We have plotted the roots for $\varepsilon = 0, 10^{-10}, 2 \times 10^{-10}, 3 \times 10^{-10}, \ldots, 10^{-9}$. Some of the roots are numbered to indicate the value of ε to which they correspond; that is, 6 means $\varepsilon = 6 \times 10^{-10}$, 3 means $\varepsilon = 3 \times 10^{-10}$, and so on. The roots starting at 11, 12, 19, and 20 move too slowly to be seen as individual dots. We conclude from this plot that very slight changes in the coefficients of a polynomial can cause drastic changes in the values of some of the roots; one must be cautious when performing numerical calculations.

In (7.3.1) E is called the energy eigenvalue and $V + W$ is called the potential. We assume that $V(x)$ and $W(x)$ are continuous functions and that both $V(x)$ and $V(x) + W(x)$ approach ∞ as $|x| \to \infty$.

We suppose that the function $V(x) + W(x)$ is so complicated that (7.3.1) is not soluble in closed form. One can still prove from the above assumptions that nontrivial solutions $[y(x) \not\equiv 0]$ satisfying (7.3.1) and (7.3.2) exist for special discrete values of E, the allowed eigenvalues of the equation (see Sec. 1.8). On the other hand, we assume that removing the term $W(x)$ from (7.3.1) makes the equation an

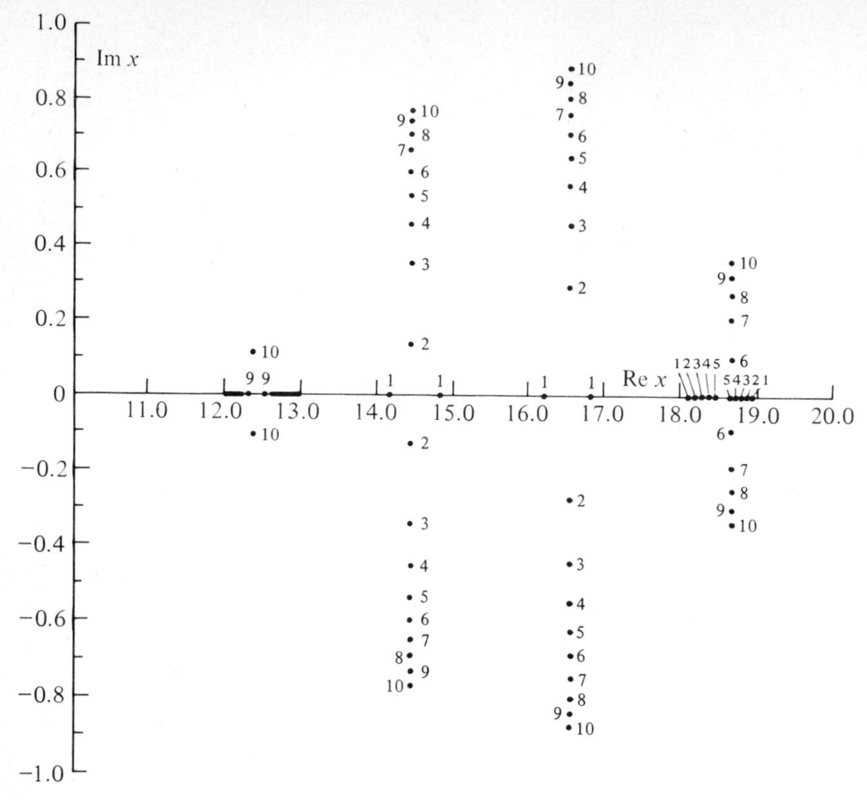

Figure 7.6 Same as in Fig. 7.5 except that the values of ε are 0, -10^{-10}, -2×10^{-10}, $-3 \times 10^{-10}, \ldots, -10^{-9}$. The roots pair up and veer off into the complex-x plane, but the pairs are not the same as in Fig. 7.5.

exactly soluble eigenvalue problem. This suggests using perturbation theory to solve the family of eigenvalue problems in which $W(x)$ is replaced by $\varepsilon W(x)$:

$$\left[-\frac{d^2}{dx^2} + V(x) + \varepsilon W(x) - E\right] y(x) = 0. \tag{7.3.3}$$

Our assumptions on the nature of $V(x)$ and $W(x)$ leave no choice about where to introduce the parameter ε if the unperturbed problem is to be exactly soluble.

Example 1 *An exactly soluble eigenvalue problem.* Several exactly soluble eigenvalue problems are given in Sec. 1.8. One such example, which is used extensively in this section, is obtained if we take $V(x) = x^2/4$. The unperturbed problem is the Schrödinger equation for the quantum-mechanical harmonic oscillator, which is just the parabolic cylinder equation

$$-y'' + \frac{x^2}{4} y - Ey = 0. \tag{7.3.4}$$

We have already shown that solutions to this equation behave like $e^{\pm x^2/4}$ as $|x| \to \infty$.

There is a discrete set of values of E for which a solution that behaves like $e^{-x^2/4}$ as $x \to \infty$ also behaves like $e^{-x^2/4}$ as $x \to -\infty$ (see Example 4 of Sec. 3.5 and Example 9 of Sec. 3.8). These values of E are

$$E = n + \tfrac{1}{2}, \qquad n = 0, 1, 2, \ldots, \tag{7.3.5}$$

and the associated eigenfunctions are parabolic cylinder functions

$$y_n(x) = D_n(x) = e^{-x^2/4}\,\mathrm{He}_n\,(x), \tag{7.3.6}$$

where $\mathrm{He}_n\,(x)$ is the Hermite polynomial of degree n: $\mathrm{He}_0\,(x) = 1$, $\mathrm{He}_1\,(x) = x$, $\mathrm{He}_2\,(x) = x^2 - 1, \ldots$.

In general, once an eigenvalue E_0 and an eigenfunction $y_0(x)$ of the unperturbed problem

$$\left[-\frac{d^2}{dx^2} + V(x) - E_0\right] y_0(x) = 0 \tag{7.3.7}$$

have been found, we may seek a perturbative solution to (7.3.3) of the form

$$E = \sum_{n=0}^{\infty} E_n \varepsilon^n, \tag{7.3.8}$$

$$y(x) = \sum_{n=0}^{\infty} y_n(x)\varepsilon^n. \tag{7.3.9}$$

Substituting (7.3.8) and (7.3.9) into (7.3.3) and comparing powers of ε gives the following sequence of equations:

$$\left[-\frac{d^2}{dx^2} + V(x) - E_0\right] y_n(x) = -Wy_{n-1}(x) + \sum_{j=1}^{n} E_j y_{n-j}(x),$$
$$n = 1, 2, 3, \ldots, \tag{7.3.10}$$

whose solutions must satisfy the boundary conditions

$$\lim_{|x|\to\infty} y_n(x) = 0, \qquad n = 1, 2, 3, \ldots. \tag{7.3.11}$$

Equation (7.3.10) is linear and inhomogeneous. The associated homogeneous equation is just the unperturbed problem and thus is soluble by assumption. However, technically speaking, only *one* of the two linearly independent solutions of the unperturbed problem (the one that satisfies the boundary conditions) is assumed known. Therefore, we proceed by the method of reduction of order (see Sec. 1.4); to wit, we substitute

$$y_n(x) = y_0(x)F_n(x), \tag{7.3.12}$$

where $F_0(x) = 1$, into (7.3.10). Simplifying the result using (7.3.7) and multiplying by the integrating factor $y_0(x)$ gives

$$\frac{d}{dx}\left[y_0^2(x)F_n'(x)\right] = y_0^2(x)\left[W(x)F_{n-1}(x) - \sum_{j=1}^{n} E_j F_{n-j}(x)\right]. \tag{7.3.13}$$

If we integrate this equation from $-\infty$ to ∞ and use $y_0^2(x)F_n'(x) = y_0(x)y_n'(x) - y_0'(x)y_n(x) \to 0$ as $|x| \to \infty$, we obtain the formula for the coefficient E_n:

$$E_n = \frac{\int_{-\infty}^{\infty} y_0(x)\left[W(x)y_{n-1}(x) - \sum_{j=1}^{n-1} E_j y_{n-j}(x)\right] dx}{\int_{-\infty}^{\infty} y_0^2(x)\, dx},$$

$$n = 1, 2, 3, \ldots, \qquad (7.3.14)$$

from which we have eliminated all reference to $F_n(x)$. [The sum on the right side of (7.3.14) is defined to be 0 when $n = 1$.]

Integrating (7.3.13) twice gives the formula for $y_n(x)$:

$$y_n(x) = y_0(x) \int_a^x \frac{dt}{y_0^2(t)} \int_{-\infty}^t ds\, y_0(s)\left[W(s)y_{n-1}(s) - \sum_{j=1}^{n} E_j y_{n-j}(s)\right],$$

$$n = 1, 2, 3, \ldots. \qquad (7.3.15)$$

Observe that in (7.3.15) a is an arbitrary number at which we choose to impose $y_n(a) = 0$. This means we have fixed the overall normalization of $y(x)$ so that $y(a) = y_0(a)$ [assuming that $y_0(a) \neq 0$]. If $y_0(t)$ vanishes between a and x, the integral in (7.3.15) seems formally divergent; however, $y_n(x)$ satisfies a differential equation (7.3.10) which has no finite singular points. Thus, it is possible to define $y_n(x)$ everywhere as a finite expression (see Prob. 7.24).

Equations (7.3.14) and (7.3.15) together constitute an iterative procedure for calculating the coefficients in the perturbation series for E and $y(x)$. Once the coefficients $E_0, E_1, \ldots, E_{n-1}, y_0, y_1, \ldots, y_{n-1}$ are known, (7.3.14) gives E_n, and once E_n has been calculated (7.3.15) gives y_n. The remaining question is whether or not these perturbation series are convergent.

Example 2 *A regular perturbative eigenvalue problem.* Let $V(x) = x^2/4$ and $W(x) = x$. It may be shown (Prob. 7.25) that the perturbation series for $y(x)$ is convergent for all ε and that the series for E has vanishing terms of order ε^n for $n \geq 3$. This is a regular perturbation problem.

Example 3 *A singular perturbative eigenvalue problem.* It may be shown (Prob. 7.26) that if $V(x) = x^2/4$ and $W(x) = x^4/4$, then the perturbation series for the smallest eigenvalue for positive ε is

$$E(\varepsilon) \sim \tfrac{1}{2} + \tfrac{3}{4}\varepsilon - \tfrac{21}{8}\varepsilon^2 + \tfrac{333}{16}\varepsilon^3 + \cdots, \qquad \varepsilon \to 0+. \qquad (7.3.16)$$

The terms in this series appear to be getting larger and suggest that this series may be divergent for all $\varepsilon \neq 0$. Indeed, (7.3.16) diverges for all ε because the nth term satisfies $E_n \sim -(-3)^n\Gamma(n + \frac{1}{2})\sqrt{6}/\pi^{3/2}$ $(n \to \infty)$. (This is a nontrivial result that we do not explain here.)

The divergence of the perturbation series in Example 3 indicates that the perturbation problem is singular. A simple way to observe the singular behavior is to compare $e^{-x^2/4}$, the controlling factor of the large-x behavior of the unperturbed ($\varepsilon = 0$) solution, with $e^{-x^3\sqrt{\varepsilon}/6}$, the controlling factor of the large-x behavior for $\varepsilon \neq 0$. There is an abrupt change in the nature of the solution when we pass to the limit ($\varepsilon \to 0+$). This phenomenon occurs because the perturbing term $\varepsilon x^4/4$ is not small compared with $x^2/4$ when x is large.

If the functions $V(x)$ and $W(x)$ in Example 3 were interchanged, then the resulting eigenvalue problem would be a regular perturbation problem because εx^2 is a small perturbation of x^4 for all $|x| < \infty$. However, the unperturbed problem, $(-d^2/dx^2 + x^4/4 - E_0)y_0(x) = 0$, is not soluble in closed form. Thus, it would not be possible to use (7.3.14) and (7.3.15) to compute the coefficients in the perturbation series analytically.

Also note that if the boundary conditions in Example 3 were given at $x = \pm A$, $A < \infty$, then the perturbation theory would be regular. This is because here εx^4 is a small perturbation of x^2. However, it is much more difficult to solve the unperturbed problem on a finite interval.

Thus, one is forced to accept a solution to Example 3 in the form of a divergent series. Fortunately, this series is one of many that may be summed by Padé theory to give a finite and unique result (see Sec. 8.3).

Example 4 *Another regular perturbation problem.* When $V = x^2/4$ and $W = |x|$ the perturbation problem is regular. But unlike the problem in Example 2, this perturbation series is not convergent for all ε; the series in (7.3.8) and (7.3.9) have finite radii of convergence. The significance of the finite radius of convergence is discussed in Sec. 7.5.

(D) 7.4 ASYMPTOTIC MATCHING

The purpose of this section is to introduce the notion of matched asymptotic expansions. Asymptotic matching is an important perturbative method which is used often in both boundary-layer theory (Chap. 9) and WKB theory (Chap. 10) to determine analytically the approximate global properties of the solution to a differential equation. Asymptotic matching is usually used to determine a uniform approximation to the solution of a differential equation and to find other global properties of differential equations such as eigenvalues. Asymptotic matching may also be used to develop approximations to integrals.

The principle of asymptotic matching is simple. The interval on which a boundary-value problem is posed is broken into a sequence of two or more *overlapping* subintervals. Then, on each subinterval perturbation theory is used to obtain an asymptotic approximation to the solution of the differential equation valid on that interval. Finally, the matching is done by requiring that the asymptotic approximations have the same functional form on the overlap of every pair of intervals. This gives a sequence of asymptotic approximations to the solution of the differential equation; by construction, each approximation satisfies all the boundary conditions given at various points on the interval. Thus, the end result is an approximate solution to a boundary-value problem valid over the entire interval.

Asymptotic matching bears a slight resemblance to an elementary technique for solving boundary-value problems called *patching*. Patching is helpful when the differential equation can be solved in closed form. Here is a simple example:

Example 1 *Patching.* The method of patching may be used to solve the boundary-value problem $y'' - y = e^{-|x|}$ $[y(\pm\infty) = 0]$. There are two regions to consider. When $x \le 0$, the most general solution which satisfies the boundary condition $y(-\infty) = 0$ is

$$y(x) = ae^x + \tfrac{1}{2}xe^x, \tag{7.4.1}$$

where a is a constant to be determined. When $x \geq 0$, the most general solution which satisfies $y(\infty) = 0$ is

$$y(x) = be^{-x} - \tfrac{1}{2}xe^{-x}, \tag{7.4.2}$$

where b is another constant to be determined. Solutions (7.4.1) and (7.4.2) are now patched together at the one point common to both regions, namely, $x = 0$. That is, we require that $y(x)$ and $y'(x)$ be continuous at $x = 0$. This gives a pair of equations for a and b whose solution is $a = b = -\frac{1}{2}$. Substituting these values of a and b into (7.4.1) and (7.4.2) gives the exact solution to the boundary-value problem: $y(x) = -\frac{1}{2}e^{-|x|} - \frac{1}{2}|x|e^{-|x|}$.

Matching is different from patching because an asymptotic approximation to the solution of the differential equation rather than the exact solution gets matched. Moreover, matching is done by comparing functions over an *interval* while patching is done by comparing functions and their derivatives at a *point*. In general, the length of the matching interval approaches ∞ as ε, the perturbing parameter, approaches 0. Here are several examples to introduce the techniques of asymptotic matching.

Example 2 *Asymptotic matching for a first-order differential equation.* The first-order differential equation

$$y' + (\varepsilon x^2 + 1 + 1/x^2)y = 0, \qquad y(1) = 1,$$

is exactly soluble on the interval $1 \leq x < \infty$. Nevertheless, we will use the principles of asymptotic matching to study the approximate behavior of the solution as $\varepsilon \to 0+$. When x is not too large, the term εx^2 is negligible so an approximate equation for y is

$$y_L' + (1 + 1/x^2)y_L = 0,$$

where the subscript L refers to the *left* region. The solution to this equation which satisfies $y_L(1) = 1$ is

$$y_L = e^{-x+1/x}. \tag{7.4.3}$$

When x is large, εx^2 is no longer negligible but $1/x^2$ is. Therefore, an approximate equation valid as $x \to +\infty$ is

$$y_R' + (\varepsilon x^2 + 1)y_R = 0,$$

where the subscript R refers to the *right* region. The solution to this equation is

$$y_R = ae^{-\varepsilon x^3/3 - x}, \tag{7.4.4}$$

where a is a constant.

There is a common region of validity of the two solutions (7.4.3) and (7.4.4) which enables us to determine the approximate value of a. For those values of x lying in the range

$$\varepsilon^{-1/5} \ll x \ll \varepsilon^{-1/4}, \qquad \varepsilon \to 0+, \tag{7.4.5}$$

x is so large that (7.4.3) may be approximated by $y_L(x) = e^{-x+1/x} \sim e^{-x}$ ($\varepsilon \to 0+$), but x is still small enough that (7.4.4) may be approximated by $y_R(x) = ae^{-\varepsilon x^3/3 - x} \sim ae^{-x}$ ($\varepsilon \to 0+$). In the overlap region both solutions have the *same functional dependence* on x. If both asymptotic expansions are to agree in the overlap region (7.4.5), then we must choose $a \sim 1$ ($\varepsilon \to 0+$). This is asymptotic matching; we have obtained a global approximation to the original differential equation. The condition $y(1) = 1$ translates into the condition that $a \sim 1$ ($\varepsilon \to 0+$), which completely determines the approximation to $y(x)$ as $x \to \infty$. Note that the extent of the matching region in (7.4.5) becomes infinite as $\varepsilon \to 0+$.

The matching region (7.4.5) that we have chosen is not the only possible choice. We could have matched on the interval $\varepsilon^{-1/6} \ll x \ll \varepsilon^{-1/5}$ ($\varepsilon \to 0+$) or even the interval

$\varepsilon^{-1/99} \ll x \ll \varepsilon^{-32/99}$ $(\varepsilon \to 0+)$! These regions all work because they satisfy the general matching criterion that x lie in the asymptotic interval $1 \ll x \ll \varepsilon^{-1/3}$ as $\varepsilon \to 0+$ (see Prob. 7.30).

Example 3 *Asymptotic matching in higher order.* Let us return to the differential equation in Example 2 and carry out the asymptotic matching to first order in ε. First we consider the left region. In Example 2 we merely discarded the term εx^2. Now let us seek an orderly perturbative expansion of the solution in powers of ε: $y_L = y_0(x) + \varepsilon y_1(x) + \varepsilon^2 y_2(x) + \cdots$, where $y_0(1) = 1$, $y_1(1) = 0$, $y_2(1) = 0, \ldots$. As before, $y_0 = e^{-x+1/x}$. y_1 satisfies the inhomogenous equation

$$y_1'(x) + \left(1 + \frac{1}{x^2}\right) y_1(x) = -x^2 y_0(x),$$

whose solution is

$$y_1(x) = \left(\frac{1}{3} - \frac{x^3}{3}\right) e^{-x+1/x}.$$

Hence a first-order approximation to y_L is

$$y_L = e^{-x+1/x}[1 + \varepsilon(1 - x^3)/3 + O(\varepsilon^2 x^6)], \qquad \varepsilon \to 0+, \tag{7.4.6}$$

where the form of the error term is suggested by an examination of $y_2(x)$ (see Prob. 7.30).

Next we consider the right region. A more accurate estimate of the behavior of y for large x is found by the usual technique of substituting $y = e^{S(x)}$, as explained in Chap. 3. This method normally generates only the asymptotic expansion of $y(x)$ valid as $x \to \infty$, but in this example it gives the exact solution because the differential equation for $y(x)$ is first order: $y_R(x) = ae^{-\varepsilon x^3/3 - x + 1/x}$. In the overlap region this expression is approximately

$$y_R(x) = ae^{-x+1/x}[1 - \varepsilon x^3/3 + O(\varepsilon^2 x^6)], \qquad \varepsilon \to 0+. \tag{7.4.7}$$

Comparing (7.4.6) and (7.4.7) determines a to first order in ε: $a = 1 + \varepsilon/3 + O(\varepsilon^2)$ $(\varepsilon \to 0+)$. Note that the exact value of a is $e^{\varepsilon/3}$. The method of matched asymptotic expansions has given the first two terms in the expansion of a for small ε.

Example 4 *Asymptotic matching for a second-order differential equation.* Unlike the first-order differential equation of Examples 2 and 3, the equation

$$y'' + \left(\nu + \frac{1}{2} - \frac{x^2}{4} - \varepsilon x^4\right) y = 0, \tag{7.4.8}$$

where ν is a parameter and $\varepsilon \to 0+$, does not have a closed-form solution. Nevertheless, the method of matched asymptotic expansions may be used to obtain an approximate solution to the boundary-value problem $y(0) = 1$, $y(+\infty) = 0$.

When x is so small that εx^4 is negligible compared with $x^2/4$, the original differential equation (7.4.8) may be approximated by the simpler equation $y'' + (\nu + \frac{1}{2} - x^2/4)y = 0$. This is a parabolic cylinder equation; a solution which decays exponentially for large x and satisfies $y(0) = 1$ is

$$y(x) = D_\nu(x)/D_\nu(0). \tag{7.4.9}$$

[We could also include in (7.4.9) a linearly independent solution like $D_\nu(-x)$ (when ν is nonintegral) which grows exponentially for large x, but such a solution would immediately be rejected in the course of asymptotic matching, as we will shortly see.]

When x is large and positive, we can perform a local analysis of (7.4.8). The usual procedure is to substitute $y = e^{S(x)}$ and to look for the exponentially decaying solution. The leading term in the asymptotic expansion of $y(x)$ for large x has the rather complicated form (see Prob. 7.31)

$$y(x) \sim a\left(\frac{x^2}{4} + \varepsilon x^4\right)^{-1/4} \left(\frac{\sqrt{\frac{1}{4} + \varepsilon x^2} - \frac{1}{2}}{\sqrt{\frac{1}{4} + \varepsilon x^2} + \frac{1}{2}}\right)^{\nu/2 + 1/4} \exp\left[-\frac{1}{3\varepsilon}\left(\varepsilon x^2 + \frac{1}{4}\right)^{3/2}\right], \qquad x \to +\infty. \tag{7.4.10}$$

The constant a cannot be determined by a local analysis of (7.4.8). However, asymptotic matching supplies the connection between the behavior at ∞ and the boundary condition $y(0) = 1$, and can thus be used to determine a approximately.

The asymptotic matching procedure relies on the existence of an overlap region. We will seek an overlap region where x is so large that it is valid to use (7.4.10) but where the term εx^4 is still negligible compared with $x^2/4$ so that it is valid to use (7.4.9). The interval

$$\varepsilon^{-1/4} \le x \le \varepsilon^{-1/3} \tag{7.4.11}$$

satisfies these constraints. Note that this interval becomes infinitely long as $\varepsilon \to 0+$ (see Prob. 7.32).

When x lies in this overlap region, the asymptotic approximation in (7.4.10) may be greatly simplified because εx^2 is small compared with $\frac{1}{4}$:

$$-\frac{1}{3\varepsilon}\left(\frac{1}{4} + \varepsilon x^2\right)^{3/2} \sim -\frac{1}{24\varepsilon} - \frac{1}{4}x^2,$$

$$\left(\frac{\sqrt{\frac{1}{4} + \varepsilon x^2} - \frac{1}{2}}{\sqrt{\frac{1}{4} + \varepsilon x^2} + \frac{1}{2}}\right) \sim \varepsilon x^2,$$

$$\left(\frac{x^2}{4} + \varepsilon x^4\right)^{-1/4} \sim \left(\frac{2}{x}\right)^{1/2}, \qquad \varepsilon \to 0+.$$

Thus,

$$y(x) \sim a 2^{1/2} \varepsilon^{\nu/2 + 1/4} e^{-1/24\varepsilon} x^\nu e^{-x^2/4}, \qquad \varepsilon \to 0+. \tag{7.4.12}$$

On the other hand, the expression in (7.4.9) may be replaced by its asymptotic behavior for large x when x lies in the overlap region (7.4.11) (see Chap. 3):

$$y(x) \sim \frac{1}{D_\nu(0)} x^\nu e^{-x^2/4}, \qquad \varepsilon \to 0+. \tag{7.4.13}$$

Throughout the overlap region the two asymptotic expansions (7.4.12) and (7.4.13) match; they exhibit the same dependence on x! We thus conclude that to lowest order in ε, an expression for a is

$$a = \frac{1}{D_\nu(0)} 2^{-1/2} \varepsilon^{-\nu/2 - 1/4} e^{1/24\varepsilon}.$$

Our treatment of asymptotic matching has no doubt raised several questions of principle. How do we know that an overlap region necessarily exists? If it does exist, how can we predict its size? What do we do if there is no overlap region? We postpone a discussion of these serious questions until Part IV. Our goal here is merely to introduce the mechanical aspects of asymptotic matching. With this in mind, we give three more examples which, although they are somewhat involved, clearly illustrate the depth of analytical power which the method of asymptotic matching can provide.

Example 5 *Nonlinear eigenvalue problem.* Asymptotic matching may be used to find an asymptotic approximation to the large positive eigenvalues E of the boundary-value problem

$$(x^2 - 1)y''(x) + xy'(x) + (E^2 - 2Ex)y(x) = 0, \qquad y(1) = 0,\ y(\infty) = 0. \tag{7.4.14}$$

This eigenvalue problem is different from those considered in Sec. 7.3 because the eigenvalue E appears nonlinearly.

The differential equation (7.2.3) may be recast into a somewhat more familiar form by substituting $x = \cosh s$. In terms of the new variable s (7.4.14) becomes

$$y''(s) + (E^2 - 2E \cosh s)y = 0,$$

which is called the associated Mathieu equation (Mathieu equation of imaginary argument). It becomes the ordinary Mathieu equation

$$y''(t) + (\alpha + \beta \cos t)y = 0$$

when we replace s by it.

This is an interesting but terribly impractical result! Recognizing that we are solving the Mathieu equation is by itself no progress; at best it assures us that it will be fruitless to search for a simple analytical expression for E. We thus return to (7.4.14) and attempt a perturbative solution. Let us take E to be large and positive and choose the perturbing parameter for this problem to be $1/E$. We will attempt to find an approximate formula for those eigenvalues E which are large.

When E is large, we can decompose the full interval $1 \le x < \infty$ into two smaller and overlapping regions: region I for which $1 \le x \ll E$ $(E \to +\infty)$ and region II for which $1 \ll x$ $(E \to +\infty)$. Note that the overlap region, which includes values of x lying between but far from 1 and E, becomes infinitely long as $E \to +\infty$. We will show that it is fairly easy to solve (7.4.14) in regions I and II separately. We will then require that both approximations agree in the overlap region. This requirement translates into a condition which determines the eigenvalues.

Throughout region I the term $2Ex$ is small compared with E^2. When x is of order 1, it is valid to neglect $2Ex$ compared with E^2. When we do so, the resulting differential equation

$$(x^2 - 1)y_1''(x) + xy_1'(x) + E^2 y_1(x) = 0$$

is soluble. To solve this equation we again let $x = \cosh s$ and obtain

$$y_1''(s) + E^2 y_1(s) = 0,$$

whose solutions are $y_1 = \cos Es$ and $y_1 = \sin Es$. But since

$$s = \text{arc cosh } x = \ln [x + (x^2 - 1)^{1/2}]$$

and since $y(x)$ must satisfy the boundary condition $y(1) = 0$, we have

$$y_1(x) = A \sin [E \ln (x + \sqrt{x^2 - 1})], \tag{7.4.15}$$

where A is an undetermined constant.

One should note that (7.4.15) is not quite valid throughout region I, but only when $x = 0(1)$. To understand why, one can substitute (7.4.15) into (7.4.14) and observe that terms containing E do not all cancel. A solution which uniformly satisfies (7.4.14) up to terms of order 1 for all x in region I is (see Prob. 7.35)

$$y_1(x) = A\left(1 + \frac{x}{2E}\right) \sin [E \ln (x + \sqrt{x^2 - 1}) - \sqrt{x^2 - 1}]. \tag{7.4.16}$$

Equation (7.4.16) is a higher-order perturbative approximation to the solution of (7.4.14) than is (7.4.15). In general, one can solve (7.4.14) to all orders in powers of $1/E$ as an expression of the form

$$y_1(x) = A\left[1 + \sum_{n=1}^{\infty} f_n(x)E^{-n}\right] \sin \left[E \ln (x + \sqrt{x^2 - 1}) + \sum_{n=0}^{\infty} g_n(x)E^{-n}\right] \tag{7.4.17}$$

(see Prob. 7.35), where the two series in brackets are asymptotic series in powers of $1/E$ valid as $E \to \infty$.

The matching region is the interval $1 \ll x \ll E$ as $E \to +\infty$. Since x is large on this interval, we may approximate (7.4.16) by its asymptotic expansion

$$y_{\text{overlap}}(x) = Af(x)\sin g(x),$$

$$f(x) \sim 1 + \frac{x}{2E}, \tag{7.4.18}$$

$$g(x) \sim E\ln(2x) - x, \qquad E \to \infty,\ 1 \ll x \ll E.$$

This completes the analysis of region I.

Throughout region II, the term -1 is small compared with x^2 so we may replace (7.4.14) with the approximate differential equation

$$x^2 y_{\text{II}}''(x) + xy_{\text{II}}' + (E^2 - 2Ex)y_{\text{II}} = 0. \tag{7.4.19}$$

Although the difference between (7.4.14) and (7.4.19) may seem slight, (7.4.19) is now soluble. The substitution $t = \sqrt{8Ex}$ converts (7.4.19) into

$$t^2 y''(t) + ty'(t) + (4E^2 - t^2)y(t) = 0,$$

which is a modified Bessel equation. Two linearly independent solutions are $y_{\text{II}} = I_{2iE}(t)$, $K_{2iE}(t)$.

In Example 2 of Sec. 3.5 we showed that for large positive argument $I_\nu(x)$ grows exponentially like $e^x/\sqrt{x}$ and that $K_\nu(x)$ decays exponentially like $e^{-x}/\sqrt{x}$. Thus, the most general solution of (7.4.19) which satisfies the boundary condition $y(\infty) = 0$ is

$$y_{\text{II}} = BK_{2iE}(\sqrt{8Ex}), \tag{7.4.20}$$

where B is an arbitrary multiplicative constant.

Finally, recall from Prob. 6.71 that when p and z are real and $p \gg z$, the leading behavior of $K_{ip}(z)$ is

$$\sqrt{2\pi}\,(p^2 - z^2)^{-1/4} e^{-p\pi/2} \sin\left[p\cosh^{-1}(p/z) - \sqrt{p^2 - z^2} + \pi/4\right].$$

In our case $p = 2E$ and $z = \sqrt{8Ex}$, so in the overlap region the condition $p \gg z$ actually holds. Hence, the leading behavior of $y_{\text{II}}(x)$ in the overlap region is

$$-B\left(1 + \frac{x}{2E}\right)\sqrt{\frac{\pi}{E}}\, e^{-E\pi} \sin\left(E\ln 2x - x + 2E - E\ln 4E - \frac{\pi}{4}\right),$$

$$E \to \infty,\ 1 \ll x \ll E. \tag{7.4.21}$$

This completes the analysis of region II.

We now have two asymptotic approximations (7.4.18) and (7.4.21) to $y(x)$ in the overlap region. Since they both approximate the same function, they must agree over the entire overlap region. Thus, $2E - E\ln 4E - \pi/4$ must be an integral multiple of π. This condition gives a simple approximate formula for the eigenvalues E which becomes increasingly accurate as $E \to \infty$. It states that as $n \to \infty$ the nth eigenvalue E_n satisfies the equation

$$E_n \ln 4E_n - 2E_n = (n + \tfrac{3}{4})\pi, \qquad n = 0, 1, 2, \ldots. \tag{7.4.22}$$

How accurate is this result? In Table 7.2 we compare the exact eigenvalues (obtained numerically on a computer) with the solutions of (7.4.22) for values of n ranging from 0 to 8. Observe that the percentage error does indeed decrease as n increases. But what is most remarkable is that the error is never more than 3.38 percent. This entire calculation rests on the assumption that E is large. However, it is never clear just how large is large. One might imagine that E must be a million or so before one can believe the result in (7.4.22). However, the computer calculation shows that $E = 3$ is large enough to give 3 percent errors. It is a common experience that asymptotic calculations tend to give errors which are far smaller than what one might reasonably have expected.

The asymptotic match gives the eigenfunctions as well as the eigenvalues. Demanding that (7.4.18) and (7.4.21) agree in the overlap region imposes a relation between the constants A and B: $A = B\sqrt{\pi/E}\, e^{-E\pi}(-1)^n$. Thus, the eigenfunctions are now known approximately for $1 \le x < \infty$ up to an overall multiplicative constant. Of course, without further information [such as $y(6) = 19$] the normalization of $y(x)$ cannot be determined because the boundary-value problem is homogeneous.

Approximate Evaluation of an Integral

Asymptotic matching may be used to determine asymptotically the behavior of some integrals. Consider, for example, the integral

$$F(\varepsilon) = \int_0^\infty e^{-t-\varepsilon/t}\, dt \tag{7.4.23}$$

as $\varepsilon \to 0+$.

The leading behavior of $F(\varepsilon)$ as $\varepsilon \to 0+$ is easy to find. We simply set $\varepsilon = 0$ in (7.4.23): $F(0) = \int_0^\infty e^{-t}\, dt = 1$. Even though the integrand does not approach e^{-t} uniformly near $t = 0$ as $\varepsilon \to 0+$, we will verify shortly that the leading behavior of $F(\varepsilon)$ is correctly given by

$$F(\varepsilon) \sim 1, \qquad \varepsilon \to 0+. \tag{7.4.24}$$

It is more difficult to find the corrections to this leading behavior. Differentiating $F(\varepsilon)$ gives $F'(\varepsilon) = -\int_0^\infty e^{-t-\varepsilon/t}\, dt/t$, so $\lim_{\varepsilon \to 0+} F'(\varepsilon)$ does not exist. Apparently, the perturbative expansion of $F(\varepsilon)$ for small ε has the curious property that while its first term in (7.4.24) is almost trivial to find, successive terms require some real ingenuity. It is much more common for the first term (the

Table 7.2 A comparison of the exact eigenvalues E_n of (7.4.14) obtained from computer calculations with the approximations to E_n in (7.4.22) obtained from asymptotic matching

The percentage relative error [percentage relative error = (approximate − exact)/exact] decreases as E_n increases and the error is never more than 3.38 percent. Observe that the relative error falls off roughly as $1/E$, the small parameter in the asymptotic analysis which led to the above results. The form of (7.4.22) suggests that there are logarithmic corrections to the $1/E$ behavior of the relative error

n	E_n (from computer)	E_n (from asymptotic matching)	Percentage relative error
0	3.6975	3.5724	−3.38
1	5.3723	5.2569	−2.15
2	6.8195	6.7031	−1.71
3	8.1423	8.0223	−1.47
4	9.3826	9.2583	−1.33
5	10.5625	10.4337	−1.22
6	11.6955	11.5623	−1.14
7	12.7904	12.6532	−1.07
8	13.8535	13.7128	−1.02

unperturbed problem) to require the ingenuity and the remaining terms to require nothing more than paper and patience.

We will obtain the corrections to (7.4.24) in two ways. First, we express $F(\varepsilon)$ in terms of a modified Bessel function and use this representation to quickly write down the appropriate expansion of $F(\varepsilon)$ about $\varepsilon = 0$. This method has nothing to do with matched asymptotic expansions but it rapidly establishes the answer. Second, and here is the point of this discussion, we use matched asymptotic expansions to rederive the expansion of $F(\varepsilon)$ without using any special properties of modified Bessel functions.

First, we represent $F(\varepsilon)$ as a modified Bessel function. By differentiating $F(\varepsilon)$ in (7.4.23) twice we see that $F(\varepsilon)$ satisfies the differential equation $\varepsilon F''(\varepsilon) = F(\varepsilon)$. The substitution $F(\varepsilon) = ty(t)$, where $t = 2\sqrt{\varepsilon}$ converts this equation into the modified Bessel equation of order 1:

$$t^2 y''(t) + ty'(t) - (1 + t^2)y(t) = 0.$$

Therefore,

$$F(\varepsilon) = 2A\sqrt{\varepsilon}\, I_1(2\sqrt{\varepsilon}) + 2B\sqrt{\varepsilon}\, K_1(2\sqrt{\varepsilon}),$$

where A and B are constants to be determined from boundary conditions at 0 and ∞.

To calculate A and B we use the asymptotic relations $I_1(x) \sim e^x(2\pi x)^{-1/2}$, $K_1(x) \sim e^{-x}(2x/\pi)^{-1/2}$ $(x \to +\infty)$ and $I_1(x) \sim x/2$, $K_1(x) \sim 1/x$ $(x \to 0+)$. Since the integral in (7.4.23) vanishes as $\varepsilon \to +\infty$, the first set of asymptotic relations implies that $A = 0$. Also, comparing the leading asymptotic behavior in (7.4.24) with the second set of asymptotic relations gives $B = 1$. Thus,

$$F(\varepsilon) = 2\sqrt{\varepsilon}\, K_1(2\sqrt{\varepsilon}). \tag{7.4.25}$$

Finally, we look up the behavior of $K_1(x)$ as $x \to 0+$ in the Appendix and learn that

$$F(\varepsilon) \sim 1 + \varepsilon \ln \varepsilon + \varepsilon(2\gamma - 1) + \tfrac{1}{2}\varepsilon^2 \ln \varepsilon + \varepsilon^2(\gamma - \tfrac{5}{4}) + \tfrac{1}{12}\varepsilon^3 \ln \varepsilon + (\tfrac{1}{6}\gamma - \tfrac{5}{18})\varepsilon^3 + \cdots, \qquad \varepsilon \to 0+, \tag{7.4.26}$$

where $\gamma \doteq 0.5772$ is Euler's constant.

Having established the answer, we proceed to the second and main point of this discussion: namely, an independent derivation of (7.4.26) directly from asymptotic matching. We observe that asymptotic matching may be useful here because the character of the integrand $e^{-t-\varepsilon/t}$ is very different in the two regions $t \ll 1$ $(\varepsilon \to 0+)$ and $\varepsilon \ll t$ $(\varepsilon \to 0+)$. In the *inner* region $(t \ll 1)$ it is valid to expand the integrand as

$$e^{-t-\varepsilon/t} \sim e^{-\varepsilon/t}(1 - t + \tfrac{1}{2}t^2 - \tfrac{1}{6}t^3 + \cdots), \qquad t \to 0+; \tag{7.4.27}$$

in the *outer* region $(t \gg \varepsilon)$ we have

$$e^{-t-\varepsilon/t} \sim e^{-t}\left(1 - \frac{\varepsilon}{t} + \frac{1}{2}\frac{\varepsilon^2}{t^2} - \frac{1}{6}\frac{\varepsilon^3}{t^3} + \cdots\right), \qquad \varepsilon/t \to 0+. \tag{7.4.28}$$

The terms "inner" and "outer" are borrowed from the terminology of boundary-layer theory (Chap. 9).

Since these two asymptotic regions overlap for $\varepsilon \ll t \ll 1$ ($\varepsilon \to 0+$), it is natural to write

$$F(\varepsilon) = \int_0^\delta e^{-t-\varepsilon/t}\, dt + \int_\delta^\infty e^{-t-\varepsilon/t}\, dt, \tag{7.4.29}$$

where $\delta(\varepsilon)$ is arbitrary subject to the asymptotic constraint $\varepsilon \ll \delta \ll 1$ ($\varepsilon \to 0+$).

Our plan is to calculate each of the two integrals on the right side of (7.4.29) asymptotically as $\varepsilon \to 0+$ and then to show that the two results match asymptotically; i.e., their sum depends only on ε and not on the arbitrary matching parameter δ. We will perform this match to several orders in the inner matching variable δ and the outer matching variable ε/δ. Note that in terms of the variable δ/ε, the extent of the overlap region becomes infinite as $\varepsilon \to 0+$.

Leading-Order (Zeroth-Order) Match

An approximation to the integral over the inner region $0 \le t \le \delta$ that is correct to zeroth order in δ is $\int_0^\delta e^{-t-\varepsilon/t}\, dt = O(\delta)$ ($\delta \to 0+$) because the integrand is bounded by 1 for all $t > 0$. On the other hand, a zeroth-order approximation to the integral over the outer region $\delta \le t < \infty$ is

$$\int_\delta^\infty e^{-t-\varepsilon/t}\, dt = \int_\delta^\infty e^{-t}\, dt + \int_\delta^\infty e^{-t}(e^{-\varepsilon/t} - 1)\, dt$$

$$= e^{-\delta} + O(\varepsilon/\delta), \qquad \varepsilon/\delta \to 0+,$$

because

$$\left| \int_\delta^\infty e^{-t}(e^{-\varepsilon/t} - 1)\, dt \right| \le \varepsilon \int_\delta^\infty \frac{e^{-t}}{t}\, dt \le \frac{\varepsilon}{\delta} \int_\delta^\infty e^{-t}\, dt \le \frac{\varepsilon}{\delta},$$

where we have used the inequality $|e^{-x} - 1| \le x$ for all $x \ge 0$.

Combining the contributions from the inner and outer regions gives

$$F(\varepsilon) = \int_0^\infty e^{-t-\varepsilon/t}\, dt = e^{-\delta} + O(\delta) + O(\varepsilon/\delta)$$

$$= 1 + O(\delta) + O(\varepsilon/\delta), \qquad \delta \to 0+, \varepsilon/\delta \to 0+, \tag{7.4.30}$$

because $e^{-\delta} = 1 + O(\delta)$ ($\delta \to 0+$). Note that the dependence on δ has dropped out to zeroth order (albeit in a trivial way) in both the inner expansion parameter δ and the outer expansion parameter ε/δ. We have rederived the leading behavior of $F(\varepsilon)$ in (7.4.24).

To find higher-order corrections to the leading behavior of $F(\varepsilon)$ we must match to higher order in powers of δ and ε/δ, which we now do.

First-Order Match

To find an approximation to the integral over the inner region $0 \le t \le \delta$ which is valid to first order in δ, we use the inequality $|e^{-x} - 1| \le x$ $(x \ge 0)$ to write

$$\int_0^\delta e^{-t-\varepsilon/t}\,dt = \int_0^\delta e^{-\varepsilon/t}\,dt + \int_0^\delta e^{-\varepsilon/t}(e^{-t} - 1)\,dt$$

$$= \int_0^\delta e^{-\varepsilon/t}\,dt + O(\delta^2), \qquad \delta \to 0+.$$

Setting $s = \varepsilon/t$, we obtain

$$\int_0^\delta e^{-t-\varepsilon/t}\,dt = \varepsilon \int_{\varepsilon/\delta}^\infty e^{-s}s^{-2}\,ds + O(\delta^2), \qquad \delta \to 0+.$$

The integral on the right is an incomplete gamma function. It was shown in Example 4 of Sec. 6.2 that when $N = 0, 1, 2, 3, 4, \ldots,$

$$\int_x^\infty \frac{e^{-s}}{s^{N+1}}\,ds = C_N + \frac{(-1)^{N+1}}{N!}\ln x - \sum_{\substack{n=0\\ n\ne N}}^\infty (-1)^n \frac{x^{n-N}}{n!\,(n-N)},$$

$$x \to 0+, \qquad (7.4.31)$$

where

$$C_0 = -\gamma,$$

$$C_N = \frac{(-1)^{N+1}}{N!}\left(\gamma - \sum_{n=1}^{N} \frac{1}{n}\right), \qquad N = 1, 2, 3, \ldots,$$

and $\gamma \doteqdot 0.5772$ is Euler's constant. Therefore, in the limit $\varepsilon/\delta \to 0+$, the first-order contribution to $F(\varepsilon)$ from the inner region is

$$\int_0^\delta e^{-t-\varepsilon/t}\,dt = \varepsilon[C_1 + \ln(\varepsilon/\delta) + \delta/\varepsilon + O(\varepsilon/\delta)] + O(\delta^2),$$

$$\delta \to 0+,\ \varepsilon/\delta \to 0+. \qquad (7.4.32)$$

We have retained two error terms in this expansion but we cannot yet conclude anything about the relative sizes of $O(\varepsilon^2/\delta)$ and $O(\delta^2)$.

Next, we compute a first-order approximation to the integral in the outer region $\delta \le t < \infty$. Using $|e^{-x} - 1 + x| \le x^2$ $(x \ge 0)$ and (7.4.31) with $N = 1$, we have

$$\int_\delta^\infty e^{-t-\varepsilon/t}\,dt = \int_\delta^\infty e^{-t}(1 - \varepsilon/t)\,dt + \int_\delta^\infty e^{-t}(e^{-\varepsilon/t} - 1 + \varepsilon/t)\,dt$$

$$= \int_\delta^\infty e^{-t}(1 - \varepsilon/t)\,dt + O(\varepsilon^2/\delta), \qquad \varepsilon/\delta \to 0+ \qquad (7.4.33)$$

(see Prob. 7.38). Therefore, using (7.4.31) with $N = 0$,

$$\int_\delta^\infty e^{-t-\varepsilon/t}\,dt = e^{-\delta} - \varepsilon[C_0 - \ln\delta + O(\delta)] + O(\varepsilon^2/\delta)$$

$$= 1 - \delta - \varepsilon C_0 + \varepsilon\ln\delta + O(\delta^2) + O(\varepsilon^2/\delta),$$

$$\delta \to 0+,\ \varepsilon/\delta \to 0+, \qquad (7.4.34)$$

where we neglect the error term $\varepsilon\, O(\delta)$ because $\varepsilon\delta \ll \delta^2$ $(\varepsilon/\delta \to 0+)$.

Now we combine the contributions in (7.4.32) and (7.4.34) from the inner and outer regions. Even though the parameter δ appears explicitly in these two formulas, it cancels to second order in δ and ε/δ when the formulas are added together:

$$\int_0^\infty e^{-t-\varepsilon/t}\,dt = 1 + \varepsilon\ln\varepsilon + \varepsilon(2\gamma - 1) + O(\delta^2) + O(\varepsilon^2/\delta),$$

$$\delta \to 0+,\ \varepsilon/\delta \to 0+, \qquad (7.4.35)$$

where we have used $C_0 = -\gamma$ and $C_1 = \gamma - 1$. We have now reproduced the first three terms in (7.4.26).

It is interesting to note that the original condition on δ, $\varepsilon \ll \delta \ll 1$ $(\varepsilon \to 0+)$, is not adequate to ensure that the error terms in (7.4.35) are smaller than the retained terms. The constraint on δ must be sharpened to read $\varepsilon \ll \delta \ll \varepsilon^{1/2}$ $(\varepsilon \to 0+)$. However, even though this new relation restricts δ more than $\varepsilon \ll \delta \ll 1$ $(\varepsilon \to 0+)$, the matching of the inner and outer integrals still occurs over an infinite range in terms of the matching variable δ/ε: $1 \ll \delta/\varepsilon \ll \varepsilon^{-1/2}$ $(\varepsilon \to 0+)$.

Third-Order Match

Now we use asymptotic matching to calculate the first seven terms in the series (7.4.26). We will see that the number of expansion terms in the inner and outer expansions proliferate rapidly. To reproduce (7.4.26) we have to calculate the inner integral accurate to $O(\delta^6)$ $(\delta \to 0+)$ and the outer integral to $O(\varepsilon^6/\delta^5)$! We must retain terms to this order if we are to achieve a proper match to order $\varepsilon^3 \ln \varepsilon$.

To fifth order in δ, the inner integral is

$$\int_0^\delta e^{-t-\varepsilon/t}\,dt = \int_0^\delta e^{-\varepsilon/t}[1 - t + \tfrac{1}{2}t^2 - \tfrac{1}{6}t^3 + \tfrac{1}{24}t^4]\,dt + O(\delta^6), \qquad \delta \to 0+.$$

Setting $s = \varepsilon/t$, we obtain

$$\int_0^\delta e^{-t-\varepsilon/t}\,dt = \varepsilon\int_{\varepsilon/\delta}^\infty e^{-s}\left(\frac{1}{s^2} - \frac{\varepsilon}{s^3} + \frac{\varepsilon^2}{2s^4} - \frac{\varepsilon^3}{6s^5} + \frac{\varepsilon^4}{24s^6}\right)ds + O(\delta^6), \qquad \delta \to 0+.$$

Using (7.4.31) with $N = 1, \ldots, 5$, we get

$$\int_0^\delta e^{-t-\varepsilon/t}\,dt = \varepsilon\left[C_1 + \ln(\varepsilon/\delta) + \frac{\delta}{\varepsilon} - \frac{\varepsilon}{2\delta} + \frac{\varepsilon^2}{12\delta^2} - \frac{\varepsilon^3}{72\delta^3} + \frac{\varepsilon^4}{480\delta^4} + O\left(\frac{\varepsilon^5}{\delta^5}\right)\right]$$
$$- \varepsilon^2\left[C_2 - \frac{1}{2}\ln(\varepsilon/\delta) + \frac{\delta^2}{2\varepsilon^2} - \frac{\delta}{\varepsilon} + \frac{\varepsilon}{6\delta} - \frac{\varepsilon^2}{48\delta^2} + O\left(\frac{\varepsilon^3}{\delta^3}\right)\right]$$
$$+ \frac{1}{2}\varepsilon^3\left[C_3 + \frac{1}{6}\ln(\varepsilon/\delta) + \frac{\delta^3}{3\varepsilon^3} - \frac{\delta^2}{2\varepsilon^2} + \frac{\delta}{2\varepsilon} + O\left(\frac{\varepsilon}{\delta}\right)\right]$$
$$- \frac{1}{6}\varepsilon^4\left[\frac{\delta^4}{4\varepsilon^4} - \frac{\delta^3}{3\varepsilon^3} + O\left(\frac{\delta^2}{\varepsilon^2}\right)\right]$$
$$+ \frac{1}{24}\varepsilon^5\left[\frac{\delta^5}{5\varepsilon^5} + O\left(\frac{\delta^4}{\varepsilon^4}\right)\right] + O(\delta^6), \qquad \delta \to 0+, \varepsilon/\delta \to 0+.$$

The outer integral is expanded similarly:

$$\int_\delta^\infty e^{-t-\varepsilon/t}\,dt = \int_\delta^\infty e^{-t}\left(1 - \frac{\varepsilon}{t} + \frac{\varepsilon^2}{2t^2} - \frac{\varepsilon^3}{6t^3} + \frac{\varepsilon^4}{24t^4} - \frac{\varepsilon^5}{120t^5}\right)dt + O\left(\frac{\varepsilon^6}{\delta^5}\right),$$
$$\frac{\varepsilon}{\delta} \to 0+ \qquad (7.4.36)$$

(see Prob. 7.38). Using (7.4.31) with $N = 0, \ldots, 4$ gives

$$\int_\delta^\infty e^{-t-\varepsilon/t}\,dt = e^{-\delta} - \varepsilon\left[C_0 - \ln\delta + \delta - \frac{1}{4}\delta^2 + \frac{1}{18}\delta^3 + O(\delta^4)\right]$$
$$+ \frac{1}{2}\varepsilon^2\left[C_1 + \ln\delta + \frac{1}{\delta} - \frac{1}{2}\delta + O(\delta^2)\right]$$
$$- \frac{1}{6}\varepsilon^3\left[C_2 - \frac{1}{2}\ln\delta + \frac{1}{2\delta^2} - \frac{1}{\delta} + O(\delta)\right]$$
$$+ \frac{1}{24}\varepsilon^4\left[\frac{1}{3\delta^3} - \frac{1}{2\delta^2} + O\left(\frac{1}{\delta}\right)\right]$$
$$- \frac{1}{120}\varepsilon^5\left[\frac{1}{4\delta^4} + O\left(\frac{1}{\delta^3}\right)\right] + O\left(\frac{\varepsilon^6}{\delta^5}\right),$$
$$\delta \to 0+, \ \varepsilon/\delta \to 0+.$$

The order of accuracy to which we compute the inner and outer expansions is not arbitrary; the error terms are chosen so that these expansions match through terms of order ε^3. Indeed, if we add together the inner and outer expansions, we obtain

$$F(\varepsilon) = \int_0^\infty e^{-t-\varepsilon/t}\,dt$$

$$= 1 + \varepsilon \ln \varepsilon + \varepsilon(2\gamma - 1) + \frac{1}{2}\varepsilon^2 \ln \varepsilon$$

$$+ \varepsilon^2\left(\gamma - \frac{5}{4}\right) + \frac{1}{12}\varepsilon^3 \ln \varepsilon + \varepsilon^3\left(\frac{1}{6}\gamma - \frac{5}{18}\right) + O(\delta^6)$$

$$+ O\left(\frac{\varepsilon^6}{\delta^5}\right) + O\left(\frac{\varepsilon^5}{\delta^3}\right) + O\left(\frac{\varepsilon^4}{\delta}\right) + O(\varepsilon^2\delta^2) + O(\varepsilon\delta^4) + O(\varepsilon^3\delta),$$

$$\delta \to 0+,\ \varepsilon/\delta \to 0+, \qquad (7.4.37)$$

where we have substituted the values $C_0 = -\gamma$, $C_1 = \gamma - 1$, $C_2 = -\frac{1}{2}\gamma + \frac{3}{4}$, $C_3 = \frac{1}{6}\gamma - \frac{11}{36}$. We have thus reproduced the series (7.4.26).

Note that all the error terms in (7.4.37) are negligible with respect to ε^3 if the constraint on δ is sharpened to read $\varepsilon^{3/5} \ll \delta \ll \varepsilon^{1/2}$ ($\varepsilon \to 0+$). In successively higher orders the constraint on δ becomes increasingly tight. However, in terms of the matching variable δ/ε the extent of the matching interval is always infinite.

In Fig. 7.7 we compare the series for $F(\varepsilon)$ in (7.4.30), (7.4.35), and (7.4.37) with a numerical evaluation of the integral in (7.4.23).

In the next example we use the method of matched asymptotic cxpansions to obtain higher-order terms in the expansion of a generalized Fourier integral.

Example 6 *Use of asymptotic matching to improve the predictions of stationary-phase analysis.* In this example we use asymptotic matching to find the large-x behavior of the integral

$$I(x) = \int_0^{\pi/2} e^{ix\cos t}\,dt. \qquad (7.4.38)$$

The method of stationary phase (see Example 3 of Sec. 6.5) quickly gives the leading behavior of $I(x)$:

$$\int_0^{\pi/2} e^{ix\cos t}\,dt \sim \sqrt{\frac{\pi}{2x}}\,e^{i(x-\pi/4)}, \qquad x \to +\infty. \qquad (7.4.39)$$

However, we did not explain in Sec. 6.5 how to obtain the higher-order corrections to this leading behavior.

In Sec. 6.5 we showed that the leading-order behavior is completely determined by a *local* analysis of the integrand in the neighborhood of the stationary point, which for the integral (7.4.38) lies at $t = 0$. On the other hand, higher-order corrections to the leading behavior may arise from regions in the domain of integration away from the stationary point. Therefore, some form of *global* analysis is required to obtain higher-order corrections; this example shows how asymptotic matching can be used.

As usual, the procedure consists of dividing the domain of integration into two regions: the first is a narrow region $0 \le t \le \delta$ containing the stationary point at $t = 0$; the second is the remainder of the integration interval $\delta < t \le \pi/2$. For now we say only that $\delta = \delta(x)$ is a small parameter satisfying $\delta \ll 1$ ($x \to +\infty$). Later we will impose more restrictive conditions on

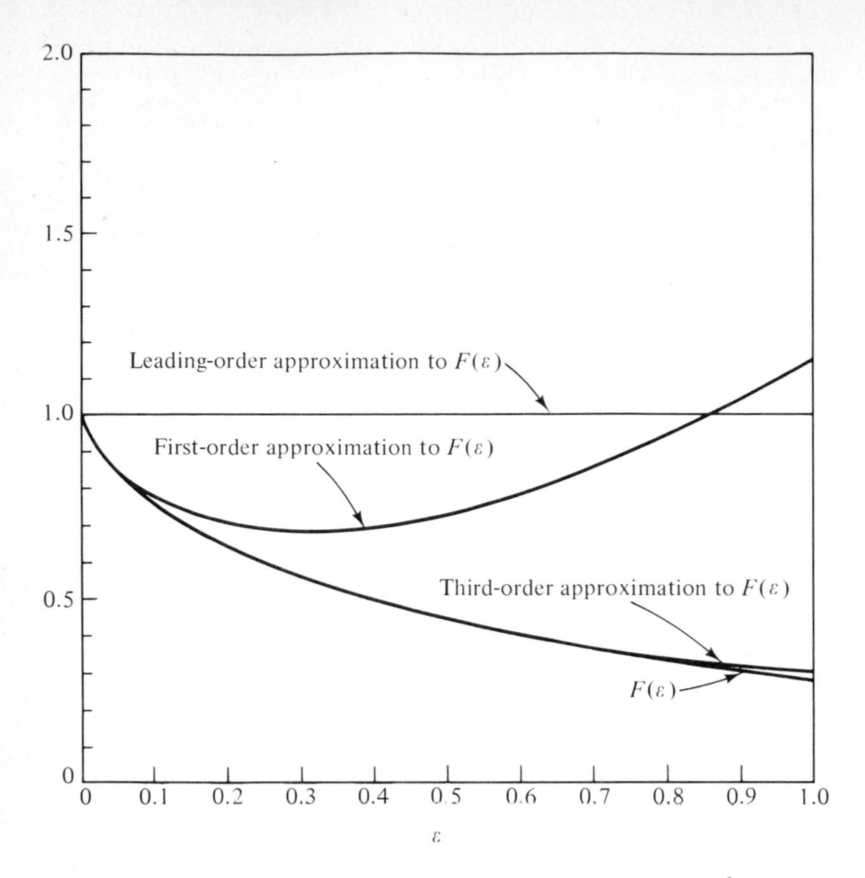

Figure 7.7 A comparison of three approximations to the integral $F(\varepsilon) = \int_0^\infty \exp(-t - \varepsilon/t)\, dt$ which were derived using asymptotic matching. The leading-order (zeroth-order) approximation in (7.4.30) is simply $F(\varepsilon) \sim 1$ $(\varepsilon \to 0+)$. The first-order and third-order approximations are given in (7.4.35) and (7.4.37). The accuracy increases rapidly with the order of the approximation.

the size of $\delta(x)$. Next, we decompose the integral $I(x)$ into two integrals $I(x) = I_1(x) + I_2(x)$, where $I_1(x) = \int_0^\delta e^{ix\cos t}\, dt$ and $I_2(x) = \int_\delta^{\pi/2} e^{ix\cos t}\, dt$. We will find asymptotic approximations to $I_1(x)$ and $I_2(x)$ as $x \to +\infty$ and $\delta(x) \to 0+$. These approximations will each depend on δ but, as we will see, their sum $I(x)$ will not depend on δ.

First, we approximate $I_1(x)$ as $x \to +\infty$. Since $\delta \ll 1$ as $x \to +\infty$, we have $\cos t = 1 - \frac{1}{2}t^2 + O(\delta^4)$ $(0 \le t \le \delta,\ \delta \to 0+)$. Therefore, $I_1(x) = e^{ix} \int_0^\delta e^{-ixt^2/2}\, dt + O(x\delta^5)$ $(x \to \infty,\ x^{1/4}\delta \to 0+)$. This result is valid if we impose the condition on δ that $x^{1/4}\delta \to 0+$ as $x \to +\infty$; we are free to impose this condition which specifies just how rapidly $\delta \to 0$ as $x \to +\infty$. To approximate the above integral further, we write $\int_0^\delta e^{-ixt^2/2}\, dt = \int_0^\infty e^{-ixt^2/2}\, dt - \int_\delta^\infty e^{-ixt^2/2}\, dt$. In both of these integrals we rotate the contour of integration by 45° in the complex-t plane. This enables us to do the first integral exactly and to approximate the second using integration by parts twice. The result for $I_1(x)$ is

$$I_1(x) = \sqrt{\frac{\pi}{2x}}\, e^{i(x-\pi/4)} + \frac{i}{x\delta} e^{ix(1-\delta^2/2)} - \frac{1}{x^2\delta^3} e^{ix(1-\delta^2/2)} - \frac{3ie^{ix(1-\delta^2/2)}}{x^3\delta^5} + O(x\delta^5) + O\left(\frac{1}{x^4\delta^7}\right),$$

$$x \to +\infty,\ x^{2/5}\delta \to 0+,\ x^{1/2}\delta \to +\infty. \tag{7.4.40}$$

To make the error incurred upon integrating by parts smaller than the smallest retained term, we have imposed two new conditions on the magnitude of δ: $x^{2/5}\delta \to 0+$ (which supplants $x^{1/4}\delta \to 0+$ because it is more stringent) and $x^{1/2}\delta \to +\infty$ as $x \to +\infty$. Observe that the conditions $x \to +\infty$, $\delta \to 0+$, $x^{2/5}\delta \to 0+$, and $x^{1/2}\delta \to +\infty$ are all satisfied if we take

$$x^{-1/2} \ll \delta \ll x^{-2/5}, \qquad x \to +\infty. \tag{7.4.41}$$

Next, we approximate $I_2(x)$ as $x \to +\infty$. Since there are no stationary points of the integrand for $\delta < t \le 1$, it is valid to integrate by parts. Three integrations by parts give

$$\begin{aligned} I_2(x) &= \left.\frac{ie^{ix\cos t}}{x \sin t}\right|_\delta^{\pi/2} - \left.\frac{e^{ix\cos t}}{x^2 \sin^3 t}\cos t\right|_\delta^{\pi/2} - \left.\frac{i(1+2\cos^2 t)e^{ix\cos t}}{x^3 \sin^5 t}\right|_\delta^{\pi/2} + O\left(\frac{1}{x^4\delta^7}\right) \\ &= \frac{i}{x} - \frac{ie^{ix\cos\delta}}{x\sin\delta} + \frac{e^{ix\cos\delta}\cos\delta}{x^2\sin^3\delta} + \frac{ie^{ix\cos\delta}(2\cos^2\delta+1)}{x^3\sin^5\delta} + O\left(\frac{1}{x^3}\right) + O\left(\frac{1}{x^4\delta^7}\right), \\ &\qquad x \to +\infty,\ x^{3/7}\delta \to +\infty. \end{aligned} \tag{7.4.42}$$

We have imposed the additional condition $x^{-3/7} \ll \delta (x \to +\infty)$ to ensure that the error term is smaller than the smallest retained term. Notice that $x^{1/2}\delta = (x^{3/7}\delta)x^{1/14} \to \infty$ when $x^{3/7}\delta \to +\infty$ and $x \to +\infty$.

Finally, we add together the asymptotic approximations to $I_1(x)$ and $I_2(x)$ in (7.4.40) and (7.4.42) and obtain

$$I(x) = \sqrt{\frac{\pi}{2x}}\, e^{i(x-\pi/4)} + \frac{i}{x} + O\left(\frac{1}{x^4\delta^7}\right) + O(x\delta^5) + O\left(\frac{1}{x^3}\right),$$
$$x \to +\infty,\ x^{2/5}\delta \to 0+,\ x^{3/7}\delta \to +\infty. \tag{7.4.43}$$

This is the answer; we have found the first *two* terms (one term beyond the leading behavior) in the expansion of $I(x)$ as $x \to +\infty$. Observe that the parameter δ cancels out of the asymptotic expansion of $I_1(x)$ and $I_2(x)$ and appears only in error terms in the final answer for $I(x)$.

The consistency of the asymptotic match depends on the error terms being smaller than the smallest retained term, i/x. There are two error terms that must be checked: $x^{-4}\delta^{-7} \ll x^{-1}$ and $x\delta^5 \ll x^{-1}$ $(x \to +\infty)$. These conditions are satisfied because it is possible to impose the asymptotic conditions $x^{-3/7} \ll \delta \ll x^{-2/5}$ $(x \to +\infty)$, which is a further refinement of the condition in (7.4.41). To stress the delicacy of this condition we rewrite it as

$$x^{-15/35} \ll \delta \ll x^{-14/35}, \qquad x \to +\infty. \tag{7.4.44}$$

The consistency of the asymptotic match depends on the existence of a parameter δ which satisfies (7.4.44). If it were not true that $x^{-15/35} \ll x^{-14/35}$ $(x \to +\infty)$, then there would have been no such δ!

We can now explain why it was necessary to integrate by parts three times even though (*a*) the final answer in (7.4.43) was determined after just one integration by parts and (*b*) further integration by parts generated terms depending on δ which cancelled when we added I_1 and I_2. Three integrations by parts were necessary to establish the consistency of asymptotic matching. If we had done just one or two integrations by parts, the final asymptotic condition on δ, instead of being consistent like that in (7.4.44), would have been inconsistent. For example, after one or two integrations by parts we would have had

$$x^{-1/3} \ll \delta \ll x^{-2/5} \quad \text{or} \quad x^{-2/5} \ll \delta \ll x^{-2/5}, \qquad x \to +\infty, \tag{7.4.45}$$

respectively, which are impossible conditions (see Prob. 7.41). In general, the result of an asymptotic match cannot be trusted until the matching scheme has been shown to be consistent. The condition in (7.4.44) is crucial because it shows that the parameter δ, which determines the location of the matching region, exists. This is the subtlety of asymptotic matching; the rest is straightforward calculation.

(TD) 7.5 MATHEMATICAL STRUCTURE OF PERTURBATIVE EIGENVALUE PROBLEMS

The series $\sum_{n=0}^{\infty} \varepsilon^n$ diverges for $|\varepsilon| \geq 1$. Naturally, this divergence reflects the singularity structure of the function $f(\varepsilon)$ that the series approximates; here, $f(\varepsilon) = 1/(1-\varepsilon)$ has a pole at $\varepsilon = 1$. Several examples of perturbative eigenvalue problems having perturbation series in the form of power series in ε were given in Sec. 7.3. In those problems, also, when the perturbation series for the eigenvalue had a finite or vanishing radius of convergence, the exact eigenvalue considered as a function of ε also had singularities in the complex-ε plane.

In this section we discuss the origin and meaning of such singularities. We will argue that the presence of singularities as well as their type is not a chance event, but is a predictable phenomenon characteristic of a broad class of perturbative eigenvalue problems.

Example 1 *Eigenvalues of a* 2×2 *matrix.* We begin by considering the simplest eigenvalue problem of all. Let A and B be real symmetric 2×2 matrices of the form

$$A = \begin{pmatrix} a & 0 \\ 0 & b \end{pmatrix}, \qquad B = \begin{pmatrix} x & z \\ z & y \end{pmatrix},$$

and consider the problem of finding the eigenvalues of $A + B$ by perturbation theory. To do this we replace B with εB and express each eigenvalue as a power series in ε. Each power series begins with the numbers a or b and is expected to be convergent for sufficiently small ε because the perturbation problem is regular.

To find the radii of convergence we solve the problem exactly. Setting the determinant of $A + \varepsilon B - I\lambda$ equal to zero gives the following formula for the two eigenvalues $\lambda_\pm$:

$$\lambda_\pm = \tfrac{1}{2}\{a + b + \varepsilon x + \varepsilon y \pm [(a - b + \varepsilon x - \varepsilon y)^2 + 4\varepsilon^2 z^2]^{1/2}\}. \tag{7.5.1}$$

$\lambda_\pm$ are analytic functions of ε except at the zeros of the square-root term. Thus, $\lambda_\pm(\varepsilon)$ have a pair of square-root branch points, symmetrically placed about the real axis, at

$$\varepsilon = \frac{a - b}{y - x \pm 2iz}. \tag{7.5.2}$$

The radius of convergence of the perturbation series is $|a - b|/[(x - y)^2 + 4z^2]^{1/2}$.

Observe that the radius of convergence vanishes when $a = b$. But a and b are the unperturbed eigenvalues, so that, if $a = b$, the unperturbed problem is degenerate. Thus, when $a = b$, the exact solution of the perturbed problem (which is nondegenerate if $\varepsilon \neq 0$) undergoes an abrupt change (the appearance of degeneracy) in the limit $\varepsilon \to 0$. The perturbation problem must therefore be singular when $a = b$ and this conclusion is consistent with the vanishing of the radius of convergence.

The noteworthy feature of this example is that the two eigenvalues in (7.5.1) are analytic continuations of each other and together they form a single two-valued function $\lambda(\varepsilon)$. $\lambda(\varepsilon)$ is defined on a two-sheeted (Riemann) surface; on the lower sheet $\lambda(\varepsilon) = \lambda_-(\varepsilon)$ and on the upper sheet $\lambda(\varepsilon) = \lambda_+(\varepsilon)$. Analytic continuation around either of the two branch points (7.5.2) exchanges the identities of the two eigenvalues because the sign of the square root in (7.5.1) changes; this phenomenon is called *level crossing*.

The existence of square-root branch-point singularities, the appearance of level crossing, and the unification of the eigenvalues into a single many-valued analytic function of ε are not just special properties of the simple problem in Example 1. Rather, these seem to be very general features of perturbative eigenvalue problems having perturbation series that diverge for sufficiently large $|\varepsilon|$. Of course, one could argue that all this analytical structure is artificial because the original problem did not involve ε. However, we are often forced to introduce a perturbation parameter ε when there is no other analytical way to make progress in computing the eigenvalues. And, when the perturbation series is divergent, the recovery of the eigenvalues depends upon a clear understanding of the analytical structure of the perturbed model, as will be shown in Chap. 8.

Next we consider a more general eigenvalue problem and show that an approximate solution displays the same basic analytic structure as the above example.

Let us reconsider (7.3.3):

$$\left[-\frac{d^2}{dx^2} + V(x) + \varepsilon W(x) - E(\varepsilon)\right] y(x) = 0 \tag{7.5.3}$$

with the boundary conditions that $y(x) \to 0$ as $|x| \to \infty$. However, instead of immmediately expanding the solution in power series in ε, let us instead represent the solution as an infinite linear combination of eigenfunctions of the unperturbed problem, which we assume are all known. We will then use the differential equation to determine the coefficients in this expansion. If we label the nth unperturbed eigenfunction and eigenvalue by the superscript (n), then

$$-\frac{d^2}{dx^2} y_0^{(n)} + V(x) y_0^{(n)} - E_0^{(n)} y_0^{(n)} = 0, \tag{7.5.4}$$

where $y_0^{(n)}(x) \to 0$ as $|x| \to \infty$.

The Schrödinger equation (7.5.4) is a Sturm-Liouville eigenvalue problem (see Sec. 1.8). Therefore, the eigenfunctions $y_0^{(n)}$ are *complete*. For these Schrödinger eigenfunctions completeness means that an arbitrary square-integrable function, such as an exact eigenfunction solution $y(x)$ of (7.5.3), can be expanded as the infinite linear combination

$$y(x) = \sum_{n=0}^{\infty} a_n y_0^{(n)}(x). \tag{7.5.5}$$

Eq. (7.5.5) is *not* a perturbation expansion because there is no perturbing parameter.

We assume that $W(x) y_0^{(n)}(x)$ may also be expanded in terms of the same eigenfunctions $y_0^{(m)}$:

$$W(x) y_0^{(n)}(x) = \sum_{m=0}^{\infty} A_m^n y_0^{(m)}(x). \tag{7.5.6}$$

Applying the differential equation (7.5.3) to (7.5.5), using (7.5.6), and equating coefficients of $y_0^{(n)}$ for each n gives an infinite matrix equation satisfied by the coefficients a_n:

$$\mathbf{M a} \equiv \begin{bmatrix} E_0^{(0)} - E + \varepsilon A_0^0 & \varepsilon A_0^1 & \varepsilon A_0^2 & \cdots \\ \varepsilon A_1^0 & E_0^{(1)} - E + \varepsilon A_1^1 & \varepsilon A_1^2 & \cdots \\ \varepsilon A_2^0 & \varepsilon A_2^1 & E_0^{(2)} - E + \varepsilon A_2^2 & \\ \vdots & \vdots & & \ddots \end{bmatrix} \begin{bmatrix} a_0 \\ a_1 \\ a_2 \\ \vdots \end{bmatrix} = 0. \tag{7.5.7}$$

If the matrix M were finite dimensional, the condition that there be nontrivial solutions $\mathbf{a}$ satisfying (7.5.7) is that $\det M = 0$ (Cramer's rule). Unfortunately, when M is infinite dimensional, as it is here, $\det M$ (called a Hill determinant) may not exist. Therefore we devise a sequence of approximations in which we truncate the matrix M. Let M_N be an $N \times N$ matrix whose entries are the same as the first N rows and columns of M. We thus approximate (7.5.7) by

$$M_N \mathbf{a}_N = 0, \tag{7.5.8}$$

where $\mathbf{a}_N = (a_0, a_1, \ldots, a_{N-1})$. We have replaced the complicated equation in (7.5.3) by a comparatively trivial sequence of matrix equations.

Note The matrix M_N can be obtained directly using the orthogonality properties of the eigenfunctions $y_0^{(n)}$. This process, called the Galerkin method, involves three steps. First, we seek an approximation to $y(x)$ in the form

$$y_{\text{approx}}(x) = \sum_{n=0}^{N-1} a_n y_0^{(n)}. \tag{7.5.9}$$

Second, we substitute $y_{\text{approx}}(x)$ into the perturbed differential equation and reexpand the result as a series of the $y_0^{(n)}$. Third, equations for the expansion coefficients a_n $(n = 0, 1, \ldots, N-1)$ are obtained by equating the coefficients of $y_0^{(n)}$ $(n = 0, 1, \ldots, N)$ to 0. The result is precisely the matrix equation (7.5.8). This so-called Galerkin procedure is very useful in numerical analysis.

The approximation (7.5.8) can also be derived in a somewhat different way using a Rayleigh-Ritz variational procedure. There, one also begins with the series (7.5.9). The coefficients a_n which give the "best fit" to $y(x)$ are determined by applying a variational principle to minimize the difference between $y(x)$ and $y_{\text{approx}}(x)$. The "best fit" is achieved when (7.5.8) is satisfied.

Next, we define $D_N(E, \varepsilon) = \det M_N$. The limit of $D_N(E, \varepsilon)$ as $N \to \infty$ may or may not exist, but we are not really interested in this limit. We are actually concerned with the behavior of the *roots* of the equation

$$D_N(E, \varepsilon) = 0. \tag{7.5.10}$$

Do the roots of this equation approach the exact eigenvalues of the differential equation (7.5.3) as $N \to \infty$? A glance at (7.5.7) shows that $D_N(E, \varepsilon)$ is an Nth-order polynomial in E and ε. Thus, given any ε we can obtain N values for E. Leaving aside all questions of rigor we will simply assume that for every value of ε these values of E do approach the correct eigenvalues of the exact problem in (7.5.3) as $N \to \infty$. We have thus replaced the complicated differential-equation eigenvalue problem (7.5.3) by a much simpler matrix eigenvalue problem very similar in structure to the one considered in Example 1.

Example 2 *Singular perturbation of the parabolic cylinder equation.* When $V = x^2/4$ and $W = x^4/4$, $D_N(E, \varepsilon)$ satisfies a five-term recursion relation (see Prob. 7.43). The results of a numerical computation of the zeros of $D_N(E, \varepsilon)$ for $\varepsilon = 1$ are given in Table 7.3. As N increases, the eigenvalues rapidly converge to the eigenvalues of the differential equation. The eigenvalues approach their limits in order of their size, the smaller ones converging more rapidly than the larger ones. This sequential (nonuniform) convergence of the eigenvalues typically occurs when infinite matrices are approximated by truncated finite matrices.

Table 7.3 Numerical calculation of the first five eigenvalues of (7.3.3) with $V(x) = x^2/4$, $W(x) = x^4/4$, and $\varepsilon = 1$

The eigenvalues are the limits as $N \to \infty$ of the zeros of $D_N(E, \varepsilon = 1)$. This table shows that as N increases zeros rapidly converge to the exact eigenvalues listed on the bottom line [obtained by Padé summation (see Chap. 8)]. The entries in the table form a checkerboard pattern with every other entry absent because the values of the zeros only change when N increases by 2. This effect is connected with the fact that the perturbed and unperturbed eigenfunctions are either even or odd functions of x

N	$E^{(0)}$	$E^{(1)}$	$E^{(2)}$	$E^{(3)}$	$E^{(4)}$
1	1.250 000				
2		5.250 000			
3	0.855 087		12.644 91		
4		3.273 837		24.226 16	
5	0.808 229		7.382 825		40.558 95
6		2.843 872		13.867 49	
7	0.805 870		5.860 713		23.373 00
8		2.752 576		10.308 12	
9	0.805 614		5.361 362		16.794 95
10		2.740 927		8.842 45	
11	0.804 698		5.215 487		13.702 25
12		2.740 828		8.240 62	
13	0.804 076		5.185 265		12.170 52
14		2.740 060		8.020 67	
15	0.803 838		5.182 772		11.435 98
16		2.738 944		7.957 72	
17	0.803 781		5.182 628		11.117 65
18		2.738 253		7.946 76	
19	0.803 774		5.181 493		11.002 51
20		2.737 979		7.946 43	
21	0.803 774		5.180 331		10.972 07
22		2.737 907		7.945 87	
23	0.803 773		5.179 657		10.967 97
24		2.737 897		7.944 56	
25	0.803 772		5.179 385		10.967 97
26		2.737 897		7.943 41	
27	0.803 771		5.179 310		10.967 08
28		2.737 896		7.942 77	
29	0.803 771		5.179 298		10.965 69
30		2.737 894		7.942 50	
∞	0.803 771	2.737 893	5.179 292	7.942 40	10.963 58

Let us examine the structure of the roots $E_N(\varepsilon)$ of $D_N(E, \varepsilon)$. Equation (7.5.10) is an implicit algebraic relation between E and ε. Therefore, the solution $E_N(\varepsilon)$ is one, or possibly several, multivalued functions (having altogether N values) and the only singularities that $E_N(\varepsilon)$ may exhibit are poles or branch points. However, from the specific form of the $D_N(E, \varepsilon)$ one may show that $E_N(\varepsilon)$ may not have poles or branch points at which $E_N(\varepsilon) = \infty$ (see Prob. 7.44). The only singularities that $E_N(\varepsilon)$ may have are branch points at which $E_N(\varepsilon)$ remains *finite*. Level crossing of the approximate eigenvalues occurs as the solutions of (7.5.10) are analytically continued around these branch points.

At a branch point of $E_N(\varepsilon)$ we expect at least two eigenvalues to become degenerate [see (7.5.1)]. Thus, at a branch point (7.5.10) must have at least a double root. The condition for a double root is

$$\frac{\partial}{\partial E} D_N(E, \varepsilon) = 0. \tag{7.5.11}$$

Since $\partial D_N/\partial E$ is a polynomial of degree $N - 1$ in both E and ε, the simultaneous solutions of (7.5.10) and (7.5.11) may yield at most $N(N - 1)$ branch points (see Prob. 7.45). These branch points typically occur as $\frac{1}{2}N(N - 1)$ complex conjugate pairs because (7.5.10) and (7.5.11) are real [see (7.5.7)].

A double root of (7.5.10) implies that $E_N(\varepsilon)$ has a square-root branch point in the ε plane. A more complicated branch point of $E_N(\varepsilon)$ would require $D_N(E, \varepsilon)$ to have a multiple root; e.g., a cube-root branch point would occur if $(\partial^2/\partial E^2)D_N(E, \varepsilon) = 0$ holds simultaneously with (7.5.10) and (7.5.11). Of course, it is not impossible for three or more simultaneous equations in two unknowns to have a solution, but it is very unlikely. The existence of a level-crossing point which is not a square-root singularity must be viewed as purely fortuitous; even if such a branch point could exist for some fixed N, it would probably disappear as soon as N is increased by 1. We conclude that in a typical $N \times N$ matrix eigenvalue problem with parameter ε there are $N(N - 1)$ square-root branch points in the ε plane.

Now let us consider what may happen to the solution of the finite matrix problem (7.5.8) as $N \to \infty$. There are four possibilities and we consider each in turn.

Possibility 1 As $N \to \infty$, the locations of the branch points stabilize, remain well separated from each other and the origin, and maintain their identities as square-root branch points. If this occurs, then the radius of convergence of the perturbation series for each eigenvalue is then nonzero and exactly equal to the distance to the nearest singularity in the complex plane at which this eigenvalue crosses with (the analytic continuation of) another eigenvalue. If possibility 1 occurs, the perturbation theory is regular.

Example 3 *Regular perturbation of the parabolic cylinder equation.* An eigenvalue equation which displays the behavior described above is

$$(d^2/dx^2 + x^2/4 + \varepsilon|x| - E)y(x) = 0, \qquad \lim_{|x|\to\infty} y(x) = 0.$$

For this differential equation the exact eigenfunctions $y(x)$ are always either even or odd under the reflection $x \to -x$. The determinant $D_N(E, \varepsilon)$ in (7.5.10) factors into a product of two determinants, $D_N(E, \varepsilon) = D_N^{(\text{even})}(E, \varepsilon) D_N^{(\text{odd})}(E, \varepsilon)$, where $D_N^{(\text{even})}(E, \varepsilon)$ contains the entries A_{2m}^{2n} and $D_N^{(\text{odd})}(E, \varepsilon)$ contains the entries A_{2m+1}^{2n+1} [see (7.5.6)]; A_{2m}^{2n+1} and A_{2m+1}^{2n} both vanish. The eigenvalues associated with even eigenfunctions and the eigenvalues associated with odd eigenfunctions are qualitatively similar, so we restrict our attention to the even eigenfunctions and their eigenvalues. The numbers A_{2m}^{2n} in equation (7.5.6) are given by (see Prob. 7.46)

$$A_{2m}^{2n} = (2/\pi)^{1/2} \frac{(-1)^{n+m+1}[2(n+m)+1](2n)!}{2^{m+n}[4(n-m)^2-1]m!\,n!}.$$

A simultaneous numerical solution of (7.5.10) and (7.5.11) gives branch points for various values of N. The locations of these branch points stabilize as N gets large (see Table 7.4). In Fig. 7.8 we plot a portion of the upper-half complex-ε plane showing the limiting values of some branch points for large N. The branch points occur in complex-conjugate pairs: each branch point in the upper-half ε plane is associated with another (not shown) in the lower-half plane. Each pair of branch points is joined by a branch cut (not shown).

What happens when the eigenvalues for this problem are analytically continued around a branch point? Contours which emerge from the origin in the ε plane, encircle a branch point, and return to the origin are indicated in Fig. 7.8. These contours all consist of sequences of line segments. The simplest contour has its corners numbered sequentially 0, 1, 2, 3, 4, 1, 0. In Fig. 7.9 a portion of the complex-E plane is plotted showing the images of this contour in the ε plane. Note that when $\varepsilon = 0$ the eigenvalues assume their unperturbed values $\frac{1}{2}, \frac{5}{2}, \frac{9}{2}, \frac{13}{2}$. [Only the eigenvalues for even eigenfunctions $y(x) = y(-x)$ are shown here; the unperturbed eigenvalues $\frac{3}{2}, \frac{7}{2}, \frac{11}{2}, \ldots$ are associated with odd functions $y(x) = -y(x)$ and behave similarly as functions of complex ε.] As the argument of each eigenvalue follows the contour in the ε plane, the eigenvalues simultaneously trace out curves in the E plane. The first two eigenvalues undergo level crossing (they exchange identities), while the other eigenvalues return to their original positions.

Figures 7.10 to 7.12 show how other pairs of levels cross when the eigenvalues are analytically continued around other branch points in Fig. 7.8. These figures are not schematic representations; they demonstrate the actual numerical behavior of the eigenvalues. The numerical error is approximately equal to the thickness of the curves.

The radius of convergence of the perturbation series for an eigenvalue increases as the size of the unperturbed eigenvalue increases because, as can be seen from Fig. 7.8, the distance to the nearest branch point at which this eigenvalue crosses with another also increases.

Table 7.4 Stabilization of the branch point "A" as $N \to \infty$

This branch point is one of five which are plotted in Fig. 7.9. The Nth approximation to a branch point is a value of ε which simultaneously solves $D_N(E, \varepsilon) = 0$ and $(\partial/\partial E) D_N(E, \varepsilon) = 0$ [see (7.5.10) and (7.5.11)]

N	Nth approximation to A
3	$-1.136 + 0.5552i$
4	$-1.209 + 0.5623i$
5	$-1.206 + 0.5741i$
6	$-1.205 + 0.5731i$
7	$-1.205 + 0.5730i$
8	$-1.205 + 0.5730i$

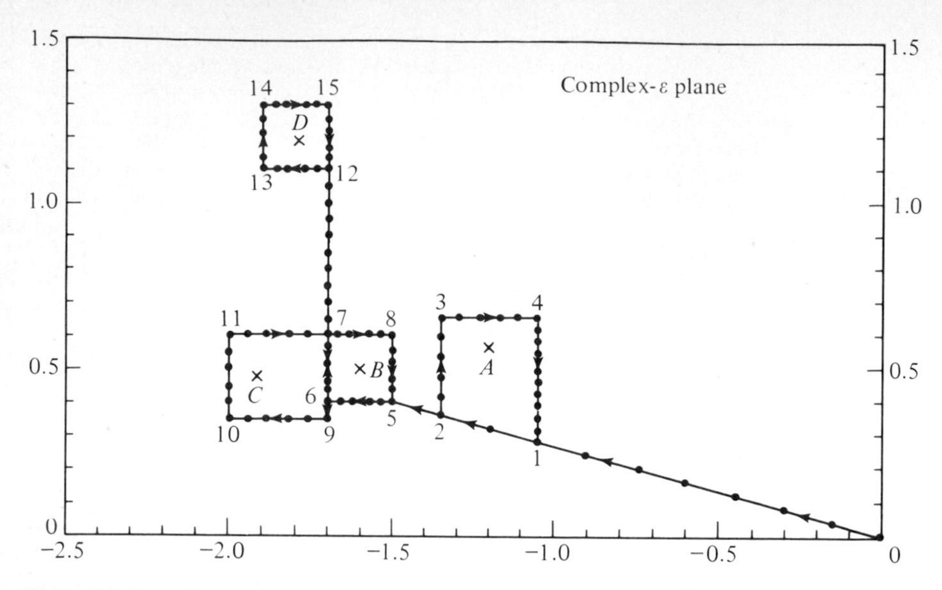

Figure 7.8 A portion of the complex-ε plane for the perturbation problem in Example 3 of Sec. 7.5. Shown are four level-crossing points A, B, C, and D. Paths made up of short line segments which start at the origin and go around the points A, B, C, and D are indicated. Particular points along these paths are labeled by the numbers 1 through 15. The images of these paths in the complex-E plane are shown in Figs. 7.9 to 7.12. Figures 7.8 through 7.12 were drawn with the help of B. Svetitsky and H. Happ.

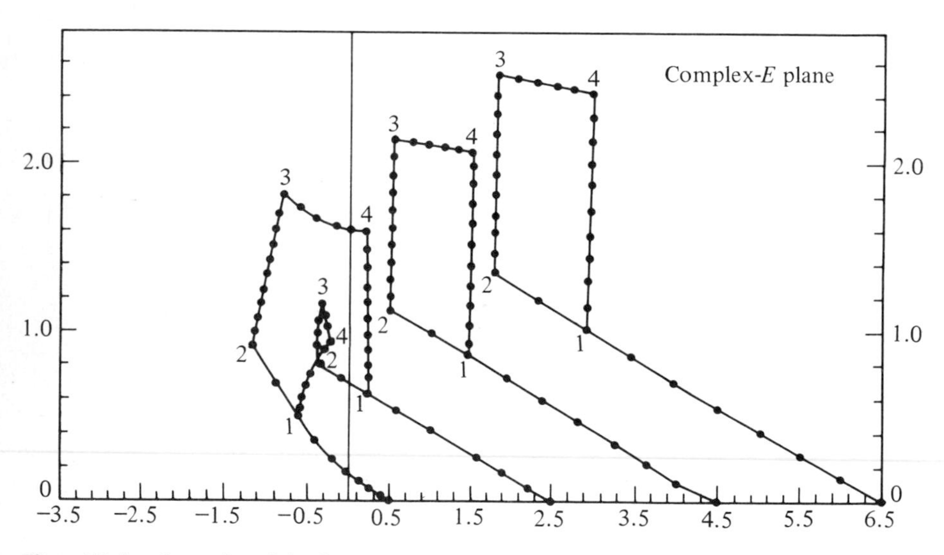

Figure 7.9 Level crossing of the first two even-parity eigenvalues in the complex-E plane. This figure shows the images of the path in the complex-ε plane that encircles the branch point A (see Fig. 7.8). For example, the images of the line segment from the origin to "1" in the ε plane are paths from 0.5, 2.5, 4.5, and 6.5 in the E plane to the four points marked "1". As we go from the origin in the ε plane to "1", "2", "3", "4", "1", and back to the origin, all of the eigenvalues in the E plane return to their original positions except for two, which exchange their identities.

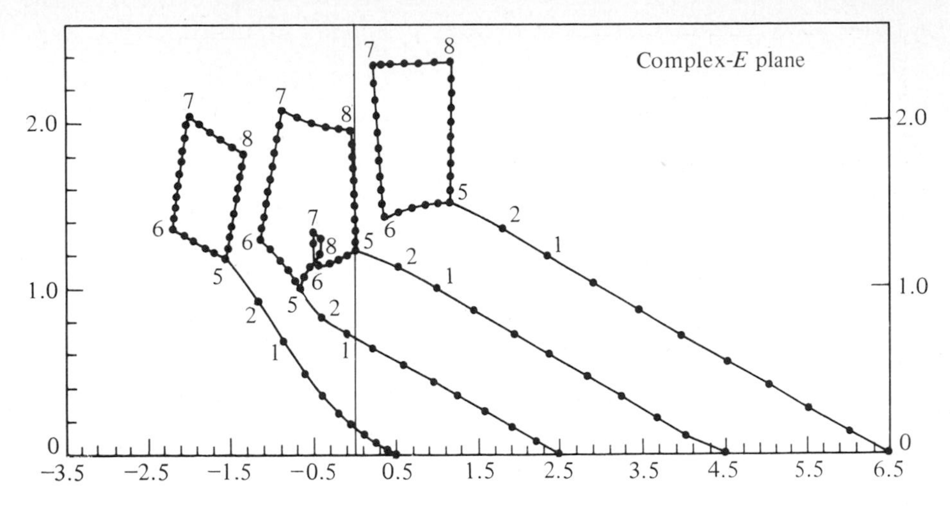

Figure 7.10 Level crossing of the second and third even-parity eigenvalues in the complex-E plane. This figure shows the images of the path in the complex-ε plane that encircles the branch point labeled "B" (see Fig. 7.8).

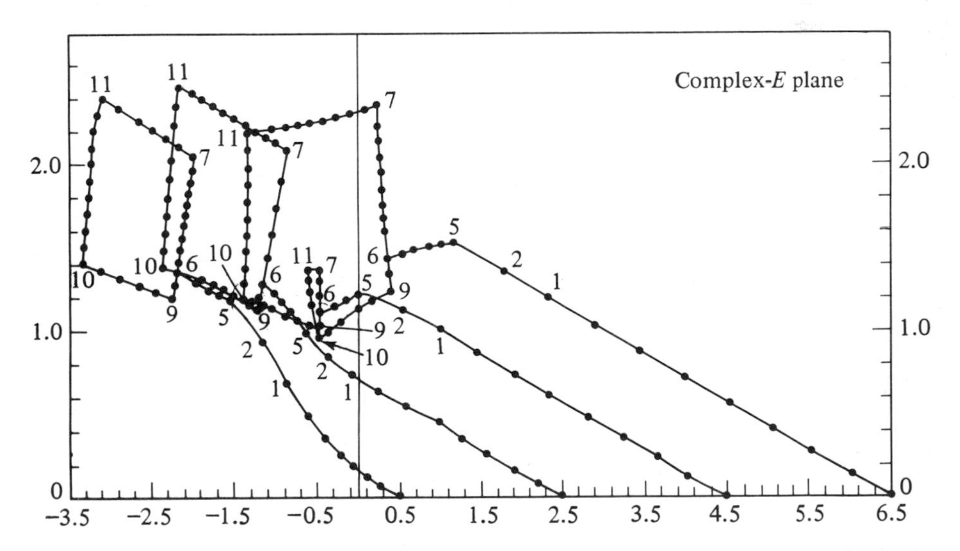

Figure 7.11 Level crossing of the third and fourth even-parity eigenvalues in the complex-E plane. This rather complicated figure shows the images of the path in the complex-ε plane that encircles the branch point labeled "C" (see Fig. 7.8).

Possibility 2 As $N \to \infty$, the branch points all move out to ∞. If this were to happen, it would mean that the eigenvalues of the original differential equation would all be entire functions of ε. Thus, the perturbation problem in question would be regular. We will give an interesting argument, based on the properties of

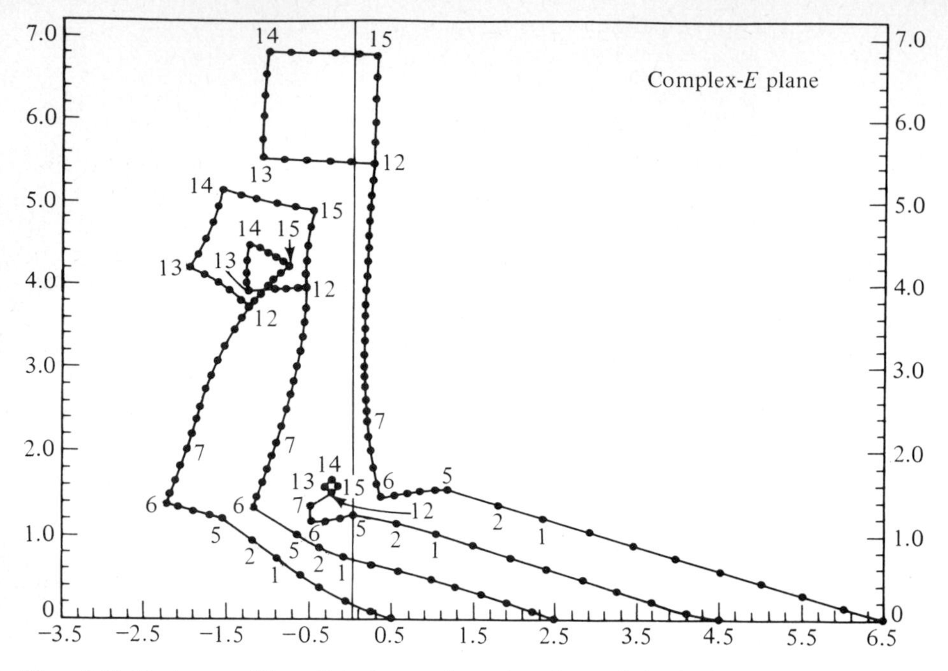

Figure 7.12 The images of the path in the complex-ε plane that encircles the branch point labeled "D" (see Fig. 7.8). The effect of traversing this path is to exchange the first two even-parity eigenvalues in the complex-E plane.

Herglotz functions, which rules out this possibility whenever W is one-signed for all x.

An analytic function is said to be *Herglotz* if $\operatorname{Im} f > 0$ when $\operatorname{Im} z > 0$, $\operatorname{Im} f = 0$ when $\operatorname{Im} z = 0$, and $\operatorname{Im} f < 0$ when $\operatorname{Im} z < 0$. For example, $f(z) = 16 + 7z$ is a Herglotz function. It is a rigorous result of complex variable theory that an entire function [$f(z)$ is *entire* if it is analytic for all $|z| < \infty$] which is Herglotz is linear [$f(z) = a + bz$]. (For the proof see Prob. 7.47.)

Next, we argue that whenever (7.5.3) is a regular perturbation problem, $E(\varepsilon)$ or $-E(\varepsilon)$ is Herglotz. If (7.5.3) is a regular perturbation problem, then for all complex ε, $\varepsilon W(x)$ becomes insignificant compared with $V(x)$ as $|x| \to \infty$. Thus, the asymptotic behavior of $y(x)$ for large $|x|$ is independent of ε. Having established this we simply multiply (7.5.3) by y^* (the complex conjugate of y) and integrate from $-\infty$ to ∞. After one integration by parts (in which the boundary term vanishes) we have

$$\int_{-\infty}^{\infty} y'(x)^* y'(x)\, dx + \int_{-\infty}^{\infty} V(x) y(x) y^*(x)\, dx$$
$$+ \varepsilon \int_{-\infty}^{\infty} W(x) y^*(x) y(x)\, dx = E \int_{-\infty}^{\infty} y^*(x) y(x)\, dx.$$

Taking the imaginary part of this equation gives

$$\text{Im } E = \frac{(\text{Im } \varepsilon) \int_{-\infty}^{\infty} W(x) y^*(x) y(x) \, dx}{\int_{-\infty}^{\infty} y^*(x) y(x) \, dx}.$$

Thus, assuming that W is one-signed, Im E has the same sign as Im ε or $-$Im ε.

The Herglotz property of $E(\varepsilon)$ explains why the branch points, which always occur in complex-conjugate pairs in the approximate theory using truncated matrices, must remain as complex-conjugate pairs in the exact theory. More importantly, this theorem almost excludes the possibility that as $N \to \infty$, the branch points all move out to ∞ leaving no singularities in the complex-ε plane. If $E(\varepsilon)$ were both Herglotz and entire, its perturbation series would have the unlikely form $E(\varepsilon) = a + b\varepsilon$. We would conclude that if the first three terms in the perturbation series for $E(\varepsilon)$ are computed and the coefficient of ε^2 is nonzero, then $E(\varepsilon)$ must almost certainly have singularities (branch points) somewhere in the finite-ε plane. (In Prob. 7.27 it is shown that when $V = x^2/4$ and $W = |x|$ the coefficient of ε^2 is nonzero. This is consistent with the existence of singularities in the ε plane in Fig. 7.8.)

This conclusion does not contradict the results of Example 2 of Sec. 7.3, where there was a nonvanishing ε^2 correction to the eigenvalues but the eigenvalues were still entire functions of ε. (Why?)

Possibility 3 Some of the singularities coalesce in the limit $N \to \infty$ to form more complicated kinds of singularities than square-root branch points. This can happen, but the kinds of singularities one might expect to find in the limit are restricted by the condition that $E(\varepsilon)$ be Herglotz. For example, poles or essential singularities cannot occur because in any neighborhood of an essential singularity or a pole it is possible to find both signs of Im E. However, complicated kinds of branch points including logarithmic branch points might well occur.

Possibility 4 As $N \to \infty$, the square-root branch points form a sequence having a limit point at $\varepsilon = 0$. Thus, in any neighborhood of the origin, however small, one can always find a branch point at which a given eigenvalue crosses with some other eigenvalue. As a result, the perturbation series for any eigenvalue has a zero radius of convergence and we have a singular perturbation theory.

Example 4 *Singular perturbation of a parabolic cylinder equation.* Possibility 4 occurs in the singular perturbation theory for which $V = x^2/4$ and $W = x^4/4$. However, it is easiest to understand how such a remarkable configuration as a converging sequence of branch points comes about by studying the *regular* perturbation problem in which the roles of V and W are reversed:

$$-y''(x) + \frac{1}{4}x^4 y(x) + \frac{\varepsilon}{4}x^2 y(x) - E(\varepsilon)y(x) = 0. \tag{7.5.12}$$

To make the connection between these two theories we perform a simple scaling transformation of the independent variable: $x = \varepsilon^{-1/4}t$. The result is the *singular* perturbation problem in which we are interested:

$$-y''(t) + \tfrac{1}{4}t^2 y(t) + \tfrac{1}{4}\delta t^4 y(t) - F(\delta)y(t) = 0, \tag{7.5.13}$$

where $\delta = \varepsilon^{-3/2}$ and $F(\delta) = \delta^{1/3}E(\delta^{-2/3})$. This scaling transformation is reminiscent of the one used in (7.2.4) to convert a singular perturbation problem to a regular one.

We can now point out four interesting features of the singular problem:

(*a*) Observe first that small values of $|\delta|$ correspond with large values of $|\varepsilon|$. Thus, since there are no level-crossing points of $E(\varepsilon)$ in the regular perturbation theory for values of $|\varepsilon|$ less than some nonzero number, it follows that $F(\delta)$, an eigenvalue of the singular perturbation theory, has no crossing points for δ lying outside some circle in the δ plane.

(*b*) The eigenvalue $F(\delta)$ has a sequence of branch points approaching the origin in the δ plane if and only if $E(\varepsilon)$ has a sequence of branch points approaching ∞ in the ε plane. One can show rigorously that, apart from a possible isolated singularity at ∞, $E(\varepsilon)$ *does* have singularities in the ε plane outside any circle $|\varepsilon| = R$. The proof (by Simon) is based on the Herglotz property of $E(\varepsilon)$ (see Prob. 7.48). Numerical calculations show that these singularities are square-root branch points.

(*c*) If we allow δ to become large, $E(\delta^{-2/3})$ approaches the small-ε (unperturbed) value of $E(\varepsilon)$. Thus, the large-δ behavior of $F(\delta)$ is

$$F(\delta) \sim E(0)\delta^{1/3}. \tag{7.5.14}$$

(*d*) Apart from any singularities of $E(\delta^{-2/3})$, $F(\delta)$ clearly has a *cube-root* singularity at the origin in the δ plane. [This is consistent with (7.5.14).] Thus, we visualize the sequence of square-root branch points in the δ plane as converging on a three-sheeted surface. The approximate locations of the branch points near the origin may be determined using WKB theory. The results are that there are actually four separate sequences of branch points which approach the origin on either side of and asymptotic to the lines $\arg \delta = \pm 3\pi/2$. This is represented schematically in Fig. 7.13.

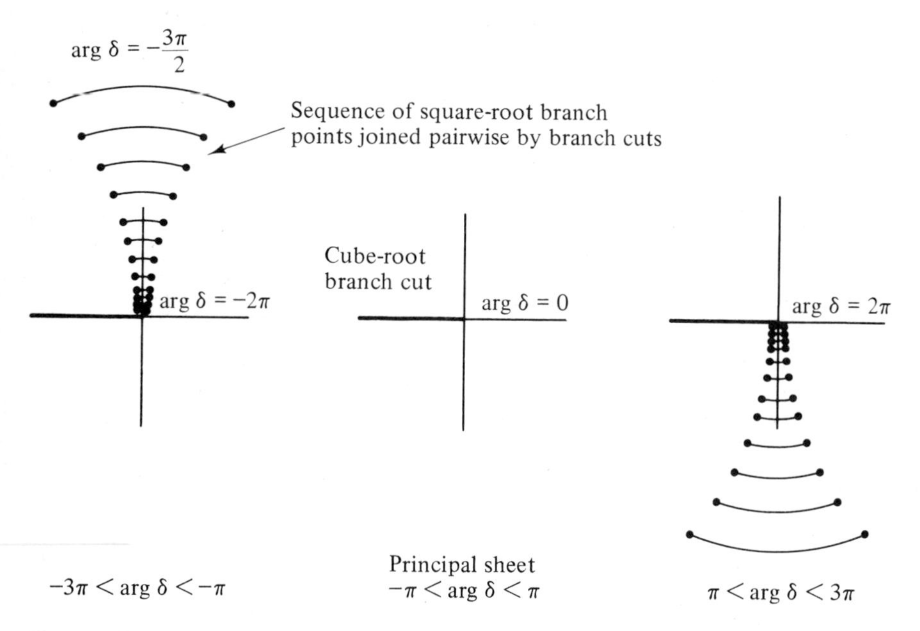

Figure 7.13 Schematic representation of a slice of the Riemann surface on which the function $F(\delta)$ in (7.5.13) is defined. The complete Riemann surface consists of an infinite number of triplets of planes; one such triplet is shown. Each triplet of planes has two pairs of sequences of square-root branch points which approach the origin in the directions $\arg \delta = -3\pi/2$ and $\arg \delta = 3\pi/2$. Each sequence of pairs of branch points is joined pairwise by branch cuts. Level crossing occurs at every branch point.

We conclude this qualitative summary of the analytic properties of perturbative eigenvalue problems with one general observation. We have seen that when we introduce a perturbing parameter ε into an eigenvalue problem having discrete and apparently unrelated eigenvalues, the eigenvalues suddenly become unified in the sense that they are now recognizable as analytic continuations of each other and therefore identifiable as branches of a single (or possibly several) analytic function(s) $E(\varepsilon)$. The discreteness of the eigenvalues for each value of ε is of course maintained, but a new and remarkable picture of what is meant by a discrete eigenvalue spectrum emerges—there is one eigenvalue for each sheet of the Riemann surface. Thus, although the introduction of the parameter ε initially may have seemed to complicate the original eigenvalue problem, it has ended up organizing and clarifying it.

PROBLEMS FOR CHAPTER 7

Sections 7.1 and 7.2

(E) **7.1** Use second-order perturbation theory to find approximations to the roots of the following equations:

(*a*) $x^2 + x + 6\varepsilon = 0$;
(*b*) $x^3 - \varepsilon x - 1 = 0$;
(*c*) $x^3 + \varepsilon x^2 - x = 0$.

(E) **7.2** Obtain a perturbative solution to the roots of $x^2 - 2.0004x + 0.9998 = 0$ and compare with the exact roots (found by solving the quadratic equation). Why does the most straightforward application of perturbation series fail?

Clue: To find the correct perturbation expansion, study the exact solution of the perturbed quadratic.

(I) **7.3** Formulate a perturbation procedure to solve the equation $(x + 1)^n = \varepsilon x$. How rapidly do the roots vary as a function of ε? Why?

(E) **7.4** Why is $x^3 - x^2 + \varepsilon = 0$ a singular perturbation problem? That is, in what sense does the exact solution undergo an abrupt change in character in the limit $\varepsilon \to 0$? Use perturbation theory to approximate the roots for small ε.

(I) **7.5** Analyze in the limit $\varepsilon \to 0$ the roots of the polynomials:

(*a*) $\varepsilon x^3 + x^2 - 2x + 1 = 0$;
(*b*) $\varepsilon x^8 - \varepsilon^2 x^6 + x - 2 = 0$;
(*c*) $\varepsilon^2 x^8 - \varepsilon x^6 + x - 2 = 0$.

(I) **7.6** Prove that the perturbation expansions (7.1.3) of the roots of the polynomial (7.1.2) in Example 1 of Sec. 7.1 converge for $|\varepsilon| < 1$.

(I) **7.7** (*a*) Show that there is one real root of $1 + (x^2 + \varepsilon)^{1/2} = e^x$ when ε is small and positive. Find a leading-order perturbative approximation to this root as $\varepsilon \to 0+$. Also find a second-order approximation.

(*b*) Find leading-order approximations to the roots of $1 + (x^n + \varepsilon)^{1/n} = e^x$ as $\varepsilon \to 0+$, where $n = 1, 2, 3, \ldots$.

(I) **7.8** Obtain a perturbative solution to $\tan \theta = 1/\theta$ when θ is large.

Clue: When θ is large, the graphs of $\tan \theta$ and θ^{-1} intersect near $\theta = N\pi$, where N is a positive integer. Thus, letting $\theta = N\pi + x$, where x is small, we have $\tan x = 1/(N\pi + x)$. Now let $x = \sum_{n=1}^{\infty} a_n \varepsilon^n$, where $\varepsilon = 1/N\pi$, and solve for the first few a_n.

(I) **7.9** If one expresses the solution y of the equation $y = x - \varepsilon \sin(2y)$ as a power series in ε, what are the first three coefficients?

(I) **7.10** Compute all of the coefficients in the perturbation series solution to the initial-value problem $y' = y + \varepsilon xy$ [$y(0) = 1$]. Show that the series converges for all values of ε. Also, compute the perturbation series indirectly by expanding the explicit exact solution in powers of ε.

(I) **7.11** Find the perturbative solution to (7.1.15).

Clue: The solution of $y_n'' = -e^{-x}y_{n-1}$ ($n = 0, \ldots$) has the form $y_n(x) = \sum_{m=0}^{n} (a_{nm} + b_{nm}x)e^{-mx}$. Find recurrence relations for a_{nm}, b_{nm}. Does the perturbation series give a uniform approximation to the solution for $0 \le x \le \infty$?

(I) **7.12** Solve perturbatively $y'' = (\sin x)y$ [$y(0) = 1$, $y'(0) = 1$].

Clue: The solution of $y_n'' = (\sin x)y_{n-1}$ ($n = 0, \ldots$) is of the form

$$y_n(x) = \sum_{m=0}^{n} (a_{nm} + b_{nm}x) \sin mx + \sum_{m=0}^{n} (c_{nm} + d_{nm}x) \cos mx.$$

Find recurrence relations for a_{nm}, b_{nm}, c_{nm}, d_{nm}. Is the resulting perturbation series uniformly valid for $0 \le x \le \infty$? Why?

(I) **7.13** Compare the perturbation methods of Example 2 of Sec. 7.1 with the techniques of local analysis. Suppose $f(x)$ in (7.1.7) is x^3 or $\sin x$. How does the second-order approximation to $y(x)$ in (7.1.12) compare with the result of computing the first few coefficients in (7.1.14)?

(I) **7.14** Convert (7.1.16) into a perturbation problem by introducing ε in the right side. Then obtain a first-order approximation to the answer. How accurate is this approximation when $\varepsilon = 1$? You may estimate the accuracy by computing the next term in the perturbation series or by solving (7.1.16) numerically on a computer.

(I) **7.15** Compare local analysis with perturbation theory for the problem $y'' = \varepsilon(a + bx)y$ [$y(0) = A$, $y'(0) = B$] with general nonzero a, b, A, B and $\varepsilon = 1$. Which contains more information, N terms of the local approximations to $y(x)$ (the Taylor series) or N terms of the perturbation series? How many terms of the Taylor series are equivalent to N terms of the perturbation series?

(D) **7.16** Solve perturbatively the boundary-value problem $y'' + y + \varepsilon f(x)y = 0$ [$y(0) = 1$, $y(\pi) = 0$] with $f(x) = O(1)$ for $0 \le x < \pi$.

Clue: Solve exactly the model problem with $f(x) \equiv 1$ to infer the form of the perturbation series that should be used. Should this problem be regarded as regular or singular?

(I) **7.17** (*a*) Explain the following paradox. We can use perturbation theory as in Example 2 of Sec. 7.1 to solve the initial-value problem $d^n y/dx^n = \varepsilon y$ [$y(0) = y_0$, $y'(0) = y_1, \ldots, y^{(n-1)}(0) = y_{n-1}$] as a power series in ε. On the other hand, solutions to $y^{(n)} = \varepsilon y$ have the form $e^{\omega\varepsilon^{1/n}x}$, where ω is an nth root of unity. Such solutions may be expanded in powers of $\varepsilon^{1/n}$. Which expansion is correct?

(*b*) Carefully contrast this perturbation problem, which is regular, with the polynomial perturbation problem $x^n = \varepsilon f(x)$, where $f(x)$ is a polynomial of degree at most $n - 1$ and $f(0) = 1$. The latter problem is singular.

(I) **7.18** Analyze the limiting behavior as $\varepsilon \to 0$ of the solution to the equation $y' = -y - \varepsilon y^2$ [$y(0) = 1$] by solving the differential equation exactly. Should this problem be classified as regular or singular for x positive? x negative? Why? Repeat the above analysis for the initial condition $y(0) = -1$.

(E) **7.19** Show that the solution to the initial-value problem $y'' + y + \varepsilon y = 0$ [$y(0) = 1$, $y'(0) = 0$] remains bounded for all real x. Obtain a first-order perturbative approximation to $y(x)$ and show that it is unbounded as $x \to \infty$. Conclude that the first-order approximation is valid only for $|x| < O(\varepsilon^{-1})$. To approximate $y(x)$ for $|x| > O(\varepsilon^{-1})$ it is best to use the multiple-scale methods discussed in Chap. 11.

(I) **7.20** (*a*) Find the asymptotic behavior of the solutions $y(x)$ to (7.2.10) as $x \to +\infty$.

(*b*) Solve (7.2.10) exactly in terms of well-known special functions.

(*c*) On the basis of the solution to part (*b*), analyze the behavior of $y(x)$ for $x = O(1)$ as $\varepsilon \to 0+$.

(*d*) How large must x be for the latter asymptotic estimate of $y(x)$ to break down?

(*e*) Show that εx is not a small perturbation in (7.2.10) unless $\varepsilon^{1/2}x \ll 1$ ($\varepsilon \to 0+$). See Prob. 11.13.

(I) **7.21** Use the regular perturbation methods explained in Example 2 of Sec. 7.1 to all orders in powers of ε. Show that when x is of order $\varepsilon^{-1/2}$, the solution to (7.2.10) satisfies $y(x) = \cos(x - \varepsilon x^2/4) + O(\varepsilon^{1/2})$ ($\varepsilon \to 0+$).

Clue: Show that when $x = O(\varepsilon^{-1/2})$

$$y(x) = \cos x \left[1 - \frac{1}{2!}\left(\frac{\varepsilon x^2}{4}\right)^2 + \frac{1}{4!}\left(\frac{\varepsilon x^2}{4}\right)^4 - \cdots\right]$$

$$+ \sin x \left[\left(\frac{\varepsilon x^2}{4}\right) - \frac{1}{3!}\left(\frac{\varepsilon x^2}{4}\right)^3 + \cdots\right] + O(\varepsilon^{1/2}), \qquad \varepsilon \to 0+.$$

(See Sec. 11.1.)

(I) **7.22** (*a*) Apply regular perturbation theory to first order in ε to estimate the effect of the εx^{19} perturbation upon the roots of (7.2.12). Show that the root at $x = k$ changes by the amount

$$(-1)^{k+1}\varepsilon \frac{k^{19}}{(k-1)!\,(20-k)!} + O(\varepsilon^2), \qquad \varepsilon \to 0.$$

(*b*) Show that the unperturbed root at $x = 16$ is most sensitive to ε. Estimate the magnitude of ε necessary to perturb each of the roots by 1 percent.

(*c*) For what value of ε does first-order perturbation theory predict "crossing" of the unperturbed roots $x = 16, 17$?

(*d*) Apply second-order perturbation theory to the roots at $x = 16, 17$. For what values of ε do the roots cross? Compare with the result of part (*c*).

(*e*) What are the radii of convergence of these perturbation series for the roots?

(D) **7.23** (*a*) Develop a perturbation theory for the roots of (7.2.12) that accounts for the fact that the roots may become complex-conjugate pairs when ε is large enough.

Clue: Seek real quadratic factors of $P(x)$ in the form $x^2 + a(\varepsilon)x + b(\varepsilon)$ where a and b are real. Carry out this calculation to first order in ε for the roots whose unperturbed values are 16 and 17. What are your predictions for these roots when $\varepsilon = -10^{-10}$, -2×10^{-10}, and -10^{-9}?

(*b*) Repeat the above calculations to second order in ε.

(*c*) What is the radius of convergence of the Taylor series expansions of $a(\varepsilon)$ and $b(\varepsilon)$ about $\varepsilon = 0$ for the unperturbed roots at 16 and 17 [$a(0) = -33$, $b(0) = 272$]?

Section 7.3

(I) **7.24** Suppose that $y_0(x)$ in (7.3.15) vanishes when $x = x_0$, making this integral expression for $y_n(x)$ formally divergent when $x > x_0 > a$. Show that it is possible to continue this integral representation for $y_n(x)$ through x_0 to give a finite expression for $y_n(x)$ for all x. Demonstrate this effect explicitly for the eigenvalue problem (7.3.3) with $V(x) = x^2/4$, $W(x) = x^4/4$, $a = 0$, $y_0(x) = (x^2 - 1)e^{-x^2/4}$, $E_0 = \frac{5}{2}$, and $n = 1$.

(I) **7.25** (*a*) Show by explicit computation that if $V(x) = x^2/4$ and $W(x) = x$ in (7.3.3), then the perturbation series for $y(x)$ is convergent for all ε and the series for $E(\varepsilon)$ has vanishing terms of order ε^n for $n \geq 3$.

(*b*) Rederive these results by showing that the effect of the perturbation is equivalent to a translation of x by 2ε and the energy level by $-\varepsilon^2$ in the unperturbed problem $V(x) = x^2/4$, $W(x) = 0$.

(I) **7.26** Show that if $V(x) = x^2/4$ and $W(x) = x^4/4$ in (7.3.3), then the perturbation series for the smallest eigenvalue is given by (7.3.16).

(D) **7.27** (*a*) Show that if $V(x) = x^2/4$ and $W(x) = |x|$ in (7.3.3), then the perturbation series for the smallest eigenvalue for positive ε begins $E(\varepsilon) = \frac{1}{2} + (2/\pi)^{1/2}\varepsilon - (\pi - 4 + 2\ln 2)\varepsilon^2/\pi + O(\varepsilon^3)$ $(\varepsilon \to 0+)$.

Clue: The expression for the coefficient of ε^2 is a tricky integral which can be evaluated in closed form by judicious use of polar coordinates.

(*b*) Show that the eigenvalues $E(\varepsilon)$ satisfy the transcendental equation

$$\frac{d}{dx} D_{\varepsilon^2 + E - 1/2}(x + 2\varepsilon)\bigg|_{x=0} = 0$$

when the eigenfunction is even in x, and

$$D_{\varepsilon^2+E-1/2}(2\varepsilon)=0$$

when the eigenfunction is odd.

Section 7.4

(E) **7.28** Use the method of patching to solve $y''(x)-y(x)=\delta(x)\ [y(\pm\infty)=0]$.

(E) **7.29** Use patching to solve $y'=y\ (y>x)$, $y'=x\ (y\le x)$, with $y(0)=0$.

(I) **7.30** (*a*) One would normally think that the upper edge of the left region in Examples 2 and 3 of Sec. 7.4 would be the largest value of x for which εx^2 is still small compared with 1 (see the differential equation). This would suggest that the left region consists of those x for which $x \ll \varepsilon^{-1/2}$ $(\varepsilon\to 0+)$. Actually, the region of validity of the left solution is $x\ll\varepsilon^{-1/3}$ $(\varepsilon\to 0+)$. Show that this is true and explain why.

(*b*) Explain why the largest possible matching region for the differential equation in Example 2 of Sec. 7.4 is $1\ll x\ll\varepsilon^{-1/3}$ $(\varepsilon\to 0+)$.

(D) **7.31** Derive (7.4.10).

(I) **7.32** Show that the largest possible matching region for Example 4 of Sec. 7.4 is $1\ll x\ll\varepsilon^{-1/2}$ $(\varepsilon\to 0+)$.

(D) **7.33** Carry out the analysis of Example 4 of Sec. 7.4 to first order in ε for $v=0$. Determine the size of the largest matching region.

(I) **7.34** Use asymptotic matching to solve the initial-value problem $y''+(v+\frac{1}{2}-\frac{1}{4}x^2-\varepsilon x^4)y=0$ $[y(0)=y'(0)=1]$. Determine the leading behavior of $y(x)$ as $x\to+\infty$.

(D) **7.35** (*a*) Show that $y_1(x)$ in (7.4.16) uniformly satisfies (7.4.14) up to terms of order 1 for all x in region I.

(*b*) More generally, show that (7.4.14) has a solution of the form (7.4.17) which is accurate throughout region I to any order in powers of $1/E$. Calculate $f_2(x)$ and $g_1(x)$.

(D) **7.36** Use matched asymptotic expansions to derive two terms in the asymptotic expansions of the following integrals as $x\to+\infty$:

(*a*) $\displaystyle\int_0^\pi e^{ixt^2}\cos t\,dt$;

(*b*) $\displaystyle\int_0^\pi e^{ixt^4}\cos t\,dt$;

(*c*) $\displaystyle\int_0^1 \exp\left(ix\sqrt{1-t^2}\right)t^{-1/4}(1-t^2)^{-1/2}\,dt$;

(*d*) $\displaystyle\int_0^1 e^{ix\ln(1+t^2)}\cos t\,dt$;

(*e*) $\displaystyle\int_0^1 e^{ix(\sinh t-t)}\sqrt{1-t^2}\,dt$.

For each of these integrals give the order of the error term after two terms in the asymptotic expansion.

(D) **7.37** Use matched asymptotic expansions to show that

$$\int_0^{\pi/4}\frac{\mathrm{Ai}\,(-x\sin t)}{\sqrt{\sin t}}\,dt\sim\sqrt{\frac{2\pi}{x}\frac{3^{-1/3}}{\Gamma(\frac{5}{6})}}-\frac{2^{11/8}\cos(2^{1/4}x/3+\pi/4)}{\sqrt{\pi}\,x^{7/4}}+O(x^{-5/2}),\qquad x\to+\infty.$$

(I) **7.38** Verify the error estimates given in (7.4.33) and (7.4.36).

Clue: Show that

$$\left|e^{-x}-\sum_{n=0}^{N}\frac{(-x)^n}{n!}\right|\le\frac{x^{N+1}}{(N+1)!}$$

for all $x\ge 0$.

(I) **7.39** Examine the $\varepsilon \to 0+$ behavior of the integral $G(\varepsilon) = \int_0^\infty \exp(-t - \varepsilon/\sqrt{t})\, dt$.

(*a*) Show that if we expand the integrand in powers of ε and keep only the first two terms we obtain $G(\varepsilon) \sim 1 - \varepsilon\sqrt{\pi}$ $(\varepsilon \to 0+)$.

(*b*) The method of matched asymptotic expansions serves to verify this behavior and to find the higher-order corrections to it. Divide the region of integration into two overlapping regions, the *inner* region $(t \ll 1)$ and the *outer* region $(t \gg \varepsilon^2)$ $(\varepsilon \to 0+)$, and decompose $G(\varepsilon)$ into inner and outer integrals: $G(\varepsilon) = \int_0^\delta \exp(-t - \varepsilon/\sqrt{t})\, dt + \int_\delta^\infty \exp(-t - \varepsilon/\sqrt{t})\, dt$, where $\delta(\varepsilon)$ satisfies $\varepsilon^2 \ll \delta \ll 1$ $(\varepsilon \to 0+)$. Using (7.4.31) show that

$$\int_0^\delta \exp(-t - \varepsilon/\sqrt{t})\, dt = \int_0^\delta \exp(-\varepsilon/\sqrt{t})\, dt + \mathrm{O}(\delta^2), \qquad \delta \to 0+,$$

$$= 2\varepsilon^2\left[\frac{\delta}{2\varepsilon^2} - \frac{\sqrt{\delta}}{\varepsilon} + \mathrm{O}\left(\ln\frac{\delta}{\varepsilon^2}\right)\right] + \mathrm{O}(\delta^2), \qquad \delta \to 0+,\ \varepsilon^2/\delta \to 0+,$$

and using (6.2.5) show that

$$\int_\delta^\infty \exp(-t - \varepsilon/\sqrt{t})\, dt = \int_\delta^\infty e^{-t}\left(1 - \frac{\varepsilon}{\sqrt{t}}\right) dt + \mathrm{O}\left(\frac{\varepsilon^2}{\delta}\right), \qquad \varepsilon^2/\delta \to 0+,$$

$$= e^{-\delta} - \varepsilon[\sqrt{\pi} - 2\sqrt{\delta} + \mathrm{O}(\delta^{3/2})] + \mathrm{O}(\varepsilon^2/\delta), \qquad \delta \to 0+,\ \varepsilon^2/\delta \to 0+.$$

Combine the inner and outer expansions to show that $G(\varepsilon) = 1 - \varepsilon\sqrt{\pi} + \mathrm{O}(\delta^2) + \mathrm{O}(\varepsilon^2/\delta)$ $(\delta \to 0+,$ $\varepsilon^2/\delta \to 0+)$. Show that the error terms are negligible compared to ε provided that $\varepsilon \ll \delta \ll \sqrt{\varepsilon}$ $(\varepsilon \to 0+)$.

(*c*) Perform a higher-order match to derive the first five terms in the expansion of $G(\varepsilon)$. In particular, show that

$$\int_0^\delta \exp(-t - \varepsilon/\sqrt{t})\, dt = \int_0^\delta \exp(-\varepsilon/\sqrt{t})(1 - t + t^2/2)\, dt + \mathrm{O}(\delta^4)$$

$$= 2\varepsilon^2\left[C_2 - \frac{1}{2}\ln\frac{\varepsilon}{\sqrt{\delta}} + \frac{\delta}{2\varepsilon^2} - \frac{\sqrt{\delta}}{\varepsilon} + \frac{\varepsilon}{6\sqrt{\delta}} + \mathrm{O}\left(\frac{\varepsilon^2}{\delta}\right)\right]$$

$$- 2\varepsilon^4\left[\frac{\delta^2}{4\varepsilon^4} - \frac{\delta^{3/2}}{3\varepsilon^3} + \frac{\delta}{4\varepsilon^2} + \mathrm{O}\left(\frac{\sqrt{\delta}}{\varepsilon}\right)\right]$$

$$+ \varepsilon^6\left[\frac{\delta^3}{6\varepsilon^6} + \mathrm{O}\left(\frac{\delta^{5/2}}{\varepsilon^5}\right)\right] + \mathrm{O}(\delta^4), \qquad \delta \to 0+,\ \varepsilon^2/\delta \to 0+,$$

and that

$$\int_\delta^\infty \exp(-t - \varepsilon/\sqrt{t})\, dt = \int_\delta^\infty e^{-t}\left(1 - \frac{\varepsilon}{\sqrt{t}} + \frac{\varepsilon^2}{2t} - \frac{\varepsilon^3}{6t^{3/2}}\right) dt + \mathrm{O}\left(\frac{\varepsilon^4}{\delta}\right)$$

$$= e^{-\delta} - \varepsilon\left[\sqrt{\pi} - 2\sqrt{\delta} + \frac{2}{3}\delta^{3/2} + \mathrm{O}(\delta^{5/2})\right]$$

$$+ \frac{1}{2}\varepsilon^2[C_0 - \ln\delta + \delta + \mathrm{O}(\delta^2)]$$

$$- \frac{1}{6}\varepsilon^3\left[-2\sqrt{\pi} + \frac{2}{\sqrt{\delta}} + \mathrm{O}(\sqrt{\delta})\right] + \mathrm{O}\left(\frac{\varepsilon^4}{\delta}\right), \qquad \delta \to 0+,\ \varepsilon^2/\delta \to 0+.$$

Combine these inner and outer expansions and substitute $C_0 = -\gamma$ and $C_2 = -\gamma/2 + \frac{3}{4}$ to show that

$$G(\varepsilon) = 1 - \varepsilon\sqrt{\pi} - \varepsilon^2\ln\varepsilon - \frac{3}{2}(\gamma - 1)\varepsilon^2 + \frac{1}{3}\sqrt{\pi}\,\varepsilon^3 + \mathrm{O}\left(\frac{\varepsilon^4}{\delta}\right) + \mathrm{O}(\varepsilon^3\sqrt{\delta})$$

$$+ \mathrm{O}(\varepsilon^2\delta^2) + \mathrm{O}(\varepsilon\delta^{5/2}) + \mathrm{O}(\delta^4), \qquad \delta \to 0+,\ \varepsilon^2/\delta \to 0+.$$

Note that the error terms are negligible compared with ε^3 in the limit $(\varepsilon \to 0+)$ provided that $\varepsilon \ll \delta \ll \varepsilon^{4/5}$ $(\varepsilon \to 0+)$.

(D) **7.40** (*a*) Use matched asymptotic expansions to show that $\int_0^\infty e^{-t-\varepsilon/t^2}\,dt \sim 1 - \sqrt{\pi\varepsilon} - \frac{1}{2}\varepsilon \ln \varepsilon - \frac{3}{2}\varepsilon(\gamma - 1) + \cdots$ $(\varepsilon \to 0+)$.

(*b*) Investigate the asymptotic behavior of the integral $\int_0^\infty e^{-t-\varepsilon/t^\alpha}\,dt$ as $\varepsilon \to 0+$ for fixed α. For which values of α does Euler's constant γ appear in this expansion?

(I) **7.41** Show how the impossible conditions (7.4.45) in Example 6 of Sec. 7.4 arise after one or two integrations by parts.

(D) **7.42** Use asymptotic matching to determine the first two terms in the asymptotic expansion as $x \to +\infty$ of $\int_0^1 e^{ixt}t^{-1/2}(1-t)^{-1/4}\,dt$.

Section 7.5

(D) **7.43** (*a*) Use the recursion property of the Hermite polynomials $\mathrm{He}_n(x)$, $\mathrm{He}_{n+1}(x) = x\,\mathrm{He}_n(x) - n\,\mathrm{He}_{n-1}(x)$, to derive the coefficients of the matrix M in (7.5.7) for $V(x) = x^2/4$ and $W(x) = x^4/4$.

(*b*) Using the result of part (*a*) derive a five-term recursion relation for $D_N(E, \varepsilon) = \det M_N$, where M_N is the $N \times N$ truncation of M.

(TI) **7.44** Show from the specific form of $D_N(E, \varepsilon)$ that $E_N(\varepsilon)$, the solution of (7.5.10), may not have poles or branch points at which $E_N(\varepsilon) = \infty$.

(I) **7.45** (*a*) If $P(x, y)$ and $Q(x, y)$ are polynomials of degree M and N, respectively, in x and y together, show that there are at most MN roots of the simultaneous equations $P(x, y) = 0$, $Q(x, y) = 0$.

Clue: Suppose that $P(x, y) = \sum_{m=0}^{M} p_{M-m}(y)x^m$, $Q(x, y) = \sum_{n=0}^{N} q_{N-n}(y)x^n$ where $p_n(y)$, $q_n(y)$ are polynomials in y of degree at most n, and consider the $(M + N)$ equations for fixed y, $x^pP(x, y) = 0$ $(p = 0, \ldots, M-1)$, $x^qQ(x, y) = 0$ $(q = 0, \ldots, N-1)$, as $(M + N)$ *linear* equations in the $(M + N)$ "unknowns" $1, x, x^2, \ldots, x^{M+N-1}$. The condition for a solution to exist is that the determinant of the coefficients of the system be zero. Show that this determinant, called Sylvester's eliminant, is a polynomial of at most degree MN in y.

(*b*) Give an example of such a pair of polynomial equations having *fewer* than MN roots.

(D) **7.46** Show that if $V(x) = x^2/4$ and $W(x) = |x|$ in (7.3.3), then A_{2m}^{2n} in (7.5.6) is given by

$$A_{2m}^{2n} = \left(\frac{2}{\pi}\right)^{1/2} \frac{(-1)^{n+m+1}[2(n+m)+1](2n)!}{2^{m+n}[4(n-m)^2 - 1]m!\,n!}.$$

(TI) **7.47** Show that if $f(z)$ is entire and Herglotz it is linear.

Clue: An entire function has a Taylor series which converges for all $|z| < \infty$. Letting $z = re^{i\theta}$ in this series gives $\operatorname{Im} f(re^{i\theta}) = \sum_{n=0}^{\infty} a_n r^n \sin(n\theta)$ $(r < \infty)$. Multiply this series by $m \sin\theta \pm \sin(m\theta)$ $(m > 1)$ and integrate from $\theta = 0$ to $\theta = \pi$. Show that $m \sin\theta \pm \sin(m\theta)$ has the same sign as $\sin\theta$, which has the same sign as $\operatorname{Im} z = r\sin\theta$. Therefore, by the Herglotz property of f we have $ma_1 r \pm a_m r^m \geq 0$ $(m > 1)$. But we may take r arbitrarily large. Thus, for this inequality to remain true we must have $a_m = 0$ $(m \geq 0)$ and $f(z) = a_0 + a_1 z$.

(TI) **7.48** Show that in addition to a possible isolated singularity at ∞, $E(\varepsilon)$ in (7.5.12) must have other singularities in the ε plane outside any circle $|\varepsilon| = R$.

Clue: Assume the contrary, namely, that $E(\varepsilon)$ is analytic except for an isolated singularity at ∞. Then

$$E(\varepsilon) = \sum_{n=-\infty}^{\infty} a_n \varepsilon^n \tag{*}$$

converges for all finite $|\varepsilon| > R$. But $E(\varepsilon)$ is Herglotz. By the same argument as in Prob. 7.47, show that $a_n = 0$ for $n \geq 2$. Using $F(\varepsilon) = \delta^{1/3}E(\delta^{-2/3})$ in (7.5.13) rewrite (*) as

$$F(\delta) = \sum_{n=-1}^{\infty} a_{-n}\delta^{(2n+1)/3} \tag{**}$$

where this series *converges* for $|\delta|$ small enough but $\delta \neq 0$. However, we already know that the perturbation series for $F(\delta)$ has the form $F(\delta) \sim \sum_{n=0}^{\infty} F_N \delta^N$ $(\delta \to 0)$. Equating coefficients of powers of δ in both series yields many contradictions. For example, the absence of terms δ^0 and δ^2 in (**) implies that F_0 and F_2 vanish but $F_0 = \frac{1}{2}$, $F_1 = \frac{3}{4}$, $F_2 = -\frac{21}{8}$ for the smallest eigenvalue [see (7.3.16)].

(I) **7.49** Show that if (7.5.13) is generalized to $-y''(t) + \frac{1}{4}t^2 y(t) + \frac{1}{4}\delta t^{2N} y(t) - F(\delta)y(t) = 0$, then the cube-root branch cut on Fig. 7.13 is replaced by an $(N + 1)$th-root branch point.

CHAPTER
EIGHT

SUMMATION OF SERIES

It is an old maxim of mine that when you have excluded the impossible, whatever remains, however improbable, must be the truth.

—Sherlock Holmes, *The Adventure of the Beryl Coronet*
Sir Arthur Conan Doyle

When perturbation methods such as those introduced in Chap. 7 are used to solve a problem, the answer emerges as an infinite series, usually involving powers of the perturbation parameter ε. In practice, only the first few terms of this series can be conveniently calculated because the iteration procedure becomes increasingly cumbersome as the order of perturbation theory increases. If the perturbation series converges rapidly, summing the few calculated terms gives a good approximation to the exact solution. However, it is more common for the series to converge slowly, if it converges at all.

Fortunately, there are powerful methods for recovering an accurate approximation to the exact answer from a few terms of a slowly convergent or divergent perturbation series if these terms can be supplemented by some analytic information about the answer. This chapter is a compilation of just such methods. These methods are crucial because they justify the study of perturbation theory; without them, perturbation theory would be largely ineffective.

If the methods introduced in the present chapter worked in all cases, then surely they could not work very well! The methods discussed here are powerful precisely because they do not always work. However, the difficulty with proposing methods that work only occasionally is that proofs of their validity are difficult and subject to many restrictions. In this chapter we discuss a wide variety of examples, but we prove very little. We confine most of the theory to Sec. 8.6 which treats the convergence theory of Padé approximants to Stieltjes functions.

(E) 8.1 IMPROVEMENT OF CONVERGENCE

In this section we show how to speed up the convergence of slowly converging series. An example of such a series is $\sum_{k=0}^{\infty} (-z)^k$. Although this series converges for all $|z| < 1$, the convergence is very slow as z approaches the unit circle because the limit function $(1 + z)^{-1}$ has a pole at $z = -1$. When z is near $+1$, the convergence rate is affected by the distant pole at $z = -1$. The remainder R_n after n terms of the series is $(-z)^{n+1}/(1 + z)$; R_n goes to zero as $n \to \infty$ for $|z| < 1$. Near $z = +1$ the remainder oscillates rapidly in sign (from odd to even n) and decays slowly. We will call such a term a *transient* because it resembles the transient behavior of a weakly damped harmonic oscillator, which undergoes many oscillations before coming to rest.

Another slowly converging series is the Taylor series for the function $A(z) = 1/[(z + 1)(z + 2)]$. The nth partial sum of this Taylor series is

$$A_n(z) = \sum_{k=0}^{n} (-1)^k \left(1 - \frac{1}{2^{k+1}}\right) z^k$$

$$= \frac{1}{(z + 1)(z + 2)} - \frac{(-z)^{n+1}}{z + 1} + \frac{(-z/2)^{n+1}}{z + 2}. \tag{8.1.1}$$

The poles of $A(z)$ at $z = -1$ and $z = -2$ affect the rate of convergence of $A_n(z)$ to $A(z)$. More than 1,500 terms of this series are necessary to evaluate $A(0.99)$ accurate to six decimal places. Yet the analytic structure of $A(z)$ is so simple that it would be very surprising if the first few terms did not contain enough information to compute $A(z)$ accurately. Indeed, there are several ways to accelerate the convergence of (8.1.1); one way is to perform a Shanks transformation.

The Shanks Transformation

A good way to improve the convergence rate of a sequence of partial sums (or of any sequence for that matter) is to eliminate its most pronounced transient behavior (i.e., to eliminate the term in the remainder which has the slowest decay to zero). Suppose the nth term in the sequence takes the form

$$A_n = A + \alpha q^n \tag{8.1.2}$$

with $|q| < 1$, so that $A_n \to A$ as $n \to \infty$. Here, the term αq^n is the transient. Since any member of this sequence depends on the three parameters A, α, and q, it follows that A can be determined from three terms of the sequence, say A_{n-1}, A_n, A_{n+1}: $A_{n-1} = A + \alpha q^{n-1}$, $A_n = A + \alpha q^n$, $A_{n+1} = A + \alpha q^{n+1}$. Solving this system of equations for A gives

$$A = \frac{A_{n+1} A_{n-1} - A_n^2}{A_{n+1} + A_{n-1} - 2A_n}.$$

(If the denominator vanishes, then $A_n = A$ for all n.)

This formula is exact only if the sequence A_n has just one transient of the form in (8.1.2). Nontrivial sequences may have many transients, some of which oscillate in a very irregular fashion. Nevertheless, if the *most pronounced* transient has the form αq^n, $|q| < 1$, then the nth term in the sequence takes the form $A_n = A(n) + \alpha q^n$, where for large n, $A(n)$ is a more slowly varying function of n than A_n. Let us suppose that $A(n)$ varies sufficiently slowly so that $A(n - 1)$, $A(n)$, and $A(n+1)$ are all approximately equal. Then the above discussion motivates the nonlinear transformation,

$$S(A_n) = \frac{A_{n+1}A_{n-1} - A_n^2}{A_{n+1} + A_{n-1} - 2A_n}, \tag{8.1.3}$$

investigated in depth by Shanks. This transformation creates a new sequence $S(A_n)$ which often converges more rapidly than the old sequence A_n, even if the old sequence has more than one transient (see Prob. 8.1). The sequences $S^2(A_n) = S[S(A_N)]$, $S^3(A_n) = S\{S[S(A_n)]\}$, and so on, may be even more rapidly convergent.

Example 1 *Shanks transformation of a geometric series.* The Shanks transformation applied to the sequence $A_n = \sum_{k=0}^{n} (-z)^k$ yields $S(A_n) = 1/(1+z)$, the exact sum, for every n. Note that $S(A_n) = 1/(1+z)$ even when $|z| \geq 1$ and the original sequence A_n diverges.

Example 2 *Shanks transformation of (8.1.1).* The Shanks transformation may be applied to the sequence (8.1.1). Here $S[A(z)]$ is not exactly $A(z)$ because $A_n(z)$ does not have the form in (8.1.2). Still, $S[A_n(z)]$ converges to $A(z)$ faster than $A_n(z)$ does and $S^2[A_n(z)]$ and $S^3[A_n(z)]$ converge still faster. When $z = 0.99$ the sequences of Table 8.1 are obtained. The first calculable member of the sequence $S^3[A_n(0.99)]$ is accurate to six decimal places, even though it uses information from only the first seven partial sums $A_n(0.99)$: $A_0(0.99)$, $A_1(0.99)$, ..., $A_6(0.99)$. Higher-order iterations of the Shanks transformation work even better.

Why does the Shanks transformation improve the convergence of the sequence $A_n(z)$ in (8.1.1)? Recall that $A_n(z)$ contains *two* transients:

$$A_n(z) = A(z) + \alpha q^n + \beta r^n, \tag{8.1.4}$$

where $q = -z$ and $r = -z/2$. It follows from (8.1.3) that

$$S[A_n(z)] = A(z) + \frac{\alpha\beta(q-r)^2}{\alpha r(q-1)^2 q^n + \beta q(r-1)^2 r^n} q^n r^n.$$

Table 8.1 Partial sums A_n and iterated Shanks transformations $S(A_n)$, $S^2(A_n)$, $S^3(A_n)$ of the Taylor series (8.1.1) for the function $A(z) = [(z+1)(z+2)]^{-1}$ evaluated at $z = 0.99$

The limit of A_n as $n \to \infty$ is $A(0.99) \doteqdot 0.168\ 064\ 4$, but, as is clear from this table, the convergence is slow. The iterated Shanks transformations converge much more rapidly

n	A_n	$S(A_n)$	$S^2(A_n)$	$S^3(A_n)$
1	0.500 000 0	—	—	—
1	−0.242 500 0	0.155 452 4	—	—
2	0.615 087 5	0.173 660 3	0.167 992 6	—
3	−0.294 567 8	0.165 430 9	0.168 079 6	0.168 064 2
4	0.636 009 6	0.169 336 6	0.168 060 9	0.168 064 4
5	−0.300 121 3	0.167 442 1	0.168 065 2	0.168 064 4
6	0.634 003 6	0.168 370 6	0.168 064 2	0.168 064 4
7	−0.294 420 9	0.167 913 3	0.168 064 5	0.168 064 4
10	0.617 836 9	0.168 082 7	0.168 064 4	0.168 064 4
15	−0.259 799 5	0.168 063 9	0.168 064 4	0.168 064 4
20	0.574 962 7	0.168 064 4	0.168 064 4	0.168 064 4

Since $|r| < |q|$, we may expand this expression into a series:

$$S[A_n(z)] = A(z) + \frac{\beta(q-r)^2}{(q-1)^2} \sum_{j=0}^{\infty} \frac{(-\beta)^j(r-1)^{2j}}{\alpha^j(q-1)^{2j}} \left(\frac{r^{j+1}}{q^j}\right)^{n-1}. \tag{8.1.5}$$

Thus, $S[A_n(z)]$ is the sum of an infinite number of geometrically decaying transient terms r^n, $(r^2/q)^n$, $(r^3/q^2)^n, \ldots$, each of which decays faster than the slowest decaying transient q^n of the original sequence. The sequence $S[A_n(z)]$ converges more rapidly than $A_n(z)$ because the transients in (8.1.5) decay more rapidly than those in (8.1.4). Specifically, when the Shanks transformation is applied to the series (8.1.1), the transients z^n and $(z/2)^n$ are transformed into $(z/2)^n$, $(z/4)^n$, $(z/8)^n, \ldots$, which gives a factor of 2^n improvement in the rate of convergence.

Similarly, the transient terms of the iterated Shanks transformation $S^2[A_n(z)]$ of (8.1.5) are $(r^2/q)^n$, $(r^3/q^2)^n, \ldots$. Thus, for the series (8.1.1) the leading behavior of the error in $S^2[A_n(z)]$ is $(z/4)^n$ for sufficiently large n. The convergence of $S^3[A_n(z)]$ is governed by $(z/8)^n$, and so on. The conclusions are consistent with the plots in Fig. 8.1.

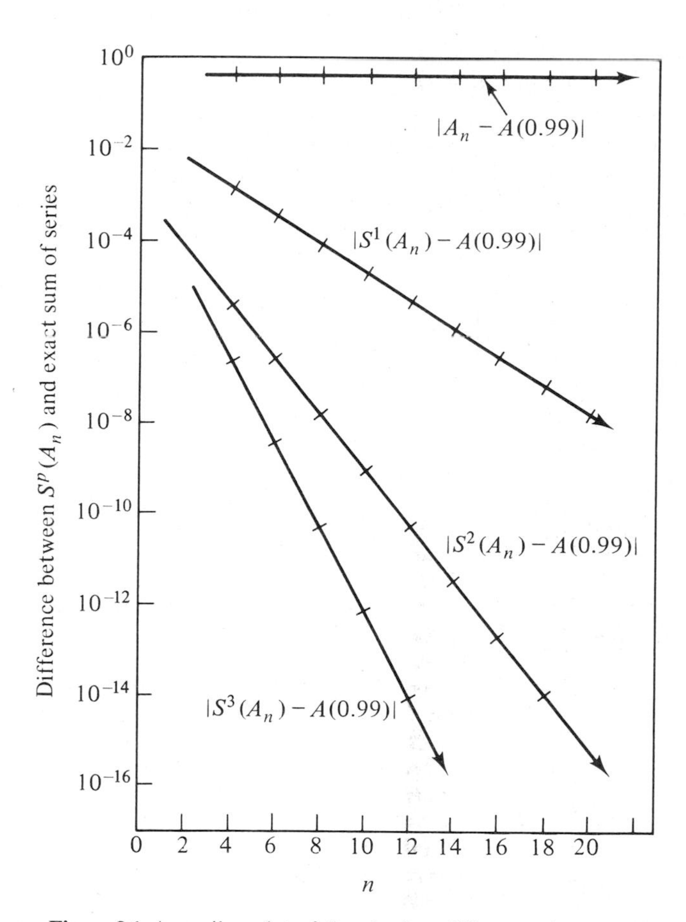

Figure 8.1 A semilog plot of the absolute differences between the iterated Shanks transformations $S^p(A_n)$ of the partial sums, $A_n = \sum_{k=0}^{n}(-z)^k(1 - 2^{-k-1})$ for $z = 0.99$, and the exact sum of the series, $A(z) = 1/[(z+1)(z+2)]$ for $z = 0.99$. As predicted in the text, the leading asymptotic behavior of this difference is proportional to $(0.99/2^p)^n$ for $S^p(A_n)$: the slope of $\ln |S^p(A_n) - A(0.99)|$ versus n is $\ln 0.99 - p \ln 2$.

A word of caution is appropriate here on the practical use of the Shanks transformation, especially when it is iterated. Computers and calculators are limited in the precision to which they can resolve numbers. For example, on an eight-digit machine $1 + 10^{-8} = 1$ because of decimal roundoff. When A_n is close to its limit, the denominator $A_{n+1} + A_{n-1} - 2A_n$ of (8.1.3) is dominated by this roundoff error, so the decimal accuracy of $A_{n+1} + A_{n-1} - 2A_n$ and thus of $S(A_n)$ is likely to be considerably less than that of A_n. This loss of accuracy becomes compounded in the computation of the iterated Shanks transformations because they converge so rapidly. High-precision arithmetic must be used for their computation. Typically, if the computer arithmetic is accurate to k digits, the results of Shanks transformations will be affected by roundoff error when convergence to only $k/2$ digits is achieved. (See Prob. 8.2.)

Example 3 *Shanks transformation of a Taylor series for a function with a branch cut.* The Shanks transformation also works well if the convergence of the original sequence is governed by branch points instead of poles. For example, the Taylor series

$$\ln(1+z) = \sum_{k=1}^{\infty} (-1)^{k+1} z^k/k \tag{8.1.6}$$

converges only for $|z| \leq 1$, $z \neq -1$, because there is a logarithmic branch point at $z = -1$. When $z = 1$, there results the very slowly converging series

$$0.693\,147\,180\,6 \doteq \ln 2 = 1 - \tfrac{1}{2} + \tfrac{1}{3} - \tfrac{1}{4} + \tfrac{1}{5} - \tfrac{1}{6} + \cdots. \tag{8.1.7}$$

About 7,000 terms of (8.1.7) are necessary to compute ln 2 with a relative accuracy of 0.01 percent. (Note that while only about 5,000 terms are necessary for the partial sum of an even number of terms to yield ln 2 accurate to four decimals, roughly 180,000 terms are necessary to get four decimal accuracy from an odd number of terms because $\ln 2 + 1/354{,}000$ differs from ln 2 in the fourth decimal place!)

A Shanks transformation of the partial sums of (8.1.7) yields the sequence given in Table 8.2 which converges to four places after 36 terms [using information from the first 37 partial sums of (8.1.7)]. The iterated Shanks transformations converge even faster as shown in Table 8.2 and Fig. 8.2; the sixth and seventh members of the second iterated Shanks sequence S^2 agree with each other and with ln 2 to four decimals, even though they only use the first eleven partial sums of (8.1.7).

To understand how the Shanks transformation improves the convergence of the series (8.1.6), we use an integral to represent the partial sum of the first n terms:

$$A_n = \sum_{k=1}^{n} (-1)^{k+1} z^k/k = \ln(1+z) - (-1)^n \int_0^z \frac{u^n}{1+u}\, du. \tag{8.1.8}$$

From (8.1.3) it follows that

$$S(A_n) = \ln(1+z) + (-1)^n \int_0^z f_n(u) u^n \, du, \tag{8.1.9}$$

where

$$f_n(u) = \frac{1}{n^2} \frac{1}{(u+1)(u+1+1/n)^2}. \tag{8.1.10}$$

Apart from an overall factor of n^{-2}, $f_n(u)$ depends only weakly on n.

From (8.1.10) we see that $S(A_n)$ converges to its limit $\ln(1+z)$ faster than A_n does, except near the branch point $z = -1$, where the convergence rate is essentially unchanged. For $|z| < 1$, the remainder in (8.1.9) is asymptotically $(-z)^{n+1}/[n(1+z)]^3$, an improvement by the factor $n^2(1+z)^2$ (see Prob. 8.3). When $z = 1$, these estimates still hold; $S(A_n)$ converges faster than A_n does by a factor of roughly $4n^2$ (in agreement with the data in Table 8.2).

Table 8.2 Improvement of convergence of the series $\ln 2 = 1 - \frac{1}{2} + \frac{1}{3} - \frac{1}{4} + \frac{1}{5} - \frac{1}{6} + \cdots \doteqdot 0.693\,147\,18$

A_n, the nth partial sum of this series, converges slowly; many thousands of terms are required to give ln 2 accurate to four decimal places. However, $S(A_n)$, $S^2(A_n)$, and $S^3(A_n)$, the iterated Shanks transformations of A_n, give very accurate approximations to ln 2 after a relatively small number of terms

n	A_n	$S(A_n)$	$S^2(A_n)$	$S^3(A_n)$
1	1.000 000	—	—	—
2	0.500 000 0	0.700 000 0	—	—
3	0.833 333 3	0.690 476 2	0.693 277 3	—
4	0.583 333 3	0.694 444 4	0.693 105 8	0.693 148 9
5	0.783 333 3	0.692 424 2	0.693 163 3	0.693 146 7
6	0.616 666 7	0.693 589 7	0.693 139 9	0.693 147 4
7	0.759 523 8	0.692 857 1	0.693 150 8	0.693 147 1
8	0.634 523 8	0.693 347 3	0.693 145 2	0.693 147 2
15	0.725 371 9	0.693 113 8	0.693 147 3	0.693 147 2
25	0.712 747 5	0.693 139 7	0.693 147 2	0.693 147 2
35	0.707 228 9	0.693 144 4	0.693 147 2	0.693 147 2

Example 4 *Shanks transformation applied to* $\sum (-1)^n/\sqrt{n}$. The convergence of the series

$$1 - \frac{1}{\sqrt{2}} + \frac{1}{\sqrt{3}} - \frac{1}{\sqrt{4}} + \cdots \doteqdot 0.604\,898\,64 \tag{8.1.11}$$

is also improved by the Shanks transformation. Shanks transformations of the partial sums are given in Table 8.3. $S(A_n)$ has already converged to three decimal places when $n = 9$.

This series is one of a general class of series $\sum_{n=0}^{\infty} a_n$ whose terms satisfy the asymptotic condition

$$\frac{a_{n+1}}{a_n} \sim c_0 + \frac{c_1}{n} + \frac{c_2}{n^2} + \cdots, \qquad n \to \infty, \tag{8.1.12}$$

Table 8.3 Improvement of convergence of the series $1 - 1/\sqrt{2} + 1/\sqrt{3} - 1/\sqrt{4} + \cdots \doteqdot 0.604\,898\,64$

A_n, the nth partial sum of this series, converges slowly, but iterated Shanks transformations give impressively accurate approximations to the exact sum for small values of n

n	A_n	$S(A_n)$	$S^2(A_n)$	$S^3(A_n)$
1	1.000 000	—	—	—
2	0.292 893 2	0.610 730 5	—	—
3	0.870 243 5	0.602 294 3	0.605 015 6	—
4	0.370 243 5	0.606 311 5	0.604 858 5	0.604 900 3
5	0.817 457 1	0.604 035 3	0.604 915 5	0.604 898 1
6	0.409 208 8	0.605 470 4	0.604 890 5	0.604 898 8
7	0.787 173 3	0.604 497 5	0.604 903 0	0.604 898 6
15	0.731 849 4	0.604 832 7	0.604 898 9	0.604 898 6
25	0.703 899 1	0.604 879 6	0.604 898 7	0.604 898 6
50	0.534 541 5	0.604 902 1	0.604 898 6	0.604 898 6
90	0.552 340 4	0.604 899 4	0.604 898 6	0.604 898 6

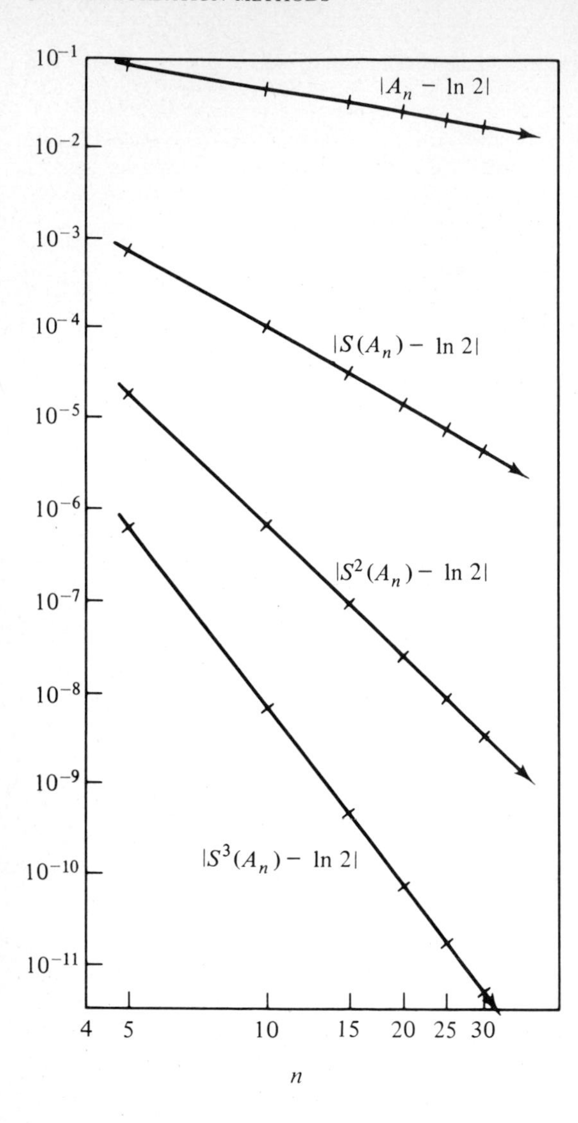

Figure 8.2 A ln-ln plot of the absolute differences between ln 2 and the iterated Shanks transformations $S^p(A_n)$ of the partial sums $A_n = \sum_{k=1}^{n} (-1)^{k+1}/k$. Equation (8.1.10) predicts that successive Shanks transformations should improve convergence by a factor of about $(4n^2)^{-1}$. This prediction is verified by the plotted curves; the slope of the curve $\ln |S^p(A_n) - \ln 2|$ versus $\ln n$ is $-2p - 1$. Points $n = 5, 10, 15, 20, 25, 30$ are labeled and shown on the graph.

where $-1 \le c_0 < 1$. Why does the Shanks transformation improve the convergence of such a series? Letting $A_n = \sum_{k=0}^{n} a_k$ represent the nth partial sum of the series, it follows from (8.1.3) that

$$S(A_n) = A_n + \frac{a_n a_{n+1}}{a_n - a_{n+1}}. \tag{8.1.13}$$

Thus,

$$S(A_n) - S(A_{n-1}) = a_n \frac{a_{n+1}/a_n - a_n/a_{n-1}}{(1 - a_{n+1}/a_n)(1 - a_n/a_{n-1})}.$$

This expression may be approximated for large n using the asymptotic condition (8.1.12). The result is that

$$S(A_n) - S(A_{n-1}) \sim -a_n \frac{kc_k}{(1 - c_0)^2 n^{k+1}}, \qquad n \to \infty,$$

if $c_0 \neq 1$ and c_k is the first nonvanishing coefficient after c_0 in (8.1.12). Thus, the convergence of $\sum_{n=0}^{\infty} a_n$ must be improved by at least a factor of n^2. For the series (8.1.11), $c_0 = -1$ and $c_1 = \frac{1}{2}$, so the convergence is improved by a factor $8n^2$, in agreement with the data in Table 8.3.

Notice that if $c_0 = -1$, we must require that $c_1 > 0$ for the series $\sum a_n$ to converge (see Prob. 8.4).

Richardson Extrapolation

There are many slowly converging series whose convergence is not improved substantially by the Shanks transformation.

Example 5 *Shanks transformation applied to* $\sum n^{-2}$. Performing a Shanks transformation on the series

$$1 + \frac{1}{2^2} + \frac{1}{3^2} + \frac{1}{4^2} + \cdots = \frac{\pi^2}{6} \doteqdot 1.644\,934\,1 \tag{8.1.14}$$

does not significantly improve its convergence. This is because $A_n = \sum_{h=1}^{n} 1/k^2 \sim \pi^2/6 - 1/n$ $(n \to \infty)$, while (8.1.13) implies that $S(A_n) = A_n + 1/(2n+1)$. Thus, the error in A_n is of order $1/n$ while that in $S(A_n)$ is of order $1/2n$, a mere factor of 2 improvement. [Note that the series in (8.1.14) does not satisfy the asymptotic condition in (8.1.12) because $c_0 = 1$.]

Roughly speaking, the slow convergence of $S(A_n)$ may be traced to the existence of a branch point singularity of the function $f(z) = \sum_{n=1}^{\infty} z^n/n^2$ at $z = 1$ (see Prob. 8.5). The branch point of $f(z)$ coincides with the point at which $f(z)$ is to be evaluated. Recall that the Shanks transformation also fails to improve the convergence of the Taylor series for $\ln(1+z)$ when z lies near the branch point at -1 (see Example 3).

When it can be shown that the nth partial sum A_n of a slowly converging series has the form

$$A_n = \sum_{k=0}^{n} a_k \sim Q_0 + Q_1 n^{-1} + Q_2 n^{-2} + Q_3 n^{-3} + \cdots, \qquad n \to \infty, \tag{8.1.15}$$

then it is appropriate to use Richardson extrapolation to calculate $Q_0 = \lim_{n\to\infty} A_n$. To use this method it is necessary to know only the form of (8.1.15), but not the individual coefficients Q_k.

The extrapolation procedure consists of first truncating the series (8.1.15) after the term $Q_N n^{-N}$ and then fitting the partial sums A_n by the resulting Nth-order polynomial in n^{-1}. To fit the polynomial we must solve a set of $(N+1)$ simultaneous equations:

$$\begin{aligned}
A_n &= Q_0 + Q_1 n^{-1} + Q_2 n^{-2} + \cdots + Q_N n^{-N}, \\
A_{n+1} &= Q_0 + Q_1 (n+1)^{-1} + Q_2 (n+1)^{-2} + \cdots + Q_N (n+1)^{-N}, \\
&\vdots \\
A_{n+N} &= Q_0 + Q_1 (n+N)^{-1} + Q_2 (n+N)^{-2} + \cdots + Q_N (n+N)^{-N}.
\end{aligned}$$

There is a nice closed-form expression for Q_0 in terms of $A_n, A_{n+1}, \ldots, A_{n+N}$ (see Prob. 8.6):

$$Q_0 = \sum_{k=0}^{N} \frac{A_{n+k}(n+k)^N(-1)^{k+N}}{k!\,(N-k)!}. \tag{8.1.16}$$

As N and n increase, the difference between this fitted value of Q_0 and $\lim_{n\to\infty} A_n$ decreases.

Example 6 *Richardson extrapolation of* $\sum_{n=1}^{\infty} n^{-2}$. The partial sums of this series may be expressed as an integral (see Prob. 8.7):

$$A_n = \sum_{k=1}^{n} \frac{1}{k^2} = \frac{\pi^2}{6} - \sum_{k=n+1}^{\infty} \frac{1}{k^2} = \frac{\pi^2}{6} - \int_0^{\infty} \frac{te^{-nt}}{e^t - 1}\, dt. \tag{8.1.17}$$

The asymptotic expansion of this integral for large n is obtained using integration by parts (see Prob. 8.7):

$$A_n \sim \frac{\pi^2}{6} - \frac{1}{n} + \frac{1}{2n^2} - \frac{1}{6n^3} + \cdots, \qquad n \to \infty. \tag{8.1.18}$$

To use Richardson extrapolation it is not necessary to know the terms in the series (8.1.18), but only that such a series exists. The numerical results of Richardson extrapolation are given in Table 8.4. Observe that the sum of the series may be calculated to extraordinary accuracy when N and n in (8.1.16) are large.

Example 7 *Richardson extrapolation of the Stirling series.* In Sec. 5.4 we used local analysis to derive the Stirling series from the difference equation for $\Gamma(n) = (n-1)!$. We showed that apart from an overall multiplicative constant C, $(n-1)! \sim Cn^{n-1/2}e^{-n}(1 + 1/12n + 1/288n^2 - \cdots)$ $(n \to \infty)$. The constant C could not be determined using local analysis methods because these methods cannot incorporate the initial condition $\Gamma(1) = 1$. However, in Chap. 6 we used an integral representation to show that $C = \sqrt{2\pi} \doteq 2.506\,628\,275$.

To verify this value for C, we carried out a Richardson extrapolation of the sequence $(n-1)!/n^{n-1/2}e^{-n}\sqrt{2\pi}$ $(n = 1, 2, 3, \ldots)$. The results are given in Table 8.5.

Example 8 *Richardson extrapolation of a recurrence relation whose limiting value is* I_0. In Sec. 5.5 we investigated recurrence relations of the form [see (5.5.5)]

$$A_n = A_{n-1} + \frac{x^2}{n(n+1)} A_{n-2}, \qquad n = 0, 1, 2, \ldots. \tag{8.1.19}$$

When A_n satisfies the initial conditions $A_{-2} = 0$, $A_{-1} = 1$, then one can show (see Prob. 8.8) that

$$\lim_{n\to\infty} A_n = I_0(2x) \tag{8.1.20}$$

and that for large n, A_n has an expansion in powers of n^{-1} (see Sec. 5.5). Thus, a novel way to calculate $I_0(2x)$ is to calculate several terms in the sequence A_n and then to determine A_∞ using Richardson extrapolation. In Table 8.6 we report the results of calculating $I_0(1)$ in this way.

Asymptotic Summation of Series

In rare cases there is an analytical expression for the difference between the Nth partial sum of a slowly converging series and its exact sum. In such cases, if it is possible to find an asymptotic expansion of this difference, then an optimal truncation of the asymptotic series may provide an accurate approximation to the sum of the series. In effect, the idea of this procedure is to accelerate the convergence of a slowly converging series by replacing it with a divergent series!

Table 8.4 Richardson extrapolation of the partial sums $\sum_{k=1}^{n} k^{-2}$ using the formula in (8.1.16)

The exact value of the sum is $\pi^2/6 \doteq 1.644\ 934\ 066\ 848$. Observe that the accuracy of the extrapolations increases with N and n. The first column ($N = 0$) is the original (nonextrapolated) sequence of partial sums. Each time N is increased by 1 there is a dramatic increase in the rate of convergence with n because the error is of order n^{-N-1} ($n \to \infty$, N fixed)

n	$N=0$	$N=1$	$N=2$	$N=3$	$N=4$	$N=5$	$N=6$
1	1.000	1.500 00	1.625 000 00	1.643 518 518 5	1.644 965 277 8	1.644 951 388 9	1.644 935 185 185
5	1.464	1.630 28	1.644 166 67	1.644 922 524 6	1.644 935 811 1	1.644 934 170 8	1.644 934 060 147
10	1.550	1.640 68	1.644 809 05	1.644 933 403 0	1.644 934 195 4	1.644 934 069 8	1.644 934 066 526
15	1.580	1.642 94	1.644 893 41	1.644 933 957 8	1.644 934 089 9	1.644 934 067 1	1.644 934 066 812
20	1.596	1.643 78	1.644 916 08	1.644 934 037 8	1.644 934 073 2	1.644 934 066 9	1.644 934 066 842
25	1.606	1.644 18	1.644 924 58	1.644 934 056 6	1.644 934 069 2	1.644 934 066 9	1.644 934 066 847

Table 8.5 Richardson extrapolation of the sequence $(n-1)!/n^{n-1/2}e^{-n}\sqrt{2\pi}$ using the formula in (8.1.16)

The Stirling formula predicts that the limit of this sequence is 1. Despite the divergence of the Sterling series, the Richardson extrapolants rapidly approach 1 as $n \to \infty$, N fixed (the error is of order n^{-N-1})

n	$N=0$	$N=1$	$N=2$	$N=3$	$N=4$	$N=5$	$N=6$
1	1.084	0.999 977	0.999 680 62	0.999 973 139 4	1.000 001 068 1	1.000 000 416 835 8	1.000 000 023 683 200
5	1.017	0.999 917	0.999 987 31	0.999 999 856 6	1.000 000 043 5	1.000 000 002 238 5	0.999 999 999 813 641
10	1.008	0.999 973	0.999 997 95	0.999 999 997 5	1.000 000 003 1	1.000 000 000 056 3	0.999 999 999 991 573
15	1.006	0.999 987	0.999 999 34	1.000 000 000 5	1.000 000 000 6	1.000 000 000 004 7	0.999 999 999 999 077
20	1.004	0.999 992	0.999 999 71	1.000 000 000 4	1.000 000 000 2	1.000 000 000 000 7	0.999 999 999 999 829
25	1.003	0.999 995	0.999 999 85	1.000 000 000 2	1.000 000 000 1	1.000 000 000 000 1	0.999 999 999 999 957

Table 8.6 Richardson extrapolation of the sequence defined by $A_n = A_{n-1} + A_{n-1}/4n(n+1)$, $A_{-2} = 0$, $A_{-1} = 1$

A_∞, the limiting value of the sequence, is $I_0(1) \doteq 1.266\,065\,877\,752$

n	$N = 0$	$N = 1$	$N = 2$	$N = 3$	$N = 4$	$N = 5$	$N = 6$
1	1.125	1.208 33	1.251 302 08	1.262 760 416 7	1.265 712 709 8	1.265 945 379 7	1.266 075 263 049
5	1.215	1.257 63	1.265 988 73	1.265 948 947 4	1.266 053 971 6	1.266 064 791 1	1.266 065 792 318
10	1.238	1.263 37	1.265 855 81	1.266 051 099 3	1.266 064 879 1	1.266 065 815 7	1.266 065 874 147
15	1.246	1.264 76	1.265 992 33	1.266 062 056 3	1.266 065 684 5	1.266 065 868 6	1.266 065 877 339
20	1.251	1.265 30	1.266 032 01	1.266 064 483 0	1.266 065 821 4	1.266 065 875 6	1.266 065 877 672
25	1.254	1.265 56	1.266 047 57	1.266 065 253 5	1.266 065 856 7	1.266 065 877 1	1.266 065 877 731

Example 9 *Computation of the Riemann zeta function.* The Riemann zeta function $\zeta(s)$ is defined by the infinite sum

$$\zeta(s) = \sum_{n=1}^{\infty} \frac{1}{n^s}, \qquad \text{Re } s > 1. \tag{8.1.21}$$

When s is near 1, this series is *very* slowly converging. About 10^{20} terms are needed to compute $\zeta(1.1)$ accurate to 1 percent!

Fortunately, there is an expression for the difference between the Nth partial sum and $\zeta(s)$ (see Prob. 8.9):

$$\zeta(s) = \sum_{n=1}^{N} \frac{1}{n^s} + \frac{1}{\Gamma(s)} \int_0^{\infty} \frac{u^{s-1}e^{-Nu}}{e^u - 1}\, du. \tag{8.1.22}$$

Watson's lemma enables us to expand the integral in (8.1.22). We need only expand $u/(e^u - 1)$ as a Taylor series in powers of u:

$$\frac{u}{e^u - 1} = 1 - \frac{1}{2}u + \sum_{n=1}^{\infty} \frac{B_{2n}u^{2n}}{n!}, \tag{8.1.23}$$

where B_n ($n = 0, 1, 2, \ldots$) are the Bernoulli numbers (see Prob. 5.18).

Substituting (8.1.23) into (8.1.22) and reversing the order of summation and integration gives

$$\zeta(s) \sim \sum_{n=1}^{N} n^{-s} + \frac{1}{(s-1)N^{s-1}} - \frac{1}{2N^s} + \frac{B_2 s}{N^{s+1}} + \frac{B_4 s(s+1)(s+2)}{2N^{s+3}} + \cdots, \qquad N \to \infty. \tag{8.1.24}$$

When N is large, an optimal truncation of (8.1.24) gives an extremely accurate approximation to $\zeta(s)$. Tables 8.7 and 8.8 summarize our numerical calculations. Note that only 38 terms in (8.1.24) are needed to compute $\zeta(1.1)$ to 26 decimal places!

The method used in Example 9 is a special case of a very general procedure for accelerating the convergence of sums of the form $\sum_{n=1}^{\infty} a_n$, where there is an explicit formula for every term a_n of the series. The procedure involves obtaining an asymptotic approximation to $\sum_{n=N}^{\infty} a_n$ valid as $N \to \infty$; this may be done using the Euler-Maclaurin summation formula (see Sec. 6.7 and Probs. 8.10 and 8.11). In Example 9 the Euler-Maclaurin summation formula (6.7.10) with $f(x) = x^{-s}$ yields (8.1.24) directly. However, this technique is only rarely useful for the summation of perturbation series because it is nearly impossible to obtain an analytic expression for all the terms in such a series. By contrast, the Shanks transformation and Richardson extrapolation only require precise knowledge of a limited number of terms of the series.

(E) 8.2 SUMMATION OF DIVERGENT SERIES

Perturbation methods commonly yield divergent series. A regular perturbation series converges only for those values of $|\varepsilon|$ less than the radius of convergence. A singular perturbation series diverges for all values of $\varepsilon \neq 0$, and even if the series is asymptotic, the value of ε may be too large to obtain much useful information.

Table 8.7 Numerical approximations to the Riemann zeta function $\zeta(s)$ for $s = 1.1$

$\zeta(s)$ is defined as the sum of the series $\sum_{n=1}^{\infty} n^{-s}$. However, since this series is very slowly convergent, we use the asymptotic approximation in (8.1.24) to reduce drastically the number of computations required to obtain accurate results. For each value of N we compute the partial sum listed in the second column of the table. Then we compute the optimal asymptotic approximation to the remainder, which is an asymptotic series in powers of $1/N$. The third column gives the optimal approximation to $\zeta(s)$ as a function of N and the last column gives the total number of terms in the partial sum and optimally truncated series for the remainder. The accuracy increases rapidly with increasing N; the result for $N = 10$ is accurate to all 27 decimal places

N	$\sum_{n=1}^{N} n^{-1.1}$	$\sum_{n=1}^{N} n^{-1.1}$ + [optimal truncation of (8.1.24)]	Number of terms in (8.1.24)
1	1.000	10.581 720 833 333 333 333 333 333 3	5
2	1.467	10.584 451 922 653 952 985 003 529 0	9
3	1.765	10.584 448 469 577 813 110 695 320 3	14
4	1.983	10.584 448 464 943 248 378 747 081 3	18
5	2.153	10.584 448 464 950 822 569 464 200 9	22
6	2.292	10.584 448 464 950 809 804 485 139 7	26
7	2.410	10.584 448 464 950 809 826 424 558 5	30
8	2.512	10.584 448 464 950 809 826 386 333 6	34
9	2.601	10.584 448 464 950 809 826 386 400 9	38
10	2.680	10.584 448 464 950 809 826 386 400 8	43

Table 8.8 Same as in Table 8.7, except that $s = 2$

When $N = 10$, $\zeta(2) = \pi^2/6$ is given correctly to 26 decimal places

N	$\sum_{n=1}^{N} n^{-2}$	$\sum_{n=1}^{N} n^{-2}$ + [optimal truncation of (8.1.24)]	Number of terms in (8.1.24)
1	1.000	1.633 333 333 333 333 333 333 333 3	5
2	1.250	1.644 949 565 408 549 783 549 783 5	9
3	1.361	1.644 934 042 800 375 673 223 020 5	13
4	1.424	1.644 934 066 887 598 150 513 176 8	17
5	1.464	1.644 934 066 848 160 013 474 414 7	21
6	1.491	1.644 934 066 848 226 329 150 352 4	26
7	1.512	1.644 934 066 848 226 436 658 707 6	30
8	1.527	1.644 934 066 848 226 436 472 088 2	34
9	1.540	1.644 934 066 848 226 436 472 415 7	38
10	1.550	1.644 934 066 848 226 436 472 415 2	42

It is discouraging to discover that a perturbation series diverges, especially if the terms in the series have been painstakingly computed. However, in this and the next three sections we show that for many divergent series it is possible to assign a sensible meaning to the "sum" of the series and even to use the first few terms to approximate this sum.

Clearly a naive summation of a divergent series by simply adding up the first N terms is silly because it gives a partial sum which gets further from the actual "sum" of the series as $N \to \infty$. By comparison, the indirect summation methods we shall introduce here again require as input the first N terms in the series, but the output is an *approximant* which converges to the "sum" of the series as $N \to \infty$.

Thus, whenever summation methods apply, they provide the reward for investing one's time in perturbative calculations; even if the perturbation series diverges and whatever the size of ε, the more terms one computes, the closer one can approximate the exact answer.

In the remainder of this chapter, we explain several methods for assigning a finite expression to, and therefore for summing, divergent series. This short section does not introduce practical methods for summing divergent perturbation series; our purpose here is merely to induce the proper frame of mind by showing that there are special kinds of divergent series whose sums can actually be defined.

Euler Summation

If a series $\sum_{n=0}^{\infty} a_n$ is algebraically divergent (the terms blow up like some power of n), then the series

$$f(x) = \sum_{n=0}^{\infty} a_n x^n \tag{8.2.1}$$

converges for all $|x| < 1$. The Euler sum S of $\sum_{n=0}^{\infty} a_n$ is *defined* as $S \equiv \lim_{x \to 1-} f(x)$ whenever the limit exists.

Example 1 *Euler summation.*

(*a*) $\sum_{n=0}^{\infty} (-1)^n$ diverges but $f(x) = \sum_{n=0}^{\infty} (-x)^n = 1/(1+x)$. Thus the Euler sum of $\sum_{n=0}^{\infty} (-1)^n$ is $\frac{1}{2}$. Notice that a Shanks transformation of this series gives the same result.

(*b*) The Euler sum of $1 + 0 - 1 + 0 + 1 + 0 - 1 \cdots$ is

$$\lim_{x \to 1-} \sum_{n=0}^{\infty} (-x^2)^n = \lim_{x \to 1-} 1/(1+x^2) = \tfrac{1}{2}.$$

Here, the Shanks transformation reproduces the original divergent series.

(*c*) The Euler sum of $\sum_{n=1}^{\infty} n$ does not exist because $\lim_{x \to 1-} f(x) = \lim_{x \to 1-} \sum_{n=1}^{\infty} nx^n = \lim_{x \to 1-} x/(1-x)^2$ does not exist.

Borel Summation

If the coefficients a_n in the series $\sum_{n=0}^{\infty} a_n$ grow faster than a power of n [$a_n \sim 2^n$ or $a_n \sim (n!)^{1/2}$, for example], then Euler summation is not applicable because $\sum_{n=0}^{\infty} a_n x^n$ diverges for x near 1. However, this power series may still have meaning as an asymptotic series.

Suppose

$$\phi(x) = \sum_{n=0}^{\infty} \frac{a_n x^n}{n!} \tag{8.2.2}$$

converges for sufficiently small x and that

$$B(x) \equiv \int_0^\infty e^{-t}\phi(xt)\,dt \tag{8.2.3}$$

exists. If we expand the integral $B(x) = \int_0^\infty e^{-t/x}\phi(t)\,dt/x$ for small x by substituting the series (8.2.2) and integrating term by term [this is justified by Watson's lemma (Sec. 6.3)], then

$$B(x) \sim \sum_{n=0}^\infty \frac{a_n}{n!}\int_0^\infty \phi(t)e^{-t/x}t^n\frac{dt}{x} = \sum_{n=0}^\infty a_n x^n, \qquad x \to 0+. \tag{8.2.4}$$

By construction, the series in (8.2.4) is asymptotic to $B(x)$ as $x \to 0+$.

The asymptotic series diverges, but since the function $B(x)$ exists it makes sense to *define* the Borel sum of $\sum_{n=0}^\infty a_n x^n$ as $B(x)$ and in particular to define the sum of $\sum_{n=0}^\infty a_n$ as $B(1)$.

Example 2 *Borel summation.* The series $\sum_{n=0}^\infty (-1)^n n!$ diverges but $\phi(x) = \sum_{n=0}^\infty (-x)^n$ converges for $|x| \le 1$ to $(1+x)^{-1}$. Thus, the Borel sum of $\sum_{n=0}^\infty (-x)^n n!$ is $B(x) = \int_0^\infty [e^{-t}/(1+xt)]\,dt$ and the Borel sum of $\sum_{n=0}^\infty (-1)^n n!$ is $B(1) = \int_0^\infty [e^{-t}/(1+t)]\,dt$.

Note that, step by step, Borel summation is the inverse of Watson's lemma. It is a technique for finding an integral representation for an asymptotic series. Of course, there are many functions $B(x)$ which have as their asymptotic series $\sum_{n=0}^\infty a_n x^n$. We postpone to Sec. 8.6 the question of *which* function $B(x)$ is picked out by Borel summation.

Generalized Borel Summation

Generalized Borel summation is an iterated version of Borel summation. The series $\sum_{n=0}^\infty a_n$ is generalized Borel summable if

$$\phi(x) = \sum_{n=0}^\infty \frac{a_n}{(n!)^k}x^n \tag{8.2.5}$$

converges for sufficiently small x and for some positive integer k. Then, when the multiple integral converges, we *define* the generalized Borel sum of $\sum_{n=0}^\infty a_n x^n$ to be

$$B(x) \equiv \int_0^\infty \cdots \int_0^\infty dt_1\,dt_2 \cdots dt_k\, e^{-t_1-t_2-\cdots-t_k}\phi(xt_1t_2\cdots t_k), \tag{8.2.6}$$

and, in particular, the sum $\sum_{n=0}^\infty a_n$ to be $B(1)$.

Example 3 *Generalized Borel summation.* The generalized Borel sum of $\sum_{n=0}^\infty (-1)^n(n!)^2x^n$ exists for positive x because if we choose $k=2$ in (8.2.5), the sum $\phi(x) = \sum_{n=0}^\infty (-1)^n x^n$ converges for $|x| < 1$ to $(1+x)^{-1}$, and the integral $B(x) = \int_0^\infty ds\, e^{-s}\int_0^\infty dt\, e^{-t}/(1+xst)$ converges.

(I) 8.3 PADÉ SUMMATION

When a power series representation of a function diverges, it indicates the presence of singularities. The divergence of the series reflects the inability of a polynomial to approximate a function adequately near a singularity. The basic idea of summation theory is to rerepresent $f(z)$, the function in question, by a convergent expression. In Euler summation this expression is the limit of a convergent series, while in Borel summation this expression is the limit of a convergent integral.

The difficulty with Euler and Borel summation is that *all* of the terms of the divergent series must be known exactly before the "sum" can be found even approximately. In realistic perturbation problems only a few terms of a perturbation series can be calculated before a state of exhaustion is reached. Therefore, a summation algorithm is needed which requires as input only a finite number of terms of a divergent series. Then, as each new term is computed, it is immediately folded in with the others to give a new and improved estimate of the exact sum of the divergent series. A well-known summation method having this property is called Padé summation.

Padé Approximants

The idea of Padé summation is to replace a power series $\sum a_n z^n$ by a sequence of rational functions (a rational function is a ratio of two polynomials) of the form

$$P_M^N(z) = \frac{\sum_{n=0}^{N} A_n z^n}{\sum_{n=0}^{M} B_n z^n}, \tag{8.3.1}$$

where we choose $B_0 = 1$ without loss of generality. We choose the remaining $(M + N + 1)$ coefficients $A_0, A_1, \ldots, A_N, B_1, B_2, \ldots, B_M$, so that the first $(M + N + 1)$ terms in the Taylor series expansion of $P_M^N(z)$ match the first $(M + N + 1)$ terms of the power series $\sum_{n=0}^{\infty} a_n z^n$. The resulting rational function $P_M^N(z)$ is called a Padé approximant.

We will see that constructing $P_M^N(z)$ is very useful. If $\sum a_n z^n$ is a power series representation of the function $f(z)$, then in many instances $P_M^N(z) \to f(z)$ as N, $M \to \infty$, even if $\sum a_n z^n$ is a divergent series. Usually we consider only the convergence of the Padé sequences $P_0^J, P_1^{1+J}, P_2^{2+J}, P_3^{3+J}, \ldots$ having $N = M + J$ with J fixed and $M \to \infty$. The special sequence $J = 0$ is called the *diagonal* sequence.

Example 1 *Computation of* $P_1^0(z)$. To compute $P_1^0(z)$ we expand this approximant in a Taylor series: $P_1^0(z) = A_0/(1 + B_1 z) = A_0 - A_0 B_1 z + O(z^2)$ $(z \to 0)$. Comparing this series with the first two terms in the power series representation of $f(z) = \sum_{n=0}^{\infty} a_n z^n$ gives two equations: $a_0 = A_0$, $a_1 = -A_0 B_1$. Thus, $P_1^0(z) = a_0/(1 - z a_1/a_0)$.

The full power series representation of a function need not be known to construct a Padé approximant—just the first $M + N + 1$ terms. Since Padé approximants involve only algebraic operations, they are more convenient for computational purposes than Borel summation, which requires one to integrate over an infinite range the analytic continuation of a function defined by a power series. In fact, the general Padé approximant can be expressed in terms of determinants.

The Padé approximant $P_M^N(z)$ is determined by a simple sequence of matrix operations. The coefficients $B_1, \ldots, B_M$ in the denominator may be computed by solving the matrix equation

$$\mathcal{A}\begin{bmatrix} B_1 \\ B_2 \\ \vdots \\ B_M \end{bmatrix} = -\begin{bmatrix} a_{N+1} \\ a_{N+2} \\ \vdots \\ a_{N+M} \end{bmatrix}, \tag{8.3.2}$$

where $\mathcal{A}$ is an $M \times M$ matrix with entries $\mathcal{A}_{ij} = a_{N+i-j}$ $(1 \le i, j \le M)$. Then the coefficients $A_0, A_1, \ldots, A_N$ in the numerator are given by

$$A_n = \sum_{j=0}^{n} a_{n-j} B_j, \qquad 0 \le n \le N, \tag{8.3.3}$$

where $B_j = 0$ for $j > M$. Equations (8.3.2) and (8.3.3) are derived by equating coefficients of $1, z, \ldots, z^{N+M}$ in the equation

$$\sum_{j=0}^{N+M} a_j z^j \sum_{k=0}^{M} B_k z^k - \sum_{n=0}^{N} A_n z^n = O(z^{N+M+1}), \qquad z \to 0, \tag{8.3.4}$$

which is just a restatement of the definition of Padé approximants.

Padé approximants often work quite well, even beyond their proven range of applicability. We postpone a discussion of the convergence theory of Padé approximants until Secs. 8.5 and 8.6. Here we simply illustrate how well Padé summation actually works by numerical examples.

Example 2 *Padé approximants for e^z.* The Taylor series for e^z is $\sum z^n/n!$. This series may be easily transformed into a sequence of Padé approximants. An efficient algorithm for doing this is discussed in Sec. 8.4. For all complex z, both the Taylor series approximants $P_0^{M+N}(z) = \sum_{n=0}^{M+N} z^n/n!$ and the Padé approximants $P_M^N(z)$ approach e^z as $M, N \to \infty$. The rates of convergence are summarized and compared in Tables 8.9 and 8.10.

Table 8.11 shows that the zeros and poles of $P_M^N(z)$ move out to ∞ as M and N increase. It is interesting to note that all the zeros of $P_M^N(z)$ lie in the left half of the complex-z plane and all the poles lie in the right half plane (see Prob. 8.61).

Example 2 shows that Padé approximants of Taylor series may elicit more information from a fixed number of terms than straightforward summation of the series; according to Table 8.9, Padé summation reduces the error after N terms by a factor proportional to 2^N. This error estimate will be derived in Example 5 of Sec. 8.5.

Table 8.9 A comparison of the convergence rates of the Taylor and Padé approximants to e^x at $x = 1$

The actual values of the approximants and their relative errors [relative error = (approximant − exact value)/exact value] are listed. Observe that (*a*) the Taylor approximants are monotone increasing while the Padé approximants have errors with the sign pattern $++--++--\cdots$, which will be explained in Sec. 8.5; and (*b*) each Padé approximant is constructed from and contains the same information as the Taylor approximant listed on the same line, but its relative error is noticeably smaller. It will be shown in Sec. 8.5 that the error in the Padé approximants is smaller than the error in the Taylor approximants by a factor proportional to 2^n as $n \to +\infty$

			Relative errors	
n	Taylor series $\sum_{k=0}^{n} \frac{1}{k!}$	Padé approximant $P_M^N(1)$	Taylor series	Padé approximant
0	1.0	$P_0^0 = 1$		
1	2.0	$P_1^0 = \infty$	−2.64 (−1)	∞
2	2.5	$P_1^1 = 3$	−8.03 (−2)	1.04 (−1)
3	2.666 67	$P_2^1 = 2.666\ 67$	−1.90 (−2)	−1.90 (−2)
4	2.708 33	$P_2^2 = 2.714\ 29$	−3.66 (−3)	−1.47 (−3)
5	2.716 67	$P_3^2 = 2.718\ 75$	−5.94 (−4)	1.72 (−4)
6	2.718 06	$P_3^3 = 2.718\ 31$	−8.32 (−5)	1.03 (−5)
7	2.718 25	$P_4^3 = 2.718\ 28$	−1.02 (−5)	−8.31 (−7)
8	2.718 28	$P_4^4 = 2.718\ 28$	−1.13 (−6)	−4.05 (−8)
9	2.718 28	$P_5^4 = 2.718\ 28$	−1.11 (−7)	2.48 (−9)
10	2.718 28	$P_5^5 = 2.718\ 28$	−1.00 (−8)	1.02 (−10)
11	2.718 28	$P_6^5 = 2.718\ 28$	−8.32 (−10)	−5.02 (−12)
12	2.718 28	$P_6^6 = 2.718\ 28$	−6.36 (−11)	−1.77 (−13)
13	2.718 28	$P_7^6 = 2.718\ 28$	−4.52 (−12)	7.31 (−15)
14	2.718 28	$P_7^7 = 2.718\ 28$	−3.00 (−13)	2.27 (−16)
15	2.718 28	$P_8^7 = 2.718\ 28$	−1.87 (−14)	−8.03 (−18)
16	2.718 28	$P_8^8 = 2.718\ 28$	−1.09 (−15)	−2.22 (−19)
17	2.718 28	$P_9^8 = 2.718\ 28$	−6.06 (−17)	6.89 (−21)
18	2.718 28	$P_9^9 = 2.718\ 28$	−3.18 (−18)	1.71 (−22)
19	2.718 28	$P_{10}^9 = 2.718\ 28$	−1.59 (−19)	−4.74 (−24)
20	2.718 28	$P_{10}^{10} = 2.718\ 28$	−7.54 (−21)	−1.07 (−25)

Example 3 *Padé approximants for* $1/\Gamma(z)$. As we saw in Sec. 5.4, the Taylor series for $1/\Gamma(z)$ converges for all z, but does so at an agonizingly slow rate; 32 terms are required to compute $1/\Gamma(4)$ accurate to about 1 percent. However, converting the Taylor series to a sequence of Padé approximants gives a substantial improvement to the rate of convergence (see Table 8.12).

Despite the nice results for the functions e^z in Example 2 and $1/\Gamma(z)$ in Example 3, there is no compelling reason to use Padé summation because the Taylor series already converge for all z and the improvement of convergence is not astounding. The real power of Padé summation is illustrated by its application to divergent series, as in the next five examples.

Example 4 *Padé approximants for* $z^{-1} \ln(1 + z)$. The Taylor series for $z^{-1} \ln(1 + z)$ converges for $|z| < 1$. However, the Padé approximants converge to $z^{-1} \ln(1 + z)$ for *all* z in the cut plane $|\arg z| < \pi$ (see Table 8.13). Loosely speaking, the reason why the Padé approximants in

Table 8.10 Same as in Table 8.9 except that the approximants are evaluated at $x = 5$

n	Taylor series $\sum_{k=0}^{n} \frac{5^k}{k!}$	Padé approximants $P_M^N(5)$	Relative errors: Taylor series	Relative errors: Padé approximants
7	128.619	$P_4^3 =$ 71.385	−1.33 (−1)	−5.19 (−1)
8	138.307	$P_4^4 =$ 128.619	−6.81 (−2)	−1.33 (−1)
9	143.689	$P_5^4 =$ 158.621	−3.18 (−2)	6.88 (−2)
10	146.381	$P_5^5 =$ 149.697	−1.37 (−2)	8.65 (−3)
11	147.604	$P_6^5 =$ 148.001	−5.45 (−3)	−2.78 (−3)
12	148.114	$P_6^6 =$ 148.362	−2.02 (−3)	−3.43 (−4)
13	148.310	$P_7^6 =$ 148.427	−6.98 (−4)	9.05 (−5)
14	148.380	$P_7^7 =$ 148.415	−2.26 (−4)	1.03 (−5)
15	148.403	$P_8^7 =$ 148.413	−6.90 (−5)	−2.28 (−6)
16	148.410	$P_8^8 =$ 148.413	−1.99 (−5)	−2.41 (−7)
17	148.412	$P_9^8 =$ 148.143	−5.42 (−6)	4.56 (−8)
18	148.413	$P_9^9 =$ 148.413	−1.40 (−6)	4.48 (−9)
19	148.413	$P_{10}^9 =$ 148.413	−3.45 (−7)	−7.43 (−10)
20	148.413	$P_{10}^{10} =$ 148.413	−8.11 (−8)	−6.81 (−11)
21	148.413	$P_{11}^{10} =$ 148.413	−1.82 (−8)	1.00 (−11)
22	148.413	$P_{11}^{11} =$ 148.413	−3.91 (−9)	8.58 (−13)
23	148.413	$P_{12}^{11} =$ 148.413	−8.07 (−10)	−1.13 (−13)
24	148.413	$P_{12}^{12} =$ 148.413	−1.60 (−10)	−9.13 (−15)
25	148.413	$P_{13}^{12} =$ 148.413	−3.05 (−11)	1.09 (−15)
26	148.413	$P_{13}^{13} =$ 148.413	−5.60 (−12)	8.29 (−17)

Table 8.11 Distance to the nearest pole in the complex-z plane of the Padé approximant $P_M^N(z)$ for the function e^z

This table can also be used to find the distance to the nearest zero of $P_M^N(z)$ because the positions of the zeros of $P_M^N(z)$ are identical to the positions of the poles of $P_N^M(-z)$ (see Prob. 8.21). Therefore, the distance to the nearest zero of $P_M^N(z)$ is the same as the distance to the nearest pole of $P_N^M(z)$. Observe that as $N, M \to \infty$ the distances to the nearest pole and the nearest zero increase. The distances to the nearest pole and the nearest zero are a measure of the size of the region in which $P_M^N(z)$ is a good approximation to e^z, which possesses neither zeros nor poles

Degree of denominator (M)	Degree of numerator (N): 0	1	2	3	4	5	6	7	8	9	10
0	—	—	—	—	—	—	—	—	—	—	—
1	1.0	2.0	3.0	4.0	5.0	6.0	7.0	8.0	9.0	10.0	11.0
2	1.4	2.5	3.5	4.5	5.5	6.5	7.5	8.5	9.5	10.5	11.5
3	1.6	2.6	3.6	4.6	5.6	6.7	7.7	8.7	9.7	10.7	11.7
4	1.9	3.0	4.0	5.0	6.0	7.1	8.1	9.1	10.1	11.1	12.1
5	2.2	3.2	4.3	5.3	6.3	7.3	8.3	9.3	10.3	11.3	12.3
6	2.5	3.6	4.6	5.6	6.7	7.7	8.7	9.7	10.7	11.7	12.7
7	2.8	3.8	4.9	5.9	6.9	7.9	8.9	9.9	10.9	12.0	13.0
8	3.1	4.2	5.2	6.2	7.3	8.3	9.3	10.3	11.3	12.3	13.3
9	3.3	4.4	5.5	6.5	7.5	8.6	9.6	10.6	11.6	12.6	13.6
10	3.6	4.7	5.8	6.9	7.9	8.9	9.9	10.9	11.9	12.9	14.0

Table 8.12 Padé summation of the Taylor series (5.4.4) for $1/\Gamma(x)$

The entries in the table were computed as follows. The truncated Taylor series for $1/[x\Gamma(x)]$ was used to construct the Padé approximants $P_N^N(x)$ and $P_{N+1}^N(x)$; the resulting Padé approximants were then inverted and multiplied by x. It is clear from the above table that the entries in the three columns are approaching their correct limits: $\Gamma(1) = 1$, $\Gamma(2) = 1$, and $\Gamma(4) = 6$. Although Padé summation enhances the convergence of the Taylor series, the effect is not dramatic: the Padé sequence requires about half the number of terms that the Taylor series requires to achieve 1 percent accuracy (see Table 5.2). The slow convergence of the Taylor series is not due to the presence of singularities [$1/\Gamma(x)$ is entire], but rather to the enormous disparity of behavior of $1/\Gamma(x)$ for positive and negative x. Padé approximants improve the convergence of series by mimicking the singularities of the function to be represented; since $1/\Gamma(x)$ has no singularities, the benefits of Padé summation are marginal

$1/[xP_N^N(x)]$

N	$x = 1$	$x = 2$	$x = 4$
1	0.787 279 606 6	0.369 614 402 3	0.176 506 593 3
2	0.993 980 121 1	0.983 777 381 1	−1.076 767 507 8
3	1.002 208 342 9	1.148 824 428 2	−0.501 000 407 3
4	0.999 993 179 2	0.996 095 931 1	1.662 143 630 6
5	0.999 992 416 9	0.995 977 024 8	1.648 553 231 8
6	1.000 000 023 8	1.000 087 294 7	9.332 982 540 2
7	0.999 999 998 8	0.999 982 355 2	4.271 068 387 3
8	1.000 000 000 0	0.999 999 507 8	5.891 861 437 4
9	1.000 000 000 0	1.000 000 001 6	6.016 004 972 1
10	1.000 000 000 0	1.000 000 000 1	6.001 556 368 7

$1/[xP_{N+1}^N(x)]$

N	$x = 1$	$x = 2$	$x = 4$
1	0.938 583 793 1	0.620 521 015 8	0.519 926 684 9
2	1.012 480 430 2	4.152 436 398 7	0.085 466 832 0
3	1.000 317 915 9	1.035 343 862 7	−1.821 772 429 8
4	0.999 989 665 5	0.995 106 544 4	1.465 422 290 2
5	0.999 999 639 8	0.999 508 399 5	3.822 815 515 8
6	1.000 000 004 4	1.000 026 327 2	7.035 967 528 8
7	1.000 000 000 0	1.000 001 311 6	6.095 643 764 4
8	1.000 000 000 0	0.999 999 880 7	5.960 840 282 9
9	1.000 000 000 0	1.000 000 001 1	6.002 325 742 5
10	1.000 000 000 0	1.000 000 000 0	6.002 625 162 0

Example 4 can provide a better description of $z^{-1} \ln(1 + z)$ than the Taylor series is that the poles of $P_M^N(z)$, which are the zeros of $1 + \sum_{m=1}^{M} B_m z^m$, have the ability to approximate the effect of the branch-cut singularity of $z^{-1} \ln(1 + z)$. The poles of $P_N^N(z)$ and $P_{N+1}^N(z)$ all lie on the negative real axis (see Table 8.14) and, as $N \to \infty$, the poles become dense and in some sense resemble the branch cut. It can be proved (see Example 4 of Sec. 8.5) that the Padé approximants converge to $z^{-1} \ln(1 + z)$ for all complex z in the cut-z plane.

Example 5 *Padé approximants for* $\sum_{n=0}^{\infty} (-z)^n n!$. The Stieltjes series $\sum_{n=0}^{\infty} (-z)^n n!$ diverges for all $z \neq 0$, but is asymptotic to the Stieltjes function $f(z) = \int_0^\infty e^{-t}\, dt/(1 + zt)$ as $z \to 0$ in the cut plane $|\arg z| < \pi$. When $|z|$ is small, the optimal asymptotic approximation provides a good numerical estimate of $f(z)$ (see Example 6 of Sec. 3.8). However, for *all* z in the cut-z plane one can

Table 8.13 Padé summation of the Taylor series for $z^{-1} \ln(1+z)$ about $z = 0$

In Sec. 8.5, it will be shown that the sequences of Padé approximants $P_N^N(z)$ and $P_{N+1}^N(z)$ converge rapidly, even beyond the circle of convergence of the Taylor series $|z| < 1$. Observe that for real positive x the Padé approximants $P_N^N(x)$ monotonically decrease and $P_{N+1}^N(x)$ monotonically increase with N to the common limit $\ln(1+x)/x$. Thus, for any N, these Padé approximants supply upper and lower bounds on $x^{-1} \ln(1+x)$

	$P_N^N(x)$		
N	$x = 0.5$	$x = 1$	$x = 2$
1	0.812 500 000 0	0.700 000 000 0	0.571 428 571 4
2	0.810 945 273 6	0.693 333 333 3	0.550 724 637 7
3	0.810 930 365 3	0.693 152 454 8	0.549 402 823 0
4	0.810 930 217 7	0.693 147 332 4	0.549 312 879 5
5	0.810 930 216 2	0.693 147 185 0	0.549 306 618 4
6	0.810 930 216 2	0.693 147 180 7	0.549 306 177 9
7	0.810 930 216 2	0.693 147 180 6	0.549 306 146 7
8	0.810 930 216 2	0.693 147 180 6	0.549 306 144 5

	$P_{N+1}^N(x)$		
N	$x = 0.5$	$x = 1$	$x = 2$
1	0.810 810 810 8	0.692 307 692 3	0.545 454 545 5
2	0.810 928 961 7	0.693 121 693 1	0.549 019 607 8
3	0.810 930 203 2	0.693 146 417 4	0.549 285 176 8
4	0.810 930 216 1	0.693 147 157 9	0.549 304 620 9
5	0.810 930 216 2	0.693 147 179 9	0.549 306 034 1
6	0.810 930 216 2	0.693 147 180 5	0.549 306 136 4
7	0.810 930 216 2	0.693 147 180 6	0.549 306 143 8
8	0.810 930 216 2	0.693 147 180 6	0.549 306 144 3

prove (see Sec. 8.6) that the Padé approximants $P_M^{M+J}(z)$ converge to $f(z)$ as $M \to \infty$, J fixed. Thus, Padé summation continues to give useful information from the terms in the asymptotic series after naive summation of the terms has failed.

The numerical convergence of the Padé approximants is examined in Table 8.15. Observe that the diagonal Padé approximants appear to be converging rapidly to the correct numerical values. In contrast to the optimal asymptotic approximation, Padé approximation gives arbitrarily accurate results as $M \to \infty$, $N - M = J$ fixed. Observe also that when z is real and positive, the convergence of $P_M^{M+J}(z)$ is monotone with M. For this series and many other Stieltjes series (see Sec. 8.6) the Padé approximants have the remarkable property that for all $J \geq -1$, as $M \to \infty$ the sequence $P_M^{M+J}(z)$ is monotone decreasing when J is even (or zero) and monotone increasing when J is odd. Thus, the Padé approximants supply their own error bounds. Asymptotic error estimates are derived from a continued function representation of $f(z)$ in Example 6 of Sec. 8.5.

Example 6 *Padé approximants for the Stirling series.* By transforming the Stirling series for $\Gamma(z)$ (see Sec. 5.4) into a sequence of Padé approximants, the applicability of the Stirling series is increased. It may now be used to compute $\Gamma(z)$ to greater accuracy than can be obtained by the

Table 8.14 Residues and poles of the Padé sequence P_N^N, P_{N+1}^N for the function $\ln(1+z)/z$

For each Padé approximant, the locations of the poles are listed; the value of the residue of each pole is given in parentheses underneath its location. For example, the residue of the pole of $P_1^1(z)$ at $z = -1.50$ is $+1.13$. Observe that all the poles of all the Padé approximants lie on the negative real axis. The numbers in this table exhibit remarkable regularities (see Prob. 8.28). For example, the sum of the locations of the poles of P_N^N is $-N(N+2)/2$, while the sum of the locations of the poles of P_{N+1}^N is $-(N+1)(N+2)$

P_1^1	−1.50						
	(1.13)						
P_2^1	−1.27	−4.73					
	(0.63)	(2.37)					
P_2^2	−1.18	−2.82					
	(0.45)	(1.44)					
P_3^2	−1.13	−2.00	−8.87				
	(0.31)	(0.89)	(2.46)				
P_3^3	−1.10	−1.69	−4.71				
	(0.24)	(0.66)	(1.55)				
P_4^3	−1.07	−1.49	−3.03	−14.40			
	(0.19)	(0.49)	(0.99)	(2.51)			
P_4^4	−1.06	−1.38	−2.40	−7.16			
	(0.15)	(0.39)	(0.75)	(1.60)			
P_5^4	−1.05	−1.30	−2.00	−4.33	−21.32		
	(0.12)	(0.31)	(0.57)	(1.04)	(2.53)		
P_5^5	−1.04	−1.25	−1.78	−3.28	10.15		
	(0.10)	(0.26)	(0.46)	(0.80)	(1.62)		
P_6^5	−1.03	−1.20	−1.61	−2.63	−5.90	−29.62	
	(0.09)	(0.22)	(0.38)	(0.61)	(1.06)	(2.54)	
P_6^6	−1.03	−1.17	−1.51	−2.27	−4.33	−13.69	
	(0.08)	(0.19)	(0.32)	(0.51)	(0.83)	(1.64)	
P_7^6	−1.03	−1.15	−1.42	−2.00	−3.37	−7.74	−39.30
	(0.07)	(0.16)	(0.27)	(0.42)	(0.64)	(1.08)	(2.54)
P_7^7	−1.02	−1.13	−1.36	−1.83	−2.84	−5.55	−17.77
	(0.06)	(0.14)	(0.24)	(0.36)	(0.53)	(0.84)	(1.65)

optimal asymptotic approximation. The results of some numerical calculations are reported in Table 8.16.

Example 7 *Padé approximants for $D_\nu(x)$.* Padé summation of the asymptotic series for $D_\nu(x)$ for large positive x (see Example 4 of Sec. 3.5) is a simple and extremely efficient means of computing $D_\nu(x)$ for $x \geq 1$ (see Table 8.17).

Padé Summation and the Generalized Shanks Transformation

There is an interesting connection between Padé summation and what we shall refer to as the *generalized* Shanks transformation. To introduce the notion of the generalized Shanks transformation, we consider the problem of improving the

Table 8.15 Padé approximants $P_N^N(x)$ and $P_{N+1}^N(x)$ for the Stieltjes series $\sum_{n=0}^{\infty} (-x)^n n!$ evaluated at $x = 1$ and $x = 10$

In contrast to these good results, the optimal asymptotic approximation to the Stieltjes series at $x = 1$ and $x = 10$ is worthless (the optimal truncation contains no terms of the series). On the other hand, $P_{14}^{14}(1)$ and $P_{15}^{14}(1)$ agree with the "sum" of the series, $\int_0^\infty e^{-t}/(1 + t)\,dt$, to better than five decimal places, while $P_{50}^{50}(1)$ and $P_{51}^{50}(1)$ agree to ten decimal places

	$x = 1$		$x = 10$	
N	$P_N^N(1)$	$P_{N+1}^N(1)$	$P_N^N(10)$	$P_{N+1}^N(10)$
0	1.0	0.5	1.0	0.090 91
1	0.666 67	0.571 43	0.523 81	0.128 63
2	0.615 38	0.588 24	0.379 73	0.149 66
3	0.602 74	0.593 30	0.314 24	0.162 95
4	0.598 80	0.595 08	0.278 47	0.171 96
5	0.597 38	0.595 78	0.256 73	0.178 36
6	0.596 82	0.596 08	0.242 56	0.183 06
7	0.596 57	0.596 21	0.232 84	0.186 60
8	0.596 46	0.596 28	0.225 93	0.189 32
9	0.596 41	0.596 31	0.220 86	0.191 45
10	0.596 38	0.596 33	0.217 06	0.193 13
50	0.596 35	0.596 35	0.201 56	0.201 39
∞	0.596 35	0.596 35	0.201 46	0.201 46

convergence of the Taylor series

$$f(z) = a_0 + a_2 z^2 + a_4 z^4 + \cdots. \tag{8.3.5}$$

The ordinary Shanks transformation fails to improve the convergence of this series because a Shanks transformation of any partial sum is identical to that partial sum. If $f_n(z)$ is a partial sum, it follows from (8.1.3) that $S[f_n(z)] = f_n(z)$ for all z! Thus, there is no way that the transformation S can be used directly to accelerate the convergence of $f_n(z)$.

The origin of this difficulty is that $f(z) = g(z^2)$, where $g(z) = \sum a_{2n} z^n$. Thus, the singularities of $f(z)$ must appear in pairs. While the transformation S in (8.1.3) was developed to eliminate the effect of one singularity in $f(z)$, it is powerless against the matched pairs of singularities of the series (8.3.5).

To improve the convergence of the series (8.3.5), we seek a generalization of the nonlinear transformation S which will eliminate two or more transient behaviors in a sequence A_n.

If A_n depends on k transient terms,

$$A_n = A + \sum_{p=1}^{k} \alpha_p q_p^n, \tag{8.3.6}$$

then A can be determined by $(2k + 1)$ members of the sequence, say A_{n-k}, $A_{n-k+1}, \ldots, A_{n+k}$. The required transformation $A = S_k(A_n)$, a kth-order Shanks

Table 8.16 Padé approximants $P_N^N(x)$ to the Stirling series for $\Gamma(x)/[(x/e)^x\sqrt{2\pi/x}] \sim 1 + 1/12x + 1/288x^2 - \cdots\ (x \to \infty)$

Notice that the sign pattern of the error in $P_N^N(x)$ is $+ - + - \cdots$. Therefore, more rapid convergence to $\Gamma(x)$ can be immediately achieved by averaging successive pairs of terms in the table. For comparison purposes, we have also listed in the table values of the optimal asymptotic approximation $\Gamma_{\text{opt}}(x)$ and the exact value of $\Gamma(x)$

	$(x/e)^x\sqrt{\frac{2\pi}{x}}\,P_N^N(1/x)$		
N	$x = 0.2$	$x = 0.5$	$x = 1.0$
0	3.326 00	1.520 35	0.922 137 01
1	5.076 52	1.796 77	1.002 322 84
2	4.171 29	1.761 14	0.999 582 32
3	4.850 31	1.777 68	1.000 113 00
4	4.329 20	1.768 90	0.999 954 49
5	4.774 21	1.774 64	1.000 020 05
6	4.397 08	1.770 75	0.999 989 13
7	4.735 03	1.773 64	1.000 005 97
8	4.435 34	1.771 46	0.999 996 23
9	4.710 81	1.773 20	1.000 002 35
10	4.460 10	1.771 80	0.999 998 37
11	4.694 19	1.772 97	1.000 001 10
12	4.477 53	1.771 99	0.999 999 18
13	4.682 03	1.772 83	1.000 000 58
14	4.490 52	1.772 11	0.999 999 55
15	4.672 69	1.772 74	1.000 000 34
$\Gamma_{\text{opt}}(x)$	4.711 83	1.762 24	0.999 499 47
$\Gamma(x)$	4.590 84	1.772 45	1.0

transform, can be expressed in terms of determinants (see Prob. 8.23) as

$$S_k(A_n) = \frac{\begin{vmatrix} A_{n-k} & \cdots & A_{n-1} & A_n \\ \Delta A_{n-k} & \cdots & \Delta A_{n-1} & \Delta A_n \\ \Delta A_{n-k+1} & \cdots & \Delta A_n & \Delta A_{n+1} \\ \cdots & \cdots & \cdots & \cdots \\ \Delta A_{n-1} & \cdots & \Delta A_{n+k-2} & \Delta A_{n+k-1} \end{vmatrix}}{\begin{vmatrix} 1 & \cdots & 1 & 1 \\ \Delta A_{n-k} & \cdots & \Delta A_{n-1} & \Delta A_n \\ \Delta A_{n-k+1} & \cdots & \Delta A_n & \Delta A_{n+1} \\ \cdots & \cdots & \cdots & \cdots \\ \Delta A_{n-1} & \cdots & \Delta A_{n+k-2} & \Delta A_{n+k-1} \end{vmatrix}}, \tag{8.3.7}$$

where $\Delta A_p = A_{p+1} - A_p$. Notice that $S_1(A_n) = S(A_n)$, as given by (8.1.3).

When S_k is applied to a sequence of partial sums whose convergence is governed by two transients [(8.1.1) is one such sequence; the two transients origi-

Table 8.17 Diagonal sequence of Padé approximants for the functions $D_\nu(x)$, $\nu = 3.5$ and $\nu = 11.5$, at $x = 1$ and $x = 2$

These Padé approximants were computed from the asymptotic expansion of $D_\nu(x)$ [see (3.5.13) and (3.5.14)] as follows. The expansion for $w(x)$ in (3.5.14) was converted to a rational function with argument x^2. The approximation to $D_\nu(x)$ was obtained by multiplying the Padé approximants by $x^\nu e^{-x^2/4}$. Observe that the Padé sequences P_N^N and P_{N+1}^N converge monotonically provided that N is large enough. The comparison between the Padé approximations in this table and the optimal asymptotic approximations in Table 3.2 is impressive. When $x = 2$, the optimal asymptotic approximations to $D_{3.5}(2)$ and $D_{11.5}(2)$ are in error by 0.3 and 0.005 percent, respectively, while the Padé approximants P_{11}^{11} are in error by about 1 part in 10^8

$\nu = 3.5$

	$x = 1$		$x = 2$	
N	P_N^N	P_{N+1}^N	P_N^N	P_{N+1}^N
4	−2.031 80	−2.039 34	−0.182 255 623	−0.182 267 983
5	−2.035 03	−2.038 24	−0.182 262 496	−0.182 265 165
6	−2.036 03	−2.037 63	−0.182 263 761	−0.182 264 459
7	−2.036 43	−2.037 28	−0.182 264 054	−0.182 264 262
8	−2.036 61	−2.037 09	−0.182 264 134	−0.182 264 201
9	−2.036 69	−2.036 97	−0.182 264 158	−0.182 264 182
10	−2.036 74	−2.036 91	−0.182 264 166	−0.182 264 175
11	−2.036 77	−2.036 87	−0.182 264 169	−0.182 264 172
Exact	−2.036 81	−2.036 81	−0.182 264 170	−0.182 264 170

$\nu = 11.5$

	$x = 1$		$x = 2$	
N	P_N^N	P_{N+1}^N	P_N^N	P_{N+1}^N
4	1244.86	865.52	1438.851 67	1366.535 33
5	−2494.96	−2926.67	1332.749 03	1332.605 32
6	−2559.00	−2623.97	1332.685 92	1332.674 88
7	−2568.73	−2585.98	1332.681 48	1332.679 94
8	−2571.19	−2577.12	1332.680 91	1332.680 61
9	−2571.99	−2574.38	1332.680 80	1332.680 73
10	−2572.30	−2573.37	1332.680 78	1332.680 76
11	−2572.43	−2572.95	1332.680 77	1332.680 77
Exact	−2572.57	−2572.57	1332.680 77	1332.680 77

nate from the two poles of $f(z)$], the result is exact: $S_k[f_n(z)] = f(z)$, for all $k \geq 2$. If the function $f(z)$ has P simple poles, then its partial sums have P transient terms (see Prob. 8.24) and S_k applied to the partial sums is exact for $k \geq P$.

These higher-order Shanks transformations S_k are closely related to Padé approximants. Indeed, this treatment could be regarded as an alternative derivation of the Padé approximants. It may be shown (see Prob. 8.25) that if $k \leq n$, then $S_k(A_n)$ is identical to the Padé approximant P_k^n of the series $A_1 + \sum_{j=1}^{\infty} (A_{j+1} - A_j)z^j$ evaluated at $z = 1$.

Generalized Padé Summation

The Padé methods that we have introduced here could be called "one-point" Padé methods because the approximants are constructed by comparing them with a power series about a particular point. However, the function in question may have been investigated in the vicinity of two or more points. For example, its large ε as well as its small ε dependences may have been determined perturbatively. One may wish to incorporate information from all these expansions in a single sequence of Padé approximants. The numerical results are sometimes impressive.

Suppose $f(z)$ has the asymptotic expansions

$$f(z) \sim \sum_{n=0}^{\infty} a_n(z - z_0)^n, \qquad z \to z_0, \tag{8.3.8a}$$

$$f(z) \sim \sum_{n=0}^{\infty} b_n(z - z_1)^n, \qquad z \to z_1, \tag{8.3.8b}$$

in the neighborhoods of the distinct points z_0 and z_1, respectively. A two-point Padé approximant to $f(z)$ is a rational function $F(z) = R_N(z)/S_M(z)$ where $S_M(0) = 1$. $R_N(z)$ and $S_M(z)$ are polynomials of degrees N and M, respectively, whose $(N + M + 1)$ arbitrary coefficients are chosen to make the first J terms $(0 \le J \le N + M + 1)$ of the Taylor series expansion of $F(z)$ about z_0 agree with (8.3.8a) and the first K terms of the Taylor series expansion of $F(z)$ about z_1 agree with (8.3.8b), where $J + K - N + M + 1$. The formulation of the general equations for the coefficients of the polynomials $R_N(z)$ and $S_M(z)$, as well as the development of efficient numerical techniques for their solution, is left to Probs. 8.26 and 8.37. Here we illustrate the method by one relatively simple example.

Example 8 *Two-point Padé approximation to an integral.* The function

$$f(z) = \frac{1}{2\sqrt{z}} e^{-z} \int_0^z \frac{e^t}{\sqrt{t}}\, dt \tag{8.3.9}$$

is the unique solution to the differential equation $2zf'(z) = -(1 + 2z)f(z) + 1$ for which $f'(0)$ is finite. This solution has power series expansions about both $z = 0$ and $z = \infty$. The series expansion about $z = 0$ is a Taylor series with an infinite radius of convergence:

$$f(z) = \sum_{n=0}^{\infty} a_n z^n, \qquad a_n = \frac{(-4)^n n!}{(2n + 1)!}. \tag{8.3.10}$$

The series expansion about $z = \infty$ is a divergent asymptotic series (see Prob. 8.27):

$$f(z) \sim \sum_{n=0}^{\infty} b_n z^{-n}, \qquad z \to +\infty,$$
$$b_n = 2(2n - 2)!\, 4^{-n}/(n - 1)!, \qquad n \ge 1, \tag{8.3.11}$$
$$b_0 = 0.$$

Table 8.18 Comparison of one-point and two-point Padé approximants for the function $f(x)$ in (8.3.9) at $x = 1$, 16, and 256

The values of the diagonal Padé sequences P_N^N are given for the two-point approximant about $x = 0$ and $x = \infty$ (see text), the one-point approximant about $x = 0$, and x^{-1} times the one-point approximant to $xf(x)$ about $x = \infty$. Observe that while the one-point approximants are very accurate in the neighborhood of the expansion point, the two-point Padé gives a better uniform approximation to $f(x)$

$x = 1$

N	Two-point Padé P_N^N	Padé P_N^N about 0	Padé P_N^N about ∞
1	0.600 000 000 0	0.523 809 523 8	1.000
2	0.548 387 096 8	0.538 245 614 0	0.600
3	0.539 748 954 0	0.538 078 546 4	0.391
4	0.537 958 185 8	0.538 079 510 2	0.087
5	0.538 045 407 2	0.538 079 506 9	−1.436
6	0.538 069 836 8	0.538 079 506 9	1.783
7	0.538 078 314 9	0.538 079 506 9	0.973
8	0.538 079 573 5	0.538 079 506 9	0.745

Exact value of $f(1) = 0.538\ 079\ 506\ 9$

$x = 16$

N	Two-point Padé P_N^N	Padé P_N^N about 0	Padé P_N^N about ∞
1	0.085 71	−0.414 41	0.032 258 064 5
2	0.038 18	0.250 90	0.032 335 329 3
3	0.033 29	−0.053 16	0.032 336 855 5
4	0.027 52	0.060 06	0.032 336 970 4
5	0.033 29	0.024 90	0.032 337 001 2
6	0.032 49	0.033 99	0.032 337 070 2
7	0.032 37	0.032 03	0.032 336 986 0
8	0.030 69	0.032 39	0.032 337 041 2
9	0.032 40	0.032 33	0.032 336 988 0
10	0.032 35	0.032 34	0.032 337 087 8

Exact value of $f(16) = 0.032\ 337\ 000\ 3$

$x = 256$

N	Two-point Padé P_N^N	Padé P_N^N about 0	Padé P_N^N about ∞
1	0.005 825 242 7	−0.651	0.001 956 947 1
2	0.001 986 947 2	0.506	0.001 956 962 3
3	0.001 957 310 9	−0.409	0.001 956 962 3
4	0.001 948 269 4	0.344	0.001 956 962 3
5	0.001 956 983 5	−0.284	0.001 956 962 3
6	0.001 956 964 4	0.242	0.001 956 962 3
7	0.001 956 962 3	−0.198	0.001 956 962 3
8	0.001 956 963 4	0.167	0.001 956 962 3

Exact value of $f(256) = 0.001\ 956\ 962\ 3$

For definiteness we discuss only the diagonal Padé sequence with $N = M$ and use as input $J = N + 1$ terms of the Taylor series (8.3.10) about $z = 0$, and $K = N$ terms of the asymptotic series (8.3.11) about $z = \infty$. To construct the Padé approximants we let $R_N(z) = \sum_{n=0}^{N} A_n z^n$ and $S_N(z) = \sum_{n=0}^{N} B_n z^n$ where $B_0 = 1$.

In order for the Taylor series expansions of $R_N(z)/S_N(z)$ and $f(z)$ to agree to $O(z^{N+1})$, it is necessary that

$$A_n = \sum_{j=0}^{n} a_{n-j} B_j, \qquad 0 \le n \le N \tag{8.3.12}$$

[see (8.3.3) and (8.3.4)].

However, (8.3.12) gives only $(N + 1)$ equations for the $(2N + 1)$ coefficients $A_0, A_1, \ldots, A_N$, $B_1, B_2, \ldots, B_N$. The other N equations are obtained by expanding $R_N(z)/S_N(z)$ in powers of $1/z$ and requiring that the coefficients of z^{-n} $(0 \le n \le N - 1)$ agree with b_n in (8.3.11). It is easy to see that this condition is equivalent to $R_N(z) - S_N(z) \sum_{n=0}^{\infty} b_n z^{-n} = O(1)$ $(z \to \infty)$ or

$$A_n = \sum_{j=N-n}^{N} B_j b_{j-n}, \qquad 1 \le n \le N. \tag{8.3.13}$$

Equations (8.3.12) and (8.3.13) determine the coefficients of the polynomials $R_N(z)$ and $S_N(z)$.

In Table 8.18 we take z real $(z = x)$ and compare the diagonal sequence of two-point Padé approximants to $f(x)$ with the diagonal sequence one-point Padé approximants obtained from the expansions (8.3.10) at $x = 0$ and (8.3.11) at $x = \infty$. It is apparent that the two-point approximant gives a more uniform approximation to $f(x)$. For small x (x smaller than 5) the two-point Padé is significantly more accurate than the ordinary one-point Padé about $x = \infty$, while it is only slightly less accurate than the one-point Padé about $x = 0$. [The two-point Padé of degree N is comparable in accuracy to the one-point $(x = 0)$ Padé of degree $N/2$.] For large x (x larger than 50) the two-point Padé is significantly more accurate than the one-point $(x = 0)$ Padé, while it is only slightly less accurate than the one-point $(x = \infty)$ Padé.

(I) 8.4 CONTINUED FRACTIONS AND PADÉ APPROXIMANTS

There are several variations on the Padé method of summing power series. One such method consists of recasting the series into *continued-fraction* instead of rational-fraction form. A continued fraction is an infinite sequence of fractions whose $(N + 1)$th member $F_N(z)$ has the form

$$F_N(z) = \cfrac{c_0}{1 + \cfrac{c_1 z}{1 + \cfrac{c_2 z}{1 + \cdots \cfrac{c_{N-1} z}{1 + c_N z}}}}. \tag{8.4.1}$$

The coefficients c_n are determined by expanding the terminated continued fraction $F_N(z)$ in a Taylor series and comparing the coefficients with those of the power series to be summed. This procedure closely resembles Padé summation because here also only algebraic operations are required.

Example 1 *Continued fractions.* For each of the following functions $f(z)$ we give the coefficients in the continued fraction representation of the formal power series of $f(z)$ about $z = 0$:

(*a*) $f(z) = e^z$: $c_0 = -c_1 = 1$, $c_{2n} = 1/(4n - 2)$, $c_{2n+1} = -1/(4n + 2)$ $(n \ge 1)$;

(b) $f(z) = \int_0^\infty dt\, e^{-t}/(1 + zt)$: $c_0 = 1$, $c_{2n-1} = c_{2n} = n$ $(n \geq 1)$;

(c) $f(z) = \int_0^\infty e^{-t}(1 + zt)^{a-1}\, dt$: $c_0 = 1$, $c_{2n-1} = n - a$, $c_{2n} = n$ $(n \geq 1)$;

(d) $f(z) = [\ln (1 + z)]/z$: $c_0 = 1$, $c_{2n-1} = n/(4n - 2)$, $c_{2n} = n/(4n + 2)$ $(n \geq 1)$;
(e) $f(z) = (1 + z)^\alpha$: $c_0 = 1$, $c_1 = -\alpha$, $c_{2n} = (n + \alpha)/(4n + 2)$, $c_{2n+1} = (n - \alpha)/(4n + 2)$ $(n \geq 1)$;
(f) $f(z) = F(a, b + 1, c + 1; z)/F(a, b, c; z)$ where $F(a, b, c; z)$ is the hypergeometric series

$$F(a, b, c; z) = \frac{\Gamma(c)}{\Gamma(a)\Gamma(b)} \sum_{n=0}^{\infty} \frac{\Gamma(a + n)\Gamma(b + n)z^n}{n!\, \Gamma(c + n)}$$

(see Sec. 3.3):

$$c_0 = 1,$$

$$c_{2n+1} = -\frac{(n + a)(n + c - b)}{(2n + c)(2n + c + 1)} \qquad n \geq 0,$$

$$c_{2n} = -\frac{(n + b)(n + c - a)}{(2n + c)(2n + c - 1)} \qquad n \geq 1.$$

For the derivation of these results see Probs. 8.29 and 8.30.

Connection between Continued Fractions and Padé Approximants

Continued fractions and Padé approximants are intimately related. In this subsection we will show how to take advantage of this relation to develop computationally efficient algorithms for summing power series by finding a simple way to represent Padé approximants as continued fractions.

We begin our discussion by considering Padé approximants of the form $P_M^M(z)$ and $P_{M+1}^M(z)$. The Padé sequence $P_0^0(z)$, $P_1^0(z)$, $P_1^1(z)$, $P_2^1(z)$, $P_2^2(z)$, $P_3^2(z)$, ... is said to be *normal* if every member of the sequence exists and no two members are identically equal. We will show that if this Padé sequence is normal, then the $(N + 1)$th term has the continued fraction representation $F_N(z)$ in (8.4.1), where the coefficients c_n $(n = 0, 1, \ldots)$ are the *same* for every term of the sequence. Thus, $P_{M+1}^M(z)$ $(M \geq 1)$ is obtained from $P_M^M(z)$ by simply replacing $c_N z$ by $c_N z/(1 + c_{N+1} z)$ where $N = 2M$, and $P_{M+1}^{M+1}(z)$ $(M \geq 0)$ is obtained from $P_{M+1}^M(z)$ by again replacing $c_N z$ by $c_N z/(1 + c_{N+1} z)$ where $N = 2M + 1$. In short, when Padé approximants are written as a ratio of polynomials, every coefficient in the rational fraction must be recomputed as we go from one member of the normal sequence to the next. Nevertheless, the entire normal sequence may be rewritten as a single continued fraction and only *one* new coefficient need be computed as we go from one member to the next.

To establish these assertions, we make two observations. First, if $N = 2M$ is even then $F_N(z)$ is the ratio of two polynomials of degree M, while if $N = 2M + 1$ is odd then $F_N(z)$ is the ratio of a polynomial of degree M divided by a polynomial of degree $M + 1$. Second, if $F_N(z)$ given by (8.4.1) is expanded in a Taylor series about $z = 0$, the coefficient of z^p $(p = 0, 1, \ldots)$ depends *only* on $c_0, c_1, c_2, \ldots, c_p$.

For example, the first four terms of the Taylor expansion of (8.4.1) are

$$c_0 - c_0 c_1 z + c_0 c_1(c_1 + c_2)z^2 - c_0 c_1(c_2 c_3 + c_2^2 + 2c_1 c_2 + c_1^2)z^3.$$

(Can you prove the above assertion for all p?)

From these observations, it follows that if $F_N(z)$ is to be a Padé approximant to $f(z) \sim \sum_{n=0}^{\infty} a_n z^n$ $(z \to 0)$, then c_p depends only on $a_0, a_1, \ldots, a_p$ for every $p \geq 0$. This shows that c_p $(p = 0, 1, \ldots)$ does not depend on N in the sequence $F_N(z)$ but only on the expansion coefficients a_n of $f(z)$. This argument is valid provided that all $c_p \neq 0$. But if $c_p = 0$ for some p, then $F_p(z) \equiv F_{p-1}(z)$ and the sequence of Padé approximants $P_0^0(z)$, $P_1^0(z)$, $P_1^1(z)$, ... is not normal.

An efficient procedure for calculating the coefficients c_p of the continued fraction (8.4.1) may be derived from the simple algebraic identity

$$\frac{N_i(z)}{D_i(z)} = \frac{N_i(0)/D_i(0)}{1 + zN_{i+1}(z)/D_{i+1}(z)}, \tag{8.4.2a}$$

where

$$N_{i+1}(z) = \left[\frac{D_i(z)}{D_i(0)} - \frac{N_i(z)}{N_i(0)}\right] \Big/ z, \tag{8.4.2b}$$

$$D_{i+1}(z) = \frac{N_i(z)}{N_i(0)}. \tag{8.4.2c}$$

We have written this identity with subscripts i because it will be applied sequentially to determine c_{i+1} once $c_0, c_1, \ldots, c_i$ are known.

From (8.4.2) it is clear that if $N_i(z)$ and $D_i(z)$ have a series expansion in powers of z (either convergent or asymptotic), then so do $N_{i+1}(z)$ and $D_{i+1}(z)$. In fact, if

$$N_i(z) \sim \sum_{n=0}^{\infty} A_n^{(i)} z^n, \qquad z \to 0, \tag{8.4.3a}$$

$$D_i(z) \sim \sum_{n=0}^{\infty} B_n^{(i)} z^n, \qquad z \to 0, \tag{8.4.3b}$$

then (8.4.2) implies that

$$A_n^{(i+1)} = \frac{B_{n+1}^{(i)}}{B_0^{(i)}} - \frac{A_{n+1}^{(i)}}{A_0^{(i)}}, \qquad n = 0, 1, \ldots, \tag{8.4.4a}$$

$$B_n^{(i+1)} = \frac{A_n^{(i)}}{A_0^{(i)}}, \qquad n = 0, 1, \ldots. \tag{8.4.4b}$$

To calculate the coefficients c_p we apply (8.4.2) to (8.4.4) iteratively for $i = 0, 1, 2, \ldots$ starting with $N_0(z) = f(z)$ and $D_0(z) = 1$, so $A_n^{(0)} = a_n$, $B_0^{(0)} = 1$, $B_n^{(0)} = 0$ $(n \geq 1)$. Comparing (8.4.2a) with (8.4.1) gives

$$c_i = \frac{A_0^{(i)}}{B_0^{(i)}} \tag{8.4.5}$$

for all i. [Note that $B_0^{(i)} = 1$ for all i by (8.4.4b).]

The calculation of $c_0, c_1, \ldots, c_N$ from (8.4.2) to (8.4.4) requires as input only the expansion coefficients $a_0, \ldots, a_N$ of the power series for $f(z)$ and involves only $N^2/2$ divisions and $N^2/2$ subtractions. The method just described is an efficient procedure for computing Padé approximants because it makes indirect use of the form of the matrix $\mathcal{A}$ in (8.3.3). A direct solution of (8.3.3) by Gaussian elimination without making use of the simple form of $\mathcal{A}$ requires on the order of N^3 operations.

Now we extend the preceding discussion by showing how to construct a continued-fraction representation for the Padé approximant $P_M^N(z)$. The $(N+1)$th member of the Padé sequence $P_0^J, P_1^J, P_1^{J+1}, P_2^{J+1}, P_2^{J+2}, \ldots$ $(J \geq 0)$ for the function $f(z) \sim \sum_{n=0}^{\infty} a_n z^n$ $(z \to 0)$ can be represented in the form

$$F_N(z) = \sum_{n=0}^{J-1} a_n z^n + \cfrac{c_0 z^J}{1 + \cfrac{c_1 z}{1 + \cfrac{c_2 z}{1 + \ddots \cfrac{c_{N-1} z}{1 + c_N z}}}}, \tag{8.4.6}$$

provided that the Padé sequence is normal (i.e., no two members of the sequence are identically equal so that all the coefficients c_p are nonzero). Any expression of the form $F_N(z)$ in (8.4.6) is the ratio of a polynomial of degree $M+J$ to a polynomial of degree M or $M+1$ $(M = 0, 1, 2, \ldots)$. Also, the coefficients of z^n in the expansion of $F_N(z)$ are a_n for $n = 0, \ldots, J-1$; the coefficients c_p affect only the coefficients of z^n for $n = J, J+1, \ldots, J+p$. Consequently, the coefficients c_p may be determined by a slight modification of the algorithm (8.4.4) and (8.4.5).

To represent the members of the Padé sequence $P_J^0, P_J^1, P_{J+1}^1, P_{J+1}^2, P_{J+2}^2, \ldots$ with $J \geq 0$ as continued fractions, we need only observe that the $P_M^N(z)$ Padé approximant to $1/f(z)$ is *identical* to the $P_N^M(z)$ Padé approximant to $f(z)$ (see Prob. 8.21). Therefore, the desired sequence of Padé approximants can be represented as the inverse of the expressions (8.4.6) with the coefficients a_n of $f(z)$ replaced by the expansion coefficients of $1/f(z)$ (the latter being easily computable in terms of a_n).

Evaluation of Continued Fractions

One may evaluate the continued fraction (8.4.1) in the obvious direct way. However, there is a nice algorithm that generates the values of $F_0(z)$, $F_1(z)$, $F_2(z), \ldots$ as a prelude to the evaluation of $F_N(z)$ at little extra cost. It is necessary to have these values to examine the convergence of the Padé approximations as $N \to \infty$. This algorithm is also a good way to develop explicit formulas for the numerator and denominator polynomials of Padé approximants from continued fractions.

The algorithm is based on the observation that if we write $F_N(z)$ in (8.4.1) in the form

$$F_N(z) = \frac{R_N(z)}{S_N(z)} \tag{8.4.7}$$

for $N = 1, 2, \ldots$, then R_N and S_N satisfy the recurrence relations

$$R_{N+1}(z) = R_N(z) + c_{N+1} z R_{N-1}(z), \tag{8.4.8a}$$

$$S_{N+1}(z) = S_N(z) + c_{N+1} z S_{N-1}(z), \tag{8.4.8b}$$

for $N = 0, 1, 2, \ldots$, where $R_{-1}(z) = 0$, $S_{-1}(z) = 1$, $R_0(z) = c_0$, $S_0(z) = 1$.

The recurrences (8.4.8) are proved by induction. They are certainly valid for $N = 0$ because they give $R_1(z) = c_0$ and $S_1(z) = 1 + c_1 z$. Assuming (8.4.8) is correct for $N < M$, we must then show that it is then correct for $N = M$. To do so, we note that when $M \geq 1$, $F_{M+1}(z)$ is obtained from $F_M(z)$ by simply replacing c_M by $c_M/(1 + c_{M+1} z)$ in all expressions. Using this modified value of c_M in (8.4.7) and (8.4.8) with $N = M - 1$, we obtain

$$F_{M+1}(z) = \frac{R_{M-1} + c_M z R_{M-2}/(1 + c_{M+1} z)}{S_{M-1} + c_M z S_{M-2}/(1 + c_{M+1} z)}$$

because, by the induction hypothesis, R_K and S_K are unchanged for $K < M$ by the change in the value of c_M. Thus,

$$\frac{R_{M+1}}{S_{M+1}} = F_{M+1}(z) = \frac{R_{M-1} + c_M z R_{M-2} + c_{M+1} z R_{M-1}}{S_{M-1} + c_M z S_{M-2} + c_{M+1} z S_{M-1}}$$

$$= \frac{R_M + c_{M+1} z R_{M-1}}{S_M + c_{M+1} z S_{M-1}},$$

which proves that (8.4.8) holds for $N = M$ and completes the proof by induction.

Continued-Function Representations

A continued fraction is a special example of a continued-function representation. A continued function is a sequence whose $(N + 1)$th term $C_N(z)$ has the form of N iterations of some function $F(z)$. For continued fractions $F(z)$ is $1/(1 + z)$ and for Taylor series $F(z)$ is $1 + z$. When $F(z) = e^z$, then $C_N(z)$ is a continued exponential:

$$C_N(z) = c_0 \exp(c_1 z \exp(c_2 z \cdots \exp(c_N z))).$$

If $F(z) = \sqrt{1 + z}$ or $\ln(1 + z)$, then we have continued roots or logarithms:

$$C_N(z) = c_0 \sqrt{1 + c_1 z \sqrt{1 + c_2 z \sqrt{1 + \cdots \sqrt{1 + c_N z}}}},$$

$$C_N(z) = c_0 \ln(1 + c_1 z \ln(1 + c_2 z \ln(1 + \cdots \ln(1 + c_N z)))).$$

For each continued function the goal is to find an efficient algorithm for computing the coefficients $c_0, c_1, c_2, \ldots, c_N$ from the first $(N + 1)$ terms of a formal power series approximation. By doing this one generates a sequence of approximants $C_0(z), C_1(z), C_2(z), \ldots$ which may converge to the original function more rapidly than the power series.

Continued functions may exhibit remarkable convergence phenomena. Some properties of continued functions are developed in Probs. 8.32 to 8.36.

(TD) 8.5 CONVERGENCE OF PADÉ APPROXIMANTS

The previous two sections emphasized the mechanics of Padé summation of power series; we carefully avoided any discussion of why Padé summation actually works. In this and the next section we give a summary of the mathematical ideas underlying Padé summation.

There are two distinct questions that must be answered to explain the amazing success of Padé approximation:

1. Why and how rapidly do the Padé approximants converge? We discuss this question in this section. Specifically, we show how to predict the rates of convergence of the Padé sequences P_N^{N+J} as $N \to \infty$ with J fixed from the continued fraction representation discussed in Sec. 8.4.
2. If the Padé sequences P_N^{N+J} do converge to a single function for all J, what is the connection between this limit function and the original power series from which the Padé approximants are generated? We postpone any discussion of this subtle and difficult question until the next section.

Digression on the Peculiar Convergence Properties of Padé Approximants

The convergence theory of Padé sequences is not a simple extension of the convergence theory of Taylor series. In fact, the regions of convergence of Taylor approximants and of the Padé approximants derived from these Taylor approximants are not, in general, closely related. The next example shows that Taylor approximants can converge where Padé approximants do not.

Example 1 *Padé approximants may diverge where Taylor approximants converge.* Consider the function $y(x) = (x + 10)/(1 - x^2) = \sum_{n=0}^{\infty} a_n x^n$, where $a_{2n} = 10$ and $a_{2n+1} = 1$. This Taylor series converges for $|x| < 1$.

The Padé approximant $P_1^N(x)$ for this series is

$$P_1^N(x) = \sum_{n=0}^{N-2} a_n x^n + \frac{a_{N-1} x^{N-1}}{1 - a_N x/a_{N-1}}.$$

Thus, if N is even, $P_1^N(x)$ has a simple pole at $x = \frac{1}{10}$. Consequently, the Padé sequence $P_1^0, P_1^1, P_1^2, P_1^3, \ldots$ does not converge to $y(x)$ throughout $|x| < 1$.

Example 1 shows that the convergence of a Padé sequence can be affected by the zeros of the denominator function. To prove that a Padé sequence converges in some region it is necessary to know that all extraneous poles of the Padé approximants (poles that do not belong to the function being approximated) move out of that region as $N \to \infty$. In Example 1 the Padé approximants P_1^{2M} all have a pole at $x = \frac{1}{10}$, which lies within the region $|x| < 1$. Similarly, extraneous zeros of a convergent Padé sequence must also move out of the region as $N \to \infty$.

Example 2 *Extraneous zeros of Padé approximants can prevent convergence.* Consider the function $z(x) = (1 - x^2)/(x + 10)$, which is the inverse of the function $y(x)$ in Example 1. The Padé approximant $P_N^1(x)$ of $z(x)$ is *identical* to the Padé approximant $P_1^N(x)$ of $y(x)$ (why?), so $P_{2M}^1(x = \frac{1}{10}) = 0$. Since $z(\frac{1}{10}) \neq 0$, the Padé sequence $P_N^1(x)$ does not converge to $z(x)$ as $N \to \infty$ throughout the region $|x| < 10$ where the Taylor series of $z(x)$ converges.

Once the distribution of poles and zeros of Padé sequences is known, it is then possible to investigate the convergence of the Padé sequence. In Prob. 8.40 we outline how this can be done for Padé approximants $P_N^N(z)$ and $P_{N+1}^N(z)$ for e^z.

The limiting behavior of sequences of Padé approximants is most easily studied using the continued-fraction formalism developed in Sec. 8.4. There we found that $F_N(z)$, the $(N+1)$th member of the Padé sequence $P_0^0(z)$, $P_1^0(z)$, $P_1^1(z), \ldots$, has the continued fraction representation in (8.4.1) and that $F_N = R_N/S_N$ can be evaluated by the recurrence relations (8.4.8). We will now apply these results to study the convergence of the Padé sequences $P_N^N(z)$ and $P_{N+1}^N(z)$ as $N \to \infty$. To determine the convergence of $P_N^{N+J}(z)$ as $N \to \infty$ for arbitrary J, one must generalize the method slightly (see Prob. 8.41).

To derive the equations which govern the convergence of the Padé sequence F_N, we multiply the two equations in (8.4.8) by S_N and R_N, respectively, and subtract the resulting equations:

$$R_{N+1}S_N - R_N S_{N+1} = -c_{N+1}z(R_N S_{N-1} - R_{N-1}S_N), \qquad N \geq 1.$$

This equation and the initial conditions $S_0 = 1$, $S_1 = 1 + c_1 z$, $R_0 = c_0$, $R_1 = c_0$ give $R_N S_{N-1} - R_{N-1} S_N = c_0 c_1 c_2 \cdots c_N(-z)^N$. Finally, we divide this last equation by $S_N S_{N-1}$ and recall that $F_N = R_N/S_N$:

$$F_N - F_{N-1} = \frac{c_0 c_1 \cdots c_N(-z)^N}{S_N(z)S_{N-1}(z)}. \tag{8.5.1}$$

This equation shows that the behavior of the denominators S_N for large N determines the convergence of the sequence F_N. The behavior of S_N for large N may be found from an asymptotic analysis of the difference equation (8.4.8*b*) if the behavior of the continued fraction coefficients c_N for large N is known.

Example 3 *Convergence of Padé approximants to* $(\sqrt{1+z} - 1)/2z$. The function $f(z) = (\sqrt{1+z} - 1)/2z$ has a Taylor series of the form

$$f(z) = \frac{1}{4} - \frac{z}{16} + \frac{z^2}{32} - \frac{5z^3}{256} + \cdots = \frac{1}{4} \sum_{n=0}^{\infty} \frac{(-1)^n \Gamma(n+\frac{1}{2})}{(n+1)!\,\Gamma(\frac{1}{2})} z^n. \tag{8.5.2}$$

How rapidly do the Padé approximants to this series converge?

To answer this question we note that $f(z)$ satisfies the algebraic equation $f(z) = \frac{1}{4}/[1 + zf(z)]$. Thus, a continued fraction expansion of $f(z)$ is easily obtained by substituting repeatedly the above equation into itself:

$$f(z) = \cfrac{\frac{1}{4}}{1 + \cfrac{z/4}{1 + zf(z)}} = \cfrac{\frac{1}{4}}{1 + \cfrac{z/4}{1 + \cfrac{z/4}{1 + zf(z)}}} = \cdots.$$

Thus, the continued fraction coefficients of $f(z)$ are simply $c_N = \frac{1}{4}$ for all N.

The convergence of the Padé sequence $F_N(x)$ for real x is determined by the behavior of the denominators S_N which satisfy the equation (8.4.8b): $S_{N+1} = S_N + \frac{1}{4}xS_{N-1}$ $(S_0 = 1,\ S_1 = 1 + x/4)$. Since this is a constant-coefficient difference equation, it may be solved in closed form (no asymptotic analysis is required); the exact solution is

$$S_N(x) = \frac{1}{\sqrt{1+x}}\left[\left(\frac{1+\sqrt{1+x}}{2}\right)^{N+2} - \left(\frac{1-\sqrt{1+x}}{2}\right)^{N+2}\right]. \tag{8.5.3}$$

However, since $|1 + \sqrt{1+x}| > |1 - \sqrt{1+x}|$ when $x > -1$, the second term on the right in (8.5.3) is subdominant with respect to the first as $N \to \infty$. Thus,

$$S_N \sim \frac{1}{\sqrt{1+x}}\left(\frac{1+\sqrt{1+x}}{2}\right)^{N+2}, \qquad N \to \infty;\ x > -1. \tag{8.5.4}$$

Substituting this result into (8.5.1) gives

$$F_N - F_{N-1} \sim B(x)\left(\frac{1-\sqrt{1+x}}{1+\sqrt{1+x}}\right)^{N+1}, \qquad N \to \infty;\ x > -1, \tag{8.5.5}$$

where $B(x) = -2(1+x)/(x + x\sqrt{1+x})$ is a function of x alone.

The goal is now to solve (8.5.5) for F_N and to show that F_N approaches a limit as $N \to \infty$. Since the left and right sides of (8.5.5) are equal except for exponentially small terms, we sum both sides from $N+1$ to ∞. The result is

$$F_N(x) \sim A(x) + \frac{\sqrt{1+x}}{x}\left(\frac{1-\sqrt{1+x}}{1+\sqrt{1+x}}\right)^{N+2}, \qquad N \to \infty;\ x > -1, \tag{8.5.6}$$

where $A(x)$ is a function of x. This equation shows that as $N \to \infty$, $F_N(x)$ tends to the limit function $A(x)$ when $x > -1$, which is what we set out to prove.

This convergence argument can be easily extended to show that the Padé sequence $P_0^0(z)$, $P_1^0(z), \cdots$ converges in the cut-z plane $|\arg(z+1)| < \pi$. One need only observe that the asymptotic relations in (8.5.4) and (8.5.5) remain valid in the cut-z plane because $|1 - \sqrt{1+z}| < |1 + \sqrt{1+z}|$ when $|\arg(z+1)| < \pi$. (Why?) Hence, $F_N(z)$ still tends to the limit function $A(z)$ as $N \to \infty$. The proof of convergence breaks down when $\arg(z+1) = \pi$ because then $|1 - \sqrt{1+z}| = |1 + \sqrt{1+z}|$ and (8.5.4) does not follow from (8.5.3). When $x < -1$, $\arg(x+1) = \pi$, and the Padé sequence $P_0^0, P_1^0, P_1^1, \ldots$ oscillates persistently and does not converge. Thus, the Padé approximants converge in the whole complex plane except for the cut from $z = -1$ to ∞ along the negative real axis; in contrast, the Taylor series for $f(z)$ [see (8.5.2)] converges only for $|z| < 1$.

We emphasize that we have proved that the Padé sequence $F_N(z)$ converges as $N \to \infty$, but we have not proved that the limit function $A(z)$ equals $f(z)$! A proof that $A(z) = f(z)$ can be based on the theory of Sec. 8.6.

Example 4 *Convergence of Padé approximants for* $z^{-1}\ln(1+z)$. The continued fraction coefficients of $z^{-1}\ln(1+z)$ are $c_0 = 1$, $c_{2n-1} = n/(4n-2)$, $c_{2n} = n/(4n+2)$ $(n \geq 1)$ (see Example 1 of Sec. 8.4). Substituting these values for c_N into the difference equation (8.4.8b) for

$S_N(z)$ and performing a local asymptotic analysis for large N gives

$$S_N(x) = C_1(z)\left(\frac{1+\sqrt{1+z}}{2}\right)^N\left[1+O\left(\frac{1}{N}\right)\right]$$

$$+ C_2(z)\left(\frac{1-\sqrt{1+z}}{2}\right)^N\left[1+O\left(\frac{1}{N}\right)\right], \qquad N \to \infty \tag{8.5.7}$$

(see Prob. 8.42). The controlling factors of the behavior of $S_N(z)$ in (8.5.8) are the same as in (8.5.3) because $c_N \sim \frac{1}{4}$ as $N \to \infty$. In general, though, when $c_N = \frac{1}{4} + O(1/N)$ as $N \to \infty$, the full leading behavior of $S_N(z)$ does not necessarily have the form in (8.5.7) because it may also contain powers of N.

If z lies in the cut-z plane $|\arg(z+1)| < \pi$, then the second term on the right side of (8.5.7) is subdominant with respect to the first as $N \to \infty$. Thus, assuming that $C_1(z) \neq 0$, we have

$$S_N = C_1(z)\left(\frac{1+\sqrt{1+z}}{2}\right)^N\left[1+O\left(\frac{1}{N}\right)\right], \qquad N \to \infty;\ |\arg(z+1)| < \pi.$$

[The initial conditions $S_0(z) = 1$, $S_1(z) = 1 + c_0 z$, and the difference equation for S_N imply that $C_1(z) \neq 0$. Why?] Therefore, (8.5.1) implies that the Padé sequence $F_N(x)$ satisfies

$$F_N - F_{N-1} \sim \frac{1+\sqrt{1+z}}{2[C_1(z)]^2}\left(\frac{1-\sqrt{1+z}}{1+\sqrt{1+z}}\right)^N, \qquad N \to \infty;\ |\arg(z+1)| < \pi. \tag{8.5.8}$$

The solution to (8.5.8) is

$$F_N(z) \sim A(z) + B(z)\left(\frac{1-\sqrt{1+z}}{1+\sqrt{1+z}}\right)^N, \qquad N \to \infty;\ |\arg(z+1)| < \pi, \tag{8.5.9}$$

for some functions $A(z)$ and $B(z)$. We conclude that the Padé sequence $P_0^0, P_1^0, P_1^1, P_2^1, \ldots$ converges geometrically as $N \to \infty$ to the limit function $A(z)$ when $|\arg(z+1)| < \pi$. In Example 1 of Sec. 8.6 it will be shown that $A(z) = z^{-1}\ln(1+z)$. If we assume that $A(z)$ is indeed $z^{-1}\ln(1+z)$, then the sign pattern of the error in the Padé series is explained; we saw in Table 8.13 that for real positive x the Padé sequence oscillates about its limit. This observation is consistent with (8.5.9).

Example 5 *Convergence of Padé approximants for e^z.* The continued fraction coefficients of e^z are $c_0 = -c_1 = 1$, $c_{2n} = 1/(4n-2)$, $c_{2n+1} = -1/(4n+2)$ $(n \geq 1)$. For these values of c_n, (8.4.8b) becomes the following system of equations:

$$S_{2M+1} - S_{2M} = -\frac{z}{4M+2}S_{2M-1}, \tag{8.5.10a}$$

$$S_{2M} - S_{2M-1} = \frac{z}{4M-2}S_{2M-2}, \qquad M \geq 1. \tag{8.5.10b}$$

We transform this system into a single difference equation of higher order by solving (8.5.10b) for S_{2M-1} and substituting the result into (8.5.10a):

$$S_{2M+2} - S_{2M} = \frac{z^2}{16M^2-4}S_{2M-2}, \qquad M \geq 1. \tag{8.5.11}$$

A local asymptotic analysis of this difference equation (see Prob. 8.39) yields

$$S_{2M}(z) = C(z)\left[1 - \frac{z^2}{16M} + O(M^{-2})\right], \qquad M \to \infty, \tag{8.5.12a}$$

for some function $C(z) \neq 0$. Combining (8.5.12a) into (8.5.10b) gives

$$S_{2M-1}(z) = C(z)\left[1 - \frac{z}{4M} - \frac{z^2}{16M} + O\left(\frac{1}{M^2}\right)\right], \qquad M \to \infty. \tag{8.5.12b}$$

Substituting (8.5.12) into (8.5.1) gives

$$F_N - F_{N-1} \sim \frac{\sigma_N z^N \sqrt{N}}{2^N N!} D(z), \qquad N \to \infty, \tag{8.5.13}$$

where $D(z)$ is a function of z (but not N) and $\sigma_N = 1, 1, -1, -1, 1, 1, -1, \ldots$ when $N = 0, 1, 2, 3, 4, 5, 6, \ldots$. By contrast, if $T_N = \sum_{n=0}^{N-1} z^n/n!$ is the sum of the first N terms of the Taylor series for e^z, then

$$T_N - T_{N-1} = \frac{z^N}{N!}. \tag{8.5.14}$$

The extra factor of 2^{-N} in (8.5.13) relative to (8.5.14) implies that the Padé sequence $F_N(z)$ converges to its limit more rapidly than the Taylor sequence $T_N(z)$ does; the difference between $F_N(z)$ and its limit is smaller than $|T_N(z) - e^z|$ by a factor of about 2^N. Indeed, we have already seen this factor of 2^N in Tables 8.9 and 8.10 which compare the rates of the convergence of Padé approximants and Taylor approximants to e^x at $x = 1$ and 5. However, while (8.5.13) proves that $\lim_{N\to\infty} F_N$ exists for all complex z, it does not prove that the limit is really e^z. This subtlety is treated in Prob. 8.40. However, assuming that $\lim_{N\to\infty} F_N(z) = e^z$, then the function σ_N in (8.5.13) explains the sign pattern of the error in $F_N(x)$ for real positive x in Tables 8.9 and 8.10, which was observed to oscillate like $+ + - - + + - - + + \cdots$.

Example 6 *Convergence of Padé approximants for Stieltjes series* $\sum (-z)^n n!$. According to Example 1 of Sec. 8.4, the continued fraction coefficients of the Stieltjes function

$$\int_0^\infty \frac{e^{-t}}{1+zt}\,dt \sim \sum_{n=0}^{\infty} (-z)^n n!, \qquad |z| \to 0, \ |\arg z| < \pi,$$

are $c_0 = c_1 = 1$, $c_{2n} = n$, $c_{2n+1} = n + 1$ $(n \geq 1)$. Therefore, (8.4.8b) becomes the system

$$S_{2N+1} - S_{2N} = (N+1)zS_{2N-1}, \tag{8.5.15a}$$

$$S_{2N} - S_{2N-1} = NzS_{2N-2}, \qquad N \geq 1. \tag{8.5.15b}$$

Consequently,

$$S_{2N+1} - [(2N+1)z + 1]S_{2N-1} + N^2 z^2 S_{2N-3} = 0. \tag{8.5.16}$$

A local analysis of (8.5.16) yields the leading behavior of $S_{2N-1}(z)$:

$$S_{2N-1}(z) \sim C(z)N!\, z^N N^{-1/4} \exp\left(2\sqrt{N/z}\right)\left[1 + \frac{\alpha(z)}{N^{1/2}}\right], \qquad N \to \infty, \ |\arg z| < \pi, \tag{8.5.17a}$$

where $C(z)$ is some function of z alone that is nonzero and $\alpha(z) = \frac{1}{12}z^{-3/2} + \frac{1}{2}z^{-1/2} - \frac{1}{4}z^{1/2}$ (see Prob. 8.43). Equation (8.5.15a) then implies that

$$S_{2N}(z) \sim C(z)N!\, z^{N+1/2} N^{1/4} \exp\left(2\sqrt{N/z}\right)\left[1 + \frac{\beta(z)}{N^{1/2}}\right], \qquad N \to \infty, \ |\arg z| < \pi, \tag{8.5.17b}$$

where $\beta(z) = \frac{1}{12}z^{-3/2} + z^{-1/2} - \frac{1}{2}z^{1/2}$.

Substituting these results into (8.5.1) gives

$$F_{2N} - F_{2N-1} \sim \frac{\exp(-4\sqrt{N/z})}{C^2(z)\sqrt{z}}\left[1 - \frac{\alpha(z)+\beta(z)}{\sqrt{N}}\right], \qquad N \to \infty,\ |\arg z| < \pi,$$

$$F_{2N+1} - F_{2N} \sim \frac{-\exp(-4\sqrt{N/z}}{C^2(z)\sqrt{z}}\left[1 - \frac{\alpha(z)+\beta(z)+z^{-1/2}}{\sqrt{N}}\right], \qquad N \to \infty,\ |\arg z| < \pi,$$

$$F_{2N-1} - F_{2N-2} \sim \frac{\exp(-4\sqrt{N/z})}{C^2(z)\sqrt{z}}\left[1 - \frac{\alpha(z)+\beta(z)-z^{-1/2}}{\sqrt{N}}\right], \qquad N \to \infty,\ |\arg z| < \pi,$$

from which we deduce that

$$F_{2N+1} - F_{2N-1} \sim \frac{\exp(-4\sqrt{N/z})}{zC^2(z)\sqrt{N}}, \qquad N \to \infty,\ |\arg z| < \pi, \tag{8.5.18a}$$

$$F_{2N} - F_{2N-2} \sim \frac{-\exp(-4\sqrt{N/z})}{zC^2(z)\sqrt{N}}, \qquad N \to \infty,\ |\arg z| < \pi. \tag{8.5.18b}$$

Equation (8.5.18) shows that the Padé approximants $P_N^N(z)$ and $P_{N+1}^N(z)$ converge exponentially in the cut-z plane $|\arg z| < \pi$ as $N \to \infty$. The formulas in (8.5.18) also verify the convergence properties of the Padé series for the Stieltjes function for Re $z = x > 0$ that were observed in Table 8.15; (8.5.18*a*) proves that the sequence $P_{N+1}^N(x)$ is monotone increasing as $N \to \infty$ and (8.5.18*b*) proves that the sequence $P_N^N(x)$ is monotone decreasing as $N \to \infty$.

(TD) 8.6 PADÉ SEQUENCES FOR STIELTJES FUNCTIONS

In Secs. 8.4 and 8.5 we explained why sequences of Padé approximants converge and we estimated their rates of convergence. In this section we address a much more profound question. Since many functions may be asymptotic to the same divergent series, which of these functions, if any, will the Padé sequence select as its limit? Moreover, if the limit function is multivalued (many of our examples involve multivalued functions), which branch of the function will be singled out as that given by the limit of the Padé sequence?

Even for simple asymptotic power series, the answers to these questions can be unexpectedly complicated. In fact, there is as yet no general theory of Padé summation for arbitrary series. However, the convergence theory for Pade approximants for the special class of Stieltjes series is relatively complete and very elegant. For Stieltjes series, we can partially answer the questions of which function the Padé sequences converge to and which branch of the function is selected.

Since these questions for Stieltjes series are tractable, it is crucial to know whether or not a given series is a series of Stieltjes before attempting to use Padé summation. Let us recall some basic definitions and results from Sec. 3.8. First, a Stieltjes series is a series of the form $\sum_{n=0}^{\infty} a_n(-z)^n$, where the coefficients a_n are the *moments* of a real *nonnegative* function $\rho(t)$:

$$a_n = \int_0^{\infty} t^n \rho(t)\, dt \ (n = 0, 1, 2, \ldots), \qquad \rho(t) \geq 0 \ (0 \leq t < \infty). \tag{8.6.1}$$

Second, we showed in Sec. 3.8 that for every Stieltjes series there is a Stieltjes function $F(z)$ to which it is asymptotic as $z \to 0$ in the cut plane $|\arg z| < \pi$:

$$F(z) = \int_0^\infty \frac{\rho(t)}{1 + zt}\, dt \sim \sum_{n=0}^\infty a_n(-z)^n, \qquad z \to 0;\ |\arg z| < \pi. \tag{8.6.2}$$

Notice that (8.6.2) implies that $F(z)$ is the Borel sum of the series $\sum a_n(-z)^n$ if the Borel sum exists; the Borel sum of $\sum a_n(-z)^n$ exists if $|a_n| \le n!\, C^n$ for some C and all n (see Sec. 8.2).

Not every series $\sum a_n(-z)^n$ is a Stieltjes series. To be a Stieltjes series the coefficients must satisfy certain inequalities (see Probs. 8.44 and 8.45). However, without knowing all the terms in a series these inequalities are impossible to verify. Here is a typical scenario. After much hard work the first few coefficients of the perturbation series have been computed and it is apparent that the series either diverges or else converges too slowly to recover useful information. One would like to use summation methods, but it is impossible to verify that the series is a Stieltjes series from the first few coefficients. Naturally, the only general advice is to use summation and hope for the best. However, in some problems it is possible to prove indirectly that a perturbation series is a Stieltjes series by showing that the exact answer to the problem is a Stieltjes function. Four properties of the exact answer must be established to verify that it is a Stieltjes function:

1. $F(z)$ is analytic in the cut plane $|\arg z| < \pi$.
2. $F(z) \to C$ as $z \to \infty$, where C is a nonnegative real constant.
3. $F(z)$ has an asymptotic series representation of the form $\sum_{n=0}^\infty a_n(-z)^n$ in the cut plane.
4. $-F(z)$ is Herglotz (see Sec. 7.5 and Prob. 8.46). (8.6.3)

In Probs. 8.47 to 8.49 we show how to reconstruct the Stieltjes density function $\rho(t)$ in (8.6.2) from $F(z)$ satisfying these properties.

Continued-Fraction Representation of Stieltjes Functions

An effective way to study limits of Padé sequences for Stieltjes functions is to use their continued-fraction representation. An important characterization of Stieltjes functions is that all their continued-fraction expansion coefficients c_n in (8.4.1) are nonnegative. This is true because if $f_n(z)$ is a Stieltjes function, then $f_{n+1}(z)$ defined by

$$f_n(z) = \frac{f_n(0)}{1 + zf_{n+1}(z)} \tag{8.6.4}$$

is also a Stieltjes function (see Prob. 8.50). Therefore, if we generate the continued-fraction representation of a Stieltjes function $F(z) \equiv f_0(z)$ by applying (8.6.4) repeatedly,

$$F(z)=\cfrac{f_0(0)}{1+zf_1(z)}=\cfrac{f_0(0)}{1+\cfrac{zf_1(0)}{1+zf_2(z)}}=\cfrac{f_0(0)}{1+\cfrac{zf_1(0)}{1+\ddots\cfrac{zf_n(0)}{1+zf_{n+1}(z)}}} \tag{8.6.5}$$

then we may conclude that the continued-fraction expansion coefficients $c_n = f_n(0)$ are all nonnegative. The function $f_n(z)$ is a Stieltjes function, so $f_n(z) = \int_0^\infty \rho_n(t)/(1 + zt)\, dt$ $[\rho_n(t) \geq 0]$ and $c_n = f_n(0) = \int_0^\infty \rho_n(t)\, dt \geq 0$. Conversely, if all the coefficients c_n are nonnegative, then there exists a Stieltjes function whose continued-fraction coefficients generated by (8.6.5) are c_n (see Prob. 8.51).

Monotone Convergence of Padé Sequences

In Sec. 8.4 it was shown that truncations of the continued fraction (8.4.1) yield the successive members of the Padé sequence $P_0^0, P_1^0, P_1^1, P_2^1, P_2^2, \ldots$. This Padé sequence has some remarkable convergence properties when all the continued-fraction coefficients are nonnegative. We shall show that if x is real and positive, then:

1. The diagonal Padé sequence $P_N^N(x)$ $(N = 0, 1, 2, \ldots)$ decreases monotonically as N increases.
2. The Padé sequence $P_{N+1}^N(x)$ $(N = 0, 1, 2, \ldots)$ increases monotonically as N increases.
3. The sequence $P_N^N(x)$ $(N = 0, 1, 2, \ldots)$ has a lower bound, while the sequence $P_{N+1}^N(x)$ $(N = 0, 1, 2, \ldots)$ has an upper bound.
4. More precisely,

$$\begin{aligned} P_1^0(x) \leq P_2^1(x) \leq P_3^2(x) \leq \cdots \leq P_{N+1}^N(x) \leq \cdots \\ \leq P_N^N(x) \leq \cdots \leq P_2^2(x) \leq P_1^1(x) \leq P_0^0(x). \end{aligned} \tag{8.6.6}$$

Properties (1) to (3) imply that $\lim_{N\to\infty} P_N^N(x)$ and $\lim_{N\to\infty} P_{N+1}^N(x)$ both exist for real positive x. Property (4) implies that $\lim_{N\to\infty} P_{N+1}^N(x) \leq \lim_{N\to\infty} P_N^N(x)$.

Finally, we will show that:

5. All Stieltjes functions $F(x)$ (there may be more than one!) whose continued-fraction coefficients are c_n satisfy

$$\lim_{N\to\infty} P_{N+1}^{N}(x) \le F(x) \le \lim_{N\to\infty} P_N^N(x). \tag{8.6.7}$$

From this property we conclude that if $\lim_{N\to\infty} P_{N+1}^N(x) = \lim_{N\to\infty} P_N^N(x)$, then there is just *one* Stieltjes function having the continued-fraction coefficients c_n and the Padé sequences converge to it when x is real and positive. Because the Padé sequences approach their limits in (8.6.7) monotonically, $P_{N+1}^N(x)$ and $P_N^N(x)$ for large N provide more than just an estimate of the Stieltjes function $F(x)$; they supply their own error bounds on the approximation. If 10 terms in the Stieltjes series are known, we may compute $P_5^4(x)$ and $P_4^4(x)$ and be sure that the exact value of $F(x)$ lies between $P_5^4(x)$ and $P_4^4(x)$. Later in this section, we discuss what happens when $\lim_{N\to\infty} P_{N+1}^N(x) \ne \lim_{N\to\infty} P_N^N(x)$.

To verify properties (1) to (5), we make use of two equations relating the numerator $R_N(x)$ and denominator $S_N(x)$ of the continued fraction $F_N(x) = R_N(x)/S_N(x)$:

$$R_N(x)S_{N-1}(x) - R_{N-1}(x)S_N(x) = c_0 c_1 c_2 \cdots c_N(-x)^N, \tag{8.6.8}$$

$$R_{N+1}(x)S_{N-1}(x) - R_{N-1}(x)S_{N+1}(x) = R_N(x)S_{N-1}(x) - R_{N-1}(x)S_N(x). \tag{8.6.9}$$

Equation (8.6.8) was derived at the beginning of Sec. 8.5; (8.6.9) follows easily from (8.4.8). Equations (8.6.8) and (8.6.9) and the positivity of the c_n imply that the sign of $R_{N+1}(x)S_{N-1}(x) - R_{N-1}(x)S_{N+1}(x)$ is $(-1)^N$ if x is positive. Thus, since $S_N > 0$ for all N [see (8.4.8)], $F_N = R_N/S_N$ satisfies $F_{2N+1}(x) \ge F_{2N-1}(x)$, $F_{2N}(x) \le F_{2N-2}(x)$, for all $N \ge 1$ and $x > 0$. But $F_{2N}(x)$ is just the Padé approximant $P_N^N(x)$ and $F_{2N+1}(x)$ is the Padé approximant $P_{N+1}^N(x)$. This proves properties (1) and (2).

To prove properties (3) to (5), it is sufficient to show that

$$P_{M+1}^M(x) \le F(x) \le P_M^M(x) \tag{8.6.10}$$

for any $M \ge 0$ and all $x > 0$. To do this, we observe from (8.6.5) that $F(x)$ is obtained from the continued fraction $F_N(x)$ by replacing $f_N(0)x$ by $f_N(0)x/[1 + xf_{N+1}(x)]$. [Recall that $f_{N+1}(x)$ is a Stieltjes function and $c_n = f_n(0)$ for all n.] Repeating the argument that leads to (8.4.8) and (8.6.8), it follows that $F(x) = R(x)/S(x)$, where

$$R(x)S_{N-1}(x) - R_{N-1}(x)S(x) = \frac{c_0 c_1 \cdots c_N}{1 + xf_{N+1}(x)}(-x)^N$$

and

$$S(x) = S_{N-1}(x) + \frac{c_N x}{1 + xf_{N+1}(x)} S_{N-2}(x) > 0.$$

Thus, the sign of $R(x)/S(x) - R_{N-1}(x)/S_{N-1}(x)$ is $(-1)^N$ when $x > 0$, proving (8.6.10). This completes the proof of properties (3) to (5).

Example 1 *Padé approximants for* $x^{-1} \ln(1+x)$. The function $\ln(1+x)/x$ is a Stieltjes function: $x^{-1}\ln(1+x) = \int_0^\infty d\rho(t)/(1+xt)$, where $\rho(t) = 1$ $(0 \le t \le 1)$ and $\rho(t) = 0$ $(t > 1)$. From Example 2 of Sec. 8.5, we know that $\lim_{N\to\infty} P_{N+1}^N(x) = \lim_{N\to\infty} P_N^N(x)$. Therefore, the Padé sequences $P_{N+1}^N(x)$ and $P_N^N(x)$ converge monotonically to $x^{-1}\ln(1+x)$ for x real and positive, as verified in Table 8.14. This proof of convergence is valid only for $x > 0$; the convergence of the Padé approximants to $\ln(1+z)/z$ in the complex-z plane is investigated in Prob. 8.59.

Example 2 *Padé approximants for* $\sum_{n=0}^\infty n!\,(-x)^n$. The series $\sum_{n=0}^\infty n!\,(-x)^n$ is asymptotic to the Stieltjes function $\int_0^\infty e^{-t}(1+xt)^{-1}\,dt$ as $x \to 0+$. Also, Example 4 of Sec. 8.5 shows that $\lim_{N\to\infty} P_{N+1}^N(x) = \lim_{N\to\infty} P_N^N(x)$. Therefore, the Padé sequences $P_{N+1}^N(x)$ and $P_N^N(x)$ converge monotonically to $\int_0^\infty e^{-t}(1+xt)^{-1}\,dt$, as verified in Table 8.16.

Convergence of Padé Sequences P_N^{N+J} for $J \neq 0, -1$

The inequalities (8.6.6) can be generalized as follows. For any $J \ge -1$ and $x > 0$, the Padé sequence $P_N^{N+J}(x)$ $(N = 0, 1, 2, \ldots)$, if generated from a Stieltjes function, is monotone increasing when J is odd and monotone decreasing when J is even. To derive this result we apply the argument used to prove (8.6.6) to the Padé sequence (8.4.6), noting that since the function being expanded is Stieltjes, c_n has the same sign for all n.

When $J < -1$, the monotonicity of $P_N^{N+J}(x)$ for $x > 0$ holds only for sufficiently large N, $N \ge N_0(x)$ where $N_0(x)$ is a positive function of x. When $J < -1$, the appearance of monotonicity is delayed because the Padé approximants P_N^{N+J} may have poles for positive x.

Example 3 *Nonmonotonic Padé sequence* P_{N+2}^N *for* $\sum n!\,(-x)^n$. A short calculation shows that the $P_{N+2}^N(x)$ Padé approximants to $\sum n!\,(-x)^n$ are

$$P_2^0(x) = \frac{1}{1 + x - x^2},$$

$$P_3^1(x) = \cfrac{1}{1 + x - \cfrac{x^2}{1+3x}},$$

$$P_4^2(x) = \cfrac{1}{1 + x - \cfrac{x^2}{1 + 3x - \cfrac{\ldots}{1 + 13x/3}}}.$$

$P_2^0(x)$ has a pole at $x_0 = \frac{1}{2} + \frac{1}{2}\sqrt{5} > 0$, $P_3^1(x)$ has no poles for $x > 0$, and $P_4^2(x)$ has a zero at $x_2 \doteq 2.33 > x_0$. Thus, $P_3^1(x) \le P_2^0(x)$ if $x < x_0$ and $P_3^1(x) \ge P_2^0(x)$ if $x > x_0$; also, $P_4^2(x) \le P_3^1(x)$ if $x < x_2$ and $P_4^2(x) \ge P_3^1(x)$ if $x > x_2$. The Padé sequence $P_{N+2}^N(x)$ (here, $J = -2$) is not monotonically decreasing for all N and $x > 0$.

Uniqueness of Limits of Padé Sequences

As mentioned above, it follows from (8.6.7) that if the Padé sequences $P_N^N(x)$ and $P_{N+1}^N(x)$ converge to the same limit as $N \to \infty$, then there is a *unique* Stieltjes function to which they converge. Conversely, if $\lim P_N^N \neq \lim P_{N+1}^N$, then there are many Stieltjes functions whose expansion coefficients are a_n in (8.6.2). A proof of this latter statement is given in Prob. 8.54. How can one know beforehand whether or not a Stieltjes series has a unique Stieltjes function as its Padé sum? Recall that a series is a Stieltjes series if there is at least one nonnegative density function $\rho(t)$ satisfying (8.6.2), so $\rho(t)$ satisfies the moment conditions (8.6.1) for all n.

The question is whether the moment conditions (8.6.1) determine the nonnegative function $\rho(t)$ uniquely. A complete solution to this problem is not known, but there are several *sufficient* (but not necessary) conditions on the moments a_n that guarantee the uniqueness of the nonnegative function $\rho(t)$. One such condition was found by Carleman. It states that if

$$\sum_{n=1}^{\infty} a_n^{-1/2n} = \infty, \tag{8.6.11}$$

then $\rho(t)$ is determined uniquely. A weak form of this result is proved in Prob. 8.55.

Convergence of Padé Approximants in the Complex Plane

Let us summarize some properties of Padé approximants to Stieltjes functions in the complex-z plane. Since Stieltjes functions are analytic in the cut plane $|\arg z| < \pi$ and since the Padé approximants $P_N^N(z)$ and $P_{N+1}^N(z)$ are Stieltjes functions, all the poles of $P_N^N(z)$ and $P_{N+1}^N(z)$ must lie on the negative real axis.

It may be shown that the Padé sequences $P_N^{N+J}(z)$ ($N = 0, 1, \ldots; J \geq -1$ fixed) converge as $N \to \infty$ in the cut-z plane: $\lim_{N\to\infty} P_N^{N+J}(z)$ exists for $J \geq -1$ fixed and $|\arg z| < \pi$. The proof of this fact is left to Prob. 8.59; the proof makes use of some properties of sequences of analytic functions and requires the convergence result (8.6.7) that the limits exist on the positive real axis.

Finally, we conclude by answering the questions posed at the beginning of this section. First, if a series is a Stieltjes series, then the Padé sequences $P_N^N(z)$ and $P_{N+1}^N(z)$ converge in the cut-z plane. The limit functions are Stieltjes functions. If the series coefficients a_n satisfy Carleman's condition (8.6.11), then the limit functions are identical and equal to the unique Stieltjes function whose asymptotic series expansion is $\sum a_n(-z^n)$ $(z \to 0+)$. Finally, if Carleman's condition is satisfied, then the branch of the Stieltjes function that is selected is the one that is real on the positive real axis.

PROBLEMS FOR CHAPTER 8

Section 8.1

(I) **8.1** Show that if $A_n \to A$ as $n \to \infty$ and $S(A_n)$, the Shanks transformation of A_n in (8.1.3), approaches S as $n \to \infty$, then $S = A$.

(I) **8.2** (*a*) If the terms in the sequence $A_1, A_2, A_3, \ldots$ are known accurately to K decimal digits, then the best approximation to $\lim_{n\to\infty} A_n$ that can be achieved by a multiply iterated Shanks transformation of the form in (8.1.3) will be accurate to about $K/2$ decimal digits. Why?

(*b*) On the other hand, if the differences $A_{n+1} - A_n$ ($n = 1, 2, 3, \ldots$) are known accurately to K decimal digits, then the multiply iterated Shanks transformation will usually determine the limit of the sequence accurately to about K decimal digits. Why?

(I) **8.3** (*a*) Show that the remainder term in (8.1.9) behaves asymptotically as $(-z)^{n+1}[n(1+z)]^{-3}$ as $n \to \infty$, $|z| < 1$.

(*b*) Discuss the behavior of the error as $z \to 1$ and as $z \to -1$.

(TI) **8.4** (*a*) Show that if $|c_0| < 1$ [see (8.1.12)], then the series $\sum a_n$ converges.

(*b*) Show that if $c_0 = -1$, then $\sum a_n$ converges if $c_1 > 0$ and diverges if $c_1 < 0$.

(E) **8.5** Show that $f(z) = \sum_{n=1}^{\infty} z^n/n^2$ has a branch point at $z = 1$.

(D) **8.6** (*a*) If we assume that (8.1.16) is valid, then applying Richardson extrapolation to the sequence 1, 1, 1, ... gives the lovely identity

$$1 = \sum_{k=0}^{N} \frac{(n+k)^N(-1)^{k+N}}{k!\,(N-k)!},$$

which holds for all $n = 0, 1, 2, \ldots$ and $N = 0, 1, 2, \ldots$. Find a direct algebraic proof of this identity which does not depend on (8.1.16).

(*b*) Use the above identity to derive (8.1.16).

(I) **8.7** (*a*) Derive (8.1.17) and (8.1.18).

(*b*) Do the Richardson extrapolants (8.1.16) of the series in (8.1.17) converge as $N \to \infty$ for fixed n?

(I) **8.8** Verify (8.1.20).

Clue: Sum both sides of (8.1.19) from $n = 1$ to ∞ and repeatedly substitute (8.1.19) to determine a Taylor series expansion of A_∞ in powers of z^2.

(I) **8.9** Verify (8.1.22).

(I) **8.10** Euler's constant γ is defined by $\gamma = \lim_{n\to\infty} (1 + \frac{1}{2} + \frac{1}{3} + \cdots + 1/n - \ln n)$. An efficient method for calculating γ is based on the Euler-Maclaurin sum formula (see Sec. 6.7). Use this sum formula to show that $1 + \frac{1}{2} + \frac{1}{3} + \cdots + 1/n \sim \ln n + \gamma + 1/2n + \sum_{k=2}^{\infty} B_k(-1)^{k+1}/kn^k$ in terms of the Bernoulli numbers B_k [see (8.1.23)]. Explain how to use this asymptotic series to compute γ to high accuracy.

(I) **8.11** Find an efficient method to evaluate the convergent sum $\sum_{n=2}^{\infty} 1/n(\ln n)^s$ for $s > 1$.

(E) **8.12** The Taylor series for e^x converges rapidly at $x = 1$, so it may be used directly to compute e. However, there is a more efficient method based on this series for calculating e. The idea is to evaluate $e^{2^{-k}}$, where k is an integer, using the Taylor series, and then to square the result k times. If it is required to evaluate e to p decimal places, determine the value of k which minimizes the number of algebraic operations. How many operations are required to calculate e to one million decimal places?

(I) **8.13** Generalize the method of the previous problem to compute accurate approximations to e^A where A is a matrix. Note that e^A is defined by its Taylor series: $e^A = I + A + A^2/2! + A^3/3! + \cdots$, where I is the identity matrix.

(E) **8.14** Use iterated Shanks transformations to accelerate the convergence of the series $\pi/4 = 1 - \frac{1}{3} + \frac{1}{5} - \frac{1}{7} + \frac{1}{9} - \cdots$. Estimate the improvement in the rate of convergence for each iteration.

(I) **8.15** A more efficient means of calculating π was used by Shanks and Wrench [*Math. Comp.*, vol. 16, p. 76 (1962)] who found the first 100,000 digits in its decimal expansion. Their method uses the formula $\pi = 24 \text{ arc tan } \frac{1}{8} + 8 \text{ arc tan } \frac{1}{57} + 4 \text{ arc tan } \frac{1}{239}$. The Taylor series for arc tan x is arc tan $x = \sum_{n=0}^{\infty} (-1)^n x^{2n+1}/(2n+1)$. Use this series both with and without the Shanks transformation to determine π and compare your results with those in Prob. 8.14. Estimate the number of operations required to calculate π to one million decimal places. Compare this method for computing π to high accuracy with the method described in Prob. 5.72.

(E) **8.16** The trapezoidal rule and Simpson's rule are two well-known methods of numerical integration. The trapezoidal rule is

$$[f(0) + 2f(1/N) + 2f(2/N) + \cdots + 2f(1 - 1/N) + f(1)]/2N \sim \int_0^1 f(x)\,dx + O(1/N^2), \qquad N \to \infty,$$

and Simpson's rule is

$$[f(0) + 4f(1/N) + 2f(2/N) + 4f(3/N) + 2f(4/N) + \cdots$$
$$+ 4f(1 - 1/N) + f(1)]/6N \sim \int_0^1 f(x)\,dx + O(1/N^4), \qquad N \to \infty,$$

where N must be an even integer in the latter formula. Show that Simpson's rule and the first Richardson extrapolation of the trapezoidal rule give identical results.

Section 8.2

(E) **8.17** Sum the series $1 - 1 + 1 - 1 + 1 - 1 \cdots$ using Euler and Borel summation and show that the results agree.

(E) **8.18** Compute the Euler sum of:
(*a*) $1 + 0 + (-1) + 0 + (-1) + 0 + 1 + 0 + 1 + 0 + (-1) + \cdots;$
(*b*) $1 + 0 + 0 + (-1) + 1 + 0 + 0 + (-1) + 1 + 0 + 0 + \cdots;$
(*c*) $1 + (-1) + 0 + 0 + 1 + (-1) + 0 + 0 + \cdots;$
(*d*) $\sum_{n=0}^{\infty} (-1)^n n^2.$

(I) **8.19** (*a*) Show that $0! - 2! + 4! - 6! + \cdots$ is not Borel summable but that $0! + 0 - 2! - 0 + 4! + 0 - 6! - 0$ is. Compute the Borel sum (Whittaker and Watson).
(*b*) Compute the generalized Borel sum of $0 - 2! + 4! - 6! + \cdots$.

(TD) **8.20** If the Euler and Borel sums of a series both exist, show that they must be equal.

Section 8.3

(E) **8.21** Let $f(z)$ have an asymptotic power series representation of the form $\sum a_n z^n$ $(z \to 0+)$ with $a_0 \neq 0$. Let $P_M^N(z)$ be the Padé approximant whose numerator has degree N and whose denominator has degree M. Show that $1/P_M^N(z)$ is the Padé approximant to $1/f(z)$ whose numerator has degree M and whose denominator has degree N.

(I) **8.22** Converting the power series $\sum_{n=0}^{\infty} (-z)^n$ into a sequence of Padé approximants is not routine because the matrix $\mathcal{A}$ in (8.3.2) has vanishing determinant. As a consequence, the coefficients $B_1, \ldots, B_M$ and also $A_0, \ldots, A_N$ are not unique. Nevertheless, the Padé approximants *are* unique. Compute $f^{(2,2)}(z)$ and $f^{(1,2)}(z)$ and show they are unique. Why does this happen in general?

(I) **8.23** Derive (8.3.7) and show that if the sequence $f_n(z)$ has P transient terms, then the kth-order Shanks transformation defined in (8.3.7) is exact for $k \geq P$.

(I) **8.24** Show that if the only singularities of $f(z)$ in the finite-z plane are P simple poles, then the remainder in the Taylor series for $f(z)$ has P transient terms.

(TD) **8.25** Show that if $k \leq n$, the kth-order Shanks transform $S_k(A_n)$ is identical to $P_k^n(1)$ where $P_k^n(z)$ is a Padé approximant of the series $A_1 + \sum_{j=1}^{\infty} (A_{j+1} - A_j)z^j$.

(I) **8.26** Formulate a set of determinantal equations for the coefficients of a two-point Padé approximant $R_N(z)/S_M(z)$ to a function $f(z)$ having the asymptotic expansions (8.3.8).
Clue: The result is analogous to (8.3.2) and (8.3.3).

(E) **8.27** Derive the asymptotic expansion (8.3.11).

(D) **8.28** Show that the poles and residues of the Padé approximants P_N^N and P_{N+1}^N to $\ln(1 + z)/z$ have the properties asserted in Table 8.14. In particular, show that the sum of the locations of the poles of P_N^N is $-N(N + 2)/2$, while the sum of the locations of the poles of P_{N+1}^N is $-(N + 1)(N + 2)$. Also show that for either P_N^N or P_{N+1}^N, if the residue of the pole at z_i is r_i then $\sum r_i/z_i^2 = \frac{1}{2}$. Can you compute $\sum r_i/z_i^n$ for all n?

Section 8.4

(I) **8.29** Derive the continued-fraction representation of $f(z) = F(a, b+1, c+1; z)/F(a, b, c; z)$ where F is the hypergeometric function

$$F(a, b, c; z) = \frac{\Gamma(c)}{\Gamma(a)\Gamma(b)} \sum_{n=0}^{\infty} \frac{\Gamma(a+n)\Gamma(b+n)}{n!\,\Gamma(c+n)} z^n$$

stated in Example 1(f) of Sec. 8.4.

Clue: It may be shown that $F(a, b, c; z)$ satisfies

$$F(a, b+1, c+1; z) = F(a, b, c; z) + \frac{a(c-b)}{c(c+1)} zF(a+1, b+1, c+2; z)$$

and $$F(a+1, b, c+1; z) = F(a, b, c; z) + \frac{b(c-a)}{c(c+1)} zF(a+1, b+1, c+2; z).$$

Use these relations to show that

$$\frac{F(a+1, b+1, c+2; z)}{F(a, b+1, c+1; z)} = 1 \bigg/ \left[1 - \frac{(b+1)(c-a+1)}{(c+1)(c+2)} z \frac{F(a+1, b+2, c+3; z)}{F(a+1, b+1, c+2; z)}\right].$$

(D) **8.30** Using the result of Prob. 8.29 derive the continued-fraction expansions of the functions e^z, $\ln(1+z)/z$, $(1+z)^\alpha$ by making special choices of a, b, c in $F(a, b, c; z)$. Check your results with Example 1 of Sec. 8.4.

Clue: See Sec. 8 of Chap. 2 of Ref. 11 or Ref. 33.

(E) **8.31** Compute the first three terms of the continued-fraction representation of the series:

(a) $\sum_{n=0}^{\infty} (n!)^2(-z)^n$;

(b) $\sum_{n=0}^{\infty} \frac{(-z)^n}{(2n)!}$;

(c) $\sum_{n=0}^{\infty} \frac{z^n}{n^2+1}$.

(I) **8.32** (a) Devise an efficient algorithm for computing the asymptotic power series coefficients g_n in $g(z) \sim \sum_{n=0}^{\infty} g_n z^n$ $(z \to 0+)$ where

(i) $g(z) = e^{f(z)}$,

(ii) $g(z) = \ln f(z)$,

(iii) $g(z) = f(z)^P$,

and $$f(z) \sim 1 + \sum_{n=1}^{\infty} f_n z^n, \qquad z \to 0+.$$

The problem is to compute g_n efficiently in terms of f_n.

Clue: When $g(z) = \ln f(z)$, $f'(z) = f(z)g'(z)$ so $(n+1)f_{n+1} = \sum_{m=0}^{n-1} (m+1)g_{m+1} f_{n-m} + (n+1)g_{n+1}$.

(b) Use the results of (a) to derive efficient algorithms for computing continued-function representations like continued exponentials, continued square roots, and continued logarithms.

(I) **8.33** Consider the series $\sum_{n=0}^{\infty} (-x)^n(2n)!/2^n n!$.

(a) Find the Borel sum of this series.

(b) Compute the Padé approximants P_N^N and P_{N+1}^N for $N = 1, 2, 3$ at $x = \frac{1}{3}$.

(c) Compute two continued-exponential, two continued-square-root, and two continued-logarithm approximations to the sum of the series at $x = \frac{1}{3}$.

(d) Compare the results obtained in parts (b) and (c) with the optimal asymptotic approximation to the sum at $x = \frac{1}{3}$ and with the Borel sum at $x = \frac{1}{3}$. Which method converges fastest?

(I) **8.34** Investigate the region of convergence of the continued exponential

$$\exp\,(z \exp\,(z \exp \ldots)).$$

(I) **8.35** Investigate the region of convergence of the continued logarithm

$$\ln\,(1 + z \ln\,(1 + z \ln\,(1 + z \ln \ldots))).$$

(I) **8.36** Compute three terms in the continued-exponential expansion of the series:

(a) $\sum_{n=0}^{\infty} (-z)^n$;

(b) $\sum_{n=0}^{\infty} n!\,(-z)^n$.

How well does the continued exponential converge compared with the Padé sequence $P_N^N(z)$?

(D) **8.37** Devise an efficient numerical method for computing two-point Padé approximants (see Sec. 8.3 and Prob. 8.26).

Clue: Try to modify the continued-fraction development of one-point Padé approximants.

(I) **8.38** Find two-point Padé approximants $P_N^N(x)$ to the functions

(a) $e^{-x} I_0(x)$,

(b) $x^{1-a} e^x \int_x^{\infty} e^{-t} t^{a-1}\, dt$,

at $x = 0, \infty$ for various N. Test the convergence of these two-point Padé approximants.

Section 8.5

(I) **8.39** Derive (8.5.12).

(TD) **8.40** (a) Show that the Padé approximants $P_N^N(z)$ and $P_{N+1}^N(z)$ to e^z converge to e^z as $N \to \infty$.

Clue: Let $F_N(z) = R_N(z)/S_N(z)$ be the Nth member of the Padé sequence $P_0^0, P_1^0, P_1^1, \ldots$. Since $S_N(z)e^z - R_N(z) = O(z^N)$ as $z \to 0$ and S_N and R_N are polynomials, Cauchy's theorem implies that

$$e^z - F_N(z) = \frac{z^N}{2\pi i S_N(z)} \oint_\Gamma \frac{S_N(t)e^t}{(t - z)t^N}\, dt,$$

where Γ is any contour on which $|z| < |t|$. Prove this. Then use (8.5.12) to show that $S_N(z) \to C(z)$ as $N \to \infty$ where $C(z)$ is some finite function. Use this result to show that $F_N(z) \to e^z$ as $N \to \infty$ provided that $C(z) \neq 0$.

(b) Show that $C(z) = e^{z/2}$ in (8.5.12).

(I) **8.41** (a) Use the continued-fraction representation of the Padé sequence $P_N^{N+J}(z)$ for arbitrary J to study the convergence of this sequence as $J \to \infty$. Generalize the result in (8.5.1.)

(b) Investigate the rate of convergence of the Padé sequence $P_N^{N+2}(z)$ to the function $1 + \frac{1}{2}z\sqrt{1 + z}$.

(I) **8.42** Derive (8.5.7).

(I) **8.43** Derive (8.5.17).

Section 8.6

(I) **8.44** Let $\sum_{n=0}^{\infty} a_n(-z)^n$ be a Stieltjes series so that there is a nonnegative function $\rho(t)$ satisfying (8.6.2) and let P_k ($k = 0, 1, \ldots, n$) be an arbitrary set of real numbers. Show that $\sum_{j=0}^{n} \sum_{k=0}^{n} a_{j+k} P_j P_k \geq 0$.

(TD) **8.45** Establish the converse of Prob. 8.44. That is, show that if all inequalities of the form $\sum a_{j+k} P_j P_k \geq 0$ are satisfied by the series coefficients a_k then $\sum a_k(-z)^n$ is a Stieltjes series. Show that in general $\rho(t)$ in (8.6.2) may be a distribution in the sense that it may behave like a δ function at discrete values of t. For example, if $a_k = \alpha^k$ then $\rho(t) = \delta(t - \alpha)$.

(I) **8.46** Show that if $F(z)$ is a Stieltjes function then $-F(z)$ is Herglotz. That is, show that $\text{Im}\,(-F) > 0$ if $\text{Im}\, z > 0$ and $\text{Im}\,(-F) < 0$ if $\text{Im}\, z < 0$.

(TD) **8.47** Prove that if $F(z)$ satisfies the four properties (8.6.3) with $C = 0$ then $F(z)$ is a Stieltjes function.

Clue: Use Cauchy's theorem to show that

$$F(z) = \frac{1}{\pi}\int_{-\infty}^{0} \frac{D(t)}{t - z}\, dt,$$

where $D(t) = \lim_{\varepsilon \to 0+} [F(t + i\varepsilon) - F(t - i\varepsilon)]/2i$. The relation between $D(t)$ and $\rho(t)$ in (8.6.2) is $\rho(t) = -D(-1/t)/\pi t$. Then use properties (3) and (4) to show that $\rho(t) \geq 0$ and that the moments of $\rho(t)$ exist.

(TD) **8.48** Generalize the construction of Prob. 8.47 to establish that properties (8.6.3) with $C > 0$ imply that $F(z)$ is a Stieltjes function.

Clue: Consider the subtracted function $F_s(z) = [F(z) - F(0)]/z$.

(E) **8.49** Use Probs. 8.47 and 8.48 to show that the density function $\rho(t)$ of a Stieltjes function $F(z)$ [see (8.6.2)] is uniquely determined by $F(z)$.

(TD) **8.50** Show that if $f(z)$ is a Stieltjes function then $g(z)$ defined by $f(z) = f(0)/[1 + zg(z)]$ is also a Stieltjes function.

Clue: Verify the properties (8.6.3).

(TI) **8.51** Show that if a continued fraction has nonnegative coefficients c_n then there exists a Stieltjes function whose continued fraction coefficients are c_n.

Clue: Use the fact that if $F(z)$ is a Stieltjes function then $1/[1 + zF(z)]$ is a Stieltjes function. Prove this result using (8.6.3). Also, show that the Padé sequence of Stieltjes functions P_N^N converges to a Stieltjes function.

(I) **8.52** (*a*) Show that if $F(x)$ is a Stieltjes function and has a Taylor series expansion about $x = 0$ with radius of convergence R then the error in the Padé approximants $P_N^N(x)$ or $P_{N+1}^N(x)$ goes to zero at least as fast as

$$\left| \frac{(1 + x/R)^{1/2} - 1}{(1 + x/R)^{1/2} + 1} \right|^{2N}.$$

(*b*) Show that the error in the Padé approximants to $F(x) = \tanh \sqrt{x}/\sqrt{x}$ is much smaller than the bound in (*a*).

(TD) **8.53** Show that if $F(x)$ is a Stieltjes function then its Padé approximants $P_M^N(x)$ satisfy the following inequalities when $x > 0$:

(*a*) $P_{N+1}^{N+1}(x) \leq P_N^{N+2}(x) \leq P_{N-1}^{N+3}(x) \leq \cdots$;

(*b*) $P_{N+1}^{N}(x) \leq P_N^{N+1}(x) \leq P_{N-1}^{N+2}(x) \leq \cdots$;

(*c*) $(d/dx)P_{N+1}^N(x) \leq F'(x) \leq (d/dx)P_N^N(x)$.

(TI) **8.54** Show that if the Padé approximants P_N^N and P_{N+1}^N to a Stieltjes series satisfy $\lim_{N\to\infty} P_N^N(x) \neq \lim_{N\to\infty} P_{N+1}^N(x)$, then there are many Stieltjes functions asymptotic to the given Stieltjes series.

(TD) **8.55** Show that if $a_n = O[(2n)!\, M^{2n}]$ $(n \to \infty)$ for some finite M, then the moment conditions (8.6.1) determine the nonnegative function $\rho(t)$ uniquely if the inequalities stated in Prob. 8.44 are satisfied. Note that if a_n satisfies this bound, then the Carleman condition (8.6.11) is satisfied.

Clue: If there are two solutions $\rho_k(t)$ $(k = 1, 2)$ to the moment equations (8.6.1), show that $\int_{-\infty}^{\infty} \sigma_k(t) \cosh (xt) = \sum_{n=0}^{\infty} a_n x^{2n}/(2n)! < \infty$ for $|x| < 1/M$, where $\sigma_k(t) = |t|\rho_k(t^2)$. Define $f_k(z) = \int_{-\infty}^{\infty} \sigma_k(t) e^{izt}\, dt$ and show that $f_k(z)$ is analytic in the strip $|\text{Im}\, z| < 1/M$. Show that $f_1(z) = f_2(z)$ by showing first that all the derivatives of f_1 equal the derivatives of f_2 at $z = 0$ and then using analytic continuation. Finally, conclude that $\sigma_1(t) = \sigma_2(t)$ because Fourier transforms are unique.

(D) **8.56** Show explicitly that if $a_n = (8n + 7)!$ then the moment equations (8.6.1) have a nonunique solution for the nonnegative function $\rho(t)$. Note that Carleman's condition (8.6.11) is violated by this sequence a_n.

Clue: Show that $\rho(t) = e^{-t^{1/8}}/8$ is one solution. Using complex variables techniques, show that the functions $\rho_a(t) = e^{-t^{1/8}}/8 + ae^{-t^{1/4}} \sin(\sqrt{3}\, t^{1/4})$ for all sufficiently small a are also solutions of (8.6.1). Show that we need only require $-0.400\,133 \cdots < a < 1.066\,27 \cdots$

(D) **8.57** Investigate the rapidity of growth of continued-fraction coefficients c_n in terms of the rapidity of growth of the coefficients a_n in $\sum a_k(-z)^k$ as $n \to \infty$. In particular, test the conjecture that c_n grows like n^k when a_n grows like $(kn)!$ as $n \to \infty$ with $k > 0$.

(D) **8.58** Show that the continued fractions $F_N(x)$ with coefficients $c_n = n^k$ $(k \geq 1)$ have the following convergence properties when $x > 0$:

(*a*) If $1 \leq k < 2$, then

$$F_N(x) - F_{N-1}(x) \sim (-1)^N A(x) \exp\left[-\frac{2N^{1-k/2}}{(2-k)\sqrt{x}}\right], \qquad N \to \infty,$$

for some nonnegative function $A(x)$.

(*b*) If $k = 2$, then $F_N(x) - F_{N-1}(x) \sim (-1)^N A(x) N^{-1/\sqrt{x}}$ $(N \to \infty)$ for some nonnegative function $A(x)$.

(*c*) If $k > 2$, then $F_N(x) - F_{N-1}(x) \sim (-1)^N A(x)$ $(N \to \infty)$ for some nonzero function $A(x)$. Thus, $F_N(x)$ does not converge as $N \to \infty$ if $k > 2$. Show that this result is consistent with Carleman's condition (8.6.11).

(TD) **8.59** Show that the Padé approximants $P_N^N(z)$ to a Stieltjes series $\sum a_n(-z)^n$ converge to an analytic function in the cut-z plane $|\arg z| < \pi$.

Clues: Use Vitali's theorem. Let $\{f_n(z)\}$ be a sequence of analytic functions in a simply connected region R, which is uniformly bounded over every finite closed domain R' lying entirely within R. If the sequence converges at an infinite set of points having at least one limit point interior to R, then the sequence converges throughout R, uniformly in finite closed domains lying entirely within R, to an analytic function. To apply this theorem, first show that for Stieltjes series each function $P_N^N(z)$ is analytic in the cut-z plane. Then show that

$$|P_N^N(z)| \leq a_0 + a_1|z|, \qquad \text{Re } z \geq 0,$$

$$|P_N^N(z)| \leq a_0 + a_0|z|/\text{Im } z, \qquad \text{Re } z < 0.$$

These bounds may be established using the representation

$$P_N^N(z) = a_0 - z \sum_{q=1}^{N} B_q/(1 + C_q z),$$

where $B_q \geq 0$, $C_q \geq 0$ for all q when $\sum a_n(-z)^n$ is a Stieltjes series. Show that $\sum_{q=1}^N B_q = -a_1$ and $\sum_{q=1}^N B_q/C_q \leq a_0$. Finally, use the fact that the Padé sequence $P_N^N(x)$ converges on the positive real axis to complete the verification of the conditions of Vitali's theorem.

(TD) **8.60** Generalize the argument of Prob. 8.59 to show that the Padé sequences $P_N^{N+J}(z)$ to a Stieltjes series converge to analytic functions in the cut-z plane as $N \to \infty$ for J fixed, $J \geq -1$.

(D) **8.61** Show that all the poles of the Padé approximants $P_M^N(z)$ to e^z lie in the right half of the complex-z plane.

Clues: Write $P_M^N = R_N/S_M$. First, consider M even. Use (8.5.10) to show that $S_{2M}(z) = C_M(z) - zD_M(z)$ where $C_0 = C_1 = 1$ and $D_0 = 0$, $D_1 = \frac{1}{2}$. Show that $D_M(z)/C_M(z)$ is a Padé approximant to a Stieltjes function of z^2. Use the Herglotz property of $-D_M/C_M$ to show that the sign of Im (D_M/C_M) is opposite to the sign of Im (z^2) so that $S_{2M} = 0$ only if Im (z^2) and Im z have the same sign. Hence, all the poles of $P_{2M}^N(z)$ must be in the right half plane. Generalize the above argument to establish the same result for $P_{2M+1}^N(z)$.

PART FOUR

GLOBAL ANALYSIS

My life is spent in one long effort to escape from the commonplaces of existence. These little problems help me to do so.

—Sherlock Holmes, *The Red-Headed League*
Sir Arthur Conan Doyle

In Part III we introduced some general perturbative methods necessary to perform global analysis. In this part we survey three specific analytical techniques of global approximation theory: boundary-layer theory, WKB theory, and multiple-scale analysis.

Boundary-layer theory and WKB theory are a collection of perturbative methods for obtaining an asymptotic approximation to the solution $y(x)$ of a differential equation whose highest derivative is multiplied by a small parameter ε. Solutions to such equations usually develop regions of rapid variation as $\varepsilon \to 0$. If the thickness of these regions approaches 0 as $\varepsilon \to 0$, they are called boundary layers, and boundary-layer theory may be used to approximate $y(x)$. If the extent of these regions remains finite as $\varepsilon \to 0$, one must use WKB theory. For linear equations boundary-layer theory is a special case of WKB theory, but boundary-layer theory also applies directly to nonlinear equations while WKB theory does not.

Multiple-scale theory is used when ordinary perturbative methods fail to give a uniformly accurate approximation to $y(x)$ for both small and large values of x. Some (although certainly not all) perturbation problems which yield to boundary-layer or WKB analysis can also be solved using multiple-scale analysis.

CHAPTER

NINE

BOUNDARY-LAYER THEORY

His career has been an extraordinary one. He is a man of good birth and excellent education, endowed by nature with a phenomenal mathematical faculty. At the age of twenty-one he wrote a treatise upon the binomial theorem, which has had a European vogue. On the strength of it he won the mathematical chair at one of our smaller universities, and had, to all appearances, a most brilliant career before him. But the man had hereditary tendencies of the most diabolical kind. A criminal strain ran in his blood, which, instead of being modified, was increased and rendered infinitely more dangerous by his extraordinary mental powers.

—Sherlock Holmes, *The Final Problem*
Sir Arthur Conan Doyle

(E) 9.1 INTRODUCTION TO BOUNDARY-LAYER THEORY

In this and the next chapter we discuss perturbative methods for solving a differential equation whose highest derivative is multiplied by the perturbing parameter ε. The most elementary of these methods is called boundary-layer theory.

A *boundary layer* is a narrow region where the solution of a differential equation changes rapidly. By definition, the thickness of a boundary layer must approach 0 as $\varepsilon \to 0$. In this chapter we will be concerned with differential equations whose solutions exhibit only *isolated* (well-separated) narrow regions of rapid variation. It is possible for a solution to a perturbation problem to undergo rapid variation over a thick region (one whose thickness does *not* vanish with ε). However, such a region is not a boundary layer. We will consider such problems in Chap. 10.

Here is a simple boundary-value problem whose solution exhibits boundary-layer structure.

Example 1 *Exactly soluble boundary-layer problem.* Consider the differential equation

$$\varepsilon y'' + (1 + \varepsilon)y' + y = 0, \qquad y(0) = 0,\ y(1) = 1. \tag{9.1.1}$$

The exact solution of this equation is

$$y(x) = \frac{e^{-x} - e^{-x/\varepsilon}}{e^{-1} - e^{-1/\varepsilon}}. \tag{9.1.2}$$

In the limit $\varepsilon \to 0+$, this solution becomes discontinuous at $x = 0$, as is shown in Fig. 9.1. For very small ε the solution $y(x)$ is slowly varying for $\varepsilon \ll x \le 1$. However, on the small interval $0 \le x \le O(\varepsilon)$ $(\varepsilon \to 0+)$ it undergoes an abrupt and rapid change. This small interval of rapid change is called a *boundary layer*. [The notation $0 \le x \le O(\varepsilon)$ means that the thickness of the boundary layer is proportional to ε as $\varepsilon \to 0+$.] The region of slow variation of $y(x)$ is called the *outer* region and the boundary-layer region is called the *inner* region.

Boundary-layer theory is a collection of perturbation methods for solving differential equations whose solutions exhibit boundary-layer structure. When the solution to a differential equation is slowly varying except in isolated boundary layers, then it may be relatively easy to obtain a leading-order approximation to that solution for small ε without directly solving the differential equation.

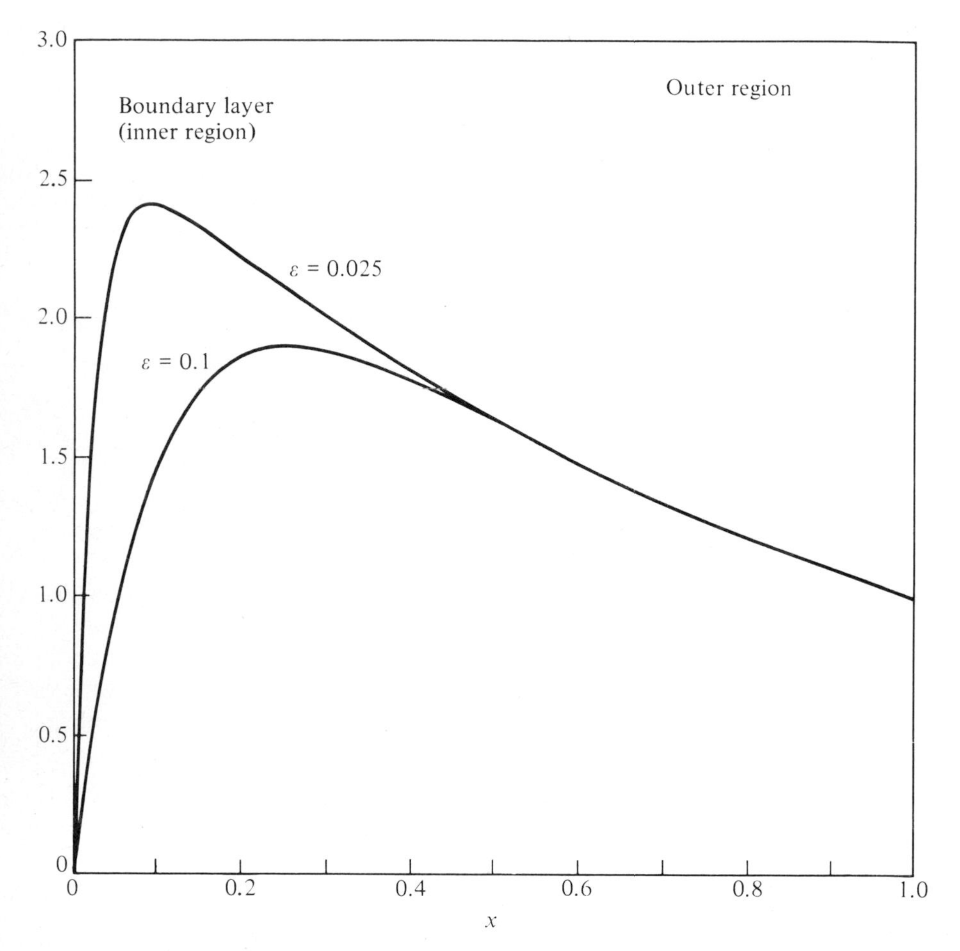

Figure 9.1 A plot of $y(x) = (e^{-x} - e^{-x/\varepsilon})/(e^{-1} - e^{-1/\varepsilon})$ $(0 \le x \le 1)$ for $\varepsilon = 0.1$ and 0.025. Note that $y(x)$ is slowly varying for $\varepsilon \ll x \le 1$ $(\varepsilon \to 0+)$. However, on the interval $0 \le x \le O(\varepsilon)$, $y(x)$ rises abruptly from 0 and becomes discontinuous in the limit $\varepsilon \to 0+$. This narrow and isolated region of rapid change is called a boundary layer.

There are two standard approximations that one makes in boundary-layer theory. In the outer region (away from a boundary layer) $y(x)$ is slowly varying, so it is valid to neglect any derivatives of $y(x)$ which are multiplied by ε. Inside a boundary layer the derivatives of $y(x)$ are large, but the boundary layer is so narrow that we may approximate the coefficient functions of the differential equation by constants. Thus, we can replace a single differential equation by a sequence of much simpler approximate equations in each of several inner and outer regions. In every region the solution of the approximate equation will contain one or more unknown constants of integration. These constants are then determined from the boundary or initial conditions using the technique of asymptotic matching which was introduced in Sec. 7.4.

The following initial-value problem illustrates these ideas.

Example 2 *First-order nonlinear boundary-layer problem.* From the initial-value problem

$$(x - \varepsilon y)y' + xy = e^{-x}, \qquad y(1) = 1/e, \tag{9.1.3}$$

we wish to determine a leading-order perturbative approximation to $y(0)$ as $\varepsilon \to 0+$.

Although this is only a first-order differential equation, it is nonlinear and is much too difficult to solve in closed form. However, in regions where y and y' are not large (such regions are called outer regions), it is valid to neglect $\varepsilon yy'$ compared with e^{-x}. Thus, in outer regions we approximate the solution to (9.1.3) by the solution to the outer equation

$$xy'_{\text{out}} + xy_{\text{out}} = e^{-x}.$$

This equation is easy to solve because it is linear. The solution which satisfies $y_{\text{out}}(1) = 1/e$ is

$$y_{\text{out}} = (1 + \ln x)e^{-x}. \tag{9.1.4}$$

Note that it is valid to impose the initial condition $y(1) = 1/e$ on $y_{\text{out}}(x)$ because $x = 1$ lies in an outer region; $x = 1$ is in an outer region because (9.1.3) implies that $y'(1) = 0$, so $y(1)$ and $y'(1)$ are of order 1 as $\varepsilon \to 0+$.

As $x \to 0+$, both $y_{\text{out}}(x)$ and $y'_{\text{out}}(x)$ become larger. Thus, near $x = 0$ the term $\varepsilon yy'$ is no longer negligible compared with e^{-x}. From the outer solution we can estimate that the thickness δ of the region in which $\varepsilon yy'$ is not small is given by

$$\delta/\ln \delta = O(\varepsilon), \qquad \varepsilon \to 0+.$$

Thus, $\delta \to 0+$ as $\varepsilon \to 0+$ [in fact, $\delta = O(\varepsilon \ln \varepsilon)$ as $\varepsilon \to 0+$] (see Prob. 9.1), and there is a boundary layer of thickness δ at $x = 0$.

In the boundary layer (the inner region), x is small so it is valid to approximate e^{-x} by 1. Furthermore, since y varies rapidly in the narrow boundary layer, we may neglect xy compared with xy'. Hence, in the inner region we approximate the solution to (9.1.3) by the solution to the inner equation

$$(x - \varepsilon y_{\text{in}})y'_{\text{in}} = 1.$$

This is a linear equation if we regard x as the dependent variable. Its solution is

$$x = \varepsilon(y_{\text{in}} + 1) + Ce^{y_{\text{in}}}, \tag{9.1.5}$$

where C is an unknown constant of integration. Since $x = 0$ is in the inner region, we may use (9.1.5) to find an approximation to $y(0)$.

C is determined by asymptotically matching the outer and inner solutions (9.1.4) and (9.1.5). Take x small but not as small as δ, say $x = O(\varepsilon^{1/2})$. Then (9.1.4) implies that $y_{\text{out}} \sim 1 + \ln x$ as $\varepsilon \to 0+$ and (9.1.5) implies that $x \sim Ce^{y_{\text{in}}}$ as $\varepsilon \to 0+$. Thus, $C = 1/e$ and a leading-order implicit equation for $y_{\text{in}}(0)$ is

$$0 = \varepsilon[y_{\text{in}}(0) + 1] + e^{y_{\text{in}}(0) - 1}. \tag{9.1.6}$$

When $\varepsilon = 0.1$ and 0.01, the numerical solutions of (9.1.6) are $y_{\text{in}}(0) \doteq -1.683$ and $y_{\text{in}}(0) \doteq -2.942$, respectively. These results compare favorably with the numerical solution to (9.1.3) which gives $y(0) \doteq -1.508$ when $\varepsilon = 0.1$ and $y(0) \doteq -2.875$ when $\varepsilon = 0.01$. For both values of ε the relative error between the perturbative and the numerical solutions for $y(0)$ is about $\frac{1}{2}\varepsilon \ln \varepsilon$. Figures 9.2 and 9.3 compare the inner and outer perturbative approximations to $y(x)$ with the numerical solution.

Boundary-layer theory can also be a very powerful tool for determining the behavior of solutions to higher-order equations.

Example 3 *Second-order linear boundary-value problem.* Let us find an approximate solution to the boundary-value problem

$$\varepsilon y''(x) + a(x)y'(x) + b(x)y(x) = 0, \qquad 0 \le x \le 1,\ y(0) = A,\ y(1) = B, \tag{9.1.7}$$

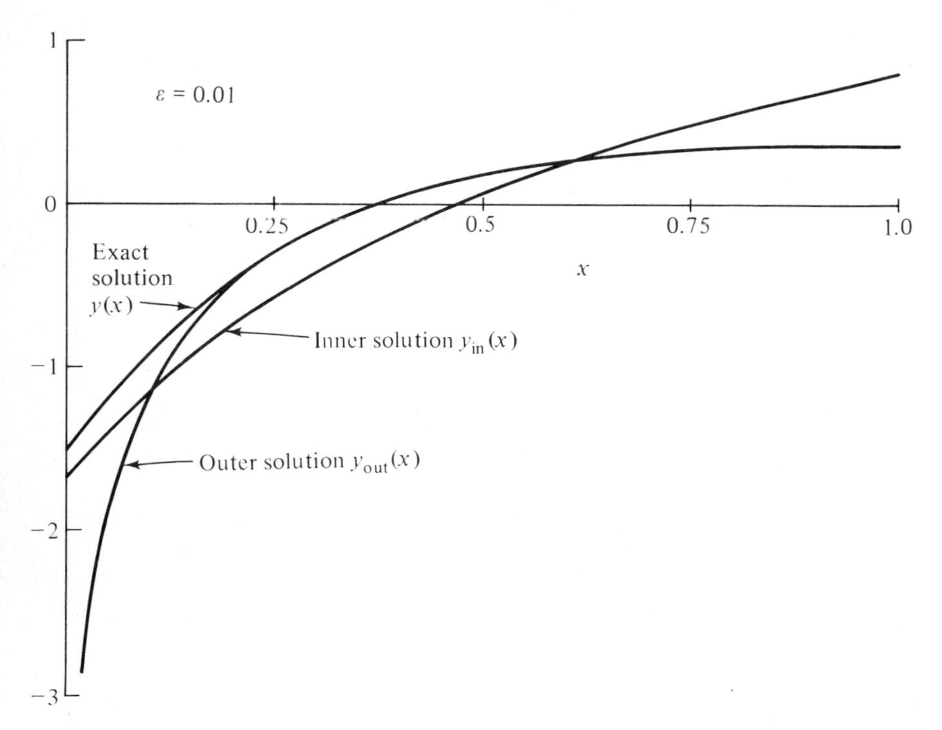

Figure 9.2 A comparison for $\varepsilon = 0.1$ of the exact solution $y(x)$ to the nonlinear differential equation (9.1.3) and the inner and outer approximations to $y(x)$ using boundary-layer theory. The integration constant in y_{out} is determined from the initial condition $y(1) = 1/e$. The integration constant in y_{in} is determined from asymptotic matching. A measure of the accuracy of the boundary-layer approximation is the magnitude of the error in the predicted value of $y(0)$. When $\varepsilon = 0.1$, $y(0) \doteq -1.508$ and $y_{\text{in}}(0) \doteq -1.683$, an error of about 10 percent.

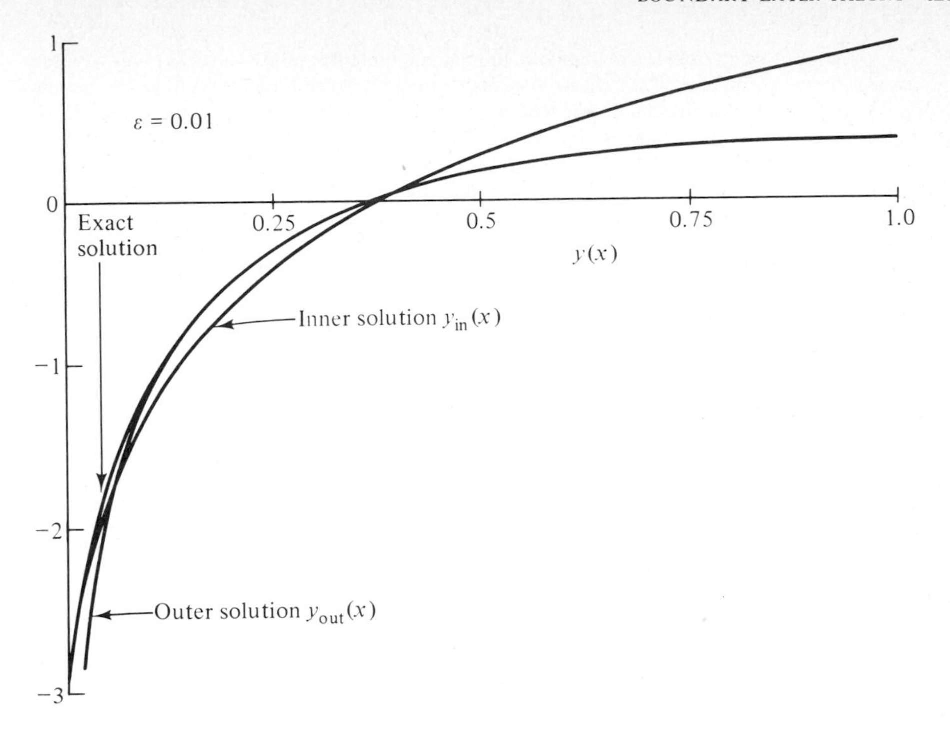

Figure 9.3 Same as Fig. 9.2 with $\varepsilon = 0.1$ replaced by $\varepsilon = 0.01$. Here, $y(0) \doteq -2.875$ and $y_{\text{in}}(0) \doteq -2.942$, an error of about 2 percent. Observe that as $\varepsilon \to 0+$, the inner and outer approximations $y_{\text{in}}(x)$ and $y_{\text{out}}(x)$ hug the exact solution $y(x)$ more closely. The error appears to be of order $\frac{1}{2}\varepsilon \ln \varepsilon$ (see Prob. 9.5).

as $\varepsilon \to 0+$. We assume for reasons to be made clear later that $a(x) \neq 0$ for $0 \le x \le 1$, and for definiteness we choose $a(x) > 0$; otherwise $a(x)$ and $b(x)$ are arbitrary continuous functions.

We shall analyze the behavior of $y(x)$ as $\varepsilon \to 0+$ by assuming that in this limit the solution $y(x)$ develops an isolated boundary layer in the neighborhood of $x = 0$ and that there are no other regions of rapid change of $y(x)$ ($\varepsilon \to 0+$). We will then justify these assumptions by showing that no other possibility is mathematically consistent.

The outer region is characterized by the absence of rapid variation of $y(x)$: $y(x)$, $y'(x)$, and $y''(x)$ are all of order 1 (assuming that A and B are finite) as $\varepsilon \to 0+$. Thus, in the outer region a good approximation to (9.1.7) is the first-order linear equation

$$a(x)y'_{\text{out}}(x) + b(x)y_{\text{out}}(x) = 0. \tag{9.1.8}$$

Observe that the outer approximation has reduced the order of the differential equation, thereby making it soluble. The solution to (9.1.8) is $y_{\text{out}}(x) = K \exp\left[\int_x^1 b(t)/a(t)\, dt\right]$, where K is an integration constant. In general, it is not possible for $y_{\text{out}}(x)$ to satisfy both boundary conditions $y(0) = A$ and $y(1) = B$. However, we have assumed that $x = 1$ lies within the outer region and that $x = 0$ does not. Thus, we should require that $y_{\text{out}}(1) = B$, but not $y_{\text{out}}(0) = A$. It follows that $K = B$:

$$y_{\text{out}}(x) = B \exp\left[\int_x^1 b(t)/a(t)\, dt\right]. \tag{9.1.9}$$

The outer solution (9.1.9) is a uniform approximation to the solution $y(x)$ as $\varepsilon \to 0+$ on the subinterval $\delta \ll x \le 1$ of $[0, 1]$, where $\delta(\varepsilon)$ is the thickness of the boundary layer. It is now becoming clear why we have assumed that $a(x_0) \ne 0$ for $0 \le x_0 \le 1$. If $a(x_0) = 0$ for some x_0 on this interval, then $y_{\text{out}}(x)$ would be singular at x_0, assuming that $b(x_0) \ne 0$. This would violate the assumption that y, y', and y'' are all of order 1.

The outer solution $y_{\text{out}}(x)$ is not valid in the neighborhood of $x = 0$ unless $y_{\text{out}}(0) = A$, in which case $y_{\text{out}}(x)$ is a uniformly valid leading-order approximation to $y(x)$ for $0 \le x \le 1$. However, since A is arbitrary, in general $y_{\text{out}}(0) \ne A$. Thus, the boundary condition $y(0) = A$ must be achieved through a boundary layer at $x = 0$. In other words, the outer solution $y_{\text{out}}(x)$ is approximately equal to $y(x)$ as x approaches 0 from above until $x = O(\delta)$. At this point $y_{\text{out}}(x)$ is approaching and already very close to $y_{\text{out}}(0)$, while the actual solution $y(x)$ rapidly veers off and approaches $y(0) = A$ (see Fig. 9.4).

To determine the behavior of $y(x)$ when $x = O(\delta)$, we may approximate the functions of $a(x)$ and $b(x)$ in the original differential equation (9.1.7) by $a(0) = \alpha \ne 0$ and $b(0) = \beta$ because δ vanishes as $\varepsilon \to 0$. Also, in the inner region, y is much smaller than y' because y is rapidly varying. Therefore, we may neglect y compared with y'. Thus, the inner approximation to (9.1.7) is the constant coefficient differential equation

$$\varepsilon y_{\text{in}}'' + \alpha y_{\text{in}}' = 0, \tag{9.1.10}$$

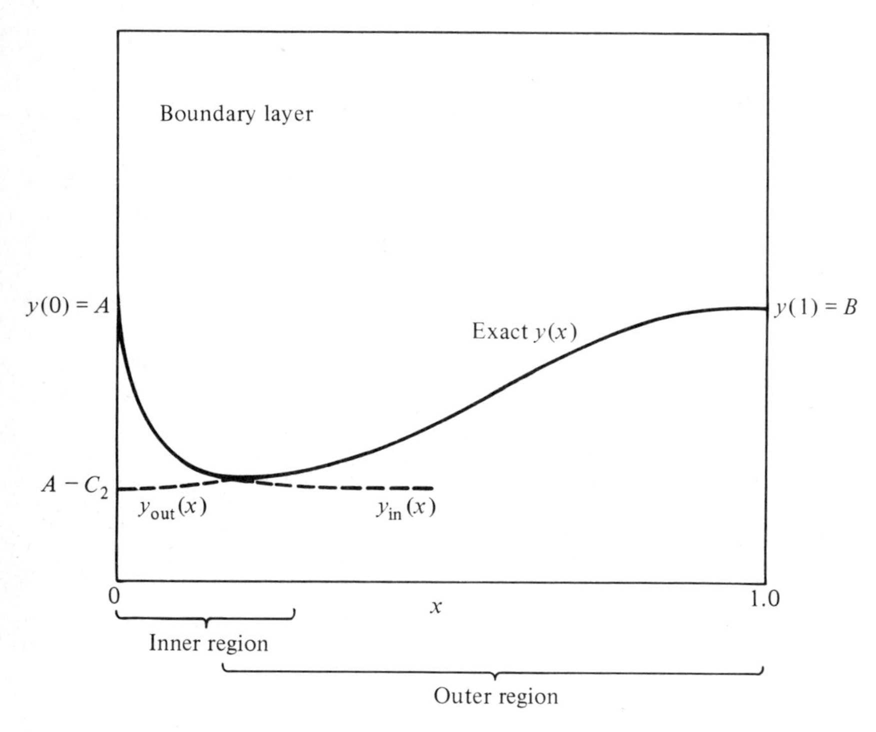

Figure 9.4 A schematic plot of the solution to the boundary-value problem $\varepsilon y''(x) + a(x)y'(x) + b(x)y(x) = 0[0 \le x \le 1;\, a(x) > 0]$ with $y(0) = A$, $y(1) = B$, in (9.1.7). The exact solution satisfies the boundary conditions $y(0) = A$ and $y(1) = B$ and has a boundary layer (region of rapid variation) of thickness $O(\varepsilon)$ at $x = 0$. The outer solution $y_{\text{out}}(x)$ is a good approximation to $y(x)$ in the outer region, but $y_{\text{out}}(0) = A - C_2$. The inner solution is a good approximation to $y(x)$ in the inner region. The asymptotic match of $y_{\text{in}}(x)$ and $y_{\text{out}}(x)$ occurs in the overlap of the inner and outer regions; in the overlap region $y_{\text{in}}(x)$ and $y_{\text{out}}(x)$ both approach the constant $A - C_2$.

which is soluble. The most general solution to (9.1.10) is

$$y_{\text{in}}(x) = C_1 + C_2 e^{-\alpha x/\varepsilon}.$$

Finally, we must require that $y_{\text{in}}(0) = y(0) = A$. Thus,

$$y_{\text{in}}(x) = A + C_2(e^{-\alpha x/\varepsilon} - 1). \tag{9.1.11}$$

The remaining constant of integration C_2 will be determined by asymptotic matching.

Since $y_{\text{in}}(x)$ in (9.1.11) varies rapidly when $x = O(\varepsilon)$, we conclude that the boundary-layer thickness δ is of order ε. The asymptotic match of the inner and outer solutions takes place between the rightmost edge of the inner region and the leftmost edge of the outer region, say for values of $x = O(\varepsilon^{1/2})$. For such values of x,

$$y_{\text{in}}(x) \sim A - C_2, \qquad \varepsilon \to 0+,$$

and

$$y_{\text{out}}(x) \sim y_{\text{out}}(0) = B \exp\left[\int_0^1 b(t)/a(t)\, dt\right], \qquad \varepsilon \to 0+.$$

Thus, if $y_{\text{in}}(x)$ and $y_{\text{out}}(x)$ are to be good approximations to $y(x)$ in the overlap of the inner and outer regions, then we must require that $C_2 = A - y_{\text{out}}(0)$.

To summarize, the boundary-layer approximation is

$$\begin{aligned} y(x) &\sim B \exp\left[\int_x^1 b(t)/a(t)\, dt\right], && 0 < x \le 1,\ \varepsilon \to 0+; \\ y(x) &\sim Ae^{-a(0)x/\varepsilon} + B(1 - e^{-a(0)x/\varepsilon}) \exp\left[\int_0^1 b(t)/a(t)\, dt\right], && x = O(\varepsilon),\ \varepsilon \to 0+. \end{aligned} \tag{9.1.12}$$

We may proceed further by combining the above two expressions into a single, *uniform* approximation y_{unif}, valid for all $0 \le x \le 1$. A suitable expression is

$$y_{\text{unif}}(x) = y_{\text{out}}(x) + y_{\text{in}}(x) - y_{\text{match}}(x),$$

where $y_{\text{match}} = A - C_2$. Hence,

$$y_{\text{unif}}(x) = B \exp\left[\int_x^1 b(t)/a(t)\, dt\right] + \left\{A - B \exp\left[\int_0^1 b(t)/a(t)\, dt\right]\right\} e^{-a(0)x/\varepsilon}. \tag{9.1.13}$$

To verify that $y_{\text{unif}}(x) \sim y(x)$ $(\varepsilon \to 0+)$, one must examine it for values of x in the inner and outer regions and check that it reduces to the two expressions in (9.1.12). Equation (9.1.13) is a uniform approximation in the sense that the difference between $y(x)$ and $y_{\text{unif}}(x)$ is uniformly $O(\varepsilon)$ $(0 \le x \le 1,\ \varepsilon \to 0+)$ (see Prob. 9.2).

We conclude this example with several observations. First, if $a(x) < 0$ throughout $[0, 1]$, then no match is possible with the boundary layer solution (9.1.11) at $x = 0$ because $y_{\text{in}}(x)$ grows exponentially with x/ε unless $C_2 = 0$. On the other hand, if the boundary layer occurs at $x = 1$, then matching is possible if $a(x) < 0$ (see Prob. 9.3). If $a(x) > 0$, it is impossible to match to a boundary layer at $x = 1$ for the same reason that a match cannot be made at $x = 0$ when $a(x) < 0$.

Second, there can be no boundary layer at an internal point x_0 $(0 < x_0 < 1)$ if $a(x_0) \ne 0$. If a boundary layer did exist at x_0, then within this narrow layer we could approximate the original differential equation (9.1.7) by $\varepsilon y''_{\text{in}} + a(x_0)y'_{\text{in}} = 0$. The general solution to this equation is

$$y_{\text{in}} = C_1 + C_2 e^{-a(x_0)(x-x_0)/\varepsilon}.$$

If $a(x_0) > 0$ (<0), then no asymptotic match is possible at the left (right) edge of the boundary layer unless $C_2 = 0$ because the approximation must remain finite. Thus, the matching conditions require that $C_2 = 0$. Hence, the outer solutions to the left and right of the boundary layer both approach the same constant C_1 as $x \to x_0$ from below and above. Thus, the outer solutions approach each other and there is no internal region of rapid change.

In summary, then, when $a(x)$ in (9.1.7) satisfies $a(x) > 0$ for $0 \le x \le 1$ the boundary layer always lies at $x = 0$ and when $a(x) < 0$ for $0 \le x \le 1$ the boundary layer always lies at $x = 1$.

This completes our heuristic introduction to boundary-layer theory. Our purpose in this section was to show how to convert difficult differential equations into easy ones by seeking approximate rather than exact solutions. However, several questions must be answered before the ideas of boundary-layer theory can really be applied with confidence. For example, how can one know *a priori* whether the solution to a differential equation has boundary-layer structure? How can one predict the locations of the boundary layers? How does one estimate δ, the thickness of the boundary layer? How can we be sure that there is an overlap region between the inner and outer regions on which to perform asymptotic matching? Is it useful to decompose a solution into its inner and outer parts if one is seeking a high-order approximation to the exact answer? These questions will be answered in the next two sections.

(E) 9.2 MATHEMATICAL STRUCTURE OF BOUNDARY LAYERS: INNER, OUTER, AND INTERMEDIATE LIMITS

Having demonstrated the power and broad applicability of boundary-layer analysis in Sec. 9.1, it is now appropriate to formalize and restate more carefully some of the rather loosely defined concepts. This section deals with the questions about boundary-layer theory that were raised at the end of the previous section.

To keep our presentation as concrete as possible we will use Example 1 of Sec. 9.1 as a model boundary-layer problem and will analyze its mathematical structure in detail. You will recall that the function

$$y(x) = \frac{e^{-x} - e^{-x/\varepsilon}}{e^{-1} - e^{-1/\varepsilon}}, \tag{9.2.1}$$

which is the exact solution of the boundary-value problem

$$\varepsilon y'' + (1 + \varepsilon)y' + y = 0, \qquad y(0) = 0,\ y(1) = 1, \tag{9.2.2}$$

has a boundary layer at $x = 0$ when $\varepsilon \to 0+$. The function $y(x)$ has two components: e^{-x}, a slowly varying function on the entire interval $[0, 1]$, and $e^{-x/\varepsilon}$, a rapidly varying function in the boundary layer $x \le O(\delta)$, where $\delta = O(\varepsilon)$ is the thickness of the boundary layer.

In boundary-layer theory we treat the solution y of the differential equation as a function of two independent variables, x and ε. The goal of the analysis is to find a global approximation to y as a function of x; this is achieved by performing a local analysis of y as $\varepsilon \to 0+$.

To explain the appearance of the boundary layer we introduce the notion of an inner and outer limit of the solution. The *outer limit* of the solution (9.2.1) is obtained by choosing a fixed x *outside* the boundary layer, that is, $\delta \ll x \le 1$, and allowing $\varepsilon \to 0+$. Thus, the outer limit is

$$y_{\text{out}}(x) \equiv \lim_{\varepsilon \to 0+} y(x) = e^{1-x}. \tag{9.2.3}$$

The difference between the outer limit of the exact solution and the exact solution itself, $|y(x) - y_{\text{out}}(x)|$, is exponentially small in the limit $\varepsilon \to 0$ when $\delta \ll x$.

Similarly, we can formally take the outer limit of the differential equation (9.2.2); the result of keeping x fixed and letting $\varepsilon \to 0+$ is simply

$$y'_{\text{out}} + y_{\text{out}} = 0, \tag{9.2.4}$$

which is satisfied by (9.2.3). Because the outer limit of (9.1.2) is a *first*-order differential equation, its solution cannot in general be required to satisfy both boundary conditions $y(0) = 0$ and $y(1) = 1$; the outer limit of (9.2.1) satisfies $y(1) = 1$ but not $y(0) = 0$.

In other words, the small-ε limit of the solution is *not* everywhere close to the solution of the unperturbed differential equation (9.2.4) [the differential equation (9.2.2) with $\varepsilon = 0$]. Thus, the problem (9.2.2) is a singular perturbation problem. The singular behavior [the appearance of a discontinuity in $y(x)$ as $\varepsilon \to 0+$] occurs because the highest-order derivative in (9.2.2) disappears when $\varepsilon = 0$.

The exact solution satisfies the boundary condition $y(0) = 0$ by developing a boundary layer in the neighborhood of $x = 0$. To examine the nature of this boundary layer, we consider the *inner limit* in which $\varepsilon \to 0+$ with $x \le O(\varepsilon)$. In this case x lies inside the boundary layer at $x = 0$. For this limit it is convenient to let $x = \varepsilon X$ with X fixed and finite. X is called an *inner* variable. X is a better variable than x to describe y in the boundary layer because, as $\varepsilon \to 0+$, y varies rapidly as a function of x but slowly as a function of X. It follows from (9.2.1) that

$$y_{\text{in}}(x) = Y_{\text{in}}(X) = \lim_{\varepsilon \to 0+} y(\varepsilon X) = e - e^{1-X}. \tag{9.2.5}$$

Similarly, the inner limit of the differential equation is obtained by rewriting (9.2.2) as

$$\frac{1}{\varepsilon}\frac{d^2 Y}{dX^2} + \left(\frac{1}{\varepsilon} + 1\right)\frac{dY}{dX} + Y = 0, \tag{9.2.6}$$

where we define $Y(X) \equiv y(x)$ and use

$$\varepsilon\frac{dy}{dx} = \frac{dY}{dX}, \qquad \varepsilon^2\frac{d^2 y}{dx^2} = \frac{d^2 Y}{dX^2}.$$

Thus, taking the inner limit, $\varepsilon \to 0+$, X fixed, gives

$$\frac{d^2 Y_{\text{in}}(X)}{dX^2} + \frac{dY_{\text{in}}(X)}{dX} = 0. \tag{9.2.7}$$

Observe that the inner limit function (9.2.5) does satisfy (9.2.7) together with the boundary condition $Y_{in}(0) = 0$.

Boundary-layer analysis is extremely useful because it allows one to construct an approximate solution to a differential equation, even when an exact answer is not attainable. This is because the inner and outer limits of an insoluble differential equation are often soluble. Once y_{in} and y_{out} have been determined, they must be asymptotically matched. This asymptotic match occurs on the overlap region which is defined by the *intermediate* limit $x \to 0$, $X = x/\varepsilon \to \infty$, $\varepsilon \to 0+$. For example, if $x = \varepsilon^{1/2}z$ with z fixed as $\varepsilon \to 0$, then an intermediate limit is obtained. A glance at (9.2.3) and (9.2.5) shows that the intermediate limits of $y_{out}(x)$ and $y_{in}(x) = Y_{in}(X)$ agree: $\lim_{x\to 0} y_{out}(x) = \lim_{X\to\infty} Y_{in}(X) = e$. This common limit verifies the asymptotic match between the inner and outer solutions. (It is *not* generally the case in boundary-layer theory that the intermediate limit is a *number* independent of x and X, as we will shortly see.) The above match condition provides the second boundary condition for the solution of (9.2.7): $Y_{in}(\infty) = e$. Observe that although the x region is finite, $0 \le x \le 1$, the size of the matching region in terms of the inner variable is infinite. As we emphasized in Chap. 7, the extent of the matching region must always be infinite.

A very subtle aspect of boundary-layer theory is the question of whether or not an overlap region for any given problem actually exists. Since one's ability to construct a matched asymptotic expansion depends on the presence of this overlap region, its existence is crucial to the solution of the problem. How did we know, for example, that the intermediate limits of y_{out} and Y_{in} would agree? That is, how did we know that the inner and outer limits of the differential equation (9.2.2) have a common region of validity?

To answer these questions we will perform a complete perturbative solution of (9.2.2) to all orders in powers of ε, and not just to leading order. First, we examine the outer solution. We seek a perturbation expansion of the outer solution of the form

$$y_{out}(x) \sim \sum_{n=0}^{\infty} y_n(x)\varepsilon^n, \qquad \varepsilon \to 0+, \tag{9.2.8}$$

and restate the boundary condition $y(1) = 1$ as

$$y_0(1) = 1, \qquad y_1(1) = 0, \qquad y_2(1) = 0, \qquad y_3(1) = 0, \qquad \ldots. \tag{9.2.9}$$

Note that $y_{out}(x)$ in (9.2.8) is *not* the same as $y_{out}(x)$ in (9.2.3); $y_{out}(x)$ in (9.2.3) is the first term $y_0(x)$ in (9.2.8).

Substituting (9.2.8) into (9.2.2) and collecting powers of ε gives a sequence of differential equations:

$$y_0' + y_0 = 0, \qquad y_0(1) = 1,$$

$$y_n' + y_n = -y_{n-1}'' - y_{n-1}', \qquad y_n(1) = 0,\ n \ge 1.$$

The solution to these equations is

$$\begin{aligned} y_0 &= e^{1-x}, \\ y_n &= 0, \qquad n \ge 1. \end{aligned} \tag{9.2.10}$$

Thus, the leading-order outer solution, $y_{\text{out}} = e^{1-x}$, is correct to *all* orders in perturbation theory. This is the reason why in the outer region, $x \gg \varepsilon$, the difference between $y(x)$ and $y_{\text{out}}(x)$ is at most exponentially small (subdominant): $|y - y_{\text{out}}| = O(\varepsilon^n)$ for *all* n as $\varepsilon \to 0+$.

Next, we perform a similar expansion of the inner solution. We assume a perturbation series of the form

$$Y_{\text{in}}(X) \sim \sum_{n=0}^{\infty} \varepsilon^n Y_n(X), \qquad \varepsilon \to 0+, \tag{9.2.11}$$

and restate the boundary condition $Y_{\text{in}}(0) = y(0) = 0$ as

$$Y_n(0) = 0, \qquad n \ge 0. \tag{9.2.12}$$

Substituting (9.2.11) into (9.2.6) gives the sequence of differential equations:

$$\begin{aligned} &Y_0'' + Y_0' = 0, && Y_0(0) = 0, \\ &Y_n'' + Y_n' = -Y_{n-1}' - Y_{n-1}, && Y_n(0) = 0, \ n \ge 1. \end{aligned}$$

These equations may be solved by means of the integrating factor e^X. The results are

$$\begin{aligned} Y_0(X) &= A_0(1 - e^{-X}), \\ Y_n(X) &= \int_0^X [A_n e^{-z} - Y_{n-1}(z)] \, dz, \qquad n \ge 1, \end{aligned} \tag{9.2.13}$$

where the A_n are arbitrary integration constants.

Does this inner solution match asymptotically, order by order in powers of ε, to $y_{\text{out}}(x)$? To see if this is so, we substitute $x = \varepsilon X$ into y_{out} in (9.2.10) and expand in powers of ε:

$$y_{\text{out}}(x) = e^{1-x} = e\left(1 - \varepsilon X + \frac{\varepsilon^2 X^2}{2!} - \frac{\varepsilon^3 X^3}{3!} + \cdots\right). \tag{9.2.14}$$

Returning to equation (9.2.13), we take X large ($X \to \infty$) and obtain $Y_0(X) \sim A_0$ ($X \to \infty$). Thus, comparing with the first term of (9.2.14), we have $A_0 = e$, as we already know. Now that Y_0 is known, we may compute Y_1 from (9.2.13):

$$Y_1(X) = (A_1 + A_0)(1 - e^{-X}) - eX.$$

Asymptotic matching with y_{out} [comparing $Y_1(X)$, when $X \to \infty$, with the second term of (9.2.14)] gives $A_1 = -e$, so $Y_1(X) = -eX$. Similarly, $Y_n(X) = e[(-1)^n/n!]X^n$. Hence the full inner expansion is

$$Y_{\text{in}}(X) = e \sum_{n=0}^{\infty} \varepsilon^n \frac{(-1)^n X^n}{n!} - e^{1-X} = e^{1-\varepsilon X} - e^{1-X}. \tag{9.2.15}$$

Evidently, the inner expansion is a valid asymptotic expansion not only for values of X inside the boundary layer [$X = O(1)$] but also for large values of X

$[X = O(\varepsilon^{-\alpha}), 0 < \alpha < 1]$. At the same time the expansion for $y_{\text{out}}(x)$ is valid for $\varepsilon \ll x \le 1$ $(\varepsilon \to 0+)$. [$y_{\text{out}}(x)$ is not valid for $x = O(\varepsilon)$ because it does not satisfy the boundary condition $y(0) = 0$; nor does it have the boundary-layer term e^{1-X} which is present in $Y_{\text{in}}(X)$.] We conclude that *to all orders in powers of* ε it is possible to match asymptotically the inner and outer expansions because they have a common region of validity: $\varepsilon \ll x \ll 1$ $(\varepsilon \to 0+)$.

We have been able to demonstrate explicitly the existence of the overlap region for this problem because it is soluble to all orders in perturbation theory. In general, however, such a calculation is too difficult. Instead, our approach will always be to assume that an overlap region exists and then to verify the consistency of this assumption by performing an asymptotic match. In the above simple boundary-value problem, we found that the size of the overlap region was independent of the order of perturbation theory. In general, however, the extent of the matching region may vary with the order of perturbation theory (see Sec. 7.4 and Example 1 of Sec. 9.3).

One final point concerns the construction of the uniform approximation to $y(x)$. The formula used in the previous section to construct a uniform approximation is $y_{\text{unif}}(x) = y_{\text{in}}(x) + y_{\text{out}}(x) - y_{\text{match}}(x)$, where $y_{\text{match}}(x)$ is the approximation to $y(x)$ in the matching region and $y_{\text{unif}}(x)$ is a uniform approximation to $y(x)$. This formula is applicable here too. For the boundary-layer solution to (9.2.2), it is easy to verify that if $y_{\text{in}}(x)$, $y_{\text{out}}(x)$, and $y_{\text{match}}(x)$ are calculated to nth order in perturbation theory, then $|y(x) - y_{\text{unif}}(x)| = O(\varepsilon^{n+1})$ $(\varepsilon \to 0+; 0 \le x \le 1)$.

The differential equation (9.2.2) is sufficiently simple that $y_{\text{unif}}(x)$ can be calculated to all orders in perturbation theory. It follows from (9.2.10) for $y_{\text{out}}(x)$, (9.2.15) for $y_{\text{in}}(x)$, and the result $y_{\text{match}}(x) = e^{1-x}$ that

$$y_{\text{unif}} = e^{1-x} - e^{1-X}$$

is the infinite-order uniform approximation to $y(x)$.

It is remarkable, however, that this expression, which is the result of summing up perturbation theory to infinite order, is actually *not* equal to $y(x)$ in (9.2.1). Thus, although the perturbation series for y_{unif} is *asymptotic* to $y(x)$ as $\varepsilon \to 0+$, the asymptotic series does not *converge* to $y(x)$ as n, the order of perturbation theory, tends to ∞; there is an exponentially small error, of order $e^{-1/\varepsilon}$, which remains undetermined. Boundary-layer theory is indeed a singular, and not a regular, perturbation theory.

Why is boundary-layer theory a singular perturbation theory? The singular nature of boundary-layer theory is intrinsic to both the inner and outer expansions. The outer expansion is singular because there is an abrupt change in the order of the differential equation when $\varepsilon = 0$. By contrast, the inner expansion is a regular perturbation expansion for finite X (see Example 2 of Sec. 7.1). However, since asymptotic matching takes place in the limit $X \to \infty$, the inner expansion is also singular (see Example 4 of Sec. 7.2). Another manifestation of the singular limit $\varepsilon \to 0$ is the location of the boundary layer in (9.2.1); when the limit $\varepsilon \to 0+$ is replaced by $\varepsilon \to 0-$, the boundary layer abruptly jumps from $x = 0$ to $x = 1$.

(E) 9.3 HIGHER-ORDER BOUNDARY-LAYER THEORY

In Secs. 9.1 and 9.2 we formulated the procedure for finding the leading-order boundary-layer approximation to the solution of an ordinary differential equation; i.e., to obtain outer and inner solutions and asymptotically match them in an overlap region. The self-consistency of boundary-layer theory depends on the success of asymptotic matching. Ordinarily, if the inner and outer solutions match to all orders in ε, then boundary-layer theory gives an asymptotic approximation to the exact solution of the differential equation. Accordingly, in Sec. 9.2 we showed how to use boundary-layer theory to all orders in powers of ε for a simple constant-coefficient differential equation.

In this section we give an example to illustrate how boundary-layer theory is used to construct higher-order approximations for more complicated equations. As we shall see, an interesting aspect of boundary-layer problems is that the size of the matching region depends on the order of perturbation theory.

Example 1 *Boundary-layer analysis of a variable-coefficient differential equation.* We wish to obtain an approximate solution to the boundary-value problem

$$\varepsilon y'' + (1 + x)y' + y = 0, \qquad y(0) = 1,\ y(1) = 1, \tag{9.3.1}$$

which is correct to order ε^3.

We seek an outer solution in the form of a perturbation series in powers of ε:

$$y_{\text{out}}(x) \sim y_0(x) + \varepsilon y_1(x) + \varepsilon^2 y_2(x) + \cdots, \qquad \varepsilon \to 0+. \tag{9.3.2}$$

Since $1 + x > 0$ for $0 \le x \le 1$, we expect a boundary layer only at $x = 0$ (see Sec. 9.1). Thus, the outer solution must satisfy the boundary condition $y_{\text{out}}(1) = 1$ and we must require that

$$y_0(1) = 1, \qquad y_n(0) = 0, \qquad n \ge 1. \tag{9.3.3}$$

Next, we substitute (9.3.2) into (9.3.1) and equate coefficients of like powers of ε. This converts (9.3.1) into a sequence of first-order inhomogeneous equations:

$$(1 + x)y_0' + y_0 = 0,\ (1 + x)y_1' + y_1 = -y_0'',\ (1 + x)y_2' + y_2 = -y_1'',\ \ldots.$$

The solutions to these equations which satisfy the boundary conditions (9.3.3) are

$$y_0(x) = 2(1 + x)^{-1},$$
$$y_1(x) = 2(1 + x)^{-3} - \tfrac{1}{2}(1 + x)^{-1},$$
$$y_2(x) = 6(1 + x)^{-5} - \tfrac{1}{2}(1 + x)^{-3} - \tfrac{1}{4}(1 + x)^{-1}.$$

Thus,

$$y_{\text{out}}(x) \sim \frac{2}{1 + x} + \varepsilon\left[\frac{2}{(1 + x)^3} - \frac{1}{2(1 + x)}\right] + \varepsilon^2\left[\frac{6}{(1 + x)^5} - \frac{1}{2(1 + x)^3} - \frac{1}{4(1 + x)}\right] + \cdots, \qquad \varepsilon \to 0+. \tag{9.3.4}$$

This completes the determination of the outer solution to second order in powers of ε.

As expected, the outer solution (9.3.4) does not satisfy the boundary condition $y(0) = 1$, so a boundary layer at $x = 0$ is necessary. As in Example 3 of Sec. 9.1, we expect the thickness δ of the boundary layer to be $O(\varepsilon)$. (In Sec. 9.4 the procedure for determining the thickness of the boundary layer is explained and examples of boundary layers having thicknesses other than ε are given.) Therefore, we introduce the inner variables $X = x/\varepsilon$ and $Y_{\text{in}}(X) \equiv y_{\text{in}}(x)$. In terms of these variables (9.3.1) becomes

$$\frac{d^2 Y_{\text{in}}}{dX^2} + (1 + \varepsilon X)\frac{dY_{\text{in}}}{dX} + \varepsilon Y_{\text{in}} = 0. \tag{9.3.5}$$

If we represent $Y_{\text{in}}(X)$ as a perturbation series in powers of ε,

$$Y_{\text{in}}(X) \sim Y_0(X) + \varepsilon Y_1(X) + \varepsilon^2 Y_2(X) + \cdots, \qquad \varepsilon \to 0+, \tag{9.3.6}$$

then the boundary condition $y(0) = 1$ translates into the sequence of boundary conditions

$$Y_0(0) = 1, \qquad Y_1(0) = 0, \qquad Y_2(0) = 0, \qquad \ldots. \tag{9.3.7}$$

Substituting (9.3.6) into (9.3.5) and equating coefficients of like powers of ε converts (9.3.5) into a sequence of second-order constant-coefficient equations:

$$\begin{aligned}
&\frac{d^2 Y_0}{dX^2} + \frac{dY_0}{dX} = 0,\\
&\frac{d^2 Y_1}{dX^2} + \frac{dY_1}{dX} = -X\frac{dY_0}{dX} - Y_0,\\
&\frac{d^2 Y_2}{dX^2} + \frac{dY_2}{dX} = -X\frac{dY_1}{dX} - Y_1.
\end{aligned} \tag{9.3.8}$$

Each of the solutions of (9.3.8) which satisfy the boundary conditions (9.3.7) have one new arbitrary constant of integration:

$$Y_0(X) = 1 + A_0(e^{-X} - 1), \tag{9.3.9}$$

$$Y_1(X) = -X + A_0(-\tfrac{1}{2}X^2 e^{-X} + X) + A_1(e^{-X} - 1), \tag{9.3.10}$$

$$\begin{aligned}
Y_2(X) = {}& X^2 - 2X + A_0(\tfrac{1}{8}X^4 e^{-X} - X^2 + 2X)\\
&+ A_1(-\tfrac{1}{2}X^2 e^{-X} + X) + A_2(e^{-X} - 1).
\end{aligned} \tag{9.3.11}$$

This completes the determination of the inner solution to second order in powers of ε.

We determine the constants $A_0, A_1, A_2, \ldots$ by asymptotically matching the inner and outer solutions. The match consists of requiring that the intermediate limits $[\varepsilon \to 0+,\ x \to 0+,\ X = x/\varepsilon \to +\infty]$ of the inner and outer solutions agree.

First, we perform a leading-order (zeroth-order in ε) match. As $x \to 0+$ in the outer solution (9.3.4), $y_{\text{out}}(x) = 2 + \mathrm{O}(\varepsilon, x)$ $(x \to 0+,\ \varepsilon \to 0+)$, where the symbol

$$\mathrm{O}(a, b, c, \ldots) \text{ means } \mathrm{O}(a) + \mathrm{O}(b) + \mathrm{O}(c) + \cdots.$$

The error term $\mathrm{O}(\varepsilon, x)$ indicates that we have neglected powers of ε higher than zero and that we have expanded the solution in a Taylor series in powers of x and have neglected all but the first term.

On the other hand, in the limit $X \to \infty$ and $\varepsilon X = x \to 0+$ the inner solution becomes

$$Y_{\text{in}}(X) = 1 - A_0 + \mathrm{O}(\varepsilon X), \qquad \varepsilon X \to 0+,\ X \to \infty, \tag{9.3.12}$$

where the correction of order εX arises from the term $\varepsilon Y_1(X)$ in the expansion of $Y_{\text{in}}(X)$. The inner and outer solutions are required to match to lowest order [that is, $y_{\text{out}}(x) \sim Y_{\text{in}}(X)$ in the intermediate limit]. Thus, $2 = 1 - A_0$ or $A_0 = -1$. This completes the leading-order match. Observe that the match occurs for values of x for which $1 \ll X = x/\varepsilon$ as well as $x \ll 1$. Thus, the size of the overlap region is determined to leading order as $\varepsilon \ll x \ll 1$ $(\varepsilon \to 0+)$.

Next, we match to first order in ε. We expand the outer solution, keeping terms of order ε and x but discarding terms of order ε^2, x^2, and εx. The result is

$$y_{\text{out}}(x) = 2 - 2x + \tfrac{3}{2}\varepsilon + \mathrm{O}(\varepsilon^2, \varepsilon x, x^2), \qquad x \to 0+,\ \varepsilon \to 0+. \tag{9.3.13}$$

We also expand the inner solution for $X \gg 1$ (x outside the boundary layer), but neglect terms of order $\varepsilon^2 X^2 = x^2$, $\varepsilon^2 X = \varepsilon x$, and ε^2. The result is

$$Y_{\text{in}}(X) = 1 - A_0 - \varepsilon X + \varepsilon A_0 X - \varepsilon A_1 + O(x^2), \qquad \varepsilon X \to 0+, \; X \to \infty. \tag{9.3.14}$$

Matching (9.3.14) with (9.3.13) gives $A_0 = -1$, which we already know, and $A_1 = -3/2$.

Observe that for a successful match to first order it is necessary to neglect terms of order ε^2, x^2, and εx, compared with x and ε. If we had retained some of these terms, we would have found that there is no way to choose the constants of integration to make (9.3.13) agree with (9.3.14). Since matching now requires that $x^2 \ll \varepsilon$, the size of the overlap region is *smaller* than it was in the leading-order match. Its extent is $\varepsilon \ll x \ll \varepsilon^{1/2}$ $(\varepsilon \to 0+)$.

To perform a second-order match, we expand the inner and outer solutions and neglect terms of order ε^3, $\varepsilon^2 x$, εx^2, and x^3. The result is

$$y_{\text{out}}(x) = 2 - 2x + 2x^2 + \varepsilon\left(\tfrac{3}{2} - \tfrac{11}{2}x\right) + \tfrac{21}{4}\varepsilon^2 + O(\varepsilon^3, \varepsilon^2 x, \varepsilon x^2, x^3),$$
$$\varepsilon \to 0+, \; x \to 0+,$$

and
$$Y_{\text{in}}(X) = 1 - A_0 - \varepsilon X + \varepsilon A_0 X - \varepsilon A_1 + \varepsilon^2 X^2 - 2\varepsilon^2 X - \varepsilon^2 A_0 X^2 + 2\varepsilon^2 A_0 X$$
$$+ \varepsilon^2 A_1 X - \varepsilon^2 A_2 + O(\varepsilon^3, \varepsilon^2 x, \varepsilon x^2, x^3),$$
$$\varepsilon X \to 0+, \; X \to \infty. \tag{9.3.15}$$

Matching requires $A_0 = -1$, $A_1 = -3/2$, $A_2 = -21/4$.

Since one must neglect x^3 compared with ε^2 to obtain a match in second order, it follows that the size of the matching region, $\varepsilon \ll x \ll \varepsilon^{2/3}$ $(\varepsilon \to 0+)$, is smaller than it was in the first-order calculation.

The process of asymptotic matching may be carried out to all orders in powers of ε (see Prob. 9.10). And as the order of perturbation theory increases, the size of the matching region continues to shrink. In nth order the common region of validity of the inner and outer expansions is $\varepsilon \ll x \ll \varepsilon^{n/(n+1)}$ $(\varepsilon \to 0+)$. However, the extent of the matching region in terms of the inner variable is still infinite as $\varepsilon \to 0+$: $1 \ll X \ll \varepsilon^{-1/(n+1)}$ $(\varepsilon \to 0+)$.

Once the boundary-layer solution is determined, one may construct a uniform approximation to the solution $y(x)$ using the formula $y_{\text{unif}}(x) = y_{\text{out}}(x) + y_{\text{in}}(x) - y_{\text{match}}(x)$, where $y_{\text{match}}(x)$ is the expansion of either the inner or outer approximations in the matching region. To third order in ε (see Prob. 9.12),

$$y_{\text{unif},3}(x) = \left(\frac{2}{1+x} - e^{-X}\right) + \varepsilon\left[\frac{2}{(1+x)^3} - \frac{1}{2(1+x)} + \left(\frac{1}{2}X^2 - \frac{3}{2}\right)e^{-X}\right]$$
$$+ \varepsilon^2\left[\frac{6}{(1+x)^5} - \frac{1}{2(1+x)^3} - \frac{1}{4(1+x)} - \left(\frac{1}{8}X^4 - \frac{3}{4}X^2 + \frac{21}{4}\right)e^{-X}\right]$$
$$+ \varepsilon^3\left[\frac{30}{(1+x)^7} - \frac{3}{2(1+x)^5} - \frac{1}{4(1+x)^3} - \frac{5}{16(1+x)}\right.$$
$$\left. + \left(\frac{1}{48}X^6 - \frac{3}{16}X^4 + \frac{21}{8}X^2 - \frac{1949}{72}\right)e^{-X}\right] + O(\varepsilon^4), \qquad \varepsilon \to 0+. \tag{9.3.16}$$

This expression is a uniform approximation to $y(x)$ over the entire region $0 \le x \le 1$:

$$|y_{\text{unif},3}(x) - y(x)| = O(\varepsilon^4), \qquad \varepsilon \to 0+. \tag{9.3.17}$$

Also, $y_{\text{unif}}(x)$ may be used to approximate the derivatives of $y(x)$:

$$|y'_{\text{unif},3}(x) - y'(x)| = O(\varepsilon^3), \qquad \varepsilon \to 0+, \tag{9.3.18}$$

$$|y''_{\text{unif},3}(x) - y''(x)| = O(\varepsilon^2), \qquad \varepsilon \to 0+, \tag{9.3.19}$$

for $0 \le x \le 1$ (see Prob. 9.13).

In Figs. 9.5 to 9.8 we show how well $y_{\text{unif}}(x)$ approximates the exact solution $y(x)$. We plot the percentage relative error [percentage relative error = $100(y_{\text{unif}} - y)/y$] for the first four uniform approximations $y_{\text{unif},\,n}$ ($n = 0, 1, 2, 3$). $y_{\text{unif},\,n}$ is the uniform approximation to $y(x)$ accurate to order ε^n; for example, $y_{\text{unif},\,1}$ is obtained from (9.3.16) by neglecting the terms containing ε^2 and ε^3. Figures 9.5 to 9.8 suggest the asymptotic nature of the boundary-layer approximation; the uniform approximation in (9.3.16) is the first four terms of a divergent asymptotic series in powers of ε (see Prob. 9.14).

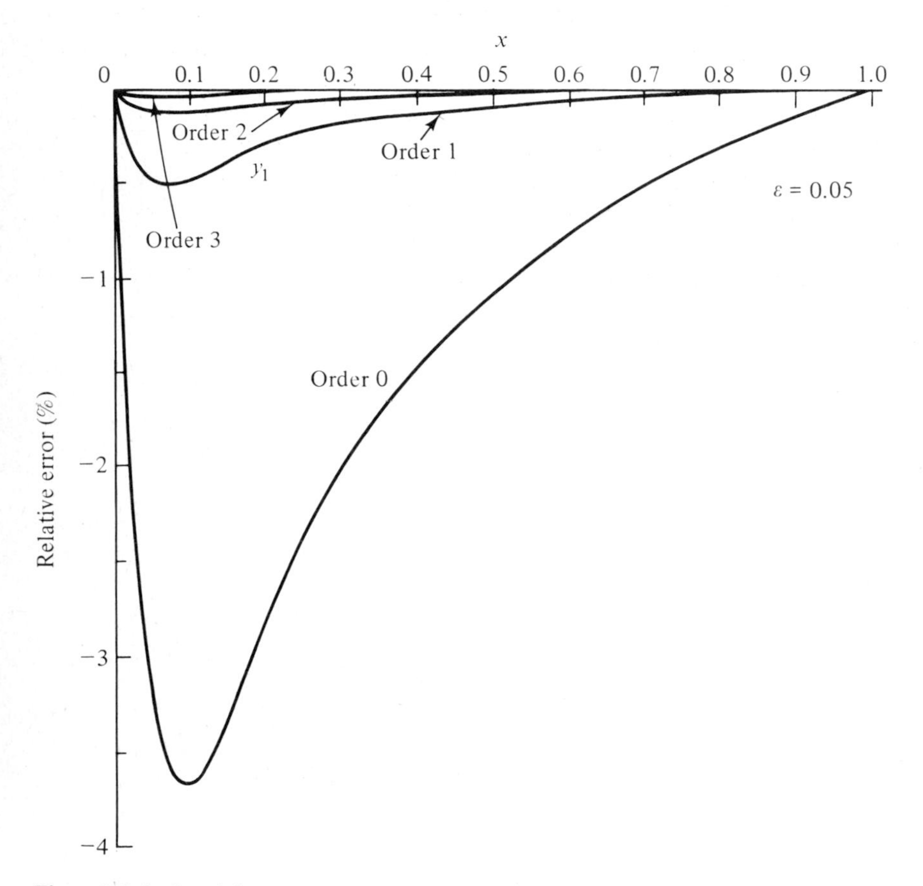

Figure 9.5 A plot of the percentage relative error between the exact solution $y(x)$ to the boundary-value problem in (9.3.1) with $\varepsilon = 0.05$ and the zeroth-order, first-order, second-order, and third-order uniform approximations to $y(x)$ obtained from boundary-layer analysis [see (9.3.16)]. The percentage relative error = $100[y_{\text{unif}}(x) - y(x)]/y(x)$. The graphs in this plot lie below the x axis because y_{unif} underestimates $y(x)$. Observe that as the order of perturbation theory increases the relative error decreases. However, for sufficiently large order the asymptotic nature of boundary-layer theory will surface and the relative error will increase with order.

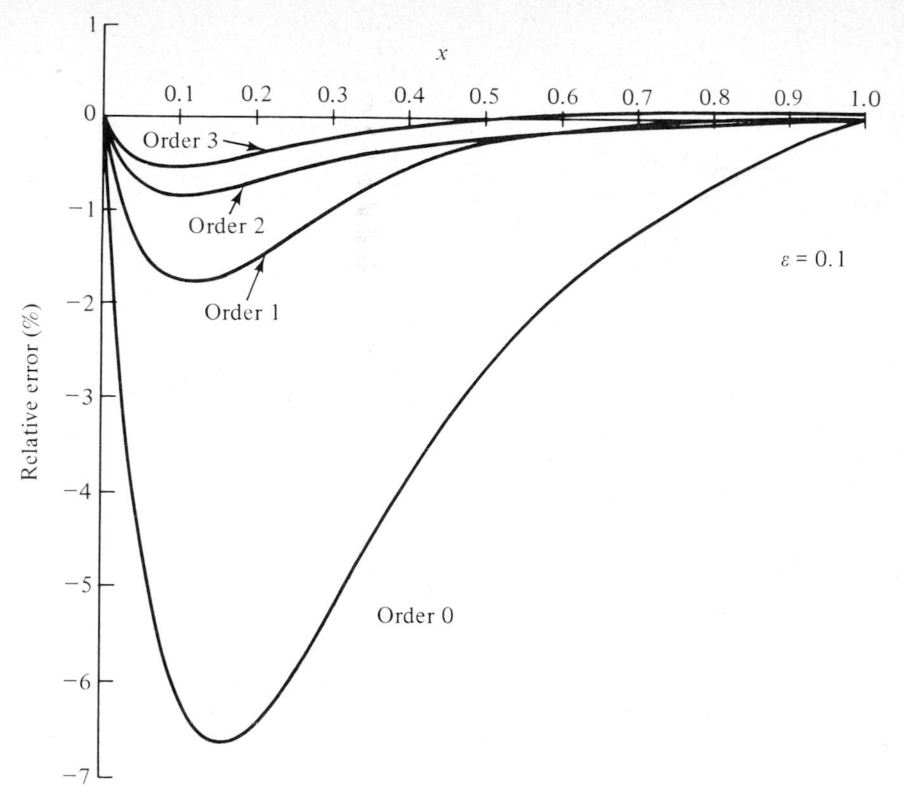

Figure 9.6 Same as in Fig. 9.5 except that $\varepsilon = 0.1$.

(I) 9.4 DISTINGUISHED LIMITS AND BOUNDARY LAYERS OF THICKNESS $\neq \varepsilon$

Until now, most of the boundary layers we have seen have had thickness $\delta = \varepsilon$. In general, however, the thickness of a boundary layer need not be of order ε as $\varepsilon \to 0+$. There are examples where $\delta = O(\varepsilon^{1/2})$, $\delta = O(\varepsilon^{2/3})$, and so on.

The determination of δ requires the notion of a *distinguished limit* which involves nothing more than a dominant-balance argument. We return to Example 3 of Sec. 9.1 to illustrate the relevant techniques. The solution of the boundary-value problem

$$\varepsilon y'' + a(x)y' + b(x)y = 0, \qquad y(0) = A,\ y(1) = B, \tag{9.4.1}$$

has a boundary layer at $x = 0$ if $a(x) > 0$ $(0 \le x \le 1)$. In the inner region we let $y(x) = Y_{\text{in}}(X)$, $X = x/\delta$, so

$$\begin{aligned} \frac{dy}{dx} &= \frac{1}{\delta}\frac{dY_{\text{in}}(X)}{dX}, \\ \frac{d^2y}{dx^2} &= \frac{1}{\delta^2}\frac{d^2Y_{\text{in}}(X)}{dX^2}. \end{aligned} \tag{9.4.2}$$

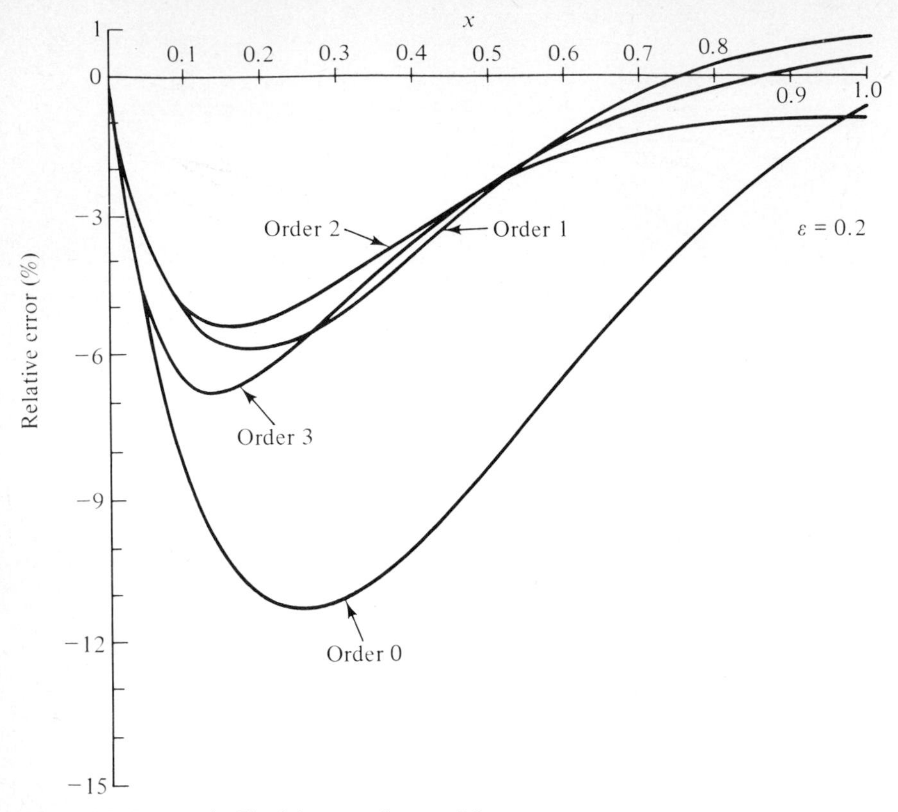

Figure 9.7 Same as in Fig. 9.5 except that $\varepsilon = 0.2$.

Thus, the differential equation (9.4.1) assumes the form

$$\frac{\varepsilon}{\delta^2}\frac{d^2 Y_{\text{in}}}{dX^2} + \frac{a(\delta X)}{\delta}\frac{dY_{\text{in}}}{dX} + b(\delta X)Y_{\text{in}} = 0. \tag{9.4.3}$$

Our task is to determine $\delta(\varepsilon)$. There are three possibilities to consider: $\delta(\varepsilon) \ll \varepsilon$, $\delta(\varepsilon) \sim \varepsilon$, and $\varepsilon \ll \delta(\varepsilon)$ as $\varepsilon \to 0$. In the first case, $\delta \ll \varepsilon$, we may approximate (9.4.3) by $d^2 Y_{\text{in}}/dX^2 = 0$. Thus, $Y_{\text{in}}(X) = A + cX$, which satisfies the boundary condition $Y_{\text{in}}(0) = A$. This inner limit does not match the outer solution because $\lim_{X\to\infty} Y_{\text{in}}(X) = \infty$ unless $c = 0$, while $y_{\text{out}}(0)$ is finite and not generally equal to A.

Similarly, $\varepsilon \ll \delta$ in (9.4.3) gives $a(0)\, dY_{\text{in}}/dX = 0$, so $Y_{\text{in}}(X) = A$ because $Y_{\text{in}}(0) = A$. Again, no match is possible if $A \neq y_{\text{out}}(0)$.

Finally, the choice $\delta = \varepsilon$ in (9.4.3) gives the leading-order equation

$$\frac{d^2 Y_{\text{in}}}{dX^2} + a(0)\frac{dY_{\text{in}}}{dX} = 0.$$

The choice $\delta = \varepsilon$ is called a *distinguished limit* because it involves a nontrivial relation (a dominant balance) between two or more terms of the equation (9.4.3);

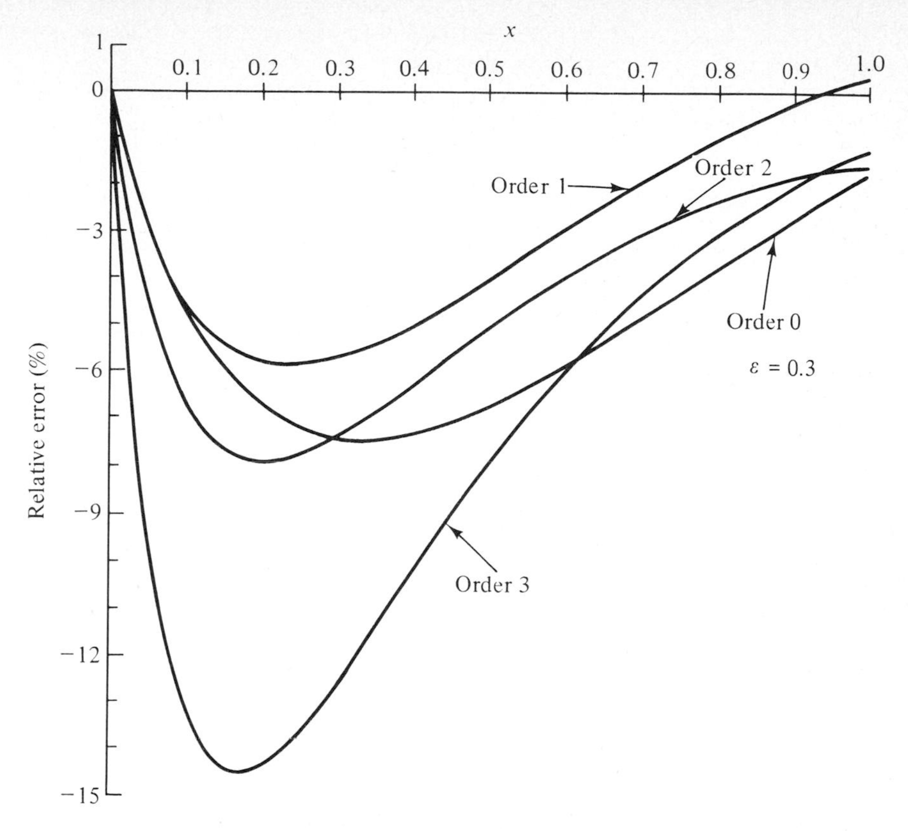

Figure 9.8 Same as in Fig. 9.5 except that $\varepsilon = 0.3$. For this value of ε the asymptotic nature of perturbation theory is evident; optimal accuracy is reached in first order and the relative error increases with the order when the order is greater than 1.

two terms are of comparable size while the third is smaller. The cases $\delta(\varepsilon) \ll \varepsilon$ and $\varepsilon \ll \delta(\varepsilon)$ as $\varepsilon \to 0$ are *undistinguished*. In general, only the distinguished limit gives a nontrivial boundary-layer structure which is asymptotically matchable to the outer solution.

In the above argument we have used the notation $\delta \sim \varepsilon$ and have neglected the possibility that $\delta \sim c\varepsilon$ where c is a constant. This is because we are only interested in the order of magnitude of the boundary-layer thickness. We will always ignore any constant of proportionality.

Example 1 *Boundary layer of thickness* $\delta = \varepsilon^{1/2}$. Consider the boundary-value problem

$$\varepsilon y'' - x^2 y' - y = 0, \qquad y(0) = y(1) = 1. \tag{9.4.4}$$

We will show that the leading-order solution to this problem has two boundary layers, one at the right boundary $x = 1$ for which $\delta = \varepsilon$ and one at the left boundary for which $\delta = \varepsilon^{1/2}$.

The leading-order outer solution satisfies $-x^2 y_0' - y_0 = 0$, so

$$y_0(x) = C_0 e^{1/x}. \tag{9.4.5}$$

Next, observe that the coefficient of y' in (9.4.4) is negative at $x = 1$. Thus, on the basis of Example 3 of Sec. 9.1 we expect that a boundary layer of thickness ε will develop at $x = 1$ as $\varepsilon \to 0+$. In terms of the inner variable $X = (1 - x)/\varepsilon$, the leading behavior in this boundary layer is governed by $d^2 Y_{0,\text{right}}/dX^2 + dY_{0,\text{right}}/dX = 0$. Hence,

$$Y_{0,\,\text{right}}(X) = A_0 + B_0 e^{-X}. \tag{9.4.6}$$

The boundary condition $y(1) = 1$ requires that $Y_{0,\,\text{right}}(0) = A_0 + B_0 = 1$ and matching to the outer solution (9.4.5) in the neighborhood of $x = 1$ requires that $A_0 = C_0 e$. Hence, $B_0 = 1 - C_0 e$. But this does not complete the solution because C_0 is still undetermined. Moreover, as $x \to 0+$ the outer solution (9.4.5) becomes infinite unless $C_0 = 0$, so it certainly cannot satisfy the boundary condition $y(0) = 1$.

Thus, we must have a boundary layer at $x = 0$ if we are to satisfy $y(0) = 1$. To determine the thickness of this layer, we use Z as the inner variable and substitute $x = Z\delta$ and (9.4.2) into (9.4.4) to obtain

$$\frac{\varepsilon}{\delta^2}\frac{d^2 Y_{\text{in, left}}}{dZ^2} - \delta Z^2 \frac{dY_{\text{in, left}}}{dZ} - Y_{\text{in, left}} = 0,$$

where $Y_{\text{in, left}}(Z) = y(x)$ in the left boundary layer. The distinguished limits are $\varepsilon/\delta^2 \sim \delta$, $\delta \sim 1$, or $\varepsilon/\delta^2 \sim 1$ as $\varepsilon \to 0+$. The first case, $\delta \sim \varepsilon^{1/3}$, is inconsistent because the undifferentiated term is dominant; the second case, $\delta \sim 1$, reproduces the outer limit. Thus, the only consistent choice of boundary-layer scale is $\delta = \varepsilon^{1/2}$. With this choice (9.4.4) becomes

$$\frac{d^2 Y_{\text{in,left}}}{dZ^2} - Y_{\text{in,left}} = \varepsilon^{1/2} Z^2 \frac{dY_{\text{in,left}}}{dZ}.$$

The leading-order inner solution $Y_{0,\text{left}}(Z)$ therefore satisfies $d^2 Y_{0,\text{left}}/dZ^2 - Y_{0,\text{left}} = 0$, whose solution is

$$Y_{0,\,\text{left}}(Z) = D_0 e^Z + E_0 e^{-Z}. \tag{9.4.7}$$

The boundary condition $y(0) = 1$ implies that $Y_{0,\,\text{left}}(0) = 1$ or that $D_0 + E_0 = 1$.

Finally, we must match $Y_{0,\,\text{left}}(Z)$ to the outer solution $y_0(x)$. However, no match is possible unless $D_0 = 0$; otherwise the inner solution grows exponentially as $Z \to \infty$. Matching also requires that $C_0 = 0$; otherwise the outer solution $y(x)$ grows exponentially as $x \to 0+$. It follows that $A_0 = C_0 = D_0 = 0$, $B_0 = E_0 = 1$.

A uniform approximation to $y(x)$ over the entire interval $0 \le x \le 1$, including both boundary layers, is given by

$$\begin{aligned} y_{\text{unif}} &= y_0 + Y_{0,\,\text{left}} + Y_{0,\,\text{right}} - y_{\text{left match}} - y_{\text{right match}} \\ &= e^{(x-1)/\varepsilon} + e^{-x/\varepsilon^{1/2}}. \end{aligned} \tag{9.4.8}$$

Thus, outside the boundary layers at $x = 0$ and $x = 1$, the solution is exponentially small. Figures 9.9 and 9.10 compare the exact solution to (9.4.4) with a plot of the leading uniform asymptotic approximation to $y(x)$ in (9.4.8) for $\varepsilon = 0.05$ and $\varepsilon = 0.025$. Observe the close agreement.

Example 2 *Higher-order treatment of $\delta = \varepsilon^{1/2}$ boundary layer.* In this example we will obtain a leading- and higher-order approximation to the solution of the singular perturbation problem

$$\varepsilon y'' + x^2 y' - y = 0, \qquad y(0) = y(1) = 1. \tag{9.4.9}$$

Note that this differential equation differs from that of the previous example by a sign change in the one-derivative term.

This problem does not quite satisfy the assumptions of the third example of Sec. 9.1. There, we assumed that $a(x) > 0$ $(0 \le x \le 1)$; in this example $a(x) = x^2$, so this inequality is violated at the left boundary. Nevertheless, $a(x) > 0$ for all other x, so the conclusions of Sec. 9.1 (that a boundary layer could not occur at $x = 1$ or at an interior point) are still valid. By elimination,

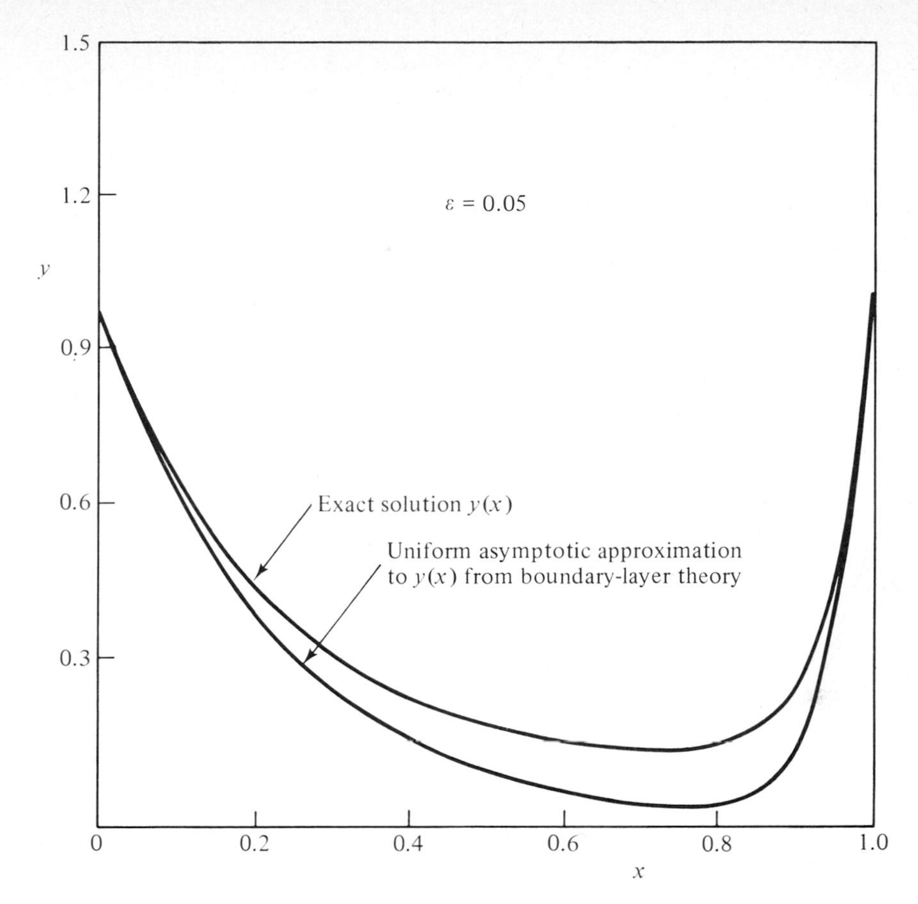

Figure 9.9 A comparison of the exact solution $y(x)$ to the boundary-value problem in (9.4.4) with the leading uniform asymptotic approximation to $y(x)$ from boundary-layer theory (9.4.8). For this plot $\varepsilon = 0.05$.

then, a boundary layer can and indeed does occur at $x = 0$. As in the previous example, the boundary layer at $x = 0$ has thickness $\delta \sim \varepsilon^{1/2}$.

In general, whenever a boundary-layer thickness is not of order ε, the matching of the inner and outer solutions is affected in a peculiar way. Recall that in the first example of Sec. 9.3, where the boundary-layer thickness was ε, the first term of the inner solution and the first term of the outer solution were matched. Next, we matched the first two terms of the inner and outer solutions and, finally, we matched the first three terms of each solution. In general, an nth-order match in this problem consists of matching the first $(n + 1)$ terms of the inner and outer solutions.

In the present example, as in the previous example, the boundary layer is comparatively thick ($\varepsilon \ll \delta = \varepsilon^{1/2}$ as $\varepsilon \to 0+$), so many more terms of the inner solution are required to describe its behavior in the matching region. Specifically, when the boundary-layer thickness is $\varepsilon^{1/2}$, $(2n - 1)$ terms of the inner expansion are required to match to n terms of the outer expansion. This and the next example will illustrate this phenomenon.

We begin the analysis by assuming an outer expansion of the form

$$y_{\text{out}}(x) \sim y_0(x) + \varepsilon y_1(x) + \cdots, \qquad \varepsilon \to 0+.$$

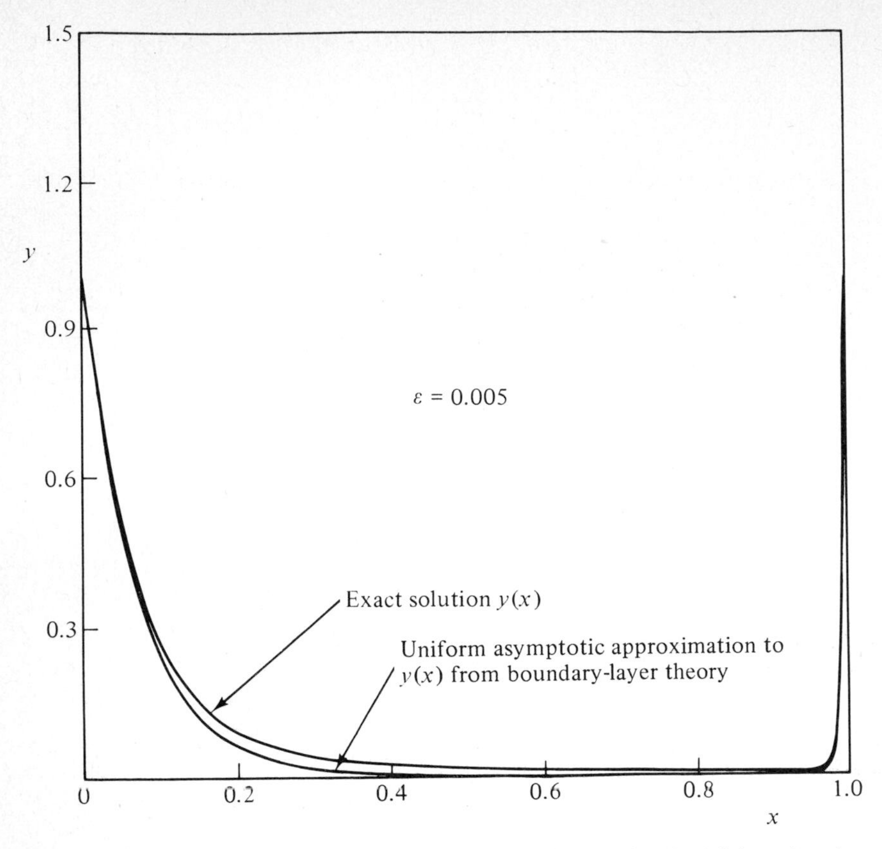

Figure 9.10 Same as in Fig. 9.9 except that $\varepsilon = 0.005$. Observe that the left boundary layer, whose thickness is of order $\varepsilon^{1/2}$, is much larger than the right boundary layer, whose thickness is of order ε.

This series leads to the sequence of equations

$$x^2 y_0' - y_0 = 0, \qquad x^2 y_n' - y_n = -y_{n-1}'', \qquad n \geq 1,$$

whose solutions are

$$\begin{aligned} y_0(x) &= C_0 e^{-1/x}, \\ y_1(x) &= C_0(\tfrac{1}{5}x^{-5} - \tfrac{1}{2}x^{-4})e^{-1/x} + C_1 e^{-1/x}, \end{aligned} \tag{9.4.10}$$

and so on. Since the coefficient of y' in (9.4.9) is positive at $x = 1$, we conclude from Example 3 of Sec. 9.1 that there cannot be a boundary layer at $x = 1$ when $\varepsilon \to 0+$. Thus, $y_{\text{out}}(1) = 1$ and the constants $C_0, C_1, \ldots$ are all determined: $C_0 = e$, $C_1 = 3e/10, \ldots$.

A boundary layer is required at $x = 0$ because the outer solution does not itself satisfy the boundary condition $y(0) = 1$. To reveal the structure of this boundary layer, we introduce the inner variable $X = x/\delta$. In terms of X, (9.4.9) becomes

$$\frac{\varepsilon}{\delta^2}\frac{d^2 Y_{\text{in}}}{dX^2} + \delta X^2 \frac{dY_{\text{in}}}{dX} - Y_{\text{in}} = 0,$$

where $y_{\text{in}}(x) = Y_{\text{in}}(X)$. As in the previous example, the only consistent distinguished limit gives $\delta = \varepsilon^{1/2}$, and with this choice (9.4.9) becomes

$$\frac{d^2 Y_{\text{in}}}{dX^2} - Y_{\text{in}} = -\varepsilon^{1/2} X^2 \frac{dY_{\text{in}}}{dX}. \tag{9.4.11}$$

Since the small parameter in this equation is $\varepsilon^{1/2}$ and not ε, the appropriate perturbation expansion of the inner solution is

$$Y_{\text{in}}(X) \sim Y_0(X) + \varepsilon^{1/2} Y_1(X) + \varepsilon Y_2(X) + \cdots, \qquad \varepsilon \to 0+. \tag{9.4.12}$$

Substituting (9.4.12) into (9.4.11) gives

$$\frac{d^2 Y_0}{dX^2} - Y_0 = 0, \qquad \frac{d^2 Y_n}{dX^2} - Y_n = -X^2 \frac{dY_{n-1}}{dX}, \qquad n \ge 1.$$

The solutions to these equations for $n = 0$, 1, and 2 have the form

$$\begin{aligned}
Y_0(X) &= A_0 e^X + B_0 e^{-X}, \\
Y_1(X) &= A_0(-\tfrac{1}{6}X^3 + \tfrac{1}{4}X^2 - \tfrac{1}{4}X)e^X \\
&\quad + B_0(-\tfrac{1}{6}X^3 - \tfrac{1}{4}X^2 - \tfrac{1}{4}X)e^{-X} + A_1 e^X + B_1 e^{-X}, \\
Y_2(X) &= A_0(\tfrac{1}{72}X^6 - \tfrac{1}{60}X^5 + \tfrac{1}{96}X^4 + \tfrac{1}{48}X^3 - \tfrac{1}{32}X^2 + \tfrac{1}{32}X)e^X \\
&\quad + B_0(\tfrac{1}{72}X^6 + \tfrac{1}{60}X^5 + \tfrac{1}{96}X^4 - \tfrac{1}{48}X^3 - \tfrac{1}{32}X^2 - \tfrac{1}{32}X)e^{-X} \\
&\quad + A_1(-\tfrac{1}{6}X^3 + \tfrac{1}{4}X^2 - \tfrac{1}{4}X)e^X + B_1(-\tfrac{1}{6}X^3 - \tfrac{1}{4}X^2 - \tfrac{1}{4}X)e^{-X} \\
&\quad + A_2 e^X + B_2 e^{-X}.
\end{aligned} \tag{9.4.13}$$

It is not possible to match the inner and outer solutions unless $A_0 = A_1 = A_2 = \cdots = 0$; exponentially growing terms are discarded because they blow up in the intermediate limit $x \to 0$, $X \to \infty$, $\varepsilon \to 0+$. The remaining constants $B_0, B_1, \ldots$ are determined by the boundary condition $Y_{\text{in}}(0) = y(0) = 1$: $B_0 = 1$, $B_1 = B_2 = \cdots = 0$. It is rather remarkable that the outer solution (9.4.10) and the inner solution (9.4.13) match asymptotically in the overlap region $X \to \infty$, $x \to 0$, $\varepsilon \to 0+$ for *all* values of the constants $C_0, C_1, \ldots, B_0, B_1, \ldots$, so long as $A_0 = A_1 = \cdots = 0$, because both the inner and outer solutions are exponentially small. There is no interaction between the outer solution and the inner solution in this problem!

To leading order, the uniform asymptotic approximation to $y(x)$ for $0 \le x \le 1$ is

$$y_{\text{unif,0}} = e^{1-1/x} + \exp(-x/\sqrt{\varepsilon}) + O(\varepsilon^{1/2}), \qquad \varepsilon \to 0+. \tag{9.4.14}$$

To first order in ε we need one additional term from the outer expansion and *two* additional terms from the inner expansion to construct the uniform expansion. The result is

$$\begin{aligned}
y_{\text{unif, 1}} &= e^{1-1/x}\left(1 - \frac{1}{2}\varepsilon x^{-4} + \frac{1}{5}\varepsilon x^{-5} + \frac{3}{10}\varepsilon\right) \\
&\quad + \exp(-x/\sqrt{\varepsilon})\left(1 - \frac{x^3}{6\varepsilon} - \frac{x^2}{4\sqrt{\varepsilon}} - \frac{x}{4} + \frac{x^6}{72\varepsilon^2} + \frac{x^5}{60\varepsilon^{3/2}}\right. \\
&\quad \left. + \frac{x^4}{96\varepsilon} - \frac{x^3}{48\sqrt{\varepsilon}} - \frac{x^2}{32} - \frac{x\sqrt{\varepsilon}}{32}\right) + O(\varepsilon^{3/2}), \qquad \varepsilon \to 0+.
\end{aligned} \tag{9.4.15}$$

In general, it is necessary to compute two new terms in the inner expansion for every new term in the outer expansion.

Figures 9.11 to 9.13 compare the above two uniform approximations to $y(x)$ with the actual numerical solution to $y(x)$ for $\varepsilon = 0.05, 0.01, 0.001$. Note that $y_{\text{unif, 1}}$ does not approximate $y(x)$ better than $y_{\text{unif, 0}}$ until ε is as small as 0.001.

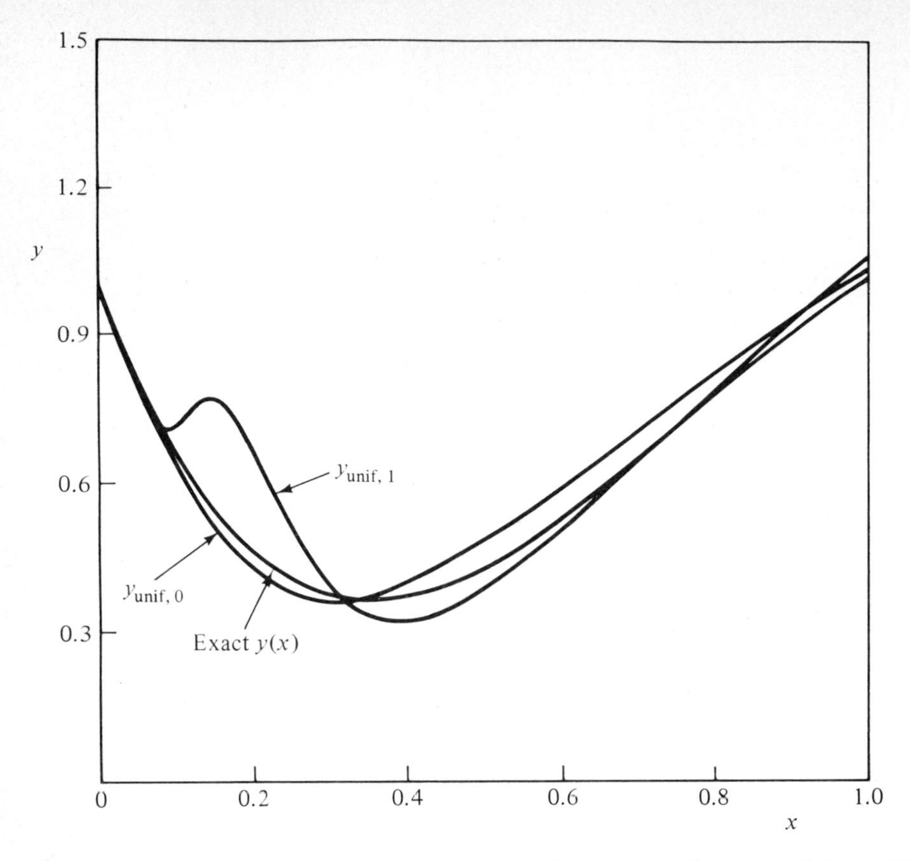

Figure 9.11 Comparison of exact and approximate solutions to $\varepsilon y'' + x^2 y' - y = 0$ [$y(0) = y(1) = 1$] for $\varepsilon = 0.05$. The two approximate solutions $y_{\text{unif},0}(x)$ and $y_{\text{unif},1}(x)$ are derived using boundary-layer theory and are given in (9.4.14) and (9.4.15). Note that even for this small value of ε, $y_{\text{unif},0}(x)$ is a better approximation to $y(x)$ than $y_{\text{unif},1}(x)$. Note that the higher-order approximation $y_{\text{unif},1}(x)$ crosses the exact solution $y(x)$ more frequently than $y_{\text{unif},0}(x)$; for very small ε, $y_{\text{unif},1}(x)$ hugs the curve $y(x)$ more closely than $y_{\text{unif},0}(x)$ (see Figs. 9.12 and 9.13).

Example 3 *Boundary-layer problem involving* ln ε. Consider the singular perturbation problem

$$\varepsilon y'' + xy' - xy = 0, \qquad y(0) = 0,\ y(1) = e. \tag{9.4.16}$$

Again, in this example there is a boundary layer at $x = 0$ whose thickness is of order $\varepsilon^{1/2}$, and not ε. However, the novelty of this example is that the inner expansion is not just a series in powers of $\varepsilon^{1/2}$. Terms containing ln ε also appear.

The outer expansion is obtained by assuming that

$$y_{\text{out}}(x) \sim y_0(x) + \varepsilon y_1(x) + \varepsilon^2 y_2(x) + \cdots, \qquad \varepsilon \to 0+. \tag{9.4.17}$$

Substituting (9.4.17) into (9.4.16) gives $y_0' - y_0 = 0$, $xy_n' - xy_n = -y_{n-1}''$ $(n \geq 1)$. A boundary layer may appear at $x = 0$ but not at $x = 1$, so the boundary condition satisfied by $y_{\text{out}}(x)$ is $y_{\text{out}}(1) = e$. The resulting outer solution is

$$y_{\text{out}}(x) \sim e^x - \varepsilon e^x \ln x + \varepsilon^2 e^x \left[\frac{1}{2}(\ln x)^2 - \frac{2}{x} + \frac{1}{2x^2} + \frac{3}{2}\right] + \cdots, \qquad \varepsilon \to 0+. \tag{9.4.18}$$

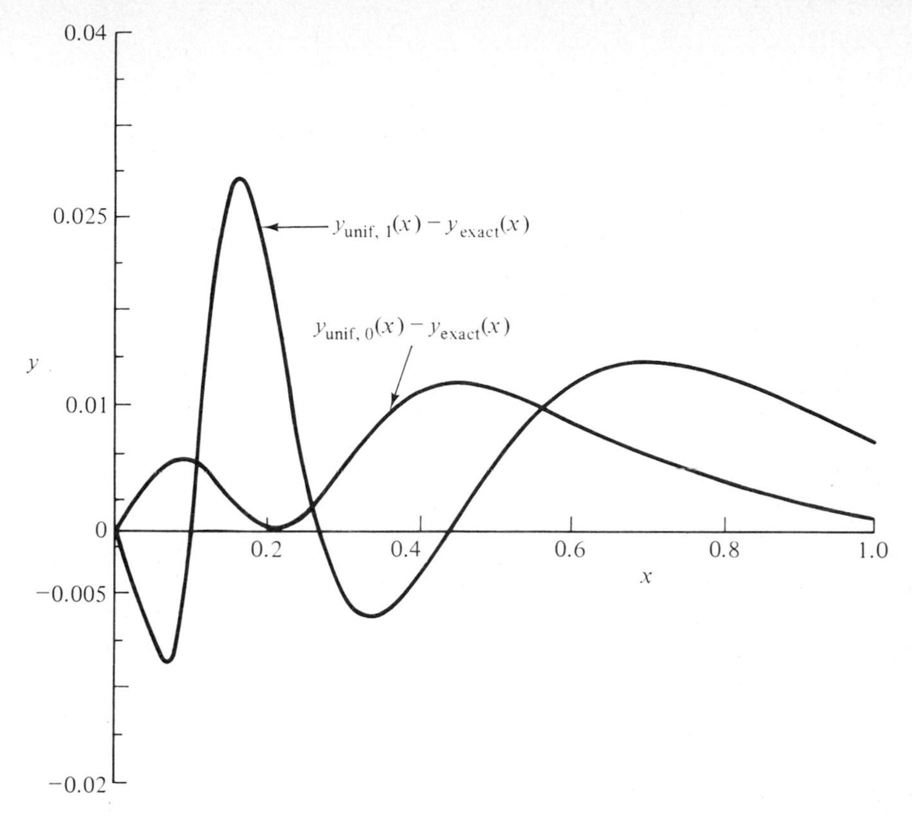

Figure 9.12 Comparison of the errors (not the relative errors) between the exact solution of $\varepsilon y'' + x^2 y' - y = 0$ $[y(0) = y(1) = 1]$ for $\varepsilon = 0.01$ and the zeroth- and first-order uniform approximations to $y_{\text{exact}}(x)$ (see Fig. 9.11). Note that $\varepsilon = 0.01$ is still too large for $y_{\text{unif},1}(x)$ to be a better approximation than $y_{\text{unif},0}(x)$.

Since $y_{\text{out}}(0) \neq 0$, there must be a boundary layer in the neighborhood of $x = 0$. Substituting (9.4.2) into (9.4.16) gives

$$\frac{\varepsilon}{\delta^2}\frac{d^2 Y_{\text{in}}}{dX^2} + X\frac{dY_{\text{in}}}{dX} - \delta X Y_{\text{in}} = 0.$$

Thus, if $\delta(\varepsilon)/\varepsilon^{1/2} \to 0$ or ∞ as $\varepsilon \to 0$, matching to the boundary condition $Y_{\text{in}}(0) = 0$ and the outer solution for $X \to \infty$ would be impossible. The distinguished limit is $\delta = \varepsilon^{1/2}$, so

$$\frac{d^2 Y_{\text{in}}}{dX^2} + X\frac{dY_{\text{in}}}{dX} - \varepsilon^{1/2} X Y_{\text{in}} = 0. \tag{9.4.19}$$

We would like to represent $Y_{\text{in}}(X)$ as a perturbation series, and in (9.4.19) the small parameter is $\varepsilon^{1/2}$ (and not ε). Thus, it would seem reasonable to assume an expansion of the form

$$Y_{\text{in}}(X) \sim Y_0(X) + \varepsilon^{1/2} Y_1(X) + \varepsilon Y_2(X) + \cdots, \qquad \varepsilon \to 0+, \tag{9.4.20}$$

where the boundary condition $y(0) = 0$ becomes $Y_n(0) = 0$ $(n \geq 0)$.

Substituting the expansion (9.4.20) into (9.4.19) gives $Y_0'' + XY_0' = 0$, whose solution is

$$Y_0(X) = A_0 \int_0^X e^{-t^2/2}\, dt, \tag{9.4.21}$$

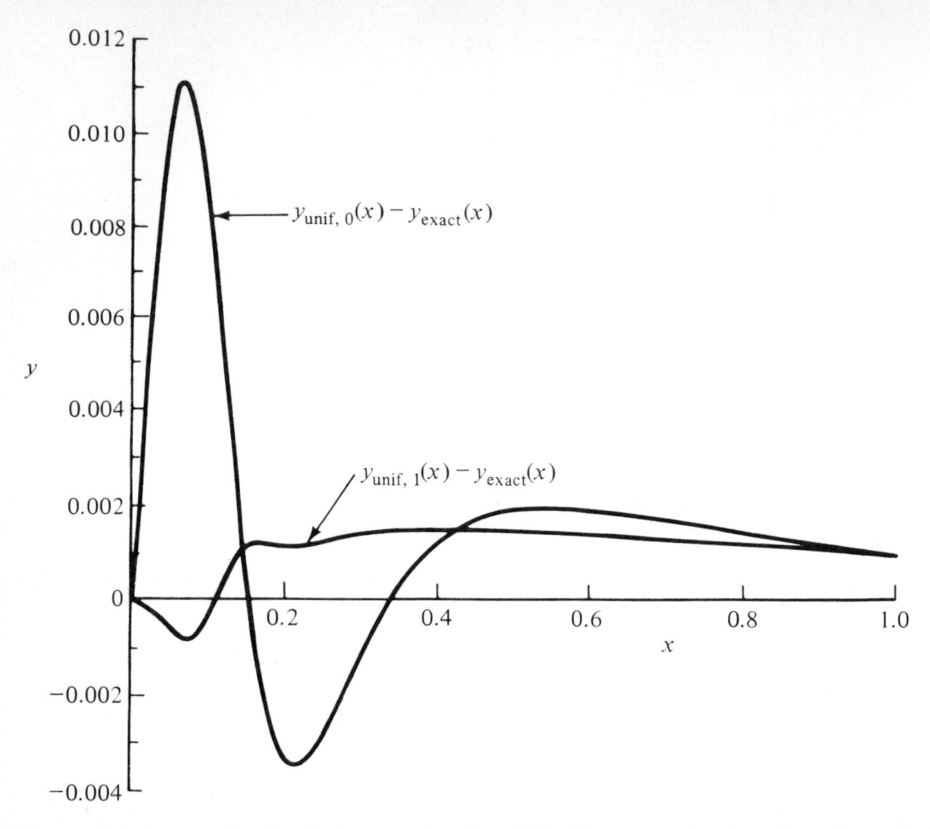

Figure 9.13 Same as in Fig. 9.12 except that $\varepsilon = 0.001$. This value of ε is sufficiently small for $y_{\text{unif, 1}}(x)$ to be a better approximation to $y_{\text{exact}}(x)$ than $y_{\text{unif, 0}}(x)$.

where A_0 is a constant to be determined by matching. Also, $Y_n'' + XY_n' = XY_{n-1}$ $(n \geq 1)$, so the relation between Y_n and Y_{n-1} is

$$Y_n(X) = A_n \int_0^X e^{-t^2/2}\, dt + \int_0^X e^{-t^2/2}\, dt \int_0^t e^{s^2/2} s Y_{n-1}(s)\, ds. \tag{9.4.22}$$

These integrals cannot be evaluated in closed form; nevertheless, we can still match the inner and outer solutions.

Matching is done by taking the intermediate limit $\varepsilon \to 0+$, $x \to 0+$, $X \to \infty$. To leading order, the outer solution (9.4.18) becomes $y_{\text{out}}(x) = 1 + O(x, \varepsilon \ln x)$, $(\varepsilon/x^2 \to 0, x \to 0+)$, while the leading-order inner solution is $Y_0(X) \sim \sqrt{\pi/2}\, A_0$ $(\varepsilon \to 0+, X \to \infty)$ with exponentially small corrections. It follows that $A_0 = \sqrt{2/\pi}$. Thus, to leading order, a uniform approximation to the solution of (9.4.16) is

$$y_{\text{unif, 0}}(x) = e^x + \sqrt{\frac{2}{\pi}} \int_0^X e^{-t^2/2}\, dt - 1 = e^x - \sqrt{\frac{2}{\pi}} \int_X^\infty e^{-t^2/2}\, dt. \tag{9.4.23}$$

A comparison between this approximation and the exact solution to (9.4.16) is given in Fig. 9.14.

To second order, the intermediate limit of the outer expansion is

$$\begin{aligned} y_{\text{out}}(x) \sim{} & 1 + x + \tfrac{1}{2}x^2 - \varepsilon \ln x + \tfrac{1}{2}\varepsilon^2 x^{-2} + \tfrac{1}{6}x^3 - \varepsilon x \ln x \\ & - \tfrac{3}{2}\varepsilon^2 x^{-1} + O(x^4, \varepsilon x^2 \ln x, \varepsilon^2 (\ln x)^2, \varepsilon^3/x^4), \qquad \varepsilon/x^2 \to 0+, x \to 0+. \end{aligned} \tag{9.4.24}$$

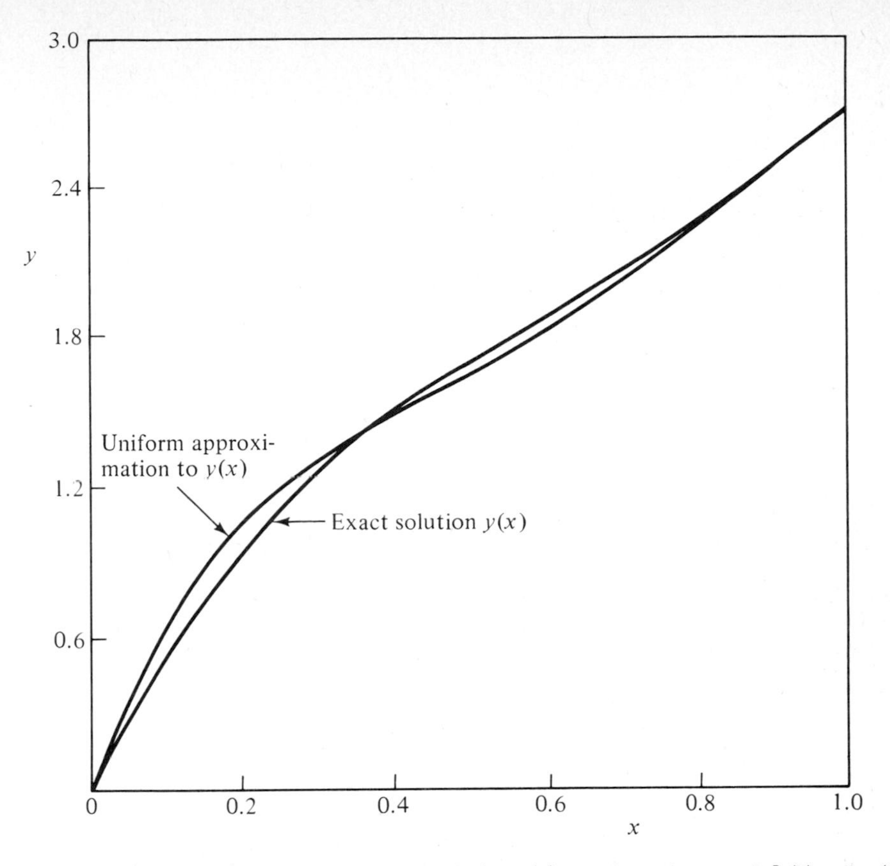

Figure 9.14 Comparison between the exact solution $y(x)$ to $\varepsilon y'' + xy' - xy = 0$ [$y(0) = 0$, $y(1) = e$] and the lowest-order uniform approximation to $y(x)$ obtained from boundary-layer analysis [see (9.4.23)]. The value of ε is 0.05.

We must also compute the intermediate limit $X \to \infty$ of the inner expansion:

$$\frac{dY_1}{dX} = A_1 e^{-X^2/2} + e^{-X^2/2} \int_0^X e^{s^2/2} s Y_0(s)\, ds \sim \sqrt{\frac{\pi}{2}} A_0, \qquad X \to \infty,$$

to within exponentially small terms. Therefore,

$$Y_1(X) \sim \sqrt{\frac{\pi}{2}} A_0 X + C_1, \qquad X \to \infty,$$

to within exponentially small terms, where C_1 is an integration constant which subsumes the constant A_1 in (9.4.22). Proceeding similarly in next order, we find that

$$Y_2(X) \sim \sqrt{\frac{\pi}{2}} A_0(\tfrac{1}{2}X^2 - \ln X + \tfrac{1}{2}X^{-2}) + C_1 X + C_2 + O(X^{-3}), \qquad X \to \infty,$$

and
$$Y_3(X) \sim \sqrt{\frac{\pi}{2}} A_0(\tfrac{1}{6}X^3 - X \ln X - \tfrac{3}{2}X^{-1}) + C_1(\tfrac{1}{2}X^2 - \ln X) + C_2 X + C_3 + O(X^{-2}), \qquad X \to \infty,$$

where C_2 and C_3 are integration constants.

The matching condition on $y_{\text{out}}(x)$ and $Y_{\text{in}}(X)$ is therefore

$$1 + x + \frac{1}{2}x^2 - \varepsilon \ln x + \frac{1}{2}\varepsilon^2 x^{-2} + \frac{1}{6}x^3 - \varepsilon x \ln x - \frac{3}{2}\varepsilon^2 x^{-1}$$

$$= \sqrt{\frac{\pi}{2}} A_0 \left(1 + \varepsilon^{1/2}X + \frac{1}{2}\varepsilon X^2 - \varepsilon \ln X + \frac{1}{2}\varepsilon X^{-2} + \frac{1}{6}\varepsilon^{3/2}X^3 - \varepsilon^{3/2}X \ln X - \frac{3}{2}\varepsilon^{3/2}X^{-1}\right)$$

$$+ \varepsilon^{1/2}C_1\left(1 + \varepsilon^{1/2}X + \frac{1}{2}\varepsilon X^2 - \varepsilon \ln X\right) + \varepsilon C_2(1 + \varepsilon^{1/2}X)$$

$$+ \varepsilon^{3/2}C_3 + O(x^4, \varepsilon x^2 \ln x, \varepsilon^2(\ln x)^2, \varepsilon^3/x^4), \qquad x \to 0+,\ \varepsilon/x^2 \to 0+, \tag{9.4.25}$$

with $X = x\varepsilon^{-1/2}$. However, matching is *not* possible here because, when the right side of (9.4.25) is rewritten in terms of $x = \varepsilon^{1/2}X$, terms of order $\varepsilon \ln \varepsilon$ appear!

Apparently, the assumption that the inner solution is a series in powers of $\varepsilon^{1/2}$ is naive. Instead, the inner expansion in (9.4.20) must be modified to read

$$Y_{\text{in}}(X) \sim Y_0(X) + \varepsilon^{1/2}Y_1(X) + \varepsilon Y_2(X) + \varepsilon \ln \varepsilon \bar{Y}_2(X) + \varepsilon^{3/2}Y_3(X)$$
$$+ \varepsilon^{3/2} \ln \varepsilon \bar{Y}_3(X) + \cdots, \qquad \varepsilon \to 0+.$$

Terms of order $\varepsilon^2 \ln \varepsilon$, $\varepsilon^2(\ln \varepsilon)^2$, and so on, must also be included in higher order.

In the intermediate matching region the additional terms contribute to the right side of (9.4.25) as

$$\varepsilon \ln \varepsilon \sqrt{\frac{\pi}{2}} \bar{A}_2(1 + \varepsilon^{1/2}X) + \varepsilon^{3/2} \ln \varepsilon \bar{C}_3,$$

where $\bar{C}_3$ is a new integration constant. Now the match in (9.2.25) can be accomplished. The matching conditions are $A_0 = \sqrt{2/\pi}$, $\bar{A}_2 = -\frac{1}{2}A_0$, $C_1 = C_2 = C_3 = \bar{C}_3 = 0$.

(I) 9.5 MISCELLANEOUS EXAMPLES OF LINEAR BOUNDARY-LAYER PROBLEMS

This section is a collection of six examples of linear differential-equation boundary-value problems which can be solved approximately using boundary-layer theory. We have selected these problems because they illustrate the broad spectrum of analysis that boundary-layer theory entails. The first two examples involve higher-order differential equations.

Example 1 *Third-order boundary-value problem.* Consider the third-order boundary-value problem

$$\varepsilon y'''(x) - y'(x) + xy(x) = 0, \qquad y(0) = y'(0) = y(1) = 1, \tag{9.5.1}$$

in the limit $\varepsilon \to 0+$. The novelty of this tricky third-order problem is that boundary layers occur at both $x = 0$ and $x = 1$.

In the outer region which contains no boundary layers, we assume an expansion of the form $y_{\text{out}}(x) \sim y_0(x) + \varepsilon y_1(x) + \varepsilon^2 y_2(x) + \cdots$ $(\varepsilon \to 0+)$. This reduces (9.5.1) to a sequence of first-order differential equations:

$$y_n' - xy_n = \begin{cases} 0, & n = 0, \\ y_{n-1}''', & n > 0. \end{cases} \tag{9.5.2}$$

We find that

$$y_0(x) = a_0 e^{x^2/2}, \tag{9.5.3}$$

$$y_1(x) = [a_0(x^4/4 + 3x^2/2) + a_1]e^{x^2/2}, \tag{9.5.4}$$

and so on.

Next, we consider the possibility of a boundary layer of thickness δ at $x = 0$. In terms of the inner variable $X = x/\delta$ with $Y_{\text{in}}(X) = y(x)$, (9.5.1) becomes

$$\frac{\varepsilon}{\delta^3}\frac{d^3 Y_{\text{in}}}{dX^3} - \frac{1}{\delta}\frac{dY_{\text{in}}}{dX} + \delta X Y_{\text{in}} = 0. \tag{9.5.5}$$

There are two consistent distinguished limits, $\delta = O(1)$ and $\delta = O(\varepsilon^{1/2})$ $(\varepsilon \to 0+)$. However, $\delta = O(1)$ reproduces the outer limit. Thus, the only consistent choice for the boundary layer thickness is $\delta = \varepsilon^{1/2}$. For this choice (9.5.5) reduces to

$$\frac{d^3 Y_{\text{in}}}{dX^3} - \frac{dY_{\text{in}}}{dX} = -\varepsilon X Y_{\text{in}}. \tag{9.5.6}$$

The next step is to approximate Y_{in} by an appropriate inner expansion. However, as we will see, the most obvious choice for an inner expansion,

$$Y_{\text{in}}(X) \sim Y_0(X) + \varepsilon Y_1(X) + \cdots \qquad \varepsilon \to 0+, \tag{9.5.7}$$

is inadequate. Substituting (9.5.7) into (9.5.6) gives

$$\frac{d^3 Y_n}{dX^3} - \frac{dY_n}{dX} = \begin{cases} 0, & n = 0, \\ -XY_{n-1}, & n > 0, \end{cases}$$

whose solutions are

$$Y_0(X) = A_0 e^X + B_0 e^{-X} + C_0, \tag{9.5.8}$$

$$\begin{aligned} Y_1(X) = {} & [-A_0(X^2/4 - 3X/4) + A_1]e^X \\ & + [-B_0(X^2/4 + 3X/4) + B_1]e^{-X} + C_0 X^2/2 + C_1, \end{aligned} \tag{9.5.9}$$

and so on. The boundary conditions $y(0) = y'(0) = 1$ become

$$Y_{\text{in}}(0) = 1, \qquad \frac{dY_{\text{in}}}{dX}(0) = \varepsilon^{1/2}. \tag{9.5.10}$$

Note that $Y_{\text{in}}(0) = 1$ implies that $A_0 + B_0 + C_0 = 1$, $A_1 + B_1 + C_1 = 0$, and so on. However, it is not possible to satisfy the condition $Y'_{\text{in}}(0) = \varepsilon^{1/2}$ for any choice of constants! Apparently, our choice of inner expansion in (9.5.7) was wrong. There is no way to represent $Y'_{\text{in}}(0) = \varepsilon^{1/2}$ by an expansion in integral powers of ε.

We therefore revise the inner expansion to read

$$Y_{\text{in}}(X) \sim Y_0(X) + \varepsilon^{1/2} Y_{1/2}(X) + \varepsilon Y_1(X) + \cdots, \qquad \varepsilon \to 0+, \tag{9.5.11}$$

where Y_0 and Y_1 are still given by (9.5.8) and (9.5.9) and

$$Y_{1/2}(X) = A_{1/2} e^X + B_{1/2} e^{-X} + C_{1/2}. \tag{9.5.12}$$

Note that the equations for Y_n with n integral and n half integral decouple because $\varepsilon^{1/2}$ does not appear explicitly in (9.5.6).

The boundary conditions (9.5.10) can now be satisfied:

$$
\begin{aligned}
Y_0(0) &= 1: & A_0 + B_0 + C_0 &= 1,\\
Y_0'(0) &= 0: & A_0 - B_0 &= 0,\\
Y_{1/2}(0) &= 0: & A_{1/2} + B_{1/2} + C_{1/2} &= 0,\\
Y_{1/2}'(0) &= 1: & A_{1/2} - B_{1/2} &= 1,\\
Y_1(0) &= 0: & A_1 + B_1 + C_1 &= 0,\\
Y_1'(0) &= 0: & A_1 - B_1 &= 0.
\end{aligned}
\tag{9.5.13}
$$

It is also necessary to require that $A_0 = A_{1/2} = A_1 = \cdots = 0$. (Otherwise, each term in the inner expansion would contain terms that grow exponentially and this would prevent the inner and outer expansions from being matched.) Combining this requirement with (9.5.13) gives $A_0 = B_0 = A_{1/2} = A_1 = B_1 = C_1 = 0$, $C_0 = C_{1/2} = 1$, $B_{1/2} = -1$. Thus, our final expression for $Y_{\text{in}}(X)$ correct to order ε is

$$Y_{\text{in}}(X) \sim 1 + \sqrt{\varepsilon}(1 - e^{-X}) + \varepsilon X^2/2 + \cdots, \qquad \varepsilon \to 0+. \tag{9.5.14}$$

In the overlap region defined by $x \to 0+$, $X \to \infty$, $\varepsilon \to 0+$, we have

$$y_{\text{out}}(x) = a_0 + a_0 x^2/2 + \varepsilon a_1 + O(x^4) + O(\varepsilon x^2) + O(\varepsilon^2), \qquad x \to 0+,\ \varepsilon \to 0+,$$

$$Y_{\text{in}}(X) = 1 + \varepsilon^{1/2} + \varepsilon X^2/2 + O(\varepsilon^{3/2} X^2), \qquad X \to \infty,\ \varepsilon^{1/2} X \to 0+$$

and again there is trouble! Matching is impossible because there is no $\varepsilon^{1/2}$ term in the expansion of $y_{\text{out}}(x)$. Apparently, it is necessary to include $\varepsilon^{n/2}$ terms in the outer expansion as well as the inner expansion. The outer expansion must be generalized to

$$y_{\text{out}}(x) \sim y_0(x) + \varepsilon^{1/2} y_{1/2}(x) + \varepsilon y_1(x) + \cdots, \qquad \varepsilon \to 0+,$$

and from (9.5.1) we have $y_{1/2}(x) = a_{1/2} e^{x^2/2}$. The new (and now correct) outer expansion in the matching region is

$$y_{\text{out}}(x) = a_0 + a_0 x^2/2 + a_{1/2}\varepsilon^{1/2} + a_1 \varepsilon + O(x^4) + O(\varepsilon^{1/2} x^2) + O(\varepsilon^{3/2}), \qquad x \to 0+,\ \varepsilon \to 0+.$$

The inner and outer expansions now match and the matching condition is $a_0 = 1$, $a_{1/2} = 1$, $a_1 = 0$. Thus, our final expression for $y_{\text{out}}(x)$ correct to order ε is

$$y_{\text{out}}(x) \sim e^{x^2/2}[1 + \varepsilon^{1/2} + \varepsilon(x^4/4 + 3x^2/2) + \cdots], \qquad \varepsilon \to 0+. \tag{9.5.15}$$

This outer solution does not satisfy the boundary condition $y(1) = 1$. Thus, there is another boundary layer at $x = 1$. Again, the only distinguished limit is $\delta = \varepsilon^{1/2}$. Thus, the appropriate inner variable is $\bar{X} = (1 - x)\varepsilon^{-1/2}$. Letting $\bar{Y}_{\text{in}}(\bar{X}) = y(x)$, (9.5.1) becomes

$$\frac{d^3 \bar{Y}_{\text{in}}}{d\bar{X}^3} - \frac{d\bar{Y}_{\text{in}}}{d\bar{X}} = \varepsilon^{1/2} \bar{Y}_{\text{in}} - \varepsilon \bar{X} \bar{Y}_{\text{in}}.$$

This suggests an inner expansion of the form

$$\bar{Y}_{\text{in}}(\bar{X}) = \bar{Y}_0(\bar{X}) + \varepsilon^{1/2} \bar{Y}_{1/2}(\bar{X}) + \varepsilon \bar{Y}_1(\bar{X}) + \cdots. \tag{9.5.16}$$

The remainder of this problem is routine. We solve for $\bar{Y}_0$, $\bar{Y}_{1/2}$, and $\bar{Y}_1$ by imposing the boundary condition $y(1) = 1$ and matching $\bar{Y}_{\text{in}}(\bar{X})$ to $y_{\text{out}}(x)$. The results are given in Prob. 9.21. By combining these results with (9.5.14) and (9.5.15), we obtain the following uniform asymptotic approximation to the solution of (9.5.1) correct to terms of order ε:

$$y_{\text{unif}}(x) = e^{x^2/2}[1 + \varepsilon^{1/2} + \varepsilon(x^4/4 + 3x^2/2)] - \sqrt{\varepsilon}\exp(-x/\sqrt{\varepsilon}) + \exp[-(1-x)/\sqrt{\varepsilon}][(\sqrt{e}-1)(\tfrac{1}{8}x^2 + \tfrac{1}{4}x - \tfrac{11}{8}) + \sqrt{\varepsilon}(-9\sqrt{e} - 3 + 3x + x\sqrt{e})/8 - \tfrac{7}{4}\varepsilon\sqrt{e}]. \tag{9.5.17}$$

In Fig. 9.15 we compare the exact solution to (9.5.1) with the uniform asymptotic approximation to $y(x)$ in (9.5.17).

Example 2 *Fourth-order boundary-value problem.* Consider the inhomogeneous fourth-order boundary-value problem

$$\varepsilon^2 \frac{d^4 y}{dx^4} - (1 + x)\frac{d^2 y}{dx^2} = 1, \qquad y(0) = y'(0) = y(1) = y'(1) = 1, \tag{9.5.18}$$

in the limit $\varepsilon \to 0+$. As in the previous example, we will see that boundary layers occur at $x = 0$ and at $x = 1$.

In the outer region the $\varepsilon^2\, d^4y/dx^4$ term is small, so we are inclined to use an outer expansion in powers of ε^2:

$$y_{\text{out}}(x) \sim y_0(x) + \varepsilon^2 y_2(x) + \cdots, \qquad \varepsilon \to 0+. \tag{9.5.19}$$

However, our experience from the previous example suggests that this choice may be naive. To determine the correct form for the outer expansion, let us first examine the boundary layers at $x = 0$ and at $x = 1$.

We take the inner limit in the neighborhood of $x = 0$ by introducing the inner variable

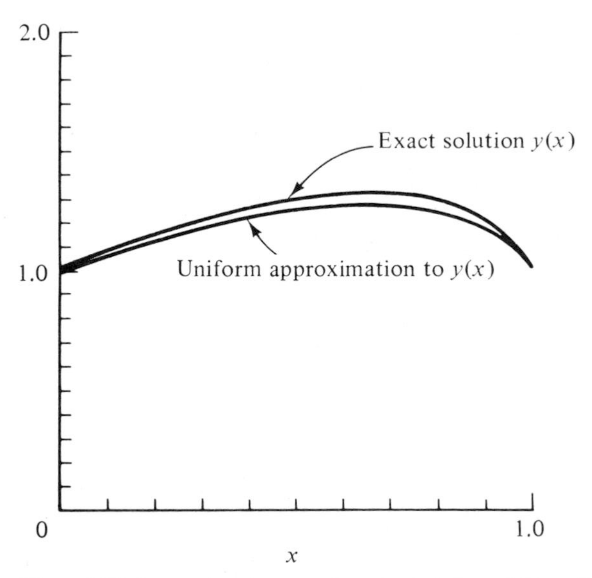

Figure 9.15 Comparison between the exact solution $y(x)$ to $\varepsilon y''' - y' + xy = 0\,[y(0) = y'(0) = y(1) = 1]$ and the uniform approximation (9.5.17) to $y(x)$ (correct to terms of order ε) obtained from boundary-layer analysis. The value of ε is 0.05. When $\varepsilon = 0.01$ the uniform approximation and the exact solution are not distinguishable on the scale of the graph.

$X = x/\delta$ and setting $Y_{in}(X) = y(x)$. In the inner region (9.5.18) becomes

$$\frac{\varepsilon^2}{\delta^4}\frac{d^2 Y_{in}}{dX^4} - \frac{1+\delta X}{\delta^2}\frac{d^2 Y_{in}}{dX^2} = 1.$$

The distinguished limits are $\delta = 1$, which reproduces the outer limit, and $\delta = \varepsilon$. Setting $\delta = \varepsilon$, we have

$$\frac{d^4 Y_{in}}{dX^4} - (1 + \varepsilon X)\frac{d^2 Y}{dX^2} = \varepsilon^2, \tag{9.5.20}$$

which suggests an inner expansion in powers of ε:

$$Y_{in}(X) \sim Y_0(X) + \varepsilon Y_1(X) + \varepsilon^2 Y_2(X) + \cdots, \qquad \varepsilon \to 0+. \tag{9.5.21}$$

We assume an inner expansion in powers of ε and not ε^2 for two reasons. First, there is a term containing ε in (9.5.20). Second, the boundary conditions at $x = 0$ when written in terms of the inner variable X read $Y_{in}(0) = 1$, $dY_{in}/dX(0) = \varepsilon$.

Substituting (9.5.21) into (9.5.20) gives the following equations and boundary conditions:

$$\frac{d^4 Y_0}{dX^4} - \frac{d^2 Y_0}{dX^2} = 0, \qquad Y_0(0) = 1,\ Y_0'(0) = 0;$$

$$\frac{d^4 Y_1}{dX^4} - \frac{d^2 Y_1}{dX^2} = X\frac{d^2 Y_0}{dX^2}, \qquad Y_1(0) = 0,\ Y_1'(0) = 1.$$

The solutions to these equations are

$$Y_0(X) = 1 + A_0(e^X - 1 - X) + B_0(e^{-X} - 1 + X),$$

$$Y_1(X) = X + \tfrac{1}{4}A_0(X^2 e^X - 5Xe^X + 5X) - \tfrac{1}{4}B_0(X^2 e^{-X} + 5Xe^{-X} - 5X)$$
$$+ A_1(e^X - 1 - X) + B_1(e^{-X} - 1 + X).$$

However, these equations simplify considerably because matching to the outer solution is impossible if terms growing like e^X as $X \to +\infty$ are present. Therefore, we must require that $A_n = 0$ for all n:

$$\begin{aligned} Y_0(X) &= 1 + B_0(e^{-X} - 1 + X), \\ Y_1(X) &= X - \tfrac{1}{4}B_0(X^2 e^{-X} + 5Xe^{-X} - 5X) + B_1(Xe^{-X} - 1 + X). \end{aligned} \tag{9.5.22}$$

In the matching region near $x = 0$, the intermediate limit is $x \to 0+$, $X = x/\varepsilon \to +\infty$. Thus, in the matching region (9.5.22) becomes

$$\begin{aligned} Y_0(X) &\sim 1 + B_0(X - 1), & X &\to +\infty, \\ Y_1(X) &\sim X + \tfrac{5}{4}B_0 X + B_1(X - 1), & X &\to +\infty, \end{aligned} \tag{9.5.23}$$

with exponentially small corrections.

A similar inner expansion can be made in the neighborhood of $x = 1$. The appropriate inner variable is $\bar{X} = (1 - x)/\varepsilon$. The first two terms of the inner solution are

$$\begin{aligned} \bar{Y}_0(\bar{X}) &= 1 + \bar{B}_0[\exp(-\sqrt{2}\,\bar{X}) - 1 + \sqrt{2}\,\bar{X}], \\ \bar{Y}_1(\bar{X}) &= -\bar{X} + \tfrac{1}{8}\bar{B}_0[\sqrt{2}\,\bar{X}^2 \exp(-\sqrt{2}\,\bar{X}) + 5\bar{X}\exp(-\sqrt{2}\,\bar{X}) - 5\bar{X}] \\ &\quad + \bar{B}_1[\exp(-\sqrt{2}\,\bar{X}) - 1 + \sqrt{2}\,\bar{X}], \end{aligned} \tag{9.5.24}$$

where the inner expansion is $\bar{Y}_{in}(\bar{X}) \sim \bar{Y}_0 + \varepsilon\bar{Y}_1 + \cdots$ $(\varepsilon \to 0+)$. In the matching region near $x = 1$, $x \to 1-$, $\bar{X} = (1 - x)/\varepsilon \to +\infty$, so

$$\begin{aligned} \bar{Y}_0(\bar{X}) &\sim 1 + \bar{B}_0(-1 + \sqrt{2}\,\bar{X}), & \bar{X} &\to +\infty, \\ \bar{Y}_1(\bar{X}) &\sim -\bar{X} - \tfrac{5}{8}\bar{B}_0\bar{X} + \bar{B}_1(-1 + \sqrt{2}\,\bar{X}), & \bar{X} &\to +\infty. \end{aligned} \tag{9.5.25}$$

In terms of the outer variable x, (9.5.23) becomes

$$Y_{\text{in}}(X) = \frac{B_0 x}{\varepsilon} + 1 - B_0 + x + \frac{5}{4} B_0 x + B_1 x - \varepsilon B_1 + O(\varepsilon^2) + O(\varepsilon x) + O(x^2),$$

$$x \to 0+,\ x/\varepsilon \to +\infty, \tag{9.5.26}$$

and (9.5.25) becomes

$$\bar{Y}_{\text{in}}(\bar{X}) = \bar{B}_0\sqrt{2}\frac{1-x}{\varepsilon} + 1 - \bar{B}_0 - (1-x) - \frac{5}{8}\bar{B}_0(1-x) + \bar{B}_1\sqrt{2}(1-x)$$

$$- \bar{B}_1 \varepsilon + O(\varepsilon^2) + O[\varepsilon(1-x)] + O[(x-1)^2],$$

$$x \to 1-,\ (1-x)/\varepsilon \to +\infty. \tag{9.5.27}$$

Now we can tell if the outer expansion in (9.5.19) is valid. If we substitute (9.5.19) into (9.5.18) and equate coefficients of like powers of ε, we obtain the leading-order outer equation $-(1+x)(d^2 y_0/dx^2) = 1$. The solution to this equation is

$$y_0(x) = -(1+x)\ln(1+x) + a_0 x + b_0.$$

In the intermediate limit the outer solution behaves near $x = 0$ like

$$y_0(x) \sim b_0 + a_0 x - x + O(x^2), \qquad x \to 0+. \tag{9.5.28}$$

Demanding that (9.5.28) match asymptotically with (9.5.26) implies that $B_0 = 0$, $b_0 = 1$, $B_1 = 0$, $a_0 = 2$, which is consistent. However, demanding that $y_{\text{out}}(x)$ also match asymptotically with (9.5.27) gives different values for the constants a_0 and b_0! Near $x = 1$ we have

$$y_0(x) \sim -2\ln 2 + a_0 + b_0 + (1-x)(1 - a_0 + \ln 2) + O[\varepsilon^2, (1-x)^2, \varepsilon(1-x)],$$

$$\varepsilon \to 0+,\ x \to 1-. \tag{9.5.29}$$

Matching (9.5.29) to (9.5.27) gives $\bar{B}_0 = 0$, $\bar{B}_1 = 0$, $b_0 = \ln 2 - 1$, $a_0 = 2 + \ln 2$.

Apparently the outer expansion (9.5.19) is wrong because it is overdetermined. Even though ε^2 seems to be the natural expansion parameter for (9.5.18), matching requires that the outer expansion should in fact be done in powers of ε: $y_{\text{out}}(x) \sim y_0(x) + \varepsilon y_1(x) + \varepsilon^2 y_2(x) + \cdots$ $(\varepsilon \to 0+)$. With this expansion $y_0(x)$ is the same as before but

$$y_1(x) = a_1 x + b_1.$$

With these additional terms in the outer expansion the matching conditions at $x = 0$ become $B_0 = 0$, $b_0 = 1$, $1 + B_1 = a_0 - 1$, $b_1 = -B_1$. Similarly, the matching conditions at $x = 1$ become $\bar{B}_0 = 0$, $1 = a_0 + b_0 - 2\ln 2$, $1 - a_0 + \ln 2 = \sqrt{2}\,\bar{B}_1 - 1$, $-\bar{B}_1 = a_1 + b_1$. Thus, a consistent match at both $x = 0$ and $x = 1$ is possible if $B_0 = \bar{B}_0 = 0$, $b_0 = 1$, $a_0 = 2\ln 2$, $B_1 = 2\ln 2 - 2$, $b_1 = 2 - 2\ln 2$, $\sqrt{2}\,\bar{B}_1 = 2 - \ln 2$, $a_1 = (\ln 2)/\sqrt{2} - \sqrt{2} - 2 - 2\ln 2$, and so on.

Notice that to lowest order the outer solution becomes

$$y_{\text{out}}(x) = -(1+x)\ln(1+x) + 2x\ln 2 + 1 + O(\varepsilon), \qquad \varepsilon \to 0+, \tag{9.5.30}$$

which satisfies the boundary conditions $y_{\text{out}}(0) = y_{\text{out}}(1) = 1$. However, $y'_{\text{out}}(0) \neq 1$ and $y'_{\text{out}}(1) \neq 1$. Apparently, boundary layers appear at $x = 0$ and at $x = 1$ to adjust the slope of the outer solution to the imposed boundary values. The values of y' have an O(1) jump across the boundary layers whose thickness is $O(\varepsilon)$. The outer solution satisfies the boundary conditions correct to $O(\varepsilon)$ because the jump in y across the boundary layers is $O(\varepsilon)$:

$$\text{Jump in } y \text{ at boundary layer} = O\left(\int_0^\varepsilon y'\,dx\right) = O(\varepsilon), \qquad \varepsilon \to 0+.$$

Similarly, we observe that the boundary layer at $x = 0$ in Example 1 has an $O(1)$ jump in y' but an $O(\varepsilon^{1/2})$ jump in y [see (9.5.15)].

In the next three examples we examine the boundary-layer structure of a singular differential equation.

Example 3 *Boundary-value problem having no boundary layers.* Consider the boundary-value problem

$$\varepsilon y'' + \frac{1}{x} y' + y = 0, \qquad y(1) = e^{-1/2},\ y(0) \text{ finite}, \tag{9.5.31}$$

in the limit $\varepsilon \to 0+$. The point $x = 0$ is a regular singular point of the differential equation (see Sec. 3.3). Frobenius' method gives the indicial exponents 0 and $1 - 1/\varepsilon$, so if $\varepsilon < 1$, the condition that $y(0)$ be finite suffices to determine the solution uniquely. (Why?) However, the condition on $y(0)$ is so weak that the solution does not exhibit a boundary layer at $x = 0$ as $\varepsilon \to 0+$, even though $1/x > 0$ for $x > 0$!

Suppose there were a boundary layer of thickness δ situated at $x = 0$. Then, we could introduce the inner variables $X = x/\delta$, $Y_{\text{in}}(X) = y(x)$ and rewrite the differential equation as

$$\frac{\varepsilon}{\delta^2}\frac{d^2 Y_{\text{in}}}{dX^2} + \frac{1}{\delta^2 X}\frac{dY_{\text{in}}}{dX} + Y_{\text{in}} = 0.$$

Observe that there is *no* distinguished limit for $\delta \ll 1$! The singularity of the differential equation at $x = 0$ ensures that the solution to (9.5.31) has no boundary layers at $x = 0$, despite first appearances. Therefore, a complete approximation of the solution to the problem on the interval $0 \le x \le 1$ is given by the outer expansion

$$y_{\text{out}}(x) \sim y_0(x) + \varepsilon y_1(x) + \varepsilon^2 y_2(x) + \cdots, \qquad \varepsilon \to 0+. \tag{9.5.32}$$

Even though there is no boundary layer at $x = 0$, (9.5.32) is not a regular perturbation expansion (see Prob. 9.24).

Substituting (9.5.32) into (9.5.31) gives

$$\frac{1}{x} y_n' + y_n = \begin{cases} 0, & n = 0, \\ -y_{n-1}'', & n > 0. \end{cases}$$

There is no boundary layer at $x = 1$ because $1/x > 0$, so we must require that $y_0(1) = e^{-1/2}$, $y_n(1) = 0$ $(n > 0)$. We obtain $y_0(x) = e^{-x^2/2}$, $y_1(x) = -\frac{1}{4}(x^2 - 1)^2 e^{-x^2/2}$, and so on. Using these equations we can predict the value of $y(0)$:

$$y(0) = 1 - \varepsilon/4 + O(\varepsilon^2), \qquad \varepsilon \to 0+. \tag{9.5.33}$$

Example 4 *Singular boundary-value problem.* Let us reexamine (9.5.31) in the limit $\varepsilon \to 0-$. With $\varepsilon < 0$ in Example 3, the effect of the coordinate singularity at $x = 0$ changes abruptly. Now, Frobenius' method gives two positive indicial exponents 0 and $1 - 1/\varepsilon$, so the condition that $y(0)$ be finite does not suffice to determine the solution. The value of $y(0)$ must be specified.

Let us change the sign of ε and pose a representative problem:

$$\varepsilon y'' - y'/x - y = 0, \qquad 0 \le x \le 1, \qquad y(0) = 1,\ y(1) = 1, \qquad \varepsilon \to 0+.$$

The boundary condition $y(0) = 1$ uniquely determines the outer solution

$$y_{\text{out}}(x) = e^{-x^2/2}[1 + \varepsilon(x^4 - 2x^2)/4 + O(\varepsilon^2)], \qquad \varepsilon \to 0+,$$

because $-1/x$ is negative for $x > 0$, so no boundary layer can exist at $x = 0$.

A boundary layer of thickness ε is required at $x = 1$. Using the matching procedures developed in this chapter, we find that the inner solution is

$$Y_{\text{in}}(X) = (1 - e^{-1/2})[1 - \tfrac{1}{2}\varepsilon(X^2 - 4X)]e^{-X} + e^{-1/2}[1 + \varepsilon X] + O(\varepsilon^2), \qquad \varepsilon \to 0+, \tag{9.5.34}$$

where $X = (1 - x)/\varepsilon$. A uniform approximation to $y(x)$ for $0 \le x \le 1$, accurate to order ε, is

$$y_{\text{unif}} = e^{-x^2/2}[1 + \tfrac{1}{4}\varepsilon(x^2 - 1)^2] + (1 - e^{-1/2})[1 - \tfrac{1}{2}\varepsilon(X^2 - 4X)]e^{-X}. \tag{9.5.35}$$

Example 5 *Imposition of boundary conditions near a singularity of the differential equation.* Suppose we change the boundary conditions in (9.5.31) slightly and formulate the new boundary-value problem

$$\varepsilon y'' + y'/x + y = 0, \qquad y(\varepsilon) = 0,\ y(1) = e^{-1/2},$$

on the restricted interval $\varepsilon \le x \le 1$. There is no coordinate singularity on the interval $\varepsilon \le x \le 1$, so boundary conditions must be imposed at both $x = \varepsilon$ and $x = 1$.

The outer solution, away from a boundary layer near $x = \varepsilon$, is given correctly to order ε^2 by

$$y_{\text{out}}(x) = e^{-x^2/2}[1 - \varepsilon(x^2 - 1)^2/4] + O(\varepsilon^2), \qquad \varepsilon \to 0+. \tag{9.5.36}$$

The interesting feature of this example is the character of the boundary layer at $x = \varepsilon$. To determine the structure of the boundary layer, we introduce the inner variables $X = (x - \varepsilon)/\delta$, $Y_{\text{in}}(X) = y(x)$. In terms of these variables (9.5.31) become

$$\frac{\varepsilon}{\delta^2}\frac{d^2 Y_{\text{in}}}{dX^2} + \frac{1}{\varepsilon + \delta X}\frac{1}{\delta}\frac{dY_{\text{in}}}{dX} + Y_{\text{in}} = 0.$$

The only distinguished limit with $\delta \ll 1$ is $\delta = \varepsilon^2$. For this choice the differential equation becomes

$$\frac{d^2 Y_{\text{in}}}{dX^2} + \frac{1}{1 + \varepsilon X}\frac{dY_{\text{in}}}{dX} + \varepsilon^3 Y_{\text{in}} = 0.$$

The solution to this equation in the narrow boundary layer at $x = \varepsilon$ must satisfy the boundary condition $y(\varepsilon) = Y_{\text{in}}(0) = 0$ and match asymptotically with (9.5.36). We find that (see Prob. 9.22)

$$Y_{\text{in}}(X) = 1 - e^{-X} - \varepsilon[(\tfrac{1}{2}X^2 + X - \tfrac{1}{4})e^{-X} + \tfrac{1}{4}] + O(\varepsilon^2), \qquad \varepsilon \to 0+. \tag{9.5.37}$$

The next example is somewhat contrived, but it illustrates the remarkable phenomenon of a boundary-layer structure inside a boundary layer.

Example 6 *Nested boundary layers.* Consider the differential equation

$$\varepsilon^3 x y'' + x^2 y' - y(x^3 + \varepsilon) = 0, \tag{9.5.38}$$

subject to the boundary conditions

$$y(0) = 1, \qquad y(1) = \sqrt{e}. \tag{9.5.39}$$

This differential equation has a regular singular point at $x = 0$ with indicial exponents 0 and 1. Therefore, both linearly independent solutions at $x = 0$ are finite and it is consistent to impose the boundary conditions in (9.5.39).

First, we determine the outer solution by assuming an outer expansion of the form $y_{\text{out}} \sim y_0 + \varepsilon y_1 + \cdots$ ($\varepsilon \to 0+$). Substituting this expansion into (9.5.38) gives the equations for y_0 and y_1:

$$x^2 y_0' - y_0 x^3 = 0, \qquad x^2 y_1' - x^3 y_1 = y_0.$$

There is no boundary layer possible at $x = 1$ because the coefficient of y' is positive. Therefore, y_0 and y_1 satisfy the boundary conditions $y_0(1) = \sqrt{e}$, $y_1(1) = 0$. Solving the above pair of differential equations and imposing the boundary conditions at $x = 1$ gives the first two terms in the outer expansion:

$$y_{\text{out}}(x) = e^{x^2/2} + \varepsilon(1 - 1/x)e^{x^2/2} + \cdots, \qquad \varepsilon \to 0+. \tag{9.5.40}$$

Even though $y_0(0) = 1$, a boundary layer is required at $x = 0$ because the second term becomes infinite at $x = 0$. To determine the thickness of the boundary layer we set $X = x/\delta$, $Y_{\text{in}}(X) = y(x)$, and obtain

$$\frac{\varepsilon^3}{\delta} X \frac{d^2 Y_{\text{in}}}{dX^2} + \delta X^2 \frac{dY_{\text{in}}}{dX} - Y(\delta^3 X^3 + \varepsilon) = 0. \tag{9.5.41}$$

There are two distinguished limits for which $\delta \ll 1$: $\delta = \varepsilon$ and $\delta = \varepsilon^2$.

We examine the case $\delta = \varepsilon$ first because the outer solution becomes large when x is of order ε. Setting $\delta = \varepsilon$ in (9.5.41) gives

$$X^2 \frac{dY_{\text{in}}}{dX} - Y_{\text{in}} = -\varepsilon X \frac{d^2 Y_{\text{in}}}{dX^2} + \varepsilon^2 X^3 Y_{\text{in}}.$$

Assuming an inner expansion of the form $Y_{\text{in}}(X) \sim Y_0(X) + \varepsilon Y_1(X) + \cdots$ $(\varepsilon \to 0+)$ gives

$$X^2 \frac{dY_0}{dX} - Y_0 = 0,$$

whose solution is

$$Y_0 = \alpha_0 e^{-1/X},$$

and

$$X^2 \frac{dY_1}{dX} - Y_1 = -X \frac{d^2 Y_0}{dX^2},$$

whose solution is

$$Y_1 = \alpha_1 e^{-1/X} + \alpha_0 \left(\frac{2}{3X^3} - \frac{1}{4X^4} \right) e^{-1/X}.$$

To determine the values of α_0 and α_1 we match $Y_{\text{in}}(X)$ to $y_{\text{out}}(x)$. Writing Y_0 and Y_1 in terms of the outer variable x and taking the intermediate limit $x \to 0$, $X \to \infty$ gives $Y_{\text{in}}(X) \sim \alpha_0(1 - \varepsilon/x) + \varepsilon\alpha_1 + O(\varepsilon^2) + O(\varepsilon^2/x^2)$ $(\varepsilon \to 0+,\ x/\varepsilon \to +\infty)$. This expansion matches asymptotically with that in (9.5.40) in the limit $x \to 0+$ and we obtain the matching conditions $\alpha_0 = \alpha_1 = 1$. Thus,

$$Y_{\text{in}}(X) = e^{-1/X} \left[1 + \varepsilon \left(1 + \frac{2}{3X^3} - \frac{1}{4X^4} \right) \right] + O(\varepsilon^2), \qquad \varepsilon \to 0+. \tag{9.5.42}$$

So far our analysis has been straightforward. However, now we observe that it is still not possible to satisfy the boundary condition $y(0) = 1$ with $Y_{\text{in}}(X)$ because $Y_{\text{in}}(X)$ vanishes exponentially as $X \to 0+$. Apparently, there is an additional boundary layer very near $x = 0$ which enables us to satisfy the boundary condition $y(0) = 1$. The thickness of this boundary layer must be $\delta = \varepsilon^2$ because this is the only other distinguished limit for the differential equation. Therefore we set $\delta = \varepsilon^2$ in (9.5.41),

$$\bar{X} \frac{d^2 \bar{Y}_{\text{in}}}{d\bar{X}^2} - \bar{Y}_{\text{in}} = -\varepsilon \bar{X}^2 \frac{d\bar{Y}_{\text{in}}}{d\bar{X}} + \varepsilon^5 \bar{X} \bar{Y}_{\text{in}},$$

and assume an inner-inner expansion of the form

$$\bar{Y}_{\text{in}} \sim \bar{Y}_0 + \varepsilon \bar{Y}_1 + \cdots, \qquad \varepsilon \to 0+.$$

$\bar{Y}_0$ satisfies

$$\frac{d^2\bar{Y}_0}{d\bar{X}^2} = \frac{\bar{Y}_0}{\bar{X}}$$

whose general solution is a linear combination of modified Bessel functions (see Prob. 9.25):

$$\bar{Y}_0(\bar{X}) = \bar{\alpha}_0\sqrt{\bar{X}}\, I_1(2\sqrt{\bar{X}}) + \bar{\beta}_0\sqrt{\bar{X}}\, K_1(2\sqrt{\bar{X}}). \tag{9.5.43}$$

Since $I_1(2\sqrt{\bar{X}})$ grows exponentially as $\bar{X} \to \infty$, we will be unable to match $Y_{\text{in}}(X)$ unless $\bar{\alpha}_0 = 0$. $\bar{\beta}_0$ is determined by the boundary condition $\bar{Y}_0(0) = 1$. We find that $\bar{\beta}_0 = 2$ (see Prob. 9.25).

Observe that $\hat{Y}_{\text{in}}(\bar{X})$ and $Y_{\text{in}}(X)$ match asymptotically in the intermediate limit $X \to 0$, $\bar{X} \to \infty$. In the matching region they both vanish exponentially.

(D) 9.6 INTERNAL BOUNDARY LAYERS

Boundary layers (localized regions of rapid change of y) may occur in the interior as well as on the edge of an interval. However, the structure of internal layers tends to be complicated, so we confine our discussion to the leading-order behavior of solutions only.

We consider the simplest second-order differential-equation boundary-value problem that can exhibit internal boundary layers:

$$\varepsilon y'' + a(x)y' + b(x)y = 0, \qquad y(-1) = A,\ y(1) = B. \tag{9.6.1}$$

We know that if $a(x) \neq 0$ for $-1 < x < 1$, then there are no internal boundary layers. However, suppose we now assume that $a(x)$ has a simple zero at $x = 0$,

$$a(x) \sim \alpha x, \qquad x \to 0,$$

and that $a(x)$ has no other zeros for $-1 \le x \le 1$. We also assume that $b(x) \sim \beta \neq 0$ $(x \to 0)$. There are two cases to consider.

Case I $\alpha > 0$. Here $a(x)$ has positive slope at $x = 0$, so $a(-1) < 0$ and $a(1) > 0$. Thus, boundary layers at either $x = +1$ or $x = -1$ are not possible.

For this problem there are *two* outer solutions, one to the left and one to the right of $x = 0$. Either outer solution $y_{\text{out}}(x)$ has an expansion of the form $y_{\text{out}}(x) = y_0(x) + \varepsilon y_1(x) + \cdots$. Thus, to lowest order we have $a(x)y_0'(x) + b(x)y = 0$. Because there are no boundary layers at $x = \pm 1$, both outer solutions are determined by the boundary conditions at $x = \pm 1$:

$$y_{0,\text{right}} = B \exp\left[\int_x^1 \frac{b(t)}{a(t)}\, dt\right], \qquad x > 0, \tag{9.6.2}$$

$$y_{0,\text{left}} = A \exp\left[-\int_{-1}^x \frac{b(t)}{a(t)}\, dt\right], \qquad x < 0. \tag{9.6.3}$$

Next we investigate the neighborhood of $x = 0$. Setting $x = \delta X$, $y(x) = Y_{\text{in}}(X)$, (9.6.1) becomes

$$\frac{\varepsilon}{\delta^2}\frac{d^2 Y_{\text{in}}}{dX^2} + \frac{a(\delta X)}{\delta}\frac{dY_{\text{in}}}{dX} + b(\delta X)Y_{\text{in}} = 0.$$

To investigate the immediate neighborhood of $x = 0$, we replace $a(x)$ by αx and $b(x)$ by β. The only distinguished limit for which $\delta \ll 1$ is $\delta = \sqrt{\varepsilon}$. Thus, to leading order we have

$$\frac{d^2 Y_0}{dX^2} + \alpha X\frac{dY_0}{dX} + \beta Y_0 = 0, \tag{9.6.4}$$

where $Y_{\text{in}} = Y_0 +$ higher-order terms.

The solution to the differential equation (9.6.4) cannot be expressed in terms of elementary functions. As we will now show, (9.6.4) is related to the parabolic cylinder equation, one of the standard equations in mathematical physics. To solve (9.6.4) we let

$$Y_0 = e^{-\alpha X^2/4}W. \tag{9.6.5}$$

$W(X)$ satisfies

$$\frac{d^2 W}{dX^2} + (\beta - \tfrac{1}{2}\alpha - \tfrac{1}{4}\alpha^2 X^2)W = 0.$$

Next, we let $\sqrt{\alpha}\,X = Z$ and obtain

$$\frac{d^2 W}{dZ^2} + \left(\frac{\beta}{\alpha} - \tfrac{1}{2} - \tfrac{1}{4}Z^2\right)W = 0,$$

which we recognize as the parabolic cylinder equation (see Sec. 3.5). Assuming that β/α is not a positive integer (we discuss this special case later), the general solution to this equation is an arbitrary linear combination of parabolic cylinder functions: $W(Z) = C_1 D_{\beta/\alpha-1}(Z) + C_2 D_{\beta/\alpha-1}(-Z)$. Thus,

$$Y_0(X) = e^{-\alpha X^2/4}[C_1 D_{\beta/\alpha-1}(X\sqrt{\alpha}) + C_2 D_{\beta/\alpha-1}(-X\sqrt{\alpha})]. \tag{9.6.6}$$

The intermediate limit is defined by $x \to 0\pm$, $X \to \pm\infty$. Does the inner solution Y_0 in (9.6.6) match asymptotically with the outer solutions in (9.6.2) and (9.6.3) in the intermediate limit? If it does, we hope that the constants C_1 and C_2 are determined by the matching condition. In the intermediate limit the arguments of the parabolic cylinder functions become large. Therefore, it is necessary to use the asymptotic behaviors of the parabolic cylinder function for large positive and negative arguments, which were determined in Sec. 3.8:

$$\begin{aligned} D_\nu(t) &\sim t^\nu e^{-t^2/4}, & t &\to +\infty, \\ D_\nu(-t) &\sim t^{-\nu-1}e^{t^2/4}\frac{\sqrt{2\pi}}{\Gamma(-\nu)}, & t &\to +\infty. \end{aligned} \tag{9.6.7}$$

Thus, as $X \to +\infty$,

$$Y_0(X) \sim C_2(X\sqrt{\alpha})^{-\beta/\alpha} \frac{\sqrt{2\pi}}{\Gamma(1-\beta/\alpha)}, \tag{9.6.8}$$

and as $X \to -\infty$,

$$Y_0(X) \sim C_1(-X\sqrt{\alpha})^{-\beta/\alpha} \frac{\sqrt{2\pi}}{\Gamma(1-\beta/\alpha)}, \tag{9.6.9}$$

where we have discarded exponentially small terms. Finally, in preparation for asymptotic matching, we replace the inner variable X in (9.6.8) by the outer variable x:

$$Y_0(X) \sim \frac{C_2\sqrt{2\pi}}{\Gamma(1-\beta/\alpha)}(\sqrt{\alpha/\varepsilon})^{-\beta/\alpha} e^{-\beta(\ln x)/\alpha}, \qquad x \to 0+, \tag{9.6.10}$$

$$Y_0(X) \sim \frac{C_1\sqrt{2\pi}}{\Gamma(1-\beta/\alpha)}(\sqrt{\alpha/\varepsilon})^{-\beta/\alpha} e^{-\beta[\ln(-x)]/\alpha}, \qquad x \to 0-. \tag{9.6.11}$$

It is now clear that (9.6.2) and (9.6.10) match as $x \to 0+$ because

$$\int_x^1 \frac{b(t)}{a(t)}\,dt \sim -\frac{\beta}{\alpha}\ln x, \qquad x \to 0+,$$

and that (9.6.3) and (9.6.11) match as $x \to 0-$ because

$$-\int_{-1}^x \frac{b(t)}{a(t)}\,dt \sim -\frac{\beta}{\alpha}\ln(-x), \qquad x \to 0-.$$

Moreover, we can determine the constants C_1 and C_2 because

$$\int_x^1 \frac{b(t)}{a(t)}\,dt + \frac{\beta}{\alpha}\ln x \sim \int_0^1 \left[\frac{b(t)}{a(t)} - \frac{\beta}{\alpha t}\right] dt, \qquad x \to 0+,$$

and

$$-\int_{-1}^x \frac{b(t)}{a(t)}\,dt + \frac{\beta}{\alpha}\ln(-x) \sim \int_0^{-1} \left[\frac{b(t)}{a(t)} - \frac{\beta}{\alpha t}\right] dt, \qquad x \to 0-,$$

where the two integrals on the right exist. Specifically,

$$C_2 = B\frac{\Gamma(1-\beta/\alpha)}{\sqrt{2\pi}}(\sqrt{\alpha/\varepsilon})^{\beta/\alpha} \exp \int_0^1 \left[\frac{b(t)}{a(t)} - \frac{\beta}{\alpha t}\right] dt, \tag{9.6.12a}$$

and

$$C_1 = \frac{A\Gamma(1-\beta/\alpha)}{\sqrt{2\pi}}(\sqrt{\alpha/\varepsilon})^{\beta/\alpha} \exp \int_0^{-1} \left[\frac{b(t)}{a(t)} - \frac{\beta}{\alpha t}\right] dt. \tag{9.6.12b}$$

The results in (9.6.2) and (9.6.3), (9.6.6), and (9.6.12) may be combined into a single uniform asymptotic approximation which is valid on the interval $-1 \le x \le 1$:

$$y_{\text{unif}}(x) = \frac{\Gamma(1-\beta/\alpha)}{\sqrt{2\pi}} (\sqrt{\alpha/\varepsilon})^{\beta/\alpha} e^{-\alpha x^2/4\varepsilon}$$
$$\times \left(A \exp\left\{ \int_x^{-1} \left[\frac{b(t)}{a(t)} - \frac{\beta}{\alpha t} \right] dt \right\} D_{\beta/\alpha - 1}(x\sqrt{\alpha/\varepsilon}) \right.$$
$$\left. + B \exp\left\{ \int_x^{1} \left[\frac{b(t)}{a(t)} - \frac{\beta}{\alpha t} \right] dt \right\} D_{\beta/\alpha - 1}(-x\sqrt{\alpha/\varepsilon}) \right). \quad (9.6.13)$$

This result is verified in Prob. 9.30.

Example 1 *Internal boundary layer, case I.* Consider the boundary-value problem

$$\varepsilon y'' + 2xy' + (1 + x^2)y = 0, \qquad y(-1) = 2,\ y(1) = 1, \quad (9.6.14)$$

in the limit $\varepsilon \to 0+$. For this problem $\alpha = 2$ and $\beta = 1$. Also,

$$\int_0^1 \left[\frac{b(t)}{a(t)} - \frac{\beta}{\alpha t} \right] dt = \int_0^{-1} \left[\frac{b(t)}{a(t)} - \frac{\beta}{\alpha t} \right] dt = \frac{1}{4}$$

and $C_1 = 2(e/2\varepsilon)^{1/4}$, $C_2 = (e/2\varepsilon)^{1/4}$. Therefore, the two outer solutions are

$$y_{0,\text{right}} = x^{-1/2} e^{(1-x^2)/4}, \quad \sqrt{\varepsilon} \ll x \le 1), \qquad y_{0,\text{left}} = 2(-x)^{-1/2} e^{(1-x^2)/4}, \ \sqrt{\varepsilon} \ll -x \le 1,$$

and the inner solution is

$$Y_0(X) = Y_0(x/\sqrt{\varepsilon}) = e^{-x^2/2\varepsilon}(e/2\varepsilon)^{1/4}[2D_{-1/2}(x\sqrt{2/\varepsilon}) + D_{-1/2}(-x\sqrt{2/\varepsilon})].$$

The outer and inner solutions may be combined to give a uniform approximation to $y(x)$:

$$y_{\text{unif}}(x) = e^{-x^2/2\varepsilon} \left(\frac{e^{(1-x^2)}}{2\varepsilon} \right)^{1/4} [2D_{-1/2}(x\sqrt{2/\varepsilon}) + D_{-1/2}(-x\sqrt{2/\varepsilon})]. \quad (9.6.15)$$

In Fig. 9.16 we compare this leading-order uniform approximation with the exact solution to (9.6.14).

CASE II $\alpha < 0$. Let us return to (9.6.1) and see what happens when $\alpha < 0$. Now $a(x)$ has negative slope at $x = 0$, so $a(-1) > 0$ and $a(1) < 0$. This implies that boundary layers may occur at both $x = -1$ and at $x = 1$. We are thus faced with the possibility of having three boundary layers at $x = -1, 0, 1$, and two outer solutions between -1 and 0, and 0 and 1! It is quite surprising that the solution to this problem is much less complicated than the solution in case I.

The proper way to approach this problem is to analyze the inner solution at $x = 0$ first. Using the same analysis as in case I we find that (see Prob. 9.30)

$$Y_0(X) = e^{-\alpha X^2/4}[C_1 D_{-\beta/\alpha}(\sqrt{-\alpha}\, X) + C_2 D_{-\beta/\alpha}(-\sqrt{-\alpha}\, X)]. \quad (9.6.16)$$

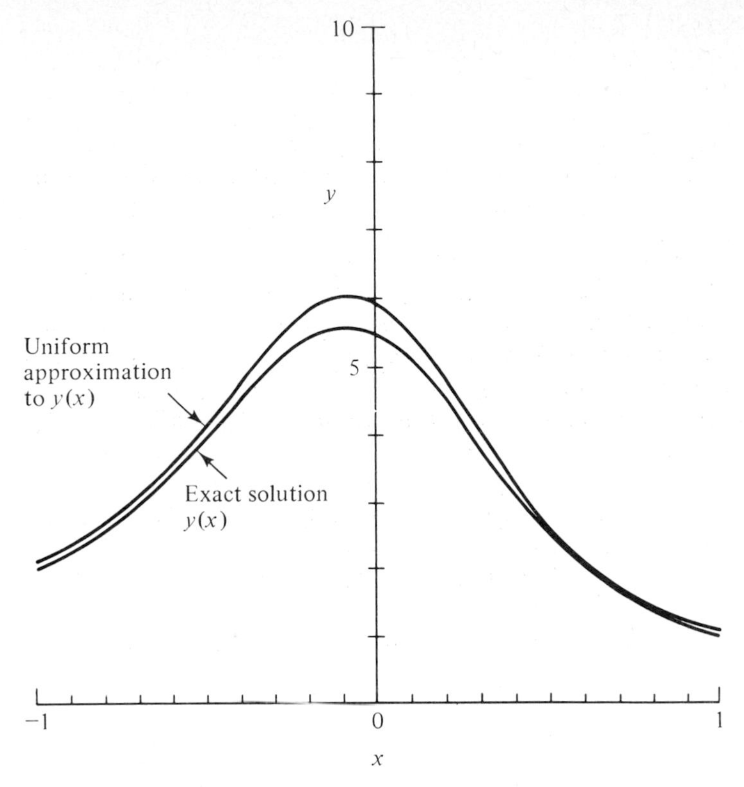

Figure 9.16 Comparison between the exact solution $y(x)$ to $\varepsilon y'' + 2xy' + (1 + x^2)y = 0$ [$y(-1) = 2$, $y(1) = 1$] and the leading-order uniform approximation (9.6.15) obtained from boundary-layer analysis. The value of ε is 0.2.

Now, α is negative, so the term $e^{-\alpha X^2/4}$ grows exponentially in both directions as $X \to \pm\infty$. (In case I it decays exponentially in both directions.) Observe that as $X \to +\infty$, $D_{-\beta/\alpha}(-\sqrt{-\alpha}\,X)$ also grows exponentially as $X \to +\infty$, assuming that $-\beta/\alpha \neq 0, 1, 2, 3, \ldots$. (We treat the integer case later.) Thus, no asymptotic match to the right outer solution is possible unless $C_2 = 0$. Similarly, as $X \to -\infty$, $D_{-\beta/\alpha}(\sqrt{-\alpha}\,X)$ grows exponentially. Thus, no asymptotic match to the left outer solution is possible unless $C_1 = 0$. We therefore obtain the very simple result that $Y_0(X) \equiv 0$. Apparently, there is *no* region of rapid change at $x = 0$ in case II.

Next, we consider the outer solutions which satisfy $a(x)y_0'(x) + b(x)y = 0$. This is a first-order homogeneous linear equation. Therefore, each outer solution is determined up to a multiplicative constant. However, requiring that the outer solutions match to $Y_0(X)$ implies that each multiplicative constant must be 0. We conclude that to leading order the solution vanishes everywhere except for boundary layers at $x = -1$ and at $x = 1$!

The leading-order boundary-layer solutions at ± 1 are easy to write down:

$$Y_{0,\text{left}} = Ae^{-a(-1)(x+1)/\varepsilon}, \tag{9.6.17}$$

$$Y_{0,\text{right}} = Be^{a(1)(1-x)/\varepsilon}. \tag{9.6.18}$$

Combining (9.6.17) and (9.6.18) we obtain the leading-order uniform asymptotic approximation

$$y_{\text{unif}}(x) = Ae^{-a(-1)(x+1)/\varepsilon} + Be^{a(1)(1-x)/\varepsilon}. \tag{9.6.19}$$

Example 2 *Case II.* Consider the boundary-value problem

$$\varepsilon y'' - 2xy' + (1 + x^2)y = 0, \qquad y(-1) = 2,\ y(1) = 1, \tag{9.6.20}$$

in the limit $\varepsilon \to 0+$. For this problem $\alpha = -2$. The leading-order uniform asymptotic approximation for this problem is

$$y_{\text{unif}} = 2e^{-2(x+1)/\varepsilon} + e^{-2(1-x)/\varepsilon}. \tag{9.6.21}$$

In Fig. 9.17 we compare the exact solution to (9.6.20) with the uniform asymptotic approximation in (9.6.21).

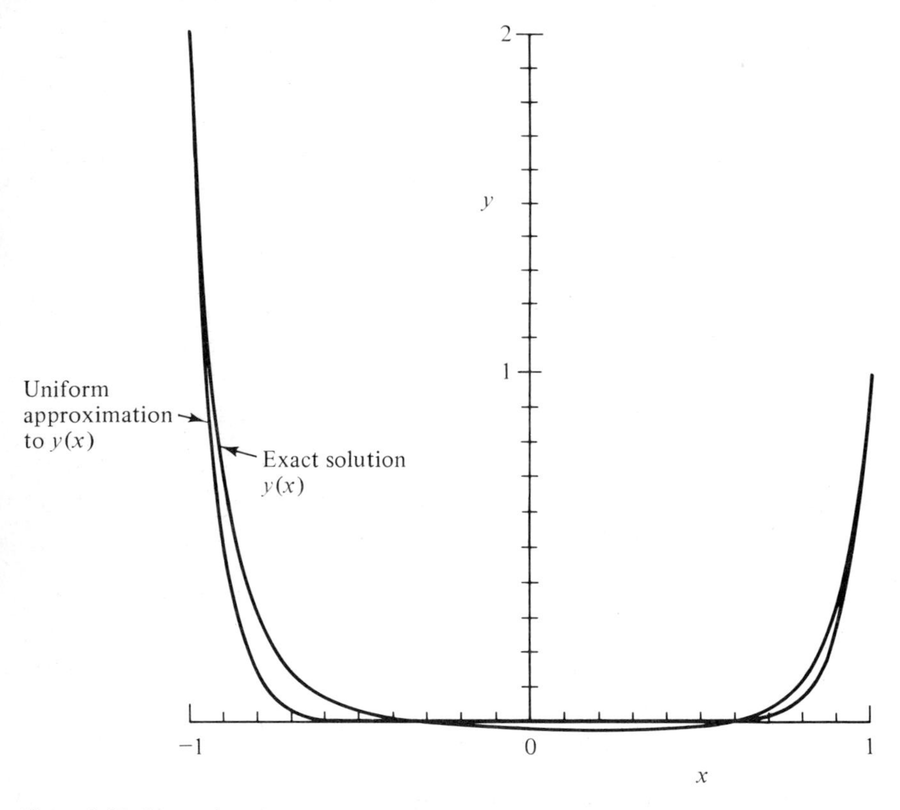

Figure 9.17 Comparison between the exact solution $y(x)$ to $\varepsilon y'' - 2xy' + (1 + x^2)y = 0$ [$y(-1) = 2$, $y(1) = 1$] and the leading-order uniform approximation (9.6.21) obtained from boundary-layer analysis. The value of ε is 0.15.

There are two exceptional cases that we have not yet discussed, namely, $\alpha > 0$ with $\beta/\alpha = 1, 2, 3, \ldots$ and $\alpha < 0$ with $\beta/\alpha = 0, -1, -2, \ldots$. In both of these cases leading-order boundary-layer analysis breaks down. Let us see why.

CASE III $\alpha > 0$, $\beta/\alpha = 1, 2, 3, \ldots$. This case is special because the indices of the parabolic cylinder functions in (9.6.6) are nonnegative integers. Recall that $D_n(Z) = \mathrm{He}_n(Z)e^{-Z^2/4}$ when $n = 0, 1, 2, \ldots$, where $\mathrm{He}_n(Z)$ is a Hermite polynomial of degree n, and that $D_n(-Z) = (-1)^n D_n(-Z)$. Therefore, $D_n(Z)$ and $D_n(-Z)$ are not linearly independent and the general leading-order inner solution $Y_0(X)$ is not given by (9.6.6). Instead, as shown in Sec. 3.8, the general solution has the form

$$Y_0(X) = e^{-\alpha X^2/4}[K_1 D_{\beta/\alpha - 1}(\sqrt{\alpha}\, X) + K_2 D_{-\beta/\alpha}(i\sqrt{\alpha}\, X)]. \quad (9.6.22)$$

From (3.7.18), the leading behavior of (9.6.22) as $X \to \pm\infty$ is

$$Y_0(X) \sim K_2(i\sqrt{\alpha}\, X)^{-\beta/\alpha}, \qquad X \to +\infty. \quad (9.6.23)$$

The contribution of the terms multiplied by K_1 is exponentially small compared to those multiplied by K_2. Therefore, the coefficient K_1 plays no role in the asymptotic matching and must remain undetermined!

No boundary layers are possible at $x = \pm 1$ when $\alpha > 0$, so we must match $y_0(X)$ to the outer solution $y_{0,\text{left}} = A \exp\left[-\int_{-1}^{x} b(t)/a(t)\, dt\right]$ $(-1 \le x < 0)$ and $y_{0,\text{right}} = B \exp\left[\int_{x}^{1} b(t)/a(t)\, dt\right]$ $(0 < x \le 1)$. Matching to $y_{0,\text{left}}$ in the intermediate limit $X \to -\infty$, $x \to 0-$ gives

$$K_2 = (i\sqrt{\alpha/\varepsilon})^{\beta/\alpha} A \exp \int_0^{-1} \left[\frac{b(t)}{a(t)} - \frac{\beta}{\alpha t}\right] dt.$$

On the other hand, matching to $y_{0,\text{right}}$ in the intermediate limit $X \to +\infty$, $x \to 0+$ gives

$$K_2 = (i\sqrt{\alpha/\varepsilon})^{\beta/\alpha} B \exp \int_0^{1} \left[\frac{b(t)}{a(t)} - \frac{\beta}{\alpha t}\right] dt.$$

Only in rare cases will the values of K_2 determined in these two different ways agree. In most cases the problem has no leading-order solution because the coefficient K_2 is *overdetermined*.

CASE IV $\alpha < 0$ with $\beta/\alpha = 0, -1, -2, \ldots$. In contrast with case III, the leading-order boundary-layer solution in case IV is *underdetermined*. Here, the general leading-order inner solution $Y_0(X)$ is not given by (9.6.16). Instead it has the form:

$$Y_0(X) = e^{-\alpha X^2/4}[K_1 D_{-\beta/\alpha}(\sqrt{-\alpha}\, X) + K_2 D_{\beta/\alpha - 1}(i\sqrt{-\alpha}\, X)]. \quad (9.6.24)$$

From (3.7.18), the leading behavior of $Y_0(X)$ as $X \to \pm\infty$ is

$$Y_0(X) \sim K_1(\sqrt{-\alpha}\, X)^{-\beta/\alpha} + K_2(i\sqrt{-\alpha}\, X)^{\beta/\alpha - 1} e^{-\alpha X^2/2}, \quad X \to \pm\infty. \quad (9.6.25)$$

Since $\alpha < 0$, matching is only possible if $K_2 = 0$, so

$$Y_0 \sim K_1(\sqrt{-\alpha/\varepsilon}\, x)^{-\beta/\alpha}, \qquad x/\sqrt{\varepsilon} \to \pm\infty. \tag{9.6.26}$$

The leading-order outer solution for $0 < x < 1$ that matches to (9.6.26) when $X \to +\infty$, $x \to 0+$ is

$$y_{0,\text{right}}(x) = K_1(x\sqrt{-\alpha/\varepsilon})^{-\beta/\alpha} \exp \int_0^x \left[\frac{\beta}{\alpha t} - \frac{b(t)}{a(t)}\right] dt, \qquad x > 0, \tag{9.6.27}$$

while the leading-order outer solution for $-1 < x < 0$ that matches to (9.6.26) when $X \to -\infty$, $x \to 0-$ is

$$y_{0,\text{left}}(x) = K_1(x\sqrt{-\alpha/\varepsilon})^{-\beta/\alpha} \exp \int_x^0 \left[\frac{b(t)}{a(t)} - \frac{\beta}{\alpha t}\right] dt, \qquad x < 0. \tag{9.6.28}$$

Since $\alpha < 0$, the outer solution (9.6.27) can be asymptotically matched as $x \to 1-$ to a boundary-layer solution; the leading-order boundary-layer approximation is

$$Y_{0,\text{right}} = Be^{a(1)(x-1)/\varepsilon} + K_1(\sqrt{-\alpha/\varepsilon})^{-\beta/\alpha}[1 - e^{a(1)(x-1)/\varepsilon}] \times \exp \int_0^1 \left[\frac{\beta}{\alpha t} - \frac{b(t)}{a(t)}\right] dt. \tag{9.6.29}$$

The outer solution (9.6.28) can also be matched as $x \to -1+$ to a boundary-layer solution at $x = -1$ because $\alpha < 0$:

$$Y_{0,\text{left}} = Ae^{\,a(-1)(x+1)/\varepsilon} + K_1(-\sqrt{-\alpha/\varepsilon})^{-\beta/\alpha}[1 - e^{-a(-1)(x+1)/\varepsilon}] \times \exp \int_{-1}^0 \left[\frac{b(t)}{a(t)} - \frac{\beta}{\alpha t}\right] dt. \tag{9.6.30}$$

The boundary-layer solutions (9.6.29) and (9.6.30), together with the outer solutions (9.6.27) and (9.6.28) and the internal layer solution (9.6.24) with $K_2 = 0$, match to leading order in ε. However, K_1 is still arbitrary, so leading-order boundary-layer theory has not determined a unique solution to the boundary-value problem!

Discussion of cases III and IV Cases III and IV have serious difficulties because $D_n(z)$ ($n = 0, 1, 2, \ldots$) decays exponentially as $z \to +\infty$ and as $z \to -\infty$. In contrast, $D_\nu(z)$ ($\nu \neq 0, 1, 2, \ldots$) decays exponentially only as $z \to +\infty$ and grows as $z \to -\infty$. Thus, in case II, for example, where $\beta/\alpha \neq 0, -1, -2, \ldots$ we could logically argue that $C_1 = C_2 = 0$ in order for a match to be possible. In case IV, where $\beta/\alpha = 0, -1, -2, \ldots$ and $\alpha < 0$ we can no longer argue that $K_1 = 0$.

The resolution of the difficulties in cases III and IV is not trivial. Sometimes it is possible to resolve these difficulties by carrying the boundary-layer analysis to higher order. A higher-order treatment may give rise to solutions in the internal boundary layer which do not decay exponentially as $X \to +\infty$ and as $X \to -\infty$.

When this happens, the higher-order analysis proceeds as in cases I and II and a unique solution can be determined (see Probs. 9.31 and 9.32). Higher-order boundary-layer analysis does not always resolve the difficulties discussed here. Often it is possible to determine a unique asymptotic solution to ambiguous internal boundary-layer problems using the methods of WKB theory (see Probs. 9.33 and 10.28).

(I) 9.7 NONLINEAR BOUNDARY-LAYER PROBLEMS

Boundary-layer analysis applies to nonlinear as well as to linear differential equations. This section is a collection of three illustrative examples. The first example is very elementary. We include this example merely to show that the techniques we have used to solve linear boundary-layer problems can apply equally well to nonlinear problcms.

Example 1 *Leading-order analysis of a nonlinear autonomous equation.* Consider the boundary-value problem

$$\varepsilon y'' + 2y' + e^y = 0, \qquad y(0) = y(1) = 0. \tag{9.7.1}$$

If e^y were a linear function of y, there would be a boundary layer at $x = 0$ (and no boundary layer at $x = 1$) because the coefficient of y' is positive. This nonlinear problem also has just one boundary layer at $x = 0$.

The outer expansion has the form

$$y_{\text{out}}(x) \sim y_0(x) + \varepsilon y_1(x) + \cdots, \qquad \varepsilon \to 0+. \tag{9.7.2}$$

Substituting (9.7.2) into (9.7.1) gives $2y_0' + e^{y_0} = 0$, whose solution is $y_0 = \ln 2/(x + c_0)$. Assuming that there is no boundary layer at $x = 1$ (a boundary layer at $x = 1$ leads to a contradiction; see Prob. 9.37), we impose the boundary condition $y_0(1) = 0$. Thus,

$$y_0(x) = \ln \frac{2}{1 + x}. \tag{9.7.3}$$

The boundary layer has thickness $\delta = \varepsilon$. (Why?) Therefore, setting $X = x/\varepsilon$, $Y_{\text{in}}(X) = y(x)$ in (9.7.1) gives

$$\frac{d^2 Y_{\text{in}}}{dX^2} + 2\frac{dY_{\text{in}}}{dX} = -\varepsilon e^{Y_{\text{in}}}.$$

Assuming an inner expansion of the form $Y_{\text{in}} = Y_0 + \varepsilon Y_1 + \cdots$ gives in leading order

$$Y_0 = A_0 + B_0 e^{-2X}. \tag{9.7.4}$$

The constants A_0 and B_0 are determined by the boundary condition at $x = 0$, $Y_0(0) = 0$, which gives $A_0 + B_0 = 0$ and the asymptotic match of (9.7.4) with (9.7.3) in the intermediate limit $x \to 0+$, $X \to +\infty$, which gives $A_0 = \ln 2$. Thus,

$$Y_0(X) = (1 - e^{-2X}) \ln 2. \tag{9.7.5}$$

We can combine (9.7.5) and (9.7.3) into a single uniform asymptotic approximation:

$$y_{\text{unif}}(x) = \ln \frac{2}{1 + x} - e^{-2x/\varepsilon} \ln 2. \tag{9.7.6}$$

A comparison between (9.7.6) and the exact solution to (9.7.1) is given in Fig. 9.18. For a higher-order treatment of (9.7.1) see Prob. 9.38.

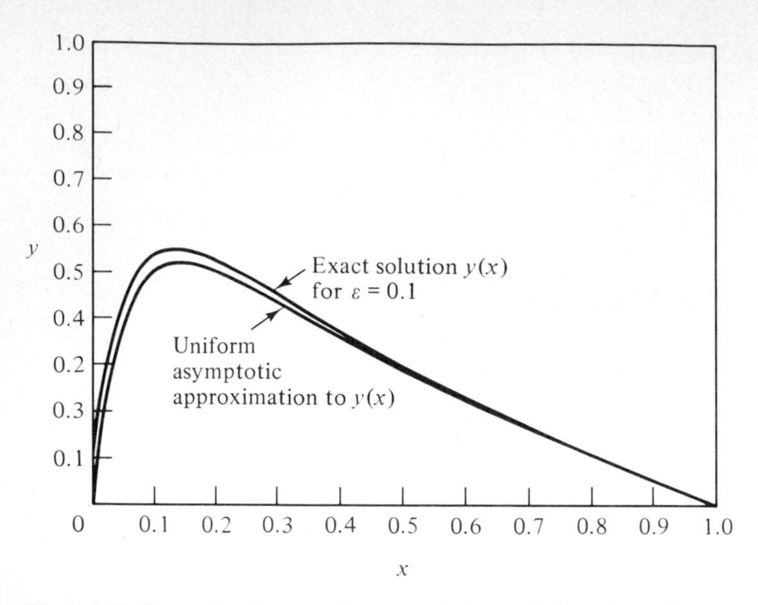

Figure 9.18 Comparison between the exact solution to the boundary-value problem $\varepsilon y'' + 2y' + e^y = 0$ $[y(0) = y(1) = 0]$ for $\varepsilon = 0.1$ in (9.7.1) and $y_{\text{unif}}(x) = \ln [2/(1 + x)] - e^{-2x/\varepsilon} \ln 2$, the leading-order uniform asymptotic approximation to $y(x)$ obtained using boundary-layer theory.

Our treatment of Example 1 was very straightforward. However, one should not be misled. Similar nonlinear boundary-layer problems can be extraordinarily difficult (see Prob. 9.39). In the next example we study a boundary-value problem that has many solutions; some can be predicted by boundary-layer theory while others are beyond the scope of boundary-layer methods.

Example 2 *Boundary-layer analysis of a nonlinear problem of Carrier.* An extremely beautiful and intricate nonlinear boundary-value problem of a type first proposed by Carrier is

$$\varepsilon y'' + 2(1 - x^2)y + y^2 = 1, \qquad y(-1) = y(1) = 0. \tag{9.7.7}$$

If we attempt a leading-order boundary-layer analysis of (9.7.7), we are immediately surprised to find that the outer equation obtained by setting $\varepsilon = 0$ is an algebraic equation rather than a differential equation:

$$y_{\text{out}}^2 + 2(1 - x^2)y_{\text{out}} - 1 = 0.$$

Because this equation is quadratic, it has two solutions

$$y_{\text{out},\pm}(x) = x^2 - 1 \pm \sqrt{1 + (1 - x^2)^2}. \tag{9.7.8}$$

In Fig. 9.19 we plot the two outer solutions. Observe that neither one satisfies the boundary conditions at $x = \pm 1$. Therefore, there must be boundary layers at $x = -1$ and at $x = +1$ which allow the boundary conditions to be satisfied. The question is, Which of the two outer solutions can be joined to inner solutions which satisfy the boundary conditions?

Let us examine the boundary layer at $x = 1$. If we substitute the inner variables $X = (1 - x)/\delta$, $Y_{\text{in}}(X) = y(x)$ into (9.7.7), we obtain in leading order

$$\frac{d^2 Y_{\text{in}}}{dX^2} + \frac{\delta^2}{\varepsilon}(Y_{\text{in}}^2 - 1) = 0.$$

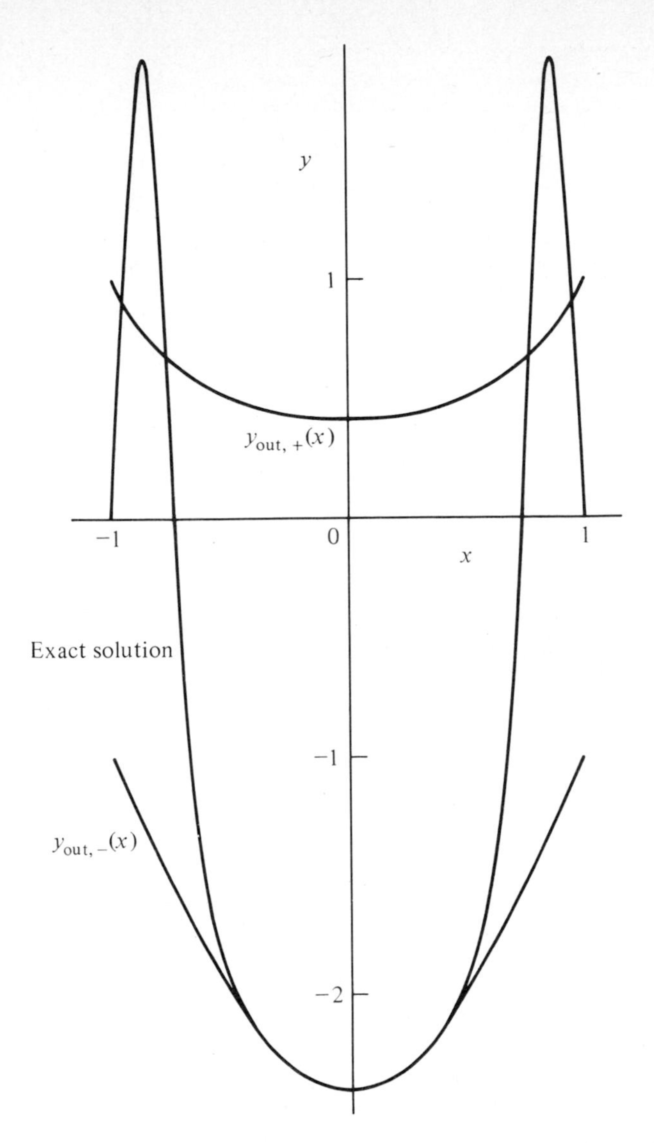

Figure 9.19 Exact solution to the nonlinear boundary-value problem in (9.7.7), $\varepsilon y'' + 2(1 - x^2)y + y^2 = 1 [y(-1) = y(1) = 0]$ for $\varepsilon = 0.01$. Also shown are the two outer approximations $y_{\text{out},\pm}$ in (9.7.8). As predicted by boundary-layer analysis, the lower outer approximation is an extremely good approximation to $y(x)$ away from the boundary layers at $x = \pm 1$. Observe that $y(x)$ has a local maximum in both boundary layers. y_{unif} in (9.7.12) with the lower choice of signs is an accurate approximation to $y(x)$ for $-1 \le x \le 1$; it predicts that the maximum value of y is 2.

Thus, the distinguished limit is $\delta = \sqrt{\varepsilon}$. The solution to the leading-order inner equation

$$\frac{d^2 Y_{\text{in}}}{dX^2} + Y_{\text{in}}^2 - 1 = 0 \tag{9.7.9}$$

must satisfy the boundary condition $Y_{\text{in}}(0) = 0$ and must match asymptotically with one (or both) of the outer solutions. That is, Y_{in} must approach either ± 1 as $X \to +\infty$.

Is it possible for Y_{in} to approach 1 as $X \to +\infty$? Suppose we let $Y_{\text{in}} = 1 + W(X)$. If $Y_{\text{in}} \to 1$, then $W(X) \to 0$ and we can replace (9.7.9) with the approximate linear equation $W'' + 2W = 0$. However, solutions to this equation oscillate as $X \to +\infty$ and do not approach 0. This simple analysis shows that it is not possible for Y_{in} to match to $y_{\text{out},+}$ in (9.7.8).

Fortunately, the same argument suggests that it *is* possible for Y_{in} to match to $y_{\text{out},-}$. Let $Y_{\text{in}} = -1 + W(X)$. Now if $Y_{\text{in}} \to -1$ then $W \to 0$ and we can replace (9.7.9) by the approximate linear equation $W'' - 2W = 0$. Since this equation has a solution which decays to 0 exponentially, it is at least consistent to assume that Y_{in} matches asymptotically with $y_{\text{out},-}$.

Having established this much, let us solve the inner equation exactly. Substituting $Y_{\text{in}} = -1 + W(X)$ into (9.7.9) gives the autonomous equation

$$W'' + W^2 - 2W = 0, \tag{9.7.10}$$

subject to the boundary conditions $W(\infty) = 0$, $W(0) = 1$. Also, since we expect W to decay exponentially as $X \to +\infty$, we may assume that $W'(\infty) = 0$. To solve (9.7.10) we multiply by $W'(X)$, integrate the equation once, and determine the integration constant by setting $X = \infty$. We obtain $\frac{1}{2}(W')^2 + \frac{1}{3}W^3 - W^2 = 0$, which is a separable first-order equation: $dW/W\sqrt{2 - 2W/3} = \pm dX$. Integrating this equation gives

$$-\sqrt{2}\tanh^{-1}\sqrt{1 - W/3} = \pm X + C.$$

The integration constant is determined by the requirement that $W = 1$ at $X = 0$. Apparently, there are *two* solutions:

$$Y_{\text{in}}(X) = -1 + 3\,\text{sech}^2\left(\pm\frac{X}{\sqrt{2}} + \tanh^{-1}\sqrt{2/3}\right). \tag{9.7.11}$$

There are also *two* inner solutions at $x = -1$ which satisfy the boundary condition $y(-1) = 0$ and match to the lower outer solution $y_{\text{out},-}$ in (9.7.8).

We can combine the outer with the two inner solutions to form a single uniform approximation valid over the entire interval $-1 \le x \le 1$:

$$\begin{aligned} y_{\text{unif}}(x) = x^2 - 1 - \sqrt{1 + (1 - x^2)^2} + 3\,\text{sech}^2\left(\pm\frac{1 - x}{\sqrt{2\varepsilon}} + \tanh^{-1}\sqrt{2/3}\right) \\ + 3\,\text{sech}^2\left(\pm\frac{1 + x}{\sqrt{2\varepsilon}} + \tanh^{-1}\sqrt{2/3}\right). \end{aligned} \tag{9.7.12}$$

Notice that the solution in (9.7.12) is not unique. There are actually four different solutions depending on the two choices of plus or minus signs in the boundary layer. For one choice of sign, $y_{\text{unif}}(x)$ in the boundary layer rapidly descends from its boundary value $y(\pm 1) = 0$ until it joins onto the outer solution $y_{\text{out},-}$. For the other choice of sign, $y_{\text{unif}}(x)$ rises rapidly until it reaches a maximum and then descends and joins onto the outer solution. It is easy to see that this maximum value of y_{unif} is 2 because the maximum value of sech is 1. It is a glorious triumph of boundary-layer theory that all four solutions actually exist and are extremely well approximated by the leading-order uniform approximation in (9.7.12)! See Figs. 9.19 to 9.21.

The analysis does not end here, however. The existence of four solutions to (9.7.7) may lead one to wonder if there are still more solutions. One may begin by asking whether there can be any internal boundary layers. We will now show that internal boundary layers are consistent.

Assume there is an internal boundary layer at $x = 0$. The thickness of such a boundary layer is $\delta = \sqrt{\varepsilon}$. (Why?) The leading-order equation is

$$Y_{\text{in}}''(X) + 2Y_{\text{in}} + Y_{\text{in}}^2 = 1. \tag{9.7.13}$$

Since $y_{\text{out},-}(0) = -1 - \sqrt{2}$, the boundary conditions on Y_{in} in (9.7.13) are $\lim_{X \to \pm\infty} Y_{\text{in}}(X) = -1 - \sqrt{2}$. The exact solution to (9.7.13) which satisfies these boundary conditions contains an arbitrary parameter A:

$$Y_{\text{in}} = 3\sqrt{2}\,\text{sech}^2\,(2^{-1/4}x/\sqrt{\varepsilon} + A) - 1 - \sqrt{2}.$$

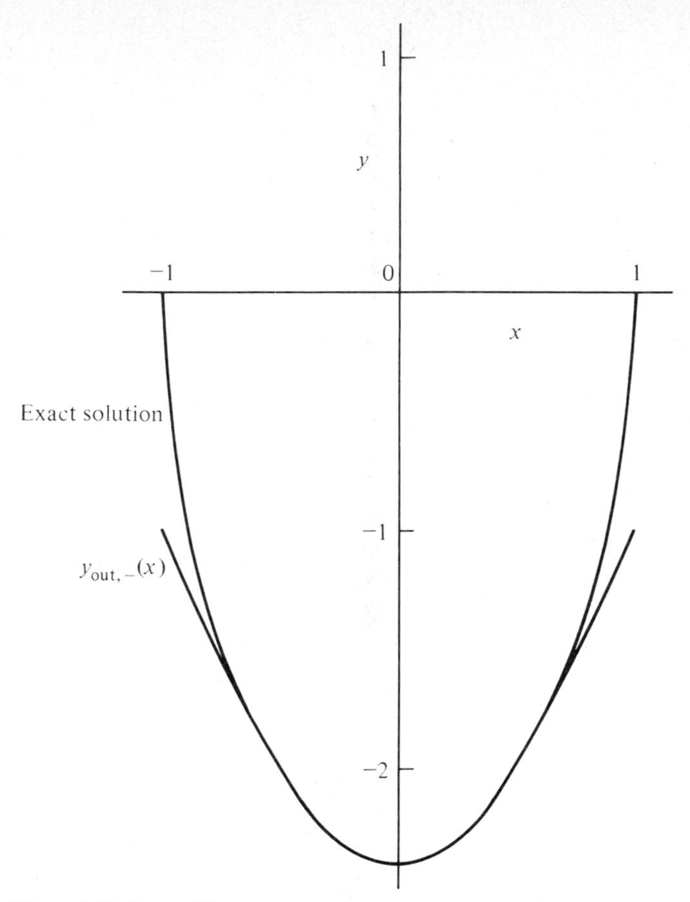

Figure 9.20 Same differential equation as in Fig. 9.19, but a different solution. y_{unif} becomes a good approximation to the plotted solution for the upper choice of signs.

Note that if $A = \pm\infty$ then there is no internal boundary-layer structure. However, for all finite values of A there is a narrow region in which Y rises abruptly to a sharp peak at which it attains a maximum value of $2\sqrt{2} - 1 \doteq 1.8$. Do you believe from this analysis that there are actual solutions to (9.7.7) having an internal boundary layer at $x = 0$? In fact, in Figs. 9.22 to 9.24 we see that for each solution in Figs. 9.19 to 9.21 there is another solution which is almost identical except that it exhibits a boundary layer at $x = 0$! What is more, the maximum in the boundary layer is close to 1.8.

Now that we have observed solutions having one internal boundary layer we may ask whether there exist solutions having multiple internal boundary layers. It is here that boundary-layer theory is no longer useful. Boundary-layer analysis shows (see Prob. 9.42) that it is consistent to have any number of internal layers at any location. However, boundary-layer theory cannot predict the number or the location of these boundary layers. In fact, for a given positive value of ε there are exactly $4(N + 1)$ solutions which have from 0 to N internal boundary layers at definite locations, where N is a finite number depending on ε. To determine N it is necessary to use some rather advanced phase-plane analysis to establish a kind of WKB quantization condition (see the References). In Figs. 9.25 to 9.28 we give some examples of solutions having several internal boundary layers. In Figs. 9.19 to 9.28 $\varepsilon = 0.01$.

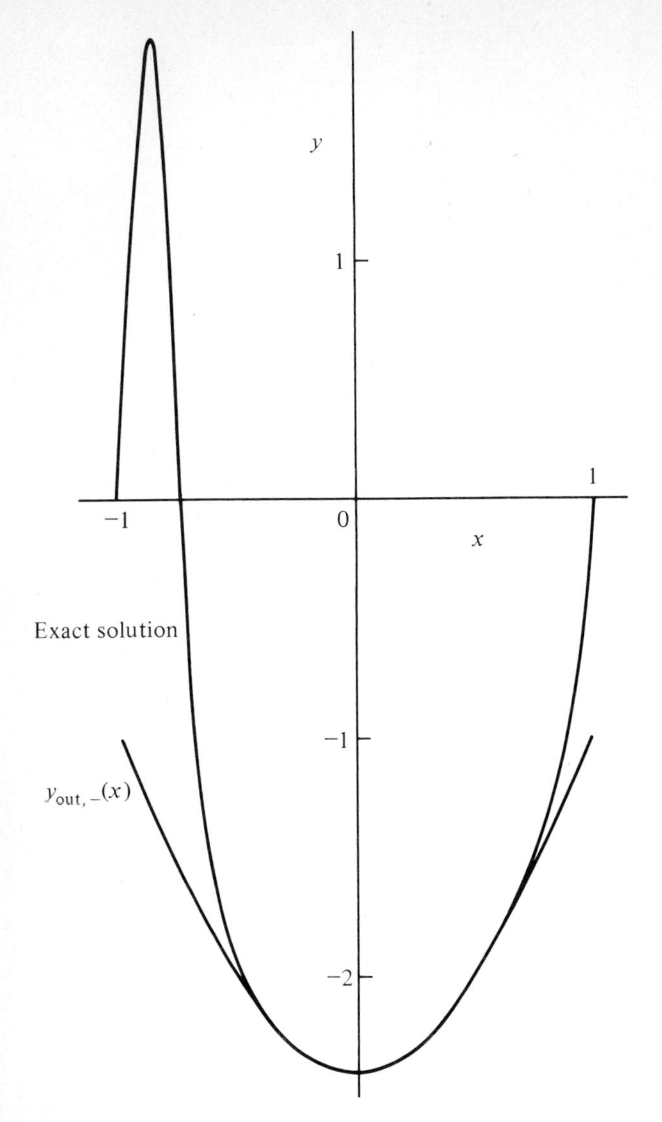

Figure 9.21 Same differential equation as in Fig. 9.19. y_{unif} in (9.7.12) is a good approximation to the plotted solution for one upper sign and one lower sign. There is also another solution which is the reflection about the y axis of the one shown here.

In the next example we use boundary-layer theory to study the approximate shape of a limit cycle in the phase plane.

Example 3 *Limit cycle of the Rayleigh oscillator.* The Rayleigh equation is

$$\varepsilon \frac{d^2y}{dt^2} - \left[\frac{dy}{dt} - \frac{1}{3}\left(\frac{dy}{dt}\right)^3\right] + y = 0. \tag{9.7.14}$$

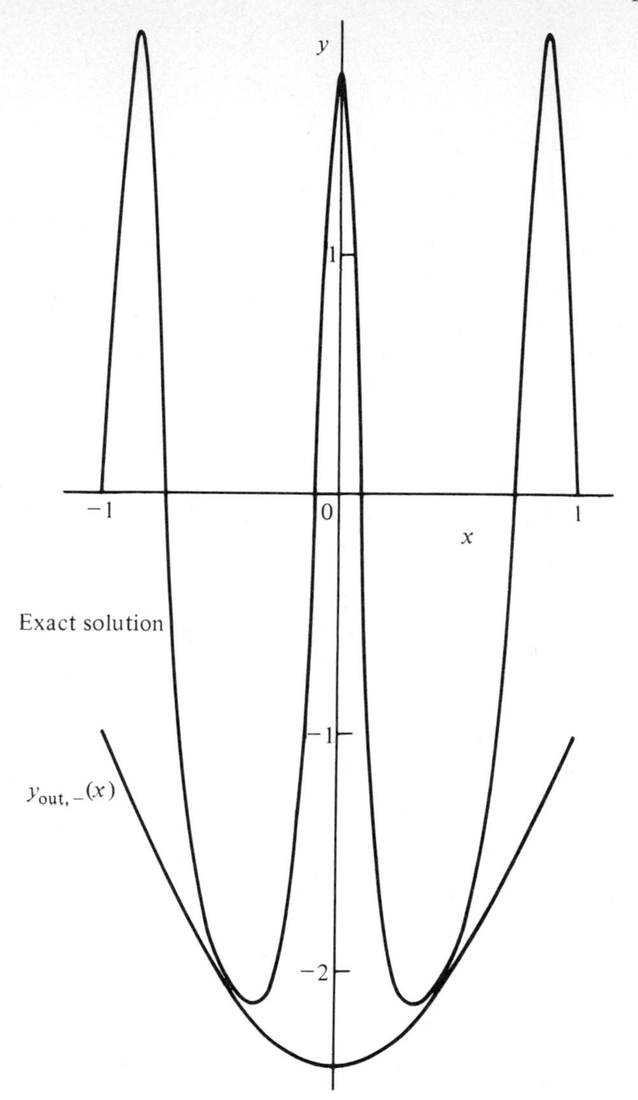

Figure 9.22 An exact solution to the boundary-value problem in (9.7.7). Apart from the internal boundary layer at $x = 0$, this solution is nearly identical to the solution in Fig. 9.19. The outer approximation $y_{\text{out},-}(x)$ in (9.7.8) is a good approximation to $y(x)$ between the boundary layers. The function $y_{\text{in}} = 3\sqrt{2}\,\text{sech}^2\,(2^{-1/4}x/\sqrt{\varepsilon}) - 1 - \sqrt{2}$ gives a good description of y in the internal boundary layer. It predicts that the maximum value of y_{in} is $2\sqrt{2} - 1 \doteqdot 1.8$, a result which is verified by this graph.

This autonomous equation can be rewritten as the system

$$\frac{dy}{dt} = z, \tag{9.7.15}$$

$$\varepsilon \frac{dz}{dt} = z - \frac{1}{3}z^3 - y, \tag{9.7.16}$$

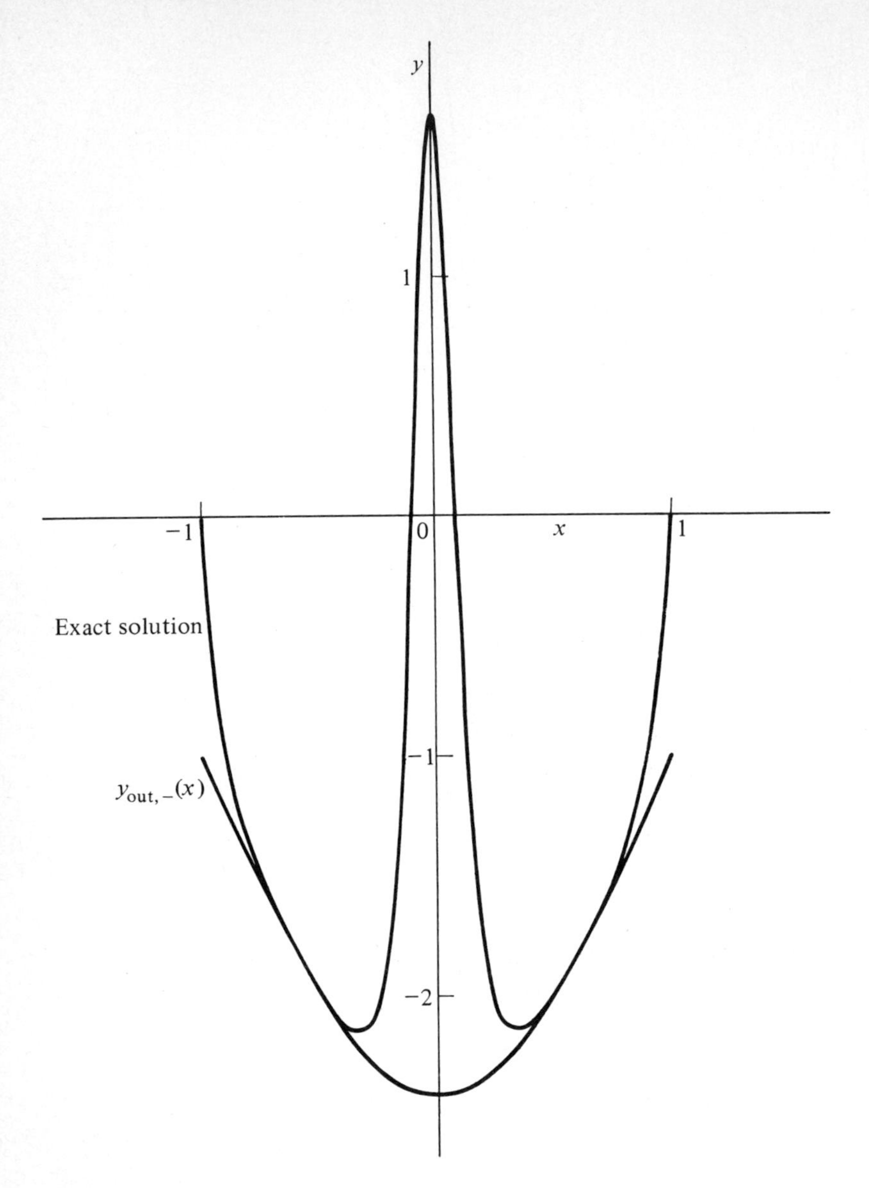

Figure 9.23 An exact solution to the boundary-value problem in (9.7.7). Apart from the internal boundary layer at $x = 0$, this solution is nearly identical to that in Fig. 9.20.

whose trajectories can be studied in the phase plane (y, z). The Rayleigh equation is an interesting model because, for any initial conditions and any $\varepsilon > 0$, $y(t)$ approaches a periodic solution as $t \to +\infty$. This periodic solution corresponds to a limit cycle in the phase plane (see Prob. 9.45). In Sec. 11.3 we study the approach to this periodic solution when ε is large using multiple-scale perturbation theory. In this example we consider the opposite limit $\varepsilon \to 0+$ and use boundary-layer theory to determine the shape of the limit cycle in the phase plane.

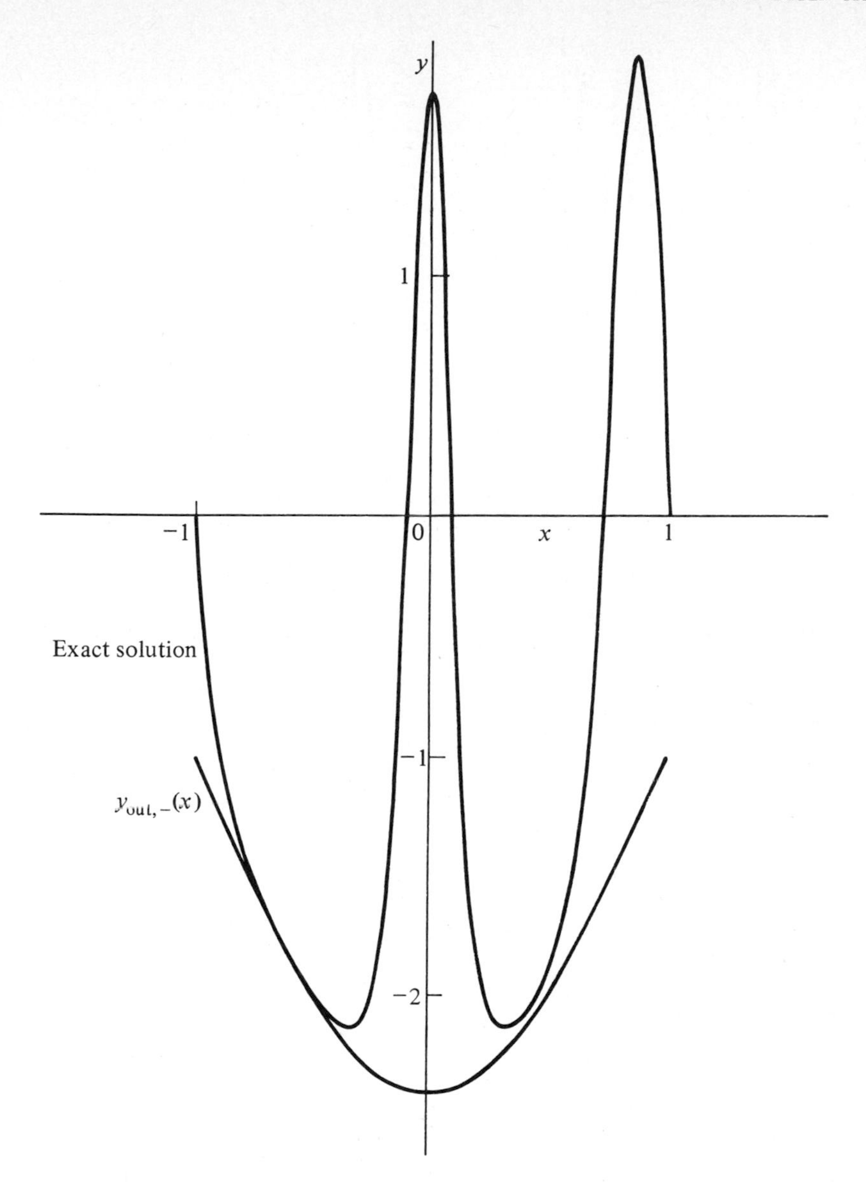

Figure 9.24 An exact solution to the boundary-value problem in (9.7.7). Apart from the internal boundary layer at $x = 0$, this solution is nearly identical to that in Fig. 9.21 reflected about the y axis.

We begin by dividing (9.7.16) by (9.7.15) to obtain

$$\varepsilon \frac{dz}{dy} = \frac{z - z^3/3 - y}{z} \tag{9.7.17}$$

and treat z as a function of the independent variable y. First, we look at the leading-order approximation to (9.7.17) as $\varepsilon \to 0+$. The outer limit is obtained by simply setting $\varepsilon = 0$. This

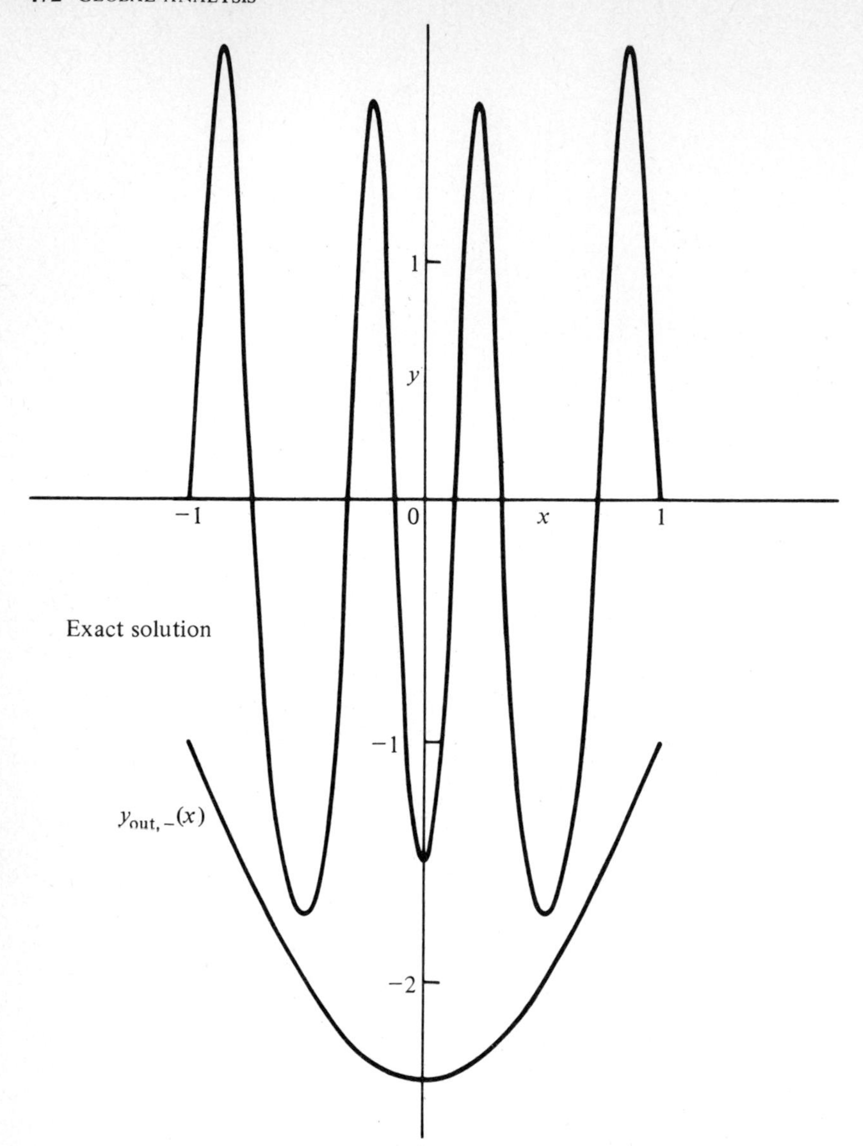

Figure 9.25 An exact solution to (9.7.7) having two internal boundary layers. Apart from the internal boundary layers, this solution is nearly identical to that in Fig. 9.19.

gives the algebraic equation

$$z_{\text{out}} - \tfrac{1}{3}z_{\text{out}}^3 = y. \tag{9.7.18}$$

This curve is plotted as a dashed line in Fig. 9.29. Observe that for $|y| < \frac{2}{3}$, there are three possible values of z_{out}. We will show that it is consistent to have a boundary-layer solution which joins the point $A(y = \frac{2}{3}, z = 1)$ to the point $B(y = \frac{2}{3}, z = -2)$ by the almost vertical line shown in Fig. 9.29 and to have a second boundary-layer solution which joins the point $C(y = -\frac{2}{3}, z = -1)$

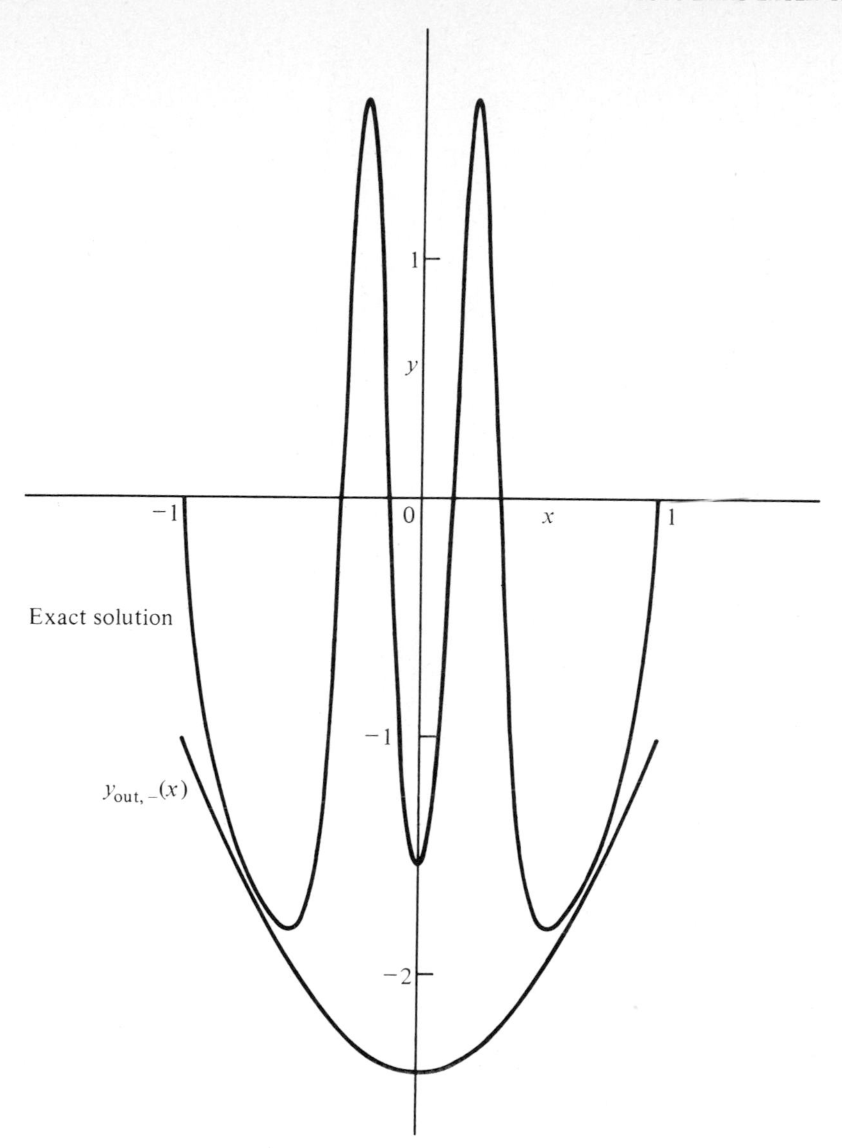

Figure 9.26 An exact solution to (9.7.7) analogous to that in Fig. 9.23 except that it has two internal boundary layers.

to the point $D(y = -\frac{2}{3}, z = 2)$. The limit cycle consists of the four segments DA and BC satisfying the outer equation (9.7.18) and AB and CD given by boundary-layer approximations.

To obtain the boundary-layer approximation joining the outer solutions DA and BC from A to B, we introduce the inner variable $Y = (y - \frac{2}{3})/\delta$ and obtain a distinguished limit $\delta = \varepsilon$:

$$\frac{dZ}{dY} = \frac{Z - Z^3/3 - \frac{2}{3} - \varepsilon Y}{Z}, \tag{9.7.19}$$

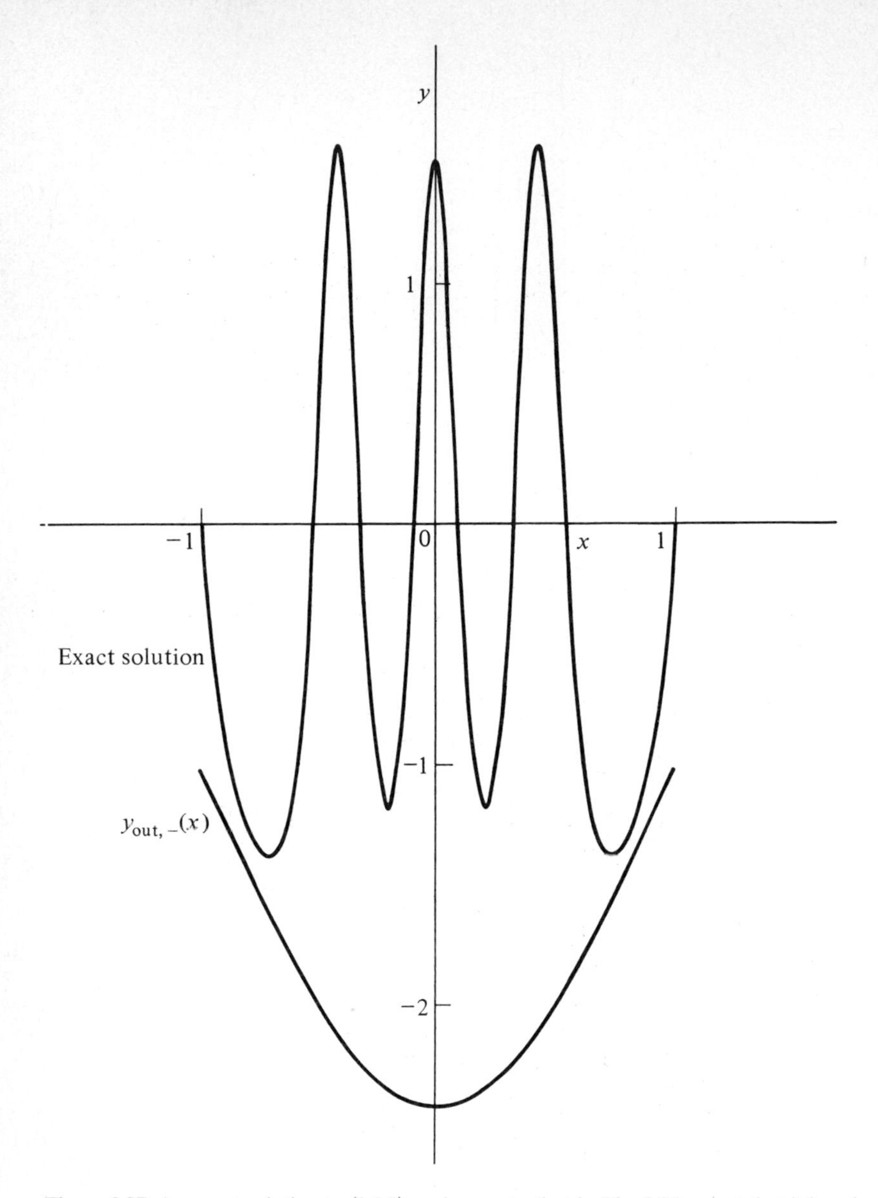

Figure 9.27 An exact solution to (9.7.7) analogous to that in Fig. 9.26 except that it has three internal boundary layers.

where $Z(Y) = z(y)$. The leading-order approximation Z_0 to Z satisfies

$$\frac{dZ_0}{dY} = \frac{Z_0 - Z_0^3/3 - \frac{2}{3}}{Z_0}. \tag{9.7.20}$$

The solution of this separable equation is

$$-\frac{2}{9}\ln|Z_0 + 2| + \frac{2}{9}\ln|Z_0 - 1| - \frac{1}{3(Z_0 - 1)} = -\frac{1}{3}(Y + c), \tag{9.7.21}$$

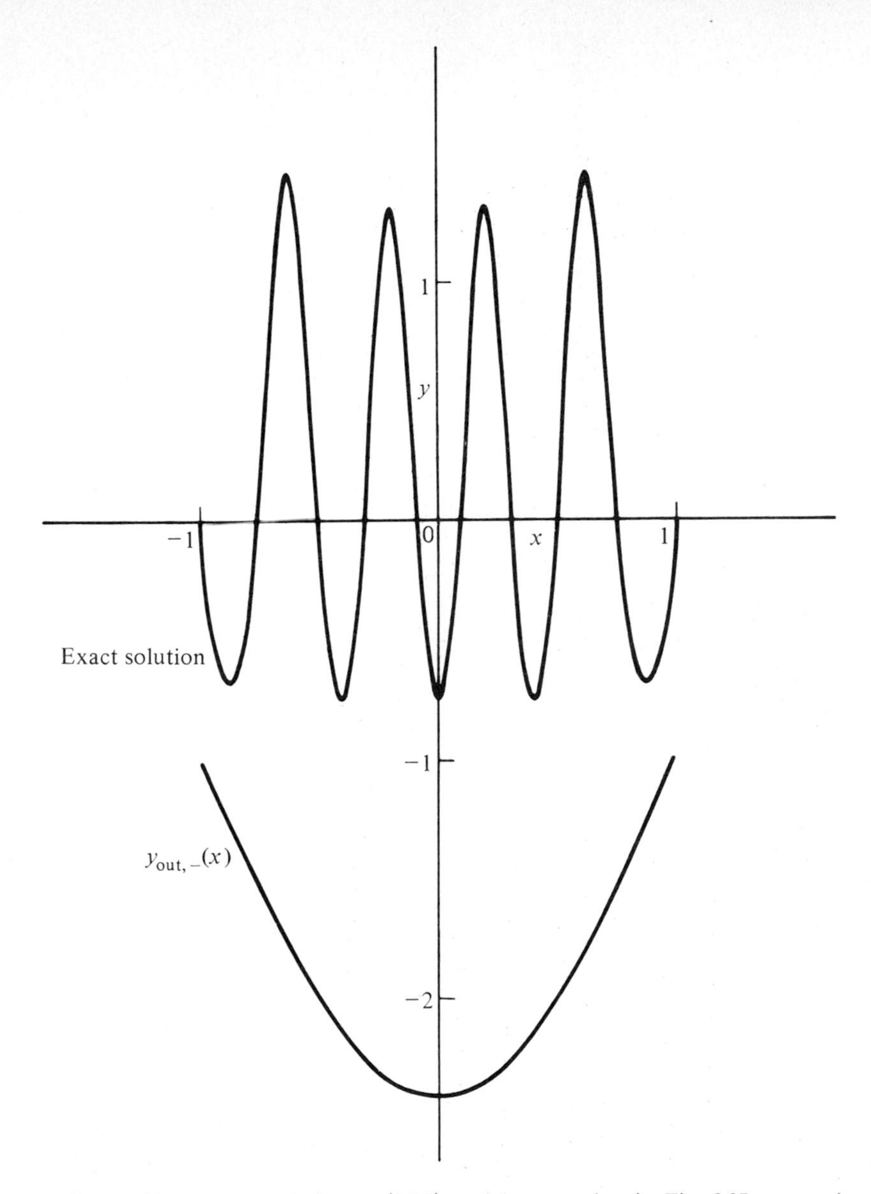

Figure 9.28 An exact solution to (9.7.7) analogous to that in Fig. 9.27 except that it has four internal boundary layers. Observe that as the number of internal boundary layers increases, the outer solution $y_{\text{out},-}(x)$ becomes a poorer approximation to $y(x)$ between boundary layers. Indeed, in this figure the regions of rapid variation of $y(x)$ are no longer localized in isolated boundary layers. Rather, individual boundary layers are so close together that $y(x)$ exhibits rapid variation on a global scale. This configuration is similar to that in Fig. 7.3. Boundary-layer analysis is useful only when the regions of rapid variation are localized.

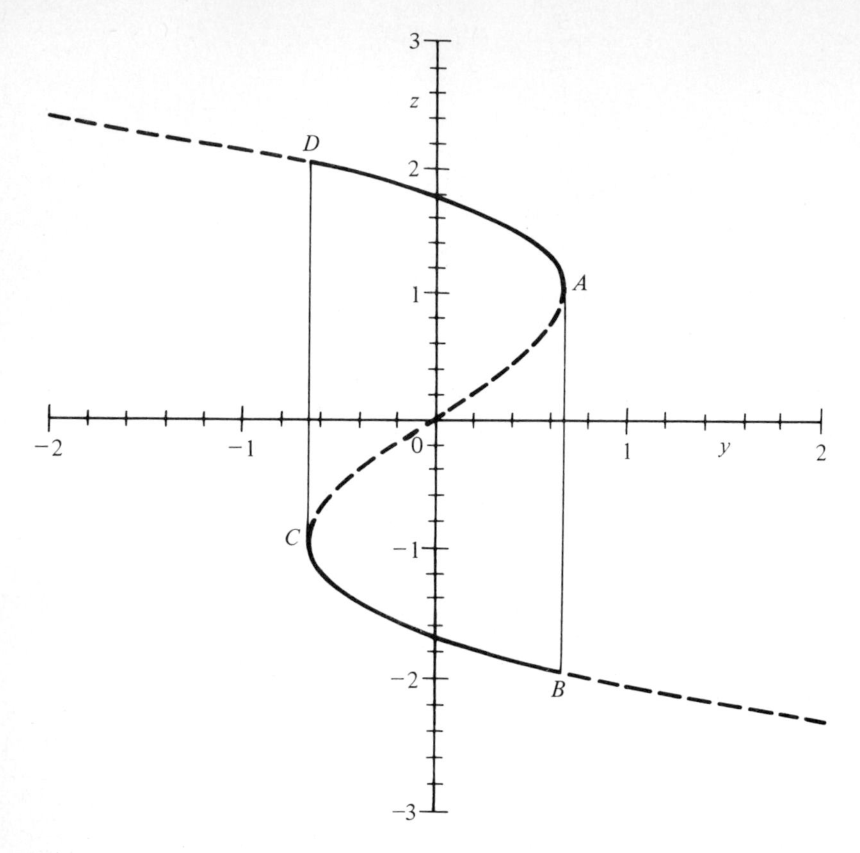

Figure 9.29 A plot of the leading-order outer solution in (9.7.18) to the Rayleigh oscillator (9.7.14). The outer solution is given by the solid lines connecting the points D to A and B to C. The dashed lines are additional segments of the graph of the algebraic functon in (9.7.18). Boundary-layer solutions connect the points A to B and C to D along the curves $y = \pm\frac{2}{3}$, respectively.

where c is a constant of integration. In the boundary-layer region this solution is a double-valued function of Y. As Z_0 decreases from 1, Y increases from $-\infty$; Y reaches a maximum at $Z_0 = 0$, $Y = 3c + \frac{2}{3}\ln 2 - 1$. As Z_0 decreases from 0 to -2, Y decreases from its maximum to $-\infty$. Asymptotic matching to lowest order is accomplished by the intermediate limits $(y, z_{\text{out}}) \to (\frac{2}{3}-, 1)$, $(Y, Z_0) \to (-\infty, 1)$ and $(y, z_{\text{out}}) \to (\frac{2}{3}-, -2)$, $(Y, Z_0) \to (-\infty, -2)$. These asymptotic matches complete the lowest-order boundary-layer analysis. Note that the constant c remains undetermined in leading order.

In the same way, it is possible to join the points C and D in Fig. 9.29 by a boundary-layer approximation; this approximation is given by (9.7.21) with the signs of Z_0 and Y reversed. The solution obtained in this way is periodic in t because the trajectory is closed in phase space. In Fig. 9.30 we plot the exact limit-cycle solution to the Rayleigh oscillator (9.7.14) in the (y, z) phase plane. Note how well leading-order boundary-layer theory predicts the shape of the limit cycle (see Fig. 9.29). In Fig. 9.31 we plot y and z versus t for the limit cycle in Fig. 9.30.

The leading-order boundary-layer solutions (9.7.18) and (9.7.21) can be used to compute a leading-order approximation to the period of the limit cycle of the Rayleigh equation as $\varepsilon \to 0+$. Since $z = dy/dt$, the period T of the limit cycle is given by

$$T = \oint \frac{dy}{z}, \tag{9.7.22}$$

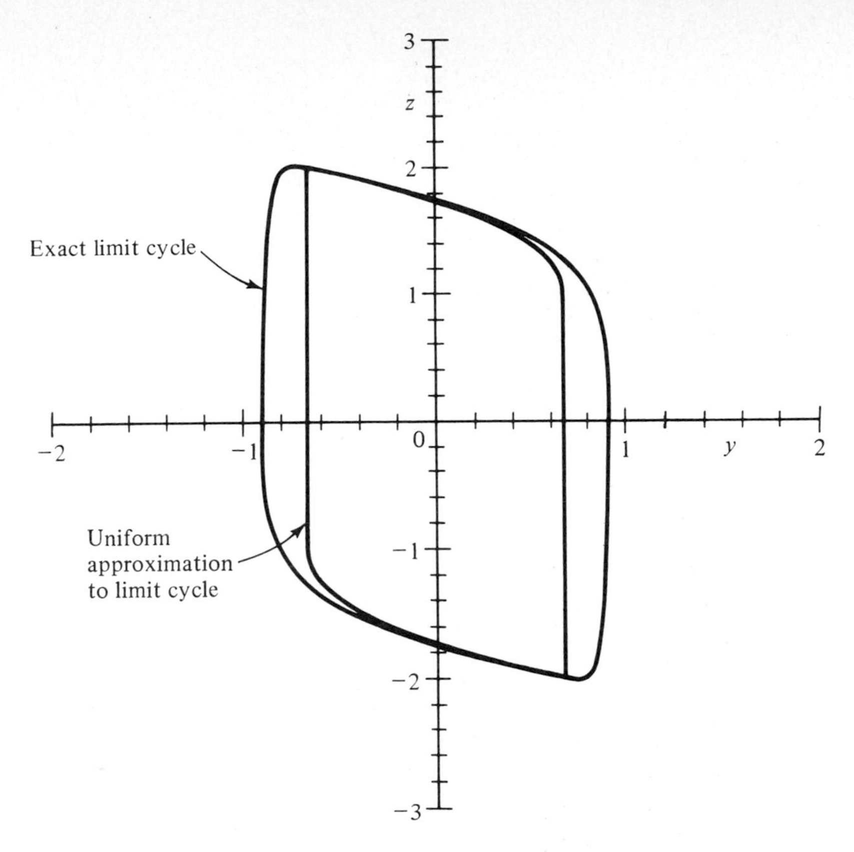

Figure 9.30 Comparison between the exact limit cycle of the Rayleigh oscillator (9.7.14) and the leading-order uniform approximation plotted in Fig. 9.29. The value of ε is 0.05.

where the integral is taken over the full limit cycle. The leading-order approximation to the period T is dominated by the contributions to the integral (9.7.22) from the outer solutions (9.7.18). The contribution to T from the boundary layer goes to 0 as $\varepsilon \to 0+$ because the width of the boundary layers in y is $O(\varepsilon)$. Therefore,

$$T \sim \int_{DA} \frac{dy}{z} + \int_{BC} \frac{dy}{z} = \int_{2}^{1} \frac{(1 - z^2)}{z} dz + \int_{-1}^{-2} \frac{(1 - z^2)}{z} dz = 3 - 2 \ln 2, \qquad \varepsilon \to 0+. \quad (9.7.23)$$

The full asymptotic expansion of the period $T(\varepsilon)$ of the limit cycles of (9.7.14) is very difficult to obtain. Dorodnicyn showed that

$$T(\varepsilon) \sim 3 - 2 \ln 2 + 3\alpha\varepsilon^{2/3} + \tfrac{1}{6}\varepsilon \ln \varepsilon + \cdots, \qquad \varepsilon \to 0+, \quad (9.7.24)$$

where α is the smallest zero of Ai $(-t)$ ($\alpha \doteqdot 2.3381$). We check the accuracy of this expansion in Table 9.1. One may wonder why the series (9.7.24) for $T(\varepsilon)$ is so complicated when the leading-order boundary-layer analysis is so simple. The reason for the complexity of (9.7.24) becomes apparent when we attempt a higher-order boundary-layer analysis.

The next-order corrections to the outer solution (9.7.18) are found by expanding $z_{\text{out}} = z_0 + z_1 + \cdots$. Here the leading-order outer approximation satisfies (9.7.18), $z_0 - \frac{1}{3}z_0^3 = y$, and from (9.7.17), the equation for z_1 is $z_0\, dz_0/dy = (1 - z_0^2)z_1$. Thus,

$$z_1 = \frac{z_0}{(1 - z_0^2)^2}. \quad (9.7.25)$$

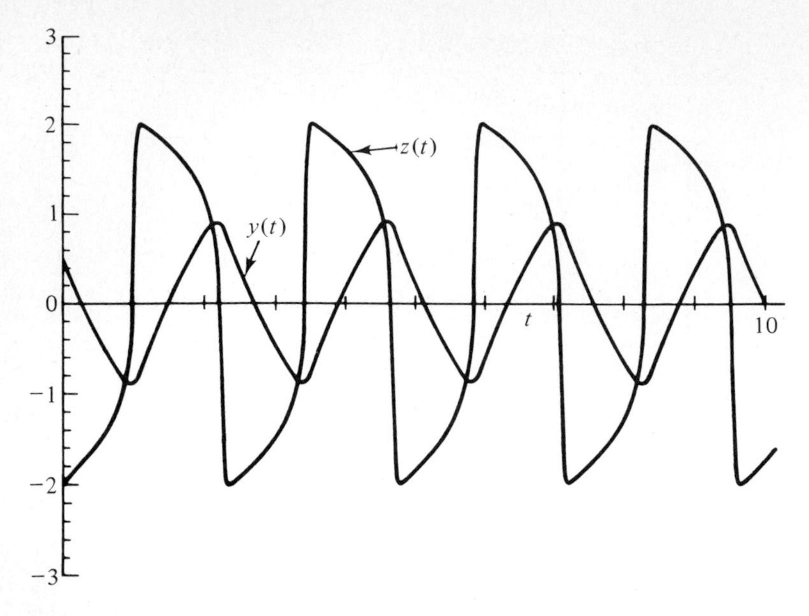

Figure 9.31 A plot of $y(t)$ and $z(t)$ versus t for the limit cycle of the Rayleigh oscillator (9.7.14) plotted in the phase plane in Fig. 9.30. The value of ε is 0.05.

To perform asymptotic matches at points A and B (see Fig. 9.29), we will need the expansions of z_{out} in the neighborhood of the points A and B:

$$z_{\text{out}} \sim 1 + \sqrt{\frac{2}{3} - y} + \frac{1}{6}\left(y - \frac{2}{3}\right) + \frac{\varepsilon}{4(\frac{2}{3} - y)} + \cdots, \qquad y \to \frac{2}{3}-, \varepsilon \to 0+, \tag{9.7.26}$$

$$z_{\text{out}} \sim -2 - \frac{1}{3}\left(y - \frac{2}{3}\right) + \cdots, \qquad y \to \frac{2}{3}-, \varepsilon \to 0+ \tag{9.7.27}$$

(see Prob. 9.46). The coefficient of $\sqrt{\frac{2}{3} - y}$ in (9.7.26) is positive because $z > z_0$ along DA; if z were less than z_0 along DA, then (9.7.25) would imply that $dz/dy > 0$ which is false.

The next-order boundary-layer approximation behaves in the neighborhood of the point A like

$$Z \sim 1 + \frac{1}{Y + c} + \frac{2 \ln |Y + c|}{3(Y + c)^2} + \frac{2 \ln 3}{3(Y + c)^2}$$
$$+ \varepsilon\left(-\frac{1}{4}Y^2 + \frac{1}{9}Y \ln Y\right) + \cdots, \qquad Y \to -\infty, \varepsilon Y^2 \to 0+, \tag{9.7.28}$$

and in the neighborhood of the point B like

$$Z \sim -2 - \tfrac{1}{3}\varepsilon Y, \qquad Y \to -\infty, \varepsilon Y \to 0+. \tag{9.7.29}$$

There is no trouble performing the asymptotic match at B; $Y = (y - \frac{2}{3})/\varepsilon$, so (9.7.27) and (9.7.29) match to first order in ε. On the other hand, (9.7.26) and (9.7.28) do *not* match at the point A; in particular, the term $\sqrt{\frac{2}{3} - y}$ in (9.7.26) is not present in (9.7.28).

This failure of boundary-layer analysis in higher order is remedied as in Example 6 of Sec. 9.5 by introducing a new scale and a new boundary layer in which the higher-order match near A can be accomplished. Surprisingly, it turns out that this new boundary-layer scale near A must be

Table 9.1 Comparison between the exact period $T(\varepsilon)$ of the Rayleigh oscillator and the asymptotic approximation (9.7.24) to $T(\varepsilon)$, $T(\varepsilon) \sim 3 - 2 \ln 2 + 3\alpha\varepsilon^{2/3} + O(\varepsilon \ln \varepsilon)$ $(\varepsilon \to 0+)$, for various values of ε

ε	Exact value of $T(\varepsilon)$	$3 - 2 \ln 2$	$3 - 2 \ln 2 + 3\alpha\varepsilon^{2/3}$
1	6.687	1.6137	8.6280
0.25	3.8155	1.6137	4.3973
0.04	2.3211	1.6137	2.4341
0.01	1.9155	1.6137	1.9393
0.0025	1.7355	1.6137	1.7429

thicker than the lowest-order boundary layer of width ε. To seek a new distinguished limit near the point A we set $y = \frac{2}{3} + \delta \bar{Y}$, $z = 1 + \eta \bar{Z}$, with $\bar{Y}$ and $\bar{Z}$ both of order 1 in the boundary layer.

Substituting into (9.7.17), we obtain

$$\frac{\varepsilon\eta}{\delta}\frac{d\bar{Z}}{d\bar{Y}} = -\frac{(\eta\bar{Z})^2 + (\eta\bar{Z})^3/3 + \delta\bar{Y}}{1 + \eta\bar{Z}}.$$

A new distinguished limit is obtained by choosing $\varepsilon\eta/\delta = \eta^2 = \delta$, so $\delta = \varepsilon^{2/3}$, $\eta = \varepsilon^{1/3}$. Thus, the thickness of the new intermediate boundary layer is $\varepsilon^{2/3}$. This explains why there is a term of order $\varepsilon^{2/3}$ in the expansion (9.7.24) for $T(\varepsilon)$. (See Prob. 9.48.)

PROBLEMS FOR CHAPTER 9

Section 9.1

(E) **9.1** Show that the solution to the asymptotic relation $\delta/\ln \delta = O(\varepsilon)$ $(\varepsilon \to 0+)$ satisfies $\delta = O(\varepsilon \ln \varepsilon)$ $(\varepsilon \to 0+)$.

(E) **9.2** Verify that $y_{\text{unif}}(x) - y(x) = O(\varepsilon)$ $(\varepsilon \to 0+;\ 0 \le x \le 1)$, where $y_{\text{unif}}(x)$ is given in (9.1.13) and $y(x)$ is the solution to (9.1.7).

(I) **9.3** (*a*) Show that if $a(x) < 0$ for $0 \le x \le 1$, then the solution to (9.1.7) has a boundary layer at $x = 1$.

(*b*) Find a uniform approximation with error $O(\varepsilon)$ to the solution (9.1.7) when $a(x) < 0$ for $0 \le x \le 1$.

(*c*) Show that if $a(x) > 0$, it is impossible to match to a boundary layer at $x = 1$.

(E) **9.4** Find leading-order uniform asymptotic approximations to the solutions of

(*a*) $\varepsilon y'' + (\cosh x)y' - y = 0$, $y(0) = y(1) = 1$ $(0 \le x \le 1)$,

(*b*) $\varepsilon y'' + (x^2 + 1)y' - x^3 y = 0$, $y(0) = y(1) = 1$ $(0 \le x \le 1)$, in the limit $\varepsilon \to 0+$.

(D) **9.5** Estimate the error between $y(0)$ and $y_{\text{in}}(0)$ in the limit as $\varepsilon \to 0+$ where $y(x)$ is the exact solution to (9.1.3) and $y_{\text{in}}(0)$ is given by (9.1.6).

(I) **9.6** Consider the initial-value problem $y' = (1 + \frac{1}{100}x^{-2})y^2 - 2y + 1$ $[y(1) = 1]$ on the interval $0 \le x \le 1$.

(*a*) Formulate this problem as a perturbation problem by introducing a suitable small parameter ε.

(*b*) Find an outer approximation correct to order ε (with errors of order ε^2). Where does this approximation break down?

(*c*) Introduce an inner variable and find the inner solution valid to order 1 (with errors of order ε). By matching to the outer solution find a uniformly valid approximation to $y(x)$ on the interval $0 \le x \le 1$. Estimate the accuracy of this approximation.

(*d*) Find the inner solution correct to order ε (with errors of order ε^2) and show that it matches to the outer solution correct to order ε.

(E) **9.7** How does the solution to (9.1.1) behave in the limit $\varepsilon \to 0-$?

Section 9.2

(I) **9.8** Use boundary-layer theory to find a uniform approximation with error of order ε^2 for the problem $\varepsilon y'' + y' + y = 0$ $[y(0) = e,\ y(1) = 1]$. Notice that there is no boundary layer in leading order, but one does appear in next order. Compare your solution with the exact solution to this problem.

(I) **9.9** Use boundary-layer methods to find an approximate solution to the initial-value problem $\varepsilon y'' + ay' + by = 0$ with $y(0) = 1$, $y'(0) = 1$, and $a > 0$. Show that the leading-order uniform approximation satisfies $y(0) = 1$ but not $y'(0) = 1$ for arbitrary b. Compare the leading-order uniform approximation with the exact solution to the problem when $a(x)$ and $b(x)$ are constants.

Section 9.3

(I) **9.10** Show that the matching region in nth-order boundary-layer theory for the problem discussed in Example 1 of Sec. 9.3 is $\varepsilon \ll x \ll \varepsilon^{n/(n+1)}$ $(\varepsilon \to 0+)$.

(I) **9.11** Obtain a uniform approximation accurate to order ε^2 as $\varepsilon \to 0+$ for the problem $\varepsilon y'' + (1 + x)^2 y' + y = 0$ $[y(0) = 1,\ y(1) = 1]$.

(I) **9.12** Verify (9.3.16).

(I) **9.13** Verify (9.3.18) and (9.3.19).

(I) **9.14** Show that the series (9.3.16) diverges like the series $\sum n!\, x^n$.

Clue: Try to find a recursion relation for the coefficient of $(1 + x)^{-2n-1}$ in the outer solution.

(I) **9.15** Find first-order uniform approximations valid as $\varepsilon \to 0+$ for the solutions of the differential equations given in Prob. 9.4.

(D) **9.16** Find second-order uniform approximations valid as $\varepsilon \to 0+$ for the solutions of the differential equations given in Prob. 9.4.

Section 9.4

(I) **9.17** For what real values of the constant α does the singular perturbation problem $\varepsilon y''(x) + y'(x) - x^\alpha y(x) = 0$, $y(0) = 1$, $y(1) = 1$ $(0 \le x \le 1)$ have a solution with a boundary layer near $x = 0$ as $\varepsilon \to 0+$?

Clue: Perform a local analysis to decide if the problem has a solution. Show that if $\alpha \le -2$, there is no solution that behaves like $y(x) \sim 1$ $(x \to 0+)$. Also show that if $-2 < \alpha < -1$, the thickness of the boundary layer at $x = 0$ is $\varepsilon^{1/(2+\alpha)}$ and that if $\alpha > -1$, the boundary layer at $x = 0$ has thickness ε.

(D) **9.18** Consider the problem (discussed by Cole) $\varepsilon y'' + \sqrt{x}\, y' - y = 0$ $[y(0) = 0,\ y(1) = e^2]$.

(*a*) Show that there is a boundary layer of thickness $\varepsilon^{2/3}$ at $x = 0$.

(*b*) Show that a leading-order uniform approximation to $y(x)$ is $y_{\text{unif},0}(x) = \exp(2\sqrt{x}) - 1 + [(\frac{3}{2})^{1/3}/\Gamma(\frac{2}{3})] \int_0^{x\varepsilon^{-2/3}} e^{-2t^{3/2}/3}\, dt$.

(*c*) Show that the next correction to the outer solution is $y_{\text{out}}(x) \sim \exp(2\sqrt{x}) + \varepsilon(-1/2x + 2/\sqrt{x} - \frac{3}{2}) \exp(2\sqrt{x}) + \cdots$ $(\varepsilon \to 0+)$.

(*d*) Find integral representations for the first four terms of the inner expansion in powers of $\varepsilon^{1/3}$. That is, calculate $Y_{\text{in}} \sim Y_0 + \varepsilon^{1/3} Y_1 + \varepsilon^{2/3} Y_2 + \varepsilon Y_3 + \cdots$ $(\varepsilon \to 0+)$.

(*e*) Show that the first two terms in the outer expansion match with the first four terms in the inner expansion.

(I) **9.19** Find a lowest-order uniform approximation to the boundary-value problem $\varepsilon y'' + y' \sin x + y \sin(2x) = 0$ $[y(0) = \pi,\ y(\pi) = 0]$.

(I) **9.20** Consider the problem $\varepsilon y'' + x^\alpha y' + y = 0$ with $y(0) = y(1) = 1$ as $\varepsilon \to 0+$. For what values of α is there a boundary layer at $x = 0$? What is the thickness of the boundary layer?

Section 9.5

(I) **9.21** Complete Example 1 of Sec. 9.5.

(*a*) Show that

$$\frac{d^3\bar{Y}_n}{d\bar{X}^3} - \frac{d\bar{Y}_n}{d\bar{X}} = \begin{cases} 0, & n = 0, \\ \bar{Y}_0, & n = \frac{1}{2}, \\ \bar{Y}_{n-1/2} - \bar{X}\bar{Y}_{n-1}, & n > \frac{1}{2}. \end{cases}$$

(*b*) Next, show that in terms of the inner variable X, the outer expansion (9.5.15) becomes $y_{\text{out}} = \sqrt{e} + \varepsilon^{1/2}\sqrt{e}(1 - \bar{X}) + \varepsilon\sqrt{e}\,\bar{X}^2 + O(\varepsilon\bar{X}, \varepsilon^{3/2}\bar{X}^3)$.

(*c*) Finally, show that the first three solutions to these equations which satisfy $y(1) = 1$ and which match asymptotically to $y_{\text{out}}(x)$ are

$$\bar{Y}_0(\bar{X}) = (1 - \sqrt{e})e^{-\bar{X}} + \sqrt{e},$$

$$\bar{Y}_{1/2}(\bar{X}) = [-\sqrt{e} - \tfrac{1}{2}(\sqrt{e} - 1)\bar{X}]e^{-\bar{X}} - \sqrt{e}\,\bar{X} + \sqrt{e}.$$

$$\bar{Y}_1(\bar{X}) = [\tfrac{1}{8}(\sqrt{e} - 1)\bar{X}^2 - \tfrac{1}{8}(\sqrt{e} + 3)\bar{X} - \tfrac{7}{4}\sqrt{e}]e^{-\bar{X}} + \sqrt{e}\,\bar{X}^2 - \sqrt{e}\,\bar{X} + \tfrac{7}{4}\sqrt{e}.$$

(*d*) Verify (9.5.17).

(I) **9.22** Verify (9.5.37).

(I) **9.23** Consider the boundary-value problems:

(*a*) $\varepsilon y'' + y'/x + y = 0$ $[y(-1) = 2e^{-1/2},\ y(1) = e^{-1/2}]$,

(*b*) $\varepsilon y'' + y'/x^2 + y = 0$ $[y(0) = 0,\ y(1) = e^{-1/3}]$,

as $\varepsilon \to 0+$. Do these problems have a solution? If so, find a leading-order approximation to the solution.

(D) **9.24** Show that the outer expansion (9.5.32) diverges.

Clue: One way to do this is to study the exact solution to (9.5.31), which is $y = e^{-1/2}x^{(\varepsilon-1)/2\varepsilon}J_{1/2-1/2\varepsilon}(x/\sqrt{\varepsilon})/J_{1/2-1/2\varepsilon}(1/\sqrt{\varepsilon})$.

(I) **9.25** (*a*) Verify (9.5.43).

(*b*) Show that $\bar{\alpha}_0 = 0$ and $\bar{\beta}_0 = 2$ in (9.5.43).

(I) **9.26** (*a*) Find a leading-order uniform approximation to the solution to the problem $\varepsilon(x + \varepsilon^2)y'' + xy' + y = 0$ $[y(0) = y(1) = 1]$ as $\varepsilon \to 0+$.

(*b*) Show that the term $\varepsilon^3 y''$ is always a small perturbation, even when $x \ll \varepsilon^2$ $(\varepsilon \to 0+)$.

(D) **9.27** Find a uniform approximation accurate to order ε for the problem $\varepsilon y'' + (1 + 2\varepsilon/x + 2\varepsilon^3/x^2)y' + 2y/x = 0$ $[y(0) = y(1) = 1]$ as $\varepsilon \to 0+$. Show that:

(*a*) $y(x) \sim \alpha + \beta e^{-2\varepsilon^2/x}$ $(x \to 0+)$, so that $y(0)$ is finite and it is appropriate to specify a nonzero value for $y(0)$;

(*b*) $y_{\text{out}}(x) \sim x^{-2} + 2\varepsilon(x^{-3} - x^{-2}) + \cdots$ $(\varepsilon \to 0+)$;

(*c*) distinguished limits are $\delta = \varepsilon$ and $\delta = \varepsilon^2$;

(*d*) inner and inner-inner expansions exist at $x = 0$ which match to each other; the inner-inner expansion satisfies the boundary condition at $x = 0$ and the inner expansion matches to $y_{\text{out}}(x)$.

(I) **9.28** Use boundary-layer theory to solve $\varepsilon y'' + a(x)y' + b(x)y = \delta(x)$ $[y(x) = 0;\ x < 0]$ to leading order in ε as $\varepsilon \to 0+$, where $a(x) > 0$ for $x \geq 0$.

Section 9.6

(I) **9.29** Find leading-order uniform approximations to the solutions of the following problems in the limit $\varepsilon \to 0+$:

(*a*) $\varepsilon y'' - 2(\tan x)y' + y = 0$ $[y(\pm 1) = 1]$;

(*b*) $\varepsilon y'' + 2(\tan x)y' - y = 0$ $[y(\pm 1) = 1]$;

(*c*) $\varepsilon y'' + (\sinh x)y' + J_0(x)y = 0$ $[y(\pm 1) = 0]$;

(*d*) $\varepsilon y'' + xy' = x\cos x$ $[y(\pm 1) = 2]$;

(*e*) $\varepsilon y'' - xy' - (3 + x)y = 0$ $[y(\pm 1) = 1]$;

(*f*) $\varepsilon y'' + (\ln x)y' - x(\ln x)y = 0$ $[y(\tfrac{1}{2}) = y(\tfrac{3}{2}) = 1]$.

(I) **9.30** (*a*) Verify (9.6.13).
(*b*) Verify (9.6.16).

(D) **9.31** Use first-order boundary-layer theory to find a uniform asymptotic approximation to the solution of $\varepsilon y'' + (x^2 - 1)y' + (x^2 - 1)^2 y = 0$ $[y(-2) = A,\ y(0) = B]$.

(*a*) This problem is an example of case IV because $\alpha = -2$, $\beta = 0$ at $x = -1$. To leading order in the internal layer near $x = -1$, let $X = (x + 1)/\sqrt{\varepsilon}$ to obtain $d^2 Y_{\text{in}}/dX^2 - 2X\, dY_{\text{in}}/dX = 0$. Show that $Y_{\text{in}} = C_1 + C_2 \int_0^X e^{t^2}\, dt$, so that matching to the outer solutions requires that $C_2 = 0$ but leaves C_1 undetermined.

(*b*) Now consider the next-order correction to the inner equation: $d^2 Y_{\text{in}}/dX^2 - 2X\, dY_{\text{in}}/dX + 4\varepsilon X^2 Y_{\text{in}} = 0$. Show that an approximate solution to this equation is $Y_{\text{in}} = e^{X^2/2}\{C_1 D_\varepsilon[X\sqrt{2}\,(1 - \varepsilon)] + C_2 D_\varepsilon[-X\sqrt{2}\,(1 - \varepsilon)]\}$. Use this result to argue that both constants C_1 and C_2 may now be determined by asymptotic matching, just as in case II.

(D) **9.32** Examples of singular perturbation problems where higher-order boundary-layer theory resolves the ambiguity of case IV of Sec. 9.6 are

(1) $\varepsilon y'' - xy' + (n + \beta x)y = 0$ $[y(-1) = A,\ y(1) = B]$,
(2) $\varepsilon y'' - xy' + (n + \beta x^2)y = 0$ $[y(-1) = A,\ y(1) = B]$,

where $\varepsilon \to 0+$, $\beta \neq 0$, and $n = 0, 1, 2, \ldots$.

(*a*) For each of these differential equations, show that the leading-order approximation to the internal layer solution $Y(X)$, where $X = x/\sqrt{\varepsilon}$, is $Y_0(X) = e^{X^2/4}[K_1 D_n(X) + K_2 D_{-n-1}(iX)]$. The difficulty with the leading-order boundary-layer analysis given in Sec. 9.6 is that when $K_2 = 0$, $Y_0(X)$ grows algebraically with X as $X \to \pm\infty$, so matching is possible for any value of K_1.

(*b*) Show that there is a higher-order approximation to $Y(X)$ that grows exponentially as $|X| \to \infty$ if either of K_1 or K_2 is not zero. In what order of perturbation theory does $Y(X)$ first grow exponentially?

(*c*) We may conclude that in order for a match to be possible, it is necessary that $K_1 = K_2 = 0$. Show that the boundary-layer analysis of these problems then reverts back to that of case II of Sec. 9.6.

(D) **9.33** Examples of case IV of Sec. 9.6 that are not resolved by higher-order boundary-layer analysis are

(1) $\varepsilon y'' - x(1 + x)y' + xy = 0$ $[y(-\frac{1}{2}) = A,\ y(\frac{1}{2}) = B]$,
(2) $\varepsilon y'' - x(1 - x^2)y' - x(x + 1)y = 0$ $[y(-\frac{1}{2}) = A,\ y(\frac{1}{2}) = B]$.

Show that higher-order corrections to the leading-order internal-layer solution $Y_0(X) = 1$ do not grow faster than algebraically as $X \to \pm\infty$. Thus, conclude that boundary-layer theory remains ambiguous to all orders in ε^n as $\varepsilon \to 0+$. The resolution of this dilemma will be given in Prob. 10.28 using WKB analysis.

(I) **9.34** Use boundary-layer techniques to find a leading-order uniform approximation to the solution of $\varepsilon y''(x) + 2xy'(x) - 4x^2 y(x) = 0$ $(-1 \le x \le 1)$, with $y(-1) = 0$, $y(+1) = e$ in the limit $\varepsilon \to 0+$.

(I) **9.35** Find leading-order uniform approximations to the solutions of the following problems:

(*a*) $\varepsilon y'' - (x + x^3)y' - 2y = 0$ $[y(-1) = A,\ y(1) = B]$,
(*b*) $\varepsilon y'' + (x + x^3)y' + 2y = 0$ $[y(-1) = A,\ y(1) = B]$,
(*c*) $\varepsilon y'' + (x^2 - 1)y' + (x^2 - 1)^2 y = 0$ $[y(0) = A,\ y(2) = B]$,

in the limit $\varepsilon \to 0+$.

(I) **9.36** Consider the boundary-value problem $\varepsilon y'' - xy' + y = 0$ $[y(-1) = -1,\ y(1) = 1]$.

(*a*) Show that there is a one-parameter family of solutions determined by boundary-layer theory.
(*b*) Solve the original differential equation exactly and show that the solution is actually unique.

Section 9.7

(E) **9.37** Show that it is inconsistent to have a boundary layer at $x = 1$ in (9.7.1) as $\varepsilon \to 0+$.

(I) **9.38** Solve (9.7.1) correct to first order in ε.

9.39 Consider the boundary-value problem $\varepsilon y'' - y' + e^y = 0$ $[y(0) = A,\ y(1) = 0]$.

(I) (*a*) Find a leading-order uniform approximation to the solution when $A < 0$.

(D) (*b*) Discuss what happens when $A \ge 0$. Is there a solution? If so, can you find it using boundary-layer theory?

(D) **9.40** Discuss the qualitative nature of the solutions to:

(*a*) $\varepsilon y'' - y' + y^2 = 0$ $[y(0) = y(1) = 1]$ in the limit $\varepsilon \to 0+$;

(*b*) $\varepsilon y'' - y' + 1/y = 0$ $[y(0) = \sqrt{2},\ y(1) = 0]$ in the limit $\varepsilon \to 0+$.

(D) **9.41** Do the boundary-value problems,

(*a*) $\varepsilon y'' - (y')^2 + e^y = 0$ $[y(0) = y(1) = 1]$,

(*b*) $\varepsilon y'' - (y')^2 + y^2 = 0$ $[y(0) = y(1) = 1]$,

have solutions when $\varepsilon \to 0+$? If they do, find a leading-order uniform approximation. Note that these problems can be solved exactly using the methods of Sec. 1.7.

Clue: Perform a boundary-layer analysis in the phase plane (y, y').

(I) **9.42** Show that it is consistent for the solution to (9.7.7) to exhibit an internal boundary layer at any value of x $(-1 < x < 1)$.

(D) **9.43** Find a leading-order approximation to the solution of $\varepsilon\, d/dx[(2 - x)^3 y\, dy/dx] - (2 - x)\, dy/dx + y = 0$, with $y(0) = y(1) = 1$ as $\varepsilon \to 0+$.

(D) **9.44** Use the approach of Example 2 of Sec. 9.7 to study the solutions of $\varepsilon y'' - xy - y^2 = 0$ $[y(1) = A,\ y(0) = 0]$ in the limit $\varepsilon \to 0+$ for various values of A.

(D) **9.45** Use phase-plane arguments to show that there exists a unique limit cycle of the Rayleigh equation (9.7.14) which attracts trajectories satisfying arbitrary initial conditions.

(I) **9.46** Derive the expansions in (9.7.26) and (9.7.27).

(D) **9.47** Show that although boundary layers are possible near any value of y with $|y| < \frac{2}{3}$, it is not possible to construct a closed trajectory of (9.7.14) except for the limit cycle discussed in Example 3 of Sec. 9.7.

(D) **9.48** Compute the term proportional to $\varepsilon^{2/3}$ in the expansion (9.7.24) for the period $T(\varepsilon)$ of the limit cycle of the Rayleigh equation.

Clue: Perform an asymptotic match of the outer solution to the intermediate layer of width $\varepsilon^{2/3}$ and then match the intermediate layer to the inner boundary layer of width ε. Show that the leading-order intermediate-layer solution satisfies a Riccati differential equation whose solution is a ratio of Airy functions. Argue that the coefficient of Bi is zero by matching to the outer solution [because the coefficient of $\sqrt{\frac{2}{3} - y}$ in (9.7.26) is positive]. Then match to the inner boundary layer as $\bar{Y} \to \alpha$, where α is the smallest solution of Ai $(-t) = 0$. Finally, compute the $\varepsilon^{2/3}$ term in (9.7.24) by suitably evaluating the integral (9.7.22) for $T(\varepsilon)$.

(D) **9.49** Perform an asymptotic boundary-layer analysis of the limit cycle of the Van der Pol equation $\varepsilon\, d^2z/dt^2 - (1 - z^2)\, dz/dt + z = 0$ by obtaining appropriate inner, outer, and intermediate expansions directly from the differential equation (not transformed to the phase plane). Show that the period $T(\varepsilon)$ of this limit cycle satisfies $T(\varepsilon) \sim 3 - 2 \ln 2 + 3\alpha\varepsilon^{2/3}$ $(\varepsilon \to 0+)$, where α is the smallest zero of Ai $(-t)$. The period of the limit cycles of the Van der Pol and Rayleigh equations are the same because the substitution $z = dy/dt$ converts the Rayleigh equation into the Van der Pol equation.

(D) **9.50** Consider the nonlinear perturbation problem $y'' + 2y'/x + \varepsilon y y' = 0$ $[y(1) = 0,\ y(+\infty) = 1]$ as $\varepsilon \to 0+$.

(*a*) Find the form of the outer solution accurate to order ε. Show that the problem is a singular perturbation problem even though ε does not multiply the highest derivative term. The problem is singular because the domain is infinite.

(*b*) Argue that there must be an "inner" expansion near $x = \infty$. Find its scale by setting $X = \delta x$ and seeking a dominant balance. Express the inner solution to order ε in terms of exponential integrals $E_n(t) = \int_t^\infty e^{-s} s^{-n}\, ds$.

(*c*) Try to perform an asymptotic match to order ε by taking the intermediate limit $x \to +\infty$, $X \to 0+$. The asymptotic expansion (6.2.11) of $E_n(t) \equiv \Gamma(1 - n, t)$ as $t \to 0+$ is helpful. Show that no match is possible.

(*d*) Argue that terms of order $\varepsilon \ln (1/\varepsilon)$ must be included in the outer expansion for matching to succeed. Introduce this intermediate-order term and show that the asymptotic match between inner and outer expansions can now be completed.

CHAPTER
TEN
WKB THEORY

When you follow two separate chains of thought, Watson, you will find some point of intersection which should approximate the truth.

—Sherlock Holmes, *The Disappearance of Lady Francis Carfax*
Sir Arthur Conan Doyle

(E) 10.1 THE EXPONENTIAL APPROXIMATION FOR DISSIPATIVE AND DISPERSIVE PHENOMENA

WKB theory is a powerful tool for obtaining a global approximation to the solution of a linear differential equation whose highest derivative is multiplied by a small parameter ε; it contains boundary-layer theory as a special case.

The WKB approximation to a solution of a differential equation has a simple structure. The exact solution may be some unknown function of overwhelming complexity; yet, the WKB approximation, order by order in powers of ε, consists of exponentials of elementary integrals of algebraic functions, and well-known special functions, such as the Airy function or parabolic cylinder function. WKB approximation is suitable for linear differential equations of any order, for initial-value and boundary-value problems, and for eigenvalue problems. It may also be used to evaluate integrals of the solution of a differential equation. The limitation of conventional WKB techniques is that they are only useful for linear equations.

Dissipative and Dispersive Phenomena

In our study of boundary-layer theory we have shown how to construct an approximate solution to a differential equation containing a small parameter ε. This construction requires one to match slowly varying outer solutions to rapidly varying inner solutions. [In perturbation theory a *slowly varying function* changes its value by $O(1)$ over an interval of size $O(1)$ as $\varepsilon \to 0+$ while a *rapidly varying function* changes its value by $O(1)$ over an interval whose size approaches 0 as $\varepsilon \to 0+$.]

An outer solution remains smooth if we allow ε to approach $0+$. But in this limit an inner solution becomes discontinuous across the boundary layer because the thickness of the boundary layer tends to 0. We thus say that the solution suffers a *local breakdown* at the boundary layer as $\varepsilon \to 0+$. A local breakdown occurs where the approximate solution is exponentially increasing or decreasing. This kind of behavior is called *dissipative* because the rapidly varying component of the solution decays exponentially (dissipates) away from the point of local breakdown. The solution of a differential equation having a strong positive or negative damping term (like ay' in $\varepsilon y'' + ay' + by = 0$) typically exhibits dissipative behavior.

Some differential equations with small parameters have solutions which exhibit a *global breakdown.* For example, the boundary-value problem

$$\varepsilon y'' + y = 0, \qquad y(0) = 0,\ y(1) = 1, \tag{10.1.1}$$

has the exact solution

$$y(x) = \frac{\sin (x/\sqrt{\varepsilon})}{\sin (1/\sqrt{\varepsilon})}, \qquad \varepsilon \neq (n\pi)^{-2}, \tag{10.1.2}$$

which becomes rapidly oscillatory for small ε (see Fig. 7.3) and discontinuous when $\varepsilon \to 0+$. The breakdown is global because it occurs throughout the finite interval $0 \leq x \leq 1$. A global breakdown is typically associated with rapidly oscillatory, or *dispersive*, behavior. A dispersive solution is wavelike with very small and slowly changing wavelengths and slowly varying amplitudes as functions of x.

Boundary-layer techniques are not powerful enough to handle dispersive phenomena. To see why, let us try to solve (10.1.1) using boundary-layer methods. Setting $\varepsilon = 0$ in (10.1.1) gives the outer solution $y_{\text{out}}(x) = 0$, which is obviously a terrible approximation to the actual solution in (10.1.2). The actual solution in Fig. 7.3 looks like a sequence of internal boundary layers with no outer solution at all. Even for this very simple problem, boundary-layer analysis is insufficient.

From our understanding of Chap. 9 we can intuit that it is the absence of a one-derivative term which causes the global breakdown of the solution to (10.1.1). In Sec. 9.6 we showed that internal boundary layers may occur in the solution of $\varepsilon y'' + a(x)y' + b(x)y = 0$ $[y(0) = A,\ y(1) = B]$ at isolated points for which $a(x) = 0$. When $a(x) \equiv 0$ on an interval, it is not surprising to find that the solution is rapidly varying on the entire interval. Fortunately, WKB theory provides a simple and general approximation method for linear differential equations which treats dissipative and dispersive phenomena equally well.

The Exponential Approximation

Dissipative and dispersive phenomena are both characterized by exponential behavior, where the exponent is real in the former case and imaginary in the latter case. Thus, for a differential equation that exhibits either or both kinds of behavior, it is natural to seek an approximate solution of the form

$$y(x) \sim A(x)e^{S(x)/\delta}, \qquad \delta \to 0+. \tag{10.1.3}$$

The *phase* $S(x)$ is assumed nonconstant and slowly varying in a breakdown region. When S is real, there is a boundary layer of thickness δ; when S is imaginary, there is a region of rapid oscillation characterized by waves having wavelength of order δ. When $S(x)$ is constant, the behavior of $y(x)$, which is characteristic of an outer solution in boundary-layer theory, is expressed by the slowly varying *amplitude* function $A(x)$.

The exponential approximation in (10.1.3) is conventionally known as a *WKB approximation*, named after Wentzel, Kramers, and Brillouin who popularized the theory. However, credit should also be given to many others including Rayleigh and Jeffreys who contributed to its early development.

Formal WKB Expansion

The exponential approximation in (10.1.3) is not in a form most suitable for deriving asymptotic approximations because the amplitude and phase functions $A(x)$ and $S(x)$ depend implicitly on δ. It is best to represent the dependences of these functions on δ explicitly by expanding $A(x)$ and $S(x)$ as series in powers of δ. We can then combine these two series in a single exponential power series of the form

$$y(x) \sim \exp\left[\frac{1}{\delta}\sum_{n=0}^{\infty}\delta^n S_n(x)\right], \qquad \delta \to 0. \tag{10.1.4}$$

This expression is the starting formula from which all WKB approximations are derived.

Example 1 *Approximate solution to a Schrödinger equation.* A second-order homogeneous linear differential equation is in Schrödinger form if the y' term is absent. The approximate solutions to the Schrödinger equation

$$\varepsilon^2 y'' = Q(x)y, \qquad Q(x) \neq 0, \tag{10.1.5}$$

are easy to find using WKB analysis when ε is small. We merely substitute (10.1.4) into (10.1.5). Differentiating (10.1.4) twice gives

$$\begin{aligned} y' &\sim \left(\frac{1}{\delta}\sum_{n=0}^{\infty}\delta^n S_n'\right)\exp\left(\frac{1}{\delta}\sum_{n=0}^{\infty}\delta^n S_n\right), & \delta \to 0, \\ y'' &\sim \left[\frac{1}{\delta^2}\left(\sum_{n=0}^{\infty}\delta^n S_n'\right)^2 + \frac{1}{\delta}\sum_{n=0}^{\infty}\delta^n S_n''\right]\exp\left(\frac{1}{\delta}\sum_{n=0}^{\infty}\delta^n S_n\right), & \delta \to 0. \end{aligned} \tag{10.1.6}$$

Next, we substitute (10.1.6) into (10.1.5) and divide off the exponential factors:

$$\frac{\varepsilon^2}{\delta^2}S_0'^2 + \frac{2\varepsilon^2}{\delta}S_0'S_1' + \frac{\varepsilon^2}{\delta}S_0'' + \cdots = Q(x). \tag{10.1.7}$$

The largest term on the left side of (10.1.7) is $\varepsilon^2 S_0'^2/\delta^2$. By dominant balance this term must have the same order of magnitude as $Q(x)$ on the right side. (Here we have used the assumption that $Q \neq 0$.) Thus, δ is proportional to ε and for simplicity we choose $\delta = \varepsilon$. As in boundary-layer theory, the small scale parameter δ is determined by a *distinguished limit* (see Sec. 9.3).

Setting $\delta = \varepsilon$ in (10.1.7) and comparing powers of ε gives a sequence of equations which determine $S_0, S_1, S_2, \ldots$:

$$S_0'^2 = Q(x), \tag{10.1.8}$$

$$2S_0'S_1' + S_0'' = 0, \tag{10.1.9}$$

$$2S_0'S_n' + S_{n-1}'' + \sum_{j=1}^{n-1} S_j'S_{n-j}' = 0, \qquad n \geq 2. \tag{10.1.10}$$

The equation for S_0 (10.1.8) is called the *eikonal* equation; its solution is

$$S_0(x) = \pm\int^x \sqrt{Q(t)}\, dt. \tag{10.1.11}$$

The equation for S_1 (10.1.9) is called the *transport* equation; its solution, apart from an additive constant, is

$$S_1(x) = -\tfrac{1}{4} \ln Q(x). \tag{10.1.12}$$

Combining (10.1.11) and (10.1.12) gives a pair of approximate solutions to the Schrödinger equation (10.1.5), one for each sign of S_0. The general solution is a linear combination of the two:

$$y(x) \sim c_1 Q^{-1/4}(x) \exp\left[\frac{1}{\varepsilon}\int_a^x dt \sqrt{Q(t)}\right] + c_2 Q^{-1/4}(x) \exp\left[-\frac{1}{\varepsilon}\int_a^x dt \sqrt{Q(t)}\right], \qquad \varepsilon \to 0, \tag{10.1.13}$$

where c_1 and c_2 are constants to be determined from initial or boundary conditions and a is an arbitrary but fixed integration point. This expression is the leading-order WKB approximation to the solution of (10.1.5); it differs from the exact solution by terms of order ε in regions where $Q(x) \neq 0$.

A more accurate approximation to $y(x)$ may be constructed from the higher terms in the WKB series. The next four terms, as computed from (10.1.10) by repeated differentiation, are

$$S_2 = \pm\int^x \left[\frac{Q''}{8Q^{3/2}} - \frac{5(Q')^2}{32Q^{5/2}}\right] dt, \tag{10.1.14}$$

$$S_3 = -\frac{Q''}{16Q^2} + \frac{5Q'^2}{64Q^3}, \tag{10.1.15}$$

$$S_4 = \pm\int^x \left[\frac{d^4Q/dx^4}{32Q^{5/2}} - \frac{7Q'Q'''}{32Q^{7/2}} - \frac{19(Q'')^2}{128Q^{7/2}} + \frac{221Q''(Q')^2}{256Q^{9/2}} - \frac{1{,}105(Q')^4}{2{,}048Q^{11/2}}\right] dt, \tag{10.1.16}$$

$$S_5 = -\frac{d^4Q/dx^4}{64Q^3} + \frac{7Q'Q'''}{64Q^4} + \frac{5(Q'')^2}{64Q^4} - \frac{113(Q')^2Q''}{256Q^5} + \frac{565(Q')^4}{2{,}048Q^6}. \tag{10.1.17}$$

A discussion of the structure and properties of these expressions is given in Sec. 10.7.

Example 2 *Solution of* $\varepsilon y'' + y = 0$ $[y(0) = 0, y(1) = 1]$ *as* $\varepsilon \to 0+$. Even though it is not possible to solve this problem using boundary-layer theory, the WKB approximation in (10.1.13), with ε replaced by $\sqrt{\varepsilon}$, leads to the exact solution. For this problem $Q(x) = -1$. Thus, (10.1.13) reduces to $y(x) = c_1 \exp(ix/\sqrt{\varepsilon}) + c_2 \exp(-ix/\sqrt{\varepsilon})$.

Imposing the boundary conditions $y(0) = 0$ and $y(1) = 1$ reproduces the exact solution in (10.1.2):

$$y(x) = \frac{\sin(x/\sqrt{\varepsilon})}{\sin(1/\sqrt{\varepsilon})}.$$

Example 3 *WKB solution of an initial-value problem.* To solve the initial-value problem $\varepsilon^2 y'' = Q(x)y$ $[y(0) = A,\ y'(0) = B]$ we set $a = 0$ in (10.1.13) and differentiate (10.1.13) to obtain simultaneous equations for c_1 and c_2:

$$[Q(0)]^{-1/4}(c_1 + c_2) = A,$$

$$-\tfrac{1}{4}Q'(0)[Q(0)]^{-5/4}(c_1 + c_2) + (c_1 - c_2)[Q(0)]^{1/4}/\varepsilon = B.$$

For example, when $A = 0$ and $B = 1$ we obtain $c_1 = -c_2 = \frac{1}{2}\varepsilon[Q(0)]^{-1/4}$. Thus, the approximate solution to this initial-value problem is

$$y(x) \sim \varepsilon[Q(x)Q(0)]^{-1/4} \sinh\left[\int_0^x dt\sqrt{Q(t)}/\varepsilon\right], \qquad \varepsilon \to 0.$$

If we now take $Q(x) = (1 + x^2)^2$, then a uniform approximation (valid for all x) to the solution of

$$\varepsilon^2 y'' = (1 + x^2)^2 y, \qquad y(0) = 0,\ y'(0) = 1, \tag{10.1.18}$$

is

$$y(x) \sim \frac{\varepsilon}{\sqrt{1 + x^2}} \sinh\left[\frac{1}{\varepsilon}(x + x^3/3)\right], \qquad \varepsilon \to 0. \tag{10.1.19}$$

In Fig. 10.1 we compare (10.1.19) with the exact solution to (10.1.18) from a computer for three values of ε. Note that ε need not be very small for (10.1.19) to be a good approximation to the exact solution (see Prob. 10.3).

Example 4 *Rederivation of a boundary-layer approximation.* Here we show that the WKB approximation contains boundary-layer theory as a special case. Consider the familiar boundary-value problem

$$\varepsilon y'' + a(x)y' + b(x)y = 0, \qquad y(0) = A,\ y(1) = B, \tag{10.1.20}$$

where we assume that $a(x) > 0$ for $0 \le x \le 1$ and $\varepsilon \to 0+$.

We begin by substituting equations (10.1.6) into (10.1.20) and neglecting terms which vanish as $\delta \to 0$:

$$\frac{\varepsilon}{\delta^2} S_0'^2 + 2\frac{\varepsilon}{\delta} S_0' S_1' + \frac{\varepsilon}{\delta} S_0'' + \frac{1}{\delta} S_0' a + S_1' a + b + \cdots = 0. \tag{10.1.21}$$

The largest of the first three terms is $\varepsilon\delta^{-2}S_0'^2$ and the largest of the next three (assuming that $a \ne 0$) is $\delta^{-1}S_0' a$. By dominant balance these two terms must be of equal magnitude, so $\varepsilon\delta^{-2} = O(\delta^{-1})$. Therefore, δ is proportional to ε and for simplicity we choose $\delta = \varepsilon$. As in Example 1, the small scale parameter δ is again determined by a *distinguished* limit. [There is another distinguished limit possible in (10.1.21); namely, $\delta = 1$. However, this limit reproduces the outer solution below. Why?]

Next we return to (10.1.21) and identify the coefficients of ε^{-1} and ε^0. The resulting equations

$$S_0'^2 + S_0' a = 0 \tag{10.1.22}$$

and

$$2S_0' S_1' + S_0'' + S_1' a + b = 0 \tag{10.1.23}$$

are easy to solve. Equation (10.1.22) yields two solutions for S_0':

$$S_0' = 0 \text{ and } S_0' = -a.$$

When $S_0' = 0$, (10.1.23) becomes $S_1' a + b = 0$. Thus, $S_1 = -\int^x [b(t)/a(t)]\, dt$, and one WKB approximation to the original differential equation in (10.1.20) is

$$y_1(x) \sim c_1 \exp\left[-\int_0^x \frac{b(t)}{a(t)}\, dt\right], \qquad \varepsilon \to 0+,$$

where c_1 is a constant which includes the term $e^{S_0/\varepsilon}$. This is the outer solution of boundary layer theory.

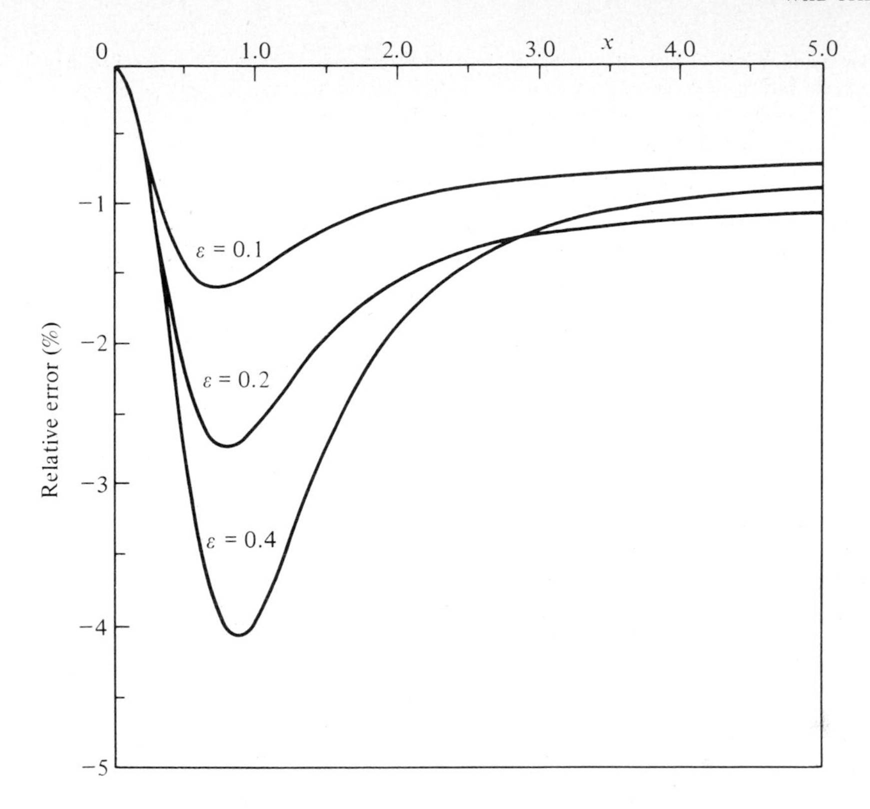

Figure 10.1 A plot of the relative error between the exact solution to the initial-value problem $\varepsilon^2 y'' = (1 + x^2)^2 y$ $[y(0) = 0, y'(0) = 1]$ in (10.1.18) and the leading-order WKB approximation to $y(x)$, $y(x) \sim \varepsilon(1 + x^2)^{-1/2} \sinh [(x + x^2/3)/\varepsilon]$ $(\varepsilon \to 0)$, in (10.1.19) for three values of ε. The relative error is defined as (WKB approximation − exact solution)/(exact solution).

When $S_0' = -a$, (10.1.23) becomes $aS_1' + a' = b$. Thus, $S_1 = -\ln a + \int_0^x [b(t)/a(t)]\, dt$, and another WKB approximation to the original differential equation in (10.1.20) is

$$y_2(x) \sim c_2 \frac{1}{a(x)} \exp \left[\int_0^x \frac{b(t)}{a(t)}\, dt - \frac{1}{\varepsilon} \int_0^x a(t)\, dt \right], \qquad \varepsilon \to 0+,$$

where c_2 is another constant. This is the inner solution of boundary-layer theory.

The general solution is a linear combination of y_1 and y_2. We must now impose the boundary conditions to determine c_1 and c_2. The boundary condition at $x = 0$ gives

$$A = c_1 + c_2/a(0) \tag{10.1.24}$$

and the boundary condition at $x = 1$ gives

$$B = c_1 \exp \left[-\int_0^1 \frac{b(t)}{a(t)}\, dt \right], \tag{10.1.25}$$

where we have neglected the exponentially small term containing $\exp [-\varepsilon^{-1} \int_0^1 a(t)\, dt]$. Solving (10.1.24) and (10.1.25) simultaneously gives

$$y(x) \sim B \exp \left[\int_x^1 \frac{b(t)}{a(t)}\, dt \right] + \frac{a(0)}{a(x)} \left[A - B \exp \int_0^1 \frac{b(t)}{a(t)}\, dt \right] \exp \left[\int_0^x \frac{b(t)}{a(t)}\, dt - \frac{1}{\varepsilon} \int_0^x a(t)\, dt \right].$$

This expression simplifies because the second term contributes only when $x = O(\varepsilon)$ $(\varepsilon \to 0+)$; it is negligible for larger values of x. Thus,

$$y(x) \sim B \exp\left[\int_x^1 \frac{b(t)}{a(t)}\,dt\right] + \left[A - B \exp \int_0^1 \frac{b(t)}{a(t)}\,dt\right] e^{-a(0)x/\varepsilon}. \tag{10.1.26}$$

Equation (10.1.26) is precisely the *uniformly* valid lowest-order boundary-layer solution that we obtained in (9.1.13). Notice that (10.1.26) was obtained without ever having to perform an asymptotic match!

Example 5 *WKB analysis of a Sturm-Liouville problem.* We know from our discussion of the Sturm-Liouville eigenvalue problem in Sec. 1.8 that the boundary-value problem

$$y''(x) + EQ(x)y(x) = 0, \qquad Q(x) > 0,\ y(0) = y(\pi) = 0, \tag{10.1.27}$$

has an infinite number of nontrivial solutions: the eigenvalues $E_1, E_2, E_3, \ldots$ are discrete, nondegenerate (eigenvalues associated with different eigenfunctions are unequal), and are all positive real numbers; the nth eigenvalue E_n is associated with the eigenfunction $y_n(x)$; eigenfunctions associated with different eigenvalues are orthogonal with respect to the weight function $Q(x)$:

$$\int_0^\pi dx\ y_n(x)y_m(x)Q(x) = 0, \qquad n \neq m. \tag{10.1.28}$$

This orthogonality property is easy to prove using integration by parts twice. See Prob. 10.7(b).

Since the differential equation and boundary conditions in (10.1.27) are homogeneous, the eigenfunctions $\{y_n\}$ are determined only up to an arbitrary multiplicative constant. It is conventional to choose the normalization of y_n so that

$$\int_0^\pi [y_n(x)]^2 Q(x)\,dx = 1. \tag{10.1.29}$$

Now the eigenfunctions form a complete orthonormal set.

WKB theory may be used to find approximate formulas for E_n and $y_n(x)$ when n is large. As the WKB theory itself will verify, E_n is approximately proportional to n^2 as $n \to \infty$; thus, the leading-order WKB approximation to the solution of $\varepsilon y''(x) + Q(x)y(x) = 0$, where $\varepsilon = 1/E_n$, is accurate for large n because $\varepsilon \to 0$ as $n \to \infty$.

The leading-order WKB approximation (10.1.13) to the general solution of $y''(x) + EQ(x)y(x) = 0$ is a linear combination of $Q^{-1/4}(x) \sin [\sqrt{E} \int_0^x \sqrt{Q(t)}\,dt]$ and $Q^{-1/4}(x) \cos [\sqrt{E} \int_0^x \sqrt{Q(t)}\,dt]$. (Recall that all the eigenvalues E are positive real numbers; also, we fix the sign of $\sqrt{E}$ to be positive.) The boundary condition $y(0) = 0$ implies that

$$y(x) \sim CQ^{-1/4}(x) \sin\left[\sqrt{E} \int_0^x \sqrt{Q(t)}\,dt\right], \qquad E \to \infty, \tag{10.1.30}$$

where C is an arbitrary normalization constant.

The boundary condition $y(\pi) = 0$ determines the eigenvalues

$$E_n \sim \left[\frac{n\pi}{\int_0^\pi \sqrt{Q(t)}\,dt}\right]^2, \qquad n \to \infty. \tag{10.1.31}$$

Next we determine the eigenfunctions. The normalization integral in (10.1.29) fixes C in (10.1.30); substituting (10.1.30) into (10.1.29) gives

$$1 \sim \int_0^\pi dx\ Q(x)C_n^2 \frac{1}{\sqrt{Q(x)}} \sin^2\left[\sqrt{E_n} \int_0^x dt\ \sqrt{Q(t)}\right], \qquad n \to \infty.$$

The change of variable $u = \sqrt{E_n} \int_0^x dt \sqrt{Q(t)}$ gives $1 \sim (C_n^2/\sqrt{E_n}) \int_0^{n\pi} du \sin^2 u$ $(n \to \infty)$, whence

$$C_n^2 \sim \frac{2}{\int_0^\pi \sqrt{Q(t)}\,dt}, \qquad n \to \infty. \tag{10.1.32}$$

Thus, the eigenfunctions are

$$y_n(x) \sim \left(\int_0^\pi \frac{\sqrt{Q(t)}}{2} dt \right)^{-1/2} Q^{-1/4}(x) \sin \left[n\pi \frac{\int_0^x \sqrt{Q(t)}\, dt}{\int_0^\pi \sqrt{Q(t)}\, dt} \right], \qquad n \to \infty. \tag{10.1.33}$$

Note that if $Q(x) \equiv 1$, then the right side of (10.1.33) reduces to $\sqrt{2/\pi} \sin(nx)$, which is the exact solution to $y'' + y = 0$ $[y(0) = y(\pi) = 0]$.

To demonstrate the accuracy of our results, we choose $Q(x) = (x + \pi)^4$. Then the approximate eigenvalues and eigenfunctions are given by

$$E_n \sim \frac{9n^2}{49\pi^4}, \qquad n \to \infty, \tag{10.1.34}$$

and

$$y_n(x) \sim \sqrt{\frac{6}{7\pi^3}} \frac{\sin[n(x^3 + 3x^2\pi + 3\pi^2 x)/7\pi^2]}{(\pi + x)}, \qquad n \to \infty. \tag{10.1.35}$$

We have checked these results numerically by computer. The comparisons between the approximate analytical and the computer solutions are given in Table 10.1 and Figs. 10.2 and 10.3.

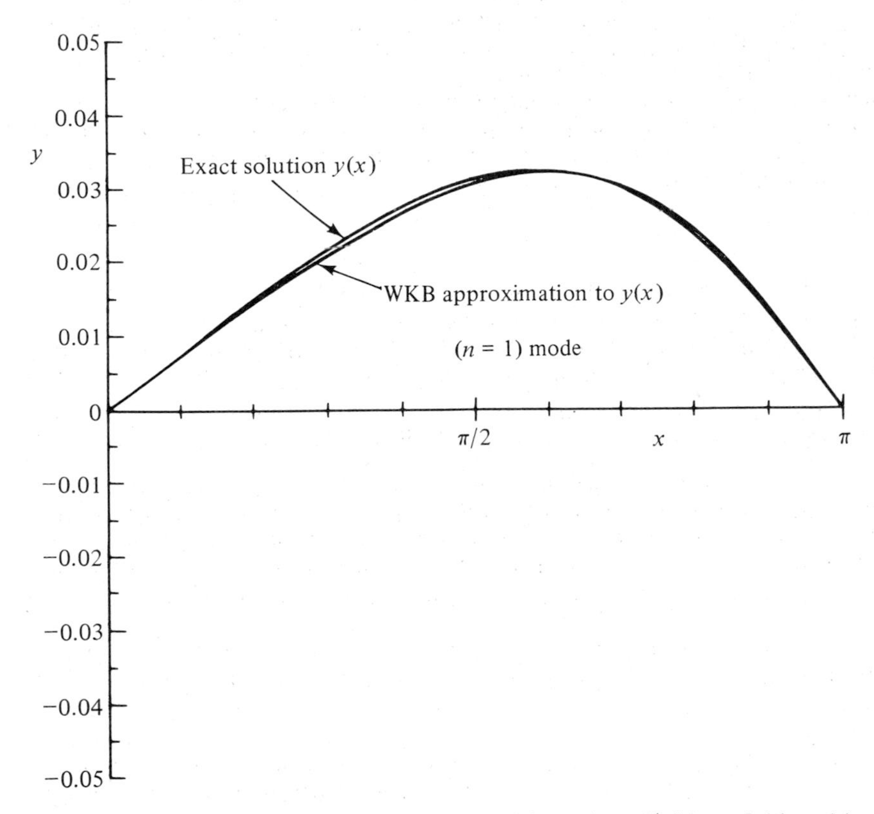

Figure 10.2 Comparison of the exact solution to $y''(x) + E_n(x + \pi)^4 y(x) = 0$ $[y(0) = y(\pi) = 0]$, with the WKB approximation to this solution as given in (10.1.35) for the lowest ($n = 1$) mode. Although WKB becomes exact as $n \to \infty$, this plot shows that even when $n = 1$ the WKB approximation is extraordinarily accurate.

Table 10.1 A comparison of the exact eigenvalues E_n of the Sturm-Liouville problem $y''(x) + E(x+\pi)^4 y(x) = 0$ $[y(0) = y(\pi) = 0]$ with the leading-order WKB prediction [see (10.1.34)] for these eigenvalues $E_n \sim 9n^2/49\pi^2$ $(n \to \infty)$

As expected, this prediction becomes more accurate as n increases. The relative error is defined as (approximate − exact)/(exact)

n	E_n(WKB)	E_n(exact)	Relative error, %
1	0.00188559	0.00174401	8.1
2	0.00754235	0.00734865	2.6
3	0.0169703	0.0167524	1.3
4	0.0301694	0.0299383	0.77
5	0.0471397	0.0469006	0.51
10	0.188559	0.188305	0.13
20	0.754235	0.753977	0.035
40	3.01694	3.01668	0.009

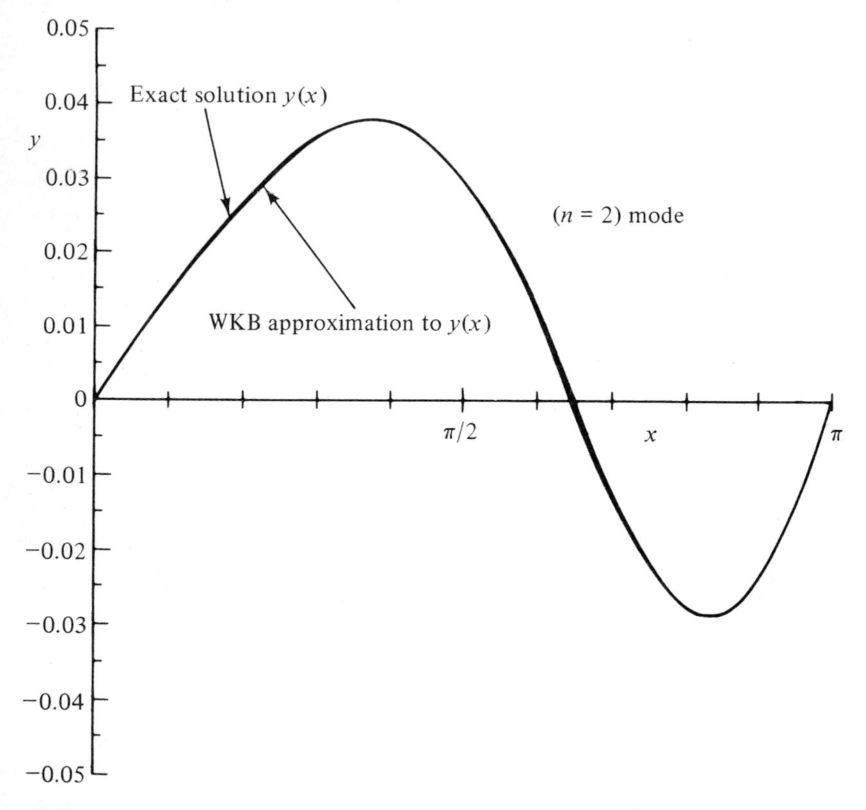

Figure 10.3 Same as in Fig. 10.2 except that $n = 2$. The exact eigenfunction and the WKB approximation are almost indistinguishable.

The preceding examples demonstrate the power and efficiency of the exponential approximation (10.1.4). However, we have already seen that rapidly varying exponentials appear naturally when one attempts to approximate solutions to linear equations. In Chap. 3 the exponential approximation was found to be useful for finding the leading behaviors of solutions near irregular singular points [see (3.4.6)]. In Chap. 6 the rapid variation of exponentials of the form (10.1.3) led to the principal ideas of approximation methods for integrals such as Laplace's method, the method of stationary phase, and the method of steepest descents. Exponentials also appear in the equations of boundary-layer theory. Thus, we are not terribly surprised when exponentials resurface in the context of WKB theory as the basis of the WKB approximation.

(E) 10.2 CONDITIONS FOR VALIDITY OF THE WKB APPROXIMATION

WKB theory is a singular perturbation theory because it is used to solve a differential equation whose highest derivative is multiplied by a small parameter (when the small parameter vanishes, the order of the differential equation changes abruptly). The singular nature of WKB theory is clearly evident in the $1/\delta$ term in the exponential approximation (10.1.4):

$$y(x) \sim \exp\left[\frac{1}{\delta}\sum_{n=0}^{\infty}\delta^n S_n(x)\right], \qquad \delta \to 0. \tag{10.2.1}$$

Unless $S_0(x) \equiv 0$, the approximation ceases to exist when $\delta = 0$. The singular nature of this approximation also surfaces in a more subtle way—the WKB series $\sum \delta^n S_n$ usually diverges. (The series converges if it truncates, but this is rare. See Prob. 10.2.) This is why we use the asymptotic notation $\sim$ rather than $=$. Nevertheless, even though the WKB series diverges, we know from the numerical examples in Sec. 10.1 that it can give an extremely accurate approximation to $y(x)$.

This section develops criteria for predicting when the WKB approximation will be useful. These criteria are quantitative; i.e., they specify how small δ must be for the WKB series in (10.2.1) to approximate $y(x)$ to some prescribed relative error.

In order that the WKB approximation (10.2.1) be valid on an interval, it is necessary that the series $\sum \delta^{n-1} S_n(x)$ be an asymptotic series in δ as $\delta \to 0$ *uniformly for all x on the interval.* This requires that the asymptotic relations

$$\begin{aligned} S_1(x) &\ll \frac{1}{\delta} S_0(x), && \delta \to 0, \\ \delta S_2(x) &\ll S_1(x), && \delta \to 0, \\ &\vdots \\ \delta^n S_{n+1}(x) &\ll \delta^{n-1} S_n(x), && \delta \to 0, \end{aligned} \tag{10.2.2}$$

hold uniformly in x. These conditions are equivalent to the requirement that each of the functions $S_{n+1}(x)/S_n(x)$ ($n = 0, 1, 2, 3, \ldots$) be bounded functions of x on the interval (although these bounds may be arbitrary functions of n). If the series $\sum \delta^{n-1} S_n(x)$ is *uniformly* asymptotic in x as $\delta \to 0$, the optimal truncation rule suggests that truncating the series before the smallest term $\delta^N S_{N+1}(x)$ gives an approximation to $\ln y$ with uniformly small relative error throughout the x interval.

However, because the WKB series appears in the *exponent* in (10.2.1), the asymptotic conditions in (10.2.2) are not sufficient to ensure that $\exp\,[\sum \delta^{n-1} S_n(x)]$ will be a good approximation to $y(x)$. For the WKB series truncated at the term $\delta^{N-1} S_N(x)$ to be a good approximation to $y(x)$, the next term must be small compared with 1 for all x in the interval of approximation:

$$\delta^N S_{N+1}(x) \ll 1, \qquad \delta \to 0. \tag{10.2.3}$$

If this relation holds, then $\exp\,[\delta^N S_{N+1}(x)] = 1 + O[\delta^N S_{N+1}(x)]$ ($\delta \to 0$). Thus, the relative error between $y(x)$ and the WKB approximation is small:

$$\frac{y(x) - \exp\,[1/\delta \sum_{n=0}^{N} \delta^n S_n(x)]}{y(x)} \sim \delta^N S_{N+1}(x), \qquad \delta \to 0.$$

Both conditions (10.2.2) and (10.2.3) must be satisfied for WKB to be useful.

Geometrical and Physical Optics

If we retain only the first term in the WKB series, we are making the approximation of *geometrical optics*: $e^{S_0(x)/\delta}$. However, while this expression may faithfully reflect the structure of $y(x)$, it does *not* constitute an asymptotic approximation to $y(x)$ because $S_1(x)$, the next term in the WKB series, is not small compared with 1 as $\delta \to 0$ (it does not depend on δ) and condition (10.2.3) is not satisfied.

The first two terms in the WKB series constitute the approximation of *physical optics*:

$$y(x) \sim e^{S_0(x)/\delta + S_1(x)}, \qquad \delta \to 0. \tag{10.2.4}$$

The relative error between y and the approximation of physical optics is of order $\delta S_2(x)$, which vanishes uniformly with δ if $S_2(x)$ is bounded. Thus, the approximation in (10.2.4) is an asymptotic relation. For example, if it is required that the physical-optics approximation be accurate to a relative error of 2 percent on an interval, we must then choose δ so small that $|\delta S_2(x)| \le 0.02$ for all x on that interval.

Usually, the approximation of physical optics expresses the leading asymptotic behavior of $y(x)$ while the approximation of geometrical optics contains just the controlling factor (the most rapidly varying component) of the leading behavior.

Example 1 *Behavior of Airy functions as $x \to +\infty$.* The Airy equation $y'' = xy$ is a Schrödinger equation with $Q(x) = x$ and $\varepsilon = 1$. Thus, from (10.1.11), (10.1.12), and (10.1.14) we have $S_0 = \pm\frac{2}{3}x^{3/2}$, $S_1 = -\frac{1}{4}\ln x$, $S_2 = \pm\frac{5}{48}x^{-3/2}$. We observe that even when $\varepsilon = 1$, the asymptotic inequalities $\varepsilon S_2 \ll S_1 \ll S_0/\varepsilon$, $\varepsilon S_2 \ll 1$ ($x \to +\infty$) hold. We conclude that for fixed ε the physical-optics approximation is valid as $x \to +\infty$. Indeed, we have just rederived the leading behaviors of solutions to the Airy equation as $x \to +\infty$ as well as the first correction to the leading behaviors [see (3.5.21)]:

$$y(x) \sim c_\pm x^{-1/4} e^{\pm 2x^{3/2}/3} \left(1 \pm \tfrac{5}{48} x^{-3/2}\right),$$

where $c_\pm$ is a constant. Note that the rapidly varying exponential factors $e^{\pm 2x^{3/2}/3}$ come from the geometrical-optics approximation.

Example 2 *Behavior of parabolic cylinder functions as* $x \to +\infty$. WKB theory also gives the large-x behavior of the solutions to the parabolic cylinder equation $y'' = (\frac{1}{4}x^2 - \nu - \frac{1}{2})y$. Here, again, even though $\varepsilon = 1$, the physical-optics approximation is valid as $x \to +\infty$ because $S_1 \ll S_0/\varepsilon$, $\varepsilon S_2 \ll S_1$, $\varepsilon S_2 \ll 1$ $(x \to +\infty)$.

The components of the physical-optics approximation are calculated from $Q(x) = \frac{1}{4}x^2 - \nu - \frac{1}{2}$ as follows:

$$S_0 = \pm \int^x \sqrt{Q(t)}\, dt \sim \pm \int^x \frac{t}{2}\left(1 - \frac{2\nu + 1}{t^2}\right) dt \sim \pm \left[\frac{x^2}{4} - \left(\nu + \frac{1}{2}\right) \ln x\right], \qquad x \to +\infty,$$

where we have used the binomial expansion and $S_1 = -\frac{1}{4} \ln Q \sim -\frac{1}{4} \ln (\frac{1}{4}x^2)$ $(x \to +\infty)$. Hence, the physical-optics approximation gives

$$y \sim c_\pm e^{S_0 + S_1} \sim \begin{cases} c_+ x^{-\nu-1} e^{x^2/4}, & x \to +\infty, \\ c_- x^{\nu} e^{-x^2/4}, & \end{cases}$$

which are the leading behaviors of solutions to the parabolic cylinder equation [see (3.5.12)]. Again, the controlling factor $e^{\pm x^2/4}$ arises from the geometrical-optics approximation.

Example 3 *Accuracy of physical optics.* How small must we make ε to be sure that the physical-optics approximation to the exponentially decaying solution of $\varepsilon^2 y'' = \sqrt{x}\, y$ is accurate to 5 percent when $x \geq 1$?

The physical-optics approximation to $y(x)$ is $y(x) \sim c x^{-1/8} e^{-4x^{5/4}/5\varepsilon}$ $(\varepsilon \to 0)$. The relative error between this asymptotic approximation and the exact solution is εS_2, which from (10.1.14) is $|\varepsilon S_2| = 9\varepsilon x^{-5/4}/160$. This equation shows that we must choose $\varepsilon \leq 0.9$ to make $|\varepsilon S_2| \leq 5$ percent for all $x > 1$.

Example 4 *Violation of criteria for validity of WKB.* Is it valid to use WKB theory to predict the large-x behavior of the solutions to

$$y''(x) = \left(\frac{\ln x}{x}\right)^2 y(x)? \tag{10.2.5}$$

First, we determine the behavior of $y(x)$ for large x without using WKB. We transform (10.2.5) by letting $s = \ln x$:

$$\frac{d^2}{ds^2} y(s) - \frac{d}{ds} y(s) = s^2 y(s).$$

The substitution $y(s) = e^{s/2} z(s)$ eliminates the dy/ds term: $(d^2/ds^2)z(s) = (s^2 + \frac{1}{4})z(s)$. Finally, the change of variable $t = \sqrt{2}\, s$ gives a parabolic cylinder equation

$$\frac{d^2}{dt^2} z(t) = \left(\frac{t^2}{4} + \frac{1}{8}\right) z(t)$$

whose solutions are $D_{-5/8}(t)$ and $D_{-5/8}(-t)$. When t is large and positive, the behavior of parabolic cylinder functions is given by [see (3.5.14)]:

$$D_\nu(t) \sim t^\nu e^{-t^2/4} \left[1 - \frac{\nu(\nu - 1)}{2^1 1!\, t^2} + \cdots\right], \qquad t \to \infty.$$

$$D_\nu(-t) \sim \frac{\sqrt{2\pi}}{\Gamma(-\nu)} t^{-\nu-1} e^{t^2/4} \left[1 + \frac{(\nu + 1)(\nu + 2)}{2^1 1!\, t^2} + \cdots\right], \qquad t \to \infty.$$

Thus, the behavior of $y(x)$ for large x is

$$y(x) \sim c_+ \sqrt{x}\,(\ln x)^{-3/8} e^{(\ln x)^2/2}[1 - \tfrac{15}{256}(\ln x)^{-2} + \cdots], \qquad x \to +\infty, \tag{10.2.6a}$$

or

$$y(x) \sim c_- \sqrt{x}\,(\ln x)^{-5/8} e^{-(\ln x)^2/2}[1 - \tfrac{65}{256}(\ln x)^{-2} + \cdots], \qquad x \to +\infty. \tag{10.2.6b}$$

Now we examine the predictions of WKB. From (10.2.5) we see that $Q(x) = (\ln x)^2 x^{-2}$. Thus,

$$S_0 = \pm \int^x \sqrt{Q(t)}\, dt = \pm \tfrac{1}{2}(\ln x)^2,$$

$$S_1 = -\tfrac{1}{4} \ln Q(x) = \tfrac{1}{2} \ln x - \tfrac{1}{2} \ln (\ln x),$$

$$S_2 = \pm \int^x \left(\frac{Q''}{8Q^{3/2}} - \frac{5Q'^2}{32Q^{5/2}} \right) dt = \pm \tfrac{1}{8} \ln (\ln x) \pm \tfrac{3}{16} (\ln x)^{-2}.$$

From these formulas, it is clear that $S_2 \ll S_1 \ll S_0$ $(x \to \infty)$, but that $1 \ll S_2$ $(x \to \infty)$. Hence, the condition in (10.2.2) is satisfied, but that in (10.2.3) with $N = 1$ is violated. Thus, while the geometrical-optics approximation e^{S_0} gives the correct controlling factors in (10.2.6), we are not surprised that the physical-optics approximation $e^{S_0+S_1}$ gives the wrong leading behavior.

Next, let us calculate S_3. From (10.1.15) we have

$$S_3(x) = -\frac{Q''}{16Q^2} + \frac{5Q'^2}{64Q^3} = \frac{3}{16}(\ln x)^{-4} - \frac{1}{16}(\ln x)^{-2}. \tag{10.2.7}$$

Here, we are gratified to find that $S_3 \ll S_2$, $S_3 \ll 1$ $(x \to +\infty)$. Thus, we expect the leading behaviors in (10.2.6) to be given by the first *three* terms in the WKB series: $y(x) \sim e^{S_0+S_1+S_2}$ $(x \to +\infty)$. Indeed, we find that

$$e^{S_0+S_1+S_2} = c_\pm\, e^{\pm(\ln x)^2/2} x^{1/2} (\ln x)^{-1/2 \pm 1/8}.$$

In Prob. 10.10 you are asked to verify that

$$e^{S_0+S_1+S_2+S_3+S_4} \tag{10.2.8}$$

reproduces the asymptotic formulas in (10.2.6).

We emphasize that in this example we have taken $\varepsilon = 1$. If we apply WKB analysis to the equation

$$\varepsilon^2 y''(x) = [(\ln x)/x]^2 y(x)$$

then the physical-optics approximation $e^{S_0/\varepsilon + S_1}$ is the leading asymptotic approximation to $y(x)$ as $\varepsilon \to 0$ for each fixed $x > 0$. However, this physical-optics approximation is not uniformly asymptotic to $y(x)$ as $\varepsilon \to 0$ for all $x > 0$. To obtain the leading behavior of $y(x)$ as $x \to \infty$ for fixed ε it is necessary to use $e^{S_0/\varepsilon + S_1 + \varepsilon S_2}$. See Prob. 10.10(*b*).

Physical Optics for Higher-Order Equations

WKB analysis is not sensitive to the order of a differential equation. It is very easy to show (see Prob. 10.11) that for the nth-order equation

$$\varepsilon^n \frac{d^n}{dx^n} y(x) = Q(x) y, \tag{10.2.9}$$

the WKB approximation has the form $e^{(S_0 + \varepsilon S_1 + \cdots)/\varepsilon}$, where

$$S_0 = \omega \int^x [Q(t)]^{1/n}\, dt, \qquad \omega^n = 1, \tag{10.2.10}$$

and
$$S_1 = \frac{1-n}{2n} \ln Q. \tag{10.2.11}$$

Compare this result with that in (3.4.28). See Prob. 10.11(*b*).

Example 5 *Behavior of solutions to $d^4y/dx^4 = xy$ for large x.* For the hyperairy equation of order 4, $d^4y/dx^4 = xy$, $n = 4$, $\varepsilon = 1$, and $Q = x$. The physical-optics approximation is

$$e^{(S_0+S_1)/\varepsilon} = cQ^{-3/8} \exp\left(\omega \int^x Q^{1/4}\, dt\right) = cx^{-3/8} \exp\left(\omega \frac{4}{5} x^{5/4}\right),$$

where c is a constant and $\omega = \pm 1$ or $\pm i$, which agrees with the leading behavior in (3.5.23).

Turning Points

Equations (10.2.10) and (10.2.11) show that the condition (10.2.2) for the validity of the WKB series on an interval is violated if $Q(x)$ vanishes on that interval. Specifically, the asymptotic relation $S_0(x)/\varepsilon \gg S_1(x) = [(1-n)/2n] \ln Q(x)$ $(\varepsilon \to 0)$ breaks down at a zero of Q because S_1 becomes singular. Points where Q vanishes are called *turning points.*

The expression "turning point" comes from the Schrödinger equation which describes a quantum mechanical particle in a potential $V(x)$:

$$\left(-\varepsilon^2 \frac{d^2}{dx^2} + V(x) - E\right) y(x) = 0.$$

$V(x)$ is the potential energy of the particle and E is the total energy of the particle. For this equation $Q(x) = V(x) - E$, so $Q(x)$ vanishes at points where $V(x) = E$. The classical orbit of a particle in the potential $V(x)$ is confined to regions where $V(x) \le E$. The particle moves until it reaches a point where $V = E$ and then it stops, turns around, and moves off in the opposite direction.

The physical-optics approximation is clearly invalid at a turning point because

$$e^{S_0/\varepsilon + S_1} = Q^{(1-n)/2n} \exp\left(\frac{\omega}{\varepsilon} \int^x Q^{1/n}\, dt\right)$$

is infinite at such points. On the other hand, the theory of linear differential equations asserts that if $Q(x)$ is analytic, then the exact solutions of the differential equation (10.2.9) are regular! We resolve this puzzle and use asymptotic matching to construct approximate solutions of differential equations with turning points in Sec. 10.4.

(E) 10.3 PATCHED ASYMPTOTIC APPROXIMATIONS: WKB SOLUTION OF INHOMOGENEOUS LINEAR EQUATIONS

From the discussion of Sec. 10.1 one might conclude that WKB theory is useful only for homogeneous linear equations; it would seem that unless a differential equation is homogeneous, it would not be possible to divide off the exponential

WKB series to obtain equations for $S_0, S_1, S_2, \ldots$. However, there is no real difficulty with an inhomogeneous equation because one can solve the associated homogeneous equations in terms of WKB approximations and then use the method of variation of parameters (see Sec. 1.5) to generate the solution of the full inhomogeneous equation. The only possible drawback with this procedure is that imposing the boundary conditions can involve evaluating cumbersome integrals. In this section we propose a simple and general method for solving the inhomogeneous Schrödinger equation in terms of a Green's function which neatly incorporates the boundary conditions at an early stage of the problem. (Green's functions are discussed in Sec. 1.5.)

We will solve the general inhomogeneous Schrödinger equation of the form

$$\varepsilon^2 y'' - Q(x)y(x) + R(x) = 0, \tag{10.3.1}$$

subject to the homogeneous boundary conditions $y(\pm\infty) = 0$. We will assume that $Q(x) > 0$ for all x (so that the homogeneous differential equation has no turning points) and that $Q(x) \gg x^{-2}$ as $x \to \pm\infty$ so that the physical-optics approximation to the solution of the homogeneous equation is valid for all x. We must also assume that the inhomogeneous term $R(x)$ is such that a solution to (10.3.1) which satisfies $y(\pm\infty) = 0$ actually exists.

One way to analyze this problem is simply to take the limit $\varepsilon \to 0+$ in (10.3.1) to obtain

$$y(x) \sim \frac{R(x)}{Q(x)}, \qquad \varepsilon \to 0+, \tag{10.3.2}$$

which is like the outer limit of boundary-layer theory. In fact, if $R(x)$ and $Q(x)$ are smooth and there are no turning points, then (10.3.2) is valid everywhere. Thus, in order that there exist a solution to (10.3.1) satisfying $y(\pm\infty) = 0$, it is necessary that $R(x) \ll Q(x)$ as $x \to \pm\infty$ (see Prob. 10.12).

The outer behavior (10.3.2) breaks down at points of discontinuity of $R(x)$ and at turning points of $Q(x)$. At such points, the solution $y(x)$ to (10.3.1) is continuous but $R(x)/Q(x)$ is discontinuous. The WKB analysis given below yields a uniformly valid approximation to $y(x)$ even in the neighborhood of discontinuities of $R(x)$. Our WKB analysis can also be extended to obtain a uniform approximation to $y(x)$ when there are turning points (see Prob. 10.24).

Our approach here consists of solving the Green's function equation

$$\varepsilon^2 \frac{\partial^2 G}{\partial x^2}(x, x') - Q(x)G(x, x') = -\delta(x - x'), \qquad G(\pm\infty, x') = 0, \tag{10.3.3}$$

and then using the Green's function $G(x, x')$ to construct the solution to (10.3.1):

$$y(x) = \int_{-\infty}^{\infty} G(x, x')R(x')\,dx'. \tag{10.3.4}$$

If we had the exact solution to (10.3.3), then (10.3.4) would constitute an exact solution to (10.3.1) because it satisfies both the differential equation and the boundary conditions. Lacking this, we propose to use the WKB physical-optics approximation to $G(x, x')$ in place of the exact Green's function. The resulting integral (10.3.4) is then asymptotic to $y(x)$ as $\varepsilon \to 0+$.

The WKB solution of (10.3.3) will require the *patching* of two WKB solutions which are valid in their respective regions. Patching is a *local* procedure because it is done at a single point $x = x'$. By contrast, asymptotic matching, which is used in the next section to construct a solution to the one-turning-point problem, is a global procedure which is performed on an interval whose length becomes infinite as $\varepsilon \to 0+$.

Construction of the Green's Function $G(x, x')$

To solve (10.3.3) we divide the x axis into two regions: region I, where $x > x'$, and region II, where $x < x'$. In each region the differential equation is homogeneous, so the WKB physical-optics approximation may be used. In region I $G \to 0$ as $x \to +\infty$; the WKB approximation to $G(x, x')$ which incorporates this condition is

$$G_{\text{I}}(x, x') = C_1[Q(x)]^{-1/4} \exp\left[-\frac{1}{\varepsilon}\int_{x'}^{x} \sqrt{Q(t)}\, dt\right], \qquad x \to x', \qquad (10.3.5)$$

where C_{I} is a constant and we have chosen the lower limit of integration to lie at x'.

In region II $G \to 0$ as $x \to -\infty$; thus,

$$G_{\text{II}}(x, x') = C_{\text{II}}[Q(x)]^{-1/4} \exp\left[-\frac{1}{\varepsilon}\int_{x}^{x'} \sqrt{Q(t)}\, dt\right], \qquad x \to x', \qquad (10.3.6)$$

where C_{II} is a second constant.

The constants C_{I} and C_{II} are determined by patching. There are two patching conditions. First, at $x = x'$, the boundary of regions I and II, we require that $G(x, x')$ be continuous: $\lim_{\eta \to 0+} [G_{\text{I}}(x' + \eta, x') - G_{\text{II}}(x' - \eta, x')] = 0$. This condition implies that $C_{\text{I}} = C_{\text{II}}$.

The second condition is derived by integrating the differential equation (10.3.3) from $x' - \eta$ to $x' + \eta$ and letting $\eta \to 0+$. We obtain

$$\lim_{\eta \to 0+}\left[\frac{\partial}{\partial x} G_{\text{I}}(x, x')\Big|_{x = x' + \eta} - \frac{\partial}{\partial x} G_{\text{II}}(x, x')\Big|_{x = x' - \eta}\right] = -\frac{1}{\varepsilon^2}.$$

[Normally, the solution to a second-order differential equation has a continuous first derivative, but the delta function in (10.3.3) gives rise to a finite discontinuity in the slope of $G(x, x')$ at $x = x'$ (a cusp).] This condition implies that

$$C_{\text{I}} = C_{\text{II}} = \frac{1}{2\varepsilon}[Q(x')]^{-1/4}.$$

$G_{\text{I}}(x, x')$ and $G_{\text{II}}(x, x')$ are now completely determined and may be combined into a single expression which is a uniformly valid approximation to the solution of (10.3.3) for all x:

$$G_{\text{unif}}(x, x') = \frac{1}{2\varepsilon}[Q(x)Q(x')]^{-1/4} \exp\left[-\frac{1}{\varepsilon}\left|\int_{x'}^{x} \sqrt{Q(t)}\, dt\right|\right]. \qquad (10.3.7)$$

For all x the relative error in this approximation is of order ε.

Example 1 *Comparison between exact and approximate Green's functions.* For $Q(x) = 1 + x^2$ the uniform approximation to $G(x, x')$ in (10.3.7) is astoundingly accurate. Equation (10.3.7) becomes

$$G_{\text{unif}}(x, x') = \frac{\exp\left(-\left|x\sqrt{x^2+1} - x'\sqrt{x'^2+1}\right|/2\varepsilon\right)}{2\varepsilon[(x^2+1)(x'^2+1)]^{1/4}} \left(\frac{x+\sqrt{x^2+1}}{x'+\sqrt{x'^2+1}}\right)^{(x'-x)/(2\varepsilon|x'-x|)} \tag{10.3.8}$$

In Figs. 10.4 and 10.5 we compare $G_{\text{unif}}(x, 0)$ with the exact numerical solution for $\varepsilon = 1$ and two values of x'. Observe how small the error is, even when ε is as large as 1.

Integrals of the Green's Function

The uniform approximation in (10.3.7) to the Green's function may be used to evaluate integrals of the Green's function. For example, to calculate $A = \int_{-\infty}^{\infty} G(x, x')\, dx$ we use integration by parts to determine the leading behav-

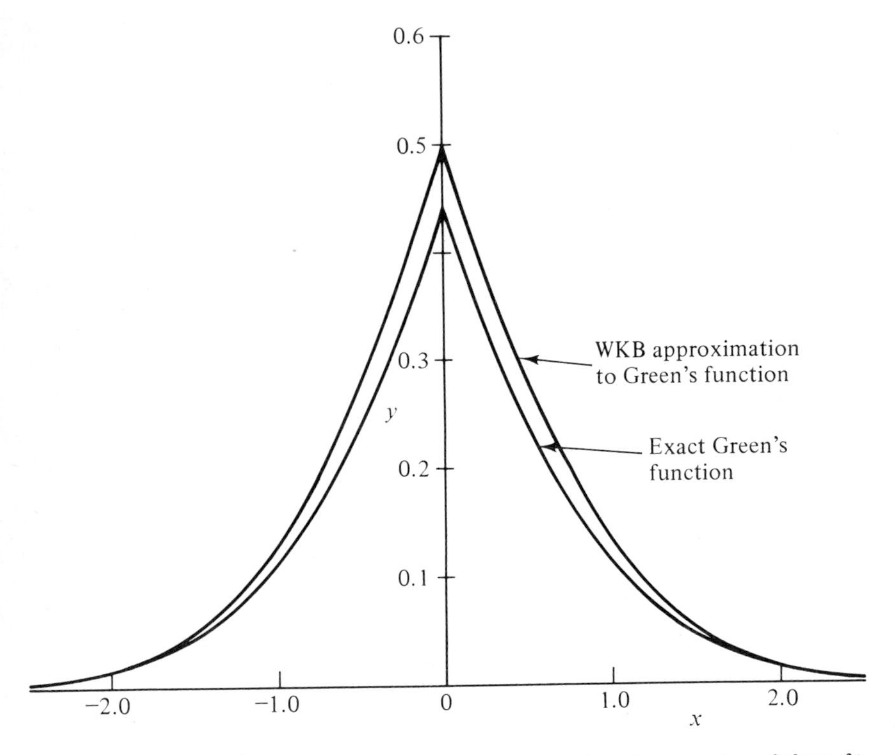

Figure 10.4 Comparison of the exact solution to the Green's function equation, $\varepsilon^2 \partial^2 G/\partial x^2(x, x') - (1 + x^2)G(x, x') = -\delta(x - x')$ $[G(\pm\infty, x') = 0]$, with $x' = 0$, $\varepsilon = 1$, with the WKB physical-optics approximation to $G(x, x')$ in (10.3.8). Observe that the error is greatest at $x = 0$, the point at which the exponent in (10.3.8) is smallest. The true value of $G(0, 0)$ is 0.443 11 . . . , while the WKB formula in (10.3.8) predicts that $G(0, 0) = (2\varepsilon)^{-1} = 0.5$. Thus, the WKB formula has a maximum error of about 5 percent.

ior of this integral. The contribution from region I is given by

$$A_{\mathrm{I}} = \left[\frac{Q(x')}{2\varepsilon}\right]^{-1/4} \int_{x'}^{\infty} dx[Q(x)]^{-1/4} \exp\left[-\frac{1}{\varepsilon}\int_{x'}^{x} \sqrt{Q(t)}\, dt\right]$$

$$= \left[\frac{Q(x')}{-2}\right]^{-1/4} \int_{x'}^{\infty} dx[Q(x)]^{-3/4} \frac{d}{dx} \exp\left[-\frac{1}{\varepsilon}\int_{x'}^{x} \sqrt{Q(t)}\, dt\right]$$

$$= \frac{1}{2Q(x')} + \mathrm{O}(\varepsilon), \qquad \varepsilon \to 0+.$$

The contribution from region II is identical. Thus, we obtain the simple result

$$\int_{-\infty}^{\infty} G(x, x')\, dx = \frac{1}{Q(x')} + \mathrm{O}(\varepsilon), \qquad \varepsilon \to 0+. \tag{10.3.9}$$

This result also follows from (10.3.4) with $R(x) = 1$ since (10.3.2) implies that the solution to $\varepsilon y'' = Q(x)y - 1$ satisfies $y \sim 1/Q(x)$ $(\varepsilon \to 0+)$.

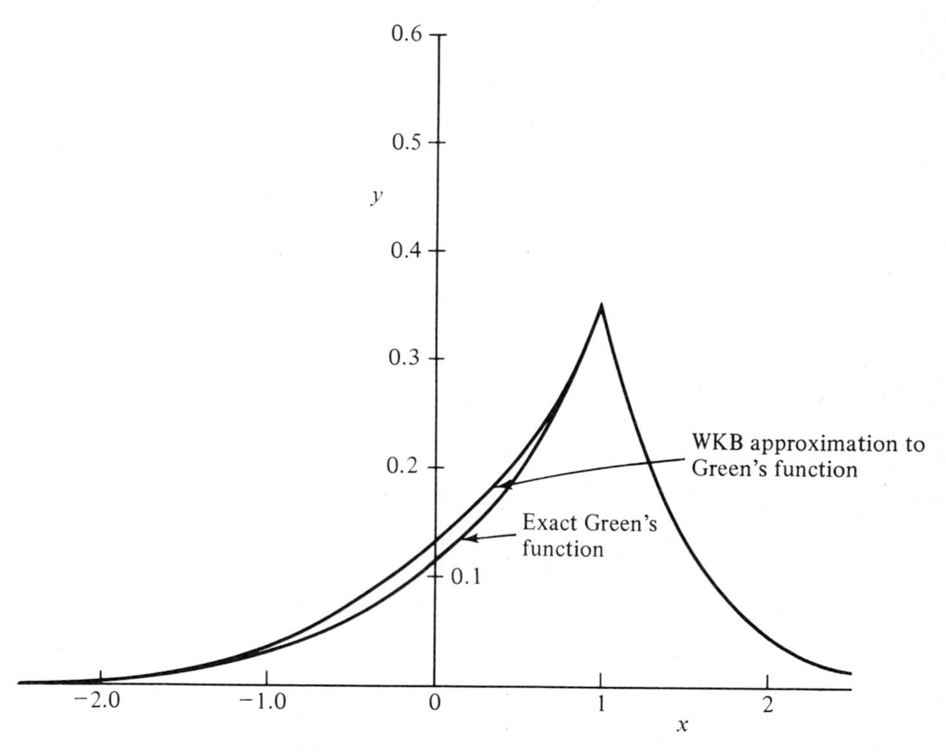

Figure 10.5 Same as in Fig. 10.4 except that $x' = 1$ instead of 0; ε is still 1. Again, the error is greatest at $x = 0$. The WKB formula in (10.3.8) predicts that when $\varepsilon = 1$, $G(1, 1) = \frac{1}{4}\sqrt{2} \doteq 0.35355$. The true value of $G(1, 1)$ is 0.349 13. ...

Example 2 *Comparison between exact and approximate integrals.* If we choose $Q(x) = 1 + x^2$ and $\varepsilon = 1$ as we did in Example 1, we conclude from (10.3.9) that $A = \int_{-\infty}^{\infty} G(x, x')\, dx = 1$ when $x' = 0$ and $A = 0.5$ when $x' = 1$. Numerical integration of the Green's function differential equation (see Figs. 10.4 and 10.5) gives the true values of A: $A = 0.623\,23$ for $x' = 0$ and $A = 0.466\,51$ for $x' = 1$. The errors are quite small considering the large size of ε.

It is just as easy to evaluate integrals of powers of G_{unif} in (10.3.7). For example (see Prob. 10.13),

$$\int_{-\infty}^{\infty} [G_{\text{unif}}(x, x')]^N\, dx = \frac{(2\varepsilon)^{1-N}}{N} [Q(x')]^{-(N+1)/2} + O(\varepsilon^{2-N}), \qquad \varepsilon \to 0+. \tag{10.3.10}$$

To solve the inhomogeneous Schrödinger equation (10.3.1), one need only apply these same integration methods to evaluate the integral in (10.3.4). At points of continuity of $R(x)$, asymptotic analysis of (10.3.4) as $\varepsilon \to 0+$ yields (10.3.2) (see Prob. 10.16).

Example 3 *Uniform approximation to the Schrödinger equation with discontinuous inhomogeneity.* If we choose $R(x)$ in (10.3.1) to be the step function

$$R(x) = \begin{cases} 1, & |x| \le 1, \\ 0, & |x| > 1, \end{cases}$$

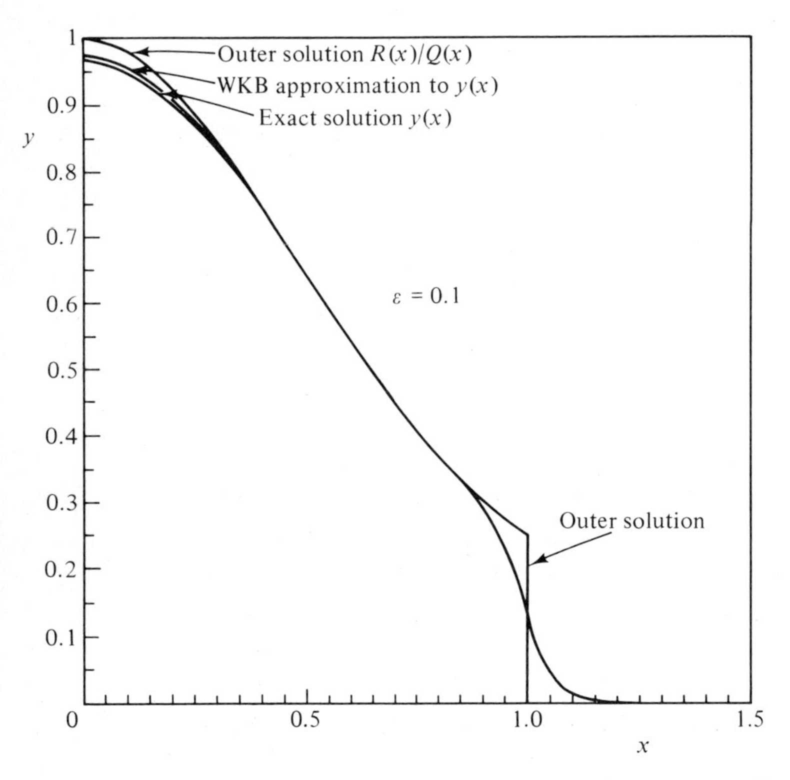

Figure 10.6 A comparison of the exact solution to (10.3.12) for $\varepsilon = 0.1$ with the leading-order WKB approximation in (10.3.11). Also plotted is the outer solution $R(x)/Q(x)$, which cuts off abruptly at $x = 1$. Observe that the WKB approximation is uniform and is especially good near the discontinuity in $R(x)$ at $x = 1$, where it is indistinguishable from the exact solution.

then we obtain $y(x) \sim \int_{-1}^{1} G_{\text{unif}}(x, x')\, dx'$ $(\varepsilon \to 0+)$. This is a nontrivial result whose accuracy we examine numerically. If we take $Q(x) = (1 + x^2)^2$, then we have a uniform approximation to $y(x)$ for all x:

$$y(x) \sim \frac{1}{2\varepsilon\sqrt{x^2+1}} \int_{-1}^{1} \frac{dx'}{\sqrt{x'^2+1}} \exp\left[-\frac{1}{\varepsilon}\left|x^3/3 + x - x'^3/3 - x'\right|\right], \qquad \varepsilon \to 0+. \tag{10.3.11}$$

Figures 10.6 and 10.7 compare the WKB prediction in (10.3.11) with the exact solution to the inhomogeneous Schrödinger equation

$$\varepsilon^2 y(x) - (1 + x^2)^2 y(x) + \begin{Bmatrix} 1, & |x| \le 1 \\ 0, & |x| > 1 \end{Bmatrix} = 0, \qquad y(\pm\infty) = 0. \tag{10.3.12}$$

In these figures, we also plot the outer approximation $R(x)/Q(x)$. Observe that this outer approximation is not uniformly valid in the neighborhood of the discontinuities of $R(x)$ at $x = \pm 1$.

A uniform approximation to $y(x)$ can also be derived using boundary-layer theory. In this case the outer solution is given by $R(x)/Q(x)$ and the inner approximation is valid at discontinuities of $R(x)$. (See Prob. 10.16.)

Example 4 *Uniform approximation to the Schrödinger equation with singular inhomogeneity.* If we choose $R(x)$ to be the singular function

$$R(x) = \begin{cases} (1 - x^2)^{-1/2}, & |x| < 1, \\ 0, & |x| \ge 1, \end{cases}$$

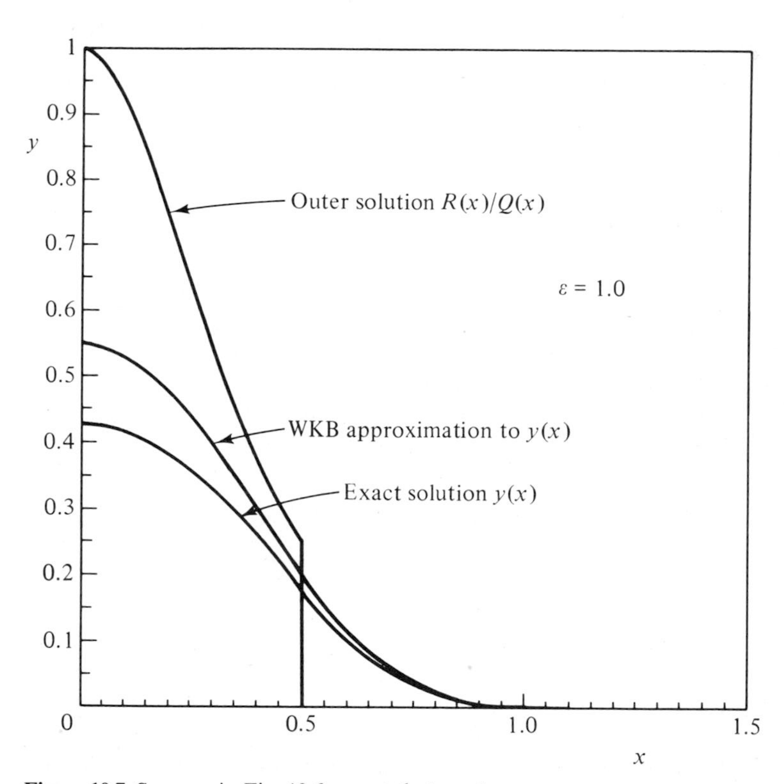

Figure 10.7 Same as in Fig. 10.6 except that $\varepsilon = 1$.

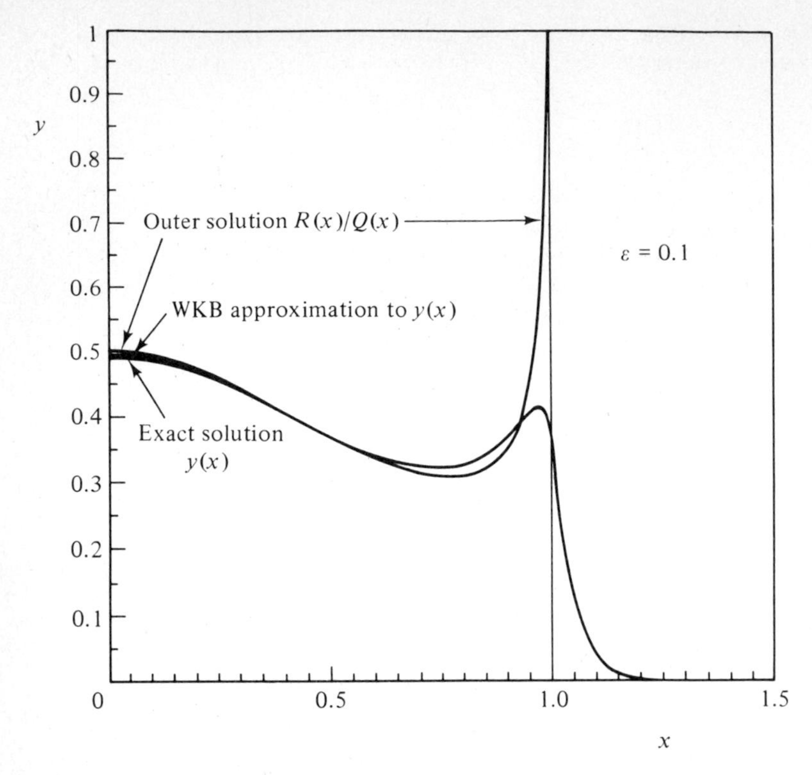

Figure 10.8 Same as in Fig. 10.6 with $\varepsilon = 0.1$ except that $R(x)$ is the singular function $(1 - x^2)^{-1/2}$ $(x < 1)$. The outer solution $R(x)/Q(x)$ becomes singular at $x = 1$, but the WKB approximation to $y(x)$ in (10.3.13) is accurate for all values of x.

and $Q(x) = (1 + x^2)^2$ as in Example 3, then a uniform approximation to $y(x)$ for all x is given by

$$y(x) \sim \frac{1}{2\varepsilon\sqrt{x^2+1}} \int_{-1}^{1} \frac{dx'}{\sqrt{1 - x'^4}} \exp\left[-\frac{1}{\varepsilon}\left|x^3/3 + x - x'^3/3 - x'\right|\right], \qquad \varepsilon \to 0+. \tag{10.3.13}$$

In Figs. 10.8 and 10.9 this WKB prediction is compared with the exact solution $y(x)$ to (10.3.1).

(I) 10.4 MATCHED ASYMPTOTIC APPROXIMATIONS: SOLUTION OF THE ONE-TURNING-POINT PROBLEM

We saw in Sec. 10.2 that the WKB exponential approximation for the Schrödinger equation is not valid in the neighborhood of a turning point. In fact, the physical-optics approximation in (10.1.13) is singular at a turning point. Nevertheless, we will see that there is a general procedure, which is based on the method of matched asymptotic expansions, for constructing a global approximation to the solution of a differential equation having turning points. The approach is very similar to that

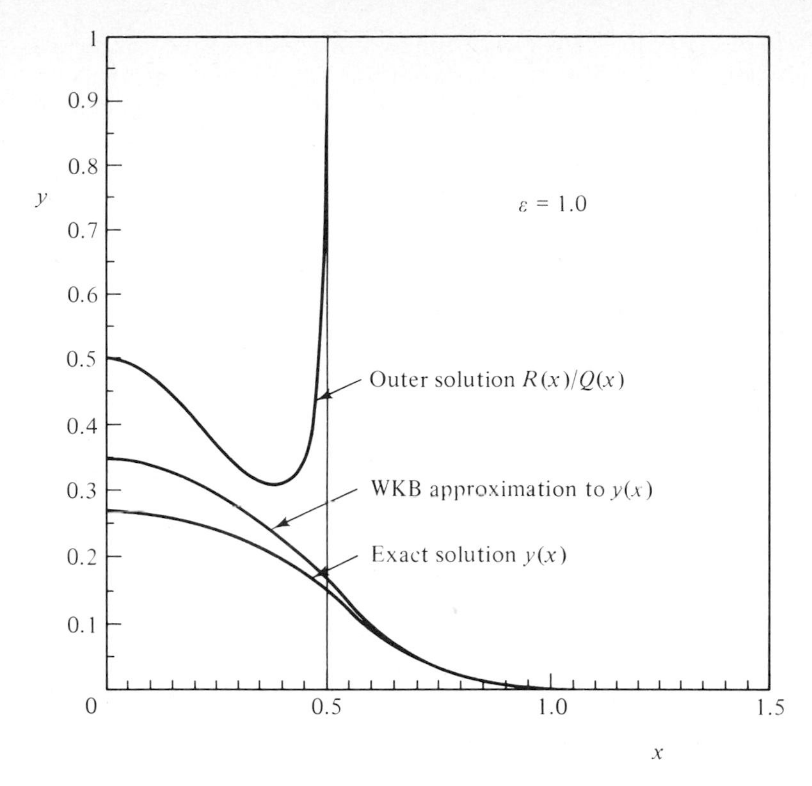

Figure 10.9 Same as in Fig. 10.8 except that $\varepsilon = 1$. Even for this large value of ε, the leading-order WKB approximation in (10.3.13) is a good estimate of $y(x)$.

used in boundary-layer theory. It consists of joining together various WKB approximations which hold in their respective regions of validity.

In this section we begin rather modestly by considering a differential equation which has just *one* turning point. Specifically, we will solve the equation

$$\varepsilon^2 y'' = Q(x)y, \qquad y(+\infty) = 0, \tag{10.4.1}$$

where $Q(x)$ is a continuous function which passes through zero just once. For simplicity, we take the turning point to lie at the origin: $Q(0) = 0$.

The Simple One-Turning-Point Problem

We begin by analyzing in detail the one-turning-point problem in which $Q(x)$ has a simple (first-order) zero: $Q(x) \sim ax$ $(x \to 0)$. For definiteness, we assume that $Q(x)$ has positive slope at $x = 0$ $(a > 0)$ and that $Q(x)$ is positive when x is positive and negative when x is negative. We also assume that $Q(x) \gg x^{-2}$ as $x \to \pm\infty$. $Q(x) = \sinh x$ and $x + x^3$ satisfy these criteria. In Probs. 10.26 and 10.29 we generalize to the case in which $Q(x)$ has a zero of order α: $Q(x) \sim ax^\alpha$ $(x \to 0)$.

Our analysis of the simple-zero one-turning-point problem proceeds as follows. We divide the x axis into three regions: region I with $x > 0$ and $x \gg \varepsilon^{2/3}$, region II with $|x| \ll 1$, and region III with $x < 0$ and $(-x) \gg \varepsilon^{2/3}$. In regions I and III the physical-optics approximation in (10.1.13) is uniformly valid. The restriction that $Q(x) \gg x^{-2}$ as $|x| \to \infty$ ensures that the physical-optics approximation is valid all the way out to $+\infty$, where we impose the boundary condition. In region II the WKB approximation is not valid because there is a turning point at $x = 0$, but we can solve the approximate differential equation

$$\varepsilon^2 y'' = axy, \tag{10.4.2}$$

valid in the neighborhood of $x = 0$, in terms of Airy functions. We show that regions I and II and regions II and III have an overlap in which both the WKB and Airy function approximations are valid. This enables us to match together asymptotically the solutions in the various regions. From this matching we obtain three formulas which together constitute a global approximation to the solution of (10.4.1). We then combine these three formulas into a single expression which is a uniformly valid approximation to $y(x)$ for all x. The global approximations to $y(x)$ are determined only up to an overall multiplicative constant because we impose only the single boundary condition $y(+\infty) = 0$. Therefore, we consider various methods for normalizing $y(x)$.

The calculation that we have just outlined begins with an analysis of the equation in region I. The physical-optics approximation to $y(x)$ in this region has the form

$$y_{\text{I}}(x) = C[Q(x)]^{-1/4} \exp\left[-\frac{1}{\varepsilon}\int_0^x \sqrt{Q(t)}\, dt\right]. \tag{10.4.3}$$

The boundary condition $y_{\text{I}}(+\infty) = 0$ has been used to eliminate the exponentially growing physical-optics solution and is explicitly satisfied by $y_{\text{I}}(x)$ in (10.4.3). We have arbitrarily chosen the lower limit of integration in (10.4.3) to lie at the turning point $x = 0$; this choice is not necessary, but it simplifies expressions appearing later in our analysis.

It is essential to determine the region of validity of the approximation (10.4.3). The two criteria for the validity of physical optics that we derived in Sec. 10.2 are $S_0/\varepsilon \gg S_1 \gg \varepsilon S_2$ $(\varepsilon \to 0+)$ and $\varepsilon S_2 \ll 1$ $(\varepsilon \to 0+)$. Because $Q(x)$ is nonzero for $x \neq 0$ and $Q(x) \gg x^{-2}$ as $|x| \to \infty$, we are assured that for x bounded away from the origin the difference between the exact solution $y(x)$ of (10.4.1) and $y_{\text{I}}(x)$ is of order ε as $\varepsilon \to 0+$. How small may x be before the physical-optics approximation $y_{\text{I}}(x)$ breaks down? When x is small, $Q(x) \sim ax$ so $S_0(x) \sim \pm\frac{2}{3}a^{1/2}x^{3/2}$ $(x \to 0+)$, $S_1(x) \sim -\frac{1}{4}\ln x$ $(x \to 0+)$, $S_2(x) \sim \pm\frac{5}{48}a^{-1/2}x^{-3/2}$ $(x \to 0+)$. Thus, the criteria for validity of the WKB physical-optics approximation are satisfied if

$$x \gg \varepsilon^{2/3}, \qquad \varepsilon \to 0+. \tag{10.4.4}$$

This relation defines the lower boundary of region I.

Next we turn to the analysis of the equation in region II. To solve the approximate differential equation (10.4.2), we make the substitution

$$t = \varepsilon^{-2/3} a^{1/3} x. \tag{10.4.5}$$

In terms of t, the differential equation for y_{II} is $d^2 y_{\text{II}}/dt^2 = t y_{\text{II}}$, which we recognize as the Airy equation. The general solution of this equation is a linear combination of Airy functions:

$$y_{\text{II}}(x) = D \text{ Ai } (\varepsilon^{-2/3} a^{1/3} x) + E \text{ Bi } (\varepsilon^{-2/3} a^{1/3} x), \tag{10.4.6}$$

where D and E are constants to be determined by asymptotic matching with $y_{\text{I}}(x)$. The approximation $y_{\text{II}}(x)$ is valid so long as

$$x \ll 1, \qquad \varepsilon \to 0+, \tag{10.4.7}$$

because it is only when x is small that we may replace $Q(x)$ by ax and thereby obtain (10.4.2) from (10.4.1). The relation in (10.4.7) defines the upper boundary of region II.

Combining (10.4.4) and (10.4.7), we observe that $y_{\text{I}}(x)$ in (10.4.3) and $y_{\text{II}}(x)$ in (10.4.6) have a common region of validity; namely, $\varepsilon^{2/3} \ll x \ll 1$ $(\varepsilon \to 0+)$. Inside this overlap region $y_{\text{I}}(x)$ and $y_{\text{II}}(x)$ are both approximate solutions to the same differential equation and therefore they must match asymptotically. However, since y_{I} and y_{II} bear so little resemblance to each other, more analysis is required to demonstrate that they actually match. We must further approximate $y_{\text{I}}(x)$ and $y_{\text{II}}(x)$ in the overlap region.

First, we consider $y_{\text{I}}(x)$. In the overlap region x is small so $Q(x)$ is approximately ax. Therefore, $[Q(x)]^{-1/4} \sim a^{-1/4} x^{-1/4}$ $(x \to 0+)$ and

$$\int_0^x \sqrt{Q(t)}\, dt \sim \tfrac{2}{3} a^{1/2} x^{3/2}, \qquad x \to 0+.$$

Hence,

$$y_{\text{I}}(x) \sim C a^{-1/4} x^{-1/4} e^{-2a^{1/2} x^{3/2}/3\varepsilon}, \qquad x \to 0+. \tag{10.4.8}$$

What is the precise region of validity of (10.4.8)? We already know that the WKB approximation is not valid unless $x \gg \varepsilon^{2/3}$ $(\varepsilon \to 0+)$. However, the upper edge of the region depends on the function $Q(x)$. Suppose, for example, that $Q(x) - ax \sim bx^2$ $(x \to 0)$. Then, a careful estimation gives

$$\begin{aligned}
\int \sqrt{Q(t)}\, dt &\sim \int_0^x \sqrt{at + bt^2}\, dt \\
&\sim \int_0^x \sqrt{at}\left(1 + \frac{bt}{2a}\right) dt \\
&\sim \frac{2}{3} a^{1/2} x^{3/2} + \frac{b}{5\sqrt{a}} x^{5/2}, \qquad x \to 0+.
\end{aligned}$$

To obtain (10.4.8) it was necessary to assume that x is sufficiently small so that $\exp\,(bx^{5/2}/5\varepsilon\sqrt{a}) \sim 1$ $(\varepsilon \to 0+)$. Hence we arrive at the condition that $x \ll \varepsilon^{2/5}$ $(\varepsilon \to 0+)$. Thus, (10.4.8) is valid in the restricted region $\varepsilon^{2/3} \ll x \ll \varepsilon^{2/5}$ $(\varepsilon \to 0+)$.

Next we consider $y_{\rm II}(x)$. In the overlap region we approximate the Airy functions by their leading asymptotic behaviors for large positive argument. The appropriate formulas are

$$\mathrm{Ai}\,(t) \sim \frac{1}{2\sqrt{\pi}} t^{-1/4} e^{-2t^{3/2}/3}, \qquad t \to +\infty,$$

$$\mathrm{Bi}\,(t) \sim \frac{1}{\sqrt{\pi}} t^{-1/4} e^{2t^{3/2}/3}, \qquad t \to +\infty.$$

These approximations may be used if the arguments of the Airy functions in (10.4.6) are large. Thus,

$$y_{\rm II}(x) \sim \frac{1}{\sqrt{\pi}} a^{-1/12} \varepsilon^{1/6} x^{-1/4} \left(\frac{1}{2} D e^{-2a^{1/2}x^{3/2}/3\varepsilon} + E e^{2a^{1/2}x^{3/2}/3\varepsilon} \right). \tag{10.4.9}$$

This result is valid if two criteria are satisfied. First, we require that $x \ll 1$ as $\varepsilon \to 0+$, so that the Airy equation (10.4.2) is a good approximation to the differential equation (10.4.1). Second, the use of the asymptotic approximations to the Airy functions requires that $t = \varepsilon^{-2/3} a^{1/3} x$ be large or equivalently that $x \gg \varepsilon^{2/3}$ as $\varepsilon \to 0+$. Thus, the region of validity of (10.4.9) is $\varepsilon^{2/3} \ll x \ll 1$ $(\varepsilon \to 0+)$.

Now observe two things. First, unlike (10.4.3) and (10.4.6), (10.4.8) and (10.4.9) have the same functional form and can therefore be matched. Second, (10.4.8) and (10.4.9) have a common region of validity over which the matching can take place:

$$\varepsilon^{2/3} \ll x \ll \varepsilon^{2/5}, \qquad \varepsilon \to 0+. \tag{10.4.10}$$

Requiring that (10.4.8) and (10.4.9) match on the overlap region (10.4.10) determines the constants D and E:

$$D = 2\sqrt{\pi}(a\varepsilon)^{-1/6} C, \tag{10.4.11a}$$

$$E = 0. \tag{10.4.11b}$$

You may recall that it was emphasized in Chaps. 7 and 9 that asymptotic matching must be performed throughout a region whose extent becomes infinite as the perturbation parameter $\varepsilon \to 0$. At first sight the overlap region (10.4.10) appears to violate this principle. However, the matching variable is not x but rather t as given by (10.4.5). In this variable the matching region is $1 \ll t \ll \varepsilon^{-4/15}$ $(\varepsilon \to 0+)$, which does indeed become infinite as $\varepsilon \to 0+$.

The problem is now half solved. We have completed the asymptotic match between regions I and II. Next we must analyze region III and match to the solution just found in region II.

The physical-optics approximation in region III is a linear combination of two rapidly oscillating WKB expressions:

$$y_{\rm III}(x) = F[-Q(x)]^{-1/4} \exp\left[\frac{i}{\varepsilon}\int_x^0 \sqrt{-Q(t)}\,dt\right] + G[-Q(x)]^{-1/4} \exp\left[-\frac{i}{\varepsilon}\int_x^0 \sqrt{-Q(t)}\,dt\right].$$

We will shortly verify that in order for this expression to match to $y_{\rm II}(x)$ in the overlap of regions II and III, the constants F and G must be chosen so that

$$y_{\rm III}(x) = 2C[-Q(x)]^{-1/4} \sin\left[\frac{1}{\varepsilon}\int_x^0 \sqrt{-Q(t)}\,dt + \frac{\pi}{4}\right]. \tag{10.4.12}$$

The result in (10.4.12) is established by comparing the asymptotic approximations to $y_{\rm III}(x)$ and to $y_{\rm II}(x)$ in the overlap of regions II and III which is $\varepsilon^{2/3} \ll (-x) \ll \varepsilon^{2/5}$ $(\varepsilon \to 0+)$. In this overlap region we may approximate $y_{\rm III}(x)$ in (10.4.12) by

$$2Ca^{-1/4}(-x)^{-1/4} \sin\left[\frac{2}{3\varepsilon}a^{1/2}(-x)^{3/2} + \frac{\pi}{4}\right].$$

Also, using the formula for the asymptotic behavior of Ai (t) for large negative argument,

$$\text{Ai}\,(t) = \frac{1}{\sqrt{\pi}}(-t)^{-1/4} \sin\phi(t), \qquad \phi(t) \sim \frac{2}{3}(-t)^{3/2} + \frac{\pi}{4}, \qquad t \to -\infty,$$

we may approximate $y_{\rm II}(x)$ in (10.4.6) with $E = 0$ by

$$D\pi^{-1/2}a^{-1/12}\varepsilon^{1/6}(-x)^{-1/4} \sin\left[\frac{2}{3\varepsilon}a^{1/2}(-x)^{3/2} + \frac{\pi}{4}\right].$$

The approximations we have just found for $y_{\rm II}(x)$ and $y_{\rm III}(x)$ in the overlap region match exactly because D and C are related by (10.4.11a). This completes the analysis of regions II and III.

In summary, we have found approximations to $y(x)$ in each of regions I, II, and III. These approximations are:

$$y_{\rm I}(x) = C[Q(x)]^{-1/4} \exp\left[-\frac{1}{\varepsilon}\int_0^x \sqrt{Q(t)}\,dt\right], \qquad x > 0,\ x \gg \varepsilon^{2/3},\ \varepsilon \to 0+; \tag{10.4.13a}$$

$$y_{\rm II}(x) = 2\sqrt{\pi}(a\varepsilon)^{-1/6}C\ \text{Ai}\,(\varepsilon^{-2/3}a^{1/3}x), \quad |x| \ll 1,\ \varepsilon \to 0+; \tag{10.4.13b}$$

$$y_{\rm III}(x) = 2C[-Q(x)]^{-1/4} \sin\left[\frac{1}{\varepsilon}\int_x^0 \sqrt{-Q(t)}\,dt + \frac{\pi}{4}\right], \qquad x < 0,\ (-x) \gg \varepsilon^{2/3},\ \varepsilon \to 0+. \tag{10.4.13c}$$

The first and third of these formulas are sometimes called *connection formulas* because they express the connection between the oscillatory and the exponentially decreasing behavior of $y(x)$ on opposite sides of the turning point. The constant C remains undetermined because we have specified only the one boundary condition $y(+\infty) = 0$. A second boundary condition is needed to determine C. For example, if we require that $y(0) = 1$, then since Ai $(0) = 3^{-2/3}/\Gamma(\frac{2}{3}) \doteq 0.355\,028\,053\,9$, we have

$$C = \tfrac{1}{2}(a\varepsilon)^{1/6}\Gamma(\tfrac{2}{3})3^{2/3}\pi^{-1/2}. \tag{10.4.14}$$

Observe the global nature of the WKB approximation; we have specified the boundary condition at $x = 0$ and at $x = \infty$ and we can predict the value of $y(-27)$, say, correct to order ε.

Uniform Asymptotic Approximation

In 1935 Langer made the amazing observation that all three formulas in (10.4.13) may be replaced by a *single* formula which is a uniformly valid approximation to $y(x)$ for all x:

$$y_{\text{unif}}(x) = 2\sqrt{\pi}\, C \left(\frac{3}{2\varepsilon} S_0\right)^{1/6} [Q(x)]^{-1/4} \text{ Ai}\left[\left(\frac{3}{2\varepsilon} S_0\right)^{2/3}\right], \tag{10.4.15}$$

where $S_0 = \int_0^x \sqrt{Q(t)}\, dt$. This result is not at all obvious. The best way to explain it is simply to verify it in all three regions. [For a derivation of the Langer formula directly from the differential equation (10.4.1) see Prob. 10.18.]

First, we consider region I, where $x \gg \varepsilon^{2/3}$. Throughout this region $3S_0(x)/2\varepsilon \gg 1$, so we may approximate Ai $[(3S_0/2\varepsilon)^{2/3}]$ by its leading asymptotic behavior:

$$\text{Ai}\left[\left(\frac{3S_0}{2\varepsilon}\right)^{2/3}\right] \sim \left(\frac{3S_0}{2\varepsilon}\right)^{-1/6} \frac{e^{-S_0(x)/\varepsilon}}{2\sqrt{\pi}}, \qquad x \gg \varepsilon^{2/3}.$$

If we substitute this expression into (10.4.15), it reduces to the first formula in (10.4.13).

In region II, where $|x| \ll 1$, the integral $S_0(x)$ may be evaluated approximately by using the first term in the Taylor series for $Q(t)$:

$$\text{Ai}\left[\left(\frac{3}{2\varepsilon} S_0\right)^{2/3}\right] \sim \text{Ai}\,(\varepsilon^{-2/3} a^{1/3} x),$$

$$\left(\frac{3}{2\varepsilon} S_0\right)^{1/6} [Q(x)]^{-1/4} \sim (a\varepsilon)^{-1/6}, \qquad |x| \ll 1,\ \varepsilon \to 0+.$$

Hence (10.4.15) reduces exactly to the second formula of (10.4.13).

In region III, where $(-x) \gg \varepsilon^{2/3}$, one must be very careful about + and − signs (see Prob. 10.19). Now,

$$S_0(x) = \int_0^x \sqrt{Q(t)}\, dt = e^{3\pi i/2} \int_x^0 \sqrt{-Q(t)}\, dt.$$

Thus, $S_0^{2/3}$ is large and *negative*, and (10.4.15) may be simplified by using the asymptotic behavior of Ai for negative argument:

$$\text{Ai}\left[\left(\frac{3}{2\varepsilon} S_0\right)^{2/3}\right] \sim \frac{1}{\sqrt{\pi}} \left[\frac{3}{2\varepsilon} \int_x^0 \sqrt{-Q(t)}\, dt\right]^{-1/6}$$

$$\times \sin\left[\frac{1}{\varepsilon} \int_x^0 \sqrt{-Q(t)}\, dt + \frac{\pi}{4}\right], \qquad \varepsilon \to 0+.$$

Also,
$$Q^{-1/4} = (-Q)^{-1/4} e^{-i\pi/4},$$
$$\left(\frac{3}{2\varepsilon} S_0\right)^{1/6} = e^{i\pi/4} \left[\frac{3}{2\varepsilon} \int_x^0 \sqrt{-Q(t)}\, dt\right]^{1/6}.$$

Thus, (10.4.15) reduces exactly to the third formula of (10.4.13).

Example 1 *Numerical comparison between exact and one-turning-point WKB solutions.* In Figs. 10.10 to 10.13 we compare the exact and uniform one-turning-point solutions in (10.4.15) to $\varepsilon^2 y''(x) = \sinh x(\cosh x)^2 y(x)$ [$y(0) = 1$, $y(+\infty) = 0$] for $\varepsilon = 0.2, 0.3, 0.5$, and 1. Note that for this choice of $Q(x)$, $a = 1$ and $\int_0^x \sqrt{Q(t)}\, dt = \frac{2}{3}(\sinh x)^{3/2}$. The agreement between the exact and the approximate solution is extremely impressive, even when ε is not small.

Directional Character of the Connection Formula

There is a subtle feature of the solution (10.4.13) to the one-turning-point problem. You will recall that in our analysis of this problem we started with the

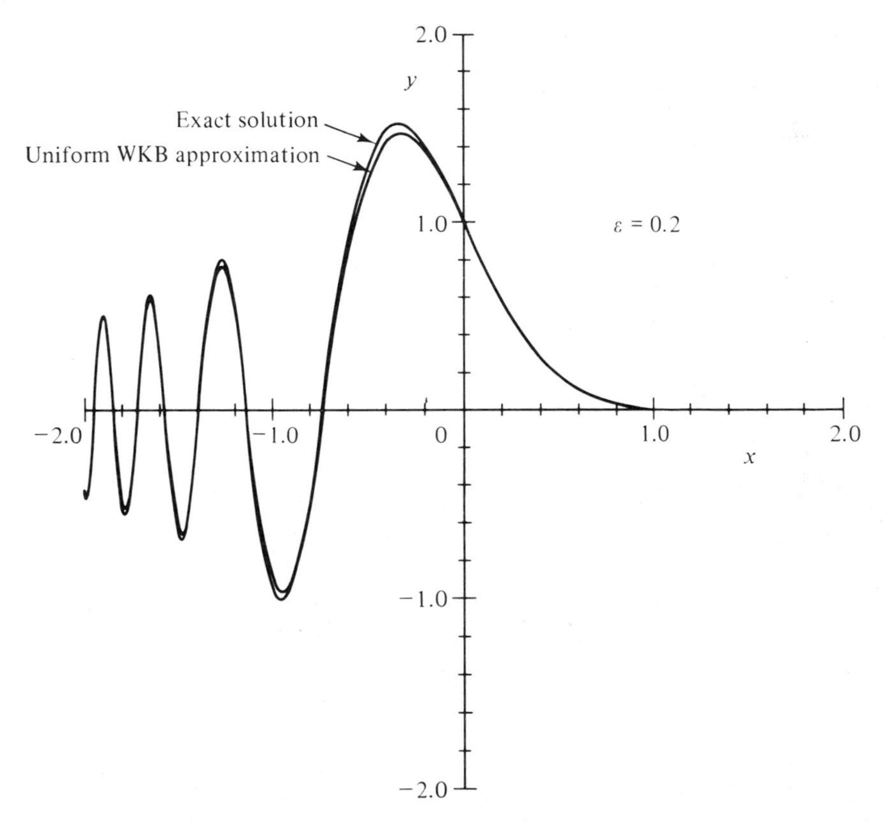

Figure 10.10 A comparison of the exact solution to $\varepsilon^2 y''(x) = \sinh x(\cosh x)^2 y(x)$ [$y(0) = 1$, $y(+\infty) = 0$], with the approximate solution from a one-turning-point WKB analysis. The WKB approximate formulas are given in (10.4.14) and (10.4.15).

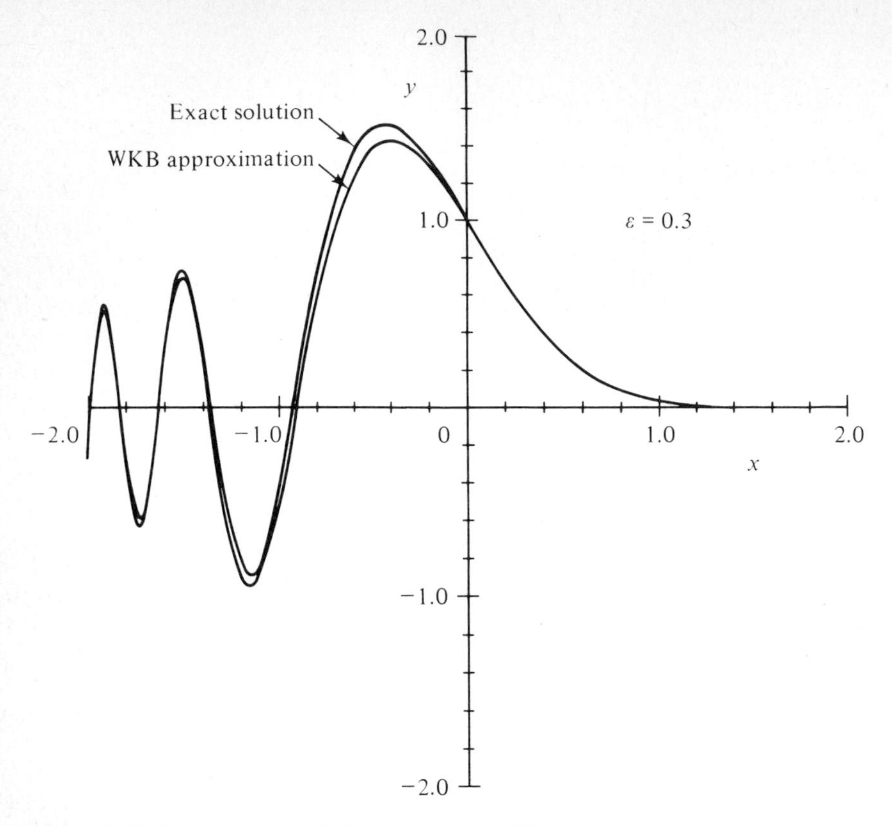

Figure 10.11 Same as in Fig. 10.10 except that $\varepsilon = 0.3$.

boundary condition $y \to 0$ as $x \to +\infty$ in region I and deduced the structure of the solution in regions II and III *in that order.* It is remarkable that the sequence in which this asymptotic analysis is carried out cannot be reversed. To wit, suppose it is given that as $x \to -\infty$ in region III, the solution to $\varepsilon^2 y'' = Q(x)y$ behaves as $2C[-Q(x)]^{-1/4} \sin [\int_x^0 \sqrt{-Q(t)}\, dt/\varepsilon + \frac{1}{4}\pi]$. One is tempted to conclude that the behavior of $y(x)$ in region I is necessarily exponentially decaying: $C[Q(x)]^{-1/4} \exp [-\int_0^x \sqrt{Q(t)}\, dt/\varepsilon]$. But this inference is wrong because the asymptotic matching through the turning-point region is only valid to leading order in ε. We may only conclude that the coefficient of the exponentially growing solution in region I vanishes to leading order in ε. We cannot be sure that the exponentially growing solution in region I is really absent unless the boundary condition $y(+\infty) = 0$ is explicitly imposed.

Apparently, the connection formula for the one-turning-point problem is directional in character. The analysis always proceeds from the region where the solution is exponentially decaying through the turning point and into the oscilla-

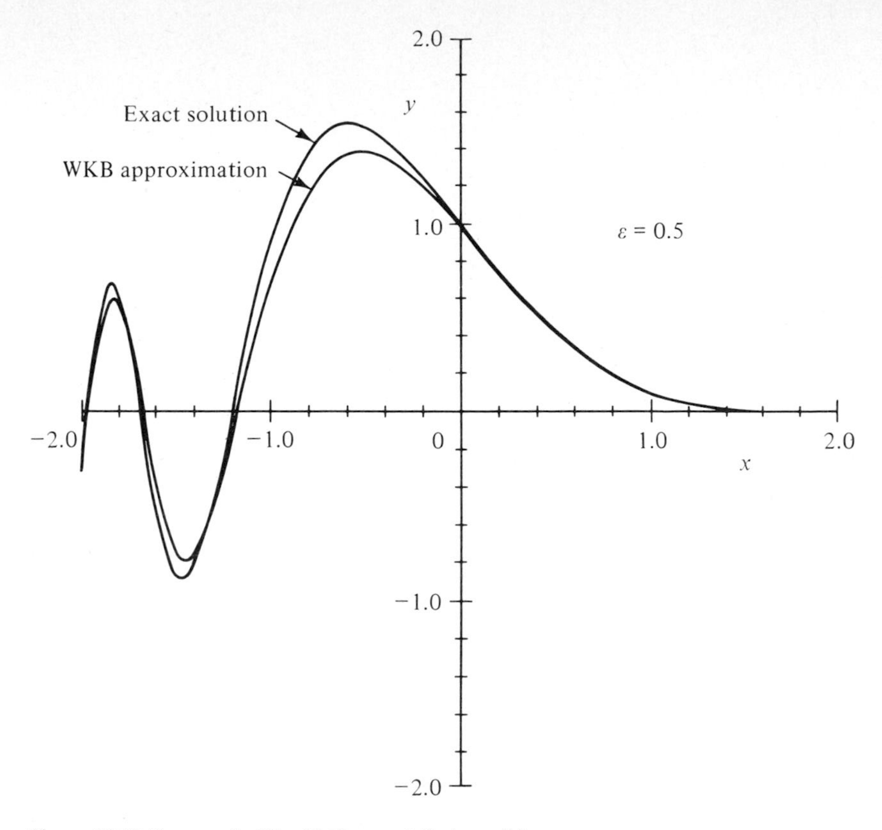

Figure 10.12 Same as in Fig. 10.10 except that $\varepsilon = 0.5$.

tory region. For this reason the descriptive notation

$$\boxed{\begin{array}{c} 2[-Q(x)]^{-1/4} \sin \left[\dfrac{1}{\varepsilon} \displaystyle\int_x^0 \sqrt{-Q(t)}\, dt + \dfrac{1}{4}\pi \right] \text{ in region III} \\ \leftarrow [Q(x)]^{-1/4} \exp \left[-\dfrac{1}{\varepsilon} \displaystyle\int_0^x \sqrt{Q(t)}\, dt \right] \text{ in region I} \end{array}} \tag{10.4.16}$$

is often used to denote the connection formula.

Normalization of the One-Turning-Point Solution

The one-turning-point solution (10.4.13) has an arbitrary multiplicative constant C because the Schrödinger equation (10.4.1) and the boundary condition $y(+\infty) = 0$ are homogeneous. In Example 1 we showed how to determine C by

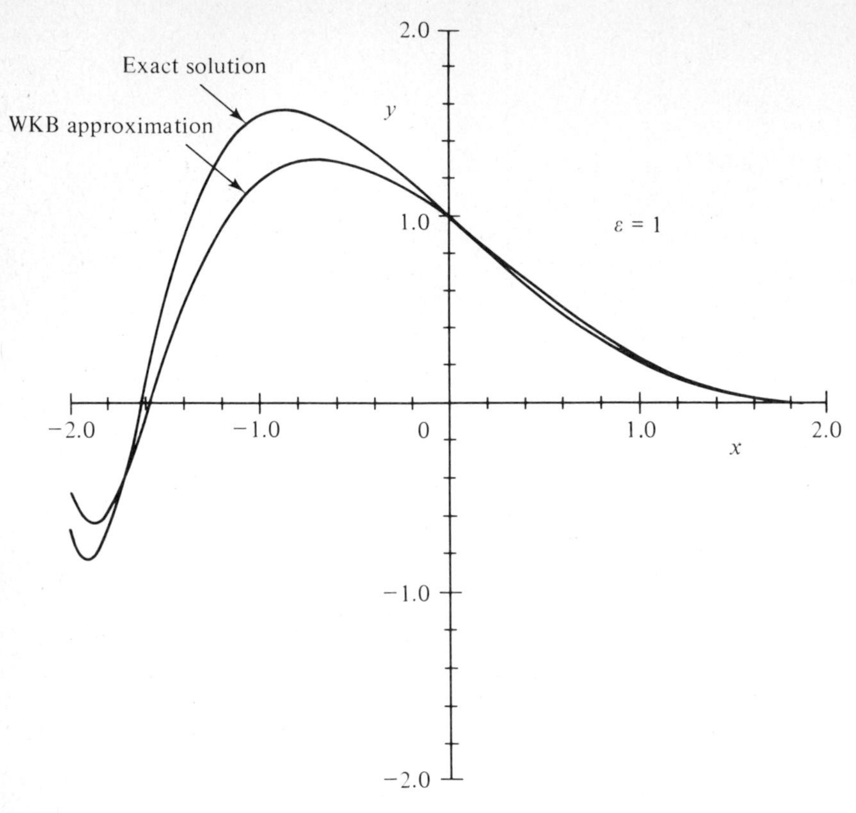

Figure 10.13 Same as in Fig. 10.10 except that $\varepsilon = 1.0$. Even for this large value of ε the agreement between the approximate and exact solutions is impressive.

imposing the additional inhomogeneous boundary condition $y(0) = 1$. Another way to determine C is to require that

$$\int_{-\infty}^{\infty} y(x)\,dx = 1 \tag{10.4.17}$$

or that

$$\int_{-\infty}^{\infty} [y(x)]^2\,dx = 1. \tag{10.4.18}$$

In contrast to the boundary condition $y(0) = 1$ which is imposed at one point, the normalization conditions (10.4.17) and (10.4.18) are global in character. The methods we shall use to evaluate these integrals are especially important because they are a prototype of the techniques for evaluating integrals of functions approximated by matched asymptotic formulas.

To evaluate the integral in (10.4.17), we introduce two arbitrary points A and B where A lies in the overlap of regions II and III and B lies in the overlap of regions I and II. Next we approximate the integral in (10.4.17) as the sum of three integrals:

$$\int_{-\infty}^{\infty} y\,dx \sim \int_{-\infty}^{A} y_{\text{III}}\,dx + \int_{A}^{B} y_{\text{II}}\,dx + \int_{B}^{\infty} y_{\text{I}}\,dx, \qquad \varepsilon \to 0+, \tag{10.4.19}$$

where y_{I}, y_{II}, y_{III} are given in (10.4.13). Since the points A and B are arbitrary and do not appear in the original integral in (10.4.17), the final answer must be completely independent of the particular choice of A and B. However, the approximations (10.4.13) are only leading-order approximations, so we expect that A and B will disappear from (10.4.19) only to leading order in ε. The cancellation of A and B in the final result is a nontrivial test of the correctness of the asymptotic approximations y_{I}, y_{II}, and y_{III}.

We evaluate $\int_B^\infty y_{\text{I}}\,dx$ using integration by parts:

$$\begin{aligned}
\int_B^\infty y_{\text{I}}\,dx &= C\int_B^\infty [Q(x)]^{-1/4} \exp\left[-\frac{1}{\varepsilon}\int_0^x \sqrt{Q(t)}\,dt\right]dx \\
&= -C\varepsilon \int_B^\infty [Q(x)]^{-3/4} \frac{d}{dx} \exp\left[-\frac{1}{\varepsilon}\int_0^x \sqrt{Q(t)}\,dt\right]dx \\
&= +C\varepsilon[Q(B)]^{-3/4} \exp\left[-\frac{1}{\varepsilon}\int_0^B \sqrt{Q(t)}\,dt\right] + \text{O}(\varepsilon^2), \qquad \varepsilon \to 0+, \\
&\sim C\varepsilon(aB)^{-3/4} \exp\left(-\frac{2\sqrt{a}}{3\varepsilon} B^{3/2}\right), \qquad \varepsilon \to 0+,
\end{aligned} \tag{10.4.20}$$

where we have retained only the boundary term after integrating by parts because a second integration by parts shows that the remaining integral is $\text{O}(\varepsilon^2)$. The last step in the calculation, where we have replaced Q by the first term in its Taylor series, is valid because B lies in the overlap region (10.4.9).

We evaluate $\int_{-\infty}^A y_{\text{III}}\,dx$ similarly. The result is

$$\int_{-\infty}^{A} y_{\text{III}}\,dx \sim 2C\varepsilon(-aA)^{-3/4} \cos\left[\frac{2\sqrt{a}}{3\varepsilon}(-A)^{3/2} + \frac{1}{4}\pi\right], \qquad \varepsilon \to 0+, \tag{10.4.21}$$

provided, of course, that the integral converges at $-\infty$. The integral converges if $Q(x) \to -\infty$ as $x \to -\infty$.

Finally, we evaluate $\int_A^B y_{\text{II}}\,dx$ by expressing it as the sum of three integrals: $\int_A^B y_{\text{II}}\,dx = \int_{-\infty}^{\infty} y_{\text{II}}\,dx - \int_{-\infty}^{A} y_{\text{II}}\,dx - \int_B^\infty y_{\text{II}}\,dx$. The first of the integrals may be done exactly using the identity

$$\int_{-\infty}^{\infty} \text{Ai}\,(t)\,dt = 1. \tag{10.4.22}$$

(See Prob. 10.20 for a derivation of this identity.) The second and third integrals may be evaluated by substituting the asymptotic behaviors of the Airy function for large negative and large positive arguments and then using integration by parts. The final result is (see Prob. 10.21)

$$\int_A^B y_{\mathrm{II}}\,dx \sim 2C\sqrt{\frac{\pi\varepsilon}{a}} - 2C\varepsilon(-aA)^{-3/4}\cos\left[\frac{2\sqrt{a}}{3\varepsilon}(-A)^{3/2}+\frac{1}{4}\pi\right]$$
$$- C\varepsilon(aB)^{-3/4}\exp\left(-\frac{2\sqrt{a}}{3\varepsilon}B^{3/2}\right), \qquad \varepsilon \to 0+. \tag{10.4.23}$$

We combine the results in (10.4.20) and (10.4.21) and (10.4.23) and are pleased to find that all reference to A and B cancels to leading order ε:

$$\int_{-\infty}^{\infty} y\,dx \sim 2C\sqrt{\frac{\pi\varepsilon}{a}}, \qquad \varepsilon \to 0+. \tag{10.4.24}$$

Example 2 *Numerical verification of (10.4.24).* Suppose that we use physical optics to solve the one-turning-point problem

$$\varepsilon^2 y'' = (x + x^3)y, \qquad y(0) = 1,\ y(+\infty) = 0. \tag{10.4.25}$$

For this choice of $Q(x)$, $a = 1$. Equation (10.4.14) implies that the leading-order solution is given by (10.4.13) with $C = \frac{1}{2}\varepsilon^{1/6}\Gamma(\frac{2}{3})3^{2/3}\pi^{-1/2}$. Thus (10.4.24) implies that

$$\begin{aligned}\int_{-\infty}^{\infty} y\,dx &\sim \varepsilon^{2/3}\Gamma(\tfrac{2}{3})3^{2/3}, \qquad \varepsilon \to 0+\\ &\doteq 2.816\,679\varepsilon^{2/3}, \qquad \varepsilon \to 0+.\end{aligned} \tag{10.4.26}$$

We have solved (10.4.25) numerically and computed $\int_{-\infty}^{\infty} y\,dx$. The results given in Table 10.2 verify the accuracy of this WKB analysis.

To evaluate the integral in (10.4.18), we again introduce two arbitrary points A and B in the overlap regions and express the integral as $\int_{-\infty}^{\infty} y^2\,dx \sim \int_{-\infty}^{A} y_{\mathrm{III}}^2\,dx + \int_A^B y_{\mathrm{II}}^2\,dx + \int_B^{\infty} y_{\mathrm{I}}^2\,dx$ ($\varepsilon \to 0+$). We evaluate $\int_B^{\infty} y_{\mathrm{I}}^2\,dx$ as before using integration by parts. The result is

$$\int_B^{\infty} y_{\mathrm{I}}^2\,dx \sim \frac{C^2\varepsilon}{2aB}\exp\left(-\frac{4\sqrt{a}}{3\varepsilon}B^{3/2}\right), \qquad \varepsilon \to 0+. \tag{10.4.27}$$

To evaluate $\int_{-\infty}^{A} y_{\mathrm{III}}^2\,dx$, we use the identity $\sin^2\theta = \frac{1}{2} - \frac{1}{2}\cos(2\theta)$ and again approximate the resulting integrals using integration by parts (see Prob. 10.22):

$$\int_{-\infty}^{A} y_{\mathrm{III}}^2\,dx \sim 2C^2\int_{-\infty}^{0}\frac{dt}{\sqrt{-Q(t)}} - \frac{4C^2}{\sqrt{a}}\sqrt{-A} - \frac{C^2\varepsilon}{aA}\cos\left[\frac{4}{3\varepsilon}\sqrt{a}(-A)^{3/2}\right],$$
$$\varepsilon \to 0+. \tag{10.4.28}$$

Finally, we evaluate $\int_A^B y_{\mathrm{II}}^2\,dx$ using a nice trick. Ai (t) satisfies the differential equation Ai$''(t) = t$ Ai (t), so

$$\frac{d}{dt}\{t[\mathrm{Ai}(t)]^2 - [\mathrm{Ai}'(t)]^2\} = [\mathrm{Ai}(t)]^2.$$

Table 10.2 Comparison between the exact value of $\int_{-\infty}^{\infty} y(x)\,dx$, where $y(x)$ satisfies $\varepsilon^2 y'' = (x + x^3)y$ $[y(0) = 1,\ y(+\infty) = 0]$ and the physical-optics approximation to this integral

$$\int_{-\infty}^{\infty} y(x)\,dx \sim \varepsilon^{2/3}\Gamma(\tfrac{2}{3})3^{2/3}, \qquad \varepsilon \to 0+,$$

as given in (10.4.26)

Observe that as ε gets smaller, the accuracy of the WKB approximation increases

ε	Exact value of $\int_{-\infty}^{\infty} y(x)\,dx$	WKB approximation to $\int_{-\infty}^{\infty} y(x)\,dx$
0.2	0.9751	0.9633
0.1	0.6136	0.6068
0.05	0.3844	0.3823
0.02	0.2079	0.2075
0.01	0.1308	0.1307
0.005	0.0823	0.0824
0.002	0.0447	0.0447

Therefore, since $y_{\text{II}}(x)$ is a constant multiple of Ai (t) with t given by (10.4.5), the integral of $[y_{\text{II}}(x)]^2$ can be evaluated in closed form in terms of Airy functions:

$$\int_A^B [y_{\text{II}}(x)]^2\,dx = 4\pi C^2 \varepsilon^{1/3} a^{-2/3}\{t[\text{Ai}\,(t)]^2 - [\text{Ai}'\,(t)]^2\}\Big|_{t = A\varepsilon^{-2/3}a^{1/3}}^{t = B\varepsilon^{-2/3}a^{1/3}}. \tag{10.4.29}$$

Naturally, we wish to approximate this expression by replacing Ai (t) and Ai′ (t) by their asymptotic expressions. However, we are surprised to find that if we use only the leading asymptotic behaviors of Ai (t) and Ai′ (t), then we obtain a vanishing result at the upper endpoint! We have emphasized repeatedly that an asymptotic calculation is wrong if the result is zero. Therefore, we must use a higher-order asymptotic approximation to Ai (t) and Ai′ (t). The appropriate formulas are

$$\text{Ai}\,(t) \sim \frac{1}{2\sqrt{\pi}} t^{-1/4} e^{-2t^{3/2}/3}\left(1 - \frac{5}{48t^{3/2}}\right), \qquad t \to \infty,$$

$$\text{Ai}'\,(t) = -\frac{1}{2\sqrt{\pi}} t^{1/4} e^{-2t^{3/2}/3}\left(1 + \frac{7}{48t^{3/2}}\right), \qquad t \to \infty.$$

We also use the higher-order asymptotic expansions of Ai (t) and Ai$'$ (t) at the lower endpoint:

$$\text{Ai}(-t) \sim \frac{1}{\sqrt{\pi}} t^{-1/4} \left[\sin\left(\frac{2}{3}t^{3/2} + \frac{\pi}{4}\right) - \cos\left(\frac{2}{3}t^{3/2} + \frac{\pi}{4}\right) \frac{5}{48t^{3/2}} \right], \qquad t \to \infty,$$

$$\text{Ai}'(-t) \sim -\frac{1}{\sqrt{\pi}} t^{1/4} \left[\cos\left(\frac{2}{3}t^{3/2} + \frac{\pi}{4}\right) - \sin\left(\frac{2}{3}t^{3/2} + \frac{\pi}{4}\right) \frac{7}{48t^{3/2}} \right], \qquad t \to \infty.$$

If these formulas are used to approximate the expression in (10.4.29), the result is

$$\int_A^B [y_{\text{II}}(x)]^2\, dx \sim \frac{4C^2}{\sqrt{a}}\sqrt{-A} - \frac{C^2\varepsilon}{aA}\cos\left[\frac{4}{3\varepsilon}\sqrt{a}\,(-A)^{3/2}\right]$$
$$- \frac{C^2\varepsilon}{2aB}\exp\left(-\frac{4\sqrt{a}}{3\varepsilon}B^{3/2}\right), \qquad \varepsilon \to 0+. \tag{10.4.30}$$

Combining the results (10.4.27), (10.4.28), (10.4.30) gives the final answer

$$\int_{-\infty}^{\infty} [y(x)]^2\, dx \sim 2C^2 \int_{-\infty}^{0} \frac{dt}{\sqrt{-Q(t)}}, \qquad \varepsilon \to 0+. \tag{10.4.31}$$

Once again, the answer is independent of A and B.

Table 10.3 Comparison between the exact value of $\int_{-\infty}^{\infty} [y(x)]^2\, dx$, where $y(x)$ satisfies $\varepsilon^2 y'' = (x + x^3)y$ $[y(0) = 1,\ y(+\infty) = 0]$ and the physical-optics approximation to this integral

$$\int_{-\infty}^{\infty} [y(x)]^2\, dx \sim \tfrac{1}{4}\varepsilon^{1/3}[\Gamma(\tfrac{2}{3})\Gamma(\tfrac{1}{4})]^2 3^{4/3}\pi^{-3/2}$$

$(\varepsilon \to 0+)$ as given in (10.4.32)

As ε decreases, the accuracy of the WKB approximation increases

ε	Exact value of $\int_{-\infty}^{\infty} [y(x)]^2\, dx$	WKB approximation to $\int_{-\infty}^{\infty} [y(x)]^2\, dx$
0.2	2.9085	2.7382
0.1	2.2308	2.1733
0.05	1.7437	1.7249
0.02	1.2751	1.2710
0.01	1.0101	1.0088
0.005	0.8011	0.8006
0.002	0.5900	0.5899

Example 3 *Numerical verification of (10.4.31).* Consider once again the differential equation in (10.4.25). For this equation, $Q(x) = x + x^3$. Thus,

$$2\int_{-\infty}^{0} \frac{dt}{\sqrt{-Q(t)}} = \left[\Gamma\left(\frac{1}{4}\right)\right]^2 \pi^{-1/2}.$$

Therefore, using C as determined in Example 2,

$$\int_{-\infty}^{\infty} [y(x)]^2 \, dx \sim \tfrac{1}{4}\varepsilon^{1/3}[\Gamma(\tfrac{2}{3})\Gamma(\tfrac{1}{4})]^2 3^{4/3}\pi^{-3/2}, \qquad \varepsilon \to 0+. \tag{10.4.32}$$

We examine the accuracy of this formula in Table 10.3 by comparing it with the integral of the numerical solution to (10.4.25).

(I) 10.5 TWO-TURNING-POINT PROBLEMS: EIGENVALUE CONDITION

In this section we show how to use the physical-optics approximation to obtain an approximate solution to the homogeneous boundary-value problem

$$\varepsilon^2 y'' = Q(x)y, \qquad y(\pm\infty) = 0, \tag{10.5.1}$$

where $Q(x)$ has two simple turning points at $x = A$ and $x = B$ with $A < B$. We also assume that $Q > 0$ if $x > B$ or $x < A$, that $Q < 0$ if $A < x < B$, and that $Q(x) \gg x^{-2}$ as $|x| \to \infty$. For most functions $Q(x)$ satisfying these conditions the only solution to (10.5.1) is $y(x) = 0$. This is because the solution to (10.5.1) which decays exponentially as $x \to +\infty$ is, in general, a mixture of growing and decaying solutions as $x \to -\infty$. We will derive an approximate constraint which must be satisfied by $Q(x)$ for the problem (10.5.1) to have nontrivial solutions. To leading order in ε this constraint is

$$\frac{1}{\varepsilon}\int_A^B \sqrt{-Q(t)}\, dt = \left(n + \frac{1}{2}\right)\pi + \mathrm{O}(\varepsilon), \qquad \varepsilon \to 0+, \tag{10.5.2}$$

where $n = 0, 1, 2, \ldots$ is a nonnegative integer.

The constraint in (10.5.2) is useful if the function Q depends on a parameter E, which we call an eigenvalue. Then (10.5.2) determines the approximate value of E correct to terms of order ε.

The derivation of (10.5.2) is done by asymptotically matching two one-turning-point solutions: the first one-turning-point solution is valid from $+\infty$ through the turning point at B and down to near the turning point at A; the second is valid from $-\infty$ through the turning point at A and up to near the turning point at B. Since these one-turning point solutions overlap in the region between the turning points at A and B, we must require that they match asymptotically. This matching condition translates into the constraint on $Q(x)$ in (10.5.2).

The one-turning-point solution that decays like

$$C_1[Q(x)]^{-1/4} \exp\left[-\frac{1}{\varepsilon}\int_B^x \sqrt{Q(t)}\, dt\right]$$

as $x \to +\infty$ behaves like

$$2C_1[-Q(x)]^{-1/4} \sin\left[\frac{1}{\varepsilon}\int_x^B \sqrt{-Q(t)}\,dt + \frac{1}{4}\pi\right] \tag{10.5.3}$$

in the region between A and B [so long as the distance between x and A or x and B is much greater than $\varepsilon^{2/3}$ (see Sec. 10.4)]. This is merely a restatement of the connection formula (10.4.16) when the turning point lies at $x = B$ instead of at $x = 0$.

The one-turning-point solution that decays like

$$C_2[Q(x)]^{-1/4} \exp\left[-\frac{1}{\varepsilon}\int_x^A \sqrt{Q(t)}\,dt\right]$$

as $x \to -\infty$ behaves like

$$2C_2[-Q(x)]^{-1/4} \sin\left[\frac{1}{\varepsilon}\int_A^x \sqrt{-Q(t)}\,dt + \frac{1}{4}\pi\right] \tag{10.5.4}$$

in the region between A and B. This result is derived in Prob. 10.30.

In order that the two physical-optics solutions in (10.5.3) and (10.5.4) match in the region between A and B, we must require that they have the same functional form. Both solutions already have identical factors of $[-Q(x)]^{-1/4}$. However, the arguments of the sine functions are not identical. To achieve the match, we rewrite (10.5.3) as

$$-2C_1[-Q(x)]^{-1/4} \sin\left[\frac{1}{\varepsilon}\int_A^x \sqrt{-Q(t)}\,dt + \frac{\pi}{4} - \left\{\frac{1}{\varepsilon}\int_A^B \sqrt{-Q(t)}\,dt + \frac{\pi}{2}\right\}\right].$$

In order that this expression be functionally identical to that in (10.5.4), it is necessary that the expression in curly brackets be an integral multiple of π. Moreover, since the expression in curly brackets is positive, it follows that we must require $(1/\varepsilon)\int_A^B \sqrt{-Q(t)}\,dt = \frac{1}{2}\pi, \frac{3}{2}\pi, \frac{5}{2}\pi, \ldots$, which is just (10.5.2). To complete the match of (10.5.3) and (10.5.4), we must choose $C_1 = (-1)^n C_2$ where n is defined in (10.5.2). This completes the derivation of (10.5.2).

The above analysis has neglected terms of order ε, namely, the higher terms in the WKB series $(\varepsilon S_2, \varepsilon^2 S_3, \ldots)$. Consequently, the constraint (10.5.2) is only accurate to terms of order ε. In Sec. 10.7 we will derive a more accurate constraint that is valid to all orders in powers of ε by taking into account the presence of higher-order terms in the WKB series.

Linear Eigenvalue Problems

We now examine a special class of eigenvalue problems in which the eigenvalue E appears linearly: $Q(x) = V(x) - E$. In the study of quantum mechanics, if $V(x)$ rises monotonically as $x \to \pm\infty$, the differential equation

$$\varepsilon^2 y'' = [V(x) - E]y(x), \qquad y(\pm\infty) = 0, \tag{10.5.5}$$

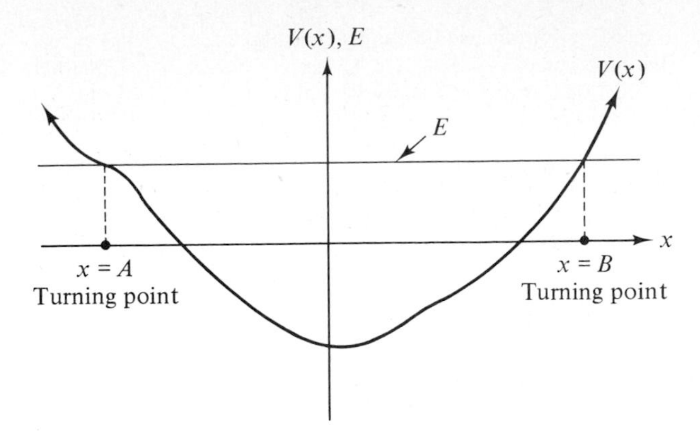

Figure 10.14 Schematic plot of the function $V(x)$ in (10.5.5). Turning points occur when $V(x) = E$. In classical mechanics if we interpret this configuration to represent a particle of energy E in a potential $V(x)$, then the particle is confined to the region between the turning points at A and B where the total energy E is greater than or equal to the potential energy $V(x)$. In classical mechanics the energy of a particle in a potential well is arbitrary so long as $E \geq V_{\min}$. In quantum mechanics E can only have special discrete values which are the eigenvalues of (10.5.5). The energy of such a particle is said to be quantized.

describes a particle of energy E confined to a potential well $V(x)$ (see Fig. 10.14). By (10.5.2) the eigenvalue E of (10.5.5) must satisfy

$$\frac{1}{\varepsilon}\int_A^B \sqrt{E - V(x)}\,dx = \left(n + \frac{1}{2}\right)\pi + O(\varepsilon), \qquad \varepsilon \to 0, \tag{10.5.6}$$

where the turning points A and B are the two solutions to the equation $V(x) - E = 0$.

We will see that if $V(x) \to \infty$ as $x \to \pm\infty$, then there are an infinite number of solutions E_n to (10.5.6) and that $E_n \to \infty$ as $n \to \infty$. In this case (10.5.6) becomes asymptotically exact as $n \to \infty$ for any *fixed* value of ε and we set $\varepsilon = 1$ as we did in our discussion of the eigenvalue problem in (10.1.27). The accuracy of the WKB approximation increases as $n \to \infty$ because, except at the turning points, $|S_0| = |\int^x \sqrt{V(t) - E}\,dt|$ increases as E increases. Thus, the conditions for the validity of WKB,

$$\frac{1}{\varepsilon}S_0 \gg S_1 \gg \varepsilon S_2 \qquad \text{and} \qquad 1 \gg \varepsilon S_2, \tag{10.5.7}$$

are satisfied either as $\varepsilon \to 0$ with E fixed or as $E \to \infty$ with ε fixed (see Prob. 10.31).

Example 1 *Eigenvalues for* $y'' = (|x| - E)y$ $[y(\pm\infty) = 0]$. For this equation the solutions of $V(x) - E = 0$ are $A = -E$ and $B = E$. Thus, the WKB eigenvalue condition becomes $\int_{-E}^{E} \sqrt{E - |x|}\,dx \sim (n + \frac{1}{2})\pi$ $(n \to \infty)$. But $\int_{-E}^{E} \sqrt{E - |x|}\,dx = \frac{4}{3}E^{3/2}$. Thus, for large n,

$$E_n \sim \left[\frac{3\pi}{4}\left(n + \frac{1}{2}\right)\right]^{2/3}, \qquad n \to \infty. \tag{10.5.8}$$

This result may be reproduced by solving the differential equation exactly. When $x > 0$, $|x| = x$ and the differential equation becomes $y'' = (x - E)y$. The exact solution to this equation is a linear combination Ai $(x - E)$ and Bi $(x - E)$. However, only Ai $(x - E)$ vanishes as $x \to +\infty$. Thus,

$$y(x) = c \text{ Ai } (x - E), \qquad x \geq 0, \tag{10.5.9}$$

where c is a constant.

When $x < 0$, $|x| = -x$ and the only solution to the differential equation $y'' = (-x - E)y$ which vanishes as $x \to -\infty$ is

$$y(x) = d \text{ Ai } (-x - E), \qquad x \leq 0, \tag{10.5.10}$$

where d is a constant.

The two solutions (10.5.9) and (10.5.10) must be patched at $x = 0$. Demanding that $y(x)$ and $y'(x)$ be continuous at $x = 0$ requires that c Ai $(-E) = d$ Ai $(-E)$ and that c Ai$'$ $(-E) = -d$ Ai$'$ $(-E)$. Thus, if $c = -d \neq 0$ then

$$\text{Ai } (-E) = 0 \tag{10.5.11}$$

and if $c = d \neq 0$ then

$$\text{Ai}' \; (-E) = 0. \tag{10.5.12}$$

The solutions of the two transcendental equations (10.5.11) and (10.5.12) comprise the complete set of eigenvalues for the differential equation. When E is large, these solutions had better agree with the WKB prediction in (10.5.8)! To check this, we replace Ai $(-E)$ and Ai$'$ $(-E)$ by their leading asymptotic expansions for large negative argument:

$$\text{Ai } (-E) \sim \frac{1}{\sqrt{\pi}} E^{-1/4} \sin\left(\frac{2}{3} E^{3/2} + \frac{\pi}{4}\right), \qquad E \to \infty,$$

has zeros whenever

$$\frac{2}{3} E^{3/2} + \frac{\pi}{4} = k\pi, \qquad k = 1, 2, 3, \ldots, \tag{10.5.13}$$

and

$$\text{Ai}' \; (-E) \sim -\frac{1}{\sqrt{\pi}} E^{1/4} \cos\left(\frac{2}{3} E^{3/2} + \frac{\pi}{4}\right)$$

has zeros whenever

$$\frac{2}{3} E^{3/2} + \frac{\pi}{4} = \left(k + \frac{1}{2}\right)\pi, \qquad k = 0, 1, 2, \ldots. \tag{10.5.14}$$

Combining (10.5.13) and (10.5.14) into a single formula gives

$$\frac{2}{3} E^{3/2} + \frac{\pi}{4} = \left(\frac{n}{2} + \frac{1}{2}\right)\pi, \qquad n = 0, 1, 2, \ldots,$$

which is equivalent to the WKB result in (10.5.8) and which is also valid when n is a large positive integer.

Example 2 *Eigenvalues of the parabolic cylinder equation.* We have seen in Example 9 of Sec. 3.8 that the eigenvalues of $(-d^2/dx^2 + x^2/4 - E)y(x) = 0$ $[y(\pm\infty) = 0]$ are exactly $E_n = n + \frac{1}{2}$ $(n = 0, 1, 2, \ldots)$. How well does WKB reproduce this result?

The turning points lie at $A = -2\sqrt{E}$ and $B = 2\sqrt{E}$. Thus, the WKB eigenvalue condition reads

$$\int_{-2\sqrt{E}}^{2\sqrt{E}} \sqrt{E - \tfrac{1}{4}x^2}\, dx \sim (n + \tfrac{1}{2})\pi, \qquad n \to \infty.$$

Upon substituting $x = 2\sqrt{E}\, t$ the above integral becomes $2E \int_{-1}^{1} dt \sqrt{1 - t^2} = E\pi$.

Thus, the WKB prediction is $E_n \sim n + \frac{1}{2}$ $(n \to \infty)$, which is not only valid as $n \to \infty$ but is *exact* for all n.

It is accidental that the leading-order (physical-optics) WKB result is exact. Indeed, the physical-optics approximation to the nth eigenfunction $y_n(x)$ is only approximate. The physical-optics approximation to the eigenvalues is exact because the corrections to E_n that result from a higher-order WKB treatment of the eigenvalue problem all happen to vanish (see Example 1 of Sec. 10.7).

Example 3 *Eigenvalues for* $y'' = (x^4 - E)y$. The turning points are at $A = -E^{1/4}$ and $B = E^{1/4}$. Thus, $\int_{-E^{1/4}}^{E^{1/4}} \sqrt{E - x^4}\, dx \sim (n + \frac{1}{2})\pi$ $(n \to \infty)$ becomes

$$E_n \sim \left[\frac{3\Gamma(\frac{3}{4})(n + \frac{1}{2})\sqrt{\pi}}{\Gamma(\frac{1}{4})}\right]^{4/3}, \qquad n \to \infty \tag{10.5.15}$$

(see Prob. 10.32).

In Table 10.4 we compare the exact eigenvalues with the WKB prediction for the eigenvalues in (10.5.15). Observe that as n increases the accuracy of the WKB prediction increases dramatically.

In Figs. 10.15 to 10.17 we compare the physical-optics approximation to $y(x)$ with the solution to the differential equation obtained by computer. Again, the accuracy increases very rapidly with n.

Table 10.4 Comparison of exact eigenvalues E_n for $y'' = (x^4 - E)y$ $[y(\pm\infty) = 0]$ and the WKB prediction for E_n in (10.5.15): $E_n \sim [3\Gamma(\frac{3}{4})(n + \frac{1}{2})\sqrt{\pi}/\Gamma(\frac{1}{4})]^{4/3}$ $(n \to \infty)$

Observe that the relative error [% relative error = 100 (WKB E_n − exact E_n)/(exact E_n)] decreases rapidly as n increases. [Note that $\Gamma(\frac{3}{4})/\Gamma(\frac{1}{4}) \doteq 0.337\,99$.]

n	Exact E_n	WKB E_n	Relative error, %
0	1.060	0.867	−18.00
2	7.456	7.414	−0.56
4	16.262	16.234	−0.17
6	26.528	26.506	−0.08
8	37.923	37.904	−0.05
10	50.256	50.240	−0.03

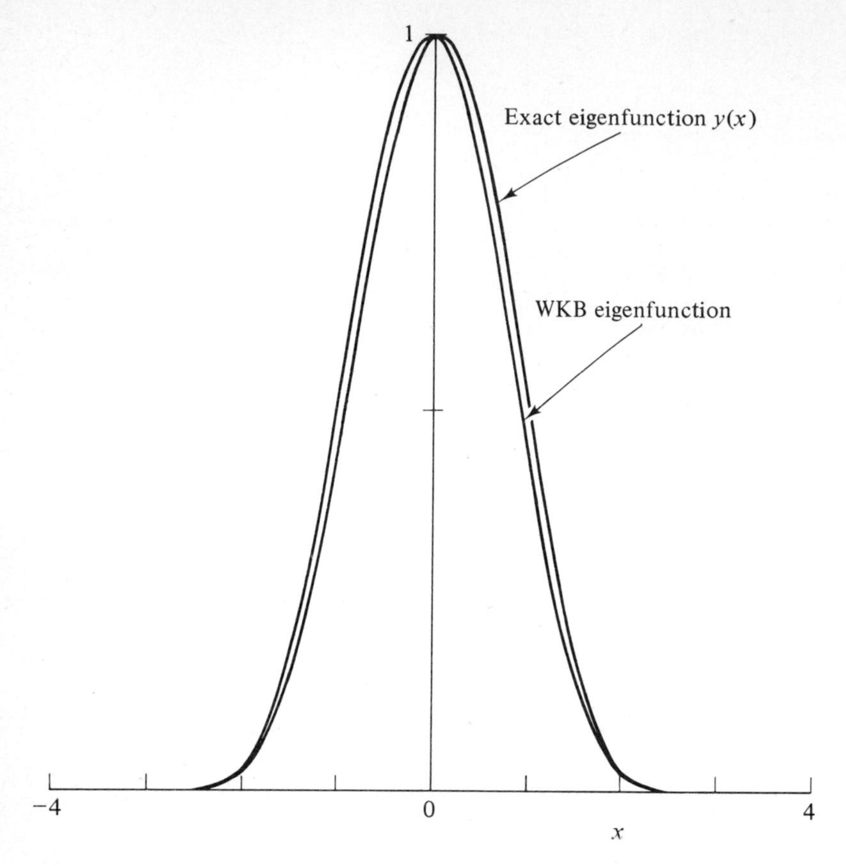

Figure 10.15 Comparison between the exact eigenfunction of $y'' = (x^4 - E)y$ $[y(\pm\infty) = 0]$, with the lowest eigenvalue $E_0(n = 0)$ and the corresponding uniform physical-optics (WKB) approximation (10.4.15) with the WKB approximation (10.5.15) to E_n. Both eigenfunctions are normalized by $y(0) = 1$. The WKB approximation is given by (10.4.15) with $Q(x) = x^4 - E_n$ for $x > 0$ and by $y(x) = y(-x)$ for $x < 0$.

(D) 10.6 TUNNELING

Tunneling is the remarkable quantum-mechanical phenomenon by which a particle passes through a potential barrier that classical mechanics predicts is impenetrable. In this section we use WKB theory to make a quantitative study of tunneling. We begin by introducing the notion of a wave.

Right-Moving and Left-Moving Waves

The phenomenon of tunneling implicitly involves motion. Thus, to describe tunneling we must begin with the time-dependent Schrödinger wave equation

$$\frac{1}{i}\frac{\partial}{\partial t}\psi(x, t) = \left[-\varepsilon^2\frac{\partial^2}{\partial x^2} + V(x)\right]\psi(x, t). \tag{10.6.1}$$

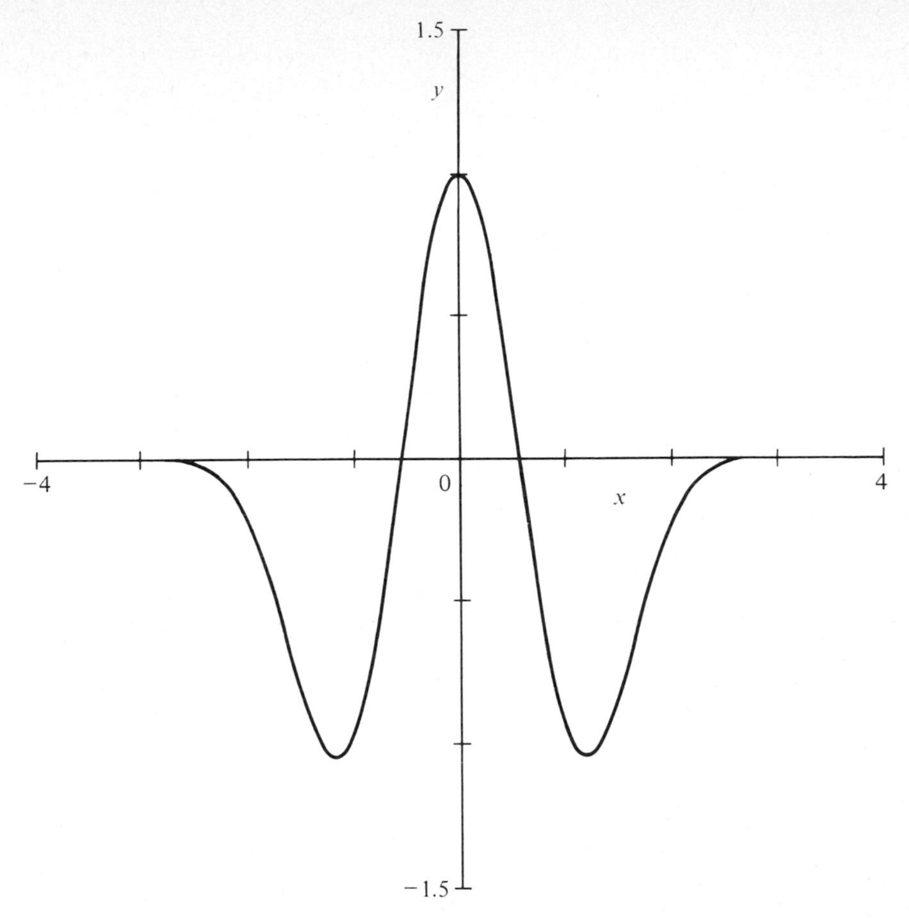

Figure 10.16 Same as Figure 10.15 except for the third lowest eigenvalue $E_2(n = 2)$ of $y'' = (x^4 - E)y$. The exact eigenfunction and the physical-optics approximation to it are not distinguishable on the scale of the plot. See Fig. 10.17 for a plot of the error $y_{\text{WKB}} - y_{\text{exact}}$.

$\psi(x, t)$ is called a wave function. Let us assume that the time dependence of $\psi(x, t)$ is purely oscillatory:

$$\psi(x, t) = y(x)e^{iEt}. \tag{10.6.2}$$

Substituting (10.6.2) into (10.6.1) gives the ordinary differential equation

$$\varepsilon^2 y'' = [V(x) - E]y(x), \tag{10.6.3}$$

which we have already examined using the WKB approximation.

In regions where $E > V(x)$ (classically allowed regions), WKB solutions to (10.6.3) are oscillatory:

$$y_{\text{WKB}}(x) = C_{\pm}[E - V(x)]^{-1/4} \exp\left[\pm\frac{i}{\varepsilon}\int^x \sqrt{E - V(t)}\, dt\right]. \tag{10.6.4}$$

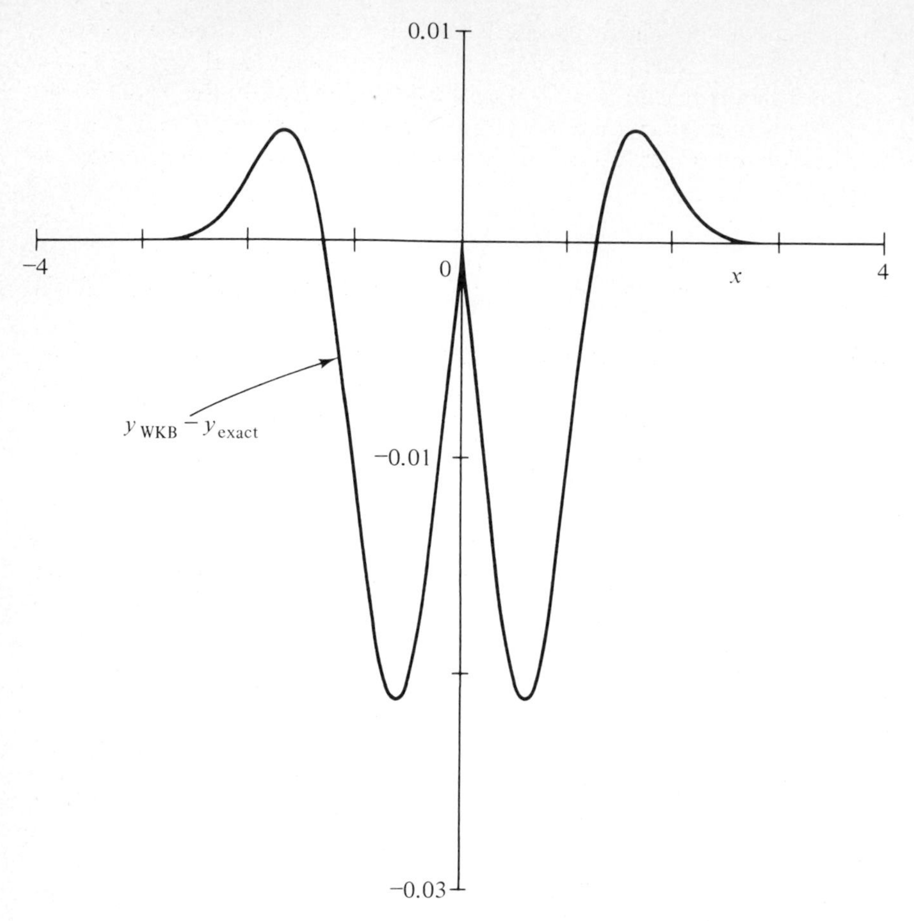

Figure 10.17 A plot of the error $y_{\text{WKB}} - y_{\text{exact}}$ in the uniform physical-optics (WKB) approximation to the third lowest eigenfunction ($n = 2$) of $y'' = (x^4 - E)y$. Both y_{exact} and y_{WKB} are normalized by $y_{\text{exact}}(0) = y_{\text{WKB}}(0) = 1$. See Fig. 10.16.

Even though the WKB approximation in (10.6.4) is time independent, we will refer to solutions having positive (negative) phase as left-moving (right-moving) waves. This is because the wave function

$$\psi_{\text{WKB}}(x, t) = C_{\pm}[E - V(x)]^{-1/4} \exp i[Et \pm (1/\varepsilon) \int^x \sqrt{E - V(t)}\, dt]$$

represents a wave which moves to the left (right) as t increases.

An Exactly Soluble Model of Tunneling

To illustrate the phenomenon of tunneling, we make a very simple choice for the potential V: $V(x) = \delta(x)$. Classically, this delta function potential confers an impulse to a particle which arrives at $x = 0$. If $E < (4\varepsilon^2)^{-1}$, a classical particle traveling toward $x = 0$ *always* bounces back (reverses its direction) when it reaches $x = 0$. This is called reflection. If $E > (4\varepsilon^2)^{-1}$, a classical particle *always* continues on when it reaches $x = 0$. This is called transmission (see Prob. 10.39).

In quantum mechanics there are well-defined probabilities T and R that a particle will undergo transmission or reflection. We now compute these probabilities exactly. To solve

$$\varepsilon^2 y'' = [\delta(x) - E]y$$

we consider two regions. When $x < 0$, we have $\varepsilon^2 y'' + Ey = 0$, whose general solution is

$$y(x) = a \exp(-ix\sqrt{E}/\varepsilon) + b \exp(+ix\sqrt{E}/\varepsilon), \qquad x < 0.$$

When $x > 0$, we have the same differential equation whose general solution is now

$$y(x) = c \exp(-ix\sqrt{E}/\varepsilon) + d \exp(+ix\sqrt{E}/\varepsilon), \qquad x > 0.$$

To observe tunneling we must choose the boundary conditions properly; we aim a monoenergetic incident beam of particles toward $x = 0$ from the left. We represent this incident beam as a right-moving wave of unit amplitude: $\exp(-ix\sqrt{E}/\varepsilon)$. There will then be a reflected (left-moving) wave for $x < 0$ of amplitude b, $b \exp(+ix\sqrt{E}/\varepsilon)$, and a transmitted (right-moving) wave for $x > 0$ of amplitude c, $c \exp(-ix\sqrt{E}/\varepsilon)$. There is *no* left-moving wave for $x > 0$.

We must patch the two solutions

$$y(x) = \exp(-ix\sqrt{E}/\varepsilon) + b \exp(+ix\sqrt{E}/\varepsilon), \qquad x < 0, \tag{10.6.5}$$

$$y(x) = c \exp(-ix\sqrt{E}/\varepsilon), \qquad x > 0, \tag{10.6.6}$$

at $x = 0$. We require that:

1. $y(x)$ be continuous at $x = 0$ and
2. $\lim_{\eta \to 0+} \varepsilon^2[y'(\eta) - y'(-\eta)] = y(0)$.

Why? (See Sec. 10.3.)

From these two conditions we obtain (see Prob. 10.38)

$$b = \frac{2\varepsilon\sqrt{E}\, i - 1}{4\varepsilon^2 E + 1}, \qquad c = \frac{2\varepsilon\sqrt{E}\,(2\varepsilon\sqrt{E} + i)}{4\varepsilon^2 E + 1}. \tag{10.6.7}$$

We define $R = |b|^2$ as the *reflection coefficient* and $T = |c|^2$ as the *transmission coefficient*. R is the probability that an incident particle of energy E will be reflected and T is the probability that the incident particle will be transmitted. We compute that

$$R = 1/(4\varepsilon^2 E + 1), \qquad T = 4\varepsilon^2 E/(4\varepsilon^2 E + 1).$$

Observe that $T + R = 1$; thus, the total probability that a particle will be reflected or transmitted is 1. Note also that $R = 0$ and $T = 1$ when $E = \infty$ and that $T = 0$ and $R = 1$ when $E = 0$. These are the only values of E for which the classical and quantum-mechanical predictions agree. Classically, $R=1$ and $T=0$ for $E<(4\varepsilon^2)^{-1}$ and $R=0$ and $T=1$ for $E>(4\varepsilon^2)^{-1}$. Thus, quantum mechanics predicts that there is a nonzero probability that a particle will penetrate (tunnel through) a potential barrier, even if its energy is smaller than the minimum energy required by classical mechanics for transmission. [What happens classically and what happens quantum mechanically when $E=(4\varepsilon^2)^{-1}$?]

WKB Description of Tunneling through Potential Barriers

Now let us take $V(x)$ in (10.6.3) to be any continuous function which vanishes as $x \to \pm\infty$ and which rises monotonically to its maximum $V_{\max}$ ($V_{\max} > E$) at $x = x_0$ as x approaches x_0 from either the left or the right side of x_0. For this potential barrier $V(x)$ there are two turning points $x = A$ and $x = B$, $A < B$, at which $V(x) = E$. Thus, there are two classically allowed regions $x < A$ and $x > B$ in which oscillatory solutions occur and a classically forbidden region $A < x < B$ in which there are exponentially growing and decaying solutions.

We will also make the technical assumption that as $x \to \pm\infty$, $V(x) \to 0$ faster than $1/x$. We then have

$$\int_B^x dt\sqrt{E-V(t)} = \sqrt{E}\int_B^x dt + \int_B^x dt[\sqrt{E-V(t)}-\sqrt{E}]$$

$$\sim x\sqrt{E} + I, \qquad x \to +\infty,$$

where $I = \int_B^\infty dt[\sqrt{E-V(t)}-\sqrt{E}] - B\sqrt{E}$ exists and the corrections to this asymptotic relation vanish as $x \to +\infty$ (see Prob. 10.40).

Similarly, we have

$$\int_x^A dt\sqrt{E-V(t)} \sim -x\sqrt{E} + J, \qquad x \to -\infty,$$

where $J = A\sqrt{E} + \int_{-\infty}^A dt[\sqrt{E-V(t)}-\sqrt{E}]$ also exists and the corrections to this asymptotic relation also vanish as $x \to -\infty$. Consequently, as $x \to \pm\infty$, the WKB approximations to $y(x)$ in the classically allowed regions $x > B$ and $x < A$ approach plane waves as $x \to +\infty$ and $x \to -\infty$:

$$y_{\mathrm{WKB}}(x) = C_\pm[E-V(x)]^{-1/4}\exp\left[\pm\frac{i}{\varepsilon}\int_B^x dt\sqrt{E-V(t)}\right]$$

$$\sim C_\pm E^{-1/4}e^{\pm iI/\varepsilon}\exp(\pm ix\sqrt{E}/\varepsilon), \qquad x \to +\infty, \tag{10.6.8}$$

$$y_{\mathrm{WKB}}(x) = D_\pm[E-V(x)]^{-1/4}\exp\left[\pm\frac{i}{\varepsilon}\int_x^A dt\sqrt{E-V(t)}\right]$$

$$\sim D_\pm E^{-1/4}e^{\pm iJ/\varepsilon}\exp(\mp ix\sqrt{E}/\varepsilon), \qquad x \to -\infty. \tag{10.6.9}$$

As in the exactly soluble model of tunneling that we discussed above, we must choose an appropriate set of boundary conditions to describe tunneling. We postulate a unit incident right-moving plane wave at $x = -\infty$. This gives rise to a right-moving transmitted plane wave at $x = +\infty$ and a left-moving reflected wave at $x = -\infty$. We formulate these boundary conditions as asymptotic relations:

$$y(x) \sim \exp(-ix\sqrt{E}/\varepsilon) + b\exp(+ix\sqrt{E}/\varepsilon), \qquad x \to -\infty, \tag{10.6.10}$$

$$y(x) \sim c\exp(-ix\sqrt{E}/\varepsilon), \qquad x \to +\infty. \tag{10.6.11}$$

The relations in (10.6.10) and (10.6.11) are the asymptotic generalizations of the exact equations in (10.6.5) and (10.6.6). The objective is to compute the constants b and c using WKB theory.

The WKB calculation requires the solution of a new kind of one-turning-point problem which reads as follows. Let $Q(0) = 0$, $Q(x) > 0$ if $x < 0$ and $Q(x) < 0$ if $x > 0$, $Q(x) \sim ax$, $a < 0$ $(x \to 0)$. If the WKB approximation to the solution of

$$\varepsilon^2 y''(x) = Q(x)y \tag{10.6.12}$$

has negative phase for positive x,

$$y_{\text{WKB}}(x) = [-Q(x)]^{-1/4}\exp\left[-\frac{i}{\varepsilon}\int_0^x \sqrt{-Q(t)}\,dt\right], \tag{10.6.13}$$

how does $y(x)$ behave for negative x?

To solve this problem, we first allow x in (10.6.13) to approach the turning point at $x = 0$ and obtain

$$y_{\text{WKB}}(x) \sim (-ax)^{-1/4}\exp\left(-\frac{2i\sqrt{-a}}{3\varepsilon}x^{3/2}\right), \qquad x \to 0+. \tag{10.6.14}$$

We know that near $x = 0$ the differential equation (10.6.12) may be approximated by

$$y'' = -ty, \qquad t = \varepsilon^{-2/3}(-a)^{1/3}x,$$

whose solution is

$$y(t) = \alpha\,\text{Ai}\,(-t) + \beta\,\text{Bi}\,(-t). \tag{10.6.15}$$

When t is large and positive, we can replace Ai $(-t)$ and Bi $(-t)$ in (10.6.15) by their asymptotic expansions

$$\text{Ai}\,(-t) \sim \frac{1}{\sqrt{\pi}}t^{-1/4}\sin\left(\frac{2}{3}t^{3/2} + \frac{\pi}{4}\right), \qquad t \to +\infty,$$

$$\text{Bi}\,(-t) \sim \frac{1}{\sqrt{\pi}}t^{-1/4}\cos\left(\frac{2}{3}t^{3/2} + \frac{\pi}{4}\right), \qquad t \to +\infty.$$

By comparing the resulting expression with that in (10.6.14), we determine α and β to leading order in the WKB approximation:

$$\beta = \sqrt{\pi}(-\varepsilon a)^{-1/6}e^{i\pi/4}, \qquad \alpha = \sqrt{\pi}(-\varepsilon a)^{-1/6}e^{-i\pi/4}. \tag{10.6.16}$$

The problem is now half solved.

Next, we allow t in (10.6.15) to be large and *negative*. The expansion of Ai $(-t)$ is negligible compared with the expansion of Bi $(-t)$ because it is exponentially small (subdominant). Using (10.6.16) and the expansion

$$\mathrm{Bi}\,(-t) \sim \frac{1}{\sqrt{\pi}}(-t)^{-1/4}e^{2(-t)^{3/2}/3}, \qquad t \to -\infty,$$

we determine that for negative x the WKB approximation to $y(x)$ is given by

$$y_{\mathrm{WKB}}(x) = [Q(x)]^{-1/4} \exp\left[\frac{i\pi}{4} + \frac{1}{\varepsilon}\int_x^0 dt\sqrt{Q(t)}\,dt\right]. \tag{10.6.17}$$

This completes the solution to the one-turning-point problem. Using the notation in (10.4.16) we summarize our result as a connection formula:

$$\begin{array}{|ll|}\hline [Q(x)]^{-1/4} \exp\left[\dfrac{i\pi}{4} + \dfrac{1}{\varepsilon}\displaystyle\int_x^0 dt\sqrt{Q(t)}\,dt\right], & x < 0 \\ \leftarrow [-Q(x)]^{-1/4} \exp\left[-\dfrac{i}{\varepsilon}\displaystyle\int_0^x \sqrt{-Q(t)}\,dt\right], & x > 0. \\ \hline\end{array} \tag{10.6.18}$$

Now we return to the solution of the two-turning-point tunneling problem for the potential $V(x)$. When $x > B$, the WKB approximation to the solution $y(x)$ of (10.6.3) which satisfies the boundary condition in (10.6.3) is [see (10.6.8)]

$$y_{\mathrm{WKB}}(x) = ce^{iI/\varepsilon}\left[\frac{E - V(x)}{E}\right]^{-1/4} \exp\left[-\frac{i}{\varepsilon}\int_B^x \sqrt{E - V(t)}\,dt\right]. \tag{10.6.19}$$

By the connection formula in (10.6.18), this expression asymptotically matches to

$$y_{\mathrm{WKB}}(x) = ce^{iI/\varepsilon}\left[\frac{V(x) - E}{E}\right]^{-1/4} e^{i\pi/4} \exp\left[\frac{1}{\varepsilon}\int_x^B dt\sqrt{V(t) - E}\right], \tag{10.6.20}$$

which is valid when $A < x < B$.

The expression in (10.6.20) may be rewritten as

$$y_{\mathrm{WKB}}(x) = ce^{iI/\varepsilon} \exp\left[\frac{1}{\varepsilon}\int_A^B dt\sqrt{V(t) - E} + \frac{i\pi}{4}\right]$$
$$\times \left[\frac{V(x) - E}{E}\right]^{-1/4} \exp\left[-\frac{1}{\varepsilon}\int_A^x dt\sqrt{V(t) - E}\right],$$

which decays exponentially as x increases toward B. We connect to the oscillatory solution which is valid for $x < A$ using the connection formula in (10.4.16) and obtain the WKB approximation

$$y_{\mathrm{WKB}}(x) = ce^{iI/\varepsilon} \exp\left[\frac{1}{\varepsilon}\int_A^B dt\sqrt{V(t) - E}\right]\left[\frac{E - V(x)}{E}\right]^{-1/4}$$
$$\times \left\{\exp\left[\frac{i}{\varepsilon}\int_x^A dt\sqrt{E - V(t)}\right] + i\exp\left[-\frac{i}{\varepsilon}\int_x^A dt\sqrt{E - V(t)}\right]\right\}, \tag{10.6.21}$$

which is valid when $x < A$.

Finally, we let $x \to -\infty$. Recall that in this limit $V(x) \to 0$ faster than $1/x$, so (10.6.21) becomes [see (10.6.9)]

$$y_{\text{WKB}}(x) \sim ce^{iI/\varepsilon} \exp\left[\frac{1}{\varepsilon}\int_A^B dt\sqrt{V(t)-E}\right]$$
$$\times [e^{iJ/\varepsilon} \exp(-i\sqrt{E}\,x/\varepsilon) + ie^{-iJ/\varepsilon} \exp(i\sqrt{E}\,x/\varepsilon)],$$
$$x \to -\infty. \qquad (10.6.22)$$

Comparing this formula with that in (10.6.8) gives expressions for the constants b and c:

$$b = ie^{-2iJ/\varepsilon},$$
$$c = \exp\left[-\frac{1}{\varepsilon}\int_A^B dt\sqrt{V(t)-E}\right] e^{-i(I+J)/\varepsilon}.$$

Thus, the reflection coefficient is

$$R = |b|^2 \sim 1, \qquad \varepsilon \to 0+, \qquad (10.6.23)$$

and the transmission coefficient is

$$T = |c|^2 \sim \exp\left[-\frac{2}{\varepsilon}\int_A^B dt\sqrt{V(t)-E}\right], \qquad \varepsilon \to 0+. \qquad (10.6.24)$$

We observe that only an exponentially small portion of the incident wave is transmitted (tunnels through the potential barrier). Notice that the leading-order WKB prediction for R and T in (10.6.23) and (10.6.24) appears to violate the constraint that $R + T = 1$, which is always exactly satisfied. Indeed, R is not 1, but differs from 1 by an exponentially small (subdominant) quantity. However, the principles of asymptotics require that we *always* disregard subdominant corrections. (Of course, we do not replace T by 0 because it is not small compared with 0!)

Scattering off the Peak of a Potential Barrier

The reflection and transmission coefficients in (10.6.23) and (10.6.24) are good asymptotic approximations only if $E < V_{\text{max}}$. What happens when $E = V_{\text{max}}$? To answer this question, we consider the simple model problem for which $V(x) = e^{-x^2/4}$, $E = 1$. Now $V_{\text{max}} = E = 1$.

A classical particle of energy $E = 1$ moving under the influence of this potential slows down as it approaches the origin. Classically, we cannot define a reflection or transmission coefficient because the particle never actually reaches the origin! The quantum-mechanical result is much more interesting, as we will now see.

The differential equation

$$\left(-\varepsilon^2 \frac{d^2}{dx^2} + e^{-x^2/4} - 1\right) y(x) = 0 \qquad (10.6.25)$$

is quite different from that in (10.6.3) where $V_{\max} > E$ because here there is just *one* real turning point which lies at $x = 0$.

To solve this equation using asymptotic matching, we divide the x axis into three regions: region I, in which $x > 0$; region II, the immediate neighborhood of $x = 0$; and region III, in which $x < 0$. For a precise asymptotic determination of the boundaries of these regions in terms of the small parameter ε, see Prob. 10.41.

We begin our analysis in region II. Since $|x|$ is small there, we may replace $e^{-x^2/4} - 1$ by $-x^2/4$, the first term in its Taylor expansion. We thereby replace (10.6.25) by the simpler differential equation

$$\left(-\varepsilon^2 \frac{d^2}{dx^2} - x^2/4\right) y_{\rm II}(x) = 0. \tag{10.6.26}$$

Note that $x = 0$ is a *second-order* turning point.

The differential equation (10.6.26) is closely related to the parabolic cylinder equation $(-d^2/dt^2 + \frac{1}{4}t^2 - \nu - \frac{1}{2})z(t) = 0$, whose general solution is $z(t) = \alpha D_\nu(t) + \beta D_\nu(-t)$ when $\nu \neq 0, 1, 2, \ldots$ (see Example 8 of Sec. 3.8). The general solution to (10.6.26) is

$$y_{\rm II}(x) = \alpha D_{-1/2}(e^{-i\pi/4}x/\sqrt{\varepsilon}) + \beta D_{-1/2}(-e^{-i\pi/4}x/\sqrt{\varepsilon}). \tag{10.6.27}$$

Now we examine the behavior of $y(x)$ in (10.6.27) as $x/\sqrt{\varepsilon} \to \pm\infty$. We use the formulas for the asymptotic behavior of $D_\nu(t)$ as $|t| \to \infty$ in the complex plane [see (3.8.22) and (3.8.24)]:

$$D_\nu(t) \sim t^\nu e^{-t^2/4}, \qquad t \to \infty;\ |\arg t| < \frac{3\pi}{4},$$

$$D_\nu(t) \sim t^\nu e^{-t^2/4} - \frac{\sqrt{2\pi}}{\Gamma(-\nu)} e^{i\pi\nu} t^{-\nu-1} e^{t^2/4}, \qquad t \to \infty;\ \frac{\pi}{4} < \arg t < \frac{5\pi}{4}.$$

Note that $\arg(e^{-i\pi/4}\varepsilon^{-1/2}x) = -\pi/4$ and $\arg(-e^{-i\pi/4}\varepsilon^{-1/2}x) = 3\pi/4$ when $x > 0$. Thus,

$$y_{\rm II}(x) \sim \varepsilon^{1/4}x^{-1/2}[(\alpha e^{i\pi/8} + \beta e^{-3i\pi/8})e^{ix^2/4\varepsilon} + \beta\sqrt{2}\, e^{i\pi/8}e^{-ix^2/4\varepsilon}],$$
$$x/\sqrt{\varepsilon} \to +\infty. \tag{10.6.28}$$

A similar expansion exists when x is large and negative:

$$y_{\rm II}(x) \sim \varepsilon^{1/4}(-x)^{-1/2}[(\alpha e^{-3i\pi/8} + \beta e^{i\pi/8})e^{ix^2/4\varepsilon} + \alpha\sqrt{2}\, e^{i\pi/8}e^{-ix^2/4\varepsilon}],$$
$$x/\sqrt{\varepsilon} \to -\infty. \tag{10.6.29}$$

Now we return to the differential equation (10.6.25) and examine it in region I where $x > 0$ and in region III where $x < 0$. In region I the WKB solution to (10.6.25) is

$$y_{\rm I} = A(1 - e^{-x^2/4})^{-1/4} \exp\left(-\frac{i}{\varepsilon}\int_0^x dt\sqrt{1 - e^{-t^2/4}}\right), \tag{10.6.30}$$

where we have included only a negative phase solution to represent only a right-going transmitted wave for positive x. Note that

$$y_{\mathrm{I}} \sim Ae^{-iI/\varepsilon}e^{-ix/\varepsilon}, \qquad x \to +\infty,$$

where $I = \int_0^\infty dt(\sqrt{1 - e^{-t^2/4}} - 1)$. We therefore impose the outgoing wave boundary condition at $+\infty$ in (10.6.11) by requiring that

$$c = Ae^{-iI/\varepsilon}. \tag{10.6.31}$$

Next we examine (10.6.30) as $x \to 0+$. When $x \to 0$, we have $1 - e^{-x^2/4} \sim x^2/4$. Thus, performing the integral in (10.6.30) we have

$$y_{\mathrm{I}} \sim A(2/x)^{1/2}e^{-ix^2/4\varepsilon}, \qquad x \to 0+.$$

This expression must match asymptotically with that in (10.6.28). Thus, we require that

$$\varepsilon^{1/4}\beta e^{i\pi/8} = A, \qquad \alpha e^{i\pi/8} + \beta e^{-3i\pi/8} = 0. \tag{10.6.32}$$

Finally, we write down the WKB solution to (10.6.25) in region III:

$$y_{\mathrm{III}} = (1 - e^{-x^2/4})^{-1/4}$$
$$\times \left[B \exp\left(\frac{i}{\varepsilon}\int_x^0 dt\sqrt{1 - e^{-t^2/4}}\right) + C \exp\left(-\frac{i}{\varepsilon}\int_x^0 dt\sqrt{1 - e^{-t^2/4}}\right)\right]. \tag{10.6.33}$$

If we allow $x \to -\infty$, we have $y_{\mathrm{III}} \sim Be^{iI/\varepsilon}e^{-ix/\varepsilon} + Ce^{-iI/\varepsilon}e^{ix/\varepsilon}$. We impose the boundary condition in (10.6.10) by requiring that

$$Be^{iI/\varepsilon} = 1, \qquad Ce^{-iI/\varepsilon} = b. \tag{10.6.34}$$

Next we match y_{III} to y_{II}. In the limit $x \to 0-$, we have

$$y_{\mathrm{III}} \sim (-2/x)^{1/2}(Be^{ix^2/4\varepsilon} + Ce^{-ix^2/4\varepsilon}), \qquad x \to 0-.$$

Comparing this expression with that in (10.6.29) gives

$$B\sqrt{2} = \varepsilon^{1/4}(\alpha e^{-3i\pi/8} + \beta e^{i\pi/8}), \qquad C = \alpha e^{i\pi/8}\varepsilon^{1/4}. \tag{10.6.35}$$

Now, we combine the algebraic equations in (10.6.31), (10.6.32), (10.6.34), and (10.6.35). We obtain the following expressions for b and c:

$$b = \frac{i}{\sqrt{2}}e^{-2iI/\varepsilon}, \qquad c = \frac{1}{\sqrt{2}}e^{-2iI/\varepsilon}.$$

Hence, the reflection and transmission coefficients are $R = |b|^2 = \frac{1}{2}$ and $T = c^2 = \frac{1}{2}$. We obtain the elegant result that half of the incident wave is reflected and half is transmitted!

In Probs. 10.43 to 10.46 we examine other aspects of the phenomenon of tunneling.

(D) 10.7 BRIEF DISCUSSION OF HIGHER-ORDER WKB APPROXIMATIONS

In this section we show how to perform a WKB approximation beyond the leading-order approximation of physical optics. We begin by constructing a higher-order solution to the one-turning-point problem discussed in Sec. 10.4. Then we use this result to obtain an eigenvalue condition which is a higher-order version of (10.5.6).

Second-Order Solution to the One-Turning-Point Problem

We follow closely the notation of Sec. 10.4. We are given the differential equation

$$\varepsilon^2 y''(x) = Q(x)y(x), \qquad y(+\infty) = 0, \tag{10.7.1}$$

where $Q(0) = 0$, $Q(x) > 0$ for $x > 0$, and $Q(x) < 0$ for $x < 0$. We assume, as we did in Sec. 10.4, that $Q(x) \gg x^{-2}$ as $|x| \to \infty$, so that the WKB approximation is valid for all x away from the turning point at $x = 0$. We also assume that

$$Q(x) = ax + bx^2 + O(x^3), \qquad x \to 0. \tag{10.7.2}$$

First, we examine region I ($x > 0$), in which the WKB approximation is valid. For a second-order calculation we must retain one term beyond the physical-optics approximation:

$$y_{\rm I}(x) = Ce^{S_0/\varepsilon + S_1 + \varepsilon S_2}, \tag{10.7.3a}$$

where

$$S_0(x) = -\int_0^x \sqrt{Q(t)}\, dt, \tag{10.7.3b}$$

$$S_1(x) = -\frac{1}{4}\ln Q(x), \tag{10.7.3c}$$

and, integrating (10.1.4) for $S_2(x)$ once by parts, we obtain

$$S_2(x) = -\frac{5}{48}\frac{Q'(x)}{Q^{3/2}(x)} - \int_\mu^x \frac{Q''(t)}{48Q^{3/2}(t)}\, dt. \tag{10.7.3d}$$

Observe that in the expression for $S_0(x)$ we have integrated from the turning point at $x = 0$. However, in the expression for $S_2(x)$ we must integrate from $\mu > 0$ to x because the integral is divergent at $x = 0$. We will treat μ as a small fixed positive number like ε^2, for example.

Next, we consider region II, the turning-point region. When $|x|$ is small, then we may replace $Q(x)$ in (10.7.1) by the first *two* terms in its Taylor expansion and obtain

$$\varepsilon^2 y_{\rm II}''(x) = (ax + bx^2)y_{\rm II}(x), \tag{10.7.4}$$

where $y_{\rm II}(x)$, as in Sec. 10.4, stands for the approximation to $y(x)$ in the turning-point region.

It is clear that by making a linear transformation of the form $x = \alpha t + \beta$ the constants α and β can be chosen so that (10.7.4) becomes a parabolic cylinder equation. However, this trick is worthless for a third-order WKB calculation. (Why?) We prefer to use a more general approach which is equally useful in all orders. We substitute $x = \varepsilon^{2/3} a^{-1/3} t$, as in (10.4.5). This converts (10.7.4) to an approximate Airy equation

$$\frac{d^2 y_{\mathrm{II}}}{dt^2} = (t + \varepsilon^{2/3} a^{-4/3} b t^2) y_{\mathrm{II}}, \tag{10.7.5}$$

which has a small correction of order $\varepsilon^{2/3}$. We can represent the approximate solution to this equation in terms of an Airy function whose argument also has small corrections of order $\varepsilon^{2/3}$:

$$\begin{aligned} y(t) \sim D(1 &+ \alpha_1 \varepsilon^{2/3} t + \alpha_2 \varepsilon^{4/3} t^2 + \cdots) \\ &\times \mathrm{Ai}\,(t + \beta_1 \varepsilon^{2/3} t^2 + \beta_2 \varepsilon^{4/3} t^3 + \cdots), \qquad \varepsilon \to 0+. \end{aligned} \tag{10.7.6}$$

Substituting (10.7.6) into (10.7.5) determines the values of the constants $\alpha_1, \alpha_2, \ldots, \beta_1, \beta_2, \ldots$. But this is a second-order calculation, so we retain only the α_1 and β_1 terms:

$$y_{\mathrm{II}}(x) \sim D\left(1 - \frac{bx}{5a}\right) \mathrm{Ai}\left[a^{1/3} \varepsilon^{-2/3}\left(x + \frac{bx^2}{5a}\right)\right]. \tag{10.7.7}$$

If we were doing a third-order calculation, we would replace $Q(x)$ by $ax + bx^2 + cx^3$ and then compute and retain the α_1, α_2, β_1, and β_2 terms in (10.7.6).

Now we must demonstrate that $y_{\mathrm{I}}(x)$ and $y_{\mathrm{II}}(x)$ match in the overlap of regions I and II and in doing so we must find the relation between the constants C and D. To perform the asymptotic match, we replace both $y_{\mathrm{I}}(x)$ in (10.7.3a) and $y_{\mathrm{II}}(x)$ in (10.7.7) by simpler functions.

First, we examine $y_{\mathrm{I}}(x)$ for small x. We make the following approximations:

$$Q^{-1/4}(x) \sim x^{-1/4} a^{-1/4}\left(1 - \frac{bx}{4a}\right), \qquad x \to 0+,$$

$$\int_0^x \sqrt{Q(t)}\, dt \sim \frac{2}{3}\sqrt{a}\, x^{3/2} + \frac{b}{5\sqrt{a}} x^{5/2}, \qquad x \to 0+,$$

$$\frac{5}{48} \frac{Q'(x)}{Q^{3/2}(x)} \sim \frac{5}{48\sqrt{a}} x^{-3/2}, \qquad x \to 0+,$$

$$\frac{1}{48} \int_\mu^x \frac{Q''(t)}{Q^{3/2}(t)}\, dt \sim \frac{1}{12} b a^{-3/2} (\mu^{-1/2} - x^{-1/2}), \qquad x \to 0+.$$

Substituting these formulas into y_{I} in (10.7.3) gives

$$y_{\text{I}}(x) \sim Ca^{-1/4}x^{-1/4}\left(1 - \frac{bx}{4a}\right)$$
$$\times \exp\left(-\frac{2}{3\varepsilon}a^{1/2}x^{3/2} - \frac{b}{5\varepsilon}a^{-1/2}x^{5/2} - \frac{5\varepsilon}{48}a^{-1/2}x^{-3/2} - \frac{b\varepsilon a^{-3/2}}{12\sqrt{\mu}}\right),$$
$$x, \varepsilon \to 0+. \qquad (10.7.8)$$

Here we are not bothering to specify the precise size of the matching region (see Prob. 10.47). We have discarded a term of the form $\varepsilon x^{-1/2}$ in (10.7.8) because it is small in the limit $\varepsilon \to 0+$; it must be included, however, in a third-order match.

Finally, we approximate $y_{\text{II}}(x)$ in (10.7.7) by expanding the Airy function for large positive argument. We take *two* terms in the expansion of Ai (t):

$$\text{Ai}\,(t) \sim \frac{1}{2\sqrt{\pi}}t^{-1/4}e^{-2t^{3/2}/3}\left(1 - \tfrac{5}{48}t^{-3/2}\right), \qquad t \to +\infty.$$

Thus, (10.7.7) gives

$$y_{\text{II}}(x) \sim D\left(1 - \frac{bx}{4a}\right)\frac{1}{2\sqrt{\pi}}\varepsilon^{1/6}a^{-1/12}x^{-1/4}\left(1 - \frac{5\varepsilon}{48}a^{-1/2}x^{-3/2}\right)$$
$$\times \exp\left(-\frac{2}{3\varepsilon}a^{1/2}x^{3/2} - \frac{b}{5\varepsilon}a^{-1/2}x^{5/2}\right), \qquad x, \varepsilon \to 0+. \qquad (10.7.9)$$

Observe that (10.7.8) and (10.7.9) match perfectly! What is more, we obtain the condition

$$C = \frac{D}{2\sqrt{\pi}}a^{1/6}\varepsilon^{1/6}\exp\left(\frac{b\varepsilon a^{-3/2}}{12\sqrt{\mu}}\right). \qquad (10.7.10)$$

The one-turning-point problem is now half done.

The next step is to write down the oscillatory WKB solution in region III and match it to $y_{\text{II}}(x)$. In Prob. 10.48 you are asked to verify that

$$y_{\text{III}}(x) = \frac{D}{\sqrt{\pi}}\varepsilon^{1/6}a^{1/6}[-Q(x)]^{-1/4}$$
$$\times \sin\left[\frac{1}{\varepsilon}\int_x^0 \sqrt{-Q(t)}\,dt - \frac{5\varepsilon Q'(x)}{48[-Q(x)]^{3/2}} + \frac{\varepsilon}{48}\int_x^{-\mu}\frac{Q''(t)}{[-Q(t)]^{3/2}}\,dt - \frac{b\varepsilon a^{-3/2}}{12\sqrt{\mu}} + \frac{\pi}{4}\right].$$
$$(10.7.11)$$

Equations (10.7.3), (10.7.7), and (10.7.11) constitute a complete second-order solution to the one-turning-point problem.

Second-Order Solution to the Two-Turning-Point Eigenvalue Problem

If we follow Sec. 10.5 and combine two second-order one-turning-point solutions, we obtain the second-order generalization of (10.5.2) (see Prob. 10.49):

$$\frac{1}{\varepsilon}\int_A^B \sqrt{-Q(t)}\,dt + \frac{\varepsilon}{48}\int_{A+\mu}^{B-\mu} \frac{Q''(t)}{[-Q(t)]^{3/2}}\,dt - \frac{b_A \varepsilon a_A^{-3/2}}{12\sqrt{\mu}} - \frac{b_B \varepsilon a_B^{-3/2}}{12\sqrt{\mu}}$$
$$= \left(n + \frac{1}{2}\right)\pi + \mathrm{O}(\varepsilon^2), \qquad \varepsilon \to 0+, \qquad (10.7.12)$$

where a_A, b_A and a_B, b_B are the expansion coefficients of $Q(x)$ in the neighborhood of A and B. If after evaluating the integrals in (10.7.2) one allows μ to tend to $0+$, one obtains a finite answer independent of μ (see Probs. 10.50 and 10.51).

If $Q(x)$ is an analytic function of x, then (10.7.12) can be replaced by a much simpler contour integral

$$\frac{1}{2i}\oint_C \left[\frac{1}{\varepsilon} S_0'(z) + \varepsilon S_2'(z)\right] dz = \left(n + \frac{1}{2}\right)\pi + \mathrm{O}(\varepsilon^2), \qquad \varepsilon \to 0+, \qquad (10.7.13)$$

where the contour C encircles the two turning points, which are connected by a branch cut on the real-z axis, and S_0 and S_2 are given in (10.7.3). Although $S_2'(z)$ is infinite at the two turning points, the integral in (10.7.13) is finite because the contour encircles the turning points without passing through them.

Complete Perturbative Solution to the Two-Turning-Point Eigenvalue Problem

Dunham discovered a lovely generalization of (10.7.13) to all orders in perturbation theory:

$$\frac{1}{2i}\oint_C \frac{1}{\varepsilon}\sum_{k=0}^{\infty} \varepsilon^k S_k'(z)\,dz \sim n\pi, \qquad \varepsilon \to 0+. \qquad (10.7.14)$$

Let us see how this formula reduces to (10.7.13). Recall that

$$S_1'(z) = -\frac{1}{4}\frac{d}{dz}\ln\,[Q(z)].$$

Thus,

$$\frac{1}{2i}\oint S_1'(z)\,dz = -\frac{1}{8i}\ln Q(z)\Bigg|_{\substack{\text{evaluated once}\\ \text{around the contour } C}} = -\frac{1}{8i}(4\pi i) = -\frac{\pi}{2}.$$

Note that evaluating $Q(z)$ once around the contour C gives $4\pi i$ because the contour encircles the two simple zeros of $Q(z)$ at the turning points A and B. This accounts for the $\pi/2$ in (10.7.13).

It is a fact that (see Prob. 10.52) $S'_{2k+1}(z)$ $(k = 1, 2, 3, \ldots)$ is a total derivative. For example [see (10.1.15)],

$$S'_3(z) = \frac{d}{dz}\left\{\frac{5[Q'(z)]^2}{64[Q(z)]^3} - \frac{Q''(z)}{16[Q(z)]^2}\right\}.$$

This becomes a single-valued function once the turning points are joined by a branch cut. Therefore, evaluating this expression once around the closed contour C gives 0. This allows us to simplify (10.7.14) to

$$\frac{1}{2i}\oint_C \frac{1}{\varepsilon}\sum_{k=0}^{\infty} \varepsilon^{2k} S'_{2k}(z)\, dz = \left(n + \frac{1}{2}\right)\pi, \qquad \varepsilon \to 0+. \tag{10.7.15}$$

Observe that only *even* orders in WKB perturbation theory contribute to a calculation of the eigenvalues.

Example 1 *Eigenvalues of the parabolic cylinder equation.* In Example 2 of Sec. 10.5 we found that the eigenvalues E of the parabolic cylinder equation $(-d^2/dx^2 + x^2/4 - E)y(x) = 0$ $[y(\pm\infty) = 0]$ are given exactly by leading-order WKB: $E = n + \frac{1}{2}$ $(n = 0, 1, 2, \ldots)$. The explanation for this surprising result is simply that all terms in (10.7.15) after the first happen to vanish upon explicit calculation (see Prob. 10.53).

Example 2 *Eigenvalues for $y'' = (x^4 - E)y$.* After much effort we have managed to evaluate the integrals and thus to calculate the first seven terms in the WKB series in (10.7.15). The series takes the form of a power series in inverse fractional powers of E:

$$E^{3/4}\sum_{n=0}^{\infty} A_{2n} E^{-3n/2} \sim \left(n + \frac{1}{2}\right)\pi, \qquad n \to \infty, \tag{10.7.16}$$

where we have set $\varepsilon = 1$ and

$$A_0 = \frac{1}{3} R\sqrt{\pi} \doteq 1.748,$$

$$A_2 = -\frac{1}{4}\frac{\sqrt{\pi}}{R} \doteq -0.1498,$$

$$A_4 = \frac{11}{3 \cdot 2^9} R\sqrt{\pi} \doteq 0.03756,$$

$$A_6 = \frac{7 \cdot 11 \cdot 61}{3 \cdot 5 \cdot 2^{11}}\frac{\sqrt{\pi}}{R} \doteq 0.09160,$$

$$A_8 = -\frac{5 \cdot 13 \cdot 17 \cdot 353}{7 \cdot 2^{19}} R\sqrt{\pi} \doteq -0.5574,$$

$$A_{10} = -\frac{11 \cdot 11 \cdot 19 \cdot 23 \cdot 1{,}009}{3 \cdot 2^{21}}\frac{\sqrt{\pi}}{R} \doteq -5.080,$$

$$A_{12} = \frac{5 \cdot 17 \cdot 29 \cdot 49{,}707{,}277}{3 \cdot 11 \cdot 2^{28}} R\sqrt{\pi} \doteq 72.54,$$

in which $R = \Gamma(\frac{1}{4})/\Gamma(\frac{3}{4}) \doteq 2.958\,675\,119\,188\,638\,892\,310\,821\,4$.

We have not been able to discover a simple formula for the terms in this series, but the series certainly looks like a typical asymptotic series. Like the Stirling series for the gamma function,

Table 10.5 Comparison of the exact eigenvalues of the x^4 potential with the 0, 2, 4, 6, 8, 10, and 12th-order WKB predictions from (10.7.16)

Observe how rapidly the maximal accuracy increases as E_n increases

E_0(exact) $\doteq$ 1.060 362 090 484 182 899 65		E_6(exact) $\doteq$ 26.528 471 183 682 518 191 81	
$(\mathrm{WKB})_0$	0.87	$(\mathrm{WKB})_0$	26.506 335 511
$(\mathrm{WKB})_2$	0.98 (1 part in 10)	$(\mathrm{WKB})_2$	26.528 512 552
$(\mathrm{WKB})_4$	0.95	$(\mathrm{WKB})_4$	26.528 471 873
$(\mathrm{WKB})_6$	0.78	$(\mathrm{WKB})_6$	26.528 471 147
$(\mathrm{WKB})_8$	1.13	$(\mathrm{WKB})_8$	26.528 471 179
$(\mathrm{WKB})_{10}$	1.40	$(\mathrm{WKB})_{10}$	26.528 471 182 (7 parts in 10^{11})
$(\mathrm{WKB})_{12}$	1.64	$(\mathrm{WKB})_{12}$	26.528 471 181
E_2(exact) $\doteq$ 7.455 697 937 986 738 392 16		E_8(exact) $\doteq$ 37.923 001 027 033 985 146 52	
$(\mathrm{WKB})_0$	7.414 0	$(\mathrm{WKB})_0$	37.904 471 845 068
$(\mathrm{WKB})_2$	7.455 8 (1 part in 10^5)	$(\mathrm{WKB})_2$	37.923 021 140 528
$(\mathrm{WKB})_4$	7.455 3	$(\mathrm{WKB})_4$	37.923 001 229 358
$(\mathrm{WKB})_6$	7.455 2	$(\mathrm{WKB})_6$	37.923 001 021 414
$(\mathrm{WKB})_8$	7.455 2	$(\mathrm{WKB})_8$	37.923 001 026 832
$(\mathrm{WKB})_{10}$	7.455 2	$(\mathrm{WKB})_{10}$	37.923 001 027 043 (7 parts in 10^{14})
$(\mathrm{WKB})_{12}$	7.455 2	$(\mathrm{WKB})_{12}$	37.923 001 027 030
E_4(exact) $\doteq$ 16.261 826 018 850 225 937 89		E_{10}(exact) $\doteq$ 50.256 254 516 682 919 039 74	
$(\mathrm{WKB})_0$	16.233 614 7	$(\mathrm{WKB})_0$	50.240 152 319 172 36
$(\mathrm{WKB})_2$	16.261 936 7	$(\mathrm{WKB})_2$	50.256 265 932 002 07
$(\mathrm{WKB})_4$	16.261 828 6 (5 parts in 10^8)	$(\mathrm{WKB})_4$	50.256 254 592 948 49
$(\mathrm{WKB})_6$	16.261 824 5	$(\mathrm{WKB})_6$	50.256 254 515 324 64
$(\mathrm{WKB})_8$	16.261 824 9	$(\mathrm{WKB})_8$	50.256 254 516 650 43
$(\mathrm{WKB})_{10}$	16.261 825 0	$(\mathrm{WKB})_{10}$	50.256 254 516 684 34
$(\mathrm{WKB})_{12}$	16.261 825 0	$(\mathrm{WKB})_{12}$	50.256 254 516 682 99 (1 part in 10^{15})

the coefficients get smaller for a while but eventually appear to grow without bound. We would therefore expect that, for any given value of n, successive approximations to the nth eigenvalue E_n, obtained by solving (10.7.16) with the series truncated after more and more terms, should improve to some maximal accuracy and then become worse. Moreover, since E_n increases with n, more terms in the series should be required to reach maximal accuracy as n increases and the accuracy should also increase with n. This is precisely what happens (see Table 10.5). The rate at which the accuracy increases is particularly impressive. The error of the most accurate WKB approximations are indicated in parentheses.

PROBLEMS FOR CHAPTER 10

Section 10.1

(I) **10.1** Derive equations (10.1.14), (10.1.15), (10.1.16), and (10.1.17).

(E) **10.2** Show that, for the Schrödinger equation $\varepsilon^2 y'' = Q(x)y$, if $S_2' = 0$ then $S_n' = 0$ for $n \ge 2$. Deduce that the most general function $Q(x)$ for which the equation $\varepsilon^2 y'' = Q(x)y$ has the physical-optics approximation as its exact solution is $Q(x) = (c_1 x + c_2)^{-4}$.

(E) **10.3** Estimate how small ε must be for the approximation in (10.1.9) to be accurate to a relative error of ≤ 1 percent when $x \ge 1$.

(I) **10.4** Using the asymptotic methods of Chap. 6, evaluate the integral $\int_0^1 [y(x)]^n\, dx$ to leading order in powers of ε, where $y(x)$ is given in (10.1.19).

(I) **10.5** Use WKB theory to obtain the solution to $\varepsilon y'' + a(x)y' + b(x)y = 0$ $[a(x) > 0,\ y(0) = A,\ y(1) = B]$ correct to order ε.

(I) **10.6** Use second-order WKB theory to derive a formula which is more accurate than (10.3.31) for the nth eigenvalue of the Sturm-Liouville problem in (10.1.27). Let $Q(x) = (x + \pi)^4$ and compare your formula with the values of E_n in Table 10.1.

(I) **10.7.** (*a*) Show that the eigenvalues of (10.1.27) are nondegenerate. That is, show that all eigenfunctions having the same eigenvalue E are proportional to each other.

(*b*) Show that distinct eigenfunctions are orthogonal in the sense that the integral $\int_0^\pi Q(x)y_n(x)y_m(x) = 0$ when $n \neq m$.

(*c*) The Sturm-Liouville eigenfunctions $y_n(x)$ for the boundary-value problem (10.1.27) form a complete orthonormal set and may therefore be used to expand functions on the interval $(0, \pi)$ into Fourier series. The Fourier expansion of the function $f(x)$ has the form $\sum_{n=0}^{\infty} a_n y_n(x)$. Show that the Fourier coefficients are given by $a_n = \int_0^\pi dx\, f(x)y_n(x)Q(x)$.

(*d*) Let $f(x)$ be continuous and $f(0) \neq 0$. Find the leading behavior of a_n as $n \to \infty$ from (10.1.33). Is the resulting series differentiable?

Section 10.2

(E) **10.8** Consider the equations $\varepsilon^2 y''(x) = (\sin x)y$, $\varepsilon^2 y''(x) = (\sin x^2)y$, $\varepsilon^2 y''(x) = [1 + (\sin x)^2]y$. For which fixed values of x is the WKB physical-optics approximation a good approximation to $y(x)$ as $\varepsilon \to 0$. Is WKB accurate as $x \to \infty$.

(E) **10.9** For the following equations estimate how small ε must be for $\exp[(1/\varepsilon)S_0(x) + S_1(x) + \varepsilon S_2(x)]$ to be accurate to a relative error of less than 0.1 percent for all $x \geq 0$: $\varepsilon^2 y'' = Q(x)y$; (*a*) $Q(x) = \cosh x$, (*b*) $Q(x) = 1 + x^2$, (*c*) $Q(x) = 1 + x^4$.

(I) **10.10** (*a*) Verify that (10.2.8) reproduces the asymptotic formulas in (10.2.6).

(*b*) Show that the leading behavior of solutions to $\varepsilon^2 y'' = x^{-2}(\ln x)^2 y$ as $x \to +\infty$ is given by (10.2.6) with $(\ln x)^{-3/8}$ and $(\ln x)^{-5/8}$ replaced by $(\ln x)^{-1/2+\varepsilon/8}$ and $(\ln x)^{-1/2-\varepsilon/8}$, respectively. Use these results to demonstrate that physical optics is valid as $\varepsilon \to 0$ with $x > 0$ fixed, but that the higher-order approximation (10.2.8) must be used to find the leading behavior of $y(x)$ as $x \to +\infty$ with $\varepsilon > 0$ fixed.

(I) **10.11** (*a*) Derive the physical-optics approximation to the solutions of $\varepsilon^n y^{(n)} = Q(x)y$. Show that $S_0(x)$ and $S_1(x)$ are given correctly by (10.2.10) and (10.2.11).

(*b*) Show that if $Q(x)$ is sufficiently smooth and that $|x^n Q(x)| \to \infty$ as $x \to +\infty$, then physical optics gives the correct leading behavior of solutions to (10.2.9) as $x \to +\infty$ with ε fixed. This justifies the result stated after (3.4.28).

Section 10.3

(E) **10.12** Show that for (10.3.1) to have a solution $y(x)$ for which $y(\pm\infty) = 0$ it is necessary that $R(x) \ll Q(x)$ as $x \to \pm\infty$.

(I) **10.13** Verify (10.3.9) and (10.3.10).

(D) **10.14** Solve the Green's function equation $\varepsilon^2\, \partial^2 G/\partial x^2 - Q(x)G = -\delta(x - x')$ to one order beyond physical optics. That is, include S_0, S_1, and S_2 in the WKB series. Evaluate $\int_{-\infty}^{\infty} G(x, x')\, dx$ and $\int_{-\infty}^{\infty} [G(x, x')]^2\, dx$ correct to order ε.

(I) **10.15** Prove that $\int_{-\infty}^{\infty} [G_{\text{unif}}(x, x')]^3 (x - x')^2\, dx' = [Q(x)]^3/54 + O(\varepsilon)$ $(\varepsilon \to 0+)$, where G_{unif} is given in (10.3.7).

(I) **10.16** Show how to use the notions of boundary-layer theory to derive the approximate solution (10.3.11) to the Schrödinger equation (10.3.12). Show that boundary layers (localized regions of rapid change) occur at $x = \pm 1$. Find the thickness of the boundary layers. Match inner and outer solutions to derive (10.3.11).

(I) **10.17** Consider the fourth-order Green's function equation $\varepsilon^4 d^4y/dx^4 + Q(x)y = \delta(x)$ [$Q(x) > 0$, $y(\pm\infty) = 0$]. Use physical optics to derive a uniform asymptotic approximation to $y(x)$ for all x. What is $y(0)$? Evaluate $\int_{-\infty}^{\infty} y(x)\, dx$.

Section 10.4

(D) **10.18** Deduce Langer's uniform approximation (10.4.15) to the solution of the Schrödinger equation $\varepsilon^2 y'' = Q(x)y$ directly from the differential equation.

Clue: First introduce a new dependent variable which is some function $f(x)$ times the old dependent variable $y(x)$. Then show that $f(x)$ may be chosen such that for *all* x there is a new independent variable for which the equation may be approximated by the Airy equation.

(I) **10.19** Prove that Langer's formula in (10.4.15) reduces to (10.4.13*c*) in region III.

(I) **10.20** From the integral representation for Ai (x) in Prob. 6.75 prove (10.4.22). Specifically, show that $\int_0^\infty \mathrm{Ai}\,(x)\, dx = \frac{1}{3}$, $\int_{-\infty}^0 \mathrm{Ai}\,(x)\, dx = \frac{2}{3}$.

(I) **10.21** Verify (10.4.23).

(I) **10.22** Verify (10.4.28).

(I) **10.23** Solve $\varepsilon^2 y''(x) = Q(x)y(x)$, where $Q(x)$ is even, $Q(x)$ vanishes just once at $x = 0$, and $Q(x) \sim a|x|$ near $x = 0$ $(a > 0)$. Find that solution y which vanishes as $x \to \infty$.

(D) **10.24** Solve the one-turning-point Green's function equation $y''(x) - Q(x)y(x) = -\delta(x - x')$, where $Q(x)$ has a simple zero at $x = 0$, $Q(x) > 0$ if $x > 0$, $Q(x) < 0$ if $x < 0$, and $Q(x) \sim ax (x \to 0)$. $y(x)$ is required to satisfy $y(\infty) = 0$ and $y(0) = 1$. Compute $y(x')$ and $\int_{-\infty}^{\infty} y(x)\, dx$.

(E) **10.25** Use WKB theory to approximate the Bessel function $J_\nu(x)$ for $\nu, x > 0$.

Clue: The Bessel equation is $y'' + y'/x + (1 - \nu^2/x^2)y = 0$. Let $x = e^z$, remember that $J_\nu(0) = 0$ when $\nu > 0$, and use Langer's formula (10.4.15).

(D) **10.26** (*a*) Consider the one-turning-point problem $\varepsilon^2 y''(x) = Q(x)y(x)$, where $Q(x)$ vanishes just once at $x = 0$ and $Q(x) \sim a^2x^2$ as $x \to 0$, $a > 0$. Find a complete physical optics WKB approximation to $y(x)$ for that solution which approaches 0 as $x \to +\infty$ and is normalized by $y(0) = 1$. *Note:* Your answer should consist of three formulas valid in each of three regions. Combine your three formulas, à la Langer, to obtain a *single* formula which is a uniformly valid approximation to $y(x)$ for $-\infty < x < \infty$.

Clue: The final answer is

$$y_{\mathrm{unif}}(x) = \left[\frac{4aS(x)}{\pi^2 Q(x)}\right]^{1/4} \Gamma\left(\frac{3}{4}\right) D_{-1/2}[2\varepsilon^{-1/2} S^{1/2}(x)],$$

where $S(x) = \int_0^x \sqrt{Q(t)}\, dt$.

(*b*) Show that to leading order in powers of ε, $\int_0^\infty x[y(x)]^2\, dx \sim [\Gamma(\frac{3}{4})/\Gamma(\frac{1}{4})]^2 2\varepsilon/a$ $(\varepsilon \to 0+)$.

(TI) **10.27** Suppose we attempt to derive the one-turning-point connection formula using WKB in the complex plane. Let $Q(0) = 0$ and write down the WKB approximation

$$y_{\mathrm{WKB}}(z) = C[Q(z)]^{-1/4} \exp\left[-\frac{1}{\varepsilon}\int_0^z \sqrt{Q(t)}\, dt\right],$$

which is valid when z is real and positive. Then analytically continue this expression to negative z along a path which does not pass through $z = 0$ and which goes around the turning point at $z = 0$ in a counterclockwise sense in the upper half plane. We fail to derive (10.4.16). Next take a path which goes around the turning point in a clockwise sense in the lower half plane. Again we fail to derive (10.4.16). Explain the breakdown of these analytic continuations of the WKB solution in terms of the Stokes phenomenon.

(I) **10.28** Use WKB analysis to show that uniform leading-order approximations to the solutions of the differential equations of Prob. 9.33 as $\varepsilon \to 0+$ are

(*a*) $y_{\mathrm{unif},0}(x) = 2A(x + 1) + (B - 3A)e^{(6x-3)/(8\varepsilon)}$;

(*b*) $y_{\mathrm{unif},0}(x) = \dfrac{3A - B}{4}(1 - x) + \dfrac{9B - 3A}{8} e^{(6x-3)/(16\varepsilon)} + \dfrac{3B - A}{8} e^{-(6x+3)/(16\varepsilon)}$.

Explain why there is a boundary layer only at $x = \frac{1}{2}$ in (*a*) while there are boundary layers near both $x = -\frac{1}{2}$ and $x = \frac{1}{2}$ in (*b*). Observe that the approximations $y_{\text{unif},0}(x)$ are *not* exponentially small when $|x| < \frac{1}{2}$ even though these differential equations correspond to case IV of Sec. 9.6. Higher-order corrections do *not* force the solution to be exponentially small within $|x| < \frac{1}{2}$.

Clue: Observe that the outer solutions $1 + x$ and $1 - x$ to the problems are also exact solutions to the differential equations. You need only find a linearly independent solution which must grow exponentially fast away from the internal layer at $x = 0$.

(D) **10.29** Derive the connection formula for a turning point which is a $(2n + 1)$-fold zero, going from a classically forbidden to a classically allowed region. Assume that $y(x) \to 0$ as $x \to \infty$ in the classically forbidden region.

Section 10.5

(I) **10.30** Derive (10.5.4).

(I) **10.31** Show that the eigenvalue condition in (10.5.6) is a valid asymptotic relation as $\varepsilon \to 0+$ with E fixed or as $E \to +\infty$ with ε fixed.

(E) **10.32** Verify (10.5.15).

(E) **10.33** Show that the physical-optics approximation to the eigenvalues E of the equation $-y'' + (x^{2K} - E)y = 0$ $[y(\pm\infty) = 0]$, with $K = 1, 2, 3, 4, \ldots$, is

$$E_n \sim \left\{\frac{(n + \frac{1}{2})\sqrt{\pi}\,\Gamma[(3 + 1/K)/2]}{\Gamma[(2 + 1/K)/2]}\right\}^{2K/(K+1)}, \qquad n \to \infty.$$

(I) **10.34** Show that the WKB physical-optics formula for the eigenvalues of the equation $-\varepsilon^2 y'' + [V(x) - E]y = 0$ $[y(0) = 0, y(+\infty) = 0]$, where $V(0) = 0$, $V(+\infty) = +\infty$, $V(x)$ rises monotonically as x increases from 0, is $(1/\varepsilon)\int_0^{x_0} \sqrt{E - V(x)}\, dx = (n - \frac{1}{4})\pi + O(\varepsilon)$ $(\varepsilon \to 0+)$, where $E - V(x_0) = 0$ and $n = 1, 2, 3, \ldots$.

(E) **10.35** The gravitational potential rises linearly with x. Use the result in Prob. 10.34 to find the eigenvalues in a gravitational potential well.

(I) **10.36** Consider the eigenvalue problem $y'' + E(\cos x)y = 0$ $[y(\pm\pi) = 0]$. Find an approximation to E which is valid for large values of E.

(D) **10.37** Find a physical-optics approximation to the eigenvalues E of $d^4y/dx^4 = [E - V(x)]y(x)$ $[y(\pm\infty) = 0]$, where $V(\pm\infty) = \infty$. Check your result by using it to find the eigenvalues of $d^4y/dx^4 = (E - x^2)y(x)$ $[y(\pm\infty) = 0]$, which can be solved using a Fourier transform.

Clue: It will help if you solve Prob. 6.83 first.

Section 10.6

(E) **10.38** Verify (10.6.7).

(I) **10.39** Prove that a classical particle incident on a delta-function potential $V(x) = \delta(x)$ bounces back if $E < (4\varepsilon^2)^{-1}$ and continues on if $E > (4\varepsilon^2)^{-1}$.

(E) **10.40** Show that the corrections to the asymptotic relation $\int_B^x dt\sqrt{E - V(t)} \sim x\sqrt{E} + I$ $(x \to +\infty)$, where $I = \int_B^\infty dt[\sqrt{E - V(t)} - \sqrt{E}] - B\sqrt{E}$, vanish as $x \to +\infty$ if $V(x) \to 0$ faster that $1/x$.

(I) **10.41** Find precise asymptotic estimates of the boundaries of regions I, II, and III for (10.6.25).

(I) **10.42** If $e^{-x^2/4}$ in (10.6.25) were replaced by $1/(1 + x^2)$, how would the leading-order WKB predictions for R and T change?

(D) **10.43** Suppose $e^{-x^2/4}$ in (10.6.25) were replaced by $e^{-x^2/4}/2$ or by $1/(2 + 2x^2)$. In classical mechanics all incident particles penetrate the potential bump at $x = 0$. However, in quantum mechanics, there is an exponentially small reflection coefficient R. Find a physical-optics approximation to R.

Clue: It is necessary to find the connection formula for a turning point in the complex plane. See the discussion of this problem in Ref. 18.

(D) **10.44** In this problem we investigate the quantum-mechanical phenomenon of resonance.

(*a*) Let $V(x)$ in (10.6.3) be two delta functions at $x = 0$ and $x = 1$: $V(x) = \delta(x) + \delta(x - 1)$. Solve (10.6.3) for this $V(x)$ exactly and show that there is an infinite number of discrete energies for which $V(x)$ becomes transparent ($T = 1$, $R = 0$). Compute these energies.

(*b*) Let $V(x)$ in (10.6.3) be $x^2 - x^4$. Use WKB theory to find a physical-optics approximation for the resonant energies E_n.

(D) **10.45** Suppose we wish to calculate the eigenvalues of a double potential well separated by a potential hill (like the letter W). Then there are two cases to consider:

(*a*) If each of the wells has a different shape, then to a good approximation the eigenvalue spectrum (for those eigenvalues below the peak of the hill) is the union of the spectra of each of the two wells separately. Corrections are exponentially small (subdominant). Explain why.

(*b*) Suppose the two wells are identical. For this problem take $V(x) = x^4 - x^2$. Now the eigenvalue spectrum consists of almost degenerate pairs of eigenvalues. Use WKB theory to calculate the splitting between pairs of eigenvalues.

Clue: For the above $V(x)$ assume that eigenfunctions are either even or odd functions of x.

(D) **10.46** In this problem we investigate the quantum-mechanical phenomenon of radioactive decay. Radioactive decay, the tunneling of a wavefunction out of a confined region, is clearly a time-dependent phenomenon. Therefore we must return to the time-dependent Schrödinger equation (10.6.1). We represent the time dependence as in (10.6.2). The decay constant is the imaginary part of E, which must be positive. Establish the following results:

(*a*) Define a probability density $\rho(x, t)$ and a probability current $j(x, t)$ by $\rho(x, t) = \psi^*\psi/\varepsilon$, $j(x, t) = i\varepsilon[\psi^*(\partial\psi/\partial x) - \psi(\partial\psi^*/\partial x)]$. Show that as a consequence of (10.6.1), j and ρ satisfy a local conservation law: $\partial j/\partial x + \partial\rho/\partial t = 0$.

(*b*) Consider a potential V which looks like an upside-down letter W. For example, take $V(x) = x^2 - x^4$. Show that in general

$$\text{Im } E = \frac{j(x_2, t) - j(x_1, t)}{2\int_{x_1}^{x_2} \rho(x, t)\, dx},$$

where x_1 and x_2 are points to the left and right of and outside the potential well. What is the connection between the sign of Im E and the direction of flow of probability current? Does Im E vary with time?

(*c*) Impose outgoing wave boundary conditions and find a physical-optics approximation to Im E. What happens to Im E if we impose incoming wave boundary conditions? Explain.

Section 10.7

(I) **10.47** What is the size of the region in which $y_{\text{I}}(x)$ in (10.7.8) and $y_{\text{II}}(x)$ in (10.7.9) match asymptotically?

(I) **10.48** Verify (10.7.11).

(D) **10.49** Derive (10.7.12) by combining two second-order one-turning-point WKB approximations.

(I) **10.50** Evaluate (10.7.12) for $Q(x) = x^2/4 - E$.

Clue: Show that $A = 2\sqrt{E}$ and $B = -2\sqrt{E}$; $a_A = a_B = \sqrt{E}$; $b_A = b_B = \frac{1}{4}$. Then show that for small μ, the left side of (10.7.12) reduces to $E\pi - \sqrt{\mu}\, E^{-3/4}/64$. Thus, in the limit $\mu \to 0+$, we recover the physical-optics result $E \sim n + \frac{1}{2}$ $(n \to \infty)$.

(I) **10.51** Evaluate (10.7.12) for $Q(x) = x^4 - E$. Show that in the limit $\mu \to 0+$, the left side of (10.7.12) becomes

$$\frac{E^{3/4}\Gamma(\frac{1}{4})\sqrt{\pi}}{3\Gamma(\frac{3}{4})} - \frac{E^{-3/4}\Gamma(\frac{3}{4})\sqrt{\pi}}{4\Gamma(\frac{1}{4})}.$$

Conclude that

$$E \sim \left[3\left(n + \frac{1}{2}\right)\sqrt{\pi}\,\Gamma\left(\frac{3}{4}\right)\Big/\Gamma\left(\frac{1}{4}\right)\right]^{4/3}\left[1 + \frac{1}{81(n + \frac{1}{2})^2\pi}\right]$$

as $n \to \infty$. Can you understand why the numerical results in Table 10.5 improve so dramatically with increasing n?

(TD) **10.52** Formulate a proof that S'_{2k+1} $(k = 1, 2, 3, \ldots)$ is a total derivative. Check your result by explicitly calculating S'_3, S'_5, and S'_7 for arbitrary Q.

(D) **10.53** Calculate explicitly the first three terms in the series (10.7.15) for $Q(x) = \frac{1}{4}x^2 - E$ and show that the only nonvanishing term is the first. Explain why *every* term vanishes except the first.

CHAPTER

ELEVEN

MULTIPLE-SCALE ANALYSIS

And here—ah, now, this really is something a little recherché.

—Sherlock Holmes, *The Musgrave Ritual*
Sir Arthur Conan Doyle

(E) 11.1 RESONANCE AND SECULAR BEHAVIOR

Multiple-scale analysis is a very general collection of perturbation techniques that embodies the ideas of both boundary-layer theory and WKB theory. Multiple-scale analysis is particularly useful for constructing uniformly valid approximations to solutions of perturbation problems.

In this section we show how nonuniformity can appear in a regular perturbation expansion as a result of resonant interactions between consecutive orders of perturbation theory. To illustrate, we examine a simple perturbation problem, show how resonances occur and lead to a nonuniformly valid perturbation expansion, and finally show how to interpret and eliminate these nonuniformities. The formal development of multiple-scale analysis is postponed to Sec. 11.2.

Resonance

The phenomenon of resonance is nicely illustrated by the differential equation

$$\frac{d^2}{dt^2} y(t) + y(t) = \cos(\omega t). \tag{11.1.1}$$

This equation represents a harmonic oscillator of natural frequency 1 which is driven by a periodic external force of frequency ω. The general solution to this equation for $|\omega| \neq 1$ has the form

$$y(t) = A \cos t + B \sin t + \frac{\cos(\omega t)}{1 - \omega^2}, \qquad |\omega| \neq 1. \tag{11.1.2}$$

Observe that for all $|\omega| \neq 1$ the solution remains bounded for all t. If $|\omega|$ is close to 1, the amplitude of oscillation becomes large because the system absorbs large amounts of energy from the external force. Nevertheless, the amplitude of the system is still bounded when $|\omega| \neq 1$ because the system is oscillating out of phase with the driving force.

The solution in (11.1.2) is incorrect when $|\omega| = 1$. The correct solution has an amplitude which grows with t:

$$y(t) = A \cos t + B \sin t + \tfrac{1}{2}t \sin t, \qquad |\omega| = 1. \tag{11.1.3}$$

The amplitude of oscillation of this solution is unbounded as $t \to \infty$ because the oscillator continually absorbs energy from the periodic external force. This system is in *resonance* with the external force.

The term $\frac{1}{2}t \sin t$, whose amplitude grows with t, is said to be a *secular* term. The secular term $\frac{1}{2}t \sin t$ has appeared because the inhomogeneity $\cos t$ in (11.1.1) with $|\omega| = 1$ is itself a solution of the homogeneous equation associated with (11.1.1): $d^2y/dt^2 + y = 0$. In general, secular terms always appear whenever the inhomogeneous term is itself a solution of the associated homogeneous *constant-coefficient* differential equation. A secular term always grows more rapidly than the corresponding solution of the homogeneous equation by at least a factor of t.

Example 1 *Appearance of secular terms.*

(*a*) The solution to the differential equation $d^2y/dt^2 - y = e^{-t}$ has a secular term because e^{-t} satisfies the associated homogeneous equation. The general solution is $y(t) = Ae^{-t} + Be^t - \frac{1}{2}te^{-t}$. The particular solution $-\frac{1}{2}te^{-t}$ is secular relative to the homogeneous solution Ae^{-t}; we must regard the term $-\frac{1}{2}te^{-t}$ as secular even though it is negligible compared with the homogeneous solution Be^t as $t \to \infty$.

(*b*) The solution to the differential equation $d^2y/dt^2 - 2dy/dt + y = e^t$ has a secular term because e^t satisfies the associated homogeneous equation. The general solution is $y(t) = Ae^t + Bte^t + \frac{1}{2}t^2e^t$. In this case, the particular solution $\frac{1}{2}t^2e^t$ is secular with respect to all solutions of the associated homogeneous equation.

Nonuniformity of Regular Perturbation Expansions

The appearance of secular terms signals the nonuniform validity of perturbation expansions for large t. The nonlinear oscillator equation (Duffing's equation)

$$\frac{d^2y}{dt^2} + y + \varepsilon y^3 = 0, \qquad y(0) = 1,\ y'(0) = 0, \tag{11.1.4}$$

provides a good illustration of what we mean by nonuniformity. A perturbative solution of this equation is obtained by expanding $y(t)$ as a power series in ε:

$$y(t) = \sum_{n=0}^{\infty} \varepsilon^n y_n(t), \tag{11.1.5}$$

where $y_0(0) = 1$, $y_0'(0) = 0$, $y_n(0) = y_n'(0) = 0$ $(n \geq 1)$. Substituting (11.1.5) into the differential equation (11.1.4) and equating coefficients of like powers of ε gives a sequence of linear differential equations of which all but the first are inhomogeneous:

$$y_0'' + y_0 = 0, \tag{11.1.6a}$$

$$y_1'' + y_1 = -y_0^3, \tag{11.1.6b}$$

and so on.

The solution to (11.1.6*a*) which satisfies $y_0(0) = 1$, $y_0'(0) = 0$ is

$$y_0(t) = \cos t.$$

To solve (11.1.6*b*) we invoke the trigonometric identity $\cos^3 t = \frac{1}{4}\cos 3t + \frac{3}{4}\cos t$ to rewrite the inhomogeneous term. The formulas in (11.1.2)–(11.1.3) then provide the general solution to (11.1.6*b*):

$$y_1(t) = A\cos t + B\sin t + \tfrac{1}{32}\cos 3t - \tfrac{3}{8}t\sin t;$$

the particular solution satisfying $y_1(0) = y_1'(0) = 0$ is

$$y_1(t) = \tfrac{1}{32}\cos 3t - \tfrac{1}{32}\cos t - \tfrac{3}{8}t\sin t.$$

We observe that $y_1(t)$ contains a secular term. This secularity necessarily occurs because $\cos^3 t$ contains a component, $\frac{3}{4}\cos t$, whose frequency equals the natural frequency of the unperturbed oscillator.

In summary, the first-order perturbative solution to (11.1.4) is

$$y(t) = \cos t + \varepsilon[\tfrac{1}{32}\cos 3t - \tfrac{1}{32}\cos t - \tfrac{3}{8}t\sin t] + O(\varepsilon^2), \qquad \varepsilon \to 0+. \qquad (11.1.7)$$

We emphasize that the term $O(\varepsilon^2)$ in the above expression means that for *fixed* t the error between $y(t)$ and $y_0(t) + \varepsilon y_1(t)$ is at most of order ε^2 as $\varepsilon \to 0+$. The nonuniformity of this result surfaces if we consider large values of t—specifically, values of t of order $1/\varepsilon$ or larger as $\varepsilon \to 0+$. For such large values of t, the secular term in $y_1(t)$ suggests that the amplitude of oscillation grows with t. However, as we will now show, the exact solution $y(t)$ remains bounded for all t.

Boundedness of the Solution to (11.1.4)

To show that the solution to (11.1.4) is bounded for all t, we construct an integral of the differential equation. Multiplying (11.1.4) by the integrating factor dy/dt converts each term in the differential equation to an exact derivative:

$$\frac{d}{dt}\left[\frac{1}{2}\left(\frac{dy}{dt}\right)^2 + \frac{1}{2}y^2 + \frac{1}{4}\varepsilon y^4\right] = 0.$$

Thus,

$$\frac{1}{2}\left(\frac{dy}{dt}\right)^2 + \frac{1}{2}y^2 + \frac{1}{4}\varepsilon y^4 = C, \qquad (11.1.8)$$

where C is a constant. Since $y(0) = 1$ and $y'(0) = 0$, $C = \frac{1}{2} + \frac{1}{4}\varepsilon$. When $\varepsilon > 0$, the integral in (11.1.8) shows that $\frac{1}{2}y^2 \le C$ for all t. Therefore, $|y(t)|$ is bounded for all t by $\sqrt{1 + \varepsilon/2}$.

The argument just given is frequently used in applied mathematics to prove boundedness of solutions to both ordinary and partial differential equations. The integral in (11.1.8) is called an *energy* integral. Equation (11.1.8) may be interpreted graphically as a closed bounded orbit in the phase plane whose axes are labeled by y and dy/dt (see Fig. 11.1).

Perturbative Construction of a Bounded Solution to (11.1.4)

We have arrived at an apparent paradox; we have shown that the exact solution $y(t)$ to (11.1.4) is bounded for all t but that the first-order perturbative solution in (11.1.7) is secular (grows with t for large t). The resolution of this paradox lies in

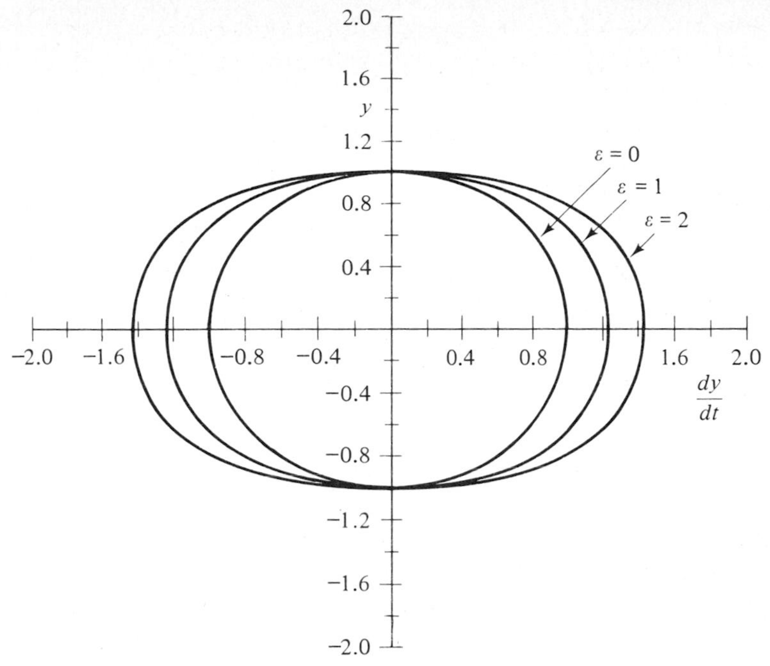

Figure 11.1 A phase-plane plot (y versus dy/dt) of solutions to Duffing's equation $d^2y/dt^2 + y + \varepsilon y^3 = 0$ [$y(0) = 1$, $y'(0) = 0$] for $\varepsilon = 0$, 1, and 2. The orbits shown are constant-energy curves [see (11.1.8)] which satisfy $(dy/dt)^2 + y^2 + \varepsilon y^4/2 = 1 + \varepsilon/2$.

the summation of the perturbation series (11.1.5). We know that the problem (11.1.4) is a regular perturbation problem as $\varepsilon \to 0+$ for fixed t (see Sec. 7.2). Therefore, the series (11.1.5) converges to the solution $y(t)$ for each t. We conclude that although order by order each term in the perturbation expansion may be secular, the secularity must disappear when the series is summed.

To illustrate how summing a perturbation series can eliminate secularity, consider the perturbation series

$$1 - \varepsilon t + \tfrac{1}{2}\varepsilon^2 t^2 - \tfrac{1}{6}\varepsilon^3 t^3 + \cdots + \varepsilon^n t^n[(-1)^n/n!] + \cdots, \qquad \varepsilon \to 0+.$$

Each term in this series is secular when t is of order $1/\varepsilon$ or larger. Nevertheless, the sum of the series $e^{-\varepsilon t}$ is bounded for all positive t!

We will now examine the more complicated perturbation series (11.1.5) and show that the sum of the most secular terms in each order in perturbation theory is actually not secular. We will show, using an inductive argument, that the most secular term in $y_n(t)$ has the form

$$A_n t^n e^{it} + A_n^* t^n e^{-it}, \tag{11.1.9}$$

where * denotes complex conjugation. There are less secular terms in $y_n(t)$ which grow like t^k ($k < n$), but we ignore such terms for now.

The final result of our calculations will be

$$A_n = \frac{1}{2}\frac{1}{n!}\left(\frac{3i}{8}\right)^n. \tag{11.1.10}$$

Using this formula for A_n we see that the sum of the most secular terms in the perturbation series (11.1.5) is a cosine function:

$$\sum_{n=0}^{\infty} \frac{1}{2} \varepsilon^n t^n \left[\frac{1}{n!}\left(\frac{3i}{8}\right)^n e^{it} + \frac{1}{n!}\left(-\frac{3i}{8}\right)^n e^{-it}\right] = \cos\left[t\left(1 + \frac{3}{8}\varepsilon\right)\right]. \quad (11.1.11)$$

Observe that this expression is not secular; it remains bounded for all t.

The expression (11.1.11) is a much better approximation to the exact solution $y(t)$ than $y_0(t) = \cos t$ because it is a good approximation to $y(t)$ for $0 \le t = O(1/\varepsilon)$. The difference between $y(t)$ and $\cos t$ is small so long as $0 \le t \ll 1/\varepsilon$ ($\varepsilon \to 0+$), while $\cos [t(1 + \frac{3}{8}\varepsilon)]$ is an accurate approximation to $y(t)$ over a much larger range of t. These assertions are explained as follows. In order that $y_0(t)$ be a good approximation to $y(t)$, it is necessary that $\varepsilon^n y_n(t) \ll y_0(t)$ ($\varepsilon \to 0+$) for all $n \ge 1$; this is true if $0 \le t \ll 1/\varepsilon$. On the other hand, the terms that we ignored in deriving (11.1.11) all have the form

$$\varepsilon[A\varepsilon^k(\varepsilon t)^l e^{imt} + A^*\varepsilon^k(\varepsilon t)^l e^{-imt}],$$

where k, l, m are nonnegative integers. Therefore, when $t = O(1/\varepsilon)$, each of these ignored terms is in fact negligible compared to at least one of the secular terms included in (11.1.11). We accept without proof the nontrivial result that the sum of all these small terms is still small. The higher-order terms are analyzed in Probs. 11.5 to 11.7.

We interpret the formula in (11.1.11) to mean that the cubic anharmonic term in (11.1.4) causes a shift in the frequency of the harmonic oscillator $y'' + y = 0$ from 1 to $1 + \frac{3}{8}\varepsilon$. This small frequency shift causes a phase shift which becomes noticeable when t is of order $1/\varepsilon$ (see Figs. 11.2 to 11.4 in Sec. 11.2).

Inductive Derivation of (11.1.10)

Comparing the first-order perturbation theory result in (11.1.7) with (11.1.9) verifies that the coefficient of the most secular terms in zeroth and first order are given correctly by (11.1.10). To establish (11.1.10) for all n, we proceed inductively. The $(n + 2)$th equation in the sequence of equations (11.1.6) determines $y_{n+1}(t)$:

$$y''_{n+1} + y_{n+1} = -I_{n+1}, \quad (11.1.12)$$

where the inhomogeneity I_{n+1} is the coefficient of ε^n in the expansion of $[\sum_{j=0}^{\infty} \varepsilon^j y_j(t)]^3$. Thus,

$$I_{n+1} = \sum_{j+k+l=n} y_j y_k y_l. \quad (11.1.13)$$

The most secular term in $y_{n+1}(t)$ is generated by the most secular terms in $y_j(t)$ for $0 \le j \le n$ (see Prob. 11.2). If we assume that (11.1.10) is valid for $A_0, A_1, A_2, \ldots, A_n$, then the coefficient of $t^n e^{it}$ in I_{n+1} is given by

$$\frac{1}{8}\left(\frac{3}{8}\right)^n \sum_{j+k+l=n} \frac{i^{j+k-l} + i^{j+l-k} + i^{k+l-j}}{j!\,k!\,l!} = \frac{1}{8}\left(\frac{3i}{8}\right)^n \sum_{j+k+l=n} \frac{(-1)^l + (-1)^k + (-1)^j}{j!\,k!\,l!}.$$

The sum in the above expression is just three times the coefficient of x^n in the Taylor expansion of $e^x e^x e^{-x}$ (see Prob. 11.3); therefore, it has the value $3/n!$. Thus, the terms in I_{n+1} which generate the most secular terms in $y_{n+1}(t)$ are

$$\tfrac{3}{8}(\tfrac{3}{8}t)^n[i^n e^{it} + (-i)^n e^{-it}]/n!.$$

Substituting these terms into the right side of (11.1.12) and solving for $y_{n+1}(t)$ gives

$$y_{n+1}(t) = (\tfrac{3}{8}t)^{n+1}[i^{n+1}e^{it} + (-i)^{n+1}e^{-it}]/(n+1)! + \text{less secular terms.}$$

By induction, we conclude that since (11.1.10) is true for $n = 0$, it remains true for all n.

(E) 11.2 MULTIPLE-SCALE ANALYSIS

In Sec. 11.1 we showed how to eliminate the most secular contributions to perturbation theory by simply summing them to all orders in powers of ε. The method we used works well but requires a lengthy calculation which can be avoided by using the methods of multiple-scale analysis that are introduced in this section.

Once again, we consider the nonlinear oscillator problem in (11.1.4):

$$\frac{d^2y}{dt^2} + y + \varepsilon y^3 = 0, \qquad y(0) = 1,\ y'(0) = 0. \tag{11.2.1}$$

The principal result of the last section is that when t is of order $1/\varepsilon$, perturbation theory in powers of ε is invalid. Secular terms appear in all orders (except zeroth order) and violate the boundedness of the solution $y(t)$.

A shortcut for eliminating the most secular terms to all orders begins by introducing a new variable $\tau = \varepsilon t$. τ defines a long time scale because τ is not negligible when t is of order $1/\varepsilon$ or larger. Even though the exact solution $y(t)$ is a function of t alone, multiple-scale analysis seeks solutions which are functions of both variables t and τ treated as *independent* variables. We emphasize that expressing y as a function of two variables is an artifice to remove secular effects; the actual solution has t and τ related by $\tau = \varepsilon t$ so that t and τ are ultimately not independent.

The formal procedure consists of assuming a perturbation expansion of the form

$$y(t) = Y_0(t, \tau) + \varepsilon Y_1(t, \tau) + \cdots. \tag{11.2.2}$$

We use the chain rule for partial differentiation to compute derivatives of $y(t)$:

$$\frac{dy}{dt} = \left(\frac{\partial Y_0}{\partial t} + \frac{\partial Y_0}{\partial \tau}\frac{d\tau}{dt}\right) + \varepsilon\left(\frac{\partial Y_1}{\partial t} + \frac{\partial Y_1}{\partial \tau}\frac{d\tau}{dt}\right) + \cdots.$$

However, since $\tau = \varepsilon t$, $d\tau/dt = \varepsilon$. Thus,

$$\frac{dy}{dt} = \frac{\partial Y_0}{\partial t} + \varepsilon\left(\frac{\partial Y_0}{\partial \tau} + \frac{\partial Y_1}{\partial t}\right) + O(\varepsilon^2). \tag{11.2.3}$$

Also, differentiating with respect to t again gives

$$\frac{d^2y}{dt^2} = \frac{\partial^2 Y_0}{\partial t^2} + \varepsilon\left(2\frac{\partial^2 Y_0}{\partial \tau\,\partial t} + \frac{\partial^2 Y_1}{\partial t^2}\right) + O(\varepsilon^2). \tag{11.2.4}$$

Substituting (11.2.4) into (11.2.1) and collecting powers of ε gives

$$\frac{\partial^2 Y_0}{\partial t^2} + Y_0 = 0, \tag{11.2.5}$$

$$\frac{\partial^2 Y_1}{\partial t^2} + Y_1 = -Y_0^3 - 2\frac{\partial^2 Y_0}{\partial \tau\,\partial t}. \tag{11.2.6}$$

The most general real solution to (11.2.5) is

$$Y_0(t, \tau) = A(\tau)e^{it} + A^*(\tau)e^{-it}, \tag{11.2.7}$$

where $A(\tau)$ is an arbitrary complex function of τ.

$A(\tau)$ will be determined by the condition that secular terms do *not* appear in the solution to (11.2.6). From (11.2.7), the right side of (11.2.6) is

$$e^{it}\left[-3A^2A^* - 2i\frac{dA}{d\tau}\right] + e^{-it}\left[-3A(A^*)^2 + 2i\frac{dA^*}{d\tau}\right] - e^{3it}A^3 - e^{-3it}(A^*)^3.$$

Note that e^{it} and e^{-it} are solutions of the homogeneous equation $\partial^2 Y_1/\partial t^2 + Y_1 = 0$. Therefore, if the coefficients of e^{it} and e^{-it} on the right side of (11.2.6) are nonzero, then the solution $Y_1(t, \tau)$ will be secular in t. To preclude the appearance of secularity, we require that the as yet arbitrary function $A(\tau)$ satisfy

$$-3A^2A^* - 2i\frac{dA}{d\tau} = 0, \tag{11.2.8}$$

$$-3A(A^*)^2 + 2i\frac{dA^*}{d\tau} = 0. \tag{11.2.9}$$

These two complex equations do not overdetermine $A(\tau)$ because they are redundant; one is the complex conjugate of the other. If (11.2.8) and (11.2.9) are satisfied, no secularity appears in (11.2.2), at least through terms of order ε.

To solve (11.2.8) for $A(\tau)$, we represent $A(\tau)$ in polar coordinate form:

$$A(\tau) = R(\tau)e^{i\theta(\tau)}, \tag{11.2.10}$$

where R and θ are real. Substituting into (11.2.8) and equating real and imaginary parts gives

$$\frac{dR}{d\tau} = 0, \tag{11.2.11a}$$

$$\frac{d\theta}{d\tau} = \frac{3}{2}R^2. \tag{11.2.11b}$$

Therefore,

$$A(\tau) = R(0)e^{i\theta(0) + 3iR^2(0)\tau/2} \tag{11.2.12}$$

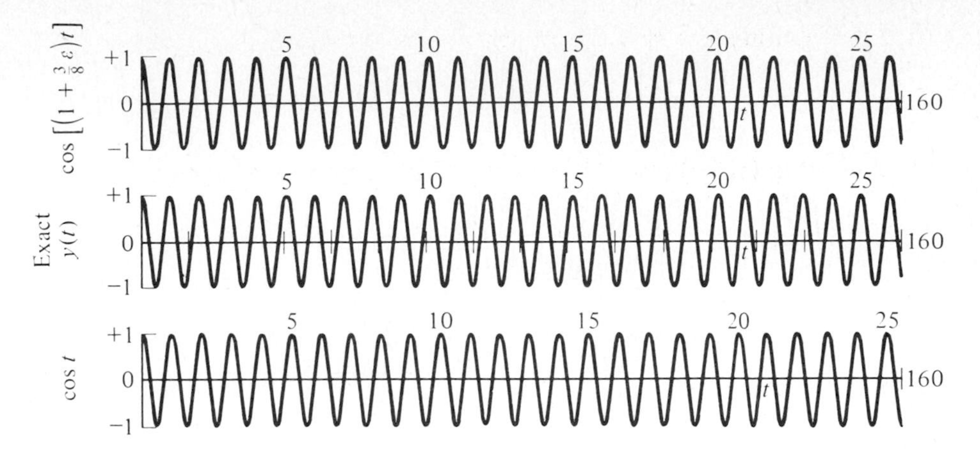

Figure 11.2 The exact solution $y(t)$ to Duffing's equation $d^2y/dt^2 + y + \varepsilon y^3 = 0$ $[y(0) = 1,\ y'(0) = 0]$ for $\varepsilon = 0.1$ (middle graph) compared with perturbative approximations to $y(t)$ (upper and lower graphs). The lower graph is a plot of $\cos t$, the first term in the regular perturbation series for $y(t)$, and the upper graph is a plot of $\cos[(1 + 3\varepsilon/8)t]$, the leading-order approximation to $y(t)$ obtained from multiple-scale methods. Both approximations, $\cos t$ and $\cos[(1 + 3\varepsilon/8)t]$, are correct up to additive terms of order ε, but $\cos t$ is not valid for large values of t; when $t = 160$, $\cos t$ is a full cycle out of phase with $y(t)$. The multiple-scale approximation closely approximates $y(t)$, even for large values of t.

and the zeroth-order solution (11.2.7) is

$$Y_0(t, \tau) = 2R(0) \cos\left[\theta(0) + \tfrac{3}{2}R^2(0)\tau + t\right]. \tag{11.2.13}$$

The initial conditions $y(0) = 1$, $y'(0) = 0$ determine $R(0)$ and $\theta(0)$. The condition $y(0) = 1$ becomes $Y_0(0, 0) = 1$, $Y_1(0, 0) = 0, \ldots$. From (11.2.3), $y'(0) = 0$ becomes $(\partial Y_0/\partial t)(0, 0) = 0$, $(\partial Y_1/\partial t)(0, 0) = -(\partial Y_0/\partial \tau)(0, 0), \ldots$. In order to satisfy these conditions, we must choose $R(0) = \frac{1}{2}$ and $\theta(0) = 0$. Therefore, the zeroth-order solution is $Y_0(t, \tau) = \cos\left[t + \frac{3}{8}\tau\right]$. Finally, since $\tau = \varepsilon t$,

$$y(t) = \cos\left[t\left(1 + \tfrac{3}{8}\varepsilon\right)\right] + O(\varepsilon), \qquad \varepsilon \to 0+,\ \varepsilon t = O(1), \tag{11.2.14}$$

and we have reproduced (11.1.11). In Figs. 11.2 to 11.4 we compare the exact solution to (11.2.1) with the approximation in (11.2.14).

A higher-order treatment of (11.2.1) is not completely straightforward. When more than two time scales are employed, there is so much freedom in the perturbation series representation that ambiguities can result (see Probs. 11.5 to 11.7).

(I) 11.3 EXAMPLES OF MULTIPLE-SCALE ANALYSIS

In this section we illustrate the formal multiple-scale technique that was developed in Sec. 11.2 by showing how to solve four elementary examples. The third and fourth of these examples are especially interesting because they show

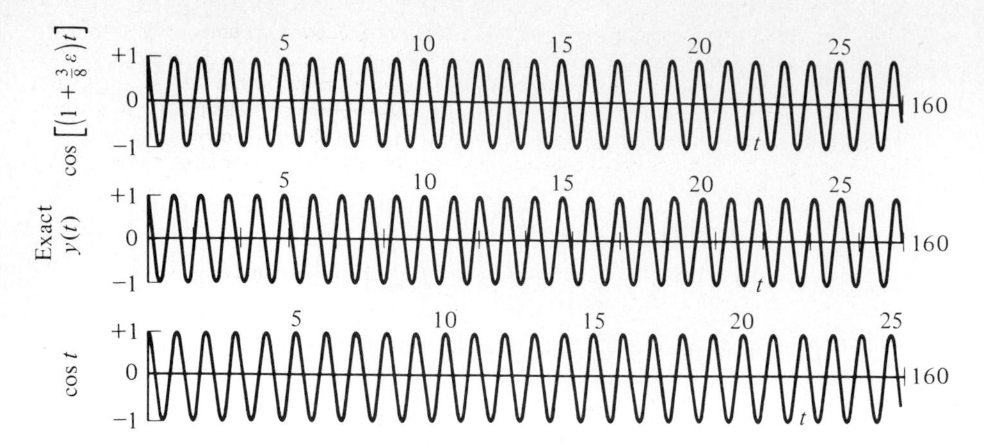

Figure 11.3 Same as in Fig. 11.2 but with $\varepsilon = 0.2$. Note that $\cos t$ is two cycles out of phase with $y(t)$ when $t = 160$.

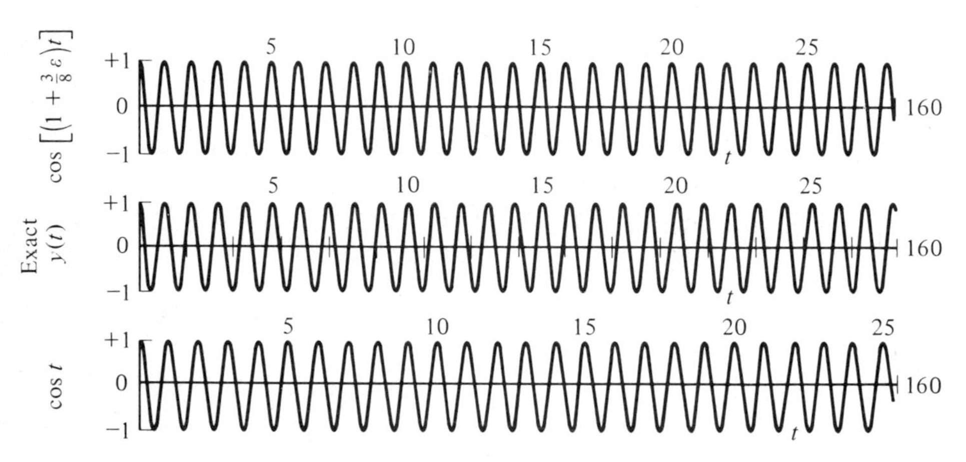

Figure 11.4 Same as in Fig. 11.2 but with $\varepsilon = 0.3$. Note that $\cos t$ is three cycles out of phase with $y(t)$ when $t = 160$.

how multiple-scale analysis can reproduce the results of boundary-layer and WKB analysis.

Example 1 *Multiple-scale analysis of a damped oscillator.* Let us consider an harmonic oscillator with a cubic damping term:

$$y'' + y + \varepsilon(y')^3 = 0, \qquad y(0) = 1,\ y'(0) = 0. \tag{11.3.1}$$

If $\varepsilon > 0$, the solution $y(t)$ must decay to 0 as $t \to \infty$. To prove this assertion, we multiply (11.3.1) by y' and construct an energy integral similar to that in (11.1.8):

$$\frac{d}{dt}\left[\frac{1}{2}(y')^2 + \frac{1}{2}y^2\right] = -\varepsilon(y')^4 \le 0. \tag{11.3.2}$$

This result shows that the energy $\frac{1}{2}(y')^2 + \frac{1}{2}y^2$ is a decreasing function of t unless $y'(t) = 0$ for all t. In Prob. 11.8 it is shown that the energy must decay to 0 as $t \to \infty$ and therefore that $y(t) \to 0$ as $t \to \infty$. [By contrast, when $\varepsilon < 0$, the energy argument just given shows that (11.3.1) represents a negatively damped system (like a self-propelled lawnmower that uses grass for fuel or a rocket with vacuum-cleaner drive that uses space dust for fuel) whose solutions grow explosively with t.]

Multiple-scale analysis may be used to study the behavior of $y(t)$ for large t. We begin by assuming a perturbation expansion for $y(t)$ in (11.3.1) of the form

$$y(t) \sim Y_0(t, \tau) + \varepsilon Y_1(t, \tau) + \cdots, \qquad \varepsilon \to 0+,$$

where $\tau = \varepsilon t$. Using (11.2.3) and (11.2.4) and equating coefficients of ε^0 and ε^1 gives two equations which correspond with (11.2.5) and (11.2.6):

$$\frac{\partial^2 Y_0}{\partial t^2} + Y_0 = 0, \tag{11.3.3}$$

$$\frac{\partial^2 Y_1}{\partial t^2} + Y_1 = -2\frac{\partial^2 Y_0}{\partial t\, \partial \tau} - \left(\frac{\partial Y_0}{\partial t}\right)^3. \tag{11.3.4}$$

The most general real solution to (11.3.3) is

$$Y_0(t, \tau) = A(\tau)e^{it} + A^*(\tau)e^{-it}. \tag{11.3.5}$$

Substituting this solution into the right side of (11.3.4) gives

$$\frac{\partial^2 Y_1}{\partial t^2} + Y_1 = -e^{it}\left[2i\frac{dA}{d\tau} + 3iA^2A^*\right] - e^{-it}\left[-2i\frac{dA^*}{d\tau} - 3i(A^*)^2A\right]$$
$$+ ie^{3it}A^3 - ie^{-3it}(A^*)^3. \tag{11.3.6}$$

Since the solutions to the homogeneous equation (11.3.3) are $e^{\pm it}$, the solution to (11.3.6) is secular unless the expressions in the square brackets vanish; in order that Y_1 not be secular, we require that $A(\tau)$ satisfy the equations

$$2i\frac{dA}{d\tau} + 3iA^2A^* = 0, \tag{11.3.7a}$$

$$-2i\frac{dA^*}{d\tau} - 3i(A^*)^2A = 0. \tag{11.3.7b}$$

To solve (11.3.7) we set $A(\tau) = R(\tau)e^{i\theta(\tau)}$, where $R(\tau)$ and $\theta(\tau)$ are real. Substituting this expression into (11.3.7) gives equations for $R(\tau)$ and $\theta(\tau)$:

$$\frac{dR}{d\tau} = -\tfrac{3}{2}R^3, \qquad \frac{d\theta}{d\tau} = 0.$$

Therefore,

$$R(\tau) = \frac{R(0)}{\sqrt{3\tau R^2(0) + 1}}, \tag{11.3.8a}$$

$$\theta(\tau) = \theta(0). \tag{11.3.8b}$$

$R(0)$ and $\theta(0)$ are determined by the initial conditions $y(0) = 1$, $y'(0) = 0$. These conditions imply that $Y_0(0, 0) = 1$, $(\partial Y_0/\partial t)(0, 0) = 0$, whence $R(0) = \frac{1}{2}$, $\theta(0) = 0$. Thus, to leading order in ε,

$$y(t) \sim \frac{\cos t}{\sqrt{1 + 3\varepsilon t/4}}, \qquad \varepsilon \to 0+,\ \varepsilon t = O(1). \tag{11.3.9}$$

This result implies that when $\varepsilon > 0$ the solution decays like $t^{-1/2}$ for large t, and that when $\varepsilon < 0$ the solution becomes infinite at a finite value of t approximately equal to $-4/3\varepsilon$. Moreover, this solution does not exhibit any phase shift (or frequency shift) to leading order in ε. These qualitative conclusions are verified numerically in Figs. 11.5 to 11.7.

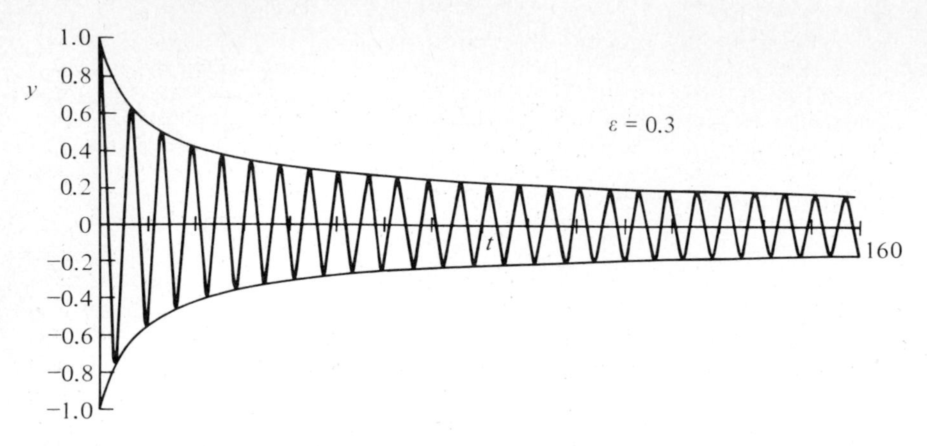

Figure 11.5 A plot of the exact solution to $y'' + y + \varepsilon(y')^3 = 0$ [$y(0) = 1$, $y'(0) = 0$] for $\varepsilon = 0.3$ [see (11.3.1)] together with a plot of the envelope $(1 + 3\varepsilon t/4)^{-1/2}$ of the leading-order multiple-scale approximation to $y(t)$ in (11.3.9). We have not plotted the full multiple-scale approximation to $y(t)$ because it is indistinguishable from the exact solution to within the thickness of the curve.

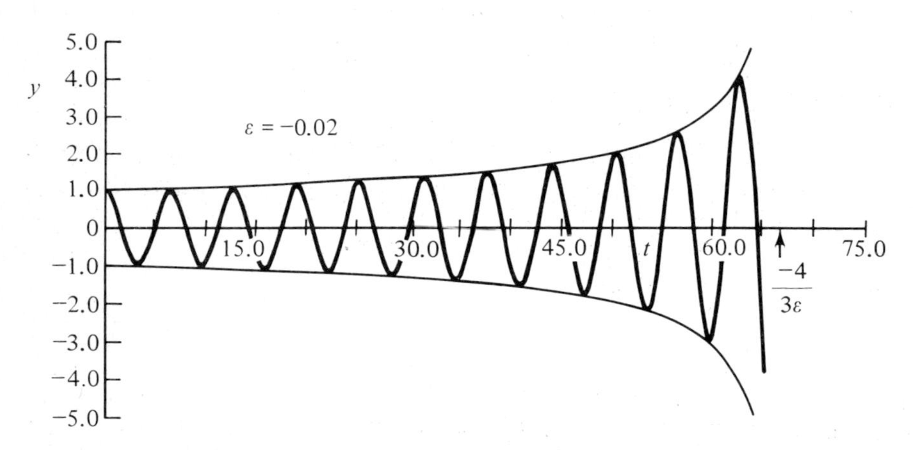

Figure 11.6 Same as in Fig. 11.5 except that $\varepsilon = -0.02$. Observe that the exact solution $y(t)$ and the multiple-scale approximation to it differ noticeably only when t is near the explosive singularity at $t = -4/3\varepsilon = 66\frac{2}{3}$.

Example 2 *Approach to a limit cycle.* The equation

$$y'' + y = \varepsilon[y' - \tfrac{1}{3}(y')^3], \qquad y(0) = 0,\ y'(0) = 2a, \tag{11.3.10}$$

known as the Rayleigh oscillator, is interesting because the solution approaches a limit cycle in the phase plane (see Sec. 4.4 and Example 3 of Sec. 9.7). Multiple-scale analysis determines the shape of this limit cycle and the rate of approach of $y(t)$ to the limit cycle.

As in Example 1, we assume a perturbation expansion for $y(t)$ in (11.3.10) of the form $y(t) \sim Y_0(t, \tau) + \varepsilon Y_1(t, \tau) + \cdots$ $(\varepsilon \to 0+)$, where $\tau = \varepsilon t$. Next we substitute (11.2.3) and (11.2.4)

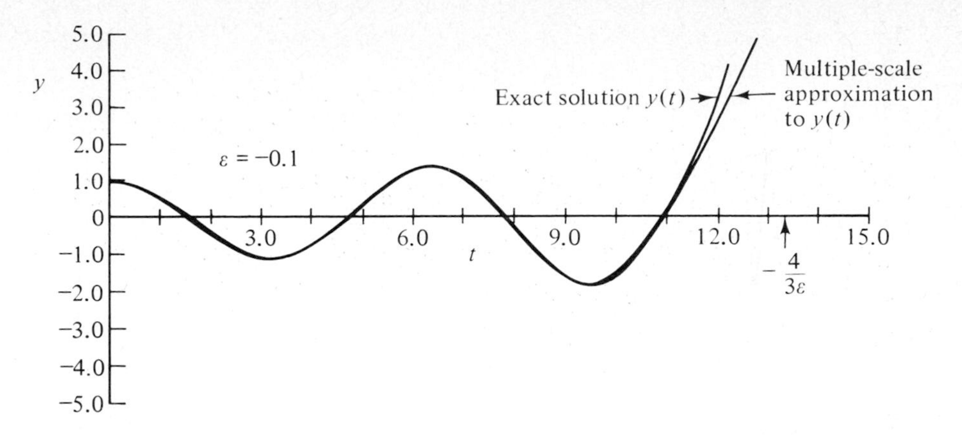

Figure 11.7 A comparison of the multiple-scale approximation and the exact solution to $y'' + y + \varepsilon(y')^3 = 0$ $[y(0) = 1, y'(0) = 0]$ for $\varepsilon = -0.1$. The approximation to $y(t)$ is extremely accurate except near the singularity at $t = -4/3\varepsilon = 13\frac{1}{3}$.

into (11.3.10) and equate coefficients of ε^0 and ε^1:

$$\frac{\partial^2 Y_0}{\partial t^2} + Y_0 = 0, \tag{11.3.11}$$

$$\frac{\partial^2 Y_1}{\partial t^2} + Y_1 = -2\frac{\partial^2 Y_0}{\partial t\, \partial \tau} + \frac{\partial Y_0}{\partial t} - \frac{1}{3}\left(\frac{\partial Y_0}{\partial t}\right)^3. \tag{11.3.12}$$

The solution to (11.3.11) is again

$$Y_0(t, \tau) = A(\tau)e^{it} + A^*(\tau)e^{-it}.$$

We substitute this expression into (11.3.12) and observe that secular terms in $Y_1(t, \tau)$ will arise unless the coefficients of $e^{\pm it}$ on the right side of (11.3.12) vanish. Thus, the conditions for the absence of secular behavior are

$$-2i\frac{dA}{d\tau} + iA - iA^2A^* = 0, \tag{11.3.13a}$$

$$2i\frac{dA^*}{d\tau} - iA^* + i(A^*)^2A = 0. \tag{11.3.13b}$$

To solve (11.3.13) we again set $A(\tau) = R(\tau)e^{i\theta(\tau)}$, where R and θ are real. The equations for R and θ are

$$2\frac{dR}{d\tau} = R - R^3, \tag{11.3.14a}$$

$$\frac{d\theta}{d\tau} = 0. \tag{11.3.14b}$$

The solutions are

$$R(\tau) = R(0)[e^{-\tau} + R^2(0)(1 - e^{-\tau})]^{-1/2}, \tag{11.3.15a}$$

$$\theta(\tau) = \theta(0). \tag{11.3.15b}$$

The initial conditions $y(0) = 0$, $y'(0) = 2a$ require that $R(0) = a$, $\theta(0) = -\frac{1}{2}\pi$. Thus, to leading order in ε, the solution to (11.3.10) is

$$y(t) \sim \frac{2a \sin t}{\sqrt{e^{-\tau} + a^2(1 - e^{-\tau})}}, \qquad \varepsilon \to 0+, \ \tau = \varepsilon t = O(1). \tag{11.3.16}$$

Observe that for all values of a, this approximate solution smoothly approaches the limit cycle $y(t) = 2 \sin t$ as $t \to \infty$. This limit cycle is represented as a circle of radius 2 in the phase plane of y and y'. If $a < 1$, the solution spirals outward to the limit cycle, and if $a > 1$, the solution spirals inward. A comparison of these asymptotic results and the numerical solution to (11.3.10) is given in Figs. 11.8 to 11.10.

Example 3 *Recovery of the WKB physical-optics approximation.* Let us consider the oscillator

$$y''(t) + \omega^2(\varepsilon t)y(t) = 0. \tag{11.3.17}$$

Note that the frequency $\omega(\varepsilon t)$ is a slowly varying function of time t.

It is easy to solve (11.3.17) using the WKB approximation. We simply introduce the new variable $\tau = \varepsilon t$ to convert (11.3.17) to standard WKB form:

$$\varepsilon^2 \frac{d^2y}{d\tau^2} + \omega^2(\tau)y = 0. \tag{11.3.18}$$

The physical-optics approximation to (11.3.18) [see (10.1.13)] is then

$$y(t) = [\omega(\tau)]^{-1/2} \exp\left[\pm i\varepsilon^{-1} \int^{\tau} \omega(s)\, ds\right]. \tag{11.3.19}$$

Now, let us rederive (11.3.19) using multiple-scale theory. The procedure requires a bit of subtlety. Suppose we naively assume that there is a linear relation $\tau = \varepsilon t$ between the appropriate long and short time scales. Then, letting $y(t) = Y_0(t, \tau) + \varepsilon Y_1(t, \tau) + \cdots$, we obtain

$$\frac{\partial^2 Y_0}{\partial t^2} + \omega^2(\tau)Y_0 = 0, \tag{11.3.20}$$

$$\frac{\partial^2 Y_1}{\partial t^2} + \omega^2(\tau)Y_1 = -2\frac{\partial^2 Y_0}{\partial t\, \partial \tau}. \tag{11.3.21}$$

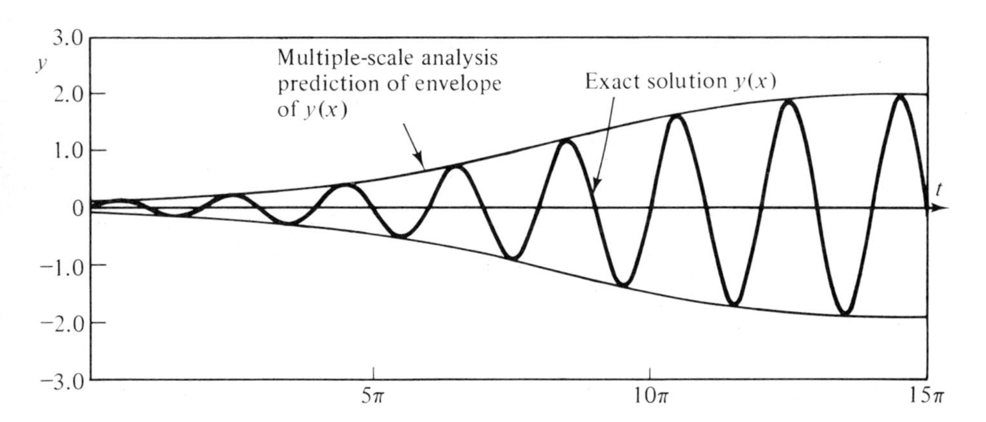

Figure 11.8 Approach to the limit cycle of the Rayleigh oscillator $y'' + y = \varepsilon[y' - \frac{1}{3}(y')^3]$ $[y(0) = 0, y'(0) = 2a]$ [see (11.3.10)], where we have taken $\varepsilon = 0.2$ and $a = 0.05$. The oscillatory curve is the numerical solution to the differential equation; the envelope is the prediction of multiple-scale analysis [see (11.3.16)]. The two curves agree to better than their thicknesses.

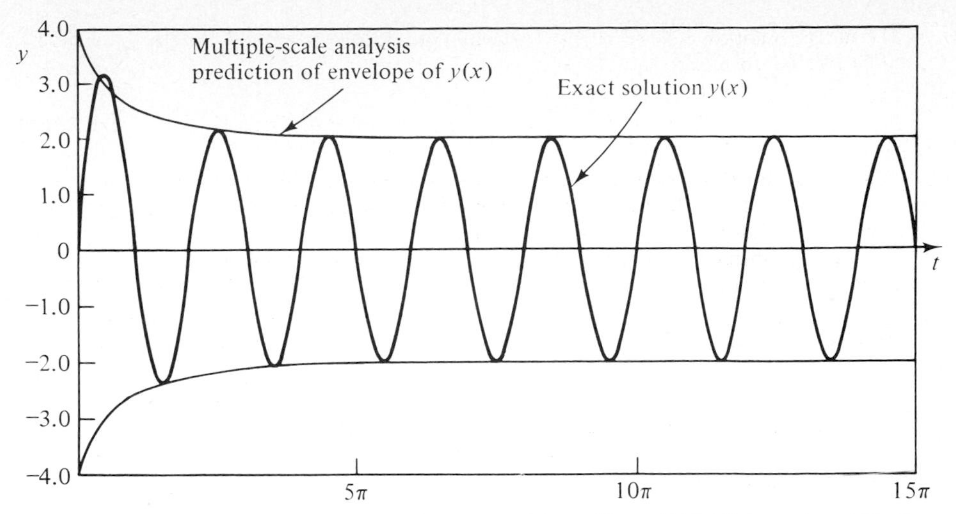

Figure 11.9 Approach to the limit cycle of the Rayleigh oscillator (11.3.10) (see Fig. 11.8). Here, $\varepsilon = 0.2$ and $a = 2.0$. Except for a small discrepancy at $t = \pi/2$ the exact and approximate solutions have nearly perfect agreement.

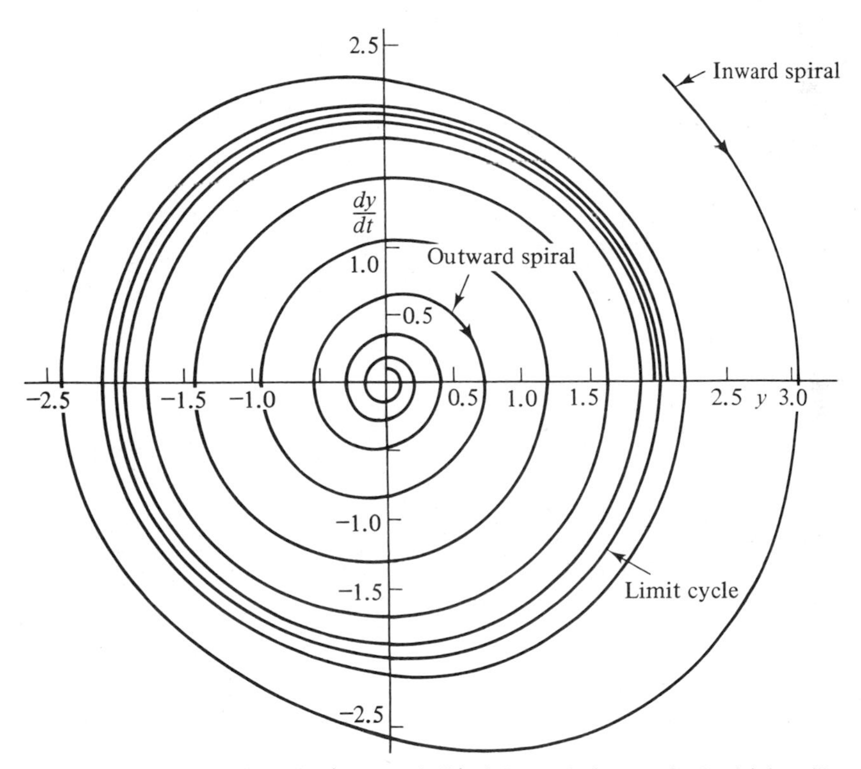

Figure 11.10 A phase-plane plot (y versus dy/dt) of three solutions to the Rayleigh oscillator (11.3.10) with $\varepsilon = 0.2$. Shown are the limit cycle solution which is approximately a circle of radius 2, the solution on Fig. 11.8 (spiraling outward toward the limit cycle), and the solution on Fig. 11.9 (spiraling inward toward the limit cycle).

The solution to (11.3.20) is $y_0 = A(\tau)e^{i\omega(\tau)t} + A^*(\tau)e^{-i\omega(\tau)t}$. Substituting this expression in the right side of (11.3.21) gives

$$\frac{\partial^2 Y_1}{\partial t^2} + \omega^2(\tau)Y_1 = -2ie^{i\omega(\tau)t}\left[\frac{d}{d\tau}(A\omega) + itA\omega\frac{d\omega}{d\tau}\right]$$
$$+ 2ie^{-i\omega(\tau)t}\left[\frac{d}{d\tau}(A^*\omega) - itA^*\omega\frac{d\omega}{d\tau}\right]. \tag{11.3.22}$$

The presence of the variable t in the square brackets implies that we cannot eliminate secularity without setting $A(\tau) \equiv 0$ (see Prob. 11.9).

This failure illustrates a crucial feature of multiple-scale perturbation methods. If the long-scale variable τ is linearly proportional to the short scale t ($\tau = \varepsilon t$), then multiple-scale methods will fail unless the frequency of the unperturbed oscillator is a constant; it must not vary even on the τ scale. Therefore, before we can apply multiple-scale methods to the oscillator (11.3.17), we must find a transformation which converts (11.3.17) to a fixed-frequency oscillator with a small perturbation term:

$$y'' + y + \varepsilon(\text{some function of } y) = 0. \tag{11.3.23}$$

With this in mind, we introduce a new time variable T:

$$T = f(t). \tag{11.3.24}$$

We will try to choose $f(t)$ to convert (11.3.17) to the form in (11.3.23). From (11.3.24) we have $d/dt = f'(t)\, d/dT$, $d^2/dt^2 = f''(t)\, d/dT + [f'(t)]^2\, d^2/dT^2$. Thus, (11.3.17) becomes

$$\frac{d^2}{dT^2}y + \frac{f''(t)}{[f'(t)]^2}\frac{d}{dT}y + \frac{\omega^2(\varepsilon t)}{[f'(t)]^2}y = 0.$$

We achieve the form in (11.3.23) if we choose $f'(t) = \omega(\varepsilon t)$. Thus,

$$T = f(t) = \int^t \omega(\varepsilon x)\, dx = \frac{1}{\varepsilon}\int^\tau \omega(s)\, ds. \tag{11.3.25}$$

In terms of T the differential equation now reads

$$\frac{d^2y}{dT^2} + y + \varepsilon\frac{\omega'(\tau)}{\omega^2(\tau)}\frac{d}{dT}y = 0. \tag{11.3.26}$$

This equation may be solved using multiple-scale methods. We expand

$$y = Y_0(T, \tau) + \varepsilon Y_1(T, \tau) + \cdots. \tag{11.3.27}$$

Using the relation $d\tau/dT = \varepsilon\, dt/dT = \varepsilon/f'(t) = \varepsilon/\omega(\tau)$, we substitute (11.3.27) into (11.3.26) and obtain, as usual, a sequence of partial differential equations:

$$\frac{\partial^2 Y_0}{\partial T^2} + Y_0 = 0, \tag{11.3.28}$$

$$\frac{\partial^2 Y_1}{\partial T^2} + Y_1 = -\frac{\omega'(\tau)}{\omega^2(\tau)}\frac{\partial Y_0}{\partial T} - \frac{2}{\omega}\frac{\partial^2 Y_0}{\partial T\, \partial\tau}. \tag{11.3.29}$$

Substituting the solution

$$Y_0 = A(\tau)e^{iT} + A^*(\tau)e^{-iT} \tag{11.3.30}$$

of (11.3.28) into the right side of (11.3.29) gives

$$\frac{\partial^2 Y_1}{\partial T^2} + Y_1 = -ie^{iT}\left[\frac{2}{\omega}\frac{dA}{d\tau} + \frac{\omega'(\tau)}{\omega^2(\tau)}A\right] + ie^{-iT}\left[\frac{2}{\omega}\frac{dA^*}{d\tau} + \frac{\omega'(\tau)}{\omega^2(\tau)}A^*\right].$$

To eliminate secularity we must require that the expressions in the square brackets vanish for all τ:

$$2\frac{dA}{d\tau} = -\frac{\omega'(\tau)}{\omega(\tau)}A,$$

$$2\frac{dA^*}{d\tau} = -\frac{\omega'(\tau)}{\omega(\tau)}A^*.$$

The solution for $A(\tau)$, apart from a multiplicative constant, is $1/\sqrt{\omega(\tau)}$. Inserting this solution into (11.3.30) gives

$$Y_0 = \frac{1}{\sqrt{\omega(\tau)}}e^{\pm iT},$$

and using the expression for T in (11.3.25) gives

$$Y_0 = \frac{1}{\sqrt{\omega(\tau)}}\exp\left[\pm\frac{i}{\varepsilon}\int^{\tau}\omega(s)\,ds\right].$$

We have reproduced the WKB result in (11.3.19).

Example 4 *Solution of a boundary-layer problem by multiple-scale perturbation theory.* Consider the elementary boundary-layer problem

$$\varepsilon y'' + ay' + by = 0, \qquad y(0) = A,\ y(1) = B,\ a > 0, \tag{11.3.31}$$

where a and b are constants. We know (see Fig. 9.4) that the solution to this problem has a boundary layer of thickness ε at $x = 0$ and is slowly varying in the range $\varepsilon \ll x \le 1$ $(\varepsilon \to 0+)$. Thus, there are two natural scales for this problem, a short scale t which describes the inner solution in the boundary layer and a long scale $x = \varepsilon t$ which describes the outer solution. Note that (11.3.31) is written in terms of the long scale. If we wish to use multiple-scale theory we must rewrite (11.3.31) in terms of the short scale t in order to eliminate secularity on the long scale:

$$\frac{d^2y}{dt^2} + a\frac{dy}{dt} + \varepsilon by = 0. \tag{11.3.32}$$

Assuming that $y(t)$ in (11.3.32) has a perturbation expansion of the form $y(t) = Y_0(t, x) + \varepsilon Y_1(t, x) + \cdots$, we obtain the following sequence of equations:

$$\varepsilon^0: \quad \frac{\partial^2 Y_0}{\partial t^2} + a\frac{\partial Y_0}{\partial t} = 0; \tag{11.3.33}$$

$$\varepsilon^1: \quad \frac{\partial^2 Y_1}{\partial t^2} + a\frac{\partial Y_1}{\partial t} = -2\frac{\partial^2 Y_0}{\partial t\,\partial x} - a\frac{\partial Y_0}{\partial x} - bY_0. \tag{11.3.34}$$

The solution to (11.3.33) has the form

$$Y_0(t, x) = A_1(x) + A_2(x)e^{-at}. \tag{11.3.35}$$

Substituting (11.3.35) into (11.3.34) gives

$$\frac{\partial^2 Y_1}{\partial t^2} + a\frac{\partial Y_1}{\partial t} = -[aA_1'(x) + bA_1(x)] + [aA_2'(x) - bA_2(x)]e^{-at}.$$

The right side of this equation is a solution to the homogeneous equation in (11.3.33) and therefore gives rise to secular terms. To eliminate the secular term that grows like t (we know from our study of boundary-layer theory that no such term is present in leading order), we set

$$aA_1'(x) + bA_1(x) = 0.$$

Thus,
$$A_1(x) = C_1 e^{-bx/a},$$
where C_1 is a constant.

Note that if $aA_2'(x) - bA_2(x) \neq 0$, then there will be a secular term of the form te^{-t} which does not occur in leading-order boundary-layer theory. It is not necessary to eliminate this secular term because it decays exponentially with increasing t.

To leading order in ε we now have

$$y(t) = C_1 e^{-bx/a} + A_2(x)e^{-at} + O(\varepsilon), \qquad \varepsilon \to 0+. \tag{11.3.36}$$

Recall that $t = x/\varepsilon$. Therefore, for all $x > 0$, it is valid to replace $A_2(x)e^{-ax/\varepsilon}$ by $A_2(0)e^{-ax/\varepsilon} + O(\varepsilon)$ $(\varepsilon \to 0+)$. Setting $A_2(0) = C_2$, (11.3.36) becomes $y(t) = C_1 e^{-bx/a} + C_2 e^{-ax/\varepsilon} + O(\varepsilon)$ $(\varepsilon \to 0+)$.

Finally, we impose the boundary conditions at $x = 0$ and $x = 1$ and obtain

$$y(t) = Be^{b/a}e^{-bx/a} + (A - Be^{b/a})e^{-ax/\varepsilon} + O(\varepsilon), \qquad \varepsilon \to 0+,$$

which agrees with the uniform leading-order boundary-layer solution in (9.1.13).

If a and b in (11.3.31) vary with x, it is necessary to perform a transformation of variable like that in Example 3 before one can use multiple-scale perturbation theory (see Prob. 11.12).

(I) 11.4 THE MATHIEU EQUATION AND STABILITY

The Mathieu equation

$$\frac{d^2y}{dt^2} + (a + 2\varepsilon \cos t)y = 0, \tag{11.4.1}$$

in which a and ε are parameters, is an example of a differential equation whose coefficients are periodic. The general theory of linear periodic differential equations, which is known as Floquet theory, predicts that there may be solutions to (11.4.1) for some values of a and ε which are unstable (grow exponentially with increasing t). As a particularly nice application of multiple-scale perturbation theory (which is valid when ε is small) we find the boundaries between the regions in the (a, ε) plane for which all solutions to the Mathieu equation are stable (remain bounded for all t) and the regions in which there are unstable solutions.

Elementary Floquet Theory

We consider here just the case of second-order linear ordinary differential equations having 2π-periodic coefficient functions. We will make use of two facts. First, since the coefficients are 2π-periodic, we know that if $y(t)$ is any solution of such an equation, so is $y(t + 2\pi)$. Second, since the equation is linear and second order, any solution $y(t)$ may be represented as a linear combination of two linearly independent solutions $y_1(t)$ and $y_2(t)$:

$$y(t) = Ay_1(t) + By_2(t). \tag{11.4.2}$$

Since the coefficients of the differential equation are 2π-periodic, $y_1(t + 2\pi)$ and $y_2(t + 2\pi)$ are also solutions, so they may be represented as linear combinations of $y_1(t)$ and $y_2(t)$:

$$y_1(t+2\pi) = \alpha y_1(t) + \beta y_2(t), \qquad y_2(t+2\pi) = \gamma y_1(t) + \delta y_2(t).$$

Thus, for $y(t)$ in (11.4.2) we have

$$\begin{aligned} y(t+2\pi) &= (A\alpha + B\gamma)y_1(t) + (A\beta + B\delta)y_2(t) \\ &= A'y_1(t) + B'y_2(t). \end{aligned} \tag{11.4.3}$$

The relation between the coefficients A and B and A' and B' in (11.4.3) involves matrix multiplication:

$$\begin{pmatrix} A' \\ B' \end{pmatrix} = \begin{pmatrix} \alpha & \gamma \\ \beta & \delta \end{pmatrix} \begin{pmatrix} A \\ B \end{pmatrix}. \tag{11.4.4}$$

Now let us choose (A, B) to be an eigenvector of the 2×2 matrix in (11.4.4). If the corresponding eigenvalue is λ, then $A' = \lambda A$ and $B' = \lambda B$ and

$$y(t+2\pi) = \lambda y(t). \tag{11.4.5}$$

Thus, if we introduce $\mu = (\ln \lambda)/2\pi$ so that $\lambda = e^{2\pi\mu}$, then we see that for all t, $y(t)$ takes the form

$$y(t) = e^{\mu t}\phi(t), \tag{11.4.6}$$

where $\phi(t)$ is a 2π-periodic function: $\phi(t+2\pi) = \phi(t)$. We say that $y(t)$ in (11.4.6) is an *unstable* solution if Re $\mu > 0$ because $y(t)$ grows exponentially with t. We say that $y(t)$ is a *stable* solution if Re $\mu \le 0$.

Stability Boundaries of the Mathieu Equation

The Mathieu equation (11.4.1) is special because it is even under the reflection $t \to -t$. Thus, if $y(t)$ is a solution of the Mathieu equation, so is $y(-t)$. Therefore, for *both* solutions $e^{\mu t}\phi(t)$ and $e^{-\mu t}\phi(-t)$ of the Mathieu equation to be stable, we must have Re $\mu = 0$.

There are well-defined regions of the (a, ε) plane for which all solutions of the Mathieu equation are stable. In Fig. 11.11 we indicate those regions (white) where all solutions are stable and those regions (cross hatched) for which there is an unstable solution. The boundaries between regions of stability and instability are called *stability boundaries*. Our main objective in this section is to find approximate expressions for the stability boundaries which are valid as $\varepsilon \to 0$.

CASE I *Perturbative investigation of stable solutions.* Here we assume that a is positive and that $a \ne n^2/4$ $(n = 0, 1, 2, 3, \ldots)$. We will show that $y(t)$ is stable for sufficiently small ε. We assume a regular perturbation expansion for $y(t)$:

$$y(t) = y_0(t) + \varepsilon y_1(t) + \varepsilon^2 y_2(t) + \cdots. \tag{11.4.7}$$

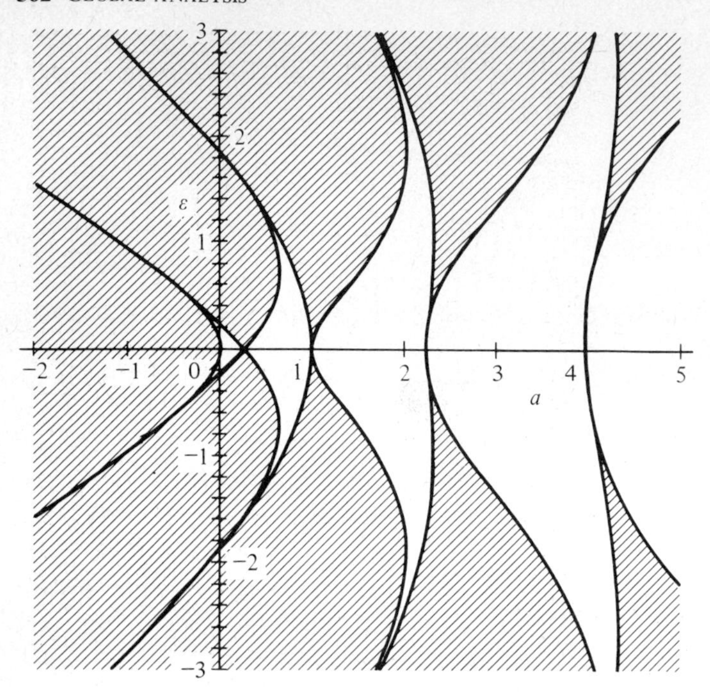

Figure 11.11 A plot of the stability boundaries of solutions to the Mathieu equation (11.4.1). In the white regions of the (a, ε) plane all solutions of the Mathieu equation are stable, while in the cross-hatched regions there is an unstable solution. When $\varepsilon = 0$, the cross-hatched regions meet the a axis at $a = n^2/4$ $(n = 0, 1, 2, \ldots)$.

Substituting $y(t)$ in (11.4.7) into the Mathieu equation (11.4.1) and comparing powers of ε gives a sequence of equations:

$$\varepsilon^0: \frac{d^2 y_0}{dt^2} + a y_0 = 0, \tag{11.4.8}$$

$$\varepsilon^1: \frac{d^2 y_1}{dt^2} + a y_1 = -2 y_0 \cos t, \tag{11.4.9}$$

$$\varepsilon^2: \frac{d^2 y_2}{dt^2} + a y_2 = -2 y_1 \cos t, \tag{11.4.10}$$

and so on.

The solution to (11.4.8) is secular (grows linearly with time) only if $a = 0$. Since we have assumed that $a > 0$, we have

$$y_0(t) = A \exp(i\sqrt{a}\, t) + \text{c.c.},$$

where c.c. stands for the complex conjugate of the exhibited terms. Substituting this result into (11.4.9), we have

$$\frac{d^2y_1}{dt^2} + ay_1(t) = -A_0 \exp\,[i(\sqrt{a}+1)t] - A_0 \exp\,[i(\sqrt{a}-1)t] + \text{c.c.}$$

Now, secular terms appear only if $\sqrt{a} \pm 1 = \pm\sqrt{a}$. But this can only happen if $a = \frac{1}{4}$.

In subsequent orders of perturbation theory the solution will be secular only if $a = 1, 9/4, 4, \ldots$ (see Prob. 11.18). But by assumption $a \neq n^2/4$. Therefore, there is no secularity. After solving to all orders in perturbation theory, we will have

$$y(t) = \exp\,(i\sqrt{a}\,t)\phi(t) + \exp\,(-i\sqrt{a}\,t)\phi^*(t),$$

where $\phi(t) = \sum \varepsilon^n A_n e^{int}$, which is a *periodic* series in t. We conclude that since (11.4.7) is a regular perturbation series, the series for $\phi(t)$ converges for sufficiently small ε to a periodic function. Thus, all solutions $y(t)$ are stable for $a > 0$, $a \neq n^2/4$ and sufficiently small ε. Figure 11.11 shows that this prediction is correct.

CASE II *Perturbative investigation of unstable solutions for a near* $\frac{1}{4}$. To investigate the behavior of solutions near $a = \frac{1}{4}$ and ε near 0, we treat a as a power series in ε:

$$a = \tfrac{1}{4} + a_1\varepsilon + a_2\varepsilon^2 + \cdots.$$

Thus, the Mathieu equation becomes

$$\frac{d^2y}{dt^2} + \left[\frac{1}{4} + (a_1 + 2\cos t)\varepsilon + \cdots\right] y = 0. \tag{11.4.11}$$

We will look for nongrowing solutions and we specifically hope to find the stability boundary (the edge of the shaded region on Fig. 11.11).

We already know that a naive perturbation expansion will yield secular terms. Thus, we will use a multiple-scale expansion:

$$y(t) = Y_0(t, \tau) + \varepsilon Y_1(t, \tau) + \cdots, \tag{11.4.12}$$

where $\tau = \varepsilon t$. Substituting (11.4.12) into (11.4.11), recalling (11.2.4), and comparing like powers of ε gives

$$\varepsilon^0\text{:}\ \frac{\partial^2 Y_0}{\partial t^2} + \frac{Y_0}{4} = 0, \tag{11.4.13}$$

$$\varepsilon^1\text{:}\ \frac{\partial^2 Y_1}{\partial t^2} + \frac{Y_1}{4} = -(a_1 + 2\cos t)Y_0 - 2\frac{\partial^2 Y_0}{\partial t\,\partial\tau}. \tag{11.4.14}$$

The solution to (11.4.13) is $Y_0 = A(\tau)e^{it/2} + A^*(\tau)e^{-it/2}$. Substituting this result into (11.4.14) gives

$$\partial^2 Y_1/\partial t^2 + Y_1/4 = -[a_1 A(\tau) + iA'(\tau) + A^*(\tau)]e^{it/2} - A(\tau)e^{3it/2} + \text{c.c.}$$

To eliminate the terms which cause Y_1 to exhibit secular behavior, we take

$$iA'(\tau) = -a_1 A(\tau) - A^*(\tau), \qquad -iA^{*\prime}(\tau) = -a_1 A^*(\tau) - A(\tau).$$

This system becomes simpler if we decompose $A(\tau)$ into its real and imaginary parts:

$$A(\tau) = B(\tau) + iC(\tau).$$

The equations for $B(\tau)$ and $C(\tau)$ are

$$B'(\tau) = (-a_1 + 1)C(\tau), \qquad C'(\tau) = (a_1 + 1)B(\tau).$$

Thus, the equation for $B(\tau)$ is

$$B''(\tau) = (1 - a_1^2)B(\tau),$$

and $B(\tau)$ has solutions of the form

$$B(\tau) = K \exp\left(\pm\sqrt{1 - a_1^2}\,\tau\right), \tag{11.4.15}$$

where K is a constant.

Instability (solutions growing exponentially with τ) occurs if $\sqrt{1 - a_1^2}$ is real. Thus, $|a_1| < 1$ gives unstable solutions and $|a_1| > 1$ gives stable solutions. We conclude that near $\varepsilon = 0$, the stability boundary is the pair of straight lines

$$a = \tfrac{1}{4} \pm \varepsilon + O(\varepsilon^2), \qquad \varepsilon \to 0, \tag{11.4.16}$$

which intersect the a axis at 45° angles. This conclusion is verified in Fig. 11.11.

Higher-Order Corrections to the Stability Boundary near $a = \frac{1}{4}$

We now set $a_1 = 1$ and pursue our analysis to higher order to determine the location of the stability boundary more precisely than in (11.4.16). This analysis is particularly interesting because when $a_1 = 1$ there is apparently a new time scale for the problem. To see why, suppose we set $a_1 = 1 + a_2\varepsilon$ in (11.4.15). Then $B(\tau)$ becomes approximately $K \exp\left(\pm\sqrt{-2a_2\varepsilon}\,\tau\right) = K \exp\left(\sqrt{-2a_2}\,\varepsilon^{3/2}t\right)$, which suggests that we must introduce a new time scale $\sigma = \varepsilon^{3/2}t$.

We therefore substitute

$$a = \tfrac{1}{4} + \varepsilon + a_2\varepsilon^2 \tag{11.4.17}$$

into the Mathieu equation (11.4.11), set $\sigma = \varepsilon^{3/2}t$, and expand

$$y = Y_0(t, \sigma) + \varepsilon^{1/2}Y_1(t, \sigma) + \varepsilon Y_2(t, \sigma) + \varepsilon^{3/2}Y_3(t, \sigma) + \varepsilon^2 Y_4(t, \sigma) + \cdots. \tag{11.4.18}$$

We have expanded y in powers of $\varepsilon^{1/2}$ rather than ε as in (11.4.12) because σ will inject powers of $\sqrt{\varepsilon}$ into the perturbation series. Note that it is necessary to go to fourth order in powers of $\sqrt{\varepsilon}$ to determine a_2!

Next, we substitute

$$\frac{d^2y}{dt^2}=\frac{\partial^2 Y_0}{\partial t^2}+\varepsilon^{1/2}\frac{\partial^2 Y_1}{\partial t^2}+\varepsilon\frac{\partial^2 Y_2}{\partial t^2}$$

$$+\varepsilon^{3/2}\left(\frac{\partial^2 Y_3}{\partial t^2}+2\frac{\partial^2 Y_0}{\partial t\,\partial\sigma}\right)+\varepsilon^2\left(\frac{\partial^2 Y_4}{\partial t^2}+2\frac{\partial^2 Y_1}{\partial t\,\partial\sigma}\right)+\cdots$$

into the Mathieu equation (11.4.11) and equate powers of $\varepsilon^{1/2}$:

$$\begin{aligned}
\varepsilon^0&:\frac{\partial^2 Y_0}{\partial t^2}+\frac{1}{4}Y_0=0, \qquad \text{so } Y_0=A_0(\sigma)e^{it/2}+\text{c.c.};\\
\varepsilon^{1/2}&:\frac{\partial^2 Y_1}{\partial t^2}+\frac{1}{4}Y_1=0, \qquad \text{so } Y_1=A_1(\sigma)e^{it/2}+\text{c.c.};\\
\varepsilon^1&:\frac{\partial^2 Y_2}{\partial t^2}+\frac{1}{4}Y_2=-(1+2\cos t)Y_0\\
&\qquad =-A_0e^{it/2}-A_0e^{3it/2}-A_0^*e^{it/2}+\text{c.c.}
\end{aligned} \tag{11.4.19}$$

To remove the secularity on this level we take $A_0=-A_0^*$, so that $A_0(\sigma)=iB(\sigma)$ with B real. Now we can solve for Y_2:

$$Y_2(t,\sigma)=A_2(\sigma)e^{it/2}+\tfrac{1}{2}A_0(\sigma)e^{3it/2}+\text{c.c.} \tag{11.4.20}$$

Equating coefficients of $\varepsilon^{3/2}$ gives

$$\begin{aligned}
\varepsilon^{3/2}&:\frac{\partial^2 Y_3}{\partial t^2}+\frac{1}{4}Y_3=-2\frac{\partial^2 Y_0}{\partial t\,\partial\sigma}-(1+2\cos t)Y_1\\
&\qquad =-i\frac{dA_0}{d\sigma}e^{it/2}-A_1e^{it/2}-A_1^*e^{it/2}-A_1e^{3it/2}+\text{c.c.}
\end{aligned}$$

Eliminating secularity on this level gives $i(dA_0/d\sigma)=-A_1-A_1^*=-dB/d\sigma$, so

$$\frac{dB}{d\sigma}=A_1+A_1^*.$$

Finally, using (11.4.19) and (11.4.20), we equate coefficients of ε^2:

$$\begin{aligned}
\varepsilon^2:\frac{\partial^2 Y_4}{\partial t^2}+\frac{1}{4}Y_4&=-i\frac{dA_1}{d\sigma}e^{it/2}-A_2e^{it/2}-A_2^*e^{it/2}-A_2e^{3it/2}-\frac{1}{2}A_0e^{3it/2}\\
&\quad -\frac{1}{2}A_0e^{5it/2}-\frac{1}{2}A_0e^{it/2}-a_2A_0e^{it/2}+\text{c.c.}
\end{aligned}$$

Setting all terms which can give rise to secularity equal to zero gives

$$-i\frac{dA_1}{d\sigma} = A_2 + A_2^* + \tfrac{1}{2}iB + a_2 iB, \qquad i\frac{dA_1^*}{d\sigma} = A_2^* + A_2 - \tfrac{1}{2}iB - a_2 iB.$$

From these equations we have

$$-i\frac{d(A_1 + A_1^*)}{d\sigma} = iB + 2a_2 iB.$$

Letting $A_1 + A_1^* = C$, C real, gives

$$\frac{dB}{d\sigma} = C, \qquad -\frac{dC}{d\sigma} = (1 + 2a_2)B.$$

Finally, eliminating C gives

$$\frac{d^2B}{d\sigma^2} = (2a_2 + 1)B,$$

whose solution is

$$B(\sigma) = (\text{constant}) \exp\left(\pm \sigma\sqrt{2a_2 + 1}\right).$$

We conclude that we have stability when $a_2 < -\frac{1}{2}$ and instability when $a_2 > -\frac{1}{2}$. The higher-order stability boundary is thus given by

$$a(\varepsilon) = \tfrac{1}{4} + \varepsilon - \tfrac{1}{2}\varepsilon^2 + O(\varepsilon^3), \qquad \varepsilon \to 0. \tag{11.4.21}$$

For further analysis of the Mathieu equation see Prob. 11.19.

PROBLEMS FOR CHAPTER 11

Section 11.1

(I) **11.1** The pendulum of a grandfather clock swings to a maximum angle of 5° from the vertical. How many seconds does the clock gain or lose each day if the clock is adjusted to keep perfect time when the angular swing is 2° from the vertical?

(E) **11.2** Show that the most secular term in y_{n+1} in (11.1.5) arises from the most secular term in y_n.

(E) **11.3** Show that the coefficient of x^n in the Taylor expansion of $e^{x}e^{x}e^{-x}$ is $\sum_{j+k+l=n} (-1)^l/(j!\, k!\, l!)$.

Section 11.2

(E) **11.4** There is an alternative to the method discussed in Sec. 11.2 to eliminate secular terms. In the *method of averaging* we consider the integral $I = \int_0^{2\pi} Y_0(\partial^2 Y_1/\partial t^2 + Y_1)\, dt$, taken over the short time-scale period of oscillation of (11.2.1). Y_0 and Y_1 are defined in (11.2.2). Throughout this integration the long time scale τ remains fixed, and thus Y_0 and Y_1 should be periodic in t.

(*a*) Show that if Y_0 and Y_1 are periodic in t and (11.2.2) is uniformly valid, then $I = 0$.

(*b*) Use (11.2.6) and the requirement that $I = 0$ to derive (11.2.8) and (11.2.9).

(I) **11.5** We know that the solution to Duffing's equation (11.2.1) $d^2y/dt^2 + y + \varepsilon y^3 = 0$ [$y(0) = 1$, $y'(0) = 0$] has the form $y = \cos(\omega t) + O(\varepsilon)$ ($\varepsilon \to 0+$), where $\omega^2 - 1 = a\varepsilon + b\varepsilon^2 + \cdots$ represents the frequency shift caused by the εy^3 term. Let us rewrite Duffing's equation as $d^2y/dt^2 + \omega^2 y + \varepsilon y^3 - a\varepsilon y - b\varepsilon^2 y \cdots = 0$.

(*a*) Adjust a so that no secularity appears to first order in the perturbation expansion of y. (Treat ω^2 as a parameter in this calculation.)

(*b*) Adjust b so that no secularity appears to second order in the perturbation expansion of y.

(*c*) From your determination of a and b, compute ω_1 and ω_2 in the expansion of ω: $\omega = 1 + \omega_1 \varepsilon + \omega_2 \varepsilon^2 + \cdots$. Show that $\omega_1 = \frac{3}{8}$ and $\omega_2 = -\frac{21}{256}$.

(E) **11.6** Rederive the result in Prob. 11.5 by integrating the separable differential equation in (11.1.8) over one period T: $\int dy/\sqrt{1 - y^2 + \varepsilon(1 - y^4)/2} = \int dt = T$. The relation between the frequency ω and the period T is $\omega T = 2\pi$. Show that $\omega = \pi/2I$, where $I = \int_0^{\pi/2} d\theta/\sqrt{1 + \varepsilon(1 + \sin^2 \theta)/2}$, and expand I as a power series in ε.

(D) **11.7** Perform a multiple-scale analysis of the Duffing equation (11.2.1) to second order in ε. That is, take three terms in the expansion (11.2.2): $y(t) = Y_0(t, \tau, \sigma) + \varepsilon Y_1(t, \tau, \sigma) + \varepsilon^2 Y_2(t, \tau, \sigma) + \cdots$, where $\tau = \varepsilon t$, $\sigma = \varepsilon^2 t$.

(*a*) Derive the partial differential equations satisfied by Y_0, Y_1, Y_2.

(*b*) Show that first-order multiple-scale analysis gives $Y_0 = A(\tau, \sigma)e^{it} + \text{c.c.}$, where $A(\tau, \sigma) = R(\sigma)e^{i[3R^2(\sigma)\tau/2 + \theta(\sigma)]}$. Also show that $Y_1 = \frac{1}{8}R^3(\sigma)e^{3i[t + 3R^2(\sigma)\tau/2 + \theta(\sigma)]} + B(\tau, \sigma)e^{it} + \text{c.c.}$

(*c*) Show that second-order multiple-scale analysis does *not* determine $R(\sigma)$ and $\theta(\sigma)$ uniquely. However, if it is assumed that B is a constant, then it is possible to reproduce the second-order results cited in Prob. 11.5 and rederived in Prob. 11.6. Can this assumption be weakened?

(*d*) Show that third-order multiple-scale analysis does not remove the ambiguities encountered in second order.

Section 11.3

(E) **11.8** Use the energy integral (11.3.2) to show that all solutions to (11.3.1) decay to 0 as $t \to +\infty$.

(E) **11.9** Show that demanding that secular terms on the right side of (11.3.22) vanish leads to the conclusion that $A(\tau) = 0$.

(I) **11.10** Consider the nonlinear oscillator $d^2y/dt^2 + \omega^2(\varepsilon t)y + \varepsilon y^3 = 0$ $[y(0) = 1,\ y'(0) = 0]$. Use multiple-scale perturbation theory to find an approximation to $y(t)$ which is valid on the εt time scale.

(I) **11.11** The Van der Pol equation is given by $d^2y/dt^2 + y - \varepsilon(1 - y^2)\, dy/dt = 0$. For arbitrary initial conditions the solution to this equation approaches a limit cycle. Find the approach to this limit cycle using multiple-scale perturbation theory.

(D) **11.12** Consider the boundary-layer problem $\varepsilon y''(x) + a(x)y'(x) + b(x)y = 0$ $[y(0) = A,\ y(1) = B,\ a(x) > 0]$. Show that naive multiple-scale perturbation theory, in which the short time scale is $t = x/\varepsilon$ and the long time scale is x, breaks down. Find a suitable transformation for which the method of multiple scales does work and reproduce the result in (9.1.13).

(D) **11.13** Before solving this problem read Example 4 of Sec. 7.2. Now consider an oscillator governed by the equation $\ddot{y} + y - \varepsilon t y = 0$. Assume that $y(0) = 1$ and that $\dot{y}(0)$ has been chosen so that $y \to 0$ as $t \to \infty$.

(*a*) Find the leading asymptotic behavior of $y(t)$ for large positive values of t.

(*b*) Using regular perturbation theory, obtain an approximation to y valid to first order in powers of ε. How large may t be before secular behavior appears.

(*c*) Use multiple-scale theory to eliminate this lowest-order secular behavior in y.

Clue: To do this you must consider three time scales: t, $T_1 = \sqrt{\varepsilon}\, t$, and $T_2 = \varepsilon t$. Then consider the differential equation $\ddot{y} + y - \sqrt{\varepsilon}\, T_1 y = 0$ and try a perturbation series of the form $y(t) = Y_0(t, T_1, T_2) + \varepsilon Y_1(t, T_1, T_2) + \cdots$.

Show that for times $t = O(\varepsilon^{-1/2})$ the effect of the perturbation term is to make the frequency of the oscillator time dependent. Find the frequency of the oscillator to first order in powers of ε.

(*d*) Find the exact solution to the differential equation and use it to verify the result of part (*c*).

(I) **11.14** Use multiple-scale perturbation theory to find a leading-order approximation to $\ddot{y} + y + \varepsilon \dot{y} y^2 = 0$ $[y(0) = 1,\ \dot{y}(0) = 0,\ \varepsilon > 0]$.

(I) **11.15** Consider the following perturbation problem: $d^2u/dt^2 + u = \varepsilon u^2$ $(\varepsilon \to 0)$ with initial conditions $u(0) = 2$, $u'(0) = 0$.

(*a*) In what order of ε does a secularity first appear in the regular perturbation solution for $u(t)$? Find $u(t)$ to that order in ε in this expansion.

(*b*) Introduce a suitable long time scale to eliminate the secularity found in part (*a*).

(*c*) Find the zeroth-order solution valid on the long time scale. Find the amplitude change and frequency shift.

(I) **11.16** Cheng and Wu considered the following simple differential equation that illustrates the limitations of WKB and multiple-scale perturbation theories: $y'' + e^{-\varepsilon x}y = 0$ $[y(0) = 0,\ y'(0) = 1]$.

(*a*) Show that the exact solution is

$$y(x) = \frac{\pi}{\varepsilon}\left[Y_0\left(\frac{2}{\varepsilon}\right)J_0\left(\frac{2}{\varepsilon}e^{-\varepsilon x/2}\right) - J_0\left(\frac{2}{\varepsilon}\right)Y_0\left(\frac{2}{\varepsilon}e^{-\varepsilon x/2}\right)\right].$$

(*b*) Show that leading-order WKB and multiple-scale (MS) analysis give the approximation

$$y_{\text{WKB}}(x) = y_{\text{MS}}(x) = e^{\varepsilon x/4}\sin\left[\frac{2}{\varepsilon}(1 - e^{-\varepsilon x/2})\right].$$

(*c*) Show that the error in y_{WKB} and y_{MS} is small only if $\varepsilon e^{\varepsilon x/2} \ll 1$ $(\varepsilon \to 0+)$.

(*d*) Show that the reason for the breakdown of the approximations as $x \to +\infty$ is that there is a turning point of the differential equation at ∞. Argue that when $x = O[(1/\varepsilon)\ln(1/\varepsilon)]$ the character of the solution changes. How does it change?

Section 11.4

(TI) **11.17** Prove the following result in Floquet theory. If the eigenvalues of the 2×2 matrix in (11.4.4) are degenerate, then there are solutions $y_1(t)$ and $y_2(t)$ with the properties that $y_1(t) = e^{\mu t}\phi(t)$, $y_2(t) = [t\phi(t) + \psi(t)]e^{\mu t}$, where $\phi(t)$ and $\psi(t)$ are 2π-periodic.

(I) **11.18** Show that the perturbation expansion (11.4.7) has secular terms if and only if $a = n^2/4$ $(n = 0, 1, 2, \ldots)$. When $a = n^2/4$, secularity appears in nth-order perturbation theory.

(D) **11.19** Find the next term in the expansion (11.4.21). Note that when $a_2 = -\frac{1}{2}$, the appropriate long time scale is $\varepsilon^2 t$.

(I) **11.20** It might seem that $A_0 = -A_0^*$ in (11.4.19) gives only one solution to the Mathieu equation (11.4.1). Can you find a second linearly independent solution in the context of multiple-scale analysis?

(D) **11.21** (*a*) Consider a pendulum of length L whose pivot point is oscillating up and down a distance l with frequency ω according to $l\cos(\omega t)$. Show that if the pendulum undergoes a small angular displacement θt, then $\theta(\omega t)$ satisfies the Mathieu equation (11.4.1) in which $s = \omega t$, $2\varepsilon = l/L$, and $a = \pm g/(\omega^2 L)$ if the pendulum is hanging downward (upward).

(*b*) Explain the physical meaning of the instabilities for a near $n^2/4$ when the pendulum is hanging downward. (See Fig. 11.11.) This is a parametric amplifier.

(*c*) Show from Fig. 11.11 that for certain ranges of ω the pendulum undergoes stable oscillation when it is hanging upward. Build a gadget to demonstrate this phenomenon.

USEFUL FORMULAS

The world is full of obvious things which nobody by any chance ever observes.

—Sherlock Holmes, *The Hound of the Baskervilles*
Sir Arthur Conan Doyle

AIRY FUNCTIONS

1. Differential equation:

$$y'' = xy.$$

Solutions are linear combinations of Ai (x) and Bi (x).

2. Taylor series:

$$\text{Ai}\,(x) = 3^{-2/3} \sum_{n=0}^{\infty} \frac{x^{3n}}{9^n n!\,\Gamma(n+\frac{2}{3})} - 3^{-4/3} \sum_{n=0}^{\infty} \frac{x^{3n+1}}{9^n n!\,\Gamma(n+\frac{4}{3})},$$

$$\text{Bi}\,(x) = 3^{-1/6} \sum_{n=0}^{\infty} \frac{x^{3n}}{9^n n!\,\Gamma(n+\frac{2}{3})} + 3^{-5/6} \sum_{n=0}^{\infty} \frac{x^{3n+1}}{9^n n!\,\Gamma(n+\frac{4}{3})},$$

$$\text{Ai}\,(0) = \text{Bi}\,(0)/\sqrt{3} = 3^{-2/3}/\Gamma(\tfrac{2}{3}) \doteq 0.355\,028,$$

$$\text{Ai}'\,(0) = -\text{Bi}'\,(0)/\sqrt{3} = -3^{-1/3}/\Gamma(\tfrac{1}{3}) \doteq -0.258\,819.$$

3. Functional relations:

$$\text{Ai}\,(z) + \omega\,\text{Ai}\,(\omega z) + \omega^2\,\text{Ai}\,(\omega^2 z) = 0,$$

$$\text{Bi}\,(z) = i\omega\,\text{Ai}\,(\omega z) - i\omega^2\,\text{Ai}\,(\omega^2 z),$$

where $\omega = e^{-2i\pi/3}$.

4. Relation to Bessel functions:

$$\text{Ai}\,(z) = \pi^{-1}\sqrt{z/3}\,K_{1/3}(2z^{3/2}/3),$$

$$\text{Bi}\,(z) = \sqrt{z/3}\,[I_{-1/3}(2z^{3/2}/3) + I_{1/3}(2z^{3/2}/3)].$$

5. Asymptotic expansions:

$$\mathrm{Ai}\,(z) \sim \tfrac{1}{2}\pi^{-1/2} z^{-1/4} e^{-2z^{3/2}/3} \sum_{n=0}^{\infty} (-1)^n c_n z^{-3n/2}, \quad z \to \infty;\ |\arg z| < \pi,$$

$$\mathrm{Bi}\,(z) \sim \pi^{-1/2} z^{-1/4} e^{2z^{3/2}/3} \sum_{n=0}^{\infty} c_n z^{-3n/2}, \quad z \to \infty;\ |\arg z| < \tfrac{1}{3}\pi,$$

$$\mathrm{Ai}\,(z) = w_1(z) \sin\left[\frac{2}{3}(-z)^{3/2} + \frac{\pi}{4}\right] - w_2(z) \cos\left[\frac{2}{3}(-z)^{3/2} + \frac{\pi}{4}\right],$$

$$\mathrm{Bi}\,(z) = w_2(z) \sin\left[\frac{2}{3}(-z)^{3/2} + \frac{\pi}{4}\right] + w_1(z) \cos\left[\frac{2}{3}(-z)^{3/2} + \frac{\pi}{4}\right],$$

$$w_1(z) \sim \pi^{-1/2}(-z)^{-1/4} \sum_{n=0}^{\infty} c_{2n} z^{-3n}, \quad z \to \infty;\ \frac{\pi}{3} < \arg z < \frac{5\pi}{3},$$

$$w_2(z) \sim \pi^{-1/2}(-z)^{-7/4} \sum_{n=0}^{\infty} c_{2n+1} z^{-3n}, \quad z \to \infty;\ \frac{\pi}{3} < \arg z < \frac{5\pi}{3},$$

$$c_n = \frac{(2n+1)(2n+3)\cdots(6n-1)}{144^n n!}$$

$$= \frac{1}{2\pi}\left(\frac{3}{4}\right)^n \frac{\Gamma(n+\frac{5}{6})\Gamma(n+\frac{1}{6})}{n!}, \quad c_0 = 1.$$

6. Integral representations:

$$\mathrm{Ai}\,(x) = \frac{1}{\pi}\int_0^{\infty} \cos\left(\frac{1}{3}t^3 + xt\right) dt,$$

$$\mathrm{Bi}\,(x) = \frac{1}{\pi}\int_0^{\infty} \left[e^{-t^3/3 + xt} + \sin\left(\frac{1}{3}t^3 + xt\right)\right] dt.$$

MODIFIED BESSEL FUNCTIONS

1. Differential equation:

$$x^2 y'' + xy' - (x^2 + \nu^2)y = 0.$$

Solutions are linear combinations of $I_\nu(x)$ and $K_\nu(x)$.

2. Frobenius series:

$$I_\nu(x) = \left(\frac{1}{2}x\right)^\nu \sum_{n=0}^{\infty} \frac{(\frac{1}{4}x^2)^n}{n!\,\Gamma(n+\nu+1)},$$

$$K_\nu(x) = \pi\frac{I_{-\nu}(x) - I_\nu(x)}{2\sin\nu\pi}, \text{ if } \nu \text{ is nonintegral,}$$

$$K_n(x) = (-1)^{n+1}\left[\ln\left(\frac{x}{2}\right) + \gamma\right] I_n(x) + \frac{1}{2}\left(\frac{x}{2}\right)^{-n} \sum_{k=0}^{n-1} \frac{(n-k-1)!}{k!}\left(-\frac{x^2}{4}\right)^k$$

$$+ (-1)^n \left(\frac{x}{2}\right)^n \sum_{k=0}^{\infty} \left[1 + \frac{1}{2} + \frac{1}{3} + \cdots + \frac{1}{k} + \frac{1}{2(k+1)} + \frac{1}{2(k+2)} + \cdots + \frac{1}{2(n+k)}\right] \frac{(\frac{1}{4}x^2)^k}{k!\,(n+k)!},$$

where the sum $\sum_{k=0}^{n-1}$ is absent when $n = 0$.

3. Functional relations:

$$I_\nu(z) = \pm \frac{i}{\pi} K_\nu(ze^{\pm i\pi}) \mp \frac{ie^{\mp i\nu\pi}}{\pi} K_\nu(z),$$

$$K_\nu(z) = 2\cos(\pi\nu) K_\nu(ze^{\pm i\pi}) - K_\nu(ze^{\pm 2i\pi}).$$

4. Asymptotic expansions:

$$K_\nu(z) \sim \left(\frac{\pi}{2z}\right)^{1/2} e^{-z} \sum_{n=0}^{\infty} c_n z^{-n}, \qquad z \to \infty;\ |\arg z| < \frac{3}{2}\pi,$$

$$I_\nu(z) \sim (2\pi z)^{-1/2} e^z \sum_{n=0}^{\infty} (-1)^n c_n z^{-n} + i(2\pi z)^{-1/2} e^{-z+\nu\pi i} \sum_{n=0}^{\infty} c_n z^{-n},$$

$$z \to \infty;\ -\frac{1}{2}\pi < \arg z < \frac{3}{2}\pi,$$

$$c_n = \frac{(4\nu^2 - 1^2)(4\nu^2 - 3^2)(4\nu^2 - 5^2)\cdots(4\nu^2 - (2n-1)^2)}{8^n n!}, \qquad c_0 = 1.$$

5. Integral representations:

$$I_\nu(z) = \frac{1}{\pi}\int_0^\pi e^{z\cos t}\cos(\nu t)\,dt - \frac{\sin(\nu\pi)}{\pi}\int_0^\infty e^{-z\cosh t - \nu t}\,dt, \quad |\arg z| < \frac{1}{2}\pi,$$

$$K_\nu(z) = \int_0^\infty e^{-z\cosh t}\cosh(\nu t)\,dt, \qquad |\arg z| < \frac{1}{2}\pi.$$

6. Difference equations [$y_\nu(x)$ is either $I_\nu(x)$ or $K_\nu(x)$]:

$$y_{\nu-1}(x) - y_{\nu+1}(x) = \frac{2\nu}{x} y_\nu(x),$$

$$2y'_\nu(x) = y_{\nu-1}(x) + y_{\nu+1}(x),$$

$$I'_0(x) = I_1(x),$$

$$K'_0(x) = -K_1(x).$$

7. Generating function:

$$e^{z\cos t} = I_0(z) + 2\sum_{k=1}^{\infty} I_k(z)\cos(kt).$$

BESSEL FUNCTIONS

1. Differential equation:

$$x^2y'' + xy' + (x^2 - \nu^2)y = 0.$$

Solutions are linear combinations of $J_\nu(x)$ and $Y_\nu(x)$.

2. Frobenius series:

$$J_\nu(x) = (\tfrac{1}{2}x)^\nu \sum_{n=0}^{\infty} \frac{(-\tfrac{1}{4}x^2)^n}{n!\,\Gamma(n+\nu+1)},$$

$$Y_\nu(x) = \frac{J_\nu(x)\cos(\nu\pi) - J_{-\nu}(x)}{\sin(\nu\pi)}, \quad \text{if } \nu \text{ is nonintegral,}$$

$$Y_n(x) = \frac{2}{\pi}[\ln(\tfrac{1}{2}x) + \gamma]J_n(x) - \frac{1}{\pi}(\tfrac{1}{2}x)^{-n}\sum_{k=0}^{n-1}\frac{(n-k-1)!}{k!}(\tfrac{1}{4}x^2)^k$$

$$-\frac{1}{\pi}(\tfrac{1}{2}x)^n \sum_{k=0}^{\infty}\left[1 + \tfrac{1}{2} + \tfrac{1}{3} + \cdots + \frac{1}{k} + \frac{1}{2(k+1)} + \frac{1}{2(k+2)} + \cdots + \frac{1}{2(n+k)}\right]\frac{(-\tfrac{1}{4}x^2)^k}{k!\,(n+k)!},$$

where the sum $\sum_{k=0}^{n-1}$ is absent if $n = 0$.

3. Functional relations:

$$J_\nu(z) = e^{i\nu\pi/2} I_\nu(ze^{-i\pi/2}), \qquad -\frac{\pi}{2} < \arg z \le \pi,$$

$$Y_\nu(z) = ie^{i\nu\pi/2} I_\nu(ze^{-i\pi/2}) - \frac{2}{\pi}e^{-i\nu\pi/2}K_\nu(ze^{-i\pi/2}), \qquad -\frac{\pi}{2} < \arg z \le \pi.$$

4. Asymptotic expansions:

$$J_\nu(z) = w_1(z)\left(\frac{2}{\pi z}\right)^{1/2}\cos(z - \tfrac{1}{2}\nu\pi - \tfrac{1}{4}\pi) - w_2(z)\left(\frac{2}{\pi z}\right)^{1/2}\sin(z - \tfrac{1}{2}\nu\pi - \tfrac{1}{4}\pi),$$

$$Y_\nu(z) = w_2(z)\left(\frac{2}{\pi z}\right)^{1/2}\cos(z - \tfrac{1}{2}\nu\pi - \tfrac{1}{4}\pi) + w_1(z)\left(\frac{2}{\pi z}\right)^{1/2}\sin(z - \tfrac{1}{2}\nu\pi - \tfrac{1}{4}\pi),$$

$$w_1(z) \sim \sum_{n=0}^{\infty}(-1)^n c_{2n} z^{-2n}, \qquad z \to \infty;\ |\arg z| < \pi,$$

$$w_2(z) \sim \sum_{n=0}^{\infty}(-1)^n c_{2n+1} z^{-2n-1}, \qquad z \to \infty;\ |\arg z| < \pi,$$

$$c_n = \frac{(4\nu^2 - 1^2)(4\nu^2 - 3^2)\cdots(4\nu^2 - (2n-1)^2)}{8^n n!}, \qquad c_0 = 1.$$

5. Integral representations:

$$J_\nu(z) = \frac{1}{\pi}\int_0^\pi \cos(z\sin t - \nu t)\,dt - \frac{\sin(\nu\pi)}{\pi}\int_0^\infty e^{-z\sinh t - \nu t}\,dt, \qquad |\arg z| < \tfrac{1}{2}\pi,$$

$$Y_\nu(z) = \frac{1}{\pi}\int_0^\pi \sin(z\sin t - \nu t)\,dt - \frac{1}{\pi}\int_0^\infty [e^{\nu t} + e^{-\nu t}\cos(\nu\pi)]e^{-z\sinh t}\,dt,$$

$$|\arg z| < \tfrac{1}{2}\pi.$$

6. Difference equations [$y_\nu(x)$ is either $J_\nu(x)$ or $Y_\nu(x)$]:

$$y_{\nu-1}(x) + y_{\nu+1}(x) = \frac{2\nu}{x}\,y_\nu(x),$$

$$2y'_\nu(x) = y_{\nu-1}(x) - y_{\nu+1}(x),$$

$$J'_0(x) = -J_1(x),$$

$$Y'_0(x) = -Y_1(x).$$

7. Generating function:

$$e^{z(t-1/t)/2} = \sum_{k=-\infty}^{\infty} t^k J_k(z).$$

8. Other differential equations:

(*a*) $$y'' + a^2x^{k-2}y = 0,$$

$$y = \sqrt{x}[\alpha J_{1/k}(2ax^{k/2}/k) + \beta Y_{1/k}(2ax^{k/2}/k)].$$

(*b*) $$\frac{d^{2n}y}{dx^{2n}} = (-a^2)^n x^{-n} y,$$

$$y = x^{n/2}[\alpha J_n(2a\omega x^{1/2}) + \beta Y_n(2a\omega x^{1/2})],$$

where $\omega^n = 1$.

PARABOLIC CYLINDER FUNCTIONS

1. Differential equation:

$$y'' + (\nu + \tfrac{1}{2} - \tfrac{1}{4}x^2)y = 0.$$

Solutions are $D_\nu(\pm x)$ and $D_{-\nu-1}(\pm ix)$. Only two of these functions are linearly independent.

2. Taylor series:

$$D_\nu(x) = \frac{\pi^{1/2}2^{\nu/2}}{\Gamma(\frac{1}{2} - \frac{1}{2}\nu)}\sum_{n=0}^{\infty}\frac{a_{2n}x^{2n}}{(2n)!} - \frac{\pi^{1/2}2^{(\nu+1)/2}}{\Gamma(-\frac{1}{2}\nu)}\sum_{n=0}^{\infty}\frac{a_{2n+1}x^{2n+1}}{(2n+1)!},$$

where $a_0 = a_1 = 1$ and $a_{n+2} = -(\nu + \frac{1}{2})a_n + \frac{1}{4}n(n-1)a_{n-2}$.

$$D_\nu(0) = \pi^{1/2} 2^{\nu/2}/\Gamma(\tfrac{1}{2} - \tfrac{1}{2}\nu).$$

$$D'_\nu(0) = -\pi^{1/2} 2^{(\nu+1)/2}/\Gamma(-\tfrac{1}{2}\nu).$$

3. Functional relation:

$$D_\nu(z) = e^{i\nu\pi} D_\nu(-z) + \frac{(2\pi)^{1/2}}{\Gamma(-\nu)} e^{i(\nu+1)\pi/2} D_{-\nu-1}(-iz).$$

4. Asymptotic expansions:

$$D_\nu(z) \sim z^\nu e^{-z^2/4} \sum_{n=0}^{\infty} (-1)^n c_n z^{-2n}, \qquad z \to \infty;\ |\arg z| < \tfrac{3}{4}\pi,$$

$$D_\nu(z) \sim z^\nu e^{-z^2/4} \sum_{n=0}^{\infty} (-1)^n c_n z^{-2n} - \frac{(2\pi)^{1/2}}{\Gamma(-\nu)} e^{i\pi\nu} z^{-\nu-1} e^{z^2/4} \sum_{n=0}^{\infty} d_n z^{-2n},$$

$$z \to \infty;\ \frac{1}{4}\pi < \arg z < \frac{5}{4}\pi,$$

$$c_n = \frac{\nu(\nu-1)\cdots(\nu-2n+1)}{2^n n!}, \qquad c_0 = 1,$$

$$d_n = \frac{(\nu+1)(\nu+2)\cdots(\nu+2n)}{2^n n!}, \qquad d_0 = 1.$$

5. Integral representation:

$$D_\nu(x) = \sqrt{\frac{2}{\pi}}\, e^{x^2/4} \int_0^\infty e^{-t^2/2} t^\nu \cos(xt - \nu\pi/2)\, dt, \qquad \operatorname{Re} \nu > -1.$$

6. Difference equations:

$$xD_\nu(x) = D_{\nu+1}(x) + (\nu + \tfrac{1}{2}) D_{\nu-1}(x),$$

$$D'_\nu(x) = -\tfrac{1}{2} x D_\nu(x) + (\nu + \tfrac{1}{2}) D_{\nu-1}(x).$$

7. Relation to Hermite polynomials:

$$D_n(x) = \mathrm{He}_n(x) e^{-x^2/4}.$$

GAMMA AND PSI FUNCTIONS

1. Integral representation:

$$\Gamma(z) = \int_0^\infty t^{z-1} e^{-t}\, dt, \qquad \operatorname{Re} z > 0.$$

2. Difference equation:

$$\Gamma(x+1) = x\Gamma(x).$$

3. Special values:

$$\Gamma(0) = 1,\ \Gamma(\tfrac{1}{2}) = \sqrt{\pi}, \qquad \Gamma(n+1) = n!.$$

4. Stirling's asymptotic formula:

$$\Gamma(z) \sim (z/e)^z \sqrt{2\pi/z}\left[1 + \frac{1}{12z} + \frac{1}{288z^2} - \frac{139}{51840z^3} - \cdots\right],$$

$$z \to \infty;\ |\arg z| < \pi.$$

5. Other formulas:

$$\Gamma(z)\Gamma(1-z) = \pi/\sin(\pi z),$$

$$\Gamma(2z) = \tfrac{1}{2}\pi^{-1/2}4^z\Gamma(z)\Gamma(z+\tfrac{1}{2}),$$

$$\int_0^1 t^{x-1}(1-t)^{y-1}\,dt = \Gamma(x)\Gamma(y)/\Gamma(x+y), \qquad \operatorname{Re} x > 0, \operatorname{Re} y > 0.$$

6. Psi function:

$$\psi(z) = \Gamma'(z)/\Gamma(z).$$

7. Difference equation:

$$\psi(z+1) = \psi(z) + \frac{1}{z}.$$

8. Special values:

$$\psi(1) = -\gamma, \qquad \psi(n+1) = -\gamma + \sum_{k=1}^{n} 1/k,$$

where $\gamma \doteqdot 0.5772$ is Euler's constant.

9. Taylor series:

$$\psi(1+z) = -\gamma - \sum_{n=2}^{\infty} \zeta(n)(-z)^{n-1},$$

where $\zeta(n) = \sum_{k=1}^{\infty} k^{-n}$ is the Riemann zeta function.

10. Asymptotic expansion:

$$\psi(z) \sim \ln z - \frac{1}{2z} - \frac{1}{12z^2} + \frac{1}{120z^4} - \frac{1}{252z^6} + \cdots, \qquad z \to \infty;\ |\arg z| < \pi.$$

EXPONENTIAL INTEGRALS

1. Integral representation:

$$E_n(z) = \int_1^{\infty} \frac{e^{-zt}}{t^n}\,dt, \qquad \operatorname{Re} z > 0.$$

2. Series expansion:

$$E_n(z) = \frac{(-z)^{n-1}}{(n-1)!}[-\ln z + \psi(n)] - \sum_{\substack{m=0 \\ m \neq n-1}}^{\infty} \frac{(-z)^m}{(m-n+1)m!}, \qquad |\arg z| < \pi,$$

where $\psi(n)$ is the psi function.

3. Recurrence relations:

$$E_{n+1}(z) = \frac{1}{n}[e^{-z} - zE_n(z)],$$

$$\frac{dE_n(z)}{dz} = -E_{n-1}(z).$$

4. Asymptotic expansions:

$$E_n(z) \sim \frac{e^{-z}}{z}\left[1 - \frac{n}{z} + \frac{n(n+1)}{z^2} - \cdots\right], \qquad z \to \infty;\ |\arg z| < \frac{3}{2}\pi.$$

REFERENCES

What one man can invent another can discover.

—Sherlock Holmes, *The Adventure of the Dancing Men*
Sir Arthur Conan Doyle

Chapter 1

Some good general texts on differential equations:

1. Birkoff, G., and Rota, G. C., *Ordinary Differential Equations*, Ginn and Company, Boston, 1962.
2. Boyce, W. E., and DiPrima, R. C., *Elementary Differential Equations*, John Wiley and Sons, Inc., New York, 1969.
3. Carrier, G., and Pearson, C. E., *Ordinary Differential Equations*, Blaisdell Publishing Company, Waltham, Mass., 1968.

Advanced texts on differential equations:

4. Coddington, E. A., and Levinson, N., *Theory of Ordinary Differential Equations*, McGraw-Hill Book Company, Inc., New York, 1955.
5. Ince, E. L., *Ordinary Differential Equations*, Dover Publications, Inc., New York, 1956.

Green's function, δ functions, transform methods, and eigenfunction expansions are discussed by:

6. Courant, R., and Hilbert, D., *Methods of Mathematical Physics*, vol. 1, Interscience Publishers, New York, 1953.
7. Jeffreys, H., and Jeffreys, B. S., *Methods of Mathematical Physics*, 3d ed., Cambridge University Press, London, 1956.
8. Lighthill, M. J., *Introduction to Fourier Analysis and Generalized Functions*, Cambridge University Press, London, 1958.
9. Morse, P. M., and Feshbach, H., *Methods of Theoretical Physics*, pt. I, McGraw-Hill Book Company, Inc., New York, 1953.

Useful reference works on the special functions of mathematical physics and applied mathematics:

10. Abramowitz, M., and Stegan, I. A., *Handbook of Mathematical Functions*, Dover Publications, Inc., New York, 1964.
11. Erdelyi, A., Magnus, W., Oberhettinger, F., and Tricomi, F. G., *Higher Transcedental Functions*, 3 vols, McGraw-Hill Book Company, Inc., New York, 1953.
12. Whittaker, E. T., and Watson, G. N., *A Course of Modern Analysis*, 4th ed., Cambridge University Press, Cambridge, 1927.

Chapter 2

General references on difference equations:

13. Hildebrand, F. B., *Methods of Applied Mathematics*, chap. 3, Prentice-Hall, Inc., Englewood Cliffs, N.J., 1952.
14. Milne-Thomson, L. M., *The Calculus of Finite Differences*, Macmillan and Co., London, 1953.

Also see Ref. 3.

Chapter 3

For a general discussion see:

15. Erdelyi, A., *Asymptotic Expansions*, Dover Publications, Inc., New York, 1956.
16. de Bruijn, N. G., *Asymptotic Methods in Analysis*, North-Holland Publishing Company, Amsterdam, 1958.
17. Carrier, G. F., Krook, M., and Pearson, C. E., *Functions of a Complex Variable*, McGraw-Hill Book Company, Inc., New York, 1966.
18. Heading, J., *An Introduction to Phase-Integral Methods*, Methuen and Co., Ltd., London, 1962.
19. Wasow, W. A., *Asymptotic Expansions for Ordinary Differential Equations*, John Wiley and Sons, Inc., New York, 1965.

Also see Refs. 4 and 5.

For asymptotic expansions of the special functions of mathematical physics and applied mathematics see Refs. 10 to 12 and:

20. Watson, G. N., *A Treatise on the Theory of Bessel Functions*, 2d ed., Cambridge University Press, Cambridge, 1944.

Chapter 4

For a general discussion see Ref. 5 and:

21. Davis, H. T., *Introduction to Nonlinear Differential and Integral Equations*, Dover Publications, Inc., New York, 1962.

For a discussion of the Thomas-Fermi equation see:

22. Messiah, A., *Quantum Mechanics*, vol. II, John Wiley and Sons, Inc., New York, 1962.

For a discussion of phase-plane analysis see Refs. 2 and 4. For an advanced discussion see:

23. Arnold, V. I., and Avez, A., *Ergodic Problems of Classical Mechanics*, W. A. Benjamin, Inc., New York, 1968.

Chapter 5

For a general discussion of asymptotic methods see Ref. 14. The Stirling series is discussed in Ref. 12. For a discussion of nonlinear recursion relations see Ref. 16 and:

24. Stein, P. R., and Ulam, S., "Lectures on Nonlinear Algebraic Transformations," in A. O. Barut (Ed.), *Studies in Mathematical Physics*, Reidel, Dordrecht, Holland, 1970.

Chapter 6

Some general texts on the asymptotic expansion of integrals:

25. Bleistein, N., and Handelsman, R. A., *Asymptotic Expansions of Integrals*, Holt, Rinehart and Winston, New York, 1975.
26. Copson, E. T., *Asymptotic Expansions*, Cambridge University Press, Cambridge, 1967.
27. Olver, F. W. J., *Asymptotics and Special Functions*, Academic Press, Inc., New York, 1974.

See also Refs. 15 to 17.

Integral representations of special functions are given in Refs. 10 to 12 and:

28. Gradshteyn, I. S., and Ryzhik, I. W., *Tables of Integrals, Series, and Products*, 4th ed., Academic Press, Inc., New York, 1965.

Chapter 7

For general discussions of perturbation methods see Ref. 16 and:

29. Nayfeh, A. H., *Perturbation Methods*, John Wiley and Sons, Inc., New York, 1973.
30. Merzbacher, E., *Quantum Mechanics*, John Wiley and Sons, Inc., New York, 1970.

Perturbation methods for eigenvalue problems are discussed in Ref. 22 (vol. I) and:

31. Landau, L. D., and Lifshitz, E. M., *Quantum Mechanics, Non-Relativistic Theory*, Pergamon Press, London, 1965.

See also:

32. Reed, M., and Simon, B., *Methods of Modern Mathematical Physics*, vols. I and II, Academic Press, Inc., New York, 1972.

Chapter 8

33. Baker, G. A., *Essentials of Padé Approximants*, Academic Press, Inc., New York, 1975.
34. Hardy, G. H., *Divergent Series*, Oxford University Press, Oxford, 1956.
35. Knopp, K., *Theory and Application of Infinite Series*, Hafner Publishing Company, New York, 1947.

Chapter 9

For a general discussion of boundary-layer methods see Refs. 3, 29, and:
36. Cole, J. D., *Perturbation Methods in Applied Mathematics*, Ginn and Company, Boston, 1968.
37. Kaplun, S., *Fluid Mechanics and Singular Perturbations*, P. A. Lagerstrom, L. N. Howard, and C. S. Liu (Eds.), Academic Press, Inc., New York, 1967.
38. O'Malley, R. E., *Introduction to Singular Perturbations*, Academic Press, Inc., New York, 1974.
39. Van Dyke, M., *Perturbation Methods in Fluid Mechanics*, Academic Press, Inc., New York, 1964.

Chapter 10

See Refs. 9, 18, 22 (vol. I), 30, 31, and:
40. Fröman, N., and Fröman, P. O., *JWKB Approximation*, North-Holland Publishing Company, Amsterdam, 1965.

Chapter 11

See Refs. 29, 36, 39, and:
41. Eckhaus, W., *Studies in Nonlinear Stability Theory*, Springer-Verlag, New York, 1965.
42. Stoker, J. J., *Nonlinear Vibrations*, Interscience Publishers, New York, 1950.

Quite simple, my dear Watson.

—Sherlock Holmes, *The Adventure of the Retired Colourman*
Sir Arthur Conan Doyle

INDEX

f refers to figures
p refers to problems
t refers to tables